拥有智能化信息化平台、智能方法和产品、大数据和云计算、物联网和自动化、数字化设计等五位一体的智能化能力体系

面向钢铁行业提供智能制造整体解决方案和产品体系

拥有CISDigital™工业互联网云平台，提供包括智能生产管控、企业运营及外部生态协同服务的一站式云平台

率先打造多个智能制造标杆项目

业内首个钢铁智慧中心(宝武韶钢)
实现生产高效协同和低成本炼铁

业内首套智能原料场(宝武湛江)
实现原料场生产的完全自动化、高度智能化和管控一体化

业内首套智能热轧钢卷库(宝武八钢)
实现行车无人化和库管智慧化

业内首套智能铁水运输系统(宝武湛江)
实现铁水运输全流程全天候智慧化无人化运行

中冶赛迪微信号

中冶赛迪集团有限公司
地址：重庆市渝中区双钢路1号，400013
电话：+86-23-63850431
传真：+86-23-63548888
网址：www.cisdigroup.com.cn

壳动工业 油内智外

| 全能顾问　　高效运维
| 智慧预测　　协力创新

Shell
LubeMaster

壳智汇　智能润滑管家

全能管家 智享润滑

壳牌工业润滑油

壳动工业潜能

LubeSensor
实时数据分析

LubeExpert
专业的设备运营咨询

LubeAnalyst
专业润滑监测、诊断与建议

LubeCoach
润滑知识培训课程

LubeChat
壳牌小贝24*7人工智能客服

中国金属学会 编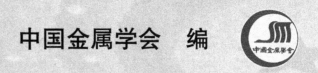

第十二届
中国钢铁年会论文集
（摘要）

Proceedings of the 12th CSM Steel Congress

北　京
冶金工业出版社
2019

内 容 简 介

本论文集共收录论文 700 余篇,约 850 万字。全书内容包括炼铁与原料、炼钢与连铸、轧制与热处理、表面与涂镀、金属材料深加工、汽车用钢、海工用钢、电工钢、能源用钢、轨道用钢、建筑用钢、机械用钢、非晶合金、粉末冶金、冶金能源与资源利用、冶金环保、钢铁材料表征与评价、冶金设备与工程技术、冶金自动化与智能化、建筑诊治等,全面反映了近两年来我国及世界钢铁行业科研、生产、管理等方面的最新成果,是一本内容全面、新颖,具有较高学术水平的专业论文集。本书可供钢铁行业的科研人员、管理人员、工程技术人员、高校师生等学习参考。

第十二届中国钢铁年会论文集以纸质图书和电子版方式出版。纸质图书收录论文摘要(含特邀报告),电子版收录论文全文。本书为摘要集。

图书在版编目(CIP)数据

第十二届中国钢铁年会论文集:摘要/中国金属学会编. —北京:冶金工业出版社,2019.10
ISBN 978-7-5024-8292-3

Ⅰ.①第… Ⅱ.①中… Ⅲ.①钢铁工业—学术会议—文集 Ⅳ.①TF4-53

中国版本图书馆 CIP 数据核字(2019)第 209512 号

出 版 人 谭学余
地　　址　北京市东城区嵩祝院北巷 39 号　邮编　100009　电话　(010)64027926
网　　址　www.cnmip.com.cn　电子信箱　yjcbs@cnmip.com.cn
责任编辑　李培禄　美术编辑　彭子赫　版式设计　孙跃红
责任校对　王永欣　责任印制　牛晓波

ISBN 978-7-5024-8292-3

冶金工业出版社出版发行;各地新华书店经销;北京虎彩文化传播有限公司印刷
2019 年 10 月第 1 版,2019 年 10 月第 1 次印刷
210mm×297mm;38.75 印张;2 彩页;1203 千字;564 页
120.00 元

冶金工业出版社　投稿电话　(010)64027932　投稿信箱　tougao@cnmip.com.cn
冶金工业出版社营销中心　电话　(010)64044283　传真　(010)64027893
冶金工业出版社天猫旗舰店　yjgycbs.tmall.com
　　　　(本书如有印装质量问题,本社营销中心负责退换)

第十二届中国钢铁年会
组委会

名誉主席 殷瑞钰 翁宇庆

年会主席 干勇

执行主席 赵沛

委　　员（以姓氏笔画为序）

于勇　王新江　左良　曲阳

沈彬　张少明　张欣欣　陈德荣

赵沛　赵继　赵民革　姚林

年会秘书长 王新江

年会秘书处 中国金属学会学术工作部

《第十二届中国钢铁年会论文集》
编委会

主　任　赵　沛

副主任　王新江

编　委　（以姓氏笔画为序）

王一德	王天义	王立新	王国栋	王昭东	王海舟
王崇愚	王　喆	王新东	王新华	毛新平	龙红明
田志凌	白晨光	冯光宏	曲选辉	朱　荣	朱　静
朱仁良	刘　玠	刘　浏	刘正东	刘宏民	刘国栋
米振莉	许家彦	孙彦广	杜　涛	杜　斌	李　卫
李文秀	李红霞	李依依	李鹤林	杨　勇	杨天钧
杨春政	杨晓东	杨景玲	张　杰	张　跃	张万山
张丕军	张寿荣	张启富	张建良	张春霞	张统一
张晓刚	张晓军	张福明	陈　杰	陈　卓	陈其安
陈思联	尚成嘉	岳清瑞	周少雄	周国治	周清跃
郑　云	郑文华	赵希超	赵振锐	胡正寰	洪及鄙
贾成厂	柴天佑	钱　刚	徐安军	徐金梧	殷国茂
唐　荻	唐立新	黄　导	常好诵	康永林	董　瀚
韩国瑞	曾加庆	温燕明	谢英秀	谢建新	鲍　磊
蔡九菊	穆怀明				

前　言

第十二届中国钢铁年会于 2019 年 10 月 15~16 日在北京召开。中华人民共和国成立 70 年以来，作为社会经济支撑产业和制造业强国建设的材料基础，我国钢铁工业发展取得了辉煌的成就和举世瞩目的国际影响，但仍面临环境、资源、能源、质量、成本、品种、服务等多方面的严峻挑战，迫切需要深入实施创新驱动发展战略，实现高质量发展。本届年会以"**创新引领，高质量发展，建设钢铁强国**"为主题，围绕钢铁生产全流程的基础理论、工艺技术、产品设计、制造和应用技术，研讨钢铁领域科技创新的方向和路径，促进钢铁行业向绿色化、智能化、品牌化发展，加快钢铁强国建设。

本届年会的征文工作得到全国冶金及材料领域的专家、学者和科技人员的积极支持，共收到 951 篇投稿论文和 80 篇特邀报告摘要，经专家评审并根据作者意愿，共有 700 余篇论文结集出版。内容包括：炼铁与原料、炼钢与连铸、轧制与热处理、表面与涂镀、金属材料深加工、汽车用钢、海工用钢、电工钢、能源用钢、轨道用钢、建筑用钢、机械用钢、非晶合金、粉末冶金、冶金能源与资源利用、冶金环保、钢铁材料表征与评价、冶金设备与工程技术、冶金自动化与智能化、建筑诊治等。年会论文全文和部分摘要以电子版方式出版，录用论文摘要以纸质版方式出版。

由于论文集编辑、出版时间较紧，疏漏与错误之处，恳请读者批评指正。

中国金属学会
2019 年 10 月

目　　录

大会特邀报告

国际标准化发展趋势和中国钢铁高质量发展 ·· 张晓刚　I-1
Development Trend of International Standardization and High-quality Development of Chinese
　　Steel Industry ·· Zhang Xiaogang
坚持绿色发展道路，打造世界一流企业 ··· 陈德荣　I-1
Persist in Green Development, Build a World-Class Enterprise ······················· Chen Derong
日本制铁针对全球环境问题的应对之策 ··· Inoue Akihiko　I-2
Efforts of Nippon Steel Corporation for Global Environmental Problems ········· Inoue Akihiko
韩国钢铁产品技术最新进展 ·· Kim Sung-Joon　I-3
Recent Progress of Steel Product Technology in Korea ······························ Kim Sung-Joon
钢铁材料品种开发及其应用进展 ·· 田志凌　I-4
Development and Application of Steel Product Research in Central Iron and Steel Institute ······ Tian Zhiling

分会场特邀报告

我国高炉使用高比例球团生产技术经济分析 ············· 沙永志，马丁·戈德斯，宋阳升　II-1
Technical and Economic Analysis of Blast Furnace Operation with High Pellet Ratio in China
　　·· Sha Yongzhi, Maarten Geerdes, Sunny Song
首钢炼铁技术进步与展望 ··· 张福明　II-1
Progress and Expectation of Ironmaking Technology in Shougang ······················ Zhang Fuming
炼铁过程炉料结构智能优化系统 ··········· 张玉柱，王宝祥，陈　伟，武鹏飞，陈　颖，马劲红　II-2
Intelligent Optimization System of Burden Structure in Ironmaking Process
　　························· Zhang Yuzhu, Wang Baoxiang, Chen Wei, Wu Pengfei, Chen Ying, Ma Jinhong
高炉本体长寿设计中的若干认识与建议 ··· 汤清华　II-3
The Opinion and Advice about the Design of BF Proper's Long Campaign ············· Tang Qinghua
钢铁造块节能减排与智能控制技术 ·· 范晓慧　II-4
我国焦化行业运行及对未来焦炭需求预测 ·· 郑文华　II-4
China's Coking Industry Operation and Coke Demand Forecast for the Future ······ Zheng Wenhua
以可视化、模式识别为核心的高炉智能控制系统的开发及应用 ································ 陈令坤　II-5
Development and Application of Artificial Intelligence(AI) Control System Based on the Technologies of
　　Visualization and Pattern Recognition ·· Chen Lingkun
高铁车辆轮轴用钢大型非金属夹杂物和宏观偏析控制技术研究
　　·· 王新华，姜　敏，于会香，汪开忠，孙　维　II-5
Control of Large-sized Inclusions and Macro Segregation in Stells for High-speed Train Wheels and Axles
　　·· Wang Xinhua, Jiang Min, Yu Huixiang, Wang Kaizhong, Sun Wei
特殊钢特种冶金技术和生产流程的发展趋势 ·· 姜周华　II-6

炼钢底喷粉工艺的关键技术及实践 ······ 朱　荣　II-6
Key Technology and Practice of Bottom Powder Injection in Steelmaking　Zhu Rong
92Si 桥梁缆索用钢生产过程中非金属夹杂物的演变规律 ······ 罗　锋，王　聪　II-7
Evolution of Non-metallic Inclusions in Production of 92Si Steel for Bridge Cable　Luo Feng, Wang Cong
铁水旋涌脱硫扒渣工艺开发及应用 ······ 万雪峰　II-8
电弧炉炼钢工程的流程设计与装备选型 ······ 潘宏涛　II-8
Process Design and Equipment Selection of EAF Steelmaking Project　Pan Hongtao
高品质高氮钢加压冶金理论、制备技术及品种开发 ······ 李花兵，姜周华，冯　浩，朱红春　II-8
Pressurized Metallurgy Theory, Manufacturing Technologies and Variety Development of High Nitrogen
　　Steels　Li Huabing, Jiang Zhouhua, Feng Hao, Zhu Hongchun
洁净钢用 CA6 基耐火材料的基础研究 ······ 侯新梅，王恩会，陈俊红，李　斌，周国治　II-9
Preparation and Application of CA6-based New Functional Refractories
　　　　　　　Hou Xinmei, Wang Enhui, Chen Junhong, Li Bin, Zhou Kuozhi
钢中复合夹杂物/钢基体的电势差与电偶腐蚀的关系
　　······ 侯延辉，刘林利，李光强，李腾飞，刘　昱　II-10
The Correlation between Potential Difference and Galvanic Corrosion of Composite Inclusions/Steel Matrix
　　in Steel　Hou Yanhui, Liu Linli, Li Guangqiang, Li Tengfei, Liu Yu
转炉吹炼后期碳含量连续预报模型算法研究
　　······ 林文辉，孙建坤，刘　青，焦树强，周凯啸，刘　敏，苏　醒　II-11
Study on Algorithm of Continuous Carbon Prediction Model in the End Blowing Stage of BOF Converter
　　　　　　　Lin Wenhui, Sun Jiankun, Liu Qing, Jiao Shuqiang, Zhou Kaixiao, Liu Min, Su Xing
X80 管线钢非金属夹杂物控制研究 ······ 李战军，初仁生，刘金刚，郝　宁　II-12
Study on Control of Non-metallic Inclusions in X80 Pipeline Steel
　　　　　　　Li Zhanjun, Chu Rensheng, Liu Jingang, Hao Ning
高温板坯全连续淬火新技术开发及应用 ······ 朱苗勇，蔡兆镇，刘志远，王重君，赵佳伟　II-13
Development and Application of New Continuously Quenching Technology for High Temperature Slab
　　　　　　　Zhu Miaoyong, Cai Zhaozhen, Liu Zhiyuan, Wang Chongjun, Zhao Jiawei
板坯连铸过程流动、传热及夹杂物分布的数值模拟研究 ······ 张立峰　II-13
Mathematical Modeling on the Fluid Flow, Heat Transfer, and Inclusion Distribution during the
　　Continuous Casting Slab Strand　Zhang Lifeng
连续铸钢设备设计理论研究 ······ 杨拉道　II-14
Research on Design Theory of Continuous Casting Steel Equipment　Yang Ladao
高表面品质汽车板连铸浸入式水口结构优化工艺研究 ······ 李海波　II-15
Optimization of SEN Structures for Slab Continuous Casting High Surface Quality IF Steels　Li Haibo
改善重轨钢连铸坯均质化与致密化的基础研究与应用实践 ······ 陈　永　II-15
Fundamental Study and Application of the Improvement of Solute Homogenization and
　　Compactness of Continuously Cast Rail Steel　Chen Yong
轧钢过程节能减排技术进展及应用实例分析 ······ 康永林　II-16
Progress and Application of Energy Saving and Emission Reduction Technology in Steel Rolling Process
　　　　　　　Kang Yonglin
宽带钢热轧智能产线关键技术研究进展 ······ 何安瑞　II-17
Development of Key Technologies of Smart Production Line for Hot Wide Strip　He Anrui
轧机振动控制研究及推广应用 ······ 闫晓强　II-18

Research and Application of Rolling Mill Vibration Control ……………………… Yan Xiaoqiang

I&Q&P 工艺中配分时的竞争机制及变形协调机理的研究 …………………………… 于　浩　II-19

首钢超大型水电压力钢管用高强度钢板的开发 ……………………………………… 邹　扬　II-19
Development of High Strength Steel Plate for Large Hydroelectric Penstock of Shougang

热连轧板"唇印结疤"成因分析与对策 ………………………………………………… 郭　斌　II-20
Formation Analyze and Countermeasure of Lip Print Scab Defect on Hot Rolled Strip ……… Guo Bin

超超临界锅炉用新型耐热无缝管 C-HRA-5 的开发 …… 方旭东，包汉生，李　阳，徐芳泓，夏　焱　II-21
Development of Model Heat Resisting Seamless Tube C-HRA-5 for Ultra-Supercritical Power Plant
　Boilerand ……………………………………… Fang Xudong, Bao Hansheng, Li Yang, Xu Fanghong, Xia Yan

热轧钢筋分段气雾冷却工艺研究 ………………… 王卫卫，李大亮，周秉国，李志新，徐晓林，杨　杰　II-22
Study on Piecewise Spray Evaporation Cooling Technology of Hot-rolled Reinforced Bar
　……………………………………… Wang Weiwei, Li Daliang, Zhou Bingguo, Li Zhixin, Xu Xiaolin, Yang Jie

我国锌与锌合金镀层钢板发展 ………………………………………………………… 张启富　II-22

冷热复合成形技术与结构轻量化应用 ………………………………………………… 韩静涛　II-23

1500MPa 液压成形管件扭力梁开发及疲劳开裂问题改善 ……… 蒋浩民，逯若东，苏海波，徐小华　II-23
Development of 1500MPa Tubular Torsion Beam with Hydroforming and Improvement of Fatigue Cracking
　……………………………………………………………… Jiang Haomin, Lu Ruodong, Su Haibo, Xu Xiaohua

定制辊压成形技术与中国制造业升级 …………………………………………… 晏培杰，毕若凌　II-24
Customized Roll Forming Technology and China's Manufacturing Upgrading ……… Yan Peijie, Bi Ruoling

帽形件的多道次链模成形技术研究 …… 张智恒，钱　朕，张学广，何广忠，邹天下，李大永，丁士超　II-25
Study on Multi-pass Chain-die Forming of Hat-shaped Parts
　　Zhang Zhiheng, Qian Zhen, Zhang Xueguang, He Guangzhong, Zou Tianxia, Li Dayong, Ding Shichao

基于压溃试验增强成形性双相钢吸能特性分析
　………………………………………… 张　伟，李春光，林兴明，杨建炜，刘立现，刘华赛　II-26
Analysis of Energy Absorption Characteristics of Dual Phase Steel with High Formability based on
　Drop Test ……………… Zhang Wei, Li Chunguang, Lin Xingming, Yang Jianwei, Liu Lixian, Liu Huasai

逆转变奥氏体对高强塑积中锰钢(0.1C-5Mn)氢脆敏感性的影响
　………………………………………………… 惠卫军，张永健，王存宇，曹文全，董　瀚　II-27
Effect of Reverted Austenite Fraction on Hydrogen Embrittlement of TRIP-aided Medium Mn
　Steel (0.1C-5Mn) ………………………… Hui Weijun, Zhang Yongjian, Wang Cunyu, Cao Wenquan, Dong Han

基于 CALPHAD 方法的马氏体转变驱动力计算和 Ms 预测 ……………… 罗　群，陈双林，李　谦　II-28
The Calculation of Driving Force for Martensitic Transformation and Ms based on CALPHAD Approach

超高强度汽车钢剪切边裂纹敏感性研究 ……………………………………………… 王存宇　II-28
Study on the Sensitivity of Shear Edge Crack of Ultra-High Strength Automobile Steel

超高强度钢塑性成形与热处理一体化工艺探索 …………………… 金学军，龚　煜，李　伟　II-29
Integrating Deformation and Quenching-Partitioning-Tempering Novel Treatment and Development of
　Advanced Steels ……………………………………………………… Jin Xuejun, Gong Yu, Li Wei

首钢汽车用先进高强钢研发进展及应用实践 ………………………………………… 韩　赟　II-30
Research, Development and Application of Shougang AHSS for Automobile ……………… Han Yun

相图计算技术在先进汽车钢板材料设计中的应用研究 ……… 何燕霖，李　麟，郑伟森，鲁晓刚　II-30
Study on Application of CALPHAD Technology in Material Design of Advanced Automobile Steel Plate
　……………………………………………………………… He Yanlin, Li Lin, Zheng Weisen, Lu Xiaogang

汽车用高强钢动态和断裂力学特性的测试分析及仿真应用 …………………………… 吴海龙　II-31

微观组织对低合金结构钢耐蚀性影响的关键基础研究 ………………………………… 李晓刚，刘 超 II-31
Key Fundamental Aspects in the Microstructural Regulation of Corrosion Resistant Low Alloy Structural Steels
………………………………………………………………………………… Li Xiaogang, Liu Chao

铁磁应变玻璃的特征与磁致伸缩性能 ……………………………………………………… 杨 森 II-32

中国电工钢发展变化及新时代需求的研究 ………………………………………………… 陈 卓 II-33
Research on the Development Change and the Demand of the Electrical Steel in China in the New Era
……………………………………………………………………………………………… Chen Zhuo

新能源车发展新趋势及高端电机磁材料应用技术 ……………………………………… 裴瑞琳 II-33
Development of EV Car and Application Technology of Magnetic Materials in High-End Electric Machines
……………………………………………………………………………………………… Pei Ruilin

高效环保变压器用高性能取向硅钢制备技术
……………………………………… 龚 坚，马家骥，孙茂林，司良英，黎先浩，王现辉 II-34

"绿色"驱动 创新引领 助力新能源用钢高质量发展 ……………………… 钱 刚，许晓红 II-34
High-level Development of New Energy Steel Moved Forward by "Green" and Innovation

金属构筑成形热加工过程界面演化 …………………… 孙明月，徐 斌，李殿中，李依依 II-35
Interface Evolution during Metal Additive Forging Process …… Sun Mingyue, Xu Bin, Li Dianzhong, Li Yiyi

宝钢三代和四代核电关键设备用钢开发与工程应用进展 ………………………………… 张汉谦 II-36

增材制造技术及其在高性能大型金属构件制造领域的应用 …………………… 吉玲康，胡美娟 II-36
Additive Manufacturing Technology and Its Application in Manufacturing of Large Metal Components
……………………………………………………………………………… Ji Lingkang, Hu Meijuan

低内应力高强韧厚规格管线钢带的生产工艺研究
……………………………………… 缪成亮，李 飞，程 政，朱腾威，牛 涛，徐海卫 II-37
Research on Production Process of Large Thickness Low Internal Stress Pipeline Steel Strip with Good
   Strength and Toughness …… Miao Chengliang, Li Fei, Cheng Zheng, Zhu Tengwei, Niu Tao, Xu Haiwei

兴澄特钢高性能能源用钢的研发和产业化 …… 许晓红，叶建军，韩 雷，徐 君，邱文军，邓云飞 II-38
R&D and Industrialization of High Performance Energy Steel in Xingcheng Special Steel
…………………………………………… Xu Xiaohong, Ye Jianjun, Han Lei, Xu Jun, Qiu Wenjun, Deng Yunfei

重载铁路用贝氏体钢轨的研究进展 …………… 高古辉，白秉哲，桂晓露，谭谆礼，翁宇庆 II-39
Research Progress of Bainitic Rail Steel for Heavy Haul Railway
…………………………………………… Gao Guhui, Bai Bingzhe, Gui Xiaolu, Tan Zhunli, Weng Yuqing

高速重载铁路高锰钢辙叉制造技术创新 ……………… 陈 晨，马 华，王 琳，张福成 II-39
Technological Innovation in Manufacturing Hadfield Steel Frogs for High-Speed and Heavy-Haul Railway
…………………………………………………… Chen Chen, Ma Hua, Wang Lin, Zhang Fucheng

高强抗震耐蚀耐火钢组织性能研究 …………… 王学敏，丛菁华，李江文，尚成嘉 II-41
The Study on the Microstructure and Properties of High-strength Anti-seismic Corrosion-resistant
   Fire-resistant Steel …………………………… Wang Xuemin, Cong Jinghua, Li Jiangwen, Shang Chengjia

多元微合金化耐火钢研究进展 ………………………… 李昭东，王 鑫，杨忠民，雍岐龙 II-41
Research Progress on Multi-microalloyed Fire-resistant Steels
…………………………………………… Li Zhaodong, Wang Xin, Yang Zhongmin, Yong Qilong

高钛高耐磨钢的物理冶金学基础及应用 ……………… 孙新军，梁小凯，刘罗锦，许 帅 II-42
Physical Metallurgy and Application of High Ti High Wear-resistant Steel
…………………………………………… Sun Xinjun, Liang Xiaokai, Liu Luojin, Xu Shuai

研发材料的熵调控模式 ……………………………………………………………………… 汪卫华 II-43
Developing New Alloys by Entropy Modulation

高饱和磁感强度铁基软磁非晶合金研究进展 .... 姚可夫 II-43
Research Progress of Fe-based Amorphous Alloys with High Saturation Flux Density .... Yao Kefu

基于非晶的纳米结构及其功能特性 .... 沈 军 II-44
Nanostructures Grown from Amorphous Metals and Their Functional Properties .... Shen Jun

铁基非晶纳米晶软磁合金的成分设计与性能研究 .... 吕 旷，李育洛，窦正旭，惠希东 II-45
Composition Design and Properties of Fe-based Amorphous and Nanocrystalline Alloys
.... Lv Kuang, Li Yuluo, Dou Zhengxu, Hui Xidong

快速凝固亚稳合金催化材料的制备与性能 .... 张海峰 II-46

"金属-类金属"型高熵块体非晶合金的制备及性能 .... 张 伟 II-46
Fabrication and Properties of Metal-metalloid Type High Entropy Bulk Metallic Glasses .... Zhang Wei

铁磁性非晶合金结构不均匀性调控及由此带来的 GFA 及力学性能变化 .... 沈宝龙 II-47
Structural Heterogeneity Modulation of Ferromagnetic Metallic Glasses and Effects on Glass Forming
    Ability and Mechanical Properties .... Shen Baolong

FeSiB 三元非晶合金熔体原子扩散实验研究 .... 张 博 II-48
Diffusion Experiments of FeSiB Glass Forming Melts .... Zhang Bo

锆基非晶合金对大鼠骨组织的影响 .... 王 刚 II-49
Implantation of a Zr-based Metallic Glass in the Bone Tissue of SD Rat .... Wang Gang

高熵非晶合金的异常玻璃转变行为 .... 刘雄军 II-50
Unusual Glass Transition Behaviors of in High-entropy Bulk Metallic Glass .... Liu Xiongjun

调控非晶合金结构提高其塑性变形能力 .... 王 丽 II-51
Tuning the Structure of Metallic Glasses and Improving Their Plasticity .... Wang Li

金属玻璃熔体异常黏度降低的根源研究 .... 胡丽娜 II-52
Probing into the Abnormal Viscosity Drop during Cooling of Metallic Glass Forming Melts .... Hu Lina

基于磁性非晶合金的金属-半导体-金属转变行为研究 .... 陈 娜 II-53
Metal-Semiconductor-Metal Transitions in Amorphous Ferromagnetic Alloys .... Chen Na

块体非晶合金材料产业化技术开发及应用介绍 .... 李扬德，张 涛，高 宽，朱旭光 II-54
The Development of Industrialization Technology and the Introduce of Application in Bulk Amorphous
    Alloys .... Li Yangde, Zhang Tao, Gao Kuan, Zhu Xuguang

先进热管理材料及技术的研究与应用进展 .... 郭 宏 II-54
Progress in Research and Application of Advanced Thermal Management Materials and Technologies
.... Guo Hong

粉末冶金技术在航空发动机高温合金涡轮盘制备中的应用 .... 田高峰 II-55
Application of Powder Metallurgy Technology in Superalloy Turbine Disk of Aeroengine .... Tian Gaofeng

钢铁工业的资源消耗及废物减排的方向与途径 .... 蔡九菊 II-56
Resources Consumption and Wastes Emission Reduction of the Iron and Steel Industry .... Cai Jiuju

钢铁冶金工程清洁高效燃烧技术现状与展望 .... 张福明 II-57

真空冶金技术在资源综合利用领域的应用
.... 徐宝强，杨 斌，戴卫平，刘大春，汤文通，李一夫，蒋文龙 II-57

钢铁行业固废处理及资源化利用 .... 王天义 II-58

钢铁水污染全过程控制技术与应用 .... 曹宏斌 II-58

环境承载力与钢铁行业绿色发展 .... 杨晓东 II-58
Environmental Carrying Capacity and Green Development of the Steel Industry .... Yang Xiaodong

钢铁烧结烟气全流程减排技术 .... 魏进超 II-59
Flue Gas Full-process Emission Reduction Technology for Sintering of Iron and Steel .... Wei Jinchao

钢铁企业超低排放技术路线与实践 ······ 朱晓华 II-60

大型钢铁企业从内陆搬迁至沿海的分析与实践 ······ 吴礼云 II-60
Analysis and Practice of Major Steel Industry Transfer from Inland to Coastal Areas ······ Wu Liyun

激光诱导击穿光谱技术在钢铁在线分类中的初步应用
······ 刘 佳，沈学静，史孝侠，崔飞鹏，徐 鹏，李晓鹏 II-61
Application of Laser Induced Breakdown Spectroscopy in Online Sorting of Steel
······ Liu Jia, Shen Xuejing, Shi Xiaoxia, Cui Feipeng, Xu Peng, Li Xiaopeng

中国材料与试验团体标准体系（CSTM） ······ 陈 鸣，杨植岗，李梦琪，王 矛，王 洋，王 蓬 II-62
The Introduction of Chinese Society for Testing & Materials (CSTM)

无人天车与智能库管技术的研究与应用 ······ 王晓晨 II-62

中冶京诚板带处理新技术新装备 ······ 王建兵 II-63

冶金车间智能化建设研究与实践 ······ 赵振锐，甄景燕，尹国强 II-63
Research and Practice on Intelligent Construction of Metallurgical Workshop
······ Zhao Zhenrui, Zhen Jingyan, Yin Guoqiang

机器视觉技术及在钢铁生产中的应用 ······ 徐 科，周 鹏，贺 笛，邓能辉 II-63
Brief Introduction of Machine Vision and Its Application in Steel Production

轧钢过程异常工况诊断与容错控制 ······ 彭开香 II-64
Fault Diagnosis and Fault Tolerant Control of Abnormal Conditions for Industrial Processes of Steel Rolling
······ Peng Kaixiang

长材生产线智能轧钢技术的研发与应用 ······ 房继虹 II-65
Development and Application of Intelligent Technology of Rolling-steel for Long Product Line
······ Fang Jihong

既有工业建筑结构诊治与性能提升关键技术研究 ······ 常好诵 II-66
Key Technology Research on Existing Industrial Building Structure Diagnosis and Performance Improvement
······ Chang Haosong

锈损钢结构承载性能退化规律研究 ······ 徐善华 II-66
Research on Degradation Law of Bearing Capacity of Steel Structures with Corrosion Damage ······ Xu Shanhua

1 炼铁与原料

焦炉炉门砖应用现状与发展趋势 ······ 武 吉，庞克亮，刘冬杰，王 超，朱庆庙，蔡秋野 1-1
The Application State and Development Trend of Coke Oven Door Brick
······ Wu Ji, Pang Keliang, Liu Dongjie, Wang Chao, Zhu Qingmiao, Cai Qiuye

澳洲FMG粉矿烧结性能研究 ······ 王纪元，邱兰欣，阎丽娟 1-1
Search on Sintering Properties of Australian FMG Powder Ore ······ Wang Jiyuan, Di Lanxin, Yan Lijuan

无精粉的哈扬粉、国王混合粉的烧结性能比较研究 ······ 邱兰欣 1-2
Technical Analysis on the Addition of HIY Fines and FMG Blend Fines at Sintering Process ······ Di Lanxin

宝钢1号高炉稳定炉况生产实践 ······ 王 波，陈永明，宋文刚，王士彬 1-3
Production Practice of Stable Furnace Conditions for Baosteel No.1 Blast Furnace
······ Wang Bo, Chen Yongming, Song Wengang, Wang Shibin

脱硫废液中硫代硫酸铵的再利用研究 ······ 贾楠楠，芦 参，张 展，张大奎，姚 君，李冈亿 1-3
Reuse of Ammonium Thiosulfate in Desulfurization Waste Liquid
······ Jia Nannan, Lu Shen, Zhang Zhan, Zhang Dakui, Yao Jun, Li Gangyi

利用酸洗废水制备絮凝剂的工艺研究 ······ 张 展，张大奎，姚 君，贾楠楠，张馨予，王晓楠 1-4

Research on Preparation of Polyferric Chloride by Pickling Wastewater
　　　　　　　　　　　Zhang Zhan, Zhang Dakui, Yao Jun, Jia Nannan, Zhang Xinyu, Wang Xiaonan

梅钢四号烧结机余热锅炉改造实践……………………………………黄　辉，周康军，邱海雨　1-5
Retrofit Practice of Waste Heat Boiler of No. Fourth Sintering Machine in Mei Steel
　　　　　　　　　　　　　　　　　　　　　　　　Huang Hui, Zhou Kangjun, Qiu Haiyu

提高炼焦加热速度对焦炭性能的影响
　　　　　　　　　　　王　超，庞克亮，杜庆海，朱庆庙，武　吉，蔡秋野，刘福军　1-5
Influence of Increasing Coking Heating Rate on Coke Performance
　　　　　　　　　　Wang Chao, Pang Keliang, Du Qinghai, Zhu Qingmiao, Wu Ji, Cai Qiuye, Liu Fujun

焦炉放散点火系统的改进与应用………………朱庆庙，庞克亮，甘秀石，王　超，武　吉　1-6
Application and Improvement of Ignition Bleeding System on Coke Oven
　　　　　　　　　　　　　　　　Zhu Qingmiao, Pang Keliang, Gan Xiushi, Wang Chao, Wu Ji

4000m³ 高炉炉缸炉底耐材性能的试验分析……………………刘英才，张晓萍，李小静　1-7
Experimental Analysis of Hearth Bottom Refractory Performance in 4000m³ BF
　　　　　　　　　　　　　　　　　　　　　　　Liu Yingcai, Zhang Xiaoping, Li Xiaojing

高炉返矿喷吹工艺探讨……………………………………………………汤楚雄，闫朝付　1-7
Discussion on Injection of Under-screen Ore into Blast Furnaces　Tang Chuxiong, Yan Chaofu

烧结喷加气体燃料均匀性的可视化技术研究………………………武　轶，李小静，张晓萍　1-8
Research on Visualization Technology of Gas Fuel Injection Uniformity on Sintering Process
　　　　　　　　　　　　　　　　　　　　　　　　　Wu Yi, Li Xiaojing, Zhang Xiaoping

硫铵单元振动流化干燥床内件的改造……………………………………肖红春，赵　飞　1-9
Modification of Inner Parts in Vibration Fluidized Dryer of Ammonium Sulfate Unit
　　　　　　　　　　　　　　　　　　　　　　　　　　　　　　Xiao Hongchun, Zhao Fei

7m 焦炉边火道热工管理优化的实践………赵　华，周　鹏，马银华，党　平，代　成，彭　磊　1-9
剩余氨水蒸氨工艺设备系统优化的生产实践………赵　华，赵恒波，周　鹏，武　斌，何亦光，代　成　1-10
三种熄焦车防碰撞技术在鞍钢鲅鱼圈的应用………赵　华，马银华，于庆泉，魏世明，何亦光，李　志　1-10
高炉槽下称量系统误差补偿控制算法的优化…………………………………………陈　明　1-10
Optimization of Error Compensation Control Algorithms for Weighing System under Blast Furnace Groove
　　　　　　　　　　　　　　　　　　　　　　　　　　　　　　　　　　　　Chen Ming

大型高炉炉缸长寿数值解析与操作对策………………………………王训富，毛晓明　1-11
Numerical Analysis and Operational Countermeasures for Hearth Long Life Campaign of Large Blast Furnace
　　　　　　　　　　　　　　　　　　　　　　　　　　　　　　Wang Xunfu, Mao Xiaoming

鞍钢利用镁基粘结剂生产镁质球团的工业试验
　　　　　　　　　　　孙　彬，杨熙鹏，李建军，高宏庄，李林春，曾　宇，李晓春　1-12
Industrial Test of Magnesium Pellets Produced by Binder base on Magnesium at Anshan Iron and Steel
　　Company　　Sun Bin, Yang Xipeng, Li Jianjun, Gao Hongzhuang, Li Linchun, Zeng Yu, Li Xiaochun

140t/h 干熄焦一次除尘入口膨胀节在线改为砌砖结构技术
　　　　　　　　　　　　　　　　　张　雷，王君敏，洪耀明，陈艳伟，王　健　1-12
On-line Replacement of Expansion Joint of Primary Dust Removal Inlet for 140t/h CDQ to Brick-laying
　　Structure Technology　　Zhang Lei, Wang Junmin, Hong Yaoming, Chen Yanwei, Wang Jian

提高烧结矿海砂配入量的实验研究………………………………马　杰，程功金，薛向欣　1-13
Study on Improving the Ratio of Sea Sand in Sinter　　Ma Jie, Cheng Gongjin, Xue Xiangxin

鞍钢 5 号高炉稳定顺行和提产攻关实践………赵东明，张延辉，刘振宇，黄　勇，朱建伟，姜　喆　1-14

Practice of Keeping Stable and Smooth Running and Raising Productivity on No. 5 Blast Furnace at Ansteel
　　　　　　　　　　Zhao Dongming, Zhang Yanhui, Liu Zhenyu, Huang Yong, Zhu Jianwei, Jiang Zhe

高配比海砂矿制备氧化球团的试验研究 ………………………… 邢振兴，程功金，杨　合，薛向欣　1-14
Experimental Research on Preparation of Oxidized Pellets with High Proportion Sea Sand Mine
　　　　　　　　　　　　　　　Xing Zhenxing, Cheng Gongjin, Yang He, Xue Xiangxin

球团直径对碳热还原过程传热传质的影响 ……………… 刘程宏，丁学勇，代婧鑫，唐昭辉，董　越　1-15
Effect of Pellet Diameter on Heat and Mass Transfer in Carbothermal Reduction Process
　　　　　　　　　　　Liu Chenghong, Ding Xueyong, Dai Jingxin, Tang Zhaohui, Dong Yue

高炉冶炼专家系统的研究现状与发展趋势
…………………………………… 车玉满，郭天永，孙　鹏，姜　喆，姚　硕，费　静，刘炳南　1-16
Current Status and Development on Expert System of Blast Furnace
　　　　　　　　　Che Yuman, Guo Tianyong, Sun Peng, Jiang Zhe, Yao Shuo, Fei Jing, Liu Bingnan

高比例配用镜铁矿烧结技术研究 ………………………………… 史先菊，李光强，李　军，刘代飞　1-16
Study on Sintering Technology of Specularite with High Ratio　　Shi Xianju, Li Guangqiang, Li Jun, Liu Daifei

生产大型高炉用焦炭配煤炼焦工艺选择 ………………………………………… 胡德生，孙维周　1-17
Selection of Coal Blending and Cokemaking Technology for Large Blast Furnace Used Coke
　　　　　　　　　　　　　　　　　　　　　　　　　　　　　　　Hu Desheng, Sun Weizhou

湛江钢铁 1BF 稳定热负荷生产实践 …………………………………………………… 王志宇，张永新　1-18

炉顶可视化技术在朝阳钢铁 2600m³ 高炉的应用
…………………………………… 王光伟，胡德顺，刘喜亮，金　遥，关　爽，马秀生　1-18
Application of Top Visualization Technology in Chaoyang Iron and Steel 2600m³ Blast Furnace
　　　　　　　　　Wang Guangwei, Hu Deshun, Liu Xiliang, Jin Yao, Guan Shuang, Ma Xiusheng

朝阳钢铁 1#高炉使用修复风口实践 ………………… 吕宝栋，李泽安，王振东，李　男，刘喜亮，王渐灵　1-19
Practice of Repairing Tuyere in Chaoyang Iron and Steel No.1 BF
　　　　　　　　　　Lv Baodong, Li Zean, Wang Zhendong, Li Nan, Liu Xiliang, Wang Jianling

朝阳钢铁铁前系统科学管理高效生产实践 ………………………………… 胡德顺，王光伟，刘喜亮　1-19
High Efficiency Production Practice of Chaoyang Iron and Steel Iron Front System Scientific Management
　　　　　　　　　　　　　　　　　　　　　　　　　Hu Deshun, Wang Guangwei, Liu Xiliang

朝阳钢铁高炉炉缸管理实践 ………………………………… 张洪宇，王光伟，胡德顺，刘喜亮　1-20
Practice of the Management of the Hearth of Chaoyang Steel
　　　　　　　　　　　　　　　Zhang Hongyu, Wang Guangwei, Hu Deshun, Liu Xiliang

冶金石灰硅钙硫含量的测定-波长色散 X 射线荧光光谱法
………………………………………… 胡　涛，崔　隽，沈　克，曾　莹，关　晖　1-21
Metallurgical Lime Determination of Silicon Calcium and Sulfur Content Wavelength Dispersive X-ray
　Fluorescence Spectrometry ……………………… Hu Tao, Cui Juan, Shen Ke, Zeng Ying, Guan Hui

6m 焦炉配用高硫煤的生产实践 ……………………………………………………………… 夏雷雷　1-21
2500 高炉循环水系统冬季送水实践 …………………………………………………………… 付　涛　1-22
A 高炉更换冷却壁操作实践 ………………………………………………………… 李刚义，黄银春　1-22
浅谈 V 形锥流量计在测量八钢焦化煤气介质的应用 ………………………………… 苗　钧，肖先锋　1-22
八钢 2500m³ 高炉顶压控制及优化方案 …………………………………………………………… 李　杰　1-23
焦化除尘灰回配炼焦试验初步研究 …………………………………………………… 夏雷雷，王雪超　1-23
八钢 2500 高炉经济指标改善途径思考 ……………………………………………………………… 王宗乐　1-23
八钢 A 高炉长周期检修开炉达产实践 ……………………………………………… 黄银春，马　锐　1-24

八钢干熄焦生产与优化操作实践 ··· 刘智江 1-24
Practice of Production and Optimization of Dry Coke Quenching in Bayi Iron & Steel Co. ······ Liu Zhijiang

八钢 C 高炉钒钛冶炼理论与实践 ··· 刘永想，高小雷 1-24
The Theory and Practice of the V-Ti Smelting in the C-Blast Furnace of the Eighth Steel
··· Liu Yongxiang, Gao Xiaolei

八钢高炉高锌负荷操作 ··· 张文庆，高小雷 1-25
八钢焦化 VOC 气体治理新技术的浅析 ··· 杜生强 1-25
Analysis on the New Technology of Coking VOC Gas Treatment in Bayi Iron and Steel Co. ······ Du Shengqiang

八钢欧冶炉配加高锌烧结矿生产实践 ··· 李维浩 1-26
八钢焦化干熄焦红焦装入系统技术改进 ··· 刘明钊 1-26
非高炉炼铁技术的发展现状 ··· 田宝山 1-27
Development Status of Non-blast Furnace Ironmaking Technology ····················· Tian Baoshan

八钢 B 高炉炉缸侧壁温度升高的原因分析及治理实践 ······················· 舒 艺，刘晓勇 1-27
斜脱硫熔硫釜清液冷却器的设计与应用 ··· 戴 杰 1-28
Design and Application of Clean Liquid Cooler for Inclined Desulfurization Fusion Tank ······ Dai Jie

基于"一带一路"沿线烧结烟气脱硫脱硝对策探讨 ······················· 冶 飞，李维浩 1-28
Discussed for Countermeasures of Desulfurization and Denitration in Sintering Plant Based on One Belt and
One Road Along the Steel Project Construction ································ Ye Fei, Li Weihao

炉前开口机旋转拖链优化设计 ··· 李海云 1-29
欧冶炉绿色炼铁生产实践 ··· 陈若平 1-29
Production Practice of Green Ironmaking in the OY Smelter ························· Chen Ruoping

欧冶炉开炉降硅降焦比冶炼实践对比分析 ································· 田宝山，陈若平 1-30
欧冶炉冷填洗涤器液位计设计及操作实践 ··· 季书民 1-30
Design and Operation Practice of Liquid Level Meter for Cold Fill Scrubber in Europe Smelter
  in the Bayisteel OY ·· Ji Shumin

浅析焦炉集气管压力的几种调节手段 ··· 占国保 1-31
Analysis of Coke Oven Gas Pressure of Adjusting Means ·························· Zhan Guobao

烧结机尾余热锅炉系统改造现状及趋势研究 ··· 马 猛 1-31
烧结系统提高工业废水用量降低工序成本实践 ······················· 谭哲湘，臧疆文，彭 丹 1-31
Practice of Increasing Industrial Wastewater Consumption and Reducing Process Cost in Sintering System
··· Tan Zhexiang, Zang Jiangwen, Peng Dan

一种新型振动式清扫器在生产实践中的运用 ······························· 荆 斌，张忠德 1-32
鞍钢炼铁总厂 11 号高炉 3#热风炉长寿攻关 ························· 张荣军，徐吉林，王 志 1-32
Long Life Tackling of No.3 Hot Blast Furnace of No.11 Blast Furnace in Angang Iron and Steel General
  Plant ··· Zhang Rongjun, Xu Jilin, Wang Zhi

高炉控料线停炉操作实践 ················· 王彩永，任艳军，何少松，冯凯民，袁兆锋，金亚丽 1-33
Operation Practice of Cutting off the Charging Line of Blast Furnace
··· Wang Caiyong, Ren Yanjun, He Shaosong, Feng Kaimin, Yuan Zhaofeng, Jin Yali

湘钢降低球团矿膨润土用量研究 ··················· 汤 勇，刘敏媛，周云花，侯盼盼，严永红 1-34
Experimental Study on Minimizing Bentonite Mixture Proportion of Pelletizing in Xiangtan
  Iron & Steel Group ·············· Tang Yong, Liu Minyuan, Zhou Yunhua, Hou Panpan, Yan Yonghong

某非洲粉矿烧结性能研究 ··································· 汤 勇，刘敏媛，周云花 1-34
Study on Sintering Properties of Afican Iron Ore Fines ········· Tang Yong, Liu Minyuan, Zhou Yunhua

| 大型高炉中心漏斗和边缘平台实验研究 ········ 赵华涛，杜 屏，卢 瑜，张明星，朱 华 | 1-35 |

Experimental Research of Burden Layer Distribution in Large Blast Furnace

　　　　　　　　　　　　　　　Zhao Huatao, Du Ping, Lu Yu, Zhang Mingxing, Zhu Hua

| 输送带硫化工艺的优化及创新 ·· 陈小波 | 1-36 |
| 水平管中高密相煤粉气力输送的模拟研究 ······ 代婧鑫，丁学勇，徐海法，刘程宏，唐昭辉，董 越 | 1-36 |

Numerical Simulation on Pneumatic Conveying of Dense Phase Pulverized Coal in Horizontal Pipe

　　　　　　　　　　Dai Jingxin, Ding Xueyong, Xu Haifa, Liu Chenghong, Tang Zhaohui, Dong Yue

| 宝钢炼焦环保技术现状及探讨 ·· 余国普，张福行 | 1-37 |

The Present State of Coking Environmental Protection Techniques and Discussion on Baosteel

　　　　　　　　　　　　　　　　　　　　　　　　　　　　　Yu Guopu, Zhang Fuhang

| 烧结机机尾除尘超低排放关键技术及工程应用 ································· 臧疆文 | 1-38 |

Key Technology and Engineering Application of Ultra-low Emission of Sintering Machine Tail Dust Removal

　　　　　　　　　　　　　　　　　　　　　　　　　　　　　　　　Zang Jiangwen

| 烧结机柔性传动持续周期地摆动原因分析 ······································· 马 云 | 1-38 |
| 某尾矿库膏体尾矿固结性研究 ······················ 毛新福，刘 威，吴 迪，李 新 | 1-39 |

Study on Consolidation of Paste Tailings in One Tailings Pond　　Mao Xinfu, Liu Wei, Wu Di, Li Xin

| 炼铁创新引领和高质量发展 ·································· 刘文权，宋文刚，吴记全 | 1-39 |

Ironmaking Innovation Leads to High Quality Development　　Liu Wenquan, Song Wengang, Wu Jiquan

| 欧冶炉全氧风口喷吹顶煤气的工业实践 ··· 田 果 | 1-40 |

Industrial Practice of Injecting Top Gas into Total Oxygen Tuyere of European Smelting Furnace

　　　　　　　　　　　　　　　　　　　　　　　　　　　　　　　　　　Tian Guo

电感耦合等离子体发射光谱法测定硅铁中的硼 ······························· 王聪磊	1-41
鞍钢集团朝阳钢铁增建 4#热风炉工程设计实践	
·················· 徐吉林，杜春松，王光伟，李勇勤，敖永和，张式宝	1-41

Engineering Design Practice of Additional (No.4) Hot Blast Stove Project in Chaoyang Iron and Steel Company

of Ansteel ········· Xu Jilin, Du Chunsong, Wang Guangwei, Li Yongqin, Ao Yonghe, Zhang Shibao

| 硅铁还原回收低品位钒钛磁铁矿有价元素研究 ···················· 高子先，程功金，薛向欣 | 1-42 |

Ferrosilicon as Reduction Reagent on Recoveries of Valuable Elements of Low-grade Vanadia-titania

　　Magnetite　　　　　　　　　　　　　　　　Gao Zixian, Cheng Gongjin, Xue Xiangxin

| 粘结剂对铁焦强度的影响研究 ················ 徐润生，邓书良，陈绍鹏，王 炜，黄晓明 | 1-43 |

Effect of Binder on the Drum Strength of Ferro-coke

　　　　　　　　　　　　　　　Xu Runsheng, Deng Shuliang, Chen Shaopeng, Wang Wei, Huang Xiaoming

| 锌对焦炭冶金性能的影响研究 ···················· 王 炜，陈柏文，徐润生，昝日安 | 1-43 |

Study on the Effect of Zinc on the Metallurgical Properties of the Coke

　　　　　　　　　　　　　　　　　　　　Wang Wei, Chen Bowen, Xu Runsheng, Zan Rian

| 烧结矿还原对料层内焦炭消耗程度的影响 ········ 王 炜，王浩翔，徐润生，昝日安，杨代伟 | 1-44 |

Effect of Sinter Reduction on Coke Consumption in Material Layer

　　　　　　　　　　　　　　　　Wang Wei, Wang Haoxiang, Xu Runsheng, Zan Rian, Yang Daiwei

| 煤胶质层指数测定影响因素分析 ························ 乌木提江，吴 琨，张启发 | 1-45 |

Analysis of Factors Affecting the Measurement of Coal Colloid Layer Index

　　　　　　　　　　　　　　　　　　　　　　　　　Wumu Tijiang, Wu Kun, Zhang Qifa

| 添加废塑料对含碳压块热成型和焙烧还原的影响 | |
| ················· 孟庆民，龙红明，春铁军，魏汝飞，王 平，李家新 | 1-45 |

Waste Plastic Addition on the Thermoforming and Roasting Reduction of Carbon-bearing Iron Ore Briquette
······ Meng Qingmin, Long Hongming, Chun Tiejun, Wei Rufei, Wang Ping, Li Jiaxin

高炉铜冷却壁长寿技术分析 ············ 张 勇，龚卫民，贾国利，赵满祥，焦克新，余雪峰 1-46

汉钢 2#高炉炉缸侧壁温度升高护炉实践 ··· 毛洁成，王纪民 1-46

欧冶炉氧气风口喷吹煤气的结构变化 ·· 田 果 1-47
Industrial Practice of Injecting Top Gas into Total Oxygen Tuyere of European Smelting Furnace
······ Tian Guo

巴西南部高硅粉矿在湛江钢铁的试验研究和工业使用实践 ······················ 章苇玲，牛长胜 1-47
Experimental Research and Industrial Use Practice of Southern Brazil's High Silica Iron Ore
······ Zhang Weiling, Niu Changsheng

风口流场与温度场分布的数值模拟研究 ············ 许海法，孙成烽，廖哲晗，寇明银，徐 健 1-48
Numerical Simulation of Flow and Temperature Distributions inside the Furnace Tuyere
······ Xu Haifa, Sun Chengfeng, Liao Zhehan, Kou Mingyin, Xu Jian

高炉风口焦炭质量特性研究 ·· 张文成，张小勇，郑明东 1-49
Study on Quality Characteristics of Coke in Blast Furnace Tuyere
······ Zhang Wencheng, Zhang Xiaoyong, Zheng Mingdong

在线激光料面探测系统在鞍钢高炉的应用
······ 高征铠，戴建华，赵东铭，张洪宇，李建军，胡德顺，高 泰 1-50
Application of On-line Laser Burden Surface Detection System on Blast Furnaces of Ansteel Group
······ Gao Zhengkai, Dai Jianhua, Zhao Dongming, Zhang Hongyu, Li Jianjun, Hu Deshun, Gao Tai

铁精矿粉配比对球团制备和冶金性能的影响 ········ 杨子超，储满生，柳政根，唐 珏，高立华，葛 杨 1-51
Effect of Iron Concentrate Mixing Ratio on Pellet Preparation and Metallurgical Properties
······ Yang Zichao, Chu Mansheng, Liu Zhenggen, Tang Jue, Gao Lihua, Ge Yang

武钢一号高炉提高煤气利用率生产实践 ··· 龚 臣，唐少波 1-51

针对近年国内高炉大型化的思考 ·· 覃开伟，王中华，王 涛 1-52
Consideration on Recent Trend of Large-Scale Blast Furnace ····· Qin Kaiwei, Wang Zhonghua, Wang Tao

钛、钒、铬对氧化球团三维显微矿相结构的影响 ··············· 唐 珏，储满生，欧 聪，柳政根 1-52
Effect of Titanium, Vanadium and Chromium on Three-dimensional Microscopic Phase Structure of
    Oxidized Pellet ·· Tang Jue, Chu Mansheng, Ou Cong, Liu Zhenggen

宣钢 1#高炉炉役后期指标提升实践 ············ 路 鹏，裴生谦，王洪余，吕志敏，褚润林，王 斌 1-53
The Practice of Improving the Indices at Post Campaign in Xuansteel's No. 1 BF
······ Lu Peng, Pei Shengqian, Wang Hongyu, Lv Zhimin, Chu Runlin, Wang Bin

烧结烟气 CO、NO_x 及二噁英的协同处理技术 ··· 申明强 1-54
Co-treatment of CO, NO_x and Dioxins in Sintering Flue Gas ····························· Shen Mingqiang

7.63m 焦炉湿熄焦焦炭水分的控制措施 ··· 赵松靖，张爱平 1-54

通过荷载效应分析高炉炉壳上涨及解决实践 ································ 卢维强，张俊杰，段国建 1-55

新常态下的焦炭质量预测模型的研究与应用 ··· 宋林娜，张 磊 1-55
Research and Application of Coke Quality Prediction Model in the New Normal State
······ Song Linna, Zhang Lei

炉料性能及锌富集对高炉波动影响的分析 ············ 肖志新，王齐武，陈令坤，郑华伟，刘栋梁 1-56
Analysis for the Effect of Charge Performance and Zinc Enrichment on Blast Furnace
······ Xiao Zhixin, Wang Qiwu, Chen Lingkun, Zheng Huawei, Liu Dongliang

鞍钢超厚料层双层预烧结新工艺研究与工业试验
······ 周明顺，王义栋，赵东明，朱建伟，姜 涛，钟 强 1-56

Research and Industrial-scale Pilot on Double-layer Pre-sintering New Process for Ultra-thick Bed Layer
·················· Zhou Mingshun, Wang Yidong, Zhao Dongming, Zhu Jianwei, Jiang Tao, Zhong Qiang

八钢烧结配加除尘灰工艺影响及分析 ·················· 王雪超，秦　斌 1-58

斗轮减速机润滑装置的改进 ·················· 王明江 1-58

斗轮机斗轮轴承防尘密封装置的改进 ·················· 王明江，邬显敏 1-58

富氧干馏技术制备富氢还原气体试验研究 ·················· 张小蕊，邹　冲，赵俊学，马　成，李　宝，吴　浩 1-58
Experimental Study on Preparation of Hydrogen-rich Reducing Gas by Oxygen-enriched Dry Distillation
·················· Zhang Xiaorui, Zou Chong, Zhao Junxue, Ma Cheng, Li Bao, Wu Hao

大型钢铁企业焦炉煤气净化技术应用前景探讨 ·················· 吴江伟 1-59
Discussion on the Application Prospect of COG Purification System in Iron and Steel Complex
·················· Wu Jiangwei

铝酸盐型高炉渣的物理化学性质 ·················· 庞正德，严志明，吕学伟，蒋宇阳，凌家伟 1-60
Physicochemical Properties of Alumina Based Slag for Blast Furnace
·················· Pang Zhengde, Yan Zhiming, Lv Xuewei, Jiang Yuyang, Ling Jiawei

安钢1#烧结机热风烧结技术的研制与应用 ·················· 王军锋，周东锋，李发展 1-61
Development and Application of Hot Air Sintering Technology for 1# Sintering Machine at Angang
·················· Wang Junfeng, Zhou Dongfeng, Li Fazhan

碱金属碳酸盐对焦炭气化反应的催化动力学研究
·················· 程可扬，方云鹏，吴　亮，祝宇涵，朱荣锦，胡义卿，张生富 1-61
Kinetics of Coke Gasification Reaction under the Catalysis of Alkali Carbonates
·················· Cheng Keyang, Fang Yunpeng, Wu Liang, Zhu Yuhan, Zhu Rongjin, Hu Yiqing, Zhang Shengfu

电磁作用对白云鄂博铌精矿渣金分离的影响 ·················· 文　明，赵增武，李保卫 1-62
Effect of Electromagnetic Action on Separation of Metal from Slag of Bayan Obo Niobium Concentrate
·················· Wen Ming, Zhao Zengwu, Li Baowei

高炉炉缸结构与炉衬选型 ·················· 卢正东，陈令坤，张正东，向武国，顾华志，李承志 1-63
Hearth Structure and Lining Selection of Blast Furnace
·················· Lu Zhengdong, Chen Lingkun, Zhang Zhengdong, Xiang Wuguo, Gu Huazhi, Li Chengzhi

焦粉粒度对烧结减排的影响 ·················· 刘敏媛，严永红，汤　勇，刘守文 1-64

新型碳铁复合炉料冷压制备工艺的实验研究 ·················· 鲍继伟，储满生，韩　冬，曹来更，柳政根，唐　珏 1-64
Experimental Study on the Cold Pressing Preparation Process of New Ironmaking Burden-iron Carbon Agglomerates ·················· Bao Jiwei, Chu Mansheng, Han Dong, Cao Laigeng, Liu Zhenggen, Tang Jue

新疆低质煤热解过程中煤的拉曼结构特征研究 ·················· 姚利，朱子宗，武强，孙灿，宗楷 1-65
Study on the Evolution Characteristics of Coal's Roman Structure in Pyrolysis Process of Low-quality Coal in Xinjiang ·················· Yao Li, Zhu Zizong, Wu Qiang, Sun Can, Zong Kai

高炉煤气高效、高值和创新利用技术 ·················· 刘文权，吴记全 1-66
High Efficiency, High Value and Innovative Utilization Technology of Blast Furnace Gas
·················· Liu Wenquan, Wu Jiquan

根据实用性分类配煤提高M40指标的方法 ·················· 周俊兰，张明星，杜　屏 1-66
Method to Promote M40 Index According to Coal Blending Based on Applied Classification
·················· Zhou Junlan, Zhang Mingxing, Du Ping

韶钢7号高炉护炉操作实践 ·················· 陈生利，匡洪锋，陈国忠 1-67
Operation Practice of Furnace Maintenance of BF #7 at SGIS
·················· Chen Shengli, Kuang Hongfeng, Chen Guozhong

韶钢2200m³高炉无计划长期停炉恢复炉况实践 ·················· 齐万兵，匡洪锋，曹旭博 1-68

Practice of Recovering Furnace Conditions after Long Term Unscheduled Shutdown for a 2200m³ BF at SGIS
　　　　　　　　　　　　　　　　　　　　　　　　Qi Wanbing, Kuang Hongfeng, Cao Xubo

韶钢六号高炉高生矿比生产实践 …………………………………… 李国权，王　振，柏德春　1-68
Production Practice of High Rawore Ratio for BF #6 at SGIS　　Li Guoquan, Wang Zhen, Bai Dechun

巨型高炉炉缸死料柱焦炭研究 ………… 牛　群，陈艳波，郑朋超，程树森，徐文轩，于兆斌，马元明　1-69
Study on Deadman Coke of a Huge Industrial Blast Furnace
　　……… Niu Qun, Chen Yanbo, Zheng Pengchao, Cheng Shusen, Xu Wenxuan, Yu Zhaobin, Ma Yuanming

热风炉焊接残余应力周向分布与消除
　　…………………………… 陈　辉，孙　健，王　伟，梁海龙，熊　军，王长水，王　建，要志超　1-70
Circumferential Distribution and Elimination of Welding Residual Stress for Hot Blast Stove
　　……… Chen Hui, Sun Jian, Wang Wei, Liang Hailong, Xiong Jun, Wang Changshui, Wang Jian, Yao Zhichao

并罐高炉无钟炉顶设备结构对布料偏析的影响
　　………………………………………… 徐文轩，高永会，谢建军，程树森，牛　群，梅亚光　1-71
Influence of Bell-less Top Equipment Structure on the Segregation of Burden Distribution in Blast Furnace
　　with Parallel Hopper　……　Xu Wenxuan, Gao Yonghui, Xie Jianjun, Cheng Shusen, Niu Qun, Mei Yaguang

真空碳酸钾法脱硫工艺优化和改进 ………………………………………… 杨志军，杨　涛　1-72
Optimization and Improvement of Vacuum Potassium Carbonate Desulfurization Process
　　　　　　　　　　　　　　　　　　　　　　　　　　　　　　　　Yang Zhijun, Yang Tao

装炉煤细度对焦化生产的影响研究 ………………… 王永亮，郭　飞，何　波，陈　康，嵇长发　1-72
Study on the Influence of Fineness of Charging Coal on Coking Production
　　　　　　　　　　　　　　　　　　　　　Wang Yongliang, Guo Fei, He Bo, Chen Kang, Ji Changfa

宝钢湛江钢铁高炉长寿技术设计与应用 ………………… 周　琦，贾海宁，苏　威，沙华玮　1-73
Design & Practice of Long Service-life of Blast Furnaces in Baosteel Zhanjiang Iron & Steel Works
　　　　　　　　　　　　　　　　　　　　　　　　　Zhou Qi, Jia Haining, Su Wei, Sha Huawei

湛钢烧结节能环保生产实践 ……………………………………………………… 李　明　1-74
Pratice of Energysaving and Environmental of Zhansteel Sintering Process ………………… Li Ming

韶钢烧结适宜焦粉粒度组成的技术研究 ……………………………………… 毛爱香，蓝伯洋　1-74
Study on Sintering Technology of Suitable Particle Size Composition of Coke Powder at SGIS
　　　　　　　　　　　　　　　　　　　　　　　　　　　　　　　Mao Aixiang, Lan Boyang

BLOOM LAKE 精粉烧结试验研究和生产实践 ………… 欧阳希，毛爱香，余　骏，黄承芳　1-75
Research and Production Practice of BLOOM LAKE Fine Powder Sintering Trial
　　　　　　　　　　　　　　　　　　　　　　　　Ouyang Xi, Mao Aixiang, Yu Jun, Huang Chengfang

要努力提高高炉炼铁喷煤比 ………………………………………………………… 王维兴　1-75
Strive to Increase the Coal Injection Ratio of Blast Furnace …………………………… Wang Weixing

降低高炉燃料比的作用 ……………………………………………………………… 王维兴　1-76
Effect on Reducing the Fuel Ratio of Blast Furnace ……………………………………… Wang Weixing

基于烧结基础特性的烧结优化配矿模型 ………… 潘禹竹，柳政根，储满生，高立华，唐　珏　1-77
Ore-matching Optimization Model based on the Basic Characteristics of Sintering
　　　　　　　　　　　　　　　　　　　　Pan Yuzhu, Liu Zhenggen, Chu Mansheng, Gao Lihua, Tang Jue

超高品位铁精矿气基直接还原的可行性研究 ………… 黄欧瑾，应自伟，吴宇航，林兴良，李靖宇　1-77
Discussion on Feasibility of Production of Direct Reduction Iron by Usingultra-high Grade Iron Concentrate
　　　　　　　　　　　　　　　　　Huang Oujin, Ying Ziwei, Wu Yuhang, Lin Xingliang, Li Jingyu

高炉冶炼高钒高铬型钒钛磁铁矿资源综合利用关键技术 ……… 薛向欣，程功金，杨　合，段培宁　1-78

Key Technology of Resource Comprehensive Utilization for High-Chromium and High-Vanadium Vanadium-Titanium Magnetite Smelting in the Blast Furnace······Xue Xiangxin, Cheng Gongjin, Yang He, Duan Peining

辽西高钒高钛型磁铁矿资源综合利用关键技术
··········程功金，薛向欣，高子先，杨 合，邢振兴，马 杰，刘学志　1-79

Key Technology of Resource Comprehensive Utilization for Liaoxi High-Vanadium and High-Titanium Magnetite······Cheng Gongjin, Xue Xiangxin, Gao Zixian, Yang He, Xing Zhenxing, Ma Jie, Liu Xuezhi

SCOPE21 工艺现状及应用推广存在的问题··········黄世平，Alex Wong，刘守显，牛 虎　1-80

Current Status of SCOPE21 Process and its Problems in Application
··········Huang Shiping, Alex Wong, Liu Shouxian, Niu Hu

无气隙热阻球墨铸铁冷却壁的研制··········宗燕兵，张学栋，周传禄，辛虹霓，万奇林　1-80

Development of Nodular Cast Iron Cooling Stave without Air Gap Thermal Resistance
··········Zong Yanbing, Zhang Xuedong, Zhou Chuanlu, Xin Hongni, Wan Qilin

鞍钢 4#高炉热风围管吊挂设计总结··········赵晓峰，张 帅，张维巍　1-81

Summary of Hanging Design of Hot Blast Pipe in No. 4 BF at Ansteel
··········Zhao Xiaofeng, Zhang Shuai, Zhang Weiwei

2　炼钢与连铸

2.1　炼　钢

迁钢 210t 转炉经济炉龄探索··········孙 亮，刘风刚，刘珍童，毕泽阳，赵艳宇　2-1

Exploration on Economic Age of 210t Converter in Qian Steel
··········Sun Liang, Liu Fenggang, Liu Zhentong, Bi Zeyang, Zhao Yanyu

RH 无铬耐火材料的应用与改进··········吕志勇，邢维义，于海岐，苏小利　2-1

炼钢生产模式比较及应用··········董金刚　2-2

Comparison and Application of Steelmaking Production Patterns··········Dong Jingang

超低碳钢中的硫化物析出行为··········张 峰　2-3

Precipitation Behavior of Sulfide Inclusions in Ultra-low Carbon Concentration Steel Sheets······Zhang Feng

迁钢 210t 转炉自动出钢技术的开发与应用··········江腾飞，朱 良，成天兵　2-4

Development and Application of Automatic Tapping Technology for 210t Converter at Qiangang
··········Jiang Tengfei, Zhu Liang, Cheng Tianbing

鞍钢提高转炉废钢比的生产实践··········李伟东，何海龙，李 冰，李 泊　2-4

Production Practice of Improving Scrap Ratio of Converter of Anshan Iron and Steel Group Coporation
··········Li Weidong, He Hailong, Li Bing, Li Bo

镁质熔剂在转炉炼钢中的应用研究··········李伟东，何海龙，孙振宇，李 冰，李 泊　2-5

Application of Magnesium Flux in Converter Steelmaking
··········Li Weidong, He Hailong, Sun Zhenyu, Li Bing, Li Bo

90t 顶吹转炉冶炼超低磷钢工艺开发
··········尚世震，臧绍双，李 泊，陶功捷，王 爽，高立超，潘瑞宝　2-5

Process Development of Smelting Molten Ultra-low Phosphor Steel by 90t Oxygen Top-Blown Converter
··········Shang Shizhen, Zang Shaoshuang, Li Bo, Tao Gongjie, Wang Shuang, Gao Lichao, Pan Ruibao

鞍钢 KR 法脱硫浸入深度研究与应用··········宋吉锁，何海龙，刘鹏飞，孙 群，李伟东，乔冠男　2-6

Development of Model with Deep Immersion for Desulphurization by KR Method in Ansteel and Application of the Model··········Song Jisuo, He Hailong, Liu Pengfei, Sun Qun, Li Weidong, Qiao Guannan

260t 转炉轻烧镁球冶炼脱磷生产实践 ……………………… 李玉德，李叶忠，朱国强，查松妍，齐志宇　2-7
Production Practice of Light Burning Magnesium Ball Smelting Dephosphorization in 260t Converter
　　　　　　　　　　　　　　　　　　　　Li Yude, Li Yezhong, Zhu Guoqiang, Zha Songyan, Qi Zhiyu

鞍钢 260t 转炉 IF 钢顶渣改质工艺研究与应用
　　　　　　　　　　　　　　　李　冰，李　泊，朱国强，齐志宇，何文英，孙振宇，高立超　2-7
Research and Application of IF Steel Ladle Top Slag Improving Process on Ansteel 260t Conventer
　　　　　　　　　　　　　　Li Bing, Li Bo, Zhu Guoqiang, Qi Zhiyu, He Wenying, Sun Zhenyu, Gao Lichao

转炉单渣留渣高效冶炼技术的研究与应用 ……… 刘忠建，王忠刚，高志滨，赵立峰，公　斌，张　丽　2-8
Research and Application of High Efficiency Smelting Technology with Single Slag Retaining in Converter
　　　　　　　　　　　　　　　　Liu Zhongjian, Wang Zhonggang, Gao Zhibin, Zhao Lifeng, Gong Bin, Zhang Li

莱钢钢包全程智能吹氩工艺开发与应用 ………………………… 薛　志，王忠刚，高志滨，丁洪周　2-9
Development and Application of Intelligent Argon Blowing Process for Ladle in Laigang
　　　　　　　　　　　　　　　　　　　　　　　Xue Zhi, Wang Zhonggang, Gao Zhibin, Ding Hongzhou

脱磷剂在转炉深脱磷中的作用研究 ………………………………………………………………… 尚　游　2-9
Study on the Role of Dephosphorizing Agent in Deep Dephosphorization of Converter ……… Shang You

KR 法铁水预脱硅工艺研究与应用 ………………………… 王　强，王忠刚，高志滨，薛　志，尚　游　2-10
Research and Application of Hot Metal Desilication with KR Technology
　　　　　　　　　　　　　　　　　　　　　Wang Qiang, Wang Zhonggang, Gao Zhibin, Xue Zhi, Shang You

转炉底吹供气元件选择与维护讨论 ………………………………………………………………… 王　东　2-11
Discussion on Selection and Maintenance of Bottom Blowing Components in Converter ……… Wang Dong

转炉自动化炼钢控制技术研究 ……………… 万雪峰，曹　东，王丽娟，马　勇，赵　亮，高学中　2-11
Research of Converter Automation Steelmaking Control Technology
　　　　　　　　　　　　　　　　　Wan Xuefeng, Cao Dong, Wang Lijuan, Ma Yong, Zhao Liang, Gao Xuezhong

高碳铬轴承钢氮化钛的控制 ……………………………………………… 李广帮，魏崇一，廖相巍　2-12
Control of Titanium Nitride Inclusion in High Carbon Chromium Bearing Steel
　　　　　　　　　　　　　　　　　　　　　　　　　　　　　Li Guangbang, Wei Chongyi, Liao Xiangwei

钢包底吹智能控制技术研究与应用 ………………………………………………………………… 李卫东　2-13
Research and Application of Intelligent Control Technology for Ladle Bottom Blowing ……… Li Weidong

重轨钢质量控制实践 …………………… 王　宁，常宏伟，金纪勇，张　锐，孙振宇，李　旭　2-14
Quality Control Practice of Heavy Rail Steel
　　　　　　　　　　　　　　　　　Wang Ning, Chang Hongwei, Jin Jiyong, Zhang Rui, Sun Zhenyu, Li Xu

基于图像分析的钢包底吹搅拌在线监视与自动控制 ………… 张才贵，郭海滨，孙玉军，徐建雄　2-14
Online Monitoring and Automatic Control of Ladle Bottom Stirring by Using Image Analysis
　　　　　　　　　　　　　　　　　　　　　　　Zhang Caigui, Guo Haibin, Sun Yujun, Xu Jianxiong

低氧含硫易切削钢的工艺优化 ………………………………… 刘　峰，马玉强，莫　丹，司焕庆　2-15
Process Optimization of Reducing Oxygen Content in Free-cutting Steel
　　　　　　　　　　　　　　　　　　　　　　　Liu Feng, Ma Yuqiang, Mo Dan, Si Huanqing

分钢种精炼窄成分渣系工艺技术开发与应用 ………………… 王　键，刘文凭，郭庆军，王玉春　2-16
Development and Application of Process Technology for Refining Narrow Composition Slag System of Steel
　　　　　　　　　　　　　　　　　　　　　　　Wang Jian, Liu Wenping, Guo Qingjun, Wang Yuchun

基于 SOM-RELM 的 LF 炉钢水温度预测
　　　　　　　　　　　　信自成，张江山，张军国，路博勋，李军明，郑　瑾，刘　青　2-16
Temperature Prediction of Molten Steel in a Ladle Furnace based on SOM-RELM Model
　　　　　　Xin Zicheng, Zhang Jiangshan, Zhang Junguo, Lu Boxun, Li Junming, Zheng Jin, Liu Qing

莱钢转炉炉体维护技术研究与实践 ……………………… 王　键，张昭平，任科社，郭庆军，杨普庆　2-17
Research and Practice on the Maintenance Technology of Furnace Body of Laigang Converter
　　　　　　　　　　　　　　　　　Wang Jian, Zhang Zhaoping, Ren Keshe, Guo Qingjun, Yang Puqing

转炉动态精准控制技术研究与应用 …………………… 郭伟达，王　键，张昭平，任科社，杨普庆　2-18
Research and Application of Converter Dynamic Precise Control Technology
　　　　　　　　　　　　　　　　　Guo Weida, Wang Jian, Zhang Zhaoping, Ren Keshe, Yang Puqing

供异型钢高裂纹敏感铸坯质量控制研究 ………………… 郭　达，刘文凭，张昭平，杨普庆，郭庆军　2-19
For Special-shaped Steel High Crack Sensitive Cast Blank Quality Control Research
　　　　　　　　　　　　　　　　　Guo Da, Liu Wenping, Zhang Zhaoping, Yang Puqing, Guo Qingjun

转炉高效低成本冶炼集成技术研究与应用 …………………………… 任科社，郭庆军，杨普庆，张昭平　2-19
Research and Application of Converter Integrated Technology with High Efficiency and Low Cost
　　　　　　　　　　　　　　　　　Ren Keshe, Guo Qingjun, Yang Puqing, Zhang Zhaoping

钛微合金化钢精炼工艺研究与应用 …………………………………………… 王　键，刘文凭，任科社　2-20
Research and Application of Titanium Microalloying Steel Refining Process
　　　　　　　　　　　　　　　　　Wang Jian, Liu Wenping, Ren Keshe

耐火材料对切割丝用钢洁净度的影响 ………………… 李　阳，杜高锋，杜沣庭，武锦涛，王　硕　2-20
Influence of Refractory on Cleanliness of Steel for Cutting Wire
　　　　　　　　　　　　　　　　　Li Yang, Du Gaofeng, Du Fengting, Wu Jintao, Wang Shuo

复吹转炉底枪布置对熔池搅拌混匀的影响 …… 孙建月，杨晓江，杜文斌，周泉林，张　全，钟良才　2-21
Influence of Bottom Tuyere Configurations on Bath Stirring and Mixing in a Combined Blown Converter
　　　　　　　　　　　　　　　　　Sun Jianyue, Yang Xiaojiang, Du Wenbin, Zhou Quanlin, Zhang Quan, Zhong Liangcai

电感耦合等离子体发射光谱法测定锰铁中 P 含量的测量不确定度评定
　　　　　　　　　　　　　　　　　………………… 李玉洁，陈高莉，崔　隽，陈晓燕，刘丽荣　2-22
Evaluation of Uncertainty for Determination of P in Manganese-iron Alloy by ICP Method
　　　　　　　　　　　　　　　　　Li Yujie, Chen Gaoli, Cui Jun, Chen Xiaoyan, Liu Lirong

烟气分析智能系统在转炉冶炼中的应用 ……………… 郭伟达，王　键，杨普庆，任科社，张昭平　2-23
Application of Smoke Analysis Intelligent System in Converter Smelting
　　　　　　　　　　　　　　　　　Guo Weida, Wang Jian, Yang Puqing, Ren Keshe, Zhang Zhaoping

气体[N]产生废次降的控制实践 ……………………………………………………… 陈代明，王永胜　2-23
Control Practice of Waste Inferior Degrade Production due to Gas[N] …… Chen Daiming, Wang Yongsheng

我国转炉氧枪及喷头的发展现状 …………………………………………………………………… 冯　超　2-24
Development Status of Converter Oxygen Lance and Nozzle in China ………………………… Feng Chao

ER70S-G 焊丝钢冶炼技术探讨 ……………………………………………… 赵晓锋，张志强，王建忠　2-24
Discussion on Welding Wire Steel Smelting Technology of ER70S-G
　　　　　　　　　　　　　　　　　Zhao Xiaofeng, Zhang Zhiqiang, Wang Jianzhong

宣钢 120t 转炉高效化生产实践 ………………………………………………………… 张利江，张明海　2-25
Practice of High Efficiency Production of 120t Converter in Xuanhua Steel
　　　　　　　　　　　　　　　　………………………………………… Zhang Lijiang, Zhang Minghai

干式除尘转炉自动炼钢高拉碳工艺研究 ……………………………… 张立君，王金龙，王宏斌，张明海　2-26
The Study on the Technology of High Carbon Steel Automatic Dry Dust Converter
　　　　　　　　　　　　　　　　　Zhang Lijun, Wang Jinlong, Wang Hongbin, Zhang Minghai

邯钢 260t 转炉顶底复吹过程数值模拟研究及应用 ……………………… 范　佳，李太全，高福彬，靖振权　2-26
Research and Application of Numerical Simulation on Top-bottom Combined Blowing Process of 260 ton
　　Converter in Han-Steel ……………………………………… Fan Jia, Li Taiquan, Gao Fubin, Jing Zhenquan

攀钢半钢冶炼转炉脱磷工艺技术研究 ………………………… 陈 均，曾建华，梁新腾，喻 林 2-27
Research on Dephosphorization Technology by Semi-steel Steelmaking Converter of Pangang
　　　　　　　　　　　　　　　　　　　　　Chen Jun, Zeng Jianhua, Liang Xinteng, Yu Lin

老厂房 150t 转炉铁水一罐到底工艺研究与应用 ………… 田云生，郭永谦，孙 拓，张远强，李志广 2-28
Research and Application of Hot Metal Get the Ladle Done at One Go for 150 t Converter in Old Factory
　　Building　　　　　　　　　Tian Yunsheng, Guo Yongqian, Sun Tuo, Zhang Yuanqiang, Li Zhiguang

一种低碳高磷系列钢造渣冶炼方法 ……………………………………………………………… 佟 迎 2-29
A Slag Smelting Method for Low Carbon and High Phosphorus Steel Series　　　　　　　　Tong Ying

韶钢 120t 铁水包加废钢提升废钢比生产实践 …………………………………………………… 佟 迎 2-29
Production Practice of Raising Scrap Ratio by Adding Scrap Steel to 120t Ladle at Shaoguan Iron and
　　Steel Co.　　　　　　　　　　　　　　　　　　　　　　　　　　　　　　　　　　Tong Ying

冷轧基料冶炼工艺优化与实践 ……………………………………… 吕圣会，李海洋，王克忠 2-30
Optimization Practice of Smelting Process of Base Material for Cold Rolling
　　　　　　　　　　　　　　　　　　　　　　　　　　Lv Shenghui, Li Haiyang, Wang Kezhong

铝基脱磷剂对高磷锰硅合金还原脱磷的影响 ……………… 孙 灿，朱子宗，宗 楷，焦万宜 2-31
Effect of Al-based Dephosphorization Agent on Reduction and Dephosphorization for High Phosphorus
　　MnSi Alloys　　　　　　　　　　　　　　　　Sun Can, Zhu Zizong, Zong Kai, Jiao Wanyi

X80 管线钢非金属夹杂物控制研究 ………………………… 李战军，初仁生，刘金刚，郝 宁 2-31
Study on Control of Non-metallic Inclusions in X80 Pipeline Steel
　　　　　　　　　　　　　　　　　　　　　　　Li Zhanjun, Chu Rensheng, Liu Jingang, Hao Ning

应用极值分析标准评估轴承钢中大尺寸夹杂物分布特征
　　　　　　　　　　　　　　　　 郭洛方，高永彬，陈殿清，赵东记，张 杰，雷 富 2-32
Size Distribution Characteristics of Large Size Inclusions in Bearing Steel by Extreme Value Analysis
　　Standard　　　　　　　Guo Luofang, Gao Yongbin, Chen Dianqing, Zhao Dongji, Zhang Jie, Lei Fu

热送热装在青岛特钢的实践与应用 ………………………… 张军卫，刘 澄，刘政鹏，钟 浩 2-33
Practice and Application of Hot Delivery and Hot Loading in Qingdao Special Steel
　　　　　　　　　　　　　　　　　　　　　Zhang Junwei, Liu Cheng, Liu Zhengpeng, Zhong Hao

钢包底吹过程物理模拟研究 ………………………… 支保宁，赵定国，张印棠，王书桓，张福君 2-34
Physical Simulation of Ladle Bottom Blowing Process
　　　　　　　　　　　　　　Zhi Baoning, Zhao Dingguo, Zhang Yintang, Wang Shuhuan, Zhang Fujun

韶钢铁水一罐制改造的探索与实践 ………………………………… 黄纯旭，陈 贝，邓长付 2-34
Exploration and Practice of One-pot System Reform of SGIS …… Huang Chunxu, Chen Bei, Deng Changfu

550A 磨球钢圆钢断裂原因分析 ………………………… 叶德新，邓湘斌，曾令宇，冯杰斌 2-35
Analysis of Fractures on Grinding Ball Steel 550A　　Ye Dexin, Deng Xiangbin, Zeng Lingyu, Feng Jiebin

20 钢连铸圆管坯的研发和生产 …………………………………… 李祥才，张 虎，柯家祥 2-36
Research and Production of 20 Steel for Continuous Casting Round Tube Blank
　　　　　　　　　　　　　　　　　　　　　　　　　　　　　Li Xiangcai, Zhang Hu, Ke Jiaxiang

承钢 150t 转炉分钢种高效脱磷方法与低成本控制
　　　　　　　　　　　　　　　　 韩德文，仇 军，胡凤伟，郭盈伟，张文彪，高建国 2-36
High Efficiency Dephosphorization Method and Low Cost Control for 150t Converter of Bearing Steel
　　　　　　　　　Han Dewen, Qiu Jun, Hu Fengwei, Guo Yingwei, Zhang Wenbiao, Gao Jianguo

转炉冶炼低硫钢的控硫方法 …………………………………………………… 赵 科，邓长付 2-37
Method of Sulphur Control for Converter Smelting of Low Sulphur Steel　　　Zhao Ke, Deng Changfu

钢中复合夹杂物/钢基体的电势差与电偶腐蚀的关系 …… 侯延辉，刘林利，李光强，李腾飞，刘 昱 2-37

The Correlation between Potential Difference and Galvanic Corrosion of Composite Inclusions/steel
　　Matrix in Steel ··· Hou Yanhui, Liu Linli, Li Guangqiang, Li Tengfei, Liu Yu
CaF$_2$ 和 Na$_2$O 对钢渣含磷相析出过程的影响 ·· 王达志，包燕平，王　敏　2-38
Effects of CaF$_2$ and Na$_2$O on the Precipitation Process of Phosphorus-containing Phase in Steel Slag
　　·· Wang Dazhi, Bao Yanping, Wang Min
煤粉中氮含量的检验 ·· 崔　隽，张巧燕，沈　克，胡　涛，李玉洁，陈高莉　2-39
Inspection of Nitrogen Content in Pulverized Coal
　　·· Cui Jun, Zhang Qiaoyan, Shen Ke, Hu Tao, Li Yujie, Chen Gaoli
硅脱氧弹簧钢 55SiCr 炼钢过程氧化物夹杂变化研究 ············ 孟耀青，赵铮铮，赵昊乾，吕海瑶，赵　垒　2-40
Study on the Change of Oxide Inclusions during Si-killed Spring Steel 55SiCr Steelmaking Process
　　·· Meng Yaoqing, Zhao Zhengzheng, Zhao Haoqian, Lv Haiyao, Zhao Lei
转炉留渣双渣工艺前期脱磷热力学及实践 ·· 孙学刚　2-40
Thermodynamics and Practice of SGRS Process during BOF ·· Sun Xuegang
八钢汽车大梁钢（B510L）钢中夹杂物探讨 ·· 李立民　2-41
Discussion on Inclusions in Eight Steel Big Beam Steel(B510L) ··· Li Limin
炉底厚度维护对转炉炉衬寿命的影响 ·· 韩东亚　2-42
Effect of Bottom Thickness Maintenance on Converter Lining Life ··· Han Dongya
钛强化低合金高强钢夹杂物的控制研究 ············ 吾　塔，李立民，吴　军，卜志胜，丁　寅，刘军威　2-42
Study on Inclusion Control of Titanium Strengthened Low Alloy High Strength Steel
　　··· Wu Ta, Li Limin, Wu Jun, Bu Zhisheng, Ding Yin, Liu Junwei

2.2　连　铸

薄板坯 9SiCr 合金工具钢高温热塑性研究
　　·· 李具中，朱万军，王春锋，齐江华，孙宜强，蔡　珍，邱　晨　2-43
High-temperature Thermoplastic Study of 9SiCr of CSP
　　····················· Li Juzhong, Zhu Wanjun, Wang Chunfeng, Qi Jianghua, Sun Yiqiang, Cai Zhen, Qiu Chen
连铸生产模式比较及应用 ·· 董金刚　2-44
Comparison and Application of Continuous Casting Production Patterns ·· Dong Jingang
260t 钢包水口尺寸对卷渣高度的水模拟研究 ··· 彭春霖，张晓光，郭庆涛，贾吉祥，柴明亮，杨　骥　2-45
Water Simulation Study on Critical Height of Vortex from the Different Nozzle Diameter of the 260t Steel
　　Ladle ················ Peng Chunlin, Zhang Xiaoguang, Guo Qingtao, Jia Jixiang, Chai Mingliang, Yang Ji
高铝双相钢连铸 T 坯质量控制及改进研究 ··· 胡署名，常文杰，亢占英，王迎春，杨启宇，吕宪雨　2-45
Study on Quality Control and Improvement of Continuous Casting T Slab of High Aluminum Dual-phase Steel
　　···························· Hu Shuming, Chang Wenjie, Kang Zhanying, Wang Yingchun, Yang Qiyu, Lv Xianyu
外加电场对浸入式水口内壁结瘤行为的影响研究
　　·· 田　晨，翟晓毅，肖国华，徐　斌，孙江波，李海斌，袁　磊，于景坤　2-46
Effect of Clogging Behavior on the Submerged Entry Nozzle by External Electric Field
　　···················· Tian Chen, Zhai Xiaoyi, Xiao Guohua, Xu Bin, Sun Jiangbo, Li Haibin, Yuan Lei, Yu Jingkun
异型坯单点浇注条件下结晶器控制模型研究与应用
　　·· 公　斌，王忠刚，卢　波，赵立峰，张　丽，刘忠建　2-47
Research and Application of Mold Control Model under Single Point Casting of Blank
　　··································· Gong Bin, Wang Zhonggang, Lu Bo, Zhao Lifeng, Zhang Li, Liu Zhongjian
异型坯 C345 工程用钢的研究与开发 ··················· 张　丽，王忠钢，赵立锋，公　斌，刘忠建　2-48

Research and Development of C345 Engineering Steel for H-beam
　　　　　　　　　　　　　　　　Zhang Li, Wang Zhonggang, Zhao Lifeng, Gong Bin, Liu Zhongjian

莱钢宽厚板线轧材折叠黑线缺陷原因分析及对策
　　　　　　　　　　　　　　　　赵立峰，王忠刚，卢　波，张　丽，公　斌，刘忠建　　2-48
Analysis and Countermeasure of Folding Black Line Defect in Wide and Heavy Plate Rolling in Laigang
　　　　　　　　　　　　　　　　Zhao Lifeng, Wang Zhonggang, Lu Bo, Zhang Li, Gong Bin, Liu Zhongjian

鞍钢电磁冶金技术应用及展望　　　　　郭庆涛，贾吉祥，唐雪峰，彭春霖，康　磊，廖相巍　　2-49
Application and Prospect of the Electromagnetic Metallurgy in Ansteel
　　　　　　　　　　　　　　　　Guo Qingtao, Jia Jixiang, Tang Xuefeng, Peng Chunlin, Kang Lei, Liao Xiangwei

静态凝固末端强冷在八钢连铸机上的应用　　　　　　　　　　　秦　军，卜志胜，陈晓山　　2-50
Effect of Static Solidification Ends Strong Cooling on Bayi Steel Slab Caster
　　　　　　　　　　　　　　　　Qin Jun, Bu Zhisheng, Chen Xiaoshan

45#钢圆钢表面纵裂纹成因分析及改进措施　　　　　　　　　　　　　　　　　　宋德山　　2-50
Cause Analysis and Improvement Measures for Longitudinal Cracks on Surface of 45# Roung Steel Bars
　　　　　　　　　　　　　　　　Song Deshan

矩形连铸坯夹杂物控制技术的研究与应用　　　王　键，王玉春，郭　达，刘文凭，谭学样　　2-51
Research and Application of Control Technology for Rectangular Continuous Casting Inclusions
　　　　　　　　　　　　　　　　Wang Jian, Wang Yuchun, Guo Da, Liu Wenping, Tan Xueyang

IF钢Al-Ca-O系细微夹杂物控制研究与实践　　李俊伟，赵　元，袁少江，杨成威，代碧波　　2-51
Control and Practice on Al-Ca-O Tiny Inclusion in Interstitial Free Steel
　　　　　　　　　　　　　　　　Li Junwei, Zhao Yuan, Yuan Shaojiang, Yang Chengwei, Dai Bibo

铌微合金化小方坯生产实践　　　　　　　　郭伟达，王　键，郭　达，任科社，谭学样　　2-52
The Production Practice of Small Square Blanks in Micro-alloying
　　　　　　　　　　　　　　　　Guo Weida, Wang Jian, Guo Da, Ren Keshe, Tan Xueyang

浸入式水口穿孔原因分析及对策　　　汪　雷，龚志翔，李　新，王　洛，王俊北，樊明宇　　2-53
Analysis on Origin of Perforation of Submerged Nozzle and Counter Measure
　　　　　　　　　　　　　　　　Wang Lei, Gong Zhixiang, Li Xin, Wang Luo, Wang Junbei, Fan Mingyu

连铸塞棒中间包冶金集成技术研究与实践　　　　　　　王　键，谭学样，郭　达，王玉春　　2-53
Continuous Casting Stopper Integrated Package Metallurgy Research and Practice
　　　　　　　　　　　　　　　　Wang Jian, Tan Xueyang, Guo Da, Wang Yuchun

高裂纹敏感性钢铸坯表面质量的研究与控制　　王　键，王玉春，郭　达，刘文凭，谭学样　　2-54
Research and Control on the Surface Quality of High Crack Sensitivity Steel Cast Blanks
　　　　　　　　　　　　　　　　Wang Jian, Wang Yuchun, Guo Da, Liu Wenping, Tan Xueyang

六流方坯连铸中间包内型结构优化　　　　　　　　　　　夏振东，胡　增，胡　睿，岳　强　　2-55
Optimization of Six-Strand Tundish Configuration for Billet Continuous Casting
　　　　　　　　　　　　　　　　Xia Zhendong, Hu Zeng, Hu Rui, Yue Qiang

板坯连铸过程结晶器内钢液面异常波动的机理　　苏志坚，江中块，陈　进，徐承乾，范　围　　2-55
Study on Mechanism of Large Amplitude Mold Level Fluctuation in Slab Continuous Casting Process
　　　　　　　　　　　　　　　　Su Zhijian, Jiang Zhongkuai, Chen Jin, Xu Chengqian, Fan Wei

IF钢在双辊薄带连铸过程中微观结构演变规律的研究　　张华龙，张同生，王万林，吕培生　　2-56
Study on the Evolution of Microstructure of IF Steel during Twin-roll Strip Continuous Casting
　　　　　　　　　　　　　　　　Zhang Hualong, Zhang Tongsheng, Wang Wanlin, Lv Peisheng

钎具钢轧后角部裂纹形成机理研究　　　　　　张建康，王万林，周乐君，陈俊宇，薛利文　　2-57

Study on the Formation Mechanism of Drill Steel Corner Crack after Rolling
······ Zhang Jiankang, Wang Wanlin, Zhou Lejun, Chen Junyu, Xue Liwen

QP980 钢亚快速凝固研究 ············ 徐慧，王万林，曾杰，路程，朱晨阳，吕培生 2-58
Effect of Cooling Rate on Microstructure and Properties of QP980 Steel by using Rroplet Solidification
　Technique ············ Xu Hui, Wang Wanlin, Zeng Jie, Lu Cheng, Zhu Chenyang, Lv Peisheng

板坯连铸过程流动、传热及夹杂物分布的数值模拟研究 ············ 张立峰，陈威 2-59
Mathematical Modeling on the Fluid Flow, Heat Transfer, and Inclusion Distribution during the Continuous
　Casting Slab Strand ············ Zhang Lifeng, Chen Wei

含硼钢板坯角横裂缺陷的成因与工艺控制 ············ 江中块，夏兆东，蔡兆镇 2-59
Cause of Formation Transverse Corner Crack and Process in Boron Steel CC Slab
············ Jiang Zhongkuai, Xia Zhaodong, Cai Zhaozhen

电磁搅拌下板坯连铸结晶器内的瞬态三相流动 ············ 安文，刘中秋，李向龙，吴存友，李宝宽 2-60
Three-phase Flow in Slab Mould under Electromagnetic Stirring
············ An Wen, Liu Zhongqiu, Li Xianglong, Wu Cunyou, Li Baokuan

CSP 薄板坯表面纵裂缺陷原因分析与控制 ············ 巩彦坤，张志克，杨学雨，王鹏 2-61
Analysis and Control of Longitudinal Crack on CSP Thin Slab Surface
············ Gong Yankun, Zhang Zhike, Yang Xueyu, Wang Peng

新一代连铸结晶器电液直驱伺服缸智能振动 ············ 刘玉，蔡春扬，王永猛，彭晓华，龙灏，李新有 2-61
New Generation Continuous Casting Mould Oscillation by Electro-hydraulic Direct Drive Servo Cylinder
············ Liu Yu, Cai Chunyang, Wang Yongmeng, Peng Xiaohua, Long Hao, Li Xinyou

基于高频超声检测的铸坯内部缺陷三维重构可视化
············ 丁恒，李雪，王柱，卫广运，刘宏强，黎敏 2-62
3D Reconstruction Visualization for Internal Defects of Slab based on High Frequency Ultrasonic Testing
············ Ding Heng, Li Xue, Wang Zhu, Wei Guangyun, Liu Hongqiang, Li Min

大圆坯连铸机轻压下仿真计算 ············ 曹学欠，陈杰 2-63
Finite Element Analysis of Process of Soft Reduction for Round Bloom Continuous Casting
············ Cao Xueqian, Chen Jie

连铸坯热芯大压下技术轧制规程实验研究 ············ 李睿昊，李海军，李天祥，李双江，郭子强 2-64
Experimental Research on Rolling Procedure of Hot-Core Heavy Reduction Rolling Tech for Continuous
　Casting Billet ············ Li Ruihao, Li Haijun, Li Tianxiang, Li Shuangjiang, Guo Ziqiang

板坯连铸结晶器窄面锥度对铸坯传热行为的影响
············ 牛振宇，蔡兆镇，刘志远，王重君，陈长芳，朱苗勇 2-64
Thermal Behavior of Slab under Different Mold Tapers during Continuous Casting
············ Niu Zhenyu, Cai Zhaozhen, Liu Zhiyuan, Wang Chongjun, Chen Changfang, Zhu Miaoyong

巨能特钢 2 号连铸机的技术特点和应用 ············ 樊伟亮，陈杰，曹学欠 2-65
The Technological Characteristic and Application of No.2 Bloom Caster in Shandong Juneng Special Steel
　Co., Ltd. ············ Fan Weiliang, Chen Jie, Cao Xueqian

基于 CA 模型的多元合金微观组织模拟 ············ 高晓晗，孟祥宁，崔磊，朱苗勇 2-66
Numerical Simulation of Microstructure Evolution based on Multiple Alloy CA Model
············ Gao Xiaohan, Meng Xiangning, Cui Lei, Zhu Miaoyong

低碳钢角部纵裂缺陷分析及控制实践 ············ 秦伟，刘红军 2-66
Analysis and Control Practice of Longitudinal Corner Cracks in Low Carbon Steel ············ QinWei, Liu Hongjun

高效自清洗喷嘴在板坯连铸机上的应用 ············ 吕宪雨，张威东，赵晓波 2-67

Application of Efficient Self-cleaning Nozzle in Slab Caster ……… Lv Xianyu, Zhang Weidong, Zhao Xiaobo

含 B_2O_3 无氟保护渣的微观结构和黏度特性研究
…………………………………… 吴 婷, 钟 磊, Shama Sadaf, 廖直友, 王海川, 王万林   2-68
Study on Microstructure and Viscosity of B_2O_3-Containing and Fluoride-free Mold Fluxes
…………………………………… Wu Ting, Zhong Lei, Shama Sadaf, Liao Zhiyou, Wang Haichuan, Wang Wanlin

基于热丝法的保护渣溶解 Al_2O_3 动力学研究新方法 ………………… 陈富杭, 唐 萍, 文光华, 谷少鹏   2-69
A Novel Method of Research on Dissolution Kinetics of Alumina in Mold Flux based on Hot Thermocouple
Technique ……………………………………………… Chen Fuhang, Tang Ping, Wen Guanghua, Gu Shaopeng

异型坯连铸技术专利分析 ……………………………………………… 高 仲, 王 颖, 陈卫强, 陈 杰   2-69
Analysis of the Patents of Beam Blank Continuous Casting Technology
…………………………………………………………… Gao Zhong, Wang Ying, Chen Weiqiang, Chen Jie

重轨钢凝固末端机械压下对溶质偏析行为的影响
…………………………………… 祭 程, 关 锐, 朱苗勇, 吴国荣, 陈天明, 李红光   2-70
Effect of Mechanical Reduction Technology in the Heavy Rail Bloom Continuous Casting on the Solute
Segregation …………… Ji Cheng, Guan Rui, Zhu Miaoyong, Wu Guorong, Chen Tianming, Li Hongguang

大圆坯凝固末端电磁搅拌流动行为数值模拟 ………… 郭壮群, 罗 森, 张文杰, 王卫领, 朱苗勇   2-71
Numerical Simulation of Fluid Flow in Liquid Core of Strand during Round Bloom Continuous Casting with
Final Electromagnetic Stirring ……… Guo Zhuangqun, Luo Sen, Zhang Wenjie, Wang Weiling, Zhu Miaoyong

旋流情况下晶粒移动行为研究 ……………………………… 王 鹏, 罗 森, 刘光光, 王卫领, 朱苗勇   2-72
Study on the Motion of Dendrite under Swirling
…………………………………… Wang Peng, Luo Sen, Liu Guangguang, Wang Weiling, Zhu Miaoyong

连铸结晶器电磁制动下导电壁面对湍流运动的影响 ……………………… 吴颖东, 刘中秋, 李宝宽   2-73
Effect of an Electrically-Conducting Wall on Turbulent Flow in a Continuous-Casting Mold with an
Electromagnetic Brake ……………………………………… Wu Yingdong, Liu Zhongqiu, Li Baokuan

微合金钢高温板坯全连续淬火新技术开发及应用 ………… 蔡兆镇, 刘志远, 王重君, 赵佳伟, 朱苗勇   2-73
Development and Application of New Continuously Quenching Technology for High Temperature
Micro-alloyed Steel Slab ……… Cai Zhaozhen, Liu Zhiyuan, Wang Chongjun, Zhao Jiawei, Zhu Miaoyong

IF 钢 Al-Ca-O 系细微夹杂物控制研究与实践 ………… 李俊伟, 赵 元, 袁少江, 杨成威, 代碧波   2-74
Control and Practice on Al-Ca-O Tiny Inclusion in Interstitial Free Steel
…………………………………………… Li Junwei, Zhao Yuan, Yuan Shaojiang, Yang Chengwei, Dai Bibo

连铸结晶器非正弦振动装置试验研究 ……………………………………… 张兴中, 周 超, 任素波   2-75
Experimental Investigation on Non-sinusoidal Oscillator of Continuous Casting Mold
…………………………………………………………………… Zhang Xingzhong, Zhou Chao, Ren Subo

连铸坯凸型辊压下工艺技术发展及其应用 ………… 逯志方, 赵昊乾, 田新中, 孟耀青, 王晓英   2-75
The Development and Application of Convex Roll Reduction Technology in Continuous Casting Bloom
…………………………………… Lu Zhifang, Zhao Haoqian, Tian Xinzhong, Meng Yaoqing, Wang Xiaoying

3 轧制与热处理

高强抗震螺纹钢连铸连轧短流程轧机研究与应用 ……………… 戴江波, 王保元, 刘 宏, 张业华   3-1
Application and Research of Short Flow Rolling Mill for the High Strength Seismic Thread Steel
Continuous Casting and Rolling ………………………… Dai Jiangbo, Wang Baoyuan, Liu Hong, Zhang Yehua

热处理工艺对 9%Ni 钢低温韧性影响的研究 ……………………… 杜 林, 朱莹光, 侯家平, 张宏亮   3-1

Study on the Influence of Heat Treatment Process on Cryogenic Toughness of 9%Ni Steel
······ Du Lin, Zhu Yingguang, Hou Jiaping, Zhang Hongliang

先进压水堆核电站用 SA-738Gr.B 特厚钢板的开发
······ 胡海洋，孙殿东，胡昕明，王　爽，颜秉宇，欧阳鑫 3-2
Development of SA-738Gr.B Extra-heavy Plate for Advanced PWR Nuclear Power Station
······ Hu Haiyang, Sun Diandong, Hu Xinming, Wang Shuang, Yan Bingyu, Ouyang Xin

20NCD14-7 钢最佳热处理工艺研究 ······ 王　爽，孙殿东，胡海洋，颜秉宇 3-3
Research on Heat-treatment Process for 20NCD14-7 Steel
······ Wang Shuang, Sun Diandong, Hu Haiyang, Yan Bingyu

热连轧板"唇印结疤"成因分析与对策 ······ 郭　斌 3-3
Formation Analyze and Countermeasure of Lip Print Scab Defect on Hot Rolled Strip ······ Guo Bin

EDC 线材表面黄锈的分析和解决 ······ 尹　一，郭大勇，刘　祥，徐　曦，安绘竹 3-4
Analysis and Solution of Surface Yellow Rust of EDC Wire Rod
······ Yin Yi, Guo Dayong, Liu Xiang, Xu Xi, An Huizhu

厚板轧机关于水梁印问题的分析及优化 ······ 陈国锋 3-5
Analysis and Improvement for Water Beam Mark during Heavy Plate Rolling ······ Chen Guofeng

湛江厚板 MULPIC 系统头尾遮挡技术的应用及优化 ······ 陈国锋 3-6
The Application and Optimization for Head Tail Masking of Zhanjiang Heavy Plate Mill MULPIC Cooling System ······ Chen Guofeng

极限宽厚比淬火钢板研发与应用 ······ 韩　钧，付天亮，王昭东，王国栋 3-6
Development and Application of Quenching Steel Plate with Ultimate Width-thickness Ratio
······ Han Jun, Fu Tianliang, Wang Zhaodong, Wang Guodong

基于余弦定理的钢种相似度评估应用研究 ······ 王金涛，张志超，杨　军，单旭沂 3-7
An Comparability Evaluation Method about Steel Grade based on Cosine Theorem
······ Wang Jintao, Zhang Zhichao, Yang Jun, Shan Xuyi

退火温度对 170MPa 级 IF 钢组织和性能影响 ······ 王占业，杨　平，李　进 3-8
Effect of Annealing Temperature on Microstructure and Properties of 170MPa IF Steel
······ Wang Zhanye, Yang Ping, Li Jin

冷连轧机组生产薄规格带钢划伤问题的分析与控制 ······ 王　静，孙荣生，辛利峰 3-8
The Analysis and Control of the Scratch Defect of the Thin Strip Produced by TCM
······ Wang Jing, Sun Rongsheng, Xin Lifeng

耐候桥梁钢板 Q345qNH/Q370qNH 的试制开发 ······ 向　华，王敬忠，秦　军，陈晓山，赵　虎 3-9
Development of Weather-resistant Bridge Steel Q345qNH/Q370qNH
······ Xiang Hua, Wang Jingzhong, Qin Jun, Chen Xiaoshan, Zhao Hu

12MnNiVR 钢板拉伸试样分层开裂原因分析 ······ 欧阳鑫，胡昕明，王　储，胡海洋，孙殿东，李广龙 3-10
Analysis in Delamination and Cracking of 12MnNiVR Steel Tensile Sample
······ Ouyang Xin, Hu Xinming, Wang Chu, Hu Haiyang, Sun Diandong, Li Guanglong

IF 系列钢再结晶退火时选择性氧化的研究 ······ 卢秉仲，周宏伟 3-10
Research of Selective Oxidation in IF-Series Steel during Annealing ······ Lu Bingzhong, Zhou Hongwei

鞍钢冷轧连退低碳钢性能影响因素研究 ······ 郭洪宇，王　越，付　薇，刘英明 3-11

Q235B 焊管开裂原因分析 ······ 张爱梅 3-11
Failure Analysis of Q235B Steel Welded Pipe ······ Zhang Aimei

平整工作辊崩边原因分析及对策 ······ 曹七华，方　俊，刘文杰，唐　华 3-12
Analysis and Solution of Work Roll Side Damage for SPM ······ Cao Qihua, Fang Jun, Liu Wenjie, Tang Hua

热轧标签打印焊挂系统 ··· 窦 刚，蔡 炜 3-12

17CrNiMo6 钢的等温正火工艺研究 ··· 相 楠 3-13
Study on Isothermal Normalizing of 17CrNiMo6 ································ Xiang Nan

一种冷连轧机轧制油的选择方法 ·· 贾生晖，赵利明 3-13
Choice Method of Rolling Oil for Tandem Cold Rolling Mill ············ Jia Shenghui, Zhao Liming

低淬火开裂的锚具钢开发 ·· 陈定乾 3-14
The Development of Anchor Steel with Low Quenching Cracking ··················· Chen Dingqian

B、Ti 元素对 SPHC 酸洗板屈服平台的影响研究 ············· 李 霞，岳重祥，李 冉，李化龙 3-15
Effects of B, Ti Elements on the YPE of SPHC P.O. Plate ······ Li Xia, Yue Chongxiang, Li Ran, Li Hualong

加热炉自动燃烧控制模型的研发与应用 ··················· 李小新，向永光，亓鲁刚 3-15
Research and Application of Automatic Combustion Control Model for Heating Furnace
·· Li Xiaoxin, Xiang Yongguang, Qi Lugang

GCr15 线材网状碳化物控冷工艺研究 ·· 熊钟铃 3-16

冷轧板表面斑迹控制技术研究 ······················· 张建波，刘 坤，张 涛，吴首民 3-16
Surface Cleanliness Control Technology for Cold Rolling Steel
·· Zhang Jianbo, Liu Kun, Zhang Tao, Wu Shoumin

38MnSiVS 非调质钢的组织性能预测模型 ··················· 卢世康，陈雨来，余 伟 3-17
Prediction Model of Microstructure and Properties of 38MnSiVS Non-quenched and Tempered Steel
·· Lu Shikang, Chen Yulai, Yu Wei

高碳钢盘条 72A 表面氧化铁皮控制研究 ········ 王海宾，于 聪，贾建平，曹光明，沈俊杰，刘振宇 3-18
Study on Controlling of Surface Oxide Scale of High Carbon Steel Wire Rod 72A
·· Wang Haibin, Yu Cong, Jia Jianping, Cao Guangming, Shen Junjie, Liu Zhenyu

16mm 厚度 Q690CFD 高强煤机钢弯曲不合原因分析 ································ 薛如锋 3-19
Analysis of the Causes of Bending Misalignment of Q690CFD High Strength Coal Machine Steel with 16mm
Thickness ··· Xue Rufeng

连退在线平整轧制力异常原因分析及对策 ············· 孙超凡，王雅晴，方 圆，张宝来，胡小明 3-19
Cause Analysis and Control Strategy of Online Temper Rolling Force Abnormity in Continuous Annealing Line
·· Sun Chaofan, Wang Yaqing, Fang Yuan, Zhang Baolai, Hu Xiaoming

连续退火炉炉辊结瘤类缺陷原因分析及控制措施 ··············· 李 军，李 源，杨麒冰，周诗正 3-20
Cause Analysis and Control Measures of Nodulation Defects in Rollers of Continuous Annealing Furnace
·· Li Jun, Li Yuan, Yang Qibing, Zhou Shizheng

JMatPro 在核电 SA738 Gr.B 钢热处理工艺设计中的应用
·· 张舒展，李艳梅，杨梦奇，姜在伟，叶其斌 3-21
Application of JMatPro in Heat Treatment Process Design of SA738 Gr. B Steel in Nuclear Power Plant
·· Zhang Shuzhan, Li Yanmei, Yang Mengqi, Jiang Zaiwei, Ye Qibin

U 型钢板桩轧制翘曲的有限元分析 ··············· 杨 洋，刘 凯，苏 磊，吴功军，刘 杨 3-21
Finite Element Analysis of U-sheet Pile Rolling Warpage
·· Yang Yang, Liu Kai, Su Lei, Wu Gongjun, Liu Yang

轧钢车间降尘工艺及设备综述 ······················· 徐言东，韩 爽，王占坡，程知松 3-22
Summary of Dust Suppression Process and Equipment of the Steeling Rolling Plant
·· Xu Yandong, Han Shuang, Wang Zhanpo, Cheng Zhisong

冷轧处理线卷取机带头定位精度改进 ·· 黄海生，胡剑斌 3-23

以高端钢板桩产品为定位的钢板桩系列产品研发 ········ 王君珂，杨 洋，苏 磊，吴功军，刘 杨 3-23

Research and Development of Steel Sheet Pile Series Products Oriented by High-end Steel Sheet Pile Products
　　　　　　　　　　　　　　　　　　　　　　　　Wang Junke, Yang Yang, Su Lei, Wu Gongjun, Liu Yang

宽厚板 3.5m 线镰刀弯刮框的原因分析及预防措施 ……………… 谢富强，邹星禄，何明涛，乔　坤　3-23
Reasons Analysis and Preventive Measures of 3.5m Wire Sickle Bending Scraper for Wide Plate
　　　　　　　　　　　　　　　　　　　　　　　　Xie Fuqiang, Zou Xinglu, He Mingtao, Qiao Kun

浅谈 3.5m 超快速冷却设备与板形控制 ………………………… 谢富强，何明涛，邹星禄，乔　坤　3-24
Discussion on 3.5m Ultra-fast Cooling Equipment and Plate Control
　　　　　　　　　　　　　　　　　　　　　　　　Xie Fuqiang, He Mingtao, Zou Xinglu, Qiao Kun

型钢轧辊磨损规律分析与研究 ……………………………………………… 郭　平，张明海，谢海深　3-25
Research on Wear of Roll for Section Steel ……………………… Guo Ping, Zhang Minghai, Xie Haishen

柔性支撑技术在矫直机换辊中的应用 ……………………… 栗增杰，李艳辉，李光伟，白世军　3-25
The Application of Type of Flexible Support in Roll Change of PPL
　　　　　　　　　　　　　　　　　　　　　　　　Li Zengjie, Li Yanhui, Li Guangwei, Bai Shijun

钢管控制冷却工艺的试验研究 …………………………………………… 洪　汛，邸　军，谷大伟　3-26
Experimental Study on Controlled Cooling Process of Steel Tubes …… Hong Xun, Di Jun, Gu Dawei

单机架可逆冷轧机的厚控系统升级与优化技术 ……………………… 陈跃华，吴有生，王志军　3-27
Upgrade and Optimization Technology of RCM AGC System …… Chen Yuehua, Wu Yousheng, Wang Zhijun

基于图像识别技术的加热炉自动上料系统 ………………………… 蔡　炜，叶理德，吉　青，祝兵权　3-27

Ti-IF 钢铁素体轧制实践及工艺探讨 ………………………… 蔡　珍，梁　文，汪水泽，何龙义　3-28
Practice and Discussion of Ferrite Rolling in Ti-IF Steel …… Cai Zhen, Liang Wen, Wang Shuize, He Longyi

无缝钢管某 PQF 连轧机组扩大产品尺寸范围初步探讨 …… 李琳琳，郭海明，张　尧，张卫东，杨永海　3-28
Preliminary Discussion on the Range of Product Size of a PQF Continuous Rolling Mill for Seamless Steel
　　Tubes …………………… Li Linlin, Guo Haiming, Zhang Yao, Zhang Weidong, Yang Yonghai

优化 177PQF 脱管机孔型 ………………………… 郭海明，张　尧，李琳琳，张卫东，杨永海　3-29
Optimization of Extractor Caliber in 177 PQF Seamless Tube Unit
　　　　　　　　　　　　　　　　　　Guo Haiming, Zhang Yao, Li Linlin, Zhang Weidong, Yang Yonghai

京唐公司热轧检查线生产效率提升的研究与实践 ………… 伊成志，曹艳生，肖　楠，崔惠民　3-30
Research and Practice on Working Efficiency Improvement of Hot Rolling Inspection Line of Jingtang
　　Company ………………………… Yi Chengzhi, Cao Yansheng, Xiao Nan, Cui Huimin

攀钢热连轧普碳钢低温轧制技术研究与应用 …… 肖　利，陈　永，方淑芳，张中平，刘　勇，文永才　3-30
Pan Steel Hot Strip Plain Carbon Steel Low Temperature Rolling Technology Research and Application
　　　　　　　　　　　Xiao Li, Chen Yong, Fang Shufang, Zhang Zhongping, Liu Yong, Wen Yongcai

低温环境服役大壁厚管件 X80 钢板的研发 …… 张志军，邓建军，龙　杰，高　雅，祝　鹏，程海林　3-31
Research and Development of X80 Steel Plate for Large Wall Thickness Pipe Fitting in Low Temperature
　　Service ………………… Zhang Zhijun, Deng Jianjun, Long Jie, Gao Ya, Zhu Peng, Cheng Hailin

U 型钢孔型设计 ……………………………………………… 安华荣，王　伟，左　岩，卜俊男　3-32
Design Method of U-shaped Steel for Coal Roadway Support
　　　　　　　　　　　　　　　　　　　　　　　　　An Huarong, Wang Wei, Zuo Yan, Bu Junnan

热连轧带钢温度对轧后板形影响的研究 …………………………… 杨立庆，饶　静，张文兴　3-32
The Research of the Influence of Strip Temperature on Strip Shape after Hot Rolling
　　　　　　　　　　　　　　　　　　　　　　　　　　　　Yang Liqing, Rao Jing, Zhang Wenxing

低碳钢工艺设计对冷板组织的影响研究 …………………………………………………… 李　雯　3-33
Effect of Low Carbon Steel Procession on Microstructure of Cold Rolling Steel …………… Li Wen

推拉式酸洗机组工艺段技术改造 ………………………… 陈　普，侯元新，吕圣才，刘军燕　3-34

Technical Reform of Process Section of Push-pull Pickling Unit
　　　　　　　　　　　　　　　　　　　　　　　Chen Pu, Hou Yuanxin, Lv Shengcai, Liu Junyan

韶钢中棒线轧钢工艺技术创新及优化改造 ………………………………………… 周小兵，李学保　3-34
Technological Innovation and Optimum Reform of Medium Bar Rolling in Shaoguan Iron and Steel Co., Ltd.
　　　　　　　　　　　　　　　　　　　　　　　Zhou Xiaobing, Li Xuebao

针对特殊钢棒材控轧控冷工艺的温度场优化 ……………… 张华鑫，程知松，余 伟，徐言东　3-35
Temperature Field Optimization for Controlled Rolling and Controlled Cooling Process of Special Steel Bars
　　　　　　　　　　　　　　　　Zhang Huaxin, Cheng Zhisong, Yu Wei, Xu Yandong

帘线钢珠光体片间距的测定 ……………………………………… 李珊珊，王 涛，赵 磊　3-36
Determination of Pearlitic Plate Spacing in Steel Cord　　　Li Shanshan, Wang Tao, Zhao Lei

时效处理对大形变珠光体钢丝扭转性能的影响 …………………… 王 猛，黄 波，胡乃志　3-36
Effect of Annealing Treatment on Torsion Property of Heavily Cold Drawn Pearlitic Steel Wire
　　　　　　　　　　　　　　　　　　　　　　　Wang Meng, Huang Bo, Hu Naizhi

高速线材智能夹送辊控制系统的研究应用 ………………………………… 谭秋生，王正洁　3-37
Study on the Automatic Control System of Intelligent Pinch Roll for High Speed Wire Mill
　　　　　　　　　　　　　　　　　　　　　　　Tan Qiusheng, Wang Zhengjie

基于 800xA 平台的 ABB 线材减定径生产线 ……………………… 王廷雷，谭秋生，王正洁　3-38
ABB Wire Reducing and Sizing Production Line based on the Platform of 800xA
　　　　　　　　　　　　　　　　　　　Wang Tinglei, Tan Qiusheng, Wang Zhengjie

斯太尔摩风冷线工艺优化对焊接用钢 ER50-6 组织的影响 ………………… 郝文权，钟 浩　3-39
Effect of Technological Optimization of Stelmore Air Cooling Line on Microstructure of Welding Steel ER50-6
　　　　　　　　　　　　　　　　　　　　　　　Hao Wenquan, Zhong Hao

水冷式加热炉节能措施的探讨分析 ………………………………… 陈海明，胡学民，查安鸿　3-39
Discussion and Analysis of Energy-saving Measures for Water-cooling Reheating Furnace
　　　　　　　　　　　　　　　　　　　　　　　Chen Haiming, Hu Xuemin, Zha Anhong

大棒 1#加热炉二级优化控制系统研究与开发 …………………… 蒋国强，张宝华，陈建洲　3-40
Research and Development of L2 Optimizing Control System for Reheating Furnace #1 of Large-size Bar
　Rolling Line　　　　　　　　　　　　Jiang Guoqiang, Zhang Baohua, Chen Jianzhou

低碳高含硫钢易切削钢轧制技术应用 ……………………………… 李学保，周小兵，潘泽林　3-41
Application of Rolling Technology for Low Carbon and High Sulphur Free Cutting Steel
　　　　　　　　　　　　　　　　　　　　　　　Li Xuebao, Zhou Xiaobing, Pan Zelin

17CrNiMo6 圆钢表面裂纹成因分析与对策 ………………………… 吴学兴，岳 峰，钟芳华　3-41
Analysis and Countermeasure on the Formation of Surface Stress Cracks on Steel Rods 17CrNiMo6
　　　　　　　　　　　　　　　　　　　　　　　Wu Xuexing, Yue Feng, Zhong Fanghua

超厚特种钢板连续辊式淬火装备技术与应用 ……………… 付天亮，王昭东，邓想涛，王国栋　3-42
Technology and Application of Continuous Roller Quenching Equipment for Ultra Heavy Special Steel Plate
　　　　　　　　　　　　　　　　Fu Tianliang, Wang Zhaodong, Deng Xiangtao, Wang Guodong

鞍钢 5500mm 线异型断面钢板生产技术研究 …… 王若钢，李靖年，丛津功，李新玲，姚 震，韩 旭　3-43
Study on Production Technology of 5500mm Profile Steel Plate in Angang
　　　　　　Wang Ruogang, Li Jingnian, Cong Jingong, Li Xinling, Yao Zhen, Han Xu

测厚仪精度影响因素浅析 ………………… 常福刚，赵立军，丛津功，王若钢，翟忠军，安宇恒　3-44
Elementary Analysis on the Influencing Factors of Thickness Meter Accuracy
　　　　　　Chang Fugang, Zhao Lijun, Cong Jingong, Wang Ruogang, Zhai Zhongjun, An Yuheng

超宽 X80M 管线钢板冷矫板形的研究 …………… 周 强，王亮亮，姚 震，王若钢，丛津功，李新玲　3-44

Study on Cold Correction of Ultra Wide X80M Pipeline Steel Plate
　　　　　…………… Zhou Qiang, Wang Liangliang, Yao Zhen, Wang Ruogang, Cong Jingong, Li Xinling

基于 AMEsim 软件研究轧机液压辊缝控制系统震荡问题
　　　　　………………………… 张秀超，赵立军，肖争光，朱丽娜，丛津功，王若钢　3-45
Based on AMEsim Software Research the Shaking Problem of Rolling Mill Hydraulic Gap Control System
　　　　　…………… Zhang Xiuchao, Zhao Lijun, Xiao Zhengguang, Zhu Lina, Cong Jingong, Wang Ruogang

汽包钢蓄热式室式炉生产轧制裂纹分析
　　　　　………………………… 韩　旭，刘长江，王若钢，丛津功，段江涛，罗　军，李新玲　3-46
Analysis of Rolling Cracks in Regenerative Chamber Furnace of Drum Steel
　　　　　…………… Han Xu, Liu Changjiang, Wang Ruogang, Cong Jingong, Duan Jiangtao, Luo Jun, Li Xinling

延长炉底辊挂腊周期的实践 ………………… 王亮亮，周　强，李新玲，王若钢，丛津功，姚　震　3-46
Practice of Prolonging Wax Hanging Period of Bottom Roller
　　　　　…………… Wang Liangliang, Zhou Qiang, Li Xinling, Wang Ruogang, Cong Jingong, Yao Zhen

变角度刀梁式静电涂油机在高速薄带材的使用 ………… 任予昌，彭　强，张东方，龚　艺，黄秋华　3-47
Variable Angle Statical Electricity Coating Oil Machine Used in High Speed Line
　　　　　…………… Ren Yuchang, Peng Qiang, Zhang Dongfang, Gong Yi, Huang Qiuhua

SKP 振动痕分析研讨 …………………………………………………… 罗明辉，朱璧学，李子武　3-47

冷轧酸轧机组激光焊机焊缝断带分析 …………………………………… 吉学军，李保卫，张核新　3-48
Analysis on Weld Strip Breakage of Laser Welder for Pickling and Cold Rolling Continuous Production Line
　　　　　……………………………………………………… Ji Xuejun, Li Baowei, Zhang Hexin

连退镀锡基板斑迹类缺陷研究 …………………………………………… 吉学军，胡小明，张文亮　3-48
Study on Residual Patches Defects about Continuous Annealing Raw Tin Coil
　　　　　……………………………………………………… Ji Xuejun, Hu Xiaoming, Zhang Wenliang

冷轧带钢连续退火炉绿色制造探索与实践 …………………………… 何建锋，王　鲁，李庆胜　3-49
Green Manufacturing Exploration and Practice of Cold Rolling Strip Continuous Annealing Furnace
　　　　　……………………………………………………… He Jianfeng, Wang Lu, Li Qingsheng

基于 Python 的变形抗力预测系统的实现与应用 ………………………………………… 彭　诚　3-49
Implementation and Application of Python based on Deformation Resistance Prediction System
　　　　　……………………………………………………………………………… Peng Cheng

双机架四辊 CVC 平整机辊形配置优化与应用
　　　　　……………………… 文　杰，于　孟，莫志英，王永强，张宝来，刘学良，李永新　3-50
Optimization and Application of Roll Contour Configuration on Double Stands 4-h CVC Temper Mill
　　　　　…………… Wen Jie, Yu Meng, Mo Zhiying, Wang Yongqiang, Zhang Baolai, Liu Xueliang, Li Yongxin

基于焊缝 3D 检测系统应用的超高强钢焊接工艺优化研究 ……………………… 朱健华，何建锋　3-51
Research on Welding Process Optimization of Ultra-high Strength Steel based on the Application of Welding
　　Seam 3D Detection System ………………………………………… Zhu Jianhua, He Jianfeng

钢种成分和工艺对连退 DC01 产品力学性能的影响 …………… 王业科，辜蕾钢，刘显军，杨　薇　3-52
Effect of Steel Composition and Process on Mechanical Properties of DC01 Product in CAL
　　　　　……………………………………………… Wang Yeke, Gu Leigang, Liu Xianjun, Yang Wei

剪切断面质量对焊接的影响及控制 ……………………………………… 张　志，黄爱军，夏大斌　3-52
Influence of Longitudinal Section Quality on Welding and Its Control ……… Zhang Zhi, Huang Aijun, Xia Dabin

首钢智新电工钢连退机组入口和出口张力辊装置方案的设计与分析
　　　　　…………………………………………………… 张　伟，何云飞，张乐峰，孟祥军　3-53

The Design and Analysis of the Entry and Exit Bridle Roll in the Silicon Steel Continuous Annealing Line
············ Zhang Wei, He Yunfei, Zhang Lefeng, Meng Xiangjun

酸洗工艺参数对热轧板酸洗效果影响的研究 ············ 张 鹏，孙 力，王 峰，任振远，刘连喜 3-54
Effect of Pickling Process Parameters on Pickling Effect of Hot Rolled Sheet
············ Zhang Peng, Sun Li, Wang Feng, Ren Zhenyuan, Liu Lianxi

浅谈冷轧薄宽规格产品炉内擦伤的控制与研究 ············ 李国栋 3-54
Control and Research on Furnace Scratch in Cold Rolled Thin and Wide Specification Products
············ Li Guodong

冷连轧过程中宽度变化时的头尾板形优化策略 ············ 任延庆 3-55
Head-to-tail Flatness Optimization Strategy for Width Variation during Cold Rolling ············ Ren Yanqing

酸洗轧机联合机组清洗效果对汽车板表面质量影响 ············ 张彦雨，李 宁 3-56
The Effect of Picking Mill Combination on the Surface of Automobile Plate ············ Zhang Yanyu, Li Ning

本钢冷轧 1#酸轧机组厚度不合控制优化 ············ 张 勇 3-56
The Thickness Deviation Control Optimization of Benxi Steel Cold Rolling of 1 # Pickling Line Tandem Cold Rolling Mill ············ Zhang Yong

卡罗塞尔卷取机挫伤缺陷分析及控制方案 ············ 田永强，周三保，陈 俊，刘 挺，张 波 3-57

N 含量超标对冷轧低合金高强钢 280VK 力学性能影响浅析
············ 狄彦军，赵小龙，罗晓阳，赵占彪，王 强 3-57
Analysis on the Effect of N Content Exceeding the Standard on the Mechanical Properties of Cold Rolled High Strength Low Alloy Steel 280VK
············ Di Yanjun, Zhao Xiaolong, Luo Xiaoyang, Zhao Zhanbiao, Wang Qiang

基于 CSP 流程的镀铝锌锌花不均缺陷分析 ············ 孙朝勇 3-58
Analysis of Spangles Inhomogeneous Defect for Galvalume in CSP Process ············ Sun Chaoyong

酒钢 CSP 线 600MPa 级热轧双相钢生产工艺研究 ············ 杨 华，蔺理平，王云平，成洋洋，刘 靖 3-59
Process Study of 600MPa Grade Hot-rolled Dual Phase Steel in CSP Production Line in JISCO
············ Yang Hua, Lin Liping, Wang Yunping, Cheng Yangyang, Liu Jing

酒钢冷连轧机二级模型的工艺优化应用 ············ 周文宾 3-60
Process Optimization Application of Two - level Model of Cold Tandem Mill at JISCO ············ Zhou Wenbin

冷轧带钢乳化液斑迹研究及控制措施 ············ 翁 星 3-60
Study on the Emulsion Stain Trace of the Cold-rolled Strip and Its Control Measures ············ Weng Xing

浅谈"一贯质量管理" ············ 赵占彪 3-61
Discussion on "Consistent Quality Management" ············ Zhao Zhanbiao

热镀铝锌硅板面边厚缺陷攻关 ············ 刘海军，马维杰 3-62
Study on the Defects of the Side-thickness of the Hot-dip 55%Al-Zn-Si Line ············ Liu Haijun, Ma Weijie

轧机液压 AGC 系统故障分析 ············ 毛成海 3-62
Fault Analysis of Mill Hydraulic AGC System ············ Mao Chenghai

罩式退火炉黑斑缺陷分析及防范措施 ············ 王生东 3-63
Analysis and Preventive Measures of Black Spot Defect of Hood Type Annealing Furnace
············ Wang Shengdong

冷轧 Ti 系超低碳烘烤硬化钢的耐时效性能研究 ············ 周三保，唐梦霞，张功庭 3-63
Research on Aging Resistance of Cold-Rolling Ti-killed Ultra-Low Carbon Bake-Hardening Steel
············ Zhou Sanbao, Tang Mengxia, Zhang Gongting

冷轧平整机组异常高速甩尾问题探讨 ············ 王 勇，冯志新，文纪刚，邢顺治，刘立恒 3-64

Discuss on the Abnormal High-speed Tail out of Temper Rolling Mill
　……………………………………………… Wang Yong, Feng Zhixin, Wen Jigang, Xing Shunzhi, Liu Liheng

冷轧汽车板点状锈蚀产生机理及控制措施 …………………… 辛利峰，阮国庆，孙荣生　3-65
The Mechanism and Control Measures about Spot Rust on the Cold Rolling Automobile Strip
　……………………………………………………………… Xin Lifeng, Ruan Guoqing, Sun Rongsheng

连退立式活套双塔改单塔控制的实现 ………… 赵天鑫，张国强，王　弢，昌　亮，张国立　3-65
Realization of Double Towers Converted to Single Tower Control in Continuous Annealing Line
　………………………………………… Zhao Tianxin, Zhang Guoqiang, Wang Tao, Chang Liang, Zhang Guoli

轮盘卷取卷筒轴承拆装方法的改进 …………………………… 王桂玉，褚国嵩，张国强　3-66

酸轧联合机组生产高强钢工艺的探讨 ………………………… 王　静，孙荣生，辛利峰　3-66
The Discuss of the Technology of the High Strength Steel on the Continuous Pickling and Rolling Line
　……………………………………………………………… Wang Jing, Sun Rongsheng, Xin Lifeng

轧机轧制线的校核与调整 …………………………… 姜大鹏，王云良，张国强，王　荣　3-67
Check and Adjustment of Mill Passline ……… Jiang Dapeng, Wang Yunliang, Zhang Guoqiang, Wang Rong

轴承座润滑油含水原因分析及维护解决方案 ………………………………………… 李　军　3-68
Cause Analysis and Maintenance Solution of Bearing Chock Lubricating Oil Moisture …………… Li Jun

光整机轴承座装配轧辊轴颈工艺改进 ………………………………………………… 李　军　3-68
Modification of Roll-neck in Roll Bearing Chocks Assembly of Skin-Pass Mill ………………… Li Jun

锰含量对于 SWRCH22A 调质后组织和性能的影响 …………………… 郭晓培，阮士朋　3-69
Effect of Manganese Content on Microstructure and Properties of SWRCH22A Quenched Tempering
　…………………………………………………………………………… Guo Xiaopei, Ruan Shipeng

Q690 钢新型差温轧制工艺的有限元模拟与验证 ……………… 江　坤，蔡庆伍，余　伟　3-69
Finite Element Simulation and Verification of New Temperature Gradient Rolling Process for Q690 Steel
　…………………………………………………………………… Jiang Kun, Cai Qingwu, Yu Wei

铁素体区轧制 IF 钢组织性能 ………………………………………… 常文杲，余　伟　3-70
Microstructure and Properties of IF Steel Rolled in Ferrite Zone ………… Chang Wengao, Yu Wei

激光熔覆技术在冷轧支撑辊辊颈修复应用实践 ……………………… 林上辉，潘勋平　3-71
The Application of Laser Cladding Technology in Repairing the Neck of Backup Roll in Cold Rolling
　………………………………………………………………………… Lin Shanghui, Pan Xunping

4　表面与涂镀

热浸镀锌铝镁镀层微观组织试验研究 ………… 吕家舜，徐闻慧，杨洪刚，李　锋，苏皓璐　4-1
Experiments Investigation on Microstructure of Galvanized Zn-Al-Mg Coated Steel Sheet
　…………………………………………… Lv Jiashun, Xu Wenhui, Yang Honggang, Li Feng, Su Haolu

马钢高导电热浸镀锌产品的研究与开发 ……………… 杨兴亮，王　滕，杨　平，刘　劼　4-1
Research and Development of High Conductivity Hot Dip Galvanizing Products in Masteel
　……………………………………………………… Yang Xingliang, Wang Teng, Yang Ping, Liu Jie

电视机背板用彩涂板涂层技术研究 ………………… 施国兰，蔡　明，谷　曦，钱婷婷　4-2
Study on Coating Technology of Color Coated Plates for TV Backplane
　……………………………………………………………… Shi Guolan, Cai Ming, Gu Xi, Qian Tingting

烧结电除尘环境下碳钢防护涂层失效行为
　……… 高　鹏，陈义庆，武裕民，王佳骥，艾芳芳，李　琳，钟　彬，肖　宇，伞宏宇，苏显栋　4-3

Failure Behavior of Protective Coating on Carbon Steel under Sintering Electric Remove Condition
················ Gao Peng, Chen Yiqing, Wu Yumin, Wang Jiaji,
Ai Fangfang, Li Lin, Zhong Bin, Xiao Yu, San Hongyu, Su Xiandong

生产运营期钢结构防腐的探讨与实践 ················ 殷 栋，苏会德，吴新辉 4-3

不同纳米添加剂对微弧氧化陶瓷膜层耐腐蚀性能的影响 ········ 张 宇，刘 鑫，曾 丽，李红莉，宋仁国 4-4
Effects of Different Nano-additives on Corrosion Resistance of Micro-arc Oxidation Films
················ Zhang Yu, Liu Xin, Zeng Li, Li Hongli, Song Renguo

防氧化涂料降低弹簧脱碳层深度的生产实践 ················ 彭 超，王旭冀，朱建成 4-4

入锌锅温度对连续热镀锌 Fe-Al 抑制层的影响 ················ 李婷婷，岳重祥，李化龙 4-5
Effect of Strip-entry Temperature on the Fe-Al Inhibition Layer during Continuous Hot-dip Galvanizing Process
················ Li Tingting, Yue Chongxiang, Li Hualong

两种锌铝镁镀层板黑点缺陷分析 ················ 杜 江，许秀飞 4-5
Research of Two Kinds of Black Point Defects in Zinc-Aluminum-Magnesium Coated Sheet
················ Du Jiang, Xu Xiufei

中国钢制品及型材热浸镀锌线设计选型建议 ················ 徐言东，马树森，顾 洋 4-6
Suggestions on Design Selection of Hot Dip Galvanizing Line for Steel Products and Profiles in China
················ Xu Yandong, Ma Shusen, Gu Yang

激光熔覆技术现状及发展 ················ 周 丰，贵永亮，胡宾生，扈理想 4-7
Present Situation and Development of Laser Cladding Technology
················ Zhou Feng, Gui Yongliang, Hu Binsheng, Hu Lixiang

Zn-Al-Mg 合金镀层的微观组织及相组成分析 ···· 蔡 宁，黎 敏，赵晓非，其其格，曹建平，杨建炜 4-8
Microstructure and Phase Analysis on Zn-Al-Mg Alloy Coatings
················ Cai Ning, Li Min, Zhao Xiaofei, Qi Qige, Cao Jianping, Yang Jianwei

纯锌镀层和锌铝镁合金镀层在循环盐雾中的腐蚀行为 ········ 黎 敏，郝玉林，龙 袁，姚士聪 4-8
The Corrosion Behaviour of Zn and Zn-Al-Mg Coated Steel in the Cyclic Salt Spray Test
················ Li Min, Hao Yulin, Long Yuan, Yao Shicong

高强汽车用酸洗钢酸洗表面发黑研究与控制 ················ 肖厚念 4-9
Research and Control of Surface Blackening of Pickling Steel for High Strength Automobile ·· Xiao Hounian

Al-Si 镀层微观结构及变形能力研究 ················ 郭 健，魏焕君，耿志宇，潘文娜，杨丽娜，李 勃 4-10
Study on Microstructure and Deformation Ability of Al-Si Coating
················ Guo Jian, Wei Huanjun, Geng Zhiyu, Pan Wenna, Yang Lina, Li Bo

基板表面形貌对镀锡层覆盖能力影响的研究
················ 谢志刚，李顺祥，张青树 4-10
Study on the Influence of Substrate Surface Morphology on the Coating Coverage Rate of Tin Plating
················ Xie Zhigang, Li Shunxiang, Zhang Qingshu

宝钢无机润滑 GA 热镀锌汽车板产品的生产与应用 ················ 张 军，温乃盟，卢海峰 4-11
Production and Application of Inorganic Solid Lubricant Coated Galvannealed Automobile Steel Sheet
················ Zhang Jun, Wen Naimeng, Lu Haifeng

液中放电沉积参数对工件性能的影响 ················ 顾晓辉，何 星，张晨昀 4-12
Effect of Liquid Discharge Deposition Parameters on Workpiece Performance
················ Gu Xiaohui, He Xing, Zhang Chenyun

无铬钝化热镀锌钢板摩擦条纹的分析与改善 ················ 杨 芃，张 云 4-12
Analysis and Improvement of Friction Striation Defect on the Oiled Cr-free Passivation Hot-Dip Galvanizing Sheet
················ Yang Peng, Zhang Yun

直读光谱法测定锌合金中的铝含量 …………………………………… 黄碧芬，王君祥，李 艳　4-13
Direct Reading Spectrometric Determination of Aluminum in Zinc Alloys
　　　　　　　　　　　　　　　　　　　　　　　Huang Bifen, Wang Junxiang, Li Yan

软熔处理对锡铁合金层耐蚀性能的影响研究 ………… 彭 强，任予昌，赵 柱，张东方，刘晓峰　4-14
Study on the Effect of Soft Soluble Treatment on the Erosion Resistance of Tin Iron Alloy Layer
　　　　　　　　　　　　　　　Peng Qiang, Ren Yuchang, Zhao Zhu, Zhang Dongfang, Liu Xiaofeng

控制热镀锌汽车外板锌渣缺陷的工艺创新方法 …………………………………… 吴价宝，张雨泉　4-14

冷轧电镀锡板生产之关键技术的探讨与分析 ………………… 何云飞，周玉林，刘宏文，侯俊达　4-15
Analysis and Discussion on the Key Technology for Cold Rolled Electrolytic Tin-plate Production
　　　　　　　　　　　　　　　　　　　　He Yunfei, Zhou Yulin, Liu Hongwen, Hou Junda

一种热镀纯锌产品黑点缺陷成因及解决办法 ………………… 李鸿友，张晶晶，马 峰，艾厚波　4-15
Causes and Solutions of Black Spots Defect in Hot－Dip Galvanized Zinc Products
　　　　　　　　　　　　　　　　　　　　　Li Hongyou, Zhang Jingjing, Ma Feng, Ai Houbo

电镀锌磷化生产工艺的优化 ………………………………… 孙晨航，曹 洋，车晓宇，曹志强　4-16
Optimization of Production Process of Electroplating Zinc Phosphate
　　　　　　　　　　　　　　　　　　　　Sun Chenhang, Cao Yang, Che Xiaoyu, Cao Zhiqiang

本钢浦项热镀锌锌锅铝含量成分稳定控制 …………………………………… 王子昂，赵兴时　4-16
Stable Control of Aluminum Composition in the Hot Dip Strip Coating　　Wang Ziang, Zhao Xingshi

基于电镀锌生产线数据采集系统的分析与实现 …………………………………………… 张 勇　4-17
Analysis and Realization of Data Collection System based on Electrolytic Galvanizing Line　Zhang Yong

高强 IF 钢合金化热镀锌镀层相结构和抗粉化性能影响因素分析
　……………………………………………………………… 周诗正，李 军，张伟浩，杜小峰　4-18
Analysis of Factors Affecting Phase Microstructure and Powdering-Resistance of High Strength IF Galvannealed
　Steel Sheets　　　　　　　　　　　　　　Zhou Shizheng, Li Jun, Zhang Weihao, Du Xiaofeng

浅谈冷轧锌铝镁产品发展现状与创新实践 ………………… 王海东，严江生，陈浩杰，王骏飞　4-18
Development & Innovation Practice of ZnAlMg Coated Product
　　　　　　　　　　　　　　　　　Wang Haidong, Yan Jiangsheng, Chen Haojie, Wang Junfei

5　金属材料深加工

1006 线材制作 T 铁时表面鼓泡原因分析 …………………… 王宁涛，阮士朋，张 鹏，王利军　5-1
Analysis of the Causes of Surface Bubbling on the T-iron Made of 1006 Wire
　　　　　　　　　　　　　　　　　　　　Wang Ningtao, Ruan Shipeng, Zhang Peng, Wang Lijun

加热温度对 B 柱热冲压成形性影响的研究 ……………………………… 郭 晶，陈 宇，陈虹宇　5-1
Research on Hot Stamping Formability of B-PILLAR by Heating Temperature
　　　　　　　　　　　　　　　　　　　　　　　　　　　　　Guo Jing, Chen Yu, Chen Hongyu

热处理工艺对 Q125 套管力学性能影响的试验 ……… 解德刚，吴 红，赵 波，袁 琴，王善宝　5-2
Effect of Heat Treating on Mechanical Properties of Q125 Casing
　　　　　　　　　　　　　　　　　　Xie Degang, Wu Hong, Zhao Bo, Yuan Qin, Wang Shanbao

82A 高碳钢热变形 Arrhenius 本构模型研究 ……… 张彭磊，苏 岚，胡 磊，沈 奎，麻 晗，米振莉　5-3
Study on Hot Deformation Arrhenius Constitutive Model of 82A High Carbon Steel
　　　　　　　　　　　　　　　　Zhang Penglei, Su Lan, Hu Lei, Shen Kui, Ma Han, Mi Zhenli

镀锌家电板白痕缺陷分析与改进 ……………………………… 赵 刚，毛一标，曹 垒 5-3
Analysis and Improvement of White Mark Defect of Galvanized Electric Appliance Plate
Zhao Gang, Mao Yibiao, Cao Lei

高强钢汽车板冷轧点状缺陷分析与改善 ……………………… 赵 刚，毛一标，曹 垒 5-4
Analysis and Improvement of Point Defects in High Strength Automobile Plate Cold Rolling
Zhao Gang, Mao Yibiao, Cao Lei

冷轧普冷卷板形问题分析及改善 ……………………………… 赵 刚，毛一标，曹 垒 5-5
Analysis and Improvement of Cold Rolled Ordinary Cold Rolled Sheet
Zhao Gang, Mao Yibiao, Cao Lei

冷轧普冷卷异物压入问题分析及改善 ………………………… 赵 刚，毛一标，曹 垒 5-6
Analysis and Improvement of the Defect of Impurity Pressed in Cold Rolled Sheet
Zhao Gang, Mao Yibiao, Cao Lei

TRIP钢的各向异性性能研究 ……………………… 杨登翠，阚鑫锋，李义强，孙蓟泉 5-7
Study on Anisotropic Properties of TRIP Steel ……… Yang Dengcui, Kan Xinfeng, Li Yiqiang, Sun Jiquan

MPM连轧限动芯棒失效形式及延长寿命的方式 …………………………… 谷大伟，洪 汛 5-8
The Failure for Mandrel Retaining MPM and the Measures of Improving the Life-span
Gu Dawei, Hong Xun

微张力对大方坯初轧过程的影响 ……………………………… 张 迪，米振莉，苏 岚 5-8
Finite Element Analysis of the Micro-tension Influence on Process of Square Billet
Zhang Di, Mi Zhenli, Su Lan

NJT600高强钢的焊接金相组织分析
……………… 罗扬，马成，孙力，白丽娟，谷秀锐，孙晓冉，周士伟，刘晓龙 5-9
Microstructure Analysis of the Welded Position of NJT600 High Strength Steel
……… Luo Yang, Ma Cheng, Sun Li, Bai Lijuan, Gu Xiurui, Sun Xiaoran, Zhou Shiwei, Liu Xiaolong

高速线材斯太尔摩冷却线仿真计算模型开发 ……………… 李会健，米振莉，苏 岚 5-10
Development of Simulation Calculation Model for Stelmor Cooling Line of High Speed Wire
Li Huijian, Mi Zhenli, Su Lan

钢丝电解磷化工艺探讨及应用 …………………… 周斌斌，段建华，廖 建，彭 凯 5-10
Research and Application of Wire Electrolytic Phosphating Process
Zhou Binbin, Duan Jianhua, Liao Jian, Peng Kai

高强钢回弹控制技术 ……………… 马闻宇，杨建炜，郑学斌，张永强，王宝川，李春光 5-11
The Springback Control Technology for High Strength Steel
……… Ma Wenyu, Yang Jianwei, Zheng Xuebin, Zhang Yongqiang, Wang Baochuan, Li Chunguang

微合金化高碳盘条生产钢帘线研究 ……………… 周志嵩，孙 忍，姚海东，殷建创，苗为钢 5-12
Study on Steel Cords Produced by Microalloying High-carbon Wire Rods
Zhou Zhisong, Sun Ren, Yao Haidong, Yin Jianchuang, Miao Weigang

二十辊SUNDWIG轧机支承辊轴承外圈爆裂原因分析及对策 ……… 袁海永，李 明，周 军 5-13
Cause Analysis and Countermeasure of Outer Ring Cracking of Backup Roll Bearing of 20-high SUNDWIG Mill
Yuan Haiyong, Li Ming, Zhou Jun

钢绞线用钢SWRH82B拉拔影响因素研究 ……… 贾元海，田 鹏，张俊粉，李吉伟，王 林 5-13
Study on Influencing Factors of Drawing Steel SWRH82B for Strand
Jia Yuanhai, Tian Peng, Zhang Junfen, Li Jiwei, Wang Lin

浅谈辊道对棒材超声全截面校验信号的影响因素与克服措施 ……… 张进科，汪建军，张 浩 5-14

The Influencing Factors and Overcoming Measures of the Roller Table on the Ultrasonic Check Signal of the Whole Section of the Bar are Briefly Discussed ············ Zhang Jinke, Wang Jianjun, Zhang Hao

0Cr17Ni12Mo2 不锈钢弹簧断裂原因分析 ············ 郭燕飞，赵星明，李居强，孙克强，邢献强　5-15
Fracture Analysis of 0Cr17Ni12Mo2 Stainless Steel Spring
　　　　　　　　　　　Guo Yanfei, Zhao Xingming, Li Juqiang, Sun Keqiang, Xing Xianqiang

新能源客车超高强钢方矩管链模成形产品质量仿真研究
　　　　　　熊自柳，齐建军，孙　力，董伊康，王　健，罗　扬，丁士超，孙　勇　5-15
Research on Simulation of Chain-die Forming of New Energy Bus AHSS Rectangle Tube Product Quality
　　　　　　Xiong Ziliu, Qi Jianjun, Sun Li, Dong Yikang, Wang Jian, Luo Yang, Ding Shichao, Sun Yong

1500MPa 级超高强钢的成形特性研究 ············ 纪登鹏，连昌伟，韩　非　5-16
Research on Forming Characteristics of Ultra High Strength Steel of 1500MPa Grade
　　　　　　　　　　　Ji Dengpeng, Lian Changwei, Han Fei

超高强钢在汽车座椅骨架轻量化中的应用 ············ 徐栋恺，胡　晓，陈自凯　5-17
Application of Ultra High Strength Steels on Seat Frame Lightweight ···· Xu Dongkai, Hu Xiao, Chen Zikai

1800MPa 级别超高强钢管气胀成形特性研究 ············ 程　超，韩　非，石　磊　5-18
Experiments and Simulation on Hot Gas Formability of B1800HS ············ Cheng Chao, Han Fei, Shi Lei

热辊弯工艺在汽车前保险杠总成上的轻量化应用 ············ 张骥超，石　磊，杨智辉　5-18
Lightweight Application of Hot Roll Bending Process on Front Bumper Assembly
　　　　　　　　　　　Zhang Jichao, Shi Lei, Yang Zhihui

常化工艺对 1.5%Si 无取向硅钢组织和磁性的影响
　　　　　　　　余红春，柳金龙，毛一标，周滨新，吴圣杰，马允敏　5-19
Effect of Normalizing Process on Microstructure and Magnetic Properties of 1.5%Si Non-oriented Silicon Steel
　　　　　　　　Yu Hongchun, Liu Jinlong, Mao Yibiao, Zhou Binxin, Wu Shengjie, Ma Yunmin

ML40Cr 马达轴断裂原因分析 ············ 王冬晨，阮士朋，王　欣，杨　栋　5-20
Analysis on Cracking Reason of ML40Cr Motor Shaft
　　　　　　　　　　　Wang Dongchen, Ruan Shipeng, Wang Xin, Yang Dong

10B21 紧固件低周疲劳和高周疲劳的断口分析 ············ 张　鹏，李家杨，阮士朋，贾东涛，王宁涛　5-21
Fracture Analysis of the 10B21 Fasteners for Low Cycle Fatigue and High Cycle Fatigue
　　　　　　　　　　　Zhang Peng, Li Jiayang, Ruan Shipeng, Jia Dongtao, Wang Ningtao

6　先进钢铁材料

6.1　汽车用钢

低温环境专用汽车法兰用 00Cr12NiTi 钢的研制 ············ 王志军　6-1
Development of 00Cr12NiTi Steel for Automoblie Flange in Low Temperature Environment ··· Wang Zhijun

现代汽车板热处理工艺与热工炉设计要领探讨 ············ 许秀飞，王业科，夏强强　6-1
Discussion on Annealing Technology of Modern Automobile Sheet and Design Essentials of Annealing Furnace
　　　　　　　　　　　Xu Xiufei, Wang Yeke, Xia Qiangqiang

薄规格"以热代冷"热轧双相钢的研制与开发 ······ 王　成，张　超，刘永前，赵　强，李　波，刘　斌　6-2
Development of Hot Rolled Thin Gauge Dual Phase Steel with "Cooling with Heat"
　　　　　　Wang Cheng, Zhang Chao, Liu Yongqian, Zhao Qiang, Li Bo, Liu Bin

SAPH400 热轧汽车用钢裂纹形成原因分析
　　　　　　　　王俊雄，时晓光，孙成钱，董　毅，张　宇，刘仁东，韩楚菲　6-3

Analysis of Causes of Crack Formation of Automobile Steel SAPH400
　　……Wang Junxiong, Shi Xiaoguang, Sun Chengqian, Dong Yi, Zhang Yu, Liu Rendong, Han Chufei

500L-Z 热轧汽车用大梁钢裂纹形成原因分析
　　……孙成钱，王俊雄，董　毅，时晓光，刘仁东，张　宇，韩楚菲　6-4

Analysis of Causes of Crack Formation of Automobile Beam Steel 500L-Z
　　……Sun Chengqian, Wang Junxiong, Dong Yi, Shi Xiaoguang, Liu Rendong, Zhang Yu, Han Chufei

过程工艺对热镀锌合金化超高强钢强度的影响……李茫茫　6-4
Effect of Process Technology on Strength of Hot-dip Galvanized and Alloyed Ultra-high Strength Steel
　　……Li Mangmang

材料表面分析技术及其在锌基镀层热冲压钢开发中的应用……毕文珍，洪继要　6-5
Material Surface Analysis Technology and Its Application in the Development of Zinc-coated Press Hardened Steel……Bi Wenzhen, Hong Jiyao

卷取后温度场对低碳微合金钢组织和性能的影响……张　宇，时晓光，董　毅，孙成钱，王俊雄，刘仁东　6-6
Effect of Coiling Temperature Field on the Microstructure and Mechanical Properties of Low-carbon Micro-alloyed Steel……Zhang Yu, Shi Xiaoguang, Dong Yi, Sun Chengqian, Wang Junxiong, Liu Rendong

近年汽车用高强钢板生产和开发热点分析……郑　瑞，代云红，魏丽艳　6-6
Hot Spot Analysis of Production and Development of High Strength Steel Sheet for Automobile in Recent Years
　　……Zheng Rui, Dai Yunhong, Wei Liyan

汽车发动机连杆拉伸性能试验影响因素的探讨……刘运娜，郝彦英，刘献达　6-7
Investigation about the Influence Factors for Automotive Engine Connecting Rod Tensile Property Test Impact Property……Liu Yunna, Hao Yanying, Liu Xianda

高强汽车板 DP980 在镀锌过程中的组织转变与相变……耿志宇，郭　健，宋海武　6-8
The Microstructure Change and Transformation of High Strength Automotive Plate DP980 during Galvanizing
　　……Geng Zhiyu, Guo Jian, Song Haiwu

含 Nb 高强汽车用钢卷取前后冷却条件对组织性能的影响分析
　　……苏振军，杨建宽，曹晓恩，孔加维，李超飞　6-8
Analysis of the Effect of Cooling Conditions before and after Coiling on Microstructure and Properties of Nb-contained High Strength Automobile Steel
　　……Su Zhenjun, Yang Jiankuan, Cao Xiaoen, Kong Jiawei, Li Chaofei

轻量化耐疲劳热轧双相车轮钢 DP590 的开发……张志强，柳风林，贾改风，裴庆涛，吕德文　6-9
Development of Light Weight Fatigue Resistant Hot Rolled Dual Phase Wheel Steel DP590
　　……Zhang Zhiqiang, Liu Fenglin, Jia Gaifeng, Pei Qingtao, Lv Dewen

淬火温度对 1000MPa 级 Q&P 钢组织性能的影响……王亚东，左海霞，陈虹宇　6-10
Effect of Quenching Temperature on Microstructure and Properties of 1000 MPa Grade Q&P Steel
　　……Wang Yadong, Zuo Haixia, Chen Hongyu

等温处理过程热轧 TRIP 钢过冷奥氏体的相变行为……王晓晖，康　健，袁　国，王国栋　6-10
Transformation Behaviors of Undercooled Austenite in Hot Rolled TRIP Steels during Isothermal Treatments
　　……Wang Xiaohui, Kang Jian, Yuan Guo, Wang Guodong

热镀锌板高耐蚀性环保钝化试验研究……董学强，郭太雄，寸海红，冉长荣　6-11
Study on Environment-friendly Passivation for Hot-dip Galvanized Steel with High Corrosion Resistance
　　……Dong Xueqiang, Guo Taixiong, Cun Haihong, Ran Changrong

热轧集装箱用钢 SPA-H 薄规格的开发……田　鹏，乔　俊，梁静召，刘　伟，张振全　6-12
Development of Hot-rolled Thin Gauge SPA-H for Container Steel
　　……Tian Peng, Qiao Jun, Liang Jingzhao, Liu Wei, Zhang Zhenquan

低碳钢铁素体轧制的力学性能和织构研究 ………… 刘宏博，郝磊磊，王建功，夏银锋，田 鹏，康永林 6-13
Study on Mechanical Properties and Texture of Low Carbon Steel under Ferritic Rolling
………………………………… Liu Hongbo, Hao Leilei, Wang Jiangong, Xia Yinfeng, Tian Peng, Kang Yonglin
铌微合金化对中碳冷镦钢组织和性能的影响 ………… 王利军，李永超，王 欣，阮士朋，王宁涛 6-14
Effect of Niobium Micro-alloying on Structure and Mechanical Properties of Medium-carbon Cold-heading Steel
……………………………………………… Wang Lijun, Li Yongchao, Wang Xin, Ruan Shipeng, Wang Ningtao
高品质汽车齿轮钢关键控制技术 ……………………………………………………………… 张 永 6-14
Key Technology of High Quality Automotive Gear Steel ……………………………………… Zhang Yong

6.2 船舶及海洋工程用钢

海洋平台用 620MPa 级超高强钢的热处理工艺研究和开发
……………………………………………………………… 周 成，赵 坦，金耀辉，朱隆浩，李家安 6-15
Research and Development of Heat Treatment of 620MPa Grade High Strength Steel Plates for Offshore Platform
……………………………………………………… Zhou Cheng, Zhao Tan, Jin Yaohui, Zhu Longhao, Li Jiaan
Ti-Mg 氧化物冶金钢中夹杂物演变行为研究 ……………………… 娄号南，王丙兴，王昭东 6-16
Study on Inclusion Evolution Behaviors of Ti-Mg Oxide Metallurgy Steel
…………………………………………………………………… Lou Haonan, Wang Bingxing, Wang Zhaodong
舞钢 LNG 船用低温高锰奥氏体钢的开发
……… 莫德敏，邓建军，龙 杰，张晨光，赵燕青，庞辉勇，杨 浩，张俊凯，王少义，陈文杰 6-16
Development of Low Temperature High Manganese Austenitic Steel for LNG Marine Use in Wugang
……………………………………………………… Mo Demin, Deng Jianjun, Long Jie, Zhang Chenguang,
………………………… Zhao Yanqing, Pang Huiyong, Yang Hao, Zhang Junkai, Wang Shaoyi, Chen Wenjie
高锰钢 TRIP/TWIP 效应对组织演变与应变硬化行为的影响规律研究
……………………………………………… 严 玲，张 鹏，王晓航，李文斌，李广龙 6-17
Researched on Microstructures Evolution and Strain Hardening Behavior of TRIP/TWIP Effect of High
Manganese Steel ……………………… Yan Ling, Zhang Peng, Wang Xiaohang, Li Wenbin, Li Guanglong
极地船用大线能量焊接用高强度船板的开发 ……………… 张 朋，邓建军，龙 杰，王晓书 6-18
The Development of Large Heat Input Welding High Strength Steel Plate used for Polar Ship
……………………………………………………………… Zhang Peng, Deng Jianjun, Long Jie, Wang Xiaoshu

6.3 电工钢

浅析硅钢连续退火炉炭套质量检验 ……………… 何明生，王雄奎，张 敬，龚学成，杨 朝，周旺枝 6-19
Quality Inspection of Carbon Sleeve in Continuous Annealing Furnace for Silicon Steel Production
……………………………………… He Mingsheng, Wang Xiongkui, Zhang Jing, Gong Xuecheng, Yang Chao, Zhou Wangzhi
氧化镁颗粒度检测的研究 ……………………………………………………………………… 齐 郁 6-20
Study on the Particle Size Measurement of Magnesium Oxide ……………………………………… Qi Yu
薄板坯生产线无取向硅钢 50BW600 表面缺陷分析 …………………………………… 李毅伟，李德君 6-20
Surface Quality Control of Non-oriented Silicon Steel 50BW600 in Thin Slab Production Line
……………………………………………………………………………………………… Li Yiwei, Li Dejun
薄板坯连铸连轧工艺生产高牌号无取向硅钢研发情况 ……………………………………………… 王 媛 6-21
高牌号无取向硅钢在发电机中的应用浅研
……………………………………… 高振宇，陈春梅，李亚东，刘文鹏，罗 理，赵 健，姜福健 6-21
提高无取向硅钢焊接性能的试验研究 ……… 姜福健，高振宇，张仁波，张本尊，王 铁，张智义，李文权 6-21

Experimental Research on Ameliorating Welding Properties of Non-oriented Silicon Steel
  ·········· Jiang Fujian, Gao Zhenyu, Zhang Renbo, Zhang Benzun, Wang Tie, Zhang Zhiyi, Li Wenquan

常化工艺对 3.15% Si 硅钢冷加工性的影响 ·················· 刘文鹏，高振宇，陈春梅，李亚东   6-22
The Effect of Normalizing Parameters on the Cold Rollability of 3.15% Si Silicon Steel
  ··· Liu Wenpeng, Gao Zhenyu, Chen Chunmei, Li Yadong

C、Si、Mn 对 Hi-B 钢 γ-相影响规律的热力学研究 ········ 庞树芳，游清雷，李　莉，贾志伟，蒋奇武   6-23
Thermodynamic Study on the Effect of C、Si and Mn on the Austenite Phase of Hi-B Steel
  ··· Pang Shufang, You Qinglei, Li Li, Jia Zhiwei, Jiang Qiwu

激光刻痕对取向硅钢铁损变化的影响规律 ················ 李　莉，张　静，贾志伟，庞树芳，蒋奇武   6-23
The Effect of Laser Scribing on the Core Loss of Grain-oriented Silicon Steel
  ··· Li Li, Zhang Jing, Jia Zhiwei, Pang Shufang, Jiang Qiwu

6.4 建 筑 用 钢

浅析一种新钢筋——精轧稀螺纹钢筋的科学合理性及应用前景 ·· 曹全兴   6-24
Analysis of the Scientific Rationality and Application Prospect of Finish Rolling Thin Threar Reinforced Bar
  ·· Cao Quanxing

建筑结构用低屈服点抗震钢板 LY225 的研制开发 ········ 王晓书，邓建军，张　朋，龙　杰，赵国昌   6-25
Research and Development of LY225 Low Yield Point Anti-seismic Steel Plate for Building Structure
  ··· Wang Xiaoshu, Deng Jianjun, Zhang Peng, Long Jie, Zhao Guochang

宣钢 HRBE400 钢筋铌微合金化的探索 ··· 柳金瑞，王宏斌，张明海   6-25

氮在 T-5 CA 中的强化作用及其应用 ···················· 孙超凡，方　圆，吴志国，王雅晴，刘　伟   6-26
Reinforcement and It's Application of Nitrogen in T-5 CA Steel
  ··· Sun Chaofan, Fang Yuan, Wu Zhiguo, Wang Yaqing, Liu Wei

高层建筑用特厚高强结构钢的开发和应用 ············ 顾林豪，路士平，刘金刚，刘春明，初仁生，张苏渊   6-26
Research and Development of Ultra-thick High-strength Structural Steel for High-rise Buildings
  ··· Gu Linhao, Lu Shiping, Liu Jingang, Liu Chunming, Chu Rensheng, Zhang Suyuan

建筑结构用高强抗震耐候钢的开发与应用研究 ········ 陈振业，齐建军，孙　力，安会龙，年保国，姚纪坛   6-27
Development and Application of High Strength Earthquake Resistant Weathering Steel for Building Structure
  ··· Chen Zhenye, Qi Jianjun, Sun Li, An Huilong, Nian Baoguo, Yao Jitan

6.5 轨 道 用 钢

铝对 ML40Cr 冷镦钢的影响 ·························· 高　航，郭大勇，王宏亮，车　安，王秉喜，马立国   6-28
Influence of Aluminum on ML40Cr Cold-heading Steel
  ··· Gao Hang, Guo Dayong, Wang Hongliang, Che An, Wang Bingxi, Ma Liguo

重轨开坯孔型系统优化及有限元模拟分析 ············ 付国龙，赵利永，刘小燕，李聪颖，王世杰   6-29

基于有限单元法的重轨在线余热淬火过程数值模拟研究 ·············· 范　佳，易洪武，李钧正，韩志杰   6-29
Research on Numerical Simulation of Heavy Rail On-line Waste Heat Quenching Process based on Finite Element
  Method ·· Fan Jia, Yi Hongwu, Li Junzheng, Han Zhijie

R350HT 热处理钢轨的开发实践 ··· 王瑞敏，周剑华，费俊杰，欧阳珉路   6-30
Development of R350HT Heat-treated Rail ··················· Wang Ruimin, Zhou Jianhua, Fei Junjie, Ouyang Minlu

百米钢轨水淬工艺研究与应用 ··· 张海旺，李钧正，韩志杰   6-30
Research and Application of Water Quenching Technology for 100m Rail
  ··· Zhang Haiwang, Li Junzheng, Han Zhijie

6.6 能源用钢

L485M 钢板的组织与性能 ……………………………… 渠秀娟，应传涛，徐 烽，陈军平，惠兴伟　6-31
Microstructure and Property of L485M Plates
………………………………………… Qu Xiujuan, Ying Chuantao, Xu Feng, Chen Junping, Hui Xingwei

耐候钢板在转炉煤气柜的应用 ………………………………………………… 郭 凌，王 河，李军红　6-32

大壁厚大管径 X65M 管线钢落锤性能优化 ……………………………………… 喻赛华，史术华　6-32

高性能风电用 S420ML 宽厚钢板的研制与开发 …………… 潘中德，李 伟，胡其龙，高 飞　6-32
Research and Development of S420ML Wide & Heavy Steel Plate for High Performance Wind Power
………………………………………………………………… Pan Zhongde, Li Wei, Hu Qilong, Gao Fei

厚规格 X80 管线钢制管前后的力学性能变化分析 …………… 张豪臻，章传国，孙磊磊　6-33
Analysis on Mechanical Properties Variation of Thick-walled X80 Pipeline Steel before and after Pipe Manufacturing ……………………………………………… Zhang Haozhen, Zhang Chuanguo, Sun Leilei

高耐酸性管线钢的研制和性能研究 ………………… 孔祥磊，黄国建，黄明浩，王 杨，张英慧　6-34
Development and Properties Research of High Acid Resistance Pipeline Steel
………………………………………… Kong Xianglei, Huang Guojian, Huang Minghao, Wang Yang, Zhang Yinghui

低屈强比高强韧性海底管线钢控轧控冷工艺研究 ……………… 李冠楠，贾改风，孙 毅　6-35
Development of Low Yield Strength Ratio High Strength and Toughness Submarine Pipeline Steel
………………………………………………………………… Li Guannan, Jia Gaifeng, Sun Yi

高等级 X90 管线钢奥氏体动态再结晶规律研究 ……………… 张 海，李少坡，丁文华　6-35
Study on Dynamic Recrystallization Rule of Austenite for High Grade X90 Pipeline Steel
………………………………………………………………… Zhang Hai, Li Shaopo, Ding Wenhua

舞钢石油化工和煤化工用高性能容器钢板的生产 …………… 吴艳阳，龙 杰，李样兵　6-36
The Production of High Performance Container Steel Plate for Wusteel Petrochemical and Coal Chemical Industry ………………………………………………… Wu Yanyang, Long Jie, Li Yangbing

复合微合金化 J55 石油套管用钢生产实践 …………… 陆凤慧，邢俊芳，周少见，白宗奇　6-37
Production Practice of Compound Microalloyed J55 Oil Casing Steel
………………………………………………………………… Lu Fenghui, Xing Junfang, Zhou Shaojian, Bai Zongqi

两种低温高强钢的比较研究 ………………………………………………………… 罗 毅　6-37
Comparative Study on Two Types of Low Temperature High Strength Steels …………………… Luo Yi

6.7 工程机械用钢

500MPa 级高耐磨焊管用钢的开发 ………… 惠亚军，吴科敏，刘 锟，许克好，许斐范，令狐克志　6-38
Development of 500MPa Grade High Wear-resistant Steel for Welded Tube
………………………………………… Hui Yajun, Wu Kemin, Liu Kun, Xu Kehao, Xu Feifan, Linghu Kezhi

模拟加速腐蚀条件下耐候螺栓的腐蚀行为研究
………………… 罗志俊，王晓晨，孙齐松，徐士新，刘厚权，黄 耀，田志红　6-39
Study on Corrosion Behavior of Weathering Bolts under Simulated Accelerated Corrosion
Luo Zhijun, Wang Xiaochen, Sun Qisong, Xu Shixin, Liu Houquan, Huang Yao, Tian Zhihong

耐磨蚀钢的研制及应用 ………………………………………… 宋凤明，温东辉，杨阿娜　6-40
Development and Application of Erosion-corrosion Resistant Steel
………………………………………………………………… Song Fengming, Wen Donghui, Yang Ana

推土机铲斗用调质耐磨钢 27MnTiB 的开发 …… 秦 坤，尹绍江，霍晶晶，李 行，任树洋，赵庆宇　6-40

Development of Tempered and Wear-resistant Steel 27MnTiB for Bulldozer Bucket
　　　　　　　　　　　　　 Qin Kun, Yin Shaojiang, Huo Jingjing, Li Hang, Ren Shuyang, Zhao Qingyu

回火热处理对低碳贝氏体钢显微组织和冲击断裂过程的影响
　　　　　　　　　　　　　　　　　 黄少文，周　平，张学民，杨　恒，李长新　　6-41
Effect of Tempering Heat Treatment Process on Microstructure and Impact Fracture Process of Low Carbon Bainite Steel 　　　　　Huang Shaowen, Zhou Ping, Zhang Xuemin, Yang Heng, Li Changxin

460MPa 级工程机械用钢表面边裂分析……… 张卫攀，吕德文，孙电强，刘红艳，张瑞超，冯俊鹏　　6-42
Surface Edge Crack Analysis of Construction Machinery Steel of 460MPa Grade
　　　　　　　　　　 Zhang Weipan, Lv Dewen, Sun Dianqing, Liu Hongyan, Zhang Ruichao, Feng Junpeng

TMCP 工艺 Q550D 钢板及焊接组织和性能研究………… 孙电强，王会岭，张瑞超，王丽敏，张卫攀　　6-42
Research on Microstructure and Properties of TMCP Q550D Plate and Weld
　　　　　　　　　　　　　　 Sun Dianqiang, Wang Huiling, Zhang Ruichao, Wang Limin, Zhang Weipan

钛系热处理高强钢开发应用与强韧性机理研究………………… 何亚元，宋育来，李利巍，陆在学，薛　欢　　6-43
Development and Application of Titanium-based Heat-treated High Strength Steel and Study on Strength and
　　Toughness Mechanism ……………………………………… He Yayuan, Song Yulai, Li Liwei, Lu Zaixue, Xue Huan

25%Mn 超低温用高锰钢熔敷金属的组织与性能……………… 陈亚魁，王红鸿，孟　亮，孙　超，李东晖　　6-44
Microstructure and Properties of Deposited Metal of 25%Mn Ultra-low Temperature Steel
　　　　　　　　　　　　　　　　　　 Chen Yakui, Wang Honghong, Meng Liang, Sun Chao, Li Donghui

Nb 元素对 NM450 耐磨钢板组织和性能的影响 ……… 金　池，邓想涛，王昭东，闫强军，杨　柳，张永青　　6-44
Effect of Niobium on the Microstructure and Mechanical Properties of the NM450 Wear Resistance Steel Plate
　　　　　　　　　　　　　　　　 Jin Chi, Deng Xiangtao, Wang Zhaodong, Yan Qiangjun, Yang Liu, Zhang Yongqing

新型超级耐磨钢板研制开发与工业化应用………… 邓想涛，王昭东，付天亮，黄　龙，梁　亮，闫强军　　6-45
Development and Industrial Application of New Super Low Alloy Abrasion Resistant Steel
　　　　　　　　　　　　　　　 Deng Xiangtao, Wang Zhaodong, Fu Tianliang, Huang Long, Liang Liang, Yan Qiangjun

8.8 级紧固件用钢 SWRCH35K 盘条的研制开发……………… 汪青山，贾元海，刘效云，王　林，李吉伟　　6-46
Development of Steel SWRCH35K Wire Rod for 8.8 Grade Fastener Steel
　　　　　　　　　　　　　　　　　　　 Wang Qingshan, Jia Yuanhai, Liu Xiaoyun, Wang Lin, Li Jiwei

10.9 级紧固件用钢 ML20MnTiB 盘条的研制开发………………………… 李吉伟，贾元海，王　林，刘效云　　6-47
Development of Steel ML20MnTiB Wire Rod for 10.9 Class Fasteners
　　　　　　　　　　　　　　　　　　　　　　　　 Li Jiwei, Jia Yuanhai, Wang Lin, Liu Xiaoyun

PC 钢棒用 40Si2Mn 盘条的研制与开发……………………………… 王　林，梁　均，贾元海，李吉伟　　6-47
The Development of 40Si2Mn Wire Rod for PC Steel Bar ……… Wang Lin, Liang Jun, Jia Yuanhai, Li Jiwei

具有良好强韧性能高等级耐磨钢 NM500 生产工艺开发及应用
　　　　　　　　　　　　　　　　　　　　　　 刘红艳，王昭东，吕德文，邓想涛，姚　宙　　6-48
Development and Application of NM500 Production Process for High-grade Wear-resistant Steel with Good
　　Strength and Toughness ……………… Liu Hongyan, Wang Zhaodong, Lv Dewen, Deng Xiangtao, Yao Zhou

6.8　其他用途用钢

BN 型易切削钢中夹杂物析出规律及对性能的影响 ………………… 张　帆，任安超，夏艳花，丁礼权　　6-49
Smelting and Performance Research of BN-type Automotive Cutting Steel
　　　　　　　　　　　　　　　　　　　　　　　　　　 Zhang Fan, Ren Anchao, Xia Yanhua, Ding Liquan

抗震阻尼器用极低屈服点钢的组织与性能调控研究
　　　　　　　　　　　　　　　　　　　 李昭东，陈润农，赵　刚，张明亚，杨忠民，杨才福　　6-49

Study on the Controlling of Microstructure and Mechanical Properties of Ultra Low Yield Point Steel for Aseismic Damper ⋯ Li Zhaodong, Chen Runnong, Zhao Gang, Zhang Mingya, Yang Zhongmin, Yang Caifu

7 粉末冶金

气雾化高硅铁粉性能分析 ⋯⋯⋯⋯⋯⋯⋯⋯⋯⋯⋯⋯⋯⋯⋯⋯⋯⋯⋯⋯ 康 伟，廖相巍，贾吉祥，康 磊　7-1
Performance Analysis of High Silicon Iron Powder by Atomization Method
⋯⋯⋯⋯⋯⋯⋯⋯⋯⋯⋯⋯⋯⋯⋯⋯⋯⋯⋯⋯⋯⋯ Kang Wei, Liao Xiangwei, Jia Jixiang, Kang Lei

金属粉末做中间材热轧制备不锈钢复合板 ⋯⋯⋯ 康 磊，廖相巍，尚德礼，贾吉祥，李广帮，康 伟　7-1
Stainless Steel Composite Plate Prepared by Hot-roll Bonding with Metal Powder as Intermediate Material
⋯⋯⋯⋯⋯⋯⋯⋯⋯⋯⋯⋯⋯⋯ Kang Lei, Liao Xiangwei, Shang Deli, Jia jixiang, Li Guangbang, Kang Wei

WC-Ni 无磁硬质合金研究进展及应用 ⋯⋯⋯⋯⋯⋯⋯⋯⋯⋯ 李金普，施瑜蕾，柳学全，姜丽娟　7-2
Research Development and Application on Non-magnetic Cemented Carbide
⋯⋯⋯⋯⋯⋯⋯⋯⋯⋯⋯⋯⋯⋯⋯⋯⋯⋯⋯⋯⋯⋯⋯ Li Jinpu, Shi Yulei, Liu Xuequan, Jiang Lijuan

Fe-2Cu-0.5C 粉末烧结钢高温拉伸流变应力及预测
⋯⋯⋯⋯⋯⋯⋯⋯⋯⋯⋯⋯⋯⋯⋯⋯⋯⋯⋯⋯⋯ 李 强，郭 彪，吴 辉，敖进清，宋 欢，敖逸博　7-3
Flow Stress and Prediction of Powder Sintered Fe-2Cu-0.5C Steel during High Temperature Tensile
⋯⋯⋯⋯⋯⋯⋯⋯⋯⋯⋯⋯⋯⋯⋯⋯⋯⋯⋯ Li Qiang, Guo Biao, Wu Hui, Ao Jinqing, Song Huan, Ao Yibo

不同碳含量粉末烧结钢镦粗流变致密化行为研究
⋯⋯⋯⋯⋯⋯⋯⋯⋯⋯⋯⋯⋯⋯⋯⋯⋯⋯⋯⋯⋯ 吴 辉，郭 彪，李 强，敖进清，宋 欢，敖逸博　7-4
Study on the Flow and Densification Behaviors of Powder Sintered Steel with Different Carbon Contents during Clod Upsetting ⋯⋯⋯⋯⋯⋯⋯⋯⋯⋯ Wu Hui, Guo Biao, Li Qiang, Ao Jinqing, Song Huan, Ao Yibo

热处理对铜基粉末冶金刹车片的摩擦性能的影响 ⋯⋯⋯⋯⋯ 王 洋，胡 铮，张万昊，张 坤，魏炳忱　7-4
Effect of Heat Treatment on Friction Properties of the Cu-based Powder Metallurgy Brake Pads
⋯⋯⋯⋯⋯⋯⋯⋯⋯⋯⋯⋯⋯⋯⋯⋯⋯⋯⋯⋯⋯⋯⋯⋯ Wang Yang, Hu Zheng, Zhang Wanhao, Zhang Kun, Wei Bingchen

8 能源、环保与资源利用

8.1 能源与资源利用

热风炉热平衡软件开发与应用 ⋯⋯⋯⋯⋯⋯⋯⋯⋯⋯⋯⋯⋯⋯⋯⋯⋯⋯⋯⋯⋯⋯⋯⋯⋯ 陈冠军　8-1
Development and Application of Heat Balance Software for Hot Blast Stove ⋯⋯⋯⋯⋯⋯ Chen Guanjun

改性钢渣吸附 MTBE 污染废水的动力学研究 ⋯⋯ 刘 芳，胡绍伟，陈 鹏，王 飞，王 永，徐 伟　8-2
Dynamics Study on Adsorption of MTBE Wastewater by Modified Steel Slag
⋯⋯⋯⋯⋯⋯⋯⋯⋯⋯⋯⋯⋯⋯⋯⋯⋯⋯⋯⋯⋯⋯ Liu Fang, Hu Shaowei, Chen Peng, Wang Fei, Wang Yong, Xu Wei

苯加氢催化剂器外预硫化技术实践 ⋯⋯⋯⋯⋯⋯⋯⋯⋯⋯⋯ 陶帅江，董 毅，杜建全，孙鸿君　8-2
Practices of Ex-situ Pre-sulphurizing Process for Hydrogenation Catalysts
⋯⋯⋯⋯⋯⋯⋯⋯⋯⋯⋯⋯⋯⋯⋯⋯⋯⋯⋯⋯⋯⋯⋯⋯⋯ Tao Shuaijiang, Dong Yi, Du Jianquan, Sun Hongjun

永磁调速技术在鞍钢生活水供应中的探索与实践 ⋯⋯⋯⋯⋯⋯⋯ 闫 涛，刘 星，邸光宇，魏 彬　8-3
Exploration and Practice of Permanent Magnet Speed Regulation Technology in Life Water Supply of Ansteel
⋯⋯⋯⋯⋯⋯⋯⋯⋯⋯⋯⋯⋯⋯⋯⋯⋯⋯⋯⋯⋯⋯⋯⋯⋯⋯⋯⋯ Yan Tao, Liu Xing, Di Guangyu, Wei Bin

煤化工企业蒸汽系统节能、优化的研究与应用 ⋯⋯⋯⋯⋯⋯⋯⋯⋯ 王 震，穆春丰，李冈亿　8-4
Research and Application of Energy Saving and Optimization of Steam System in Coal Chemical Enterprises
⋯⋯⋯⋯⋯⋯⋯⋯⋯⋯⋯⋯⋯⋯⋯⋯⋯⋯⋯⋯⋯⋯⋯⋯⋯⋯⋯⋯⋯ Wang Zhen, Mu Chunfeng, Li Gangyi

冶金企业保障输电线路稳定性相关措施	刘 鹏	8-4
Metallurgical Enterprises to Ensure the Stability of Transmission Lines Related Measures	Liu Peng	
论高炉煤气干湿除尘工艺及配套TRT运行优劣	李山虎，哈 君	8-5
The Comparison of BFG Dry-type Dedusting and BXF Dedusting Technology with Related TRT	Li Shanhu, Ha Jun	
焦炉煤气制氢系统常见故障、影响及解决方案	李山虎，杨子壮，赵 冬	8-6
Problem, Effect and Solution of Hydrogen Production from Coke Oven Gas	Li Shanhu, Yang Zizhuang, Zhao Dong	
一种适用于轧钢加热炉的煤气除湿技术	徐言东，谢仲豪，刘 涛，余 伟，程知松，魏付豪	8-6
A Gas Dehumidification Technology for Steel Rolling Heating Furnace	Xu Yandong, Xie Zhonghao, Liu Tao, Yu Wei, Cheng Zhisong, Wei Fuhao	
鞍钢高炉工序CO_2排放水平及减排展望	孟凡双，李建军，曾 宇，李晓春	8-7
鞍钢低成本有效提高热风温度技术	刘德军，袁 玲，赵爱华	8-7
Low Cost Effective Hot Air Temperature Technology in Angang	Liu Dejun, Yuan Ling, Zhao Aihua	
冷轧热镀锌退火炉测试分析与节能	高 军，马光宇，刘常鹏，张天赋，赵 俣	8-8
The Energy Saving and Thermal Measurements on Annealing Furnace in Cold Rolled Hot Dip Galvanizing Line	Gao Jun, Ma Guangyu, Liu Changpeng, Zhang Tianfu, Zhao Yu	
鞍钢高炉热风炉燃烧控制系统应用实践	李建军，孟凡双，曾 宇，李晓春	8-9
Application Practice of Combustion Control System for Hot Blast Furnace of Angang	Li Jianjun, Meng Fanshuang, Zeng Yu, Li Xiaochun	
红土镍矿资源特点及矿物改性研究现状	武兵强，齐渊洪，周和敏，高建军，邹宗树	8-9
Characteristics and Research Status of Mineral Modification of Laterite Nickel Ore	Wu Bingqiang, Qi Yuanhong, Zhou Hemin, Gao Jianjun, Zou Zongshu	
鞍钢2000m^3焦炉煤气制氢项目探析	舒 畅，杨子壮	8-10
Study on the 2000m^3 COG to H_2 Project in Ansteel	Shu Chang, Yang Zizhuang	
碳纤维复合芯铝绞线在钢铁企业的应用	韩 帅，李 达	8-11
Application of Carbon Fiber Composite Core Aluminum Strand in Iron and Steel Enterprises	Han Shuai, Li Da	
厚板加热炉自动炉温设定方法初探	张敏文	8-11
Research for the Automatic Furnace Temperature Setting of Heavy Plate Mill	Zhang Minwen	
冷轧浓油废水油水分离废油的资源化分析	张国志，尹婷婷	8-12
Reclamation Analysis of Oil-Water Separation Waste Oil from Cold Rolling Heavy Oil Wastewater	Zhang Guozhi, Yin Tingting	
TRT旁通管变工况下的仿真及设计优化	潘 宏，李星星，刘丹瑶，胡学羽	8-13
Simulation and Design Optimization of TRT Bypass Pipeline under Variable Conditions	Pan Hong, Li Xingxing, Liu Danyao, Hu Xueyu	
鱼雷罐加金属化球团对铁水温度的影响	谢俊波，牛长胜，金 奕	8-13
Effect of Taking Metallized Pellets into Torpedo Tank on the Temperature of Molten Iron	Xie Junbo, Niu Changsheng, Jin Yi	
板框压滤机在含酸废水处理改进工艺中的应用	杨 跃	8-14
Application of Plate and Frame Filter Press in Improvement Process of Acid Wastewater Treatment	Yang Yue	
梅钢高炉水渣水处理工艺优化	翟玉龙	8-14

The Water Treatment Process Optimization for Blast Furnace Slag in Meishan Steel ······ Zhai Yulong

无焰富氧燃烧技术在马钢板坯加热炉上的应用 ······ 丁　毅，田　俊，范满仓，周劲军，翟　炜　8-15
Application of Flameless Oxy-fuel Burning Technology in Slab Heating Furnace of Masteel
······ Ding Yi, Tian Jun, Fan Mancang, Zhou Jinjun, Zhai Wei

CCPP 机组夏季燃料水喷射降温加湿技术及应用 ······ 薛晓金　8-16
Technology and Application of Cooling and Humidifying by Fuel Water Injection in CCPP Unit in Summert
······ Xue Xiaojin

蒸汽管网安全运行的分析与事故防治 ······ 黄显保，何　嵩，刘柏寒　8-16

新型接地选线装置的应用及案例分析 ······ 李　达，韩　帅，王林松，唐　巍，沙　金　8-17
Application and Case Analysis of New Earth Line Selection Device
······ Li Da, Han Shuai, Wang Linsong, Tang Wei, Sha Jin

焦化废水处理工艺氨氮指标的控制措施 ······ 李国庆　8-17
The Control Measures of NH_3-N Index in Coking Wastewater Treatment System ······ Li Guoqing

钢化联合企业生产能源一体化智能管控系统探讨 ······ 刘书文，欧　燕，叶理德　8-18

换热器在步进式加热炉生产中的技术进步 ······ 李国辉　8-18

酒钢矿棉加工可行性探析 ······ 赵贵清，陈亚团　8-19
Feasibility Investigation of Mineral Wool Produced in Jiusteel ······ Zhao Guiqing, Chen Yatuan

鞍钢第二发电厂 CCPP 厂用电系统经济运行方案分析 ······ 吴　猛，孙德俊，丁伟伟　8-19
Analysis on the Economic Operation Scheme of CCPP Power Plant in Angang Second Power Plant
······ Wu Meng, Sun Dejun, Ding Weiwei

钢铁厂粉尘处理及二次产物综合利用 ······ 孙利军，张　晨，张丽俊　8-20
Dust Treatment and Comprehensive Utilization of Secondary Products in Iron and Steel Works
······ Sun Lijun, Zhang Chen, Zhang Lijun

梅钢烧结矿竖冷炉内物料运动特性的模拟研究 ······ 孙俊杰，程乃良，徐　骥　8-21
Simulation of Particle Motion Characteristics in Sinter Shaft Cooler of Meisteel
······ Sun Junjie, Cheng Nailiang, Xu Ji

燃气锅炉燃烧自动控制优化 ······ 任　珲，降万勇　8-21
Optimization of Automatic Combustion Control for Gas-fired Boilers ······ Ren Hui, Jiang Wanyong

节能减排装置"高炉煤气透平机组"故障分析及解决方法 ······ 谢福成　8-22
Fault Analysis and Solution of "Top Gas Turbine Unit" for Energy Saving and Emission Reduction Device
······ Xie Fucheng

电场处理对渣金反应元素迁移的影响研究 ······ 廖直友，王博文，王海川，尹振兴　8-23
Effect of Electric Field Treatment on the Transport of Reactive Elements in Slag-metal Reaction
······ Liao Zhiyou, Wang Bowen, Wang Haichuan, Yin Zhenxing

钢铁工业能耗现状和节能潜力分析 ······ 王维兴　8-23
The Status of Energy Consumption and Analysis of Energy Saving Potential in Iron and Steel Industry
······ Wang Weixing

转炉烟气中低温余热回收工艺研究 ······ 周　涛，侯祥松，谢　建，董茂林　8-24

8.2　冶金环保

承压一体化冶金污水净化装置在连铸浊环水系统中的应用实例 ······ 李建康，李　杰，梁思懿　8-25
Application of Pressure Type Integration Metallurgy Sewage Purification Treatment Device in Turbid
　　Circulating Water System of Continuous Casting ······ Li Jiankang, Li Jie, Liang Siyi

吹扫捕集/气相色谱-质谱法和气相色谱法同步测定水中挥发性有机物和挥发性石油烃
…………………………………………………………………… 凌 冰，高巨鹏，黄 晓 8-25
Simultaneous Determination of Volatile Organic Compounds and Volatile Petroleum Hydrocarbons in Water
by Purge and Trap/Gas Chromatography-mass Spectrometry and Gas Chromatography
…………………………………………………………………… Ling Bing, Gao Jupeng, Huang Xiao

冶金除尘灰科学利用技术集成与应用 ……………………… 刘德军，张延辉，唐继忠，袁 玲 8-26
Technology Integration and Application of Scientific Utilization of Metallurgical Dust
…………………………………………………………………… Liu Dejun, Zhang Yanhui, Tang Jizhong, Yuan Ling

一种芽孢杆菌在降解焦化废水中苯酚的作用研究
…………………………………………… 王 永，袁 玲，陈 鹏，胡绍伟，王 飞，刘 芳 8-27
Study on the Role of a Bacillus in the Degradation of Phenol in Coking Wastewater
…………………………………………… Wang Yong, Yuan Ling, Chen Peng, Hu Shaowei, Wang Fei, Liu Fang

$321m^2$ 带式焙烧机头电除尘器的选型与实践 ……………………………… 金勇成，李小丽 8-27
Type Selection and Practice of $321m^2$ Belt - type Electric Dust Collector ……… Jin Yongcheng, Li Xiaoli

焦炉烟囱二氧化硫排放浓度分析与治理 ………… 张其峰，张允东，周晓锋，孔 弢，李富鑫 8-28
Analysis and Control of Sulfur Dioxide Emission from Coke Oven Chimney
…………………………………………… Zhang Qifeng, Zhang Yundong, Zhou Xiaofeng, Kong Tao, Li Fuxin

电吸附除盐技术用于梅钢再生水中试研究 ………………………… 邹正东，初 明，胡小龙 8-29
Pilot Study on Desalination of Recycled Water by Electro-absorption Technique in Meigang
…………………………………………………………………… Zou Zhengdong, Chu Ming, Hu Xiaolong

冶金物料清洁高效转运新技术的研究与探讨 …………………… 毕 琳，王得刚，徐培万 8-29
Research and Discussion on the new Technology of Clean and Efficient Transport of Metallurgical Materials
…………………………………………………………………… Bi Lin, Wang Degang, Xu Peiwan

含有氯化亚铁酸性废水出水超标的原因分析 ……………………………………… 胡金良 8-30
Causal Analysis of Acidic Waste Water Containing Ferrous Chloride Beyond-standard Discharging
…………………………………………………………………… Hu Jinliang

硅钢退火机组废气中氮氧化物含量的控制 …………… 汪剑飞，卢禹龙，范建军，毕 军，马允敏 8-31
Control of Nitrogen Oxide Content in Waste Gas of Annealing Unit of Silicon Steel
…………………………………………… Wang Jianfei, Lu Yulong, Fan Jianjun, Bi Jun, Ma Yunmin

生态环境强约束下的京津冀钢铁节能环保产业发展战略研究
…………………………………………… 汪 浩，潘崇超，刘育松，张立明，王一帆，赵喜彬 8-31
Research on the Development Strategy of Beijing-Tianjin-Hebei Steel Energy-saving and Environmental
Protection Industry under the Strong Ecological Environment
…………………………………………… Wang Hao, Pan Chongchao, Liu Yusong, Zhang Liming, Wang Yifan, Zhao Xibin

热电厂 5 号 6 号锅炉烟气在线监控系统矩阵流速仪改造的实践 ………………… 卢 亮 8-32
钢铁企业 CO_2 排放计算方法探讨 …………………………………………………… 王维兴 8-33
Discussion on the Calculation Method of CO_2 Emission in Iron and Steel Enterprises ……… Wang Weixing

9 钢铁材料表征与评价

基于大数据和 XGBoost 的热轧板带力学性能预测模型 ……… 陈金香，赵 峰，孙彦广，张 琳，尹一岚 9-1
A Prediction Model based on Big Data and XGBoost for the Mechanical Properties of Hot -rolled Strips
…………………………………………… Chen Jinxiang, Zhao Feng, Sun Yanguang, Zhang Lin, Yin Yilan

船运皮带自动取样设备选型及应用 ……………………………………………………… 罗 湘 9-2

38MnVS6 活塞销座孔内壁掉肉原因分析 ………… 李富强，林晏民，罗新中，章玉成，朱祥睿，杨明梅   9-2
Cause Analysis of Flesh out on the Inner Wall of 38MnVS6 Piston Pin Hole
                Li Fuqiang, Lin Yanmin, Luo Xinzhong, Zhang Yucheng, Zhu Xiangrui, Yang Mingmei

二安替吡啉甲烷光度法测定钛铁中钛含量 ………… 邓军华，高品，王莹，刘冬杰，李颖   9-3
Spectrophotometric Determination of Titanium Content in Titanium Ferroalloy with Diantipyrylmethane
                Deng Junhua, Gao Pin, Wang Ying, Liu Dongjie, Li Ying

SWRCH22A 冷镦钢线材拉拔断裂原因分析 ………… 李富强，罗新中，孙福猛，杨明梅，章玉成，朱祥睿   9-3
Analysis on Causes of Fracture Occurred in Drawing of SWRCH22A Cold Heading Wire Rod
                Li Fuqiang, Luo Xinzhong, Sun Fumeng, Yang Mingmei, Zhang Yucheng, Zhu Xiangrui

调质处理对油井管的微观组织和耐蚀性能的影响
                钟彬，陈义庆，高鹏，李琳，艾芳芳，伞宏宇，肖宇   9-4
Effects of Quenching and Tempering Treatment on Microstructure and Corrosion Resistance of Oil Well Pipes
                Zhong Bin, Chen Yiqing, Gao Peng, Li Lin, Ai Fangfang, San Hongyu, Xiao Yu

冷轧薄钢板 DC04 表面波纹度测试及分析 ………… 郭晶，孟庆刚，王鲲鹏   9-5
Measurement and Analysis of Surface Waviness of DC04 Cold Rolled Sheet Steel
                Guo Jing, Meng Qinggang, Wang Kunpeng

激光切割技术在制备钢铁中氮元素含量分析样品中的应用 ………… 王鹏，刘步婷，杨琳，刘俊   9-6
Application of Laser Cutting Technology in Preparing Nitrogen Content Analysis Samples in Steel
                Wang Peng, Liu Buting, Yang Lin, Liu Jun

40Cr 连接螺栓断裂失效分析 ………… 王德宝，杨峥，牟祖茂，徐辉，徐雁，宋鑫晶   9-6
Failure Analysis of 40Cr Steel Connecting Bolt
                Wang Debao, Yang Zheng, Mou Zumao, Xu Hui, Xu Yan, Song Xinjing

冷轧稀碱含油废水中挥发性有机污染物特征研究 ………… 洪涛，何晓蕾，张毅，李恩超   9-7
Study on Characteristics of VOCs in Cold Mill Dilute Alkali Oily Wastewater
                Hong Tao, He Xiaolei, Zhang Yi, Li Enchao

波长色散 X 射线荧光光谱法测定渣类化学成分 ………… 刘青桥，郭登元   9-8

原子吸收分光光度计火焰法测铅检出限结果不确定度评定 ………… 张启发，吴琨   9-8
Evaluation of Uncertainty in Results of Lead Detection Limit by Flame Atomic Absorption Spectrophotometer
                Zhang Qifa, Wu Kun

汽运铁精粉检验频次优化的探索实践 ………… 王浩，吴琨，张启发   9-9
Exploration and Practice of Frequency Optimization of Steam Transport Iron Powder Inspection
                Wang Hao, Wu Kun, Zhang Qifa

焊接工艺对耐候钢焊接接头疲劳性能的影响 ………… 邱宇，蔡宁，杨永达，王泽阳，鞠新华   9-9
Influence of Welding Processes on the Fatigue Properties of Welding Joints for Corten Steel
                Qiu Yu, Cai Ning, Yang Yongda, Wang Zeyang, Ju Xinhua

电阻应变片在高速拉伸试验中的应用 ………… 鹿宪宝，邱宇，孙博，张清水，鞠新华   9-10
Application of Resistance Strain Gage in the High Speed Tensile Test
                Lu Xianbao, Qiu Yu, Sun Bo, Zhang Qingshui, Ju Xinhua

河钢石钢钢材宏观纯净度检测技术 ………… 王殿峰，周立波，丁辉，李双居   9-11
Detection Technology of Steel Macroscopic Cleanness Rate
                Wang Dianfeng, Zhou Libo, Ding Hui, Li Shuangju

汽车外板冲压后表面轮廓变化对 2C1B 涂装质量的影响 ………… 胡燕慧，张浩，刘李斌，滕华湘，李蔚然   9-11

Effect of Surface Profile Change during Forming Process of Automobile Outer Panel on Painting Quality for 2C1B Coating Process ·········· Hu Yanhui, Zhang Hao, Liu Libin, Teng Huaxiang, Li Weiran

钢中铌原子晶界偏聚的研究现状及发展 ·· 吕俊南，魏　然，王红鸿　9-12
Research of Niobium Segregation at Grain Boundary in Steel ·········· Lv Junnan, Wei Ran, Wang Honghong

试样加工对冷轧汽车板拉伸性能检验的影响 ······ 张宏岭，叶　筱，金　鑫，甘　霖，秦　胜，江亚平　9-13
Influence of Sample Processing on Tensile Property Test of Cold Rolled Automobile Plate
　　·········· Zhang Hongling, Ye Xiao, Jin Xin, Gan Lin, Qin Sheng, Jiang Yaping

铝渣球金属铝化学分析方法探讨 ·· 张启发，吴　琨　9-14
Discussion on Chemical Analysis Method of Aluminum Slag Ball Metal Aluminum ········ Zhang Qifa, Wu Kun

比表面积分析仪校准方法的探讨 ·· 毕经亮　9-14
Discussion on Calibration Method of Specific Surface Integral Analyses ·························· Bi Jingliang

电感耦合等离子体原子发射光谱测定磷化液中 3 种主要元素 ······ 叶　筱，魏　静，熊立波，张宏岭　9-15
Determination of Three Main Elements in Phosphating Liquid by Inductively Coupled Plasma Atomic Emission Spectrometry ·········· Ye Xiao, Wei Jing, Xiong Libo, Zhang Hongling

对于低合金钢化学成分的不同测定方法比对 ······ 孙璧瑶，高宏斌，王　蓬，冯浩州，卢毓华，蔺　菲　9-16
Comparison of Different Determination Methods for Chemical Composition of Low Alloy Steel
　　·········· Sun Biyao, Gao Hongbin, Wang Peng, Feng Haozhou, Lu Yuhua, Lin Fei

ICP-AEC 法测定锰铁中铬锑铅锡砷含量 ··· 何志明，陈　溪　9-16
Determination of Chromium, Antimony, Lead, Tin and Arsenic in Ferromanganese by ICP-AEC
　　·········· He Zhiming, Chen Xi

晶粒度对奥氏体不锈钢 321 低周疲劳影响机理的研究
　　·········· 张玉刚，何国求，林　媛，佘　萌，张怀征，张文康　9-17
Research on the Influence Mechanism of Grain Size on the LCF of Austenitic Stainless Steel 321
　　·········· Zhang Yugang, He Guoqiu, Lin Yuan, She Meng, Zhang Huaizheng, Zhang Wenkang

汽车结构用钢 SAPH440 疲劳寿命预测研究 ··· 韩　丹，陈虹宇　9-18
Research on Prediction of Fatigue Life of SAPH440 Steel for Automobile Structure ··· Han Dan, Chen Hongyu

熔融制样-X 射线荧光光谱法测定磷铁中磷、硅、锰和钛
　　·········· 宋祖峰，英江霞，王忠乐，郭士光，付万云　9-19
Determination of Phosphorus, Silicon, Manganese and Titanium in Ferrophosphorus by X-ray Fluorescence Spectrometry with Fusion Sample Preparation
　　·········· Song Zufeng, Jia Jiangxia, Wang Zhongle, Guo Shiguang, Fu Wanyun

激光光谱原位分析技术对铝合金焊缝的成分分布分析 ······ 史孝侠，崔飞鹏，徐　鹏，李晓鹏，郭飞飞　9-20
Components Distribution Analysis in Aluminium Alloy Welding Joint by LIBSOPA
　　·········· Shi Xiaoxia, Cui Feipeng, Xu Peng, Li Xiaopeng, Guo Feifei

一种大气压下液体阴极辉光放电装置的研制 ······················ 史孝侠，李晓鹏，崔飞鹏，徐　鹏　9-20
Development of a Solution Cathode—Glow Discharge (SC-GD) Device at Atmospheric Pressure
　　·········· Shi Xiaoxia, Li Xiaopeng, Cui Feipeng, Xu Peng

以钛合金锻件为例的不同标准体系对比分析 ·································· 王　矛，陈　鸣　9-21
Comparison of Different Standard Systems based on Titanium Alloy Forgings ············ Wang Mao, Chen Ming

10　冶金设备与工程技术

一种开口机凿岩机油管运行装置 ······ 邓振月，陈文彬，王仲民，何润平，刘　斌，冯　伟，康大鹏　10-1

Oil Pipe Operation Device of Rock Drill Machine with Opening Machine
　　　　　　　Deng Zhenyue, Chen Wenbin, Wang Zhongmin, He Runping, Liu Bin, Feng Wei, Kang Dapeng

基于 AMESim 的双缸液压同步仿真分析 ……………………………………………… 刘秀军　10-1
Simulation Analysis of Hydraulic Synchronization System with Double Cylinders based on AIMSim
　　　　　　　　　　　　　　　　　　　　　　　　　　　　　　　Liu Xiujun

卧式活套套量设计计算 …………………………………………………………… 惠升谋　10-2
Design and Calculation of Horizontal ……………………………………… Hui Shengmou

冷轧酸洗除尘效果提升技术 ……………………………………………………… 武明明　10-3

红外吸收法测定汽车板钢中氧含量的不确定度评定 ……… 刘步婷，刘　俊，王　鹏，彭　涛　10-3
Assess of Uncertainty for Determination of Oxygen Content in Automobile Plate Steel by Infrared
　Absorption Method ……………………………………… Liu Buting, Liu Jun, Wang Peng, Peng Tao

一种冷轧带钢内径擦伤的原因分析和解决方案 ……………………………………… 李　黎　10-4
Cause Analysis and Solution of Inner Diameter Scratch of Cold Rolling Strip …………… Li Li

提高棒材 650t 冷剪机生产能力的技术改造 …………… 刘　宏，戴江波，金培革，张业华　10-4
Technical Reform of Increasing Production Capacity of 650t Bar Cold Shearing Machine
　　　　　　　　　　　　　　　　　Liu Hong, Dai Jiangbo, Jin Peige, Zhang Yehua

储煤筒仓防自燃因素分析及设计 ………………………………………………… 向甘乾　10-5
Analysis of Influencing Factors and Design for Anti-self-ingition of Coal Storage Silos …… Xiang Gan'gan

湛江钢铁厚板二切线传动系统改造 ……………………………………………… 谢清新　10-6
Cutting Line Drive System Transformation in Zhanjiang Steel Heavy Plate Plant …… Xie Qingxin

一种热轧厂 SSP 曲轴拆装技术 …………………………………………………… 曹叶飞　10-6
A Crankshaft Disassembly and Assembly Technology of SSP in Hot Rolling Mill …… Cao Yefei

混铁车电气故障诊断与处理 ……………………………………………………… 牟德新　10-7
Electrical Fault Diagnosis and Treatment of Mixed Iron Furnace Type Ladle Car …… Mou Dexin

基于定位跟踪的自动抛丸控制技术 ……………………………………………… 赵进友　10-8
Automatic Shot Blasting Control Technology based on Positioning and Tracking …… Zhao Jinyou

机旁库备件二维码的应用 ………………………………………………………… 陈敏华　10-8
Application of Two-dimensional Code for Spare Parts in Machine Side Storage …… Chen Minhua

LECO CS844 测定铌铁合金中碳元素含量 ……… 郭士光，宋祖峰，英江霞，王忠乐，张　洁，龙如成　10-9
Determination of Carbon Content in Ferroniobium by LECO CS844
　　　　　　　Guo Shiguang, Song Zufeng, Jia Jiangxia, Wang Zhongle, Zhang Jie, Long Rucheng

35kV 变压器非电量保护故障分析与改进 ……………………………………… 叶　恩　10-10
The Analysis and Improvement of Fault of Non-electric Protection for 35kV Transformer …… Ye En

无功补偿装置在湛江厚板厂的应用与优化 …………………………………… 叶　恩　10-10
The Application and Optimization of Reactive Compensation Device in Zhanjiang Heavy Plate Mill
　　　　　　　　　　　　　　　　　　　　　　　　　　　　　　　　　Ye En

鞍钢大连铸双线供钢铁路改造项目组织研究 ……… 荀　涛，刘彦栋，李新颖，李荣升，林传山　10-11
Study on Reconstruction Project of Steel Provision Double-track Railway
　　　　　　　　　　　　　Xun Tao, Liu Yandong, Li Xinying, Li Rongsheng, Lin Chuanshan

鞍钢 3200m³ 高炉煤气除尘升级改造研究与应用 …… 李　艳，卢　楠，李志远，李洪宇，袁冬海　10-12
Research and Application of Upgrading and Reforming of Gas Dust for 3200m³ Blast Furnace in Ansteel
　　　　　　　　　　　　　Li Yan, Lu Nan, Li Zhiyuan, Li Hongyu, Yuan Donghai

鞍钢西区焦炉煤气加压站变频系统改造 ……… 尚士鑫，牟柳春，杨大海，陕卫东，芦　刚　10-12

The Reform on Frequency Converter System in Coke Coal Gas Pressure Station of West Zone Anshan
　　Iron-steel Group ················· Shang Shixin, Mou Liuchun, Yang Dahai, Shan Weidong, Lu Gang

棒材生产线轧后工序的智能改造研究与技术开发
　　·································· 徐言东，张华鑫，程知松，郭新文，吴　杰，徐　平　10-13
Intelligent Transformation Research and Development for the Bar Finishing System
　　·································· Xu Yandong, Zhang Huaxin, Cheng Zhisong, Guo Xinwen, Wu Jie, Xu Ping

山钢日照2050mm热连轧工程设计中采用的先进技术 ············· 王　永，吴兆军　10-14
Advanced Technologies of 2050mm Hot Strip Rolling Line at Shansteel Rizhao
　　·································· Wang Yong, Wu Zhaojun

奥氏体不锈钢水冷隔热罩焊接裂纹分析及对策 ······· 秦书清，赵守林，殷　栋，贾士忠　10-14
Analysis and Countermeasures of Crack Welding of Austenitic Stainless Steel Water-cooled Heat Cover
　　·································· Qin Shuqing, Zhao Shoulin, Yin Dong, Jia Shizhong

高效的真空感应熔炼炉上加料装置 ·· 钱红兵　10-15
High Efficiency Overmelt Charging Chamber of VIM ·· Qian Hongbing

倾斜摄影在三维厂区可视化中的应用 ······ 左春雷，刘九阳，马亚敏，韦书剑，王海旭，王东明　10-16
The Application of Oblique Photography in Three-dimensional Visualization Plant Area
　　·································· Zuo Chunlei, Liu Jiuyang, Ma Yamin, Wei Shujian, Wang Haixu, Wang Dongming

智慧检修的思考与探索 ·································· 时元海，张鸿元，李玉凤　10-16
Thinking and Exploration of Smart Overhaul ··········· Shi Yuanhai, Zhang Hongyuan, Li Yufeng

石钢公司轴流压缩机EPU应用实践 ······················· 李　绅，曹俊卿，时永海　10-17
Application Practice of EPU in Axial Compressor of Shisteel Company
　　·································· Li Shen, Cao Junqing, Shi Yonghai

调速器因素导致游车故障的原因分析及处理方法 ································ 邓朝强　10-18
Analysis of Vehicle Failure Caused by 302Y-Z Governor and Countermeasures ······ Deng Zhaoqiang

冷轧厂镀锌线卸卷小车升降定位不准问题的解决与思考 ········· 胡增雨，冯志新，王　军　10-18

钢包渣线砖在RH精炼终点后增碳影响研究 ············ 王　洛，汪　雷，王俊北，樊明宇　10-19
Research on Carburization of Ladle Slag Lining from Refining End
　　·································· Wang Luo, Wang Lei, Wang Junbei, Fan Mingyu

机床丝杠断裂原因分析 ·· 赵　亮，樊一丁　10-19
Fracture Analysis of Machine Tool Lead Screw ··························· Zhao Liang, Fan Yiding

210t铁水罐出铁状态一体化安全校核 ······ 赵　阳，董现春，刘新垚，姜　猛，张彦丰，汪海东　10-20
Integrated Safety Check of a 210t Iron Ladle in the Tapping State
　　·································· Zhao Yang, Dong Xianchun, Liu Xinyao, Jiang Meng, Zhang Yanfeng, Wang Haidong

梅钢公司回用水厂V型滤池运行状况分析及控制技术研究 ···················· 初　明　10-21
Study on the Operation and Control Technology of V-type Filter in the Rewater Plant of Meigang Company
　　·································· Chu Ming

新型高炉旋风除尘器仿真分析 ·············· 胡　伟，耿云梅，陈玉敏，李　欣，章启夫　10-21
Simulation and Analysis of New Type Cyclone Dust Collector for Blast Furnace
　　·································· Hu Wei, Geng Yunmei, Chen Yumin, Li Xin, Zhang Qifu

两轴转向架在130t保温平车上的应用 ···················· 项克舜，严　峰，张海滨　10-22
Application of Two Axle Bogies in 130t Insulated Flat Wagon ······ Xiang Keshun, Yan Feng, Zhang Haibin

钢管超声波检测可靠性的探讨 ··························· 曾海滨，王　伟，陈　杰，李　阳　10-23
Discussion on Reliability of Ultrasonic Testing of Steel Pipe ······ Zeng Haibin, Wang Wei, Chen Jie, Li Yang

自力式调节阀原理及其在连退炉氮氢混合站的应用 ······· 张贵春，周　涛，胡剑斌，陈　璐　10-23

Principle and Application of Self-acting Pressure Regulating Valve in Continuous Annealing Furnace
　　N_2H_2 Mixing Station ·················· Zhang Guichun, Zhou Tao, Hu Jianbin, Chen Lu

最新一代热轧钢卷运输技术及其工程应用 ················· 韦富强，郑江涛，杨建立，刘天柱　10-24
Latest Generation of Hot Strip Coil Conveying Technology and Its Engineering Application
　　·················· Wei Fuqiang, Zheng Jiangtao, Yang Jianli, Liu Tianzhu

钢铁企业应用自动化立体仓库的分析与探讨 ················· 毕　琳，李洪森　10-25
Analysis and Discussion of the Application of Automated Storage and Retrieval System in Steel Works
　　·················· Bi Lin, Li Hongsen

电涡流振动位移传感器自动检定技术方案研究 ················· 郑建忠　10-25
Study on Automatic Verification Technology for Eddy-current Vibration Displacement Sensor
　　·················· Zheng Jianzhong

Φ250mm 4340 钢棒的超声波 C 扫描检测研究 ················· 万　策　10-26
Research on Ultrasonic C-scan Method for Φ250mm 4340 Steel Rod ·················· Wan Ce

油膜轴承在韶钢宽厚板轧机的应用及改进初探 ················· 陈功彬，王　萍　10-27
Application and Improvement Practice of Oil Film Bearing for Wide and Heavy Plate Rolling Mill at SGIS
　　·················· Chen Gongbin, Wang Ping

镶嵌式限位支架在真空槽合金翻板阀中的应用 ················· 何肇恒　10-27
Application of Inlaid Limit Bracket in Vacuum Tank Alloy Flap Valve ·················· He Zhaoheng

一种缓冲阀的设计和应用 ················· 肖争光，李松磊，丛津功，高轶桐　10-28
Design and Application of a New Type of Buffer Valve
　　·················· Xiao Zhengguang, Li Songlei, Cong Jingong, Gao Yitong

UCM 冷轧机力学行为及板形调控性能研究 ················· 贾生晖，陶　浩，韩国民，李洪波　10-29
Analysis of Mechanical Behavior and Flatness Control Characteristics for UCM Cold Rolling Mill
　　·················· Jia Shenghui, Tao Hao, Han Guomin, Li Hongbo

轧钢加热炉垫块的选型应用及发展趋势 ················· 赵　侯，马光宇，刘常鹏，张天赋　10-29
Selection Application and Development Trend of the Skid in Steel Rolling Heating Furnace
　　·················· Zhao Yu, Ma Guangyu, Liu Changpeng, Zhang Tianfu

冷轧带钢激光拼焊板焊缝过拉矫机力学行为仿真研究 ········· 陈　兵，张海涛，唐晓垒，李晋鹏　10-30
Simulation Study on Mechanical Behavior of Welding Seam Tension Correction Machine for Laser Welding
　　Plate of Cold Rolled Strip Steel ·················· Chen Bing, Zhang Haitao, Tang Xiaolei, Li Jinpeng

成品筛激振器的发热分析及改进实践 ················· 程明森　10-31
Heating Analysis and Improvement Practice of Fine Screen Exciter ·················· Cheng Mingsen

11　冶金自动化与智能化

绝对值编码器常用算法缺陷及解决策略 ················· 孙　抗　11-1
Common Algorithmic Deficiencies and Solutions for Absolute Encoders ·················· Sun Kang

热轧轧辊磨损及补偿控制 ················· 陈　晨，孙建林，谭耘宇　11-1
Wear and Compensation Control of Hot Rolling ·················· Chen Chen, Sun Jianlin, Tan Yunyu

钢卷库无人天车系统的开发及应用 ················· 丁　辉，邹永生，赵　平　11-2
Development and Application of Unmanned Crane System for Steel Coil Warehouse
　　·················· Ding Hui, Zou Yongsheng, Zhao Ping

基于机器学习实现冷轧产品的动态工艺调整 ················· 张振宇，彭晶晶，梁燕燕　11-3

Dynamic Process Adjustment of Cold Rolled Products based on Machine Learning
　　　　　　　　　　　　　　　　　　　　　　　　Zhang Zhenyu, Peng Jingjing, Liang Yanyan

焦炉电机车自动化应用　　孔弢，张允东，甘秀石，赵明，高薇，张其峰　11-3
Automation Application of Coke Oven Locomotive
　　　　　　　　Kong Tao, Zhang Yundong, Gan Xiushi, Zhao Ming, Gao Wei, Zhang Qifeng

MES 系统在冶金产线的应用与发展　　范海龙　11-4
Application and Development of MES System in Metallurgical Production Line　　Fan Hailong

基于 PCA 算法的高炉参数规则指标排序研究　　张秀春，宋政，周道付，范晨　11-5
Research on Index Ranking of Blast Furnace Parameter Rules based on PCA Algorithms
　　　　　　　　　　　　　　　　　　　　　Zhang Xiuchun, Song Zheng, Zhou Daofu, Fan Chen

iSCADA 系统在 CSP 厂系统迁移项目中的应用　　徐重　11-5
Application of iSCADA System in System Migration Project of CSP Plant　　Xu Zhong

Python 数据分析工具在 CSP 厂信息化开发中的应用　　徐重　11-6
Application of Python Data Analysis Tool in Information Development of CSP Plant　　Xu Zhong

炼钢数字化单元的集成实践与思考　　李立勋，杨晓江，康书广，王欣，张道良　11-7
Practice and Consideration of Steelmaking Digital Unit Integration
　　　　　　　　　　　　Li Lixun, Yang Xiaojiang, Kang Shuguang, Wang Xin, Zhang Daoliang

基于 Spark 的大数据网络日志采集分析和预警模型研究　　易可可，钱哲怡，王威　11-8
Research on Big Data Network Log Collection Analysis and Early Warning Model based on Spark
　　　　　　　　　　　　　　　　　　　　　　　　　　Yi Keke, Qian Zheyi, Wang Wei

武钢 CSP 厂 1 号 RH 真空系统 S4a、S4b 喷嘴腐蚀失效故障的分析判断及改进
　　　　　　　　　　　　　　　　　　　　　张云，王南生，李文晓，徐重　11-8
Failure Analysis and Improvement of RH Vacuum System S4a, S4b Nozzle Corrosion
　　　　　　　　　　　　　　　　　Zhang Yun, Wang Nansheng, Li Wenxiao, Xu Zhong

PRINCE2 方法在 CSP 一键式炼钢模型项目的应用　　徐重，张云　11-9
Application of PRINCE2 Method in CSP Fully Automatic Steelmaking Model Project
　　　　　　　　　　　　　　　　　　　　　　　　　　　　　　Xu Zhong, Zhang Yun

高炉炉衬监测技术发展和应用探讨　　张伟，朱建伟，张立国，任伟，李金莲　11-10
Application and Its Development for Blast Furnace Lining Thickness Detection Technology
　　　　　　　　　　　　　　　Zhang Wei, Zhu Jianwei, Zhang Liguo, Ren Wei, Li Jinlian

冷轧镀锌线减少焊缝过光整机张力波动的优化研究　　崔伦凯，郭俊明，张军，张振勇，李凤惠　11-10
The Optimization of Tension Control on the Skin Pass Mill
　　　　　　　　　　　Cui Lunkai, Guo Junming, Zhang Jun, Zhang Zhenyong, Li Fenghui

一种环境在线监测数据采集系统　　楼纬　11-11
Environment Online Monitoring Data Acquisition System　　Lou Wei

污水处理的自控系统设计　　王猛　11-12
Design of Automatic Control System for Sewage Treatment　　Wang Meng

关键皮带故障原因分析与技术改进　　朱庆庙，庞克亮，张允东，甘秀石，侯士彬，马银华　11-12
Analysis and Technical Improvement of Key Belt Fault
　　　　　　　Zhu Qingmiao, Pang Keliang, Zhang Yundong, Gan Xiushi, Hou Shibin, Ma Yinhua

焦炉机车信息化管控系统研究与实践　　于庆泉，周鹏，马富刚，崔金林，孙强　11-13
Research and Practice of Informationalized Control System for Coking Locomotive
　　　　　　　　　　　　　　Yu Qingquan, Zhou Peng, Ma Fugang, Cui Jinlin, Sun Qiang

基于产销协同处理的钢铁资源分配模型 ··· 王丽婷，刘志军　11-13
Steel Resource Allocation Model based on Production and Marketing Co Processing
·· Wang Liting, Liu Zhijun

超大规模项目全生命周期管理信息系统研究 ······································· 王晓蕾，牛春波　11-14
Research on Life Cycle Management Information System of Ultra Large-scale Project
·· Wang Xiaolei, Niu Chunbo

智能化管控信息系统的设计与应用 ····································· 李　健，陈向阳，刘　杰　11-15
Design and Application of Intelligent Information System Control ······ Li Jian, Chen Xiangyang, Liu Jie

大包回转台准确定位法在异形坯连铸机中的应用 ·· 魏晨溪　11-16
Application of the Method of Precise Positioning for Ladle Turret on Beam Blank Continuous Caster
·· Wei Chenxi

大型钢铁企业产销一体化信息系统设计与实现 ························· 徐守新，牛春波，杨金光　11-16
Design and Implementation of Integrated Information System for Sales and Production in Large Iron and
　　Steel Enterprises ·· Xu Shouxin, Niu Chunbo, Yang Jinguang

智能视频监控绊线穿越检测在河钢石钢的研究与应用 ·· 刘　霞　11-17
Research and Application of Intelligent Video Monitoring Trigger Crossing Detection in HBIS Shisteel
·· Liu Xia

轧机主电机负载平衡控制的研究应用 ··· 王　博　11-18
Research and Application of Load Balancing Control for Main Motor of Rolling Mill ······ Wang Bo

炼焦煤质量预警系统的实现和应用 ······································· 卢　瑜，杜　屏，赵华涛　11-18
Realization and Application of Coking Coal Quality Early Warning System ······ Lu Yu, Du Ping, Zhao Huatao

基于大数据分析的可视化智能运维系统在钢铁企业的应用
······································· 杨　恒，周　平，范　鹍，黄少文，李长新　11-19
Application of Visual Intelligent Operation and Maintenance System based on Big Data Analysis in Iron and
　　Steel Enterprise ············· Yang Heng, Zhou Ping, Fan Kun, Huang Shaowen, Li Changxin

钢铁流程工序界面信息一体化融合实践 ························ 谢　晖，周　平，范　鹍，王成镇，杨　恒　11-20
Practice of Integrative Information Fusion of Process Interface in Iron and Steel Process
································ Xie Hui, Zhou Ping, Fan Kun, Wang Chengzhen, Yang Heng

基于钢板轮廓精确检测的智能优化剪切方法的应用 ··················· 吴昆鹏，杨朝霖，石　杰　11-20
Application of Intelligent Optimization Shearing Method based on Accurate Detection of Steel Plate Contour
··· Wu Kunpeng, Yang Chaolin, Shi Jie

六辊轧机厚控系统精度问题分析及应对措施 ······································ 李　胤　11-21
Analysis and Countermeasure of Accuracy Problem of Thickness Control System of Six-high Mill ··· Li Yin

设备全生命周期管理系统建设与应用 ····················· 安丰涛，宋振兴，苏　威，石　江　11-22
Construction and Application of Equipment Life Cycle Management System
··· An Fengtao, Song Zhenxing, Su Wei, Shi Jiang

宣钢高炉 EPU 防喘控制系统技术应用 ··· 李蕴华　11-23
宣钢 $2500m^3$ 高炉炉顶设备最优控制 ········ 王春元，闫新卓，蔡　宇，薄　魏，倪力鑫，周建军　11-23
横河 DCS 在 $435m^2$ 烧结机自控系统中的开发及应用 ·············· 霍迎科，申存斌，宋林昊　11-23
Development and Application of YOKOGAWA DCS Automation System in the $435m^2$ Sintering Machine
··· Huo Yingke, Shen Cunbin, Song Linhao

基于模式识别的电机系统在线监测研究 ······················ 黄晨晨，郭光辉，郝春辉，张勇军　11-24
Research on Motor Online Monitoring System based on Pattern Recognition
······································· Huang Chenchen, Guo Guanghui, Hao Chunhui, Zhang Yongjun

在线实时成分检测技术在炼铁行业应用前景 ……………………………………………………… 马洪斌 11-25

加热炉二级优化控制系统研究与开发 ……………………………………… 蒋国强，张宝华，陈建洲 11-25
Research and Development of the Secondary Optimization Control System for the Heating Furnace
　　　　　　　　　　　　　　　　　　　　　　　　　Jiang Guoqiang, Zhang Baohua, Chen Jianzhou

钢铁企业数据应用创造价值的实践 ……………………………………………… 熊　鑫，王里程，白　雪 11-26
Practice of Creating Value by Data Application in Iron and Steel Enterprises
　　　　　　　　　　　　　　　　　　　　　　　　　　　　　　Xiong Xin, Wang Licheng, Bai Xue

立式连续退火炉炉辊温度模型研究 ………………………………… 李洋龙，任伟超，陈　飞，王　慧 11-26
Study on Temperature Model of Furnace Roller in Vertical Continuous Annealing Furnace
　　　　　　　　　　　　　　　　　　　　　　　　Li Yanglong, Ren Weichao, Chen Fei, Wang Hui

AI 技术在冷轧连退平整机控制模型中的应用 ……………… 袁文振，黄华钦，司华春，高新建，谢　谦 11-27
Application of AI Technology in Control of Cold Rolling Continuous Leveling Mill
　　　　　　　　　　　　　　　　　　Yuan Wenzhen, Huang Huaqin, Si Huachun, Gao Xinjian, Xie Qian

物资远程智能计量系统发展状况 …………………………………………………………………… 郑建忠 11-28
The Development Status of Material Remote Intelligent Measurement Systems …… Zheng Jianzhong

钢铁制造流程智能制造与智能设计的研究 ………………………………………………… 颉建新，张福明 11-29
Research of Intelligent Manufacturing and Intelligent Design on Iron and Steel Manufacturing Process
　　　　　　　　　　　　　　　　　　　　　　　　　　　　　　　　Xie Jianxin, Zhang Fuming

控冷装备控制精度的自动测定与提高 ……………………………………………………………… 曹育盛 11-30
Auto-evaluating and Improving the Controlling Accuracy of Control Cooling Equipment …… Cao Yusheng

韶钢原燃料汽车运输智能采样系统的设计与实现 ………………… 肖命冬，欧连发，吴红兵，石志钢 11-30
Design and Application of Remote Auto-sampling System for Trucks at SGIS
　　　　　　　　　　　　　　　　　　　　　　　Xiao Mingdong, Ou Lianfa, Wu Hongbing, Shi Zhigang

基于连续域蚁群的钢坯加热炉操作优化 …………………………… 周　毅，罗国盛，王小毛，邱奕敏 11-31
Continuous Ant Colony Optimization based Operational Optimization for Slab Reheating Furnace
　　　　　　　　　　　　　　　　　　　　　　　Zhou Yi, Luo Guosheng, Wang Xiaomao, Qiu Yimin

基于现代信息化平台的炼钢调度在沙钢的运用 …………………………………………………… 陈　超 11-32
Application of Steelmaking Scheduling based on Modern Information Platform in Shagang …… Chen Chao

全自动控制在钢包车上的应用 ……………………………………… 王雪明，黄春东，施军贤，陆网军 11-32
Full-automation Control Use in Ladle Car …… Wang Xueming, Huang Chundong, Shi Junxian, Lu Wangjun

板坯长度跟踪在 CSP 连铸机上的应用及优化 ……………………… 雷希璋，张　超，何金平，徐　重 11-33
Application and Optimization of Slab Length Tracking in CSP Caster Machine
　　　　　　　　　　　　　　　　　　　　　　　　　　Lei Xizhang, Zhang Chao, He Jinping, Xu Zhong

退休管理网上办公系统在企业中的应用 …………………………… 蔺　飞，李彩虹，杨雅娟，赵媛媛 11-34
The Application of Retirement Management Information System
　　　　　　　　　　　　　　　　　　　　　　　　Lin Fei, Li Caihong, Yang Yajuan, Zhao Yuanyuan

八钢 150t 精炼炉冶炼变压器调压绕组烧损故障分析及处理 …………………………………… 刘兴海 11-34
Analysis and Treatment of Burning Fault of Voltage Regulating Winding of Smelting Transformer in 150t
　　Refining Furnace of Bagang …………………………………………………………… Liu Xinghai

12　建 筑 诊 治

带大开间巨型框架结构抗震性能研究 ……………………………………………………… 胡光林，吴志华 12-1
The Research on the Seismic Performance of Mega-frame Structures with a Large Bay
　　　　　　　　　　　　　　　　　　　　　　　　　　　　　　　　Hu Guanglin, Wu Zhihua

提高建筑工程抗震性能的探究 ……………………………………………………………………… 杨　倩 12-1
Research on Improving Seismic Performance of Building Engineering …………………… Yang Qian

建筑抗震性能分析及提升技术 ……………………………………………………………………… 付洪坤 12-2

钢结构无损检测技术以及质量控制研究 ………………………………………………………………… 刘　冠　12-2
吊车梁辅助桁架与柱连接破坏情况分析 …………………………………………………… 王九根，郑　钧　12-3
Failure Analysis of Connection between Auxiliary Truss and Column of Crane Girder
　　………………………………………………………………………………… Wang Jiugen, Zheng Jun
某高炉出铁场除尘管道凹陷变形原因分析 ………………………… 王建强，庄继勇，郑　钧，王华丹　12-3
Deformation Cause Analysis of Dust Removal Pipeline Depression in a Blast Furnace Discharge Yard
　　………………………………………………… Wang Jianqiang, Zhuang Jiyong, Zheng Jun, Wang Huadan
某厂房抗震鉴定 …………………………………………………………………………………… 陆明磊　12-4
The Seismic Appraiser of an Workshop ……………………………………………………………… Lu Minglei
浅析建筑结构抗震技术的应用 …………………………………………………………………… 郭之亮　12-4
工业建筑钢结构疲劳损伤检测、评估及加固关键技术研究
　　…………………………………………………… 幸坤涛，赵晓青，郭小华，惠云玲，杨建平　12-5
Research on Key Technologies of Fatigue Damage Detection, Assessment and Reinforcement of Steel
　　Structures in Industrial Buildings …… Xing Kuntao, Zhao Xiaoqing, Guo Xiaohua, Hui Yunling, Yang Jianping
钢框架-中心支撑结构的结构抗震鉴定 …………………………………………………………… 郑　钧　12-6
Seismic Appraisal of Steel Frame-central Braced Structure ………………………………………… Zheng Jun

13　其　他

应用新型工业化应力退火技术研究矩形纳米晶磁芯的磁性能
　　………………………………………………… 李雪松，薛志勇，Sajad Sohrabi，汪卫华　13-1
Tailoring Magnetic Properties of Rectangular-Shaped Magnetic Cores Using a Novel Industrial-scale Stress
　　Annealing Technique …………………………… Li Xuesong, Xue Zhiyong, Sajad Sohrabi, Wang Weihua
冶金行业三维工厂设计与二维设计的分析比较 ………………………………………… 陈　馨，郭彩虹　13-2
Analysis and Comparison between 3d and 2d Design in Metallurgical Industry …… Chen Xin, Guo Caihong
草酸盐从钒渣酸浸液中除锰试验研究 …………………………………………………… 杨　晓，李道玉　13-2
Experimental Study on Removal of Manganese from Acid Leaching Solution of Vanadium Slag by Oxalate
　　……………………………………………………………………………………… Yang Xiao, Li Daoyu
Cu 表面超疏水薄膜的制备及耐腐蚀性的研究 ………………………… 汪　建，蔡　浩，张　强，龙剑平　13-3
Preparation and Corrosion Resistance of Superhydrophobic Films on Cu Foil
　　……………………………………………………… Wang Jian, Cai Hao, Zhang Qiang, Long Jianping
TMEIC 激光焊机 X 轴激光焦点系统分析与维护 ………………………………………………… 邓承龙　13-4
Analysis and Maintenance of Laser Focus System of the X-ray Axis of TMEIC Laser Welder
　　……………………………………………………………………………………………… Deng Chenglong

大会特邀报告

- ★ 大会特邀报告
- 分会场特邀报告
- 炼铁与原料
- 炼钢与连铸
- 轧制与热处理
- 表面与涂镀
- 金属材料深加工
- 先进钢铁材料
- 粉末冶金
- 能源、环保与资源利用
- 钢铁材料表征与评价
- 冶金设备与工程技术
- 冶金自动化与智能化
- 建筑诊治
- 其他

国际标准化发展趋势和中国钢铁高质量发展

张晓刚

（国际标准化组织）

摘　要：本文从国际标准化发展趋势讨论了标准化工作在中国钢铁高质量发展中的深刻意义。以标准化引领钢铁行业的科技创新和管理创新，从而实现钢铁大国向钢铁强国转变的必由之路。

Development Trend of International Standardization and High-quality Development of Chinese Steel Industry

Zhang Xiaogang

(The International Organization for Standardization)

Abstract: This article aims to illustrate the profound significance of standardization in the high quality development of Chinese steel industry complying with development trend of international standardization. Leading the technological innovation and management innovation of Chinese steel industry by standardization, so as to promote transformation of China from a large steel producer into a steel power.

坚持绿色发展道路，打造世界一流企业

——在第十二届中国钢铁年会上的主旨报告（摘要）

陈德荣

（中国宝武　党委书记、董事长）

2019 年 10 月 15 日

中国钢铁行业历经几十年的迅猛发展，粗钢产量在 2018 年首次突破 9 亿吨。虽然实现了跨越式发展，但是也存在产业集中度不高、企业竞争力不足、环保欠账多、鲜有革命性创新成果等问题。妥善处理"发展"与"生态"的关系，是当前实现中国钢铁工业高质量发展的前提，也是所有钢铁人共同构建人类命运共同体的使命担当。

中国宝武坚持创新驱动、坚持绿色发展道路，深度挖掘供应链效率和价值，创新供应链服务产品，构建价值共享的可持续发展钢铁生态圈。积极推进以"弯弓搭箭+网络型短流程钢厂"为特征的国内空间布局，主动承担行业责任，提升国内钢铁的产业集中度；稳步推进混合所有制改革，加强内部专业化整合,打造"一基五元"产业布局,持续提升竞争力;以欧冶炉、氧气高炉等低碳冶金工艺技术为重点,加快先进绿色制造工艺技术研究,全面实施废气超低排放、废水零排放、固废不出厂,引领行业绿色发展;解决城市环保之难,推动产城融合,致力绿色公益事业,助力城市生态文明建设;面向未来,多策并举,努力打造具有全球竞争力的世界一流企业,不断满足人民群众日益增长的优美生态环境需要。

Persist in Green Development, Build a World-Class Enterprise

Chen Derong

(Chairman, China Baowu)

October 15, 2019

In 2018, China's crude steel output exceeded 900 million tons for the first time after decades of its rapid development. Some problems emerged after the leap-forward development, e.g. Low industrial concentration degree, insufficient competitiveness, more environmental issues and few revolutionary innovative achievements. It is the premise of realizing the high-quality development of China's iron and steel industry to appropriately manage the relationship between "development" and "ecology" well. It is also the mission of all steelmakers to jointly build a community for shared interests for the mankind.

China Baowu insists on innovation-driven and green development path, deeply explores supply chain efficiency and its value, innovates supply chain service products, and constructs a value sharing sustainable development of steel ecosphere. China Baowu is to shape strategic spatial layout compatible with the Belt and Road Initiative and the Yangtze River Economic Belt Strategy of the nation, promote compact steel mill, undertake industrial responsibility, increase the industrial concentration degree. It steadily promotes mixed ownership reform, strengthens internal professional integration, creates an "One Base Five Fundamentals" industrial layout, continuously enhances competitiveness. China Baowu will speed up the research of advanced green manufacturing technology by focusing low-carbon metallurgical technology on oxygen blast furnace and Ouyeel furnace, comprehensively implement ultra-low emission of exhaust gas, zero discharge of waste water and solid waste out of the plant to lead the green development of steel industry; It will solve urban environmental protection difficulties, promote the integration of production and city, make efforts in green public welfare, contribute to the construction of urban ecological civilization. China Baowu will take multiple measures to face the future by striving to build a world-class enterprise with global competitiveness to continuously satisfy people with their increasing demands on the beautiful ecological environment.

日本制铁针对全球环境问题的应对之策
Efforts of Nippon Steel Corporation for Global Environmental Problems

Inoue Akihiko

(Executive Vice President, Nippon Steel Corporation)

- Short Biography of the Speaker
- NSC's Basic Environmental Policy
- NSC's Contribution to SDGs
- Japan's Reduction Target for Greenhouse Gas Emissions by 2030
- Efforts for CO_2 Reduction in the Japanese Iron and Steel Industry
- Energy-saving Investment by the Japanese Iron and Steel Industry
- New Coke Oven for energy-saving and CO_2 reduction
- Reduction of Reducing Agent Rate
- Energy-saving and CO_2 Reduction through High Performance Products
- Contribution of Eco-products
- CO_2 reduction using High performance steel products
- Overseas Transfer of Japan's Energy-saving and Environment protection Technologies
- Overview of the COURSE50 National Project
- Concept of COURSE50 Blast Furnace
- CO_2 reduction from Blast Furnace Using an Experimental Blast Furnace
- Results of Experimental Blast Furnace Operation
- Innovative CO_2 Capture Technologies
- Commercialization of the COURSE50 CO_2 Capture Technology
- COURSE50 Project Timeline
- Long-term climate change mitigation strategy by JISF: Consistency with IEA-ETP2017 2DS
- Long-term climate change mitigation strategy by JISF: super innovative technologies development
- Summary

韩国钢铁产品技术最新进展
Recent Progress of Steel Product Technology in Korea

Kim Sung-Joon

(President of Korean Institute of Metals and Materials
Graduate Institute of Ferrous Technology, POSTECH, Pohang, Korea)

Korean steel industry produced 72 million tons of crude steel in 2018 following China, India, Japan and USA. In the worldwide oversupply of steel products, the importance of research and development for the new process and products are becoming more and more important for every countries. For the last couple of years, Korean steel industries and research institutes achieved a remarkable progress in many steel products. In this presentation, some examples of those achievements will be introduced. Firstly, high Mn steels for LNG tank and oil refinery pipes were commercialized by POSCO. Secondly, stainless steels for hydrogen environment were successfully developed by POSCO and Hyundai. Finally, high strength high ductility steels for automotive parts were developed and commercialized. All these environmentally friendly products are expected to contribute to reduce global warming and build a more comfortable society.

钢铁材料品种开发及其应用进展

田志凌

(中国钢研科技集团有限公司,北京 100081)

Development and Application of Steel Product Research In Central Iron and Steel Institute

Tian Zhiling

  近年来,钢铁研究总院结合国民经济建设和国家重大工程对高性能先进钢铁材料的迫切需求,积极开展军工用钢,海洋与船舶用钢,火电、核电与石化用钢,建筑交通用钢,模具用钢以及基础件用钢的基础理论、关键技术和新产品研发工作。在基础理论和关键工艺技术方面取得一系列成果,包括高强度钢氢脆控制机理和轴承钢的长寿命控制机理,大线能量焊接用钢的第二相粒子控制技术,高强度球扁钢截面组织性能均质化控制技术,高止裂性船板韧化控制技术、热作模具钢高等向性控制技术、大型难变形合金均匀化冶炼控制技术、大锻件晶粒组织细化和均匀化的热过程控制技术等工艺技术。在此基础上开发了耐低温、抗酸、抗大变形、深海等高性能管线钢,连续油井管,超大膨胀量膨胀管,超高压压裂泵液力端用长寿命泵体材料,TiN粒子增强型耐磨钢,高性能高铁轮轴用钢,建筑用高强度耐火耐磨钢,高品质压铸系列模具钢,630℃超超临界火电用耐热材料,新型马氏体耐热钢、新一代核压力容器用钢,先进核能核岛关键装备用耐蚀合金,以及以飞机起落架、战斗部壳体用钢为代表的超高强度钢,取得了一系列丰硕的成果。同时,积极开展材料基因组,钢铁材料数据库建设,网络平台与云端服务等关键技术研究,对支撑和促进我国钢铁工业的发展和进步起到了积极的作用。

分会场特邀报告

大会特邀报告

★ 分会场特邀报告

炼铁与原料

炼钢与连铸

轧制与热处理

表面与涂镀

金属材料深加工

先进钢铁材料

粉末冶金

能源、环保与资源利用

钢铁材料表征与评价

冶金设备与工程技术

冶金自动化与智能化

建筑诊治

其他

我国高炉使用高比例球团生产技术经济分析

沙永志[1]，马丁·戈德斯[2]，宋阳升[3]

（1. 中国钢研科技集团，北京 100081；2. 戈德斯咨询公司，荷兰；3. 力拓铁矿，上海 200000）

摘　要：提高高炉球团比例正成为我国高炉炼铁发展新动向。国外大量的成功生产实践证明了高比例球团高炉冶炼的技术可行性。不同种类的球团应满足基本的质量要求标准，其中球团膨胀性能应是关注的重点。在高炉生产中，需在炉料结构、高炉炉料分布、炉热控制等方面进行优化和改进。提高球团比例的最大挑战是炼铁成本的控制。铁精矿和球团的资源供应，球团生产和使用的节能减排效果，精矿和球团的溢价水平等，都是限制球团比例升高的关键因素。

关键词：炼铁；高炉；球团

Technical and Economic Analysis of Blast Furnace Operation with High Pellet Ratio in China

Sha Yongzhi[1], Maarten Geerdes[2], Sunny Song[3]

(1. China Iron & Steel Research Institute Group, Beijing 100081, China;
2. Geerdes Advies, Holland; 3. Rio Tino Iron Ore, Shanghai 200000, China)

Abstract: Increasing pellet ratio in blast furnace burden is becoming a hot topic in China. Abundant practices in abroad have demonstrated the technical feasibility for BF operations with high pellet proportion. There are basic quality requirements and standards for different types of pellets, in which swelling behavior is a major index to be noticed. During transforming to high pellet proportion, optimization and improvements should be done on burden composition, burden distribution, and thermal control of BF. The biggest challenging for high pellet proportion is cost control in ironmaking process. The limiting factors to increase pellet proportion are supply, use values, and premium prices for both iron ore concentrates and pellets.

Key words: ironmaking; blast furnace; pellet

首钢炼铁技术进步与展望

张福明

（首钢集团有限公司，北京 100041）

摘　要：进入 21 世纪以来，为了产业结构调整和产品结构优化，首钢进行了整体搬迁。按照现代钢铁制造流程"三个功能"的工程理念，在河北唐山地区相继建成了迁钢和京唐两个现代化钢铁基地。本文介绍了首钢炼铁系统技术装备及生产技术体系，对近年来首钢炼铁技术进步进行了总结。重点阐述了烧结料面喷吹蒸汽技术、复合球团生产技术、高炉长寿铜冷却壁技术、大型高炉高比率球团冶炼技术、冶金烟气综合治理等技术的研究和应用情况。基于低碳绿色发展理念，对首钢炼铁技术未来发展趋势进行了探讨和展望。

关键词：炼铁；高炉；球团；烧结；高炉长寿；环境保护

Progress and Expectation of Ironmaking Technology in Shougang

Zhang Fuming

(Shougang Group Co., Ltd., Beijing 100041, China)

Abstract: Since the new century, the Shougang has been relocated for the structural adjustment of the industrial structure and the optimization of the product structure. According to the engineering concept of the modern iron and steel manufacturing process "Three functions", two modern steel bases of Qiangang and Jingtang are successively constructed in the region of Tangshan, Hebei province. In this paper, the technical equipments and manufacturing system are introduced, the progress of the ironmaking technology of Shougang in recent years is summarized. The development and application are expatiated and reviewed, such as the sintering bed surface steam spraying technology, the composite pellet production technology, the long service life copper cooling stave, high-proportion pellet burden operating technology of large blast furnace, and comprehensive treatment of metallurgical waste flue gas, etc. The future development trend of Shougang ironmaking technology base on the concept of low-carbon and greenization is discussed and prospected.

Key words: ironmaking; blast furnace; sintering; pelletizing; long campaign life; environmental protection

炼铁过程炉料结构智能优化系统

张玉柱[1,3]，王宝祥[1,2]，陈 伟[1,2]，武鹏飞[1,2]，陈 颖[1,2]，马劲红[1,2]

（1. 华北理工大学冶金与能源学院，河北唐山 063210；2. 河北省高品质钢连铸工程技术研究中心，河北唐山 063000；3. 现代冶金技术教育部重点实验室，河北唐山 063009）

摘 要：如何在保证高炉顺行和铁水质量的前提下配加贫矿，最大限度降低铁水成本，减少有害元素危害是目前各企业炼铁过程面临的极大难点。本研究主要应用炼铁原理、科学实验、专家知识与数学优化算法相结合的方法，建立了炼铁炉料结构优化系统，经过科学计算最终得到最优的炼铁系统炉料结构优化方案。系统由原燃料数据库、烧结配料、高炉配料以及公司成本核算四个模块组成。智能优化过程从选料开始，到烧结配料优化，再到高炉配料优化，最后利用配料计算所得数据进行公司成本核算，得到优化后利润。本炼铁炉料结构智能优化操作系统以钢铁公司烧结和高炉生产为基础，为钢铁企业解决如何智能优化配矿问题，系统操作简单，计算准确，实用性强，为最大限度降低铁水成本、满足炼铁和炼钢的成分要求、保证生产顺行、减少有害元素危害等提供可行可靠的科学理论支撑，具有较高的理论价值和应用价值。

关键词：烧结；高炉；炼铁；配料；优化；成本；有害元素

Intelligent Optimization System of Burden Structure in Ironmaking Process

Zhang Yuzhu[1,3], Wang Baoxiang[1,2], Chen Wei[1,2], Wu Pengfei[1,2], Chen Ying[1,2], Ma Jinhong[1,2]

(1. College of Metallurgy and Energy, North China University of Science and Technology, Tangshan 063210, China; 2. Hebei Engineering Research Center of High Quality Steel Continuous Casting, Tangshan 063000, China; 3. Ministry of Education Key Laboratory of Modern Metallurgy Technology, Tangshan 063009, China)

Abstract: How to use lean ore to minimize the cost of molten iron and reduce harmful elements hazards is a great difficulty faced by various enterprises in the process of ironmaking under the premise of ensuring the smooth operation of blast furnace and the quality of molten iron. In this study, the iron-making principle, scientific experiments, expert knowledge and mathematical optimization algorithm are combined to establish the optimization system of the burden structure of the iron-making system. After scientific calculation, the optimal scheme of the burden structure of the iron-making system is finally obtained. The whole system consists of four modules: raw material database, sintering burdening, blast furnace burdening, and company cost accounting. The optimization system starts from the selection of materials, optimizes the sintering burdening, optimizes the blast furnace burdening, and finally uses the calculated data of the burdening to calculate the company's cost and obtain the optimized profit. Based on sintering and blast furnace production process, this intelligent optimization system solves the problem of how to optimize ore blending for iron and steel enterprises. The system has the advantages of simple operation, accurate calculation and strong practicability. It provides feasible and reliable scientific theoretical support for minimizing the cost of molten iron, meeting the composition requirements of ironmaking and steelmaking, ensuring smooth production and reducing harmful elements. It has high theoretical value and application value.

Key words: sintering; blast furnace; ironmaking; burdening; optimize; cost; harmful elements

高炉本体长寿设计中的若干认识与建议

汤清华

（鞍钢股份公司，辽宁鞍山 114000）

摘　要：笔者通过鞍钢高炉改造与建设的实践，参与国内一些高炉事故分析与学习，结合近年来我国高炉出现的：开炉后不久炉缸炭砖温度升高，炉缸烧穿事故频繁，铜冷却壁寿命不长等现象的调查与分析，提出了结构上的一些共性问题，如：炉缸炉壳结构，炉缸环炭从热面到冷面存在水平通缝，冷却结构，炭砖使用结构及质量，铜冷却壁使用等方面的认识，与炼铁同仁讨论，以其引起重视，起到抛砖引玉的作用。

The Opinion and Advice about the Design of BF Proper's Long Campaign

Tang Qinghua

(Anshan Iron and Steel Co., Ltd., Anshan 114000, China)

Abstract: The paper is based on the practice about Ansteel BF's reconstruction and building, the analysis and study about domestic BF accidents, and the investigation & analysis about the following condition which occurred in China recently. No long after blowing-in, because of the abnormal temperature rise of carbon bricks in furnace hearth, the damage of furnace hearth had been occurred frequently. The service life of copper cold wall is not long. So some general problems about the design of BF proper in structure should be proposed. For example, the structure of furnace hearth and furnace shell, the horizontal continuous seam existed between the hot surface and the cold surface of carbon bricks in furnace hearth; cooling structure; the structure and quality of carbon bricks; copper cold wall and so on. I hope that the above problems will attract your attention.

钢铁造块节能减排与智能控制技术

范晓慧

（中南大学，湖南长沙　410083）

摘　要： 绿色化与智能化是我国钢铁工业可持续发展的重要要素。造块作为钢铁生产的第一道高温工序，其节能减排与智能控制尤为关键。为此，论文介绍了三个方面的内容：一是烧结烟气多污染物的源头、过程控制技术，二是冶金固废的高效处理和利用技术，三是烧结、球团生产的智能控制技术。

我国焦化行业运行及对未来焦炭需求预测

郑文华

（中冶焦耐工程技术有限公司，辽宁大连　116085）

摘　要： 介绍 2018 年及 2019 年上半年我国焦化行业运行现况。分析未来我国对钢铁的需求、炼铁焦比的变化和大量使用废钢铁对焦炭需求的影响，预测 2030 年我国的焦炭需求量。

关键词： 焦炭运行；预测需求量

China's Coking Industry Operation and Coke Demand Forecast for the Future

Zheng Wenhua

(ACRE Coking & Refractory Engineering Consulting Co., Ltd., Dalian 116085, China)

Abstract: Introduction of the operation status of coking industry in China in 2018 and the first half of 2019. Analyzing the future demand for steel in China, the change of coke ratio of iron smelting and the impact of large amount of waste steel on coke demand. Forecasting China's coke demand in 2030.

以可视化、模式识别为核心的高炉智能控制系统的开发及应用

陈令坤

(宝钢股份武钢有限技术中心炼铁所,湖北武汉 430081)

摘 要:国内外研究高炉智能控制的历史最少 30 年了,但研究和实际需求之间仍存在巨大的反差,究其原因是难以用技术手段跟踪高炉状态的演化,很多参数要么难以获取,要么太简单不足以反映高炉状态的变化,本研究从炉顶布料的可视化入手,采用模式识别技术对炉型、炉缸、顺行、气流等反应高炉状态变化的重要现象进行识别,可视化和模式识别的基础上构建了一个高炉上下部调剂智能控制系统并用于武钢有限 7 号高炉,尽管该高炉运行在炉役后期,冷却水管时有破损,但总体状态稳定,产量屡创新高,该系统的使用对于高炉的稳定顺行有重要作用。

关键词:高炉;可视化;布料;上下部调剂;模式识别

Development and Application of Artificial Intelligence(AI) Control System Based on the Technologies of Visualization and Pattern Recognition

Chen Lingkun

高铁车辆轮轴用钢大型非金属夹杂物和宏观偏析控制技术研究

王新华[1],姜 敏[1],于会香[1],汪开忠[2],孙 维[2]

(1. 北京科技大学冶金与生态工程学院;2. 马鞍山钢铁有限责任公司技术中心)

摘 要:作者对采用"EAF-LF-RH-CC"冶金工艺生产的高铁车轮、车轴钢中大型非金属夹杂物和宏观偏析的控制技术开展了试验研究,发现:超低氧含量(T.O<6ppm)铸坯中尺寸大于 $100\mu m$ 的夹杂物全部为在连铸过程聚合形成的簇群状夹杂物,而铸坯宏观偏析主要表现为由铸坯边部至中心正负偏析交替发生的"W"形偏析。

高铁轮轴钢 LF 精炼中,钢液中夹杂物发生由 Al_2O_3 向 $MgO\text{-}Al_2O_3$ 系、$CaO\text{-}MgO\text{-}Al_2O_3$ 系、$CaO\text{-}Al_2O_3$ 系夹杂物的顺序转变。由于固态夹杂物与钢液间表面张力大而更易去除,因此开发了更高效的二次精炼工艺,其特点主要为:(1)尽可能地缩短 LF 精炼时间,以延缓固态夹杂物向液态夹杂物的转变;(2)将 RH 精炼时间延长至 33min 以上以更充分发挥其去除夹杂物能力,使 RH 精炼后钢液中夹杂物近乎全部为在其后连铸过程不易聚合的微小液态夹杂物。

研究证实"W"形宏观偏析主要与结晶器电磁搅拌(M-EMS)和铸流区电磁搅拌(S-EMS)作用有关,通过采取措施大幅后移 S-EMS 位置并采用"弱 M-EMS /弱 S-EMS /中 F-EMS"电磁搅拌功率控制模式,解决了长期困

扰特殊钢连铸的"W"形宏观偏析问题。直径 450mm 和 600mm 连铸圆坯碳含量波动值低于 0.03%，最大偏析度（C/C_0）降低至 1.03 以下。

Control of Large-sized Inclusions and Macro Segregation in Stells for High-speed Train Wheels and Axles

Wang Xinhua[1], Jiang Min[1], Yu Huixiang[1], Wang Kaizhong[2], Sun Wei[2]

(1. School of Metallurgical and Ecological Engineering, University of Science and Technology Beijing, Beijing, China; 2. Technical Research Center, Ma Steel Co., Ltd., Maanshan, China)

Abstract: Investigation was made on control of large-sized inclusions and macro segregation in high speed train wheel and axle steels produced by EAF-LF-RH-CC steelmaking route. It was found that, in the continuously cast round blooms containing less than 6ppm T.O, the inclusions larger than 100μm were all the cluster typed inclusions formed due to aggregation of small inclusion particles in liquid steel in continuous casting and the macro segregation behaves like the letter "W", i.e. negative and positive segregations took place alternatively from edge to center of the bloom.

In LF refining of the steels, the inclusions were found to transfer from Al_2O_3 firstly to $MgO-Al_2O_3$, then to $CaO-MgO-Al_2O_3$ and finally to $CaO-Al_2O_3$ system. As the solid inclusions of Al_2O_3, $MgO-Al_2O_3$ and $CaO-MgO-Al_2O_3$ have larger contact angles and are easy to be removed, measures were taken: (1) Delay the transfer of the solid inclusions to liquid ones of $CaO-Al_2O_3$ system by shorten as much as possible the LF refining time and, (2) eliminate most solid inclusions in RH degassing by extending the RH refining time to more than 33min. With this technique, the remnant inclusions after RH were mostly small-sized liquid inclusions which were difficult to aggregate to larger inclusions in subsequent continuous casting.

The "W" typed macro segregation was proved mainly due to actions of M-EMS and S-EMS in continuous casting. By largely moving the position of S-EMS far from the meniscus and adopting the EMS power control pattern of "Week M-EMS/Week S-EMS/Middle F-EMS", the long term problem of "W" typed macro segregation in continuous casting of special steels were solved. The carbon variation range of wheel and axle steel blooms of ϕ450mm and ϕ600mm has been decreased to less than 0.03% and the maximum segregation degree of carbon (C/C_0) lowered to less than 1.03.

特殊钢特种冶金技术和生产流程的发展趋势

姜周华

（东北大学）

摘　要： 特种冶金是生产高端特殊钢和特种合金的主要手段，一直是国内外冶金技术竞争的制高点。本文首先回顾和总结了传统特种冶金技术的种类和主要生产流程。其次，概述了真空感应熔炼、真空电弧重熔、电渣重熔以及其他特种冶金技术的最新技术发展。第三，分析了特种冶金流程与转炉/电炉-精炼-连铸/模铸的普通炼钢流程的结合，实现高效和低成本生产高端特殊钢和特种合金的可行性。最后，列举了几种典型高端特殊钢和高温合金的特种冶金生产流程。

炼钢底喷粉工艺的关键技术及实践

朱　荣

（北京科技大学冶金与生态工程学院，北京　100083）

摘 要：底喷粉技术是近年炼钢领域备受关注的热点之一，具有钢铁料消耗低、钢液洁净度高、烟尘及炉渣排放少等优点。针对底吹氧及粉剂喷吹技术，本团队进行了相关理论分析、热态实验、数值及冷态模拟等基础研究，并在电弧炉及转炉进行了底喷粉炼钢工业试验，试验结果证明了底喷粉技术在炼钢领域具有广阔的发展前景。

关键词：炼钢工艺；底吹氧；粉剂喷吹

Key Technology and Practice of Bottom Powder Injection in Steelmaking

Zhu Rong

(School of Metallurgical and Ecological Engineering, University of Science and Technology Beijing, Beijing 100083, China)

Abstract: It is recently one of the hottest topics of bottom powder injection technology in the field of steelmaking, which has advantages in low consumption of iron and steel materials, high cleanliness of molten steel, less smoke and slag discharge, etc. Theoretical analysis, thermal experiment, numerical simulation and cold simulation on the technology of bottom-blowing oxygen and powder injection had been carried out by the authors' team, and the industrial test of bottom powder injection steelmaking in electric arc furnace and converter has been carried out. The test results show that the bottom powder injection technology has broad prospects in the field of steelmaking.

Key words: steelmaking process; bottom-blowing oxygen; powder injection

92Si 桥梁缆索用钢生产过程中非金属夹杂物的演变规律

罗锋，王聪

（东北大学冶金学院，辽宁沈阳 110819）

摘 要：随着我国桥梁建设的飞速发展，桥梁缆索用钢等硬线钢品种正朝着高强化、轻量化方向升级。目前我国生产的硬线钢产品普遍存在力学性能不稳定、拉拔断裂等严重问题，给后续的加工带来了很大的困难，而研究表明，钢中脆性夹杂物是导致拉断的主要原因之一。本研究对 92Si 桥梁缆索用钢工业试验生产在炉外精炼(LF+VD)、中间包、连铸坯、加热坯以及热轧盘条进行系统取样，并采用夹杂物人工统计(SEM-EDS)以及自动统计(ASPEX PSEM explorer)相结合的手段对 BOF-LF-VD-CC 工艺生产的 92Si 钢中夹杂物在各工序的成分、数量及尺寸进行了系统的研究。结果表明：92Si 钢中主要存在两类典型夹杂物，一类是转炉出钢时采用 Al/Si/Mn 复合脱氧而产生的 $MnO-SiO_2-Al_2O_3$ 系脱氧产物；另一类是渣钢相互作用形成的 $CaO-SiO_2-Al_2O_3-(MgO)$ 系夹杂；针对第一类夹杂物，结合 FactSage 热力学计算，当钢中 Si/Mn 比例为 2:1 时，能获得最佳的脱氧效果，将最终脱氧产物控制在 Fe-Al-Si-Mn-O 优势区图的液态夹杂物区域；在 VD 炉钙处理过程中，结合实际生产的成分数据，对 Ca-S、Ca-Al、Al-S 平衡进行优化计算，并在 $CaO-Al_2O_3-CaS$ 三元相图中确定了夹杂物最佳塑性变形区域。铸坯中有部分 TiN 析出及双相 Al_2O_3-MnS 夹杂生成，这与热力学计算的结果相符；加热坯中 Al_2O_3-MnS 夹杂外层的 MnS 包裹层会分解，从而使难变形的 Al_2O_3 及 TiN 均遗留在钢坯中，成为后续轧制及拉拔的应力集中源，而 MnS 夹杂能够很好地与钢基体产生协同变形。

关键词：92Si；夹杂物演变；FactSage 计算；Al/Si/Mn 脱氧；钙处理；夹杂物变形

Evolution of Non-metallic Inclusions in Production of 92Si Steel for Bridge Cable

Luo Feng, Wang Cong

(School of Metallurgy, Northeastern University, Shenyang 110819, China)

铁水旋涌脱硫扒渣工艺开发及应用

万雪峰

（鞍钢集团钢铁研究院）

摘　要："旋涌脱硫扒渣工艺"是在传统喷吹法脱硫扒渣基础上最新研发的一种铁水脱硫扒渣工艺，是铁水旋转喷吹脱硫与涌动式扒渣相结合的铁水预处理新技术。通过喷枪旋转改善铁水脱硫动力学条件，大幅度降低脱硫剂消耗。依据气泡泵原理，利用铁水自下而上涌动将脱硫渣推向扒渣口，扒渣操作简捷快速，不但缩短扒渣时间，而且降低铁水扒损。该工艺的技术指标已实现 CaO-Mg 混合喷吹降低金属 Mg≥27%、降低铁水扒损 3~10kg/t、缩短扒渣时间≥3min/罐的实绩。

电弧炉炼钢工程的流程设计与装备选型

潘宏涛

（中冶京诚工程技术有限公司）

摘　要：报告主要阐述如何开展电弧炉炼钢工程的流程设计，以及流程设计中应该注意的几个问题，同时针对如何进行电炉及炉外精炼选型给出建议。

Process Design and Equipment Selection of EAF Steelmaking Project

Pan Hongtao

高品质高氮钢加压冶金理论、制备技术及品种开发

李花兵，姜周华，冯　浩，朱红春

（东北大学冶金学院，辽宁沈阳　110819）

摘　要：作者近年来围绕高性能高氮钢加压冶金理论、制备技术及品种开发开展了系统的研究。构建了加压条件下氮溶解度模型，阐明了加压强化冷却、细化凝固组织、减轻凝固缺陷的机理，丰富和发展了加压冶金理论。成功地开发出加压感应和加压电渣单步法冶炼工艺，创新地提出了加压感应和加压电渣重熔制备高性能高氮不锈钢双联新工艺，并制备出系列高性能高氮钢材料。围绕新一代高性能航空高氮不锈轴承钢进行了初步的成分优化、热加工和热处理工艺研究，揭示了氮和氮钼协同作用对组织和性能的影响规律，获得了最佳的热加工工艺，阐明了热处理工艺对组织和性能的影响规律。

关键词：高氮钢；加压冶金；制备技术；凝固；品种开发

Pressurized Metallurgy Theory, Manufacturing Technologies and Variety Development of High Nitrogen Steels

Li Huabing, Jiang Zhouhua, Feng Hao, Zhu Hongchun

(School of Metallurgy, Northeastern University, Shenyang 110819, China)

Abstract: In recent years, the authors systematically investigated the pressurized metallurgy theory, manufacturing technologies and variety development of high nitrogen steels. The nitrogen solubility model under pressure was established. The mechanisms of enhancement of cooling, refinement of solidification structure and alleviation of solidification defects were revealed, and the pressurized metallurgy theory was enriched and developed. The single-step methods of pressurized induction melting and pressurized electroslag remelting were successfully developed, and the duplex process of pressurized induction melting and pressurized electroslag remelting was innovatively proposed, thereafter, series high performance high nitrogen steels were manufactured. The preliminary component optimization, hot deformation and heat treatment of the new generation of high performance aerospace high nitrogen stainless bearing steel were investigated. The influence of nitrogen and nitrogen-molybdenum synergism on microstructure and properties were revealed. The optimum hot deformation technology was obtained, and the effects of heat treatment processes on microstructure and properties were revealed.

Key words: high nitrogen steels; pressurized metallurgy; manufacturing technologies; solidification; variety development

洁净钢用 CA6 基耐火材料的基础研究

侯新梅，王恩会，陈俊红，李　斌，周国治

（北京科技大学钢铁共性技术协同创新中心，北京　100083）

摘　要：钢铁工业的每次重大技术进步都离不开耐火材料的支撑。随着对高品质钢要求的不断提高，传统耐火材料难以满足先进冶炼流程以及大量炉外精炼工艺的要求，已逐渐成为制约钢材品质提升的瓶颈之一。基于此，既具有优异热机械性能，即在钢水精炼过程中不"污染"或少"污染"钢水，又能一定程度上去除钢水中高熔点非金属夹杂物的新型功能化耐火材料的设计、制备及应用成为钢铁工业的进一步发展的难点。本研究团队围绕洁净钢用 CA6 基新型耐火材料展开，初步实验表明，CA6 及 $CaO-MgO-Al_2O_3$(CMA)材料兼具优异的热机械和耐渣侵性能的同时，还可以在服役过程产生低熔点精炼渣相，具备净化钢水的潜力。可以预见，上述功能化新型耐火材料有望为高品质钢的进一步发展提供有力材料支撑。

关键词：耐火材料；结构设计；可控制备；高品质钢；工业应用

Preparation and Application of CA6-based New Functional Refractories

Hou Xinmei, Wang Enhui, Chen Junhong, Li Bin, Zhou Kuozhi

(Collaborative Innovation Center of Steel Technology, University of Science and Technology Beijing, Beijing 100083, China)

Abstract: Refractories play an important role in the iron & steel industry. Confronted with the increasing demand for high-quality steels, it is difficult for traditional refractories to meet the advanced production process and large proportion of secondary refining, which become one of the bottlenecks that restrict the improvement of the steel quality. Based on this, the design, preparation and application of new functional refractories become the hottest issue for the further development in the steel industry. These refractories possess not only excellent thermo-mechanical properties, i.e. no or less contaminant for the molten steel, but also can remove the inclusions with a high melting point in the molten steel. This paper focuses on CA6 and Al_2O_3-MgO-CaO (CMA) with ternary-layer structure. Preliminary experiments show that CA6 and CMA not only possess excellent thermo-mechanical properties and high slag resistance, but also can produce refining slag phases with a low melting point. It can be predicted that CA6 based refractories are expected to provide strong support for the further development of high-quality steels.

Key words: refractory; microstructure design; controllable preparation; high-quality steels; industrial application

钢中复合夹杂物/钢基体的电势差与电偶腐蚀的关系

侯延辉[1,2]，刘林利[1,2]，李光强[1,2]，李腾飞[1,2]，刘昱[1,2]

（1. 耐火材料与冶金省部共建国家重点实验室，武汉科技大学，湖北武汉 430081；
2. 钢铁冶金及资源利用省部共建教育部重点实验室，武汉科技大学，湖北武汉 430081）

摘 要： 为了揭示高强度管线钢中典型复合夹杂物诱发点蚀的机理，本文以 Al-Ti-Mg 脱氧钢为例，采用第一性原理计算，浸泡试验，扫描电镜等研究了钢中复合夹杂物/钢基体的电势差与电偶腐蚀的关系。结果表明，在 3.5% NaCl 腐蚀环境中，MnS 夹杂物起阳极作用，优先腐蚀和溶解；$MgAl_2O_4$ 和 Al_2O_3 起阴极作用，导致铁基体的腐蚀；$MgTiO_3$ 和 $MgTi_2O_4$ 的不同端面同时起阳极和阴极的作用，因此对点蚀的影响不明显。

关键词： 夹杂物；点蚀；电偶腐蚀；第一性原理；电势差

The Correlation between Potential Difference and Galvanic Corrosion of Composite Inclusions/Steel Matrix in Steel

Hou Yanhui[1,2], Liu Linli[1,2], Li Guangqiang[1,2], Li Tengfei[1,2], Liu Yu[1,2]

(1. State Key Laboratory of Refractories and Metallurgy, Wuhan University of Science and Technology, Wuhan 430081, China; 2. Key Laboratory for Ferrous Metallurgy and Resources Utilization of Ministry of Education, Wuhan University of Science and Technology, Wuhan 430081, China)

Abstract: In order to reveal the mechanism of galvanic pitting corrosion initiation induced by typical composite inclusions, first-principles calculations, combining with immersion tests, scanning electron microscopy was used to study the correlation between electronic work function and galvanic corrosion of Al-Ti-Mg killed steel. The results show that MnS inclusions act as anodes in the electrochemical environment, preferentially corroding and dissolving; $MgAl_2O_4$ and Al_2O_3 inclusions act as cathodes, leading to the corrosion of Fe matrix; different end planes of $MgTiO_3$ and $MgTi_2O_4$ act as both anodes and cathodes, so they have little effect on pitting corrosion.

Key words: inclusions; pitting corrosion; galvanic corrosion; first-principles; potential difference

转炉吹炼后期碳含量连续预报模型算法研究

林文辉[1]，孙建坤[1]，刘　青[1]，焦树强[1]，周凯啸[1]，
刘　敏[2]，苏　醒[2]

（1. 北京科技大学钢铁冶金新技术国家重点实验室，北京　100083；
2. 新余钢铁集团有限公司，江西新余　338001）

摘　要： 本文分析了三种常用的基于烟气分析技术的转炉碳含量预报模型，即碳积分模型、指数衰减模型和三次方拟合模型，讨论了各模型的优缺点并利用工业生产的实际数据对模型预测效果进行了比较。在综合相关模型算法优点的基础上，提出了一种转炉吹炼后期熔池碳含量连续预报模型的指数修正算法。该算法首先利用历史炉次吹炼后期的脱碳氧效率数据，回归拟合得到历史平均脱碳曲线并以之作为计算参考曲线，再根据每炉实际曲线与参考曲线的偏离度，采用等距离多点连续校正的方法对计算结果及参考曲线进行同步修正，进而对吹炼后期熔池碳含量进行连续精确预报，直至吹炼结束。实践证明，本文提出的指数修正算法能够显著提高模型预报的准确率，终点碳含量预报误差在±0.02%范围内的命中率达到90%以上。

关键词： 转炉炼钢；碳含量连续预报；指数模型；脱碳氧效率；等距离多点校正

Study on Algorithm of Continuous Carbon Prediction Model in the End Blowing Stage of BOF Converter

Lin Wenhui[1], Sun Jiankun[1], Liu Qing[1], Jiao Shuqiang[1],
Zhou Kaixiao[1], Liu Min[2], Su Xing[2]

(1. State Key Laboratory of Advanced Metallurgy, University of Science and Technology Beijing, Beijing 100083, China; 2. Xinyu Iron and Steel Group Co., Ltd., Xinyu 338001, China)

Abstract: Several models for carbon content prediction of BOF process based on off-gas analysis were discussed in this paper, such as the integral model, the exponential decay model and the cubic fitting model. The advantages and disadvantages of these models were analysed respectively, then a new exponential model was established by the introduction of a correction algorithm for predicting the carbon content of the bath continuously. The proposed model involves applying the decarburization efficiency data of the historical heats firstly to obtain a average decarburization curve using the regression fitting method. The historical average curve was used as the reference curve to calculate the carbon content of current heat. According to the deviation between the reference curve and the actual curve the calculation results and parameters of the reference curve were corrected simultaneously by isometric multi-point continuous correction method.

Plant trials were carried out in a BOF converter to compare the performance of the mentioned models. The results showed that the new model exhibited better adaptability and higher accuracy than the other ones. The hit ratio of the new model reached more than 90% for the prediction of end-point carbon prediction within a tolerance of ±0.02%.

Key words: BOF process; carbon prediction; exponential decay model; decarburization efficiency; isometric multi-point continuous correction

X80 管线钢非金属夹杂物控制研究

李战军[1,2,3]，初仁生[1]，刘金刚[1]，郝 宁[1]

（1. 首钢技术研究院宽厚板所，北京 100043；2. 绿色可循环钢铁流程北京市重点实验室，北京 100043；3. 北京市能源用钢工程技术研究中心，北京 100043）

摘 要：本文研究了"铁水脱硫预处理—转炉—LF 精炼—RH 真空精炼—板坯连铸"工艺 X80 管线钢非金属夹杂物的控制，采用此工艺生产的 X80 管线钢 T[O]≤10ppm，实现了高洁净度控制；并对冶炼过程中的夹杂物转变机理进行研究，使钢坯中的夹杂物得到有效的控制：从形态上看，以球形夹杂为主；从成分上看，以高熔点的钙铝酸盐和 CaS 的复合夹杂物为主；从尺寸上看，尺寸控制在≤8μm。通过板坯轧制后，采用金相评级的方法对轧材中的非金属夹杂物进行评级，非金属夹杂物的合格率稳定控制在 99.6%以上，其中 A 类非金属夹杂物评级为 0，B 类非金属夹杂物评级≤1.5。

关键词：X80 管线钢；冶炼工艺；洁净度；非金属夹杂物

Study on Control of Non-metallic Inclusions in X80 Pipeline Steel

Li Zhanjun[1,2,3], Chu Rensheng[1], Liu Jingang[1], Hao Ning[1]

(1. Shougang Research Institute of Technical Department of Heavy and Medium Plate, Beijing 100043, China; 2. Beijing Key Laboratory of Green Recyclable Process for Iron & Steel Production Technology, Beijing 100043, China; 3. Beijing Engineering Research Center of Energy Steel, Beijing 100043, China)

Abstract: In this paper, the smelting process of "Hot metal desulphurization pretreatment-Basic oxygen furnace-LF refining-RH refining-continuous casting slab" is studied to control of non-metallic inclusions in X80 pipeline steel. The X80 pipeline steel T[O]≤10ppm produced by this process realizes high cleanliness control. The transformation mechanism of inclusions in the process of smelting was studied, so that the inclusions in the billet could be effectively controlled: spherical inclusions are dominant in morphology; high melting point calcium aluminate and CaS composite inclusions are dominant; the size is controlled ≤8μm. After slab rolling, the metallographic grading method was used to grade the non-metallic inclusions in the rolled steel. The qualified rate of non-metallic inclusions is stably controlled over 99.6%, of which the grade of A-type non-metallic inclusions is 0, and that of B-type non-metallic inclusions is less than 1.5.

Key words: X80 pipeline steel; smelting process; cleanliness; non-metallic inclusions

高温板坯全连续淬火新技术开发及应用

朱苗勇[1]，蔡兆镇[1]，刘志远[1,2]，王重君[2]，赵佳伟[1]

（1. 东北大学冶金学院，辽宁沈阳　110819；2. 唐山中厚板材有限公司，河北唐山　063610）

摘　要：研究分析了含 Nb 微合金钢高温淬火过程不同淬火温度及冷却速度下的组织结构转变及析出特点，确定了实现铸坯表层组织高塑化最佳淬火控制起始温度为 950℃、最佳冷却速度≥5℃/s。基此结合连铸坯温度场模拟计算，确定了最佳淬火位置为扇形段 12 段末。在此基础上，设计开发了连铸机在线全连续淬火装备，现场应用表明满足微合金钢连铸坯热送要求。

关键词：微合金钢；板坯；热送；表面淬火；裂纹

Development and Application of New Continuously Quenching Technology for High Temperature Slab

Zhu Miaoyong[1], Cai Zhaozhen[1], Liu Zhiyuan[1,2], Wang Chongjun[2], Zhao Jiawei[1]

(1. School of Metallurgy, Northeastern University, Shenyang 110819 China;
2. Tangshan Heavy Plate Co., Ltd., Tangshan 063610, China)

Abstract: The evolution of the microstructure of the continuous casting slab of Nb containing alloyed steel under different quenching temperatures and cooling rate were analyzed, and the optimum quenching temperature 950℃ and the optimum cooling rate ≥5°C/s were obtained. Moreover, the optimum position, where is after the end of seg. 12, of slab surface quenching was determined by the slab heat transfer simulation. Based on these, the quenching equipment was developed, and the application result showed that it can meet the need of micro-alloyed slab hot charging well.

Key words: micro-alloyed steel; slab; hot charging; surface quenching; crack

板坯连铸过程流动、传热及夹杂物分布的数值模拟研究

张立峰

（北京科技大学，北京　100083）

摘　要：连铸过程结晶器内多相流分布、传热和凝固直接影响了流场形态、液位波动及氩气泡和夹杂物被凝固坯壳捕获等现象。本研究通过建立三维数学模型研究了连铸结晶器内的多相流动、传热、凝固以及夹杂物的运动和捕获等现象。结果表明随着吹氩流量的增大，流场形态从双环流逐渐转变为复杂流和单环流，且复杂流和单环流下液位波动较大。电磁制动能够提高夹杂物在弯月面处的去除率，并且降低铸坯表层一定厚度内的夹杂物数量。夹杂物分布预测结果表明铸坯中夹杂物分布存在铸坯中心及距内弧 1/4 宽度处的条状聚集区。

Mathematical Modeling on the Fluid Flow, Heat Transfer, and Inclusion Distribution during the Continuous Casting Slab Strand

Zhang Lifeng

Abstract: The multiphase flow, heat transfer and solidification during the continuous casting process directly affect the flow pattern, surface level fluctuations and the entrapment of argon bubbles and inclusions by the solidified shell. In the current study, a three-dimensional mathematical model was established to investigate the multiphase flow, heat transfer, solidification and the transport of inclusions in the continuous casting mold. The results show that with the increase of the argon flow rate, the flow pattern gradually changed from double roll flow to complex flow and single roll flow. The surface level fluctuation under complex flow and single roll flow was larger compared with the double roll flow. The inclusion removal fraction was increased after the application of the FC-Mold. The number of inclusions in a certain thickness below the slab surface was also reduced accordingly. The inclusion distribution results indicate that two accumulation peaks of inclusions along the thickness of the slab were existed, including the centerline of the slab thickness and the 1/4 thickness from the loose side.

连续铸钢设备设计理论研究
——连铸结晶器振动理论的研究与实践

杨拉道

（中国重型机械研究院股份公司）

摘　要： 本报告总结了连铸结晶器振动系统近 20 年来的理论研究成果，连铸机常用结晶器振动设备类型，连铸机结晶器振动参数在生产应用当中的确定原则以及确定连铸机结晶器振动参数时的注意事项。为连续铸钢生产提供了理论支撑和经验指导。

Research on Design Theory of Continuous Casting Steel Equipment
—— Research and Practice of the Vibration Theory of Continuous Casting Mold

Yang Ladao

Abstract: This report summarizes the theoretical research results of continuous casting mold vibration system in the past 20 years, the type of vibration equipment commonly used by continuous casting machine, the determination principle of the vibration parameters of continuous casting mold in the production applications, and the consideration of determining the vibration parameters of continuous casting mold. It provides theoretical support and experience guidance for continuous cast steel production.

高表面品质汽车板连铸浸入式水口结构优化工艺研究

李海波

（首钢集团有限公司技术研究院，北京 100043）

摘 要：近年来首钢汽车钢板在产量和质量方面均有大幅度提升，代表钢水洁净度水平的中间包 T.O 含量均值达 15ppm，达到了较高的水平。但后道工序反应的冷轧板缺陷依然较多，主要原因是由于结晶器卷渣引起。前期调研认为水口堵塞和液面波动大是产生线状缺陷的主要原因。本文提出水口出口上部的负流区是引起水口堵塞的重要原因。基于此将水口出口面积比从 2.6 减小至 2.2，减小了负流区面积，缓解了水口堵塞。针对液面波动大的问题，水模型试验和数值模拟研究发现：提高水口角度、变换水口底部和出口形状可显著降低结晶器液面波动和卷渣发生率。提出了倾角为 20°的凹底椭圆形浸入式水口，并成功应用于首钢京唐汽车钢板连铸实践。浸入式水口结构优化后，汽车钢板的线状缺陷发生率大幅度降低。

关键词：浸入式水口；负流区；水口堵塞；液面波动

Optimization of SEN Structures for Slab Continuous Casting High Surface Quality IF Steels

Li Haibo

(Shougang Group Co., Ltd., Research Institute of Technology, Beijing 100043, China)

Abstract: In recent years, the production and quality of the cold-rolled IF steel sheets in Shougang have both significantly improved. The T.O in tundish, an index to evaluate the cleanliness of the steel, reached 15ppm. Slivers defects due to the mold powder entrapment still exists although the cleanliness of steel melt approach to a high level. It is considered that nozzle clogging and level fluctuation are the main cause for the sliver defects. The research results show that the presence of back flow zone is an important cause of the present nozzle clogging. Decreasing the ports to bore area ratio from 2.6 to 2.2 reduce the area of back flow zone and relieve the clogging. The water modeling and mathematical modeling results show that increasing the nozzle angle, changing the bottom and outlet shape can sharply decrease the level fluctuation. A nozzle with 20° angle with well bottom and oval outlet was proposed to apply in slab casters in SGJT. After applying this nozzle, the sliver defects drastically reduced.

Key words: submerged entry nozzle; back flow zone; nozzle clogging; meniscus fluctuation

改善重轨钢连铸坯均质化与致密化的基础研究与应用实践

陈 永

（攀钢集团研究院有限公司，四川攀枝花 617000）

摘　要：重轨钢连铸坯作为重轨钢生产的轧制母材，其中心偏析、中心疏松和缩孔等内部质量缺陷是造成钢轨探伤不合和钢轨断裂的主要原因，严重时危害铁路行车安全。为此，高均质化和致密化连铸坯是生产高品质钢轨的关键。本研究针对攀钢重轨钢连铸生产现状，采用数值模拟方法，建立了重轨钢连铸凝固组织生长与宏观偏析数学模型，以及连铸坯热力耦合数学模型，详细研究了连铸工艺（浇铸温度、拉速和冷却强度）对重轨钢凝固组织生长、柱状晶向等轴晶转变、等轴晶率、晶粒尺寸和溶质宏观偏析的影响规律，以及连铸设备（辊型）对重轨钢连铸坯宏观偏析和缩孔闭合度的影响规律。以此为理论基础，优化了重轨钢连铸工艺，并对拉矫辊辊型进行了改造，使重轨钢连铸坯中心等轴晶率达到35%以上，中心碳偏析控制在1.08以内，连铸坯均质化和致密化得到明显改善。

关键词：重轨钢；凝固组织；宏观偏析；缩孔；数值模拟；工业应用

Fundamental Study and Application of the Improvement of Solute Homogenization and Compactness of Continuously Cast Rail Steel

Chen Yong

(Research Institute Co., Ltd., Pangang Group, Panzhihua 617000, China)

Abstract: The continuously cast strand is the rough rolling material for rail steel production. The internal quality defects of bloom, such as center segregation, porosity, cavity, etc. are the main reasons for the rail inspection failed and rail breakage, which seriously endanger the safety of the train operation. Therefore, highly homogenous and compact bloom is the key factor to produce high quality rail steel. To achieve it, several numerical models, such as cellular automaton model for dendiritc grain growth, volume average model for marosegregation, and thermoechanic model for cavity closure, are firstly proposed to investigate the solidification microstructure evolution, solute macrosegregation, and cavity elimination of continuously cast rail steel, based on the current situation of continuous casting production of heavy rail steel in Panzhihua Iron and Steel Co., Ltd. and systematic investigation on the effect of casting operations (casting temperature, drawing speed and cooling strength) on solidification microstructure growth, columnar crystal equiaxed crystal transformation, equiaxed crystal ratio, grain size and macrosegregation of solute, and caster equipment (roller Type) on the macrosegregation and shrinkage porosity of rail steel. Based on the results, the continuous casting process of rail steel was optimized, and the roller type of the withdrawer was modified. Finally, the solute homogenization and compactness of the continuously cast bloom of rail steel produced in Panzhihua Iron and Steel Co., Ltd. are significantly improved. The equiaxed crystal ratio of continuous casting bloom is larger than 35%, and the central carbon segregation is less than 1.08.

Key words: rail steel; solidification structure; macrosegregation; shrinkage cavity; numerical simulation; industrial application

轧钢过程节能减排技术进展及应用实例分析

康永林

（北京科技大学，北京　100083）

摘　要：首先简要介绍了我国近年轧制钢材产量、品种发展情况，指出资源、能源、环境、节能减排对钢材生产及应用的迫切要求；其次，分析了钢材轧制、热处理、表面处理等技术发展趋势及与能源消耗和排放的联系；针对轧钢过程节能减排技术应用实例及效果做了具体分析，包括：（1）热轧板带钢新一代控轧控冷技术；（2）热宽带钢无头轧制/半无头轧制技术；（3）低碳、超低碳钢铁素体轧制技术；（4）热轧板带材表面氧化铁皮控制技术；（5）板带钢及型钢轧制数字化技术；（6）连铸坯凝固末端大压下技术；（7）小方坯免加热直轧技术；（8）棒材多线切分轧制技术；（9）板带钢免酸洗除氧化铁皮除锈技术；（10）轧后钢材在线热处理技术。最后，对轧钢过程节能减排技术的发展作了展望。

Progress and Application of Energy Saving and Emission Reduction Technology in Steel Rolling Process

Kang Yonglin

(School of Materials Science and Engineering, University of Science and Technology Beijing, Beijing 100083, China)

Abstract: Firstly, the production and variety development of rolling steel in China in recent years are briefly introduced, the urgent requirements of resources, energy, environment, energy saving and emission reduction for steel production and application are pointed out. Secondly, the development trend of rolling, heat treatment and surface treatment of steel and their relationship with energy saving and emission reduction are analyzed. The application examples and their effects of energy saving and emission reduction technology in steel rolling process are analyzed in detail, including: (1) new generation of controlled rolling and controlled cooling technology for hot rolled strip steel; (2) endless rolling/semi-endless rolling technology for hot rolled wide strip steel; (3) ferrite rolling technology of low-carbon and ultra-low-carbon steel; (4) oxide scale control technology on the surface of hot rolled strip steel; (5) digital technology in the plate-strip and section steel rolling; (6) heavy reduction technology at solidification end of Continuous Casting Slab; (7) direct rolling technology of free-heating for billet; (8) multi slit rolling technology for bar; (9) clean processing technology for oxide scale removal on the hot rolled strip steel; (10) on-line heat treatment technology for rolled steel. Finally, the development of energy saving and emission reduction technology in rolling process is prospected.

宽带钢热轧智能产线关键技术研究进展

何安瑞

（北京科技大学工程技术研究院，北京　100083）

摘　要： 热轧是宽带钢生产的关键工序，也是多参数耦合影响、多目标协调管控的复杂过程。提高产线的智能化程度，实现低成本、高精度、柔性化、少人化的生产，是近年来的关注热点。结合热轧生产难点、工艺流程及产品特点，围绕产品质量、生产效率、制造成本、定制化能力等核心指标，从智能感知、智能控制、智能物流、智能管控等方面介绍了基于机器视觉的非对称板形缺陷检测、基于先进算法的加热炉及轧制智能化数学模型、无人天车及智能库管系统、基于工业互联网架构的产线能源介质及工艺质量管控大数据平台等关键技术的研究进展及工业实践，为宽带钢热轧智能产线的建设提供了参考。

关键词： 热带钢轧机；智能感知；智能控制；智能物流；智能管控

Development of Key Technologies of Smart Production Line for Hot Wide Strip

He Anrui

(Institute of Engineering Technology of University of Science and Technology, Beijing 100083, China)

Abstract: Hot rolling is the key process in wide strip production, and also a complex process with multi-parameter coupling influence and multi-objective coordinated management and control. In recent years, how to improve the intelligence of production line and realize low-cost, high-precision, flexible and less-manpower production has become the hot focus. Based on the analysis of production difficulty, process and product characteristics of hot strip rolling, considering the core targets of product quality, production efficiency, manufacturing cost, customization ability, etc., development progress and industrial application of several key technologies about intelligent perception, intelligent control, intelligent logistics, intelligent management and control were introduced, including inspection technology of asymmetric shape defects based on machine vision, intelligent mathematical model of heating furnace and rolling based on advanced algorithms, unmanned crane and intelligent warehouse management system, big data platform for energy medium and process quality management and control of production line based on industrial internet architecture. All these provide the reference for the construction of smart hot rolling production line of wide strip.

Key words: hot strip mills; intelligent perception; intelligent control; intelligent logistics; intelligent management and control

轧机振动控制研究及推广应用

闫晓强

（北京科技大学机械工程学院，北京 100083）

摘 要：轧机振动是世界范围内轧制领域的一个技术难题。当轧制高强度、薄规格带钢时，轧机频繁发生强烈振动现象。由于轧制过程呈现多参数、强耦合和非线性特征，使得轧机振动问题研究变得更加复杂化。经过多年专家和学者的不懈探索和研究，取得了一些进展和成果，但还未得到诱发轧机振动机理的公认解释，成为半个多世纪以来轧制领域亟须解决的难题。

本课题从铸轧全流程的角度和视野进行研究，也是前人没有进行深入研究的盲区，经过12年承担全国22个轧机机组的振动研究，取得了突破性进展，建立了轧机机电液界多态耦合振动理论体系，总结出一套抑制轧机振动的实用措施，取得了显著的抑振效果，为企业创造了巨大的经济效益。

关键词：轧机振动；耦合振动；振动控制

Research and Application of Rolling Mill Vibration Control

Yan Xiaoqiang

(School of Mechanical Engineering, Beijing University of Science and Technology, Beijing 100083, China)

Abstract: Rolling mill vibration is a technical problem in the rolling field worldwide. When rolling high strength and thin gauge strip, the rolling mill often has strong vibration phenomenon. The rolling process is characterized by multi-parameters, strong coupling and non-linearity, which makes the study of rolling mill vibration more complicated. After years of unremitting exploration and research by experts and scholars, some progress and achievements have been made, but the mechanism of induced rolling mill vibration has not been recognized, which has become an urgent problem in the rolling field for more than half a century.

In this project, the whole process of casting and rolling is studied from the point of view and field of vision. It is also a blind area that the predecessors have not studied in depth. After 12 years of undertaking the vibration research of 22 rolling mills nationwide, breakthrough progress has been made. A theoretical system of multi-state coupling vibration in rolling mill has been established, and a set of practical measures to suppress the vibration of rolling mills have been summarized. It has achieved remarkable anti-vibration effect and created enormous economic benefits for enterprises.

Key words: rolling mill vibration; coupled vibration; vibration control

I&Q&P 工艺中配分时的竞争机制及变形协调机理的研究

于 浩

（北京科技大学，北京 100083）

摘 要： 第三代先进高强度汽车用钢兼有第一代和第二代的特点，可以满足节约资源、降低成本、汽车轻量化和提高安全性的要求，因此研发并生产具有高强塑性的第三代汽车用钢是当今的发展趋势。其中 I&Q&P (Intercritical heating, quenching and partitioning)钢已经可以进行工业化生产，而且相关研究已有很多，但仍存在一些尚未突破的机理性问题，如配分时贝氏体的转变机理、碳配分及碳化物析出对贝氏体转变的影响、不同界面上原子的偏聚机理、变形过程中各相组织之间的变形协调机理、不同复相组织对疲劳裂纹扩展的影响机理等。本文以 I&Q&P 钢为研究对象，对淬火、配分工艺过程中的几个关键、共性的科学问题进行系统研究，也力求推动淬火、配分工艺在其他领域的发展，如挤泥机绞刀、破碎机锤头、耐磨钢球、球磨机衬板等。

关键词： I&Q&P 钢；残余奥氏体；TRIP 效应；贝氏体转变；变形协调机理

首钢超大型水电压力钢管用高强度钢板的开发

邹 扬

（首钢技术研究院，京唐技术中心，北京 100043）

摘 要： 近年来随着水电行业的快速发展，水电装机容量不断增加和设计水头不断提高，对水电站压力钢管所使用的高强钢板的强度、韧性、可焊性等方面均提出了更高的要求。高可焊性的高强度钢板逐步成为目前大型水电站和抽水蓄能电站的首选钢板。首钢 4300 中厚板生产线采用优化的成分设计，通过严格的冶炼、轧制及热处理工艺控制，开发出满足水电行业设计需求高焊接性 800MPa 级钢板、150mm 规格特厚板、适应大热输入的配套焊材及高效焊接技术。800MPa 级别钢板预热温度从 150℃降低至 80℃，50kJ/cm 热输入下热影响区 KV2≥100J，焊接效率提高 1.5 倍；150mm 特厚钢板贝氏体晶粒尺寸控制在 10μm 以下，与传统工艺相比晶粒尺寸减小了 80%。

关键词：水电站；高强度贝氏体钢；焊接性

Development of High Strength Steel Plate for Large Hydroelectric Penstock of Shougang

热连轧板"唇印结疤"成因分析与对策

郭 斌

（宝钢股份中央研究院武汉分院，湖北武汉 430080）

摘 要：对热连轧板"唇印结疤"缺陷的形貌、分布以及形成原因进行了系统的分析，并提出了消除措施。"唇印结疤"宏观呈云状疤块，类似唇印，呈网状分布，发生在钢板的上表面，沿长度方向间断、无规律，有时连续分布，在宽度方向时左、时右、时中，在板宽 1/4～1/3 处居多，随着厚度增加，疤块越明显、严重；微观分析表明钢板表面"唇印结疤"是一种网状裂纹，其形成与连铸坯上的缺陷或 Cu、Ni、As 在晶界富集有关；系统调查表明连铸坯、加热坯、粗轧后中间坯表面均普遍存在由氧化铁皮下 Cu、Ni、As 富集产生的奥氏体晶界网状裂纹；钢坯除鳞时一次氧化铁皮未除净或二次氧化铁皮清除不及时，将出现裂纹中 Ni、Cu、As 富集严重或裂纹表面明显氧化，导致裂纹不能轧合；此外，除鳞喷嘴流量、重叠量大，使钢板表面冷却不均匀，局部过冷部位的裂纹也不能轧合，上述两种情况均导致钢板表面形成网状裂纹——"唇印结疤"；采取更换除鳞冷却水集管和水嘴，整改定宽机异常漏水点，严格执行加热工艺，控制 Cu、As 等残余元素含量等措施彻底消除了"唇印结疤"。

关键词：热连轧板；结疤；Cu、Ni、As；表面裂纹；除鳞

Formation Analyze and Countermeasure of Lip Print Scab Defect on Hot Rolled Strip

Guo Bin

(Wuhan Branch of Baosteel Central Research Institute, Wuhan 430080, China)

Abstract: The reasons of formation had scientifically researched for lip print scab defect on hot rolled strip, and the measure of solve scab defect had been put forward. The macroscopical features of lip print scab defect are cloudy scars, network-like distribution, similar lipstick, on the upper surface of the strip, irregularly and disjunctive distribution along the length, sometimes continuous distribution along the length, left time right and sometimes middle along width, majority on 1/4 ~1/3 of width, more obvious and serious as the thickness increases. Through microscope analysis, the results show that the lip print scab defect is surface network cracks, the cause of formation is related to the surface defect of slab or the concentration Cu, Ni, As at grain boundary. Systematic investigation shows that the austenitic reticulate crack produced by Cu, Ni, As enrichment is common below surface oxidizing iron of continuously cast slab, heated slab, roughed slab. When enrichment of Cu, Ni, As in the cracks or the oxidation of cracks is serious because one iron oxide skin has not been eliminated or two iron oxide skin is out of time by descaling, the slab surface network cracks cannot be rolled. In addition, the local overcooling cracks cannot be rolled because uneven cooling of plate surface caused by large flow rate and large overlap of water descaling sprayer. The above two aspect lead to the formation of cracks- lip print scab defect. The lip print scab defect had been thoroughly eliminated by replacing the descaling collection tube and the nozzle, by rectification of the abnormal leakage point of the fixed width machine, by strictly enforcing the heating process, by controlling the content of residual elements such as Cu, As , and so on.

Key words: hot rolled strip; scab; Cu、Ni、As; surface crack; descaling

超超临界锅炉用新型耐热无缝管 C-HRA-5 的开发

方旭东[1,2]，包汉生[3]，李　阳[1,2]，徐芳泓[1,2]，夏　焱[1,2]

（1. 先进不锈钢材料国家重点实验室（太原钢铁（集团）有限公司），山西太原　030003；
2. 山西太钢不锈钢股份有限公司技术中心，山西太原　030003；3. 钢铁研究总院，北京　100081）

摘　要： 为了满足 630~650℃超超临界电站过热器和再热器锅炉管的要求，太钢和钢铁研究总院通过成分设计、冶炼、冷热变形、热处理等技术攻关，突破关键技术瓶颈，采用"电炉+AOD（或 VIM+ESR）+径锻+挤压+冷轧"工艺，制备出 C-HRA-5 不锈钢无缝钢管。产品具有化学成分控制精确、非金属夹杂物含量低、钢质纯净、表面质量好、组织均匀、性能优异的特点，其化学成分、晶粒度、微观组织、常温力学性能、高温拉伸性能、高温持久性能等各项性能指标满足 ASME SA-213 标准的要求，抗蒸汽氧化性能和焊接性能优异，实物质量达到了国际先进水平，获得上锅、东锅、哈锅、上海成套院、钢研院的评定报告，产品取得全国锅炉压力容器标准化技术委员会的评审认证，并牵头制订了相关团体标准，具备批量供货条件，可用于超（超）临界锅炉的过热器、再热器等部件以及类似工况的受压结构件。

关键词： 超超临界锅炉；耐热钢；无缝管；C-HRA-5

Development of Model Heat Resisting Seamless Tube C-HRA-5 for Ultra-Supercritical Power Plant Boilerand

Fang Xudong[1,2], Bao Hansheng[3], Li Yang[1,2], Xu Fanghong[1,2], Xia Yan[1,2]

(1. State Key Laboratory of Advanced Stainless Steel Materials, Taiyuan Iron & Steel (Group) Co., Ltd., Taiyuan 030003, China; 2. Technology Center, Shanxi Taiyuan Stainless Steel Co., Ltd., Taiyuan 030003, China; 3. Central Iron and Steel Research Institute, Beijing 100081, China)

Abstract: In order to meet the requirements of superheater and reheater boilers in 630~650℃ ultra-supercritical power plants. The C-HRA-5 stainless steel seamless steel pipe was prepared according to the process of "EAF+AOD (or VIM+ESR)+diameter forging+extrusion+cold rolling" through the technical research of composition design, smelting, cold and hot deformation and heat treatment and breaking through the bottleneck of key technology. The products are characterized by precise chemical composition control, low content of non-metallic inclusions, pure steel, good surface quality, uniform structure and excellent performance. Its chemical composition, grain size, micro-structure, mechanical properties at room temperature, high temperature tensile properties, high temperature durability and other performance indicators meet the requirements of ASME SA-213 standard, and its steam oxidation resistance and welding performance are excellent. Its physical quality has reached the international advanced level, and has been evaluated by Shanghai Boiler Works, Ltd., Dongfang Boiler Group Co., Ltd., Harbin Boiler Co., Ltd., Shanghai Power Equipment Research Institute and

CISRI. The product has been appraised and certified by the National Technical Committee for Standardization of Boiler and Pressure Vessel, and has taken the lead in formulating relevant group standards. It can be used in superheater, reheater and other parts of ultra (ultra) critical boiler, as well as pressure structure parts under similar conditions.

Key words: ultra supercritical boiler; heat resistant steel; seamless pipe; C-HRA-5

热轧钢筋分段气雾冷却工艺研究

王卫卫[1]，李大亮[2]，周秉国[3]，李志新[4]，徐晓林[2]，杨 杰[2]

(1. 钢铁研究总院冶金工艺研究所，北京 100081；
2. 山西建龙实业有限公司，山西运城 043801；3. 金鼎钢铁集团轧钢厂，
河北武安 056300；4. 河北敬业集团有限责任公司，河北石家庄 050409)

摘 要： 本文介绍了热轧钢筋新型气雾冷却设备及工艺，对新设备和工艺下生产的钢筋进行成分-组织-性能-表面质量的分析，并对新设备的应用前景进行了讨论，为今后的热轧钢筋控冷工艺的优化提供技术参考。研究表明，气雾冷却工艺能细化边部到心部的显微组织，可以细化珠光体团块尺寸及片层间距，强化效果明显。

关键词： 热轧钢筋；气雾冷却工艺；新设备；表面质量

Study on Piecewise Spray Evaporation Cooling Technology of Hot-rolled Reinforced Bar

Wang Weiwei[1], Li Daliang[2], Zhou Bingguo[3], Li Zhixin[4], Xu Xiaolin[2], Yang Jie[2]

(1. Metallurgical Technology Institute of Central Iron & Steel Research Institute, Beijing 100081, China;
2. Shanxi Jianlong Industrial Co., Ltd., Yuncheng 043801, China; 3. Jinding Iron & Steel Co., Ltd., Wuan 056300, China; 4. Hebei Jingye Group Co., Ltd., Shijiazhuang 050409, China)

Abstract: In this paper, the application of spray evaporation cooling equipment and technology of hot-rolled reinforced bar is summarized, and the new process and equipment is explained, which about the composition-microstructure-performance-surface quality requirements of hot-rolled reinforced bar and the application prospect of the new equipment is discussed t, which provides technical reference for the optimization of the controlled cooling technology for hot-rolled steel bars in the future. The results show that the spray evaporation cooling process can refine the pearlite block size and lamellar spacing from the edge to the center, and the strengthening effect is obvious.

Key words: hot-rolled reinforced bar; spray evaporation cooling process; new equipment; surface quality

我国锌与锌合金镀层钢板发展

张启富

(中国钢研科技集团，北京 100081)

摘 要： 随着我国经济进入新常态及制造业提升，我国锌与锌合金镀层钢板的总量保持稳定，但对高端锌与锌合金

镀层钢板需求逐步增加。本文介绍了近年来我国锌及锌合金镀层钢板的发展，重点介绍了汽车用镀锌先进高强钢、Zn-Al-Mg 镀层钢板、连续 PVD 镀层、热成形钢镀层的发展及未来趋势。

冷热复合成形技术与结构轻量化应用

韩静涛

（北京科技大学，北京 100083）

摘 要：传统金属塑性加工工艺以冷加工、热加工划分，产品开发和应用受到了很大的限制，我们提出的冷热复合成形技术，很好的解决了高强、厚壁、复杂断面材料的成形技术难题，开发出多种钢材深加工新产品，使金属材料加工技术获得了更广泛的应用。

（1）汽车与运载工具轻量化。高强钢大幅降低运载设备重量，提高设备承载能力和安全性，但塑性差难成形。新开发的≥1500MPa 超高强异型钢管，2018 年起大批量应用于宇通、金龙客车车身，及轿车防撞杆、保险杠、稳定杆等结构件，将成为运载工具主要结构材料，为国内市场新增≥15 万吨/年汽车热成形钢市场；

（2）钢结构建筑/装配式住宅。钢结构技术发展瓶颈为厚壁≥6mm 构件制造。因成形部位出现厚度减薄、裂纹、残余应力等问题，占结构钢≥50%的梁柱件只能采用箱型梁/柱工艺，手工操作和焊接工作量大，质量难于控制，后期畸变和运维成本高，环境污染严重。新（角部增厚、塑性最佳、无残余应力）尖角厚壁方矩型钢管，可提高钢材贡献率≥20%，确保结构安全，住建部专家一致认为该技术将大量应用于各类构造物主辅承力结构，必将推动钢结构建筑划时代的技术进步，并为中国钢铁业新增≥4000 万吨/年应用市场。2018 年 4 月已完成第一条示范生产线建设，并于 2018 年 6 月承接了冀中能源 300 万吨/年新建大型高精钢材深加工生产基地建设任务。

（3）交通运输。钢结构桥梁为国家重点关注技术，我国钢桥比将由≤1%迅速提高到≥30%，但造成≥90%钢桥病害的传统 U 肋，成为制约"瓶颈"。端部增厚 UTU 肋，大大增强焊接接头疲劳性能，提高钢桥运行可靠性和稳定性。新产品已在≥20 座大型桥梁建设中获得应用，将为我国钢铁行业新增 2200 万吨/年市场。

1500MPa 液压成形管件扭力梁开发及疲劳开裂问题改善

蒋浩民[1,2]，逯若东[1,2]，苏海波[3]，徐小华[3]

（1. 宝山钢铁股份有限公司研究院，上海 201900；2. 汽车用钢开发与应用技术国家重点实验室（宝钢），上海 201900；3. 宝钢高新技术零部件有限公司，上海 201908）

摘 要：汽车底盘扭力梁式后悬架由于在车辆行驶过程中弯曲受扭，服役条件严苛，在开发及应用过程中容易发生疲劳开裂。管件液压成形扭力梁具有封闭变截面结构特征，结合超高强钢材料应用，可以提高扭力梁综合性能，实现结构轻量化和材料轻量化。本文以某 1500MPa 难成形管件扭力梁为目标零件，通过四序预成形及一序液压成形工艺成功开发了该零件，通过三段式密封头设计解决了管端密封漏液问题，通过原始管径由 116mm 缩小为 114mm，及调整第三序压扁角度等措施解决了侧壁凹陷缺陷。管件扭力梁通过真空炉自由油淬，脱碳层控制在 20μm 以内。针对台架及路试疲劳开裂问题，通过将激光焊管变更为高频焊管，改善成形过程中 V 形上部 R 角，0.9mm 喷丸处理等措施解决了疲劳开裂问题。

关键词：扭力梁；管件；液压成形；疲劳开裂

Development of 1500MPa Tubular Torsion Beam with Hydroforming and Improvement of Fatigue Cracking

Jiang Haomin[1,2], Lu Ruodong[1,2], Su Haibo[3], Xu Xiaohua[3]

(1. Research Institute of Baosteel, Shanghai 201900, China;
2. State Key Laboratory of Development and Application Technology of Automotive Steels (Baosteel), Shanghai 201900, China;
3. Baosteel High-tech Components Co., Ltd., Shanghai 201908, China)

Abstract: Due to the bending and torsion force of the chassis torsion beam in the process of driving, the service conditions are strict, and fatigue cracking is easy to occur in the process of development and application. The tubular torsion beam formed by hydroforming has the characteristics of closed variable section structure, which can improve the comprehensive performance of torsion beam and realize the light weight of structure and material. Taking the 1500MPa tubular torsion beam with great forming difficulty as the research object, the part was successfully developed by four steps preforming and hydroforming. The leakage problem of sealing is solved by the design of three-section sealing punch structure. The hollow defect of side wall was solved by reducing the original tube diameter from 116mm to 114mm, and by adjusting the flattening angle of the third preforming process. Vacuum furnace oil quench method is applied, and the Decarburization layer thickness is required to be less than 20 μm. The fatigue cracking problem during bench and road test was solved by changing laser tube to ERW tube, and by improving the part radius on the top of the V shape, and by 0.9mm short blasing.

Key words: torsion beam; tube; hydroforming; fatigue cracking

定制辊压成形技术与中国制造业升级

晏培杰，毕若凌

（上海宝钢型钢有限公司，上海 201999）

摘 要：定制辊压成型技术是通过对辊压型材产品进行材料、结构优化以及模块化、近成品化的整合优化设计，并通过辊压、激光焊接等多工艺集成实现柔性化自动化制造的制造工艺技术。该技术在我国制造行业整体升级，节能减排、轻量化、智能制造等先进制造理念推广进程中具有极大的应用价值和市场开发潜力。本文对定制辊压成形技术及其产品的发展从实施路线、发展难点、应用实例等方面进行了系统介绍，同时分析了该项技术的市场前景及发展方向。

关键词：定制化；辊压成形；制造业升级；应用示范

Customized Roll Forming Technology and China's Manufacturing Upgrading

Yan Peijie, Bi Ruoling

(Shanghai Baosteel Section Steel Co., Ltd., Shanghai 201999, China)

Abstract: Customized roll forming technology changes traditional roll forming profiles to materials structure optimized, modular and nearly finished products which is flexible automatic manufactured by integrated manufacturing process including the punching, roll forming and laser welding. This technology has a great application value and market development potential in the process of promote advanced manufacturing concept, such as manufacturing industry upgrading, energy conservation and emissions reduction, lightweight and intelligent manufacturing. In this paper, the design path, the difficulties in the development and application examples of the customized roll forming technology are systemic introduced. At the same time analyzed the market prospect and development direction of the technology.

Key words: customized; roll forming; manufacturing upgrading; application examples

帽形件的多道次链模成形技术研究

张智恒[1]，钱朕[2]，张学广[3]，何广忠[3]，
邹天下[4]，李大永[4]，丁士超[2]

（1. 宁波赛乐福板材成型技术有限公司，浙江宁波 315176；2. 昆士兰大学机械与矿业工程学院，澳大利亚昆士兰 4072；3. 中车长春轨道客车股份有限公司，吉林长春 130062；4. 上海交通大学机械与动力工程学院，上海 200240）

摘 要： 链模成形是近年来出现的一种新型渐进成形技术，而帽形件是一种应用于交通运输载具的重要基础零件。本文通过数值模拟与实验的方法，研究多道次链模成形帽形件的可行性。建立了多道次链模成形帽形件的有限元模型，研究了成形过程中应力、应变的分布情况及发展历程，并在多道次链模成形试验机上进行了帽形件的成形试验。成形过程中未出现圆角破裂，成形效果良好。通过将试验结果分别与仿真结果和目标形状进行对比，既验证了仿真模型的准确性，也说明了成形方案的可行性。为提高成形精度，引入了辊弯精整道次对帽形件进一步整形。虽然端部仍然存在轻微成形质量问题，但翻折角和压深的加工误差被分别控制在 1° 和 0.5 mm 以下，认为成形目标实现。

关键词： 链模成形；帽形件；有限元模拟；成形试验

Study on Multi-pass Chain-die Forming of Hat-shaped Parts

Zhang Zhiheng[1], Qian Zhen[2], Zhang Xueguang[3], He Guangzhong[3],
Zou Tianxia[4], Li Dayong[4], Ding Shichao[2]

(1. Ningbo Sairolf Metal Forming Co., Ltd., Ningbo 315176, China; 2. School of Mechanical and Mining Engineering, University of Queensland, St Lucia, QLD 4072, Australia; 3. CRRC Changchun Railway Vehicles Co., Ltd., Changchun 130062, China; 4. School of Mechanical Engineering, Shanghai Jiaotong University, Shanghai 200240, China)

Abstract: Chain-die forming is a new type of incremental forming technology in recent years, and the hat-shaped part is a kind of important basic component in transportation vehicles. By means of numerical simulation and experiment, the feasibility of multi-pass chain-die forming for hat-shaped parts is studied. The finite element model of multi-pass chain-die forming was established; the distribution and history of stress and strain were studied; forming tests were carried out on a prototype of multi-pass chain-die forming. During the forming process, no fracture happened on fillets, and the final shape was fine. By comparing test results with simulation results and target shapes, the accuracy of the simulation model and the feasibility of the forming scheme were verified. In order to improve the forming accuracy, roll forming passes were introduced to refine the hat-shaped parts preformed by chain-die forming. Although a slight forming quality problem was found at the ends, the errors of folding angle and pressing depth were controlled below 1 degree and 0.5 mm respectively, and the forming goal was achieved.

Key words: chain-die forming; hat-shaped part; finite element simulation; forming test

基于压溃试验增强成形性双相钢吸能特性分析

张 伟[1]，李春光[1]，林兴明[2]，杨建炜[1]，刘立现[1]，刘华赛[1]

（1. 首钢集团有限公司技术研究院用户技术研究所，北京 100043；
2. 北京首钢股份有限公司营销中心，北京 100043）

摘 要： 增强成形性双相钢在保证原有双相钢强度的同时，提升了材料的延伸率，适合于复杂冲压结构件。采用压溃试验，针对增强成形性双相钢 HC440/780DH 吸能特性进行对比分析。对比普通双相钢 HC440/780DP 和 HC440/780DH 在组织结构、力学性能、动态力学特性等的差异；对两种材料的成形极限进行对比，基于某复杂成形零件对两种材料的成形性能进行对比；将两种材料制作成帽形薄壁梁，基于落锤压溃试验台，在相同的冲击载荷下，对比分析两种材料的吸能特性差异。结果可知：HC440/780DH 内部组织增加了残余奥氏体，材料呈现出良好的成形性，断后延伸率和 n 值比传统双相钢提升 8%左右；HC440/780DH 的安全域度和材料减薄率比 HC440/780DP 更优，更适合于复杂零件的设计，可有效降低材料成形开裂的风险；HC440/780DH 与同级别的 HC440/780DP 表现出部分吸能特性更优，其中，吸能比和载荷比提升 2%左右；从成形性和吸能特性方面对比可知，HC440/780DH 更适合于设计复杂安全件生产。

关键词： 增强成形性双相钢；压溃试验；吸能特性；动态力学性能

Analysis of Energy Absorption Characteristics of Dual Phase Steel with High Formability based on Drop Test

Zhang Wei[1], Li Chunguang[1], Lin Xingming[2], Yang Jianwei[1],
Liu Lixian[1], Liu Huasai[1]

(1. Department of Application Technology, Shougang Co., Ltd. of Research Institute of Technology, Beijing 100043, China; 2. Beijing Shougang Co., Ltd. Marketing Center, Beijing 100043, China)

Abstract: The dual phase steel with high formability improves the material elongation, while ensuring the strength of the original dual-phase steel, which is suitable for complex stamping structural parts of the vehicle. Based on drop test, the energy absorption characteristics of the dual phase steel with high formability HC440/780DH was analyzed. The differences in the microstructure, mechanical properties and dynamic mechanical properties were compared between dual-phase steel HC440/780DP and HC440/780DH. The forming limit curves of the two materials were compared. Based on a complex shaped part of a vehicle, the forming properties of the two materials were compared. The two materials were made into hat-shaped thin-walled beams. Based on the drop test, the difference in energy absorption characteristics of the two materials was compared under the same impact load. The results show that: the internal microstructure of HC440/780DH increases the retained austenite, and the material exhibits high formability. The elongation and n value after fracture of HC440/780DH are increased by 8% compared with the traditional dual-phase steel HC440/780DP. HC440/780DH's safety domain and material thinning rate are better than HC440/780DP, which is more suitable for the design of complex parts and can effectively reduce the risk of material forming cracking. Compared with the same grade HC440/780DP, HC440/780DH shows better partial energy absorption characteristics, in which the energy absorption ratio and load ratio increased by about 2%. From the comparison of formability and energy absorption characteristics, HC440/780DH is more suitable for the design of complex safety parts of the vehicle.

Key words: dual phase steel with high formability; drop test; energy absorption characteristics; dynamic mechanical properties

逆转变奥氏体对高强塑积中锰钢(0.1C-5Mn)氢脆敏感性的影响

惠卫军[1]，张永健[1]，王存宇[2]，曹文全[2]，董 瀚[3]

（1. 北京交通大学机械与电子控制工程学院，北京 100044；2. 钢铁研究总院特殊钢所，北京 100081；3. 上海大学材料科学与工程学院，上海 200444）

摘　要： 高强塑积中锰钢主要依靠两相区退火时获得的逆转变奥氏体在变形时发生形变诱导塑性(TRIP)效应而获得高的强塑积，能否同时获得低的氢脆敏感性是其能否成功应用的一个关键。本文利用电化学充氢、慢应变速率拉伸(SSRT)及氢热分析(TDS)实验等研究了典型的热轧 0.1C-5Mn 中锰钢的氢脆断裂行为。通过改变两相区退火时间获得不同含量的逆转变奥氏体。TDS 分析结果表明，电化学充入试样中的氢主要为扩散性氢，其含量随逆转变奥氏体量几乎线性增加。随着逆转变奥氏体量的增加，实验钢拉伸变形过程中的形变诱导马氏体量增加，这使得其氢脆敏感性逐渐增加。因此，TRIP 效应对依赖于 TRIP 效应中锰钢的氢脆抗力不利，逆转变奥氏体的稳定性及含量对中锰钢氢脆断裂行为具有显著影响。

关键词： 氢脆；中锰钢；逆转变奥氏体；TRIP 效应

Effect of Reverted Austenite Fraction on Hydrogen Embrittlement of TRIP-aided Medium Mn Steel (0.1C-5Mn)

Hui Weijun[1], Zhang Yongjian[1], Wang Cunyu[2], Cao Wenquan[2], Dong Han[3]

(1. School of Mechanical, Electronic and Control Engineering, Beijing Jiaotong University, Beijing 100044, China; 2. Central Iron and Steel Research Institute, Beijing 100081, China; 3. School of Materials Science and Engineering, Shanghai University, Shanghai 200444, China)

Abstract: The present work was attempted to evaluate the influence of reverted austenite (RA) fraction on hydrogen embrittlement (HE) of a hot rolled transformation-induced plasticity (TRIP)-aided medium Mn steel (0.1C-5Mn) by using electrochemical charging, slow strain rate tensile (SSRT) test and thermal desorption spectrometry (TDS) analysis. Different volume fractions of RA (~10-30 vol%) in the tested steel sheet were obtained by changing intercritical annealing time. The result of TDS analysis demonstrated that the charged hydrogen is primarily diffusible hydrogen corresponding to low-temperature hydrogen desorption peak and its concentration increases almost linearly with an increase in the volume fraction of RA. It was found that the intercritically annealed specimen exhibits an increasing susceptibility to HE with an increase in the volume fraction of RA primarily due to the increased RA transformation to martensite during tensile deformation. It is thus regarded that the TRIP effect is harmful to the HE resistance of TRIP-aided steels, and therefore both the mechanical stability and amount of RA have a strong influence on the HE behavior of TRIP-aided medium Mn steel.

Key words: hydrogen embrittlement; medium Mn steel; reverted austenite; transformation-induced plasticity effect

基于 CALPHAD 方法的马氏体转变驱动力计算和 Ms 预测

罗 群[1]，陈双林[1,2]，李 谦[1,3,4]

（1. 省部共建高品质特殊钢国家重点实验室，上海大学，上海 200444；
2. CompuTherm, LLC, Madison, WI 53719, USA；
3. 材料基因组工程研究院，上海大学，上海 200444；
4. 上海材料基因组工程研究院，上海 200444）

摘 要：随着汽车轻量化和安全性的要求日益提升，超高强度汽车钢成为汽车结构材料中备受关注的研究方向。其中 Q&P 钢的马氏体和亚稳奥氏体双相组织调控是提高其强韧性的关键，而淬火过程的马氏体转变及马氏体转变点 Ms 的准确描述是精确控制马氏体转变的首要要素。本工作在总结前人工作的基础上，从热力学角度出发研究马氏体转变的驱动力，分为化学驱动力和非化学驱动力。其中通过化学驱动力描述同成分奥氏体转变为铁素体的 Gibbs 自由能贡献，非化学驱动力分别描述剪切应变能、膨胀能、缺陷储能对 Gibbs 自由能的贡献。Fe-C、Fe-Ni 和 Fe-C-Ni 体系的计算结果与实验结果吻合较好。将该模型与热力学计算软件相结合，通过 CALPHAD 方法可以直接计算出不同成分和温度空间的 Ms 等值线和等值面，大大提高计算效率。同时，由于非化学驱动力中量化钢的屈服强度、弹性常数等参数对 Gibbs 自由能的贡献，该模型可以定量研究奥氏体的成分、晶粒尺寸等对 Ms 的影响，对超高强度汽车钢的设计提供理论依据。

关键词：马氏体转变点；热力学计算；CALPHAD 方法；相变驱动力

The Calculation of Driving Force for Martensitic Transformation and Ms based on CALPHAD Approach

超高强度汽车钢剪切边裂纹敏感性研究

王存宇

（钢铁研究总院，北京 100081）

摘 要：机械落料是冷冲压汽车零件最常用加工方式，超高强度汽车钢的剪切落料过程中存在裂纹敏感性的现象。报告描述了超高强度钢剪切边的特征、裂纹敏感性的影响因素，并通过模拟仿真的方法对剪切边的落料过程进行预测。研究结果表明，裂纹敏感性随强度水平的升高而增加，亚稳奥氏体的相变导致汽车钢裂纹敏感性增加会降低零件成形能力，减小落料间隙可提高剪切边质量，有效降低超高强度汽车钢的裂纹敏感性。

关键词：汽车钢；剪切边；裂纹敏感性；成形

Study on the Sensitivity of Shear Edge Crack of Ultra-High Strength Automobile Steel

超高强度钢塑性成形与热处理一体化工艺探索

金学军，龚 煜，李 伟

（上海交通大学材料科学与工程学院，上海 200240）

摘 要：采用超高强钢是实现汽车轻量化兼顾安全性的必由之路，热冲压成形是高强韧汽车零件成形的关键工艺。为获得超高强韧热成形钢与良好使用性能的热冲压零件，需配合使用多种强韧化机制，本项目提出以形变QPT（DQPT）工艺为主要研究思路，探索塑性成形与热处理一体化新工艺，以获得细晶多相（亚稳奥氏体相、高强马氏体相、纳米析出相）多层级非均质组织，达到超高强韧的目标。形变QPT（DQPT）具有增加位错密度、改变相变途径（如相变临界温度降低、形核密度增加等）、优化组织形貌、调控亚稳奥氏体、促进析出、缩短热处理时间等优点，是新一代超高强韧钢的组织设计与开发应用的新手段。

Integrating Deformation and Quenching-Partitioning-Tempering Novel Treatment and Development of Advanced Steels

Jin Xuejun, Gong Yu, Li Wei

(School of Materials Science and Engineering, Shanghai Jiao Tong University, Shanghai 200240, China)

Abstract: High strength steels have seen increasing demands for the concerns of better safety and low energy consumption, especially in the automobile and clean energy industries. An innovative quenching-partitioning-tempering (QPT) heat treatment was developed in realization of complex microstructure by affordable elements and manufacturing costs.

A significant increase in elongation of a boron steel by integrated process of deformation and QPT treatment were achieved, which can be attributed to the nanoscale duplex microstructure comprising ultrafine retained austenite and martensite. The principles of QPT treatment integrated with deformation may extend its application in tailoring the multiscale microstructures of various advanced high strength steels.

Key words: high strength steels; quenching-partitioning-tempering (QPT) process; metstable austenite; nano-precipitates

首钢汽车用先进高强钢研发进展及应用实践

韩 赟

（首钢集团有限公司技术研究院，北京 100043）

摘　要：首钢汽车板自 2009 年投放市场，至今已经历十个年头，2018 年汽车板产量突破 315 万吨，位居国内第二位。随着轻量化、节能减排、碰撞安全性要求的不断提高，汽车用先进高强钢在白车身中发挥了越来越重要的作用。首钢目前已实现 1200MPa 及以下强度级别冷轧先进高强钢产品全覆盖，同时为满足用户不同成形需求，陆续开发了 DP、DH、CP、CH、MS 等系列产品。文章介绍了首钢汽车用先进高强钢产品的开发思路及研发进展，并探讨了典型钢种的应用实践及应用关键技术等。

Research, Development and Application of Shougang AHSS for Automobile

Han Yun

(Shougang Research Institute of Technology, Beijing 100043, China)

Abstract: Shougang automobile sheet has been on the market for ten years since the first coil was put in 2009. The output of Shougang automobile sheet in 2018 exceeded 3.15 million tons and ranked No.2 in China. Because of the increased requirement of lightweight, energy conservation, emission reduction and crash safety, AHSS plays more and more important role in the body in white. So far, Shougang cold rolled AHSS has covered 1200MPa grade and below. DP, DH, CP, CH and MS steels have been developed in order to satisfy the various forming requirement. In this paper, research ideas and development progress are introduced, and the typical application and relevant application technology is also discussed.

相图计算技术在先进汽车钢板材料设计中的应用研究

何燕霖，李　麟，郑伟森，鲁晓刚

（上海大学材料科学与工程学院）

摘　要：本文对相图计算技术在汽车用高性能 TRIP 钢，TWIP 钢板等材料的成分设计，可镀性预测等方面的应用展开了研究，结果表明：两相区相比例、层错能、T_0 温度以及碳元素的扩散时间等可作为材料设计的特征变量，对其进行热力学和动力学计算有助于把握成分-组织-性能相互作用的内在机理，在此基础上采用集成计算技术有望实现成分的智能优选；通过对汽车钢板热浸镀锌过程中表面氧化物形成的热力学评估，可有效预测其可镀性，为高性能镀锌板的开发和应用提供设计指导。

Study on Application of CALPHAD Technology in Material Design of Advanced Automobile Steel Plate

He Yanlin, Li Lin, Zheng Weisen, Lu Xiaogang

(School of Material Science and Engineering, Shanghai University)

Abstract: In this paper, the application of CALPHAD technology in the composition design and the prediction of galvanizability of TRIP steel, TWIP steel and other materials used in automobiles were studied. It was shown that phase composition during annealing, stack fault energy, T_0 temperature and diffusion time of carbon can be used as characteristic variable of material design. Thermodynamic and kinetic calculations about these variables were helpful to obtain the mechanism of composition-microstructure-property interaction. On this basis, it was possible to realize intelligent optimization of composition by using integrated computing technology. Moreover, the thermodynamic evaluation of the formation of surface oxides in the process of hot dip galvanization can effectively predict the galvanizability and provide design guidance for the development and application of zinc-plated sheets with higher performance.

汽车用高强钢动态和断裂力学特性的测试分析及仿真应用

吴海龙

（清华大学苏州汽车研究院）

摘　要：本报告主要介绍汽车用高强钢在汽车碰撞过程中的动态大变形和断裂失效力学行为，重点分析高强钢的动态力学性能测试技术和实现方法，设计并开展不同应力状态实验获取高强钢准确全面的力学数据并标定仿真材料卡，有利于提高整车碰撞仿真精度和加快开发进度。

Abstract: This report mainly introduces the dynamic with large deformation and failure of mechanical behavior of automotive high strength steel in the vehicle crash. The dynamic mechanical properties testing technique and realization method of high strength steel are emphatically analyzed. The accurate comprehensive mechanics data and calibration simulation material card were obtained through different stress state test of high strength steel, which can improve vehicle crash simulation accuracy.

微观组织对低合金结构钢耐蚀性影响的关键基础研究

李晓刚，刘　超

（新材料技术研究院，北京科技大学，北京　100083）

摘　要：低合金结构钢的腐蚀起源是极其复杂和多样化的。特别地，低合金钢中局部腐蚀的萌生和钢中微纳米尺度的织构密切相关，如夹杂物、显微组织结构、晶粒度、畸变和析出相等。这些不同种类的织构之间共存时，可能会因为其间的协同作用而加速局部腐蚀的萌生和发展。众所周知，这些不同的微纳米尺度织构会引起局部腐蚀的发生，问题是这些织构引起的局部腐蚀原因的本质是什么？对于夹杂物，导电性是一个关键问题，因为夹杂物是否导电决定了腐蚀机理的根本变化。不同微纳米尺度织构的尺寸也是一个重要的值得关注点。为了提高低合金钢的耐腐蚀性，低合金钢应避免"大阴极和小阳极"的微观结构设计。一些研究人员指出，晶粒取向的差异可能会采取不同的腐蚀速率。然而，需要研究更多细节以获得腐蚀过程的机制。随着夹杂物的出现，晶格畸变在钢中大量分布。晶格的程度和分布对材料的腐蚀性能影响是明显的，需要重点关注。纳米级沉淀相对局部腐蚀的影响尚未完全了解。需要大量合适的实验来验证相关因素的腐蚀效应。

关键词：低合金结构钢；局部腐蚀；夹杂物；显微组织；腐蚀机理

Key Fundamental Aspects in the Microstructural Regulation of Corrosion Resistant Low Alloy Structural Steels

Li Xiaogang, Liu Chao

(Institute of Advanced Materials & Technology, University of Science and Technology Beijing, Beijing 100083, China)

Abstract: The origin of corrosion is an extremely complex and diverse system of the low alloy steel structural steels. Especially the localized corrosion of the low alloy is closely related to micro-nano scale structure in the steel, like inclusions, microstructure, grain size and orientation, lattice distortion and precipitated phase. Synergism between these different micro-nano scale structure can also accelerate the initiation of corrosion. As well known, these different micro-nano scale structures can induced localized corrosion, the question becomes what is the nature of the localized corrosion reason induced by these structures? As for inclusions, the electrical conductivity is a key concern, because whether the inclusion is conductive or not determines the fundamental change of corrosion mechanism. The size of different micro-nano scale structures is also a key point to be regulated. In order to improve the corrosion resistance of the low alloy steel, the microstructure design of "large cathode and small anode" should be avoid in the low alloy steel. Some researchers have pointed out the difference of grain orientation may take different corrosion rate. However more details need to be investigate to get the natural mechanism of the corrosion inanition process. The lattice distortion is wildly distributed in steel along with the appearance of inclusions. The effect of lattice distribution on the corrosion inanition is obviously, and need to focus on regulation. The influence of the nano-scale precipitated phase on the localized corrosion inanition have not been fully understood. A large number of suitable experiments are needed to verify the corrosion effects of relevant factors.

Key words: low alloy steel; localized corrosion; inclusion; microstructures; corrosion mechanism

铁磁应变玻璃的特征与磁致伸缩性能

杨 森

（西安交通大学理学院，教育部物质非平衡合成与调控重点实验室，陕西西安 710049）

摘 要：材料的物相与相变一直是物理学界最关注的科学问题之一。玻璃态相变，是指从一个母相态至"长程无序但短程有序"态的转变。近十年来，我国学者在铁磁应变玻璃及团簇自旋玻璃的研究取得重大进展，发现了铁磁玻璃态与铁电玻璃态（弛豫铁电体）、铁弹玻璃态之间的物理平行性，三种玻璃都由短程无序-长程无序的关联作用而表现出相似的相变行为，同属于铁性玻璃。就铁磁玻璃而言，铁磁团簇是它的短程有序的主要表现形式。本次报告内容，将介绍我们对于铁磁性应变玻璃的最新研究进展，主要包括：铁磁应变玻璃的发现和实验证据，磁弹效应以及磁致伸缩性能等。

中国电工钢发展变化及新时代需求的研究

陈 卓

（中国金属学会电工钢分会，湖北武汉　430081）

摘　要：电工钢是电力、电器工业不可缺少的重要软磁合金，广泛应用于电力和电讯工业，用以制造发电机、电动机、变压器、互感器等设备，在电能的生成、传输与使用方面起到了关键的作用，电机设备及家用电器的能效高低很大程度取决于电工钢电磁性能的优劣等。国外企业把电工钢的制造技术和质量作为衡量一个国家钢铁产品生产和科技水平的重要标志。本文研究了我国电工钢发展变化，即产能、产量、表观消费量、进出口数量、生产与技术等方面以及新需求的利好政策、不利因素、下游行业分析、需求增长预测，并提出建议。

Research on the Development Change and the Demand of the Electrical Steel in China in the New Era

Chen Zhuo

Abstract: Electrical steel is an indispensable and important soft magnetic alloy in electric power and electric appliance industry. It is widely used in the power and telecommunications industry, used to make generators, motors, transformers, mutual inductors and other equipment. Electrical steel plays a key role in the generation, the transmission and the use of the electric energy. The energy efficiency of the electrical equipment and the household appliances depends largely on the electromagnetic performance of the electrical steel. Foreign enterprises take the manufacturing technology and the quality of the electrical steel as an important symbol to measure the production and the technological level of a country's steel products. This paper studies the development changes of the electrical steel in China, that is, the production capacity, the output, the apparent consumption, the import and export quantity, the production and technology, as well as the positive policies, the adverse factors, the downstream industry analysis, the demand growth forecast, and puts forward suggestions.

新能源车发展新趋势及高端电机磁材料应用技术

裴瑞琳

（苏州英磁新能源科技有限公司，江苏苏州　215000）

摘　要：新能源车在中国已发展5年有余，本专题将阐述未来10年国内外新能源车总体市场和技术趋势，并针对新能源车等高端电机领域对磁材料的"严苛"需求，提出未来软磁材料开发的新技术思路及创新点。

Development of EV Car and Application Technology of Magnetic Materials in High-End Electric Machines

Pei Ruilin

Abstract: There are five years development for Electric Vehicles in China. This topic presents the market and technlogy trend for EV within 10 years, and gives technology development direction and innovation topics for magnetic materials used in high-end electric machines having "strict" requirement in EV.

高效环保变压器用高性能取向硅钢制备技术

龚 坚，马家骥，孙茂林，司良英，黎先浩，王现辉

（北京首钢股份有限公司）

摘　要： 随着我国特高压电网及高能效配电网的大力发展，电网对变压器核心材料取向硅钢提出了新的需求：（1）超低损耗薄规格化；（2）优异涂层附着性；（3）低噪声。北京首钢股份有限公司与钢铁研究总院、中国电力科学研究院、沈变等全产业链联合研发，开发高效环保变压器用高性能取向硅钢。研究开发了超低损耗成套控制技术，采用抑制剂强化控制及其与二次再结晶过程精细匹配，解决了表面效应导致二次再结晶发生不稳定的难题，开发18SQGD060等3个超低铁损牌号；开发细密均质优附着底层综合控制技术，通过氧化膜细密化控制、高形核添加剂技术，产品附着性水平优异；开发了取向硅钢低噪声综合控制技术，采用脱碳退火两段式快速加热、大张力涂层均匀性控制与高叠装系数技术，开发出23SQGD080LN等低噪声牌号。产品应用于特变电工、西电集团、天威保变、ABB、西门子等国内外100多家知名变压器企业；应用于国网特高压"双百万"变压器、乌东德水电站、白鹤滩水电站、大唐托克托电厂、北京新机场等重大工程。提升了我国高磁感取向硅钢制造技术水平，保障和支撑国家能源战略可持续发展。

"绿色"驱动　创新引领　助力新能源用钢高质量发展

钱　刚，许晓红

（中信泰富特钢集团，江苏江阴　214429）

摘　要： 新能源行业的快速发展给钢铁材料行业带来新的机遇和巨大的挑战，为应对新能源汽车、风电行业等对于钢铁零部件的大型化、大厚度规格、轻量化、高强度、高速运行、大载荷、高寿命、少维护等要求，兴澄特钢应用全生产流程控制理念，开发多种高综合性能的真空连铸钢种，通过不断提高钢的纯净度来延长轴承钢寿命；通过成分配比和铸态组织控制实现高速齿轮钢的轻量化、热处理高效化；通过钢水原始夹杂物-连铸冷却制度设

计控制夹杂物形态实现零部件高效和高精度加工，降低设备运行能耗和噪声；以及高强度弹簧钢、大圆坯材料、帘线钢、大壁厚容器、管线、大厚度齿条钢等钢种开发和生产过程中的共性问题和个性问题的解决，本报告中将予以重点介绍。

关键词：新能源用钢；连铸；创新

High-level Development of New Energy Steel Moved Forward by "Green" and Innovation

金属构筑成形热加工过程界面演化

孙明月，徐 斌，李殿中，李依依

（中国科学院金属研究所沈阳材料科学国家研究中心，辽宁沈阳 110016）

摘 要：大锻件是重大装备的核心部件，其质量直接影响装备的运行可靠性。然而，当前大锻件均采用大钢锭制备，偏析、疏松、粗晶等缺陷严重，影响性能稳定。金属构筑成形工艺突破了大锻件只能采用大钢锭制造的传统思维，利用品质优、成本低的连铸坯作为基础单元，经表面加工、清洁处理、堆垛成形和真空封装后，在高温下施以保压锻造、多向锻造为特点的变形工艺，充分愈合界面，实现界面与基体完全一致的"无痕"连接，获得高质量的初始坯料。构筑成形的关键在于界面结合，本研究介绍了若干种工程材料构筑成形过程界面的组织演化和氧化物分解行为。

关键词：金属构筑成形；界面结合；界面氧化物

Interface Evolution during Metal Additive Forging Process

Sun Mingyue, Xu Bin, Li Dianzhong, Li Yiyi

(SYNL, Institute of Metal Research, Chinese Academy of Sciences, Shenyang 110016, China)

Abstract: Heavy forgings, whose quality have a direct influence on operational reliability of equipments, are key components of major equipments. However, there always exist some serious defects such as macrosegregation, center porosity and grain coarsening in present heavy forgings manufactured by large steel ingots, which may severely impair equipment performance. Jumping out of the traditional concept that a heavy forging must be made by a large steel ingot, we first put forward a novel technology of building a large high-quality forging by much better and cheaper continuous casting billets. The process is as follows, after surface machining, cleaning, the slabs are stacked in order and vacuum packaged, then the whole package is pressure forged and multiple forged under high temperature until the interfaces are welded together perfectly, and finally a high quality initial billet is manufactured. The key point of additive forging is interface bonding, thus a systematic study was performed on interface evolution during additive forging of several structural metals in the present paper.

Key words: metal additive forging; interface bonding; interfacial oxides

宝钢三代和四代核电关键设备用钢开发与工程应用进展

张汉谦

（宝山钢铁股份有限公司，上海　201000）

摘　要：自宝钢股份 5m 宽厚板产线投产以来，一直致力于我国新建三代和四代核电站核级关键设备用宽厚钢板的开发和生产。本报告介绍了宝钢股份近十年为我国已开工建设的三代和四代核电站典型核级设备开发的宽厚钢板，包括第三代压水堆 AP1000/CAP1400 核电站安全壳、设备闸门、贯穿件等用 SA738GrB 钢板，CAP1400 核电站安注箱用 SA533B/304L 轧制不锈钢复合板，华龙一号核电站安注箱用 18MND5 钢板，高温气冷堆内构件用 12Cr2Mo1R 钢板，钠冷快堆用蒸汽发生器用 SA387Gr22Cl1 钢板等。以 CAP1400 核电站安全壳、设备闸门用 SA738GrB 钢板为例，重点介绍核级设备用钢板的使用技术设计和工程应用效果。

增材制造技术及其在高性能大型金属构件制造领域的应用

吉玲康，胡美娟

（石油管材及装备材料服役行为与结构安全国家重点实验室，
中国石油集团石油管工程技术研究院，陕西西安　710077）

摘　要：本报告首先介绍了石油管材与装备领域所面临的高可靠性的技术挑战，特别是结构件的大型化和高性能要求。对金属增材制造从原理、分类、发展历史、研究和制造现状、特点等方面进行了介绍，特别介绍了增材制造技术在石油行业中的应用及其展望，并详细介绍了现有增材制造的 X80 等级大型三通的优异性能。金属增材制造方法可在大型金属构件制造领域推广应用。

关键词：增材制造；大型金属构件；三通；高性能；高韧性

Additive Manufacturing Technology and Its Application in Manufacturing of Large Metal Components

Ji Lingkang, Hu Meijuan

(State Key Laboratory for Performance and Structure Safety of Petroleum Tubular Goods and Equipment Materials, CNPC Tubular Goods Research Institute, Xi'an 710077, China)

Abstract: This paper first introduces the technical challenges of high reliability in the field of tubular goods and equipment, especially the large-scale and high performance requirements of structural parts. This paper also introduces the principle, classification, development history, research and manufacturing status and characteristics of metal additive manufacturing, especially introduces the application and prospect of additive manufacturing technology in petroleum industry. In the end of this paper, an example is given to illustrate the excellent performance of the X80 grade large tee made by the existing additive. The metal additive manufacturing technology can be popularized and applied in the field of manufacturing large-scale metal components.

Key words: additive manufacturing; large metal components; tee; high performance; high toughness

低内应力高强韧厚规格管线钢带的生产工艺研究

缪成亮[1], 李 飞[2], 程 政[1], 朱腾威[2], 牛 涛[2], 徐海卫[1]

（1. 首钢京唐钢铁联合有限责任公司，河北唐山 063200；2. 首钢技术研究院，北京 100043）

摘 要： 随着管道工程的发展，管线钢向厚壁、高强发展，在工程上，很多管线钢钢卷订单规格已接近极限。为了获得高强度高韧性，除了注重奥氏体晶粒调控外，冷却相变阶段强调提高冷速，采用较低卷取温度，以此获得细小的针状铁素体+弥散细小的MA组元，产品同时具备高强度（≥X70）、优异低温韧性，这是国内生产厚规格高强管线钢（≥X70）所采用的常规方案。但是这种工艺也将带来其他问题，由于卷取温度低，且钢带强度高、厚度大，钢卷的内应力较高，这样将导致钢卷卷取稳定性较差；钢卷在没有强约束条件下很易发生散卷；制管开卷送料难度大，如果制管设备能力不匹配将可能产生较大安全问题；高残余应力也将对制管成型产生很大影响，等等。本论文研究了厚规格高强管线钢钢卷内应力产生的主要原因，同时基于产线冷却装备，结合成分设计、轧制过程奥氏体调控工艺，提出了不同以往的钢带生产工艺，该工艺能明显降低钢卷卷取时产生的内应力，同时钢卷仍可获得理想的组织，具有良好高强度和低温韧性，很大程度上解决了厚规格高强钢卷采用常规工艺生产中所面临的一系列问题。

关键词： 低温韧性；高强度；大壁厚管线；低内应力；冷却

Research on Production Process of Large Thickness Low Internal Stress Pipeline Steel Strip with Good Strength and Toughness

Miao Chengliang[1], Li Fei[2], Cheng Zheng[1], Zhu Tengwei[2], Niu Tao[2], Xu Haiwei[1]

(1. Shougang Jingtang United Iron & Steel Co., Ltd., Tangshan 063200, China;
2. Shougang Research Institute of Technology, Beijing 100043, China)

Abstract: With the development of pipeline projects, pipeline steel has been developed to thicker wall and higher strength with better toughness. Many actual orders of pipeline steel strip have required to limit specifications. To obtain high strength and good low temperature toughness, besides the refinement controlling of austenite grain, it is emphasized that fast cooling and proper low coiling temperature should be applied during cooling stage, as a consequence, fine acicular ferrite and diffused MA constituents will be obtained, and the pipeline steel strip with large thickness may has high strength (above X70) and good low temperature toughness. This is also the conventional cooling process for the production of high-strength pipeline steel strip (above X70) in China in the past ten years.

However, this process brings the other problem in industry, due to low coiling temperature, high strength and large thickness, the internal stress of the steel coil is very strong, and it will lead to very poor coiling stability, in the absence of strong constraint of packing strap, it is easy to happen coil loosening, uncoiling and coil feeding both are difficult for pipe-making company, and there may be a big security problem, large internal stress may influence pipe quality, even pipe properties.

In this paper, it is studied that the main influenced factors for strong internal stress in high strength pipeline steel coil with large thickness. Based on a large number of lab and industrial trial, an novel production process, especially cooling process, is presented for the internal stress decreasing during coiling. Meanwhile, combining with special chemical compositions designing and austenite controlling during rolling, the steel coil can still obtain ideal microstructure, and has high strength and good low temperature toughness. It is a good solution and technology in industry to the improvement of the problems caused by the conventional production process of high strength pipeline steel with large thickness.

Key words: low temperature toughness; high strength; large thickness steel strip; low internal stress; cooling

兴澄特钢高性能能源用钢的研发和产业化

许晓红，叶建军，韩 雷，徐 君，
邱文军，邓云飞

（江阴兴澄特种钢铁有限公司，江苏江阴 214400）

摘 要：能源是国民经济发展的重要基础，随着装备大型化和技术要求的不断提高，能源的开发和利用要依靠一系列高品质钢材制造的高端装备得到保证。能源用钢因其服役条件不同，技术要求各具特色，尤其在强韧性能、抗脆化性能、抗疲劳性能和焊接性能等方面是研发的重点，且常具有大厚度、大单重等特点。本文简述了兴澄特钢在能源用特种钢板研发方面，实现了多项关键制造技术的突破和创新，成功研发出压力容器用钢、海工钢、管线钢、锅炉钢、球罐钢、水电钢、风电钢、核电钢和LNG用钢等能源用钢，产品具有优良的力学性能和使用性能，广泛应用于石油化工、油气开采和输送、低温储运等能源行业，部分产品成功替代进口，实现了能源用钢高端材料的国产化。

关键词：能源用钢；突破和创新；强韧性能；焊接性能；国产化

R&D and Industrialization of High Performance Energy Steel in Xingcheng Special Steel

Xu Xiaohong, Ye Jianjun, Han Lei, Xu Jun,
Qiu Wenjun, Deng Yunfei

(Jiangyin Xingcheng Special Steel Co., Ltd., Jiangyin 214400, China)

Abstract: Energy is an important basis for the development of the national economy. With the improvement of large-scale equipment and technical requirement, the development and utilization of energy depends on a series of high-end equipments made of high quality steels. Energy steel usually has large thickness and large unit weight. Technical requirements are different due to its different service conditions, especially in the aspects of strength, toughness, embrittlement resistance, fatigue resistance, and welding performance. This article briefly describes the research and development of energy special steel plate in Xingcheng. With the breakthrough and innovation of many key manufacturing technologies, energy steel such as pressure vessel steel, offshore steel, pipeline steel, boiler steel, spherical tank steel, hydropower steel, wind power steel, nuclear steel, and LNG steel has been successfully developed. The products have excellent mechanical properties and performances, and been widely used in petrochemical, oil and gas exploitation and transportation, low temperature storage and transportation, and other energy industries. Some products have successfully replaced the import, and realized the localization of high-end materials for energy steel.

Key words: energy steel; breakthrough and innovation; strength and toughness; weldability; localization

重载铁路用贝氏体钢轨的研究进展

高古辉，白秉哲，桂晓露，谭谆礼，翁宇庆

（北京交通大学机械与电子控制工程学院，北京 100044）

摘　要：我国大宗和长途货物的运输主要依靠货运铁路承担，铁路的货运能力直接影响着我国国民经济的发展。重载铁路是提升货运能力的最有效的途径，也是我国铁路发展的重要方向。重载铁路具有轴重大、运量大、发车密度高等特点，对钢轨材料的强韧性、耐磨性和抗疲劳性能提出了更苛刻的要求，我国现有钢轨性能已无法完全满足，成为制约我国重载铁路发展的瓶颈环节，研发新一代重载钢轨势在必行。与普通珠光体钢轨相比，贝氏体钢轨具有更高的强韧性、更优的抗接触疲劳性能和焊接性。近 20 多年来，世界各国的铁路部门、钢企、研究院校均致力于开展贝氏体钢轨的研究和应用的探索。本文在简述国内外重载钢轨研究现状的基础上，重点介绍了我国重载贝氏体钢轨的研究进展，包括贝氏体钢轨的组织性能关系、磨损与疲劳伤损分析、1380MPa 级贝氏体钢轨的研发与应用等。

关键词：贝氏体钢轨；重载铁路；抗磨损性能；抗接触疲劳性能；焊接性；组织性能关系

Research Progress of Bainitic Rail Steel for Heavy Haul Railway

Gao Guhui, Bai Bingzhe, Gui Xiaolu, Tan Zhunli, Weng Yuqing

(Material Science & Engineering Research Center, School of Mechanical, Electronic and Control Engineering, Beijing Jiaotong University, Beijing 100044, China)

Abstract: In China, the mass and long-distance goods transports rely on the freight railway. The freight capacity of railway directly affects the development of our national economy. The heavy haul railway is the most effective way to improve the freight capacity, which is also the development direction of China Railway. With increased axle load, traffic capacity and density, the heavy haul railway requires that the rail steels have more excellent strength, toughness, fatigue and wear resistance. The existing rail steels in China cannot fully meet the requirement and limits the development of heavy haul railway. Hence, to develop new rail steel is very important and imperative in China. Compared with conventional pearlite rail steels, the bainitic rail steels have higher strength, toughness, RCF resistance and weldability. In the past two decades, the rail department, steel enterprise and research institutions all over the world are devoted to the development and application of bainitic rail steels. With the brief review of the development of heavy haul rail steels, we emphatically introduce the research progress of bainitic rail steels in China, in terms of microstructure-property relationship, damage analysis of RCF and wear, and the development of 1380MPa grade bainitic rail steel.

Key words: bainitic rail steel; heavy haul railway; wear resistance; RFC resistance; weldability; microstructure-property relationship

高速重载铁路高锰钢辙叉制造技术创新

陈　晨[1]，马　华[1]，王　琳[1]，张福成[1,2]

（1. 燕山大学亚稳材料制备技术与科学国家重点实验室，河北秦皇岛 066004；
2. 燕山大学国家冷轧板带装备及工艺工程技术研究中心，河北秦皇岛 066004）

摘　要：长期以来，铁路一直是人类出行或货物运输最主要的交通方式之一。随着高速、重载铁路的快速发展，世界各国对铁路各组成零部件的服役性能也提出了更高的要求。铁路辙叉作为改变列车运行轨迹的关键零部件，其重要性不言而喻。高锰钢是应用最为广泛的铁路辙叉生产材料，因其热塑性极差，高锰钢辙叉均采用铸造生产。然而，因铸造工艺引起的铸造缺陷之间的交互作用会加剧局域应力应变集中，形成疲劳裂纹。因此，高锰钢辙叉的品质主要受控于铸造质量。在大量研究工作的基础上，我们建立了高锰钢辙叉成分和热加工技术的设计原理和准则。在此指导下，实现了高锰钢辙叉制造技术的集成创新。高锰钢加重稀土并吹氮微合金化和炉外精炼技术，提高了钢水纯净度，改善了结晶组织，降低了层错能；辙叉局部 TMCP 技术，在服役条件最为恶劣的心轨和翼轨获得了细小的奥氏体组织，铸造缩松和气孔充分焊合，夹杂物有效碎化；高频冲击辙叉预硬化技术，实现了表面的纳米孪晶硬化。综合采用上述技术，制造出了成分纯净、宏观组织致密、微观组织超细和表面孪晶化的高锰钢辙叉。该技术成果解决了铸造高锰钢辙叉寿命短且离散的技术难题，成功实现了新一代高锰钢辙叉制造技术的升级换代。

关键词：高锰钢；辙叉；合金化；形变热处理；表面硬化

Technological Innovation in Manufacturing Hadfield Steel Frogs for High-Speed and Heavy-Haul Railway

Chen Chen[1], Ma Hua[1], Wang Lin[1], Zhang Fucheng[1,2]

(1. State Key Laboratory of Metastable Materials Science and Technology,
Yanshan University, Qinhuangdao 066004, China; 2. National Engineering
Research Center for Equipment and Technology of Cold Strip Rolling,
Yanshan University, Qinhuangdao 066004, China)

Abstract: Railway has been one of the most popular transportations for a long time. As the development of high-speed and heavy-haul railway, high service performances of the component of railway are demanded throughout the world. Frogs, as key component of railway, play an important role of changing the track of trains. Hadfield steel is one of the most widely used frog-making materials, and Hadfield steel frogs are usually produced by casting because of its bad thermo-plasticity. However, interactions between casting defects will intensify local stress concentration, finally resulting in fatigue cracks. Therefore, the quality of frogs is determined by their casting quality. Based on massive research work, we established the design principles and guidelines for chemical compositions and thermo processing technologies. The integrated innovation of manufacturing technology of Hadfield steel frogs was realized. The purity of molten steel was improved through off-furnace refining and microalloying with heavy rare earth and nitrogen. The crystalline structure was refined and stacking fault energy was reduced. Fine austenitic microstructure was obtained by local thermo mechanical control processing (TMCP) at nose and wing rails, where the service conditions were the worst. Meanwhile, the casting shrinkages and gas cavities were fully welded, and the inclusions were broken effectively. The surfaces of frogs were hardened and nanotwins were prepared by high-frequency impact pre-hardening technology. Hadfield steel frog with pure chemical compositions, dense macrostructure, fine microstructure, and nanotwins were produced by using the technologies above. The problem that service life of casting Hadfield steel frog was short and discrete was solved by these technologies. A new generation of manufacturing technologies for Hadfield steel frog was successfully upgraded and replaced.

Key words: hadfield steel; frogs; alloying; TMCP; surface hardening

高强抗震耐蚀耐火钢组织性能研究

王学敏，丛菁华，李江文，尚成嘉

（北京科技大学钢铁共性技术协同创新中心，北京 100083）

摘　要：本文利用 SEM, TEM, 3D-AP 等对 460MPa 级抗震耐蚀耐火钢微观组织及力学性能进行研究。通过 TMCP 工艺调控可以得到贝氏体和铁素体双相组织，且通过对不同开冷温度的控制，得到不同体积分数的铁素体，进而调整强度级别。双相组织的存在使试验钢有较低的屈强比，因此具备优异的抗震性能。利用 OM、SEM、TEM 等手段，对组织及析出行为进行研究。试验结果显示试验钢具有很好的抵抗 600℃ 3h 保温的耐火性能。在轧后冷却过程铁素体中形成(Ti，V，Mo)C 的相间析出，并且在 600℃保温过程中保持良好的热稳定性，相间析出的尺寸仍小于 50nm，同时在此保温过程中还会有大量的过饱和析出。由于析出阻碍位错的移动，试验钢在高温时仍保持较高的强度。本研究显示由于 600℃的相间析出数量明显多于 650℃开冷试验钢，其析出强化和位错强化在高温时的贡献明显优于后者，因此 600℃开冷的试验钢耐火性能明显优于 650℃开冷试验钢。

关键词：控轧控冷工艺；相间析出；析出强化；耐火性能

The Study on the Microstructure and Properties of High-strength Anti-seismic Corrosion-resistant Fire-resistant Steel

Wang Xuemin, Cong Jinghua, Li Jiangwen, Shang Chengjia

Key words: thermo-mechanical control process(TMCP); interphase precipitation; precipitation strengthening; fire-resistance

多元微合金化耐火钢研究进展

李昭东，王　鑫，杨忠民，雍岐龙

（钢铁研究总院工程用钢研究所，北京 100081）

摘　要：随着强度级别的提高，建筑结构用钢的组织由铁素体/珠光体向铁素体/贝氏体、贝氏体、多相多尺度亚稳（M^3）组织方向发展，耐火性能调控也从单一高 Mo 微合金化为主的方式向 Nb、V、Ti、Mo 等多元微合金化方式发展。本研究比较了不同组织和微合金的耐火性能差异，提出了多元微合金纳米碳化物遇火析出增强耐火性能的新思路。针对 Q345~Q690 耐火钢，形成了差异化的合金与组织设计及其热轧/热处理技术，600℃ 3h 屈服强度不低于室温标准屈服强度的 2/3。阐明了典型多元微合金耐火钢升温-加载过程中显微组织和力学性能的变化规律，探讨了纳米碳化物高温沉淀强化、基体组织高温稳定的耐火机理。

关键词：微合金；耐火钢；纳米析出；耐火机制

Research Progress on Multi-microalloyed Fire-resistant Steels

Li Zhaodong, Wang Xin, Yang Zhongmin, Yong Qilong

(Department of Structural Steel, Central Iron and Steel Research Institute, Beijing 100081, China)

Abstract: With the increase of strength grade, the microstructure of steel for building structure was developed from ferrite/pearlite to ferrite/bainite, bainite, M^3 (multi-phase, multi-scale, meta-stable) structure. The controlling of fire resistance performance of steel was also evolved from a single high-Mo microalloying mode to a multi-microalloying mode including Nb, V, Ti, Mo and so on. In this study, the differences in fire resistance due to different microstructures and microalloying were compared. A novel idea that the fire resistance performance of steel was enhanced by multi-microalloyed nano-carbides precipitation in case of fire was proposed. For Q345~Q690 fire-resistant steels, differentiated designs of alloy and microstructure and hot rolling/heat treatment technology have been obtained. The yield strength at 600℃ after holding 3h was not lower than 2/3 of the standard yield strength at room temperature. The variation of microstructure and mechanical properties of typical multi-microalloyed fire-resistant steels during the heating-loading process were clarified. The refractory mechanism including high temperature precipitation strengthening of nano-carbide and high temperature stability of matrix structure was discussed.

Key words: microalloying; fire-resistant steel; nano-precipitation; refractory mechanism

高钛高耐磨钢的物理冶金学基础及应用

孙新军，梁小凯，刘罗锦，许　帅

（钢铁研究总院工程用钢研究所，北京　100081）

摘　要： 介绍了"十三五"国家重点研发计划项目——"煤炭采运用高耐磨性钢板及应用"在基础研究和应用方面取得的成果。针对传统耐磨钢"耐磨性-加工性"矛盾突出的问题，提出采用高Ti微合金化在钢基体中引入多尺度TiC颗粒以增强耐磨性的技术方案，实现了在不增加硬度的前提下耐磨性的大幅提高。研究发现钢中TiC颗粒呈现出"微米-亚微米-纳米"三峰分布特征，阐述了其对应的不同析出阶段及影响因素；明确了"三峰分布"TiC颗粒对耐磨钢基体组织、力学性能和耐磨性的影响规律。最后介绍了新型耐磨钢在煤矿刮板输送机上的应用情况。

关键词： 耐磨钢；高Ti微合金化；析出；耐磨性

Physical Metallurgy and Application of High Ti High Wear-resistant Steel

Sun Xinjun, Liang Xiaokai, Liu Luojin, Xu Shuai

研发材料的熵调控模式

汪卫华 [1,2]

（1. 中国科学院物理研究所，北京 100190；2. 松山湖材料实验室，广东深圳 523808）

摘 要：传统探索新材料的方法主要是通过改变和调制化学成分，或者调控结构及物相，或者调制结构缺陷来获得新材料。本报告将阐述"熵"或者"序"调制是探索材料的新的思路和途径。如非晶合金就是典型通过快速凝固或者不同元素的混合，引入"结构无序"而获得的高性能合金材料。通过改变和调制"化学序"也可获得性能独特的新材料，化学无序高熵合金就是近年来采用多组元混合引入"化学无序"获得的新型合金材料。这种通过调制材料的"序"或者"熵"的方法将会导致发现更多类似非晶的新材料，其基本的物理机制更值得关注和深入研究。

关键词：熵调控；非晶材料；序

Developing New Alloys by Entropy Modulation

高饱和磁感强度铁基软磁非晶合金研究进展

姚可夫

（清华大学材料学院，北京 100084）

摘 要：铁基软磁非晶合金自问世至今已有半个世纪。尽管铁基软磁非晶合金的饱和磁感强度 Bs 略低于传统硅钢软磁合金，但与硅钢相比，其具有更高的导磁率、更低的矫顽力和更低的铁损等优异性能，使其在输变电变压器、逆变电源、电子和通讯领域获得了广泛应用。目前，铁基非晶合金，包括以非晶合金为前驱体获得的纳米晶合金已成为一类重要的软磁材料。但随着科技发展，对软磁非晶合金的综合软磁性能，特别是饱和磁感强度 Bs 提出了更高的要求。而提高 Bs 则需要提高合金中的铁含量，但提高 Fe 含量会降低铁基合金的非晶形成能力。因此，需要研发出既具有较高非晶形成能力又具有较高 Bs 和综合软磁性能的铁基非晶合金，以满足工业领域，特别是高频工业产品和高技术领域对高性能铁基非晶合金的需求。最近 10 余年，高 Bs 铁基软磁非晶合金研发取得了显著进展。本文将综述国内外高铁含量、高饱和磁感强度 Bs 铁基非晶合金的研究进展，介绍影响铁基非晶合金饱和磁感强度的主要因素，探讨提升非晶合金饱和磁感强度和综合软磁性能的方法和途径，为铁基非晶合金研发和工程应用提供技术支持。

关键词：铁基非晶合金；软磁性能；高饱和磁感强度

Research Progress of Fe-based Amorphous Alloys with High Saturation Flux Density

Yao Kefu

(School of Materials Science and Engineering Tsinghua University, Beijing 100084, China)

Abstract: Since the Fe-based amorphous alloy was first reported 50 years ago, it has attracted lots of attention. Now Fe-based amorphous alloys have been widely applied in transformers, electric and electronic devices due to their excellent soft magnetic properties, such as higher permeability, lower coercivity and lower core loss than the traditional silicon steels, despite of that their saturation flux density (Bs) is lower than silicon steels. Nowdays, Fe-based amorphous alloys, together with the nanocrystalline alloys prepared by annealing the amorphous alloy precursors, have become a kind of important soft magnetic materials. Recently, the requirements for high performance Fe-based amorphous alloys, especially those with high saturated flux density Bs, are increasing. It is known that increasing the Bs needs to increase the Fe content in the alloys, but increasing Fe content would deteriorate their amorphous forming ability, resulting in that it is difficult to obtain Fe-based amorphous alloys with high Fe content and then high Bs. However, high technology development and high frequency products demand the Fe-based amorphous alloys with high Bs, which have stimulated materials scientists and researchers to develop high performance Fe-based amorphous alloys. Indeed, in last decade significant progress has been achieved in this aspect. Here, the research progresses on Fe-based amorphous alloys with high saturation flux density have been summarized, including the main factors affecting the high saturation flux density of the Fe-based amorphous alloys. The possible ways for enhancing the high saturation flux density and the performance of the Fe-based amorphous alloys have been discussed. It may offer new ideas and technique support for development and application of Fe-based amorphous alloys.

Key words: Fe-based amorphous alloys; soft magnetic property; high saturation flux density

基于非晶的纳米结构及其功能特性

沈 军

（深圳大学，广东深圳 518060）

摘 要： 利用水热和去合金化方法在钛锆基非晶合金的基体上制备具有纳米叶/微米剑麻形状的钛酸盐-锆酸盐纳米结构材料，其成分为 $Na_2(Ti_{0.75}Zr_{0.25})_4O_9$。纳米叶尺寸大致为长 15~20 mm，宽 200~300 nm，厚 20~30 nm。去合金化和 Ostwald 熟化是这种纳米结构材料主要的生长机制。在紫外光作用下，这种纳米结构材料具有比商业氧化钛、未掺杂钛酸盐纳米管以及一维钛酸盐-锆酸盐纳米材料更强的由甲醇水溶液制氢的能力。

关键词： 非晶合金；纳米结构；功能特性

Nanostructures Grown from Amorphous Metals and Their Functional Properties

Shen Jun

(Shenzhen University, Shenzhen 518060, China)

Abstract: A novel titanate–zirconate solid solution with controllable nanoleaf/microsisal-like three-dimensional morphology, $Na_2(Ti_{0.75}Zr_{0.25})_4O_9$, grows on the surface of a Ti-based bulk metallic glass by a combination of hydrothermal and dealloying processes. A single sisal-like bundle consists of a number of nano-sized leaves and each nanoleaf has a thickness, width and length of 20~30 nm, 200~300 nm and 15~20 mm, respectively. A modified dipole driving Ostwald ripening mechanism for the interesting architectures has been proposed based on a series of time-dependent experiments and the structure feature of titanate. Moreover, the as-synthesized nanoleaf/microsisal-like material exhibits extraordinary ability to produce hydrogen from a methanol/water solution that is higher than that of a commercial TiO_2 film, undoping titanate nanotube film and titanate–zirconate one-dimensional nanoleaf film. Possible origins of the high performance of the as-synthesized three-dimensional nanomaterials were discussed based on theoretical and experimental results.

Key words: amorphous alloy; nanostructure; functional property

铁基非晶纳米晶软磁合金的成分设计与性能研究

吕 旷，李育洛，窦正旭，惠希东

（北京科技大学新金属材料国家重点实验室，北京　100083）

摘　要： Fe 基非晶合金具有低矫顽力、高磁导率、高电阻率、低的损耗和良好的频率特性，是一种优异的节能和环保材料。Fe 基纳米晶材料中，α-Fe 纳米晶与非晶软磁相之间发生交互耦合作用，可进一步降低合金的磁各向异性和磁致伸缩系数，因而合金具有更高的饱和磁化强度和磁导率。目前铁基非晶合金及其纳米晶合金已广泛应用于电力变压器、电子和信息工业领域。本报告将介绍近年来本课题组在铁基非晶纳米晶软磁合金的研究进展。主要内容包括：高饱和磁化强度 Fe-Si-B-P-C 和 Fe-Si-B-P-C-M（M=Cu，Nb，Zr）系非晶纳米晶软磁合金的成分设计、组织结构和软磁性能，重点介绍 Si, P, C, Nb, Cu 和 Zr 元素对非晶纳米晶合金非晶形成能力、纳米晶组织、磁化强度和矫顽力等软磁性能的影响规律，从而为高饱和磁化强度非晶纳米晶软磁合金的成分设计提供参考。

关键词： 铁基非晶合金；纳米晶合金；成分；组织；高饱和磁化强度；软磁性能

Composition Design and Properties of Fe-based Amorphous and Nanocrystalline Alloys

Lv Kuang, Li Yuluo, Dou Zhengxu, Hui Xidong

(State Key Laboratory for Advanced Metals and Materials, University of Science and Technology Beijing, Beijing 100083, China)

Abstract: Fe-based amorphous alloys exhibit many desirable features, such as low coercivity, high permeability, high electrical resistivity, low loss and good frequency characteristics, which make them a kind of viable energy-saving and environmentally-friendly materials. The interactive coupling between α-Fe nanocrystalline phase and amorphous soft magnetic phase further reduces the magnetic anisotropy and magnetostrictive coefficient of Fe-based amorphous and nanocrystalline composites. Therefore, Fe-based amorphous and nanocrystalline alloys have a high saturation flux density and high magnetic permeability, and have been widely applied for electric power, electronics and information industries. In this presentation, we will introduce our recent progress in Fe-based amorphous and nanocrystalline alloys with high saturation flux density. The main contents include the compositions design, structure and soft magnetic properties of

Fe-Si-B-P-C and Fe-Si-B-P-C-M (M=Cu, Nb, Zr) high-Bs Fe-based amorphous and nanocrystalline alloys. Most works are focused on the influences of Si, P, C, Nb, Cu and Zr on the glass forming ability, microstructure of nanocrystallines, saturation flux density and coercivity of these alloys so as to provide some reference for the composition design of Fe-based amorphous and nanocrystalline alloys with high saturation flux density.

Key words: Fe-based amorphous alloy; nanocrystalline alloy; composition; microstructure; high saturation flux density; soft magnetic properties

快速凝固亚稳合金催化材料的制备与性能

张海峰

（中国科学院金属研究所，辽宁沈阳 110016）

摘 要：快速凝固亚稳合金催化材料与常规催化材料比较有如下优点：不受合金平衡相的限制，可以获得过去一般方法所不能获得的新型组分和结构；各元素分布更加均匀；表面能高，形成的活性点多，而且具有高强度、耐腐蚀性等。本文介绍了快速凝固亚稳 Ni-Al-Cr-Fe 合金和 Fe-Si-B 合金的结构特征，以及在加氢反应和染料降解脱色反应中的催化性能。结果表明：快速凝固亚稳合金催化材料不仅具有高的催化活性，而且具有优异的选择性和稳定性。

关键词：快速凝固；催化

"金属-类金属"型高熵块体非晶合金的制备及性能

张 伟

（大连理工大学，辽宁大连 116085）

摘 要：高熵合金的成分设计理念也为非晶合金的研发提供了空间。近年，新型高熵块体非晶合金体系被相继开发出来，其中大多为"金属-金属"型，研究主要涉及其玻璃形成能力和力学性能。最近，我们研制出了"金属-类金属"型的软磁性 $Fe_{25}Co_{25}Ni_{25}(B, Si)_{25}$、$Fe_{25}Co_{25}Ni_{25}(P, C, B)_{25}$ 及 $Fe_{25}Co_{25}Ni_{25}(P, C, B, Si)_{25}$。该系列合金在 0~12.5 at.% P，0~10 at.% C，5~17.5 at.% B，0~10 at.% Si 的组成范围可形成块体非晶态。其中，$Fe_{25}Co_{25}Ni_{25}(P, C, B, Si)_{25}$ 系合金具有较高玻璃形成能力，非晶形成临界直径大于 2 mm；$Fe_{25}Co_{25}Ni_{25}(P, C, B)_{25}$ 系合金具有较高过冷液体热稳定性，过冷液相温度区间达到 56K。它们高的玻璃形成能力和过冷液体热稳定性分别与其凝固和结晶化过程中 $Fe_{23}C_6$ 型化合物的析出有关。该系列非晶合金均具有优异的软磁性能和力学性能，其矫顽力、饱和磁感应强度、压缩屈服强度和塑性应变分别在 0.8~6.4 A/m、0.71~0.87 T、2817~3624 MPa 和 0.3%~3.1%的范围，最大磁导率（1kHz）达到 22000。在该系列合金的基础上，我们进而研发出了具有高热稳定性、优异耐蚀性、高强度的 FeNiCrMo(P, C, B)系高熵块体非晶合金，并对其高耐蚀性机理和结晶化行为进行了探讨。

关键词：高熵合金；块体非晶合金；磁性能；耐蚀性；结晶化行为

Fabrication and Properties of Metal-metalloid Type High Entropy Bulk Metallic Glasses

Zhang Wei

(Dalian University of Technology, Dalian 116085, China)

Abstract: Encompassing diverse behavior from metals to insulators, amorphous solids are useful in optics, electronics and catalysis. Local atomic configurations determine their electronic structure and properties, yet it poses a significant challenge to the experimental setup of a reliable structure-property correlation in amorphous materials. Here we demonstrate that a spectrum of behavior from magnetic metal to paramagnetic insulator is realized in one Co-Fe-Ta-B-O system via continuously compositional adjustability. Hence, the amorphous structure can be manipulated in a well controllable manner, leading to tunable optical, electrical and magnetic properties over a wide range. Particularly, the electric-field effect on the magnetism is obtained and varies significantly in the system. The combined optical, electrical and ferromagnetic functionalities in the amorphous system may assure them an important role in low-power-consumption electronic/spintronic devices. Furthermore, crystallization induces a semiconductor-metal transition in the amorphous ferromagnetic semiconductors based on ferromagnetic alloys. The significant resistance difference in their crystalline and amorphous states enables them to be room-temperature ferromagnetic phase change materials that could be used in new non-volatile phase change memories.

Key words: high-entropy alloy; bulk metallic glass; magnetic property; corrosion resistance; crystallization behavior

铁磁性非晶合金结构不均匀性调控及由此带来的 GFA 及力学性能变化

沈宝龙

（东南大学，江苏南京　211189）

摘　要： 本文通过组元设计和成分调整对铁磁性非晶合金的结构不均匀性进行调控，系统研究非均匀结构演化与铁磁性非晶合金非晶形成能力和力学性能的关联，进而探索制备了系列高性能铁磁性非晶合金。其中，通过调整 Co/Al 比，制备了临界直径达 8mm，最大磁熵变为 9.5J/(kg·K)的 Gd55Co17.5Al27.5 非晶合金。动力学分析结果表明，随着 Co/Al 比的降低，非晶合金的晶化驱动力降低，抑制了原子的能动性，增加了结构的不均匀性，进而大幅提升其非晶形成能力。并且通过 Tg 附近合适的弛豫处理，析出不同尺度的团簇结构，进一步提升体系非均匀性，使得合金的磁热性能也得到了改善。此外，通过在 FeMoPCB 体系中添加 Ni 和 Si 元素，制备了塑性应变 7%，屈服强度 2.82GPa，饱和磁感应强度 0.93T，矫顽力 1.9A/m 的 Fe56Ni20Mo4P11C4B4Si1 块体非晶合金。X 射线衍射结果表明该合金的原子结构松散，存在大量潜在的剪切变形区，从而在压缩变形过程中形成多重剪切带，表现为自组织临界状态下稳定的剪切带动力学；并且通过在 CoFeNiBSiNb 体系中添加适量的 Cu 元素，制备了塑性应变 5.5%，断裂强度 4.77GPa，饱和磁感应强度 0.6T，矫顽力 1.33A/m 的(Co0.7Fe0.2Ni0.1)67.7B21.9Si5.1Nb5Cu0.3 块体非晶合金。同步辐射和球差电镜结果表明其强度和塑性同时增加归因于平均原子距离的减小和原子尺度不均匀性的增加，进而导致剪切带之间强烈的交互作用，表现为复杂的塑性变形过程。

关键词： 铁磁性非晶合金；结构不均匀；GFA；力学性能

Structural Heterogeneity Modulation of Ferromagnetic Metallic Glasses and Effects on Glass Forming Ability and Mechanical Properties

Shen Baolong

(Southeast University, Nanjing 211189, China)

Abstract: In this paper, structural heterogeneity of ferromagnetic metallic glasses was modulated via compenent design and composition adjustment, and the effects of structural heterogeneity on glass formation ability (GFA) and mechanical properties was investigated in detail, then several high performance ferromagnetic metallic glasses were successfully developed. Through adjusting the Co/Al ratio in ternary GdCoAl system, a novel $Gd_{55}Co_{17.5}Al_{27.5}$ bulk metallic glass (BMG) with the critical diameter of 8mm and the maximum magnetic entropy change of 9.5J/(kg·K) was developed. Thermodynamic and kinetic investigations indicate that this BMG possesses a lower driving force for crystallization and exhibits a stronger liquid behavior due to the inhibition of atomic mobility and the increase of inhomogeneity. Furthermore, the magnetic entropy change is enhanced by appropriate annealing treatment through the formation of more complex structures comprising short-range order, medium-range order and nano-crystallized structure. In addition, through adding Ni and Si elements into FeMoPCB system, a ductile $Fe_{56}Ni_{20}Mo_4P_{11}C_4B_4Si_1$ BMG with plastic strain of 7%, yield strength of 2.82GPa, saturation magnetic flux density (Bs) of 0.93T and coercivity (Hc) of 1.9A/m and was developed. It is found that a large number of potential shear transition zone sites exist in this Fe-based BMGs due to its loose structure confirmed by radiation X-ray diffraction observation, leading to the formation of multiple shear bands, resulting in a complex deformation process follow the self-organized critical dynamics. Moreover, a $(Co_{0.7}Fe_{0.2}Ni_{0.1})_{67.7}B_{21.9}Si_{5.1}Nb_5Cu_{0.3}$ BMG with record large plasticity of 5.5%, high fracture strength of 4.77GPa, higher Bs of 0.60T, and low Hc of 1.33A/m was developed by adding proper Cu element into CoFeNiBSiNb system. The results of synchrotron X-ray diffraction measurement and high-resolution transmission electron microscopy observation indicate that the improvement of plasticity and strength is due to the reduced average atomic spacing and increased atomic-scale structural heterogeneity, leading to the strong interactions between shear bands, then resulting in the stable plastic deformation process.

Key words: ferromagnetic metallic glasses; structural heterogeneity; GFA; mechanical properties

FeSiB 三元非晶合金熔体原子扩散实验研究

张 博

（合肥工业大学，安徽合肥 230009）

摘 要： 液体金属的原子扩散系数是理解液体金属动力学行为以及动力学和结构之间的关联必需的关键参数，也是检验现有的液体动力学理论的关键物理量。同时，准确可靠的扩散数据也是液体金属凝固过程模拟的必备参数。然后由于实验测量上的困难，可靠的实验数据非常缺乏。对于高熔点的金属熔体，扩散数据更加稀少。我们采用自主研发的多层滑动剪切技术，首次在 FeSiB 非晶合金熔体中实验测量了 FeSiB 三组元之间的互扩散系数。对于该实验结果的初步分析表明，该体系中 B 和 Si 组元的添加降低了合金熔体的原子互扩散。该结果和 FeSiB 体系的非晶形成能力相吻合。

关键词： FeSiB 非晶熔体；原子扩散；多层滑动剪切技术

Diffusion Experiments of FeSiB Glass Forming Melts

Zhang Bo

(Hefei University of Technology, Hefei 230009, China)

Abstract: Atomic diffusion in liquid metals is crucial for understanding the dynamic behavior and its correlation with structure in liquid metals. It is also vertically important in the evaluation of the known theories of liquid dynamics. Meanwhile, accurate and reliable diffusion data is also basically required in the simulation and modeling of the solidification process. Due to the experimental difficulties, reliable diffusion data in liquid metal is really rare. And this situation is even worse in the high-melting-point liquid metals. In the present talk, we will present the experimental results of interdiffusion data in FeSiB glass forming melts measured by our newly developed multi-sliding cell technique. Our results indicate that the addition of elements like B and Si decreases interdiffusion in the present liquid system. To some extent, this result can explain the good glass forming tendency of the FeSiB alloy system.

Key words: FeSiB glass forming melt; atomic diffusion; multi-sliding cell technique

锆基非晶合金对大鼠骨组织的影响

王 刚

（上海大学，上海 200444）

摘 要： 当前工作将锆基非晶合金作为主要研究对象，开展了一系列有关生物相容性的系统研究。通过多种表征手段及体内、体外实验，全面对比了锆基非晶合金与生物金标准材料纯钛和PEEK（聚醚醚酮）的生物相容性差异。结果表明，相比于其他两种材料，锆基非晶合金具有更高的强度，更适宜的弹性模量和更优异的耐腐蚀性能。同时，体外细胞实验数据证明，锆基非晶合金具有更为明显的促细胞分化能力，促分化水平高于纯钛及PEEK 50%左右。再者，通过活体实验的结果分析，显示出锆基非晶合金所服役环境不仅具有更高的骨组织密度，植入体接触面还呈现了更多的血管生成量。这些优异特性可归结于其化学组成及非晶合金独特的原子结构。基于其优良的力学性能，耐腐蚀行为和生物相容性的特征，锆基非晶合金有望在生物材料领域发挥重要作用。

关键词： 非晶合金；生物相容性；体外细胞实验；体内活体实验

Implantation of a Zr-based Metallic Glass in the Bone Tissue of SD Rat

Wang Gang

(Shanghai University, Shanghai 200444, China)

Abstract: Zr-based BMG was subjected in present work for systematical object of biocompatibility comparing with the golden standard biomaterials: Cp-Ti and PEEK (ployetheretherketone). The results in terms of not only kinds of characteristic experiments but also lots of biological vitro and vivo measurements. On one hand, it exhibits Zr61Ti2Cu25Al12 BMG has higher strength, more suitable elastic modules and better corrosion resistance. On the other hand, specially, it can promote cell's differentiation larger to 50% than the control groups obviously and promote the formation of blood vessel in the living body. Meanwhile, the analysis of the Micro-CT results shows slightly higher bone contents around Zr-based BMG than other two cases at different serving time. These advantages can be attributed to the chemical composition as well as the unique atomic structure of metallic glass. Based on the better mechanical properties, corrosion behavior and biocompatibility character, it suggests Zr-based BMG could be able to act a significant role in the field of biomaterial.

Key words: metallic glass; biocompatibility; in vitro measurements; in vivo measurements

高熵非晶合金的异常玻璃转变行为

刘雄军

（北京科技大学，北京　100083）

摘　要：高熵非晶合金是兼具高熵合金多主元的成分特征和非晶合金长程无序的原子结构堆垛特性的一种新型无序合金。自从高熵非晶合金被发现以来就引起了广泛的关注。然而，高熵非晶合金的内在结构本质及其优异性能的来源仍缺乏研究。本报告将主要介绍近来我们课题组在高熵非晶合金的玻璃转变行为方面的研究工作。

（1）高熵非晶的形成能力：传统非晶合金的设计中，合金化是提高非晶形成能力（GFA）的有效方法，其实质是利用合金化多组元提高合金的复杂性（即："混乱法则"）。高熵非晶由于其多主元（至少5元）、等原子比（或近原子等比）的成分特征，具有高混合熵的特点。依据"混乱法则"，高熵效应有利于非晶结构的形成。然而，研究发现高熵非晶相对于传统非晶具有较差的GFA。同时，研究还发现高熵非晶的GFA与其热稳定性之间存在异常关系。

（2）高熵非晶合金异常玻璃转变行为：在加热过程中随着玻璃转变的发生，非晶固体转变为过冷液体，非晶态中冻结的平移自由度得到释放，因此非晶态过冷液体的比热通常将增加$3R/2$（R为气体常数）。然而，研究发现，高熵非晶合金的过冷液体具有异常小的过剩比热容，在玻璃转变过程中其比热值仅增加$R/2$，以及极其不明显的模量软化行为。进一步结构分析和计算模拟结果表明，高熵非晶合金具有更加均匀且致密的原子堆垛结构，同时高熵效应及缓慢原子扩散效应限制了原子运动，只有较少量的类液态原子发生局部的玻璃转变。

关键词：高熵非晶合金；玻璃转变；玻璃形成能力；热稳定性

Unusual Glass Transition Behaviors of in High-entropy Bulk Metallic Glass

Liu Xiongjun

(University of Science & Technology Beijing, Beijing 100083, China)

Abstract: High-entropy bulk metallic glasses (HE-BMGs) are a new kind of metallic materials, which simultaneously possess the compositional feature of multi-elements high-entropy alloys and the structural characteristics of amorphous alloys. HE-BMGs have attracted extensive attention due to their unique properties since the first HE-BMG was discovered. However, researches on the nature of atomic structures and structure-properties relationships in HE-BMGs are still lacking. In this talk, we will report our recent work on the glass transition behaviors of HE-BMGs.

(1) Poor glass-forming ability (GFA) of HE-BMGs. Alloying is an effective method to improve GFA of traditional MGs. By increasing the complexity of alloys (i.e., "Confusion Principles") is beneficial to improve GFA. In this regard,

HE-BMGs which are characterized by multiple principal elements (≥5) with equal or nearly-equal atomic ratio and high-mixing entropy are expected to possess good GFA. However, we found that HE-BMGs have worse GFA compared with respect to their traditional BMG counterparts. Moreover, we revealed an unusual relationship between the GFA and their thermal stability in HE-BMGs.

(2) Abnormal glass transition (GT) behavior of HE-BMGs: For conventional MGs, the heat capacity jump was found to be a constant value of $3R/2$ (R: gas constant) during GT due to the fact that each atom gains three translational degrees of freedom accompanying with the glass-liquid transition. However, it is found that the supercooled liquids of HE-BMGs exhibit an abnormally ultra-low heat capacity jump (about $R/2$), and a very inconspicuous modulus softening behavior. Further boson peak and structural analysis coupled with ab initio MD simulations show that the HE-BMGs have a more homogeneous and denser structure with fewer liquid-like atoms involved in GT with respect to traditional BMGs, which results in local GTs in HE-BMGs.

Key words: high entropy amorphous alloy; glass transition; glass forming ability; thermal stability

调控非晶合金结构提高其塑性变形能力

王 丽

（山东大学（威海）机电与信息工程学院，山东威海 264209）

摘 要： 通过调控单相非晶合金微观结构或在非晶基体中引入纳米级第二相等方法被广泛应用于提高非晶合金的塑性变形能力。通过分子动力学模拟方法，根据熔体团簇结构的不对称性，提出原子团簇键长偏差参数，研究了低温下非晶合金塑性变形的结构起源；探讨了单相非晶合金塑性变形时，应变局域化的结构因素及含有不同强度纳米第二相的非晶合金的变形行为。研究表明，塑性变形优先/主要发生在那些键长偏差较大的区域。单相非晶合金变形过程是否产生应变局域化，从原子结构上看，很大程度上取决于该非晶合金的结构与其 Tg 以上"类液"结构的相近程度，中程序结构的连接程度和稳定性，一定程度上也影响非晶合金变形过程应变局域化的程度。对于纳米相分离非晶合金，通过改变第二相的尺寸或体积分数，调节第二相和基体的成分关系，发生明显的韧脆转变。对于含有纳米晶体的非晶基复合材料，当晶体相无缺陷时，在变形过程中沿一定的位向发生滑移或孪晶，释放应力，有效地减缓应力陡降，减缓单一剪切带的形成；而当晶体相含有大量位错或层错等缺陷时，大量的剪切带形核，有效的阻止了单一剪切带的形成，甚至还有呈现一定的加工硬化效果。

关键词： 非晶合金；分子动力学模拟；微观结构；第二相；塑性

Tuning the Structure of Metallic Glasses and Improving Their Plasticity

Wang Li

(Shandong University, Weihai 264209, China)

Abstract: It is widely used to improve the plasticity of metallic glasses (MGs) by controlling the microstructure of single-phase or introducing nanoscale second phase into the amorphous matrix. The molecular dynamics simulations have been performed to study the structural origin of plastic deformation of MGs at low temperature using the proposed parameter of bond length deviation (BLD) of atomic clusters; to explore the structural factors of strain localization in single-phase MGs and the deformation behaviors of heterogeneous MGs with different yield strength nanoscale secondary phases. The results show that the plastic deformations have a high propensity to originate from those regions with the higher BLD clusters. Whether high strain localization occurs during the deformation of single-phase MGs depends largely on the similarity between the structure of MGs and their "liquid-like" structure above Tg. The degree and stability of medium-range order also, to a certain extent, affect the degree of strain localization in MGs during deformation. For nanophase-separated MGs, an obvious brittle to ductile transition occurs by varying the size/fraction and the chemical composition of the secondary phase. For nanocrystal-MG composites, the presence of stacking faults along a certain orientation in the defect-free nanocrystals releases most of the stress of nanocrystals and retards the formation of a mature shear band (SB). For the crystalline phases containing dislocations or stacking defects, serving as a relatively ductile secondary phase, promote the formation of a large number of plastic zones and thus retard the formation of a mature SB. Moreover, the cross-interaction of immature SBs results in a "strain-hardening".

Key words: metallic glasses; molecular dynamics simulations; nanoscale phase separation; ductility

金属玻璃熔体异常黏度降低的根源研究

胡丽娜

（山东大学，山东济南　250061）

摘　要： 在降温过程中，金属玻璃熔体的黏度通常是单调增加的。但研究发现，铜锆合金体系在一定温度区间存在异常的黏度降低的变化趋势，且与该异常动力学行为相对应，可以观察到液相线温度上存在明显的放热峰。通过精确控制冷却过程，研究证实利用异常动力学行为前后的熔体获得的非晶条带其热力学行为和微观结构表现出明显差别。通过对铜锆二元合金的热力学及动力学行为进行系统研究，已经发现该异常现象存在于 Cu_8Zr_3 和 $CuZr_2$ 的成分范围之间。结合相图，并通过高分辨电镜等试验，我们将该降温过程中发现的异常现象归因为 CuZr 二元合金熔体从高温的无序态向类金属间化合物（具有较低熔点）的有序团簇的转变。

关键词： 金属玻璃；融化遗传；粘性；相图

Probing into the Abnormal Viscosity Drop during Cooling of Metallic Glass Forming Melts

Hu Lina

(Shandong University, Jinan 250061, China)

Abstract: The viscosity of metallic glass-forming melts generally increases monotonically with temperature decreasing. However, recent reports observed an abnormal distinct viscosity drop in a certain temperature range during cooling processes in CuZr or CuZr-based melts, which has been evidenced further by a exothermic effect well above the liquidus temperature. To further understand this anomaly, metallic ribbons fabricated by liquids above and below this drop have been both studied. It has been found that thermodynamic features, as well as the microstructures, exibit distinct differences between the two types of ribbons. Such abnormal viscosity drop in CuZr binary alloys lies between compositions of Cu_8Zr_3 and $CuZr_2$ in the phase diagram. By combination with HTEM results, such abnormal viscosity decreases during cooling in CuZr metls has been attributed to the evolution from microstructures with low order-degree to intermetallic-like clusters with high local order-degree. The type of intermetalliclike clusters can be predicted by phase diagram of CuZr alloys.

Key words: metallic glasses; melt heredity; viscosity; phase diagram

基于磁性非晶合金的金属-半导体-金属转变行为研究

陈 娜

(清华大学材料学院,北京 100084)

摘 要:磁性非晶合金作为一类兼具原子结构长程无序和磁长程有序的材料,为建立和发展非晶材料和自旋电子学及器件的交叉领域方向提供了独具特色的研究对象。通过在磁性非晶合金中引入氧诱发金属-半导体转变,可以制备出成分和结构连续可调的非晶材料,实现同一材料体系磁、光、电等性能连续可调的非晶态金属、半导体、绝缘体的精准制备,为理解和揭示非晶材料领域长期存在的结构-性能关系提供新的研究思路。通过热处理晶化诱发半导体-金属转变,可以使得磁性非晶半导体转变为磁性金属。晶化前后磁性非晶薄膜的电阻率变化可达约三个数量级,有可能成为一种室温磁性相变材料,在新型非易失相变存储器件中获得应用。

关键词:磁性非晶合金;金属-半导体转变;室温磁性半导体;半导体-金属转变

Metal-Semiconductor-Metal Transitions in Amorphous Ferromagnetic Alloys

Chen Na

(Tsinghua University, Beijing 100084, China)

Abstract: Encompassing diverse behavior from metals to insulators, amorphous solids are useful in optics, electronics and catalysis. Local atomic configurations determine their electronic structure and properties, yet it poses a significant challenge to the experimental setup of a reliable structure-property correlation in amorphous materials. Here we demonstrate that a spectrum of behavior from magnetic metal to paramagnetic insulator is realized in one Co-Fe-Ta-B-O system via continuously compositional adjustability. Hence, the amorphous structure can be manipulated in a well controllable manner, leading to tunable optical, electrical and magnetic properties over a wide range. Particularly, the electric-field effect on the magnetism is obtained and varies significantly in the system. The combined optical, electrical and ferromagnetic functionalities in the amorphous system may assure them an important role in low-power-consumption electronic/spintronic devices. Furthermore, crystallization induces a semiconductor-metal transition in the amorphous ferromagnetic semiconductors based on ferromagnetic alloys. The significant resistance difference in their crystalline and amorphous states enables them to be room-temperature ferromagnetic phase change materials that could be used in new non-volatile phase change memories.

Key words: magnetic amorphous alloys; metal-semiconductor transition; room temperature magnetic semiconductor; semiconductor-metal transition

块体非晶合金材料产业化技术开发及应用介绍

李扬德，张 涛，高 宽，朱旭光

（东莞宜安科技股份有限公司，广东东莞 523662）

摘 要：块体非晶合金材料产业化涉及专利、合金成分优化、合金批量制备与回收、真空熔炼成型工艺与装备、产品机加工、切割、焊接和抛光等诸多材料加工技术等。锆基非晶合金，但作为一种全新的可以超精密真空压铸成型的高比强材料，在部分取代不锈钢和钛合金的应用领域具有明显的成型工艺和成本优势。非晶合金特有低凝固收缩、大弹性应变极限、高强度、高硬度和高耐腐蚀性特点，使其在高强精密薄壁复杂结构和弹性部件领域有着极好的应用前景。本文重点介绍液态金属在通讯器材、汽车部件和高性能工业紧固件等领域的应用研发以及关键加工技术的开发情况。

关键词：非晶合金；产业化；应用；真空压铸

The Development of Industrialization Technology and the Introduce of Application in Bulk Amorphous Alloys

Li Yangde, Zhang Tao, Gao Kuan, Zhu Xuguang

(Dongguan Eontec Co., Ltd., Dongguan 523662, China)

Abstract: The industrialization of bulk amorphous alloys involves patents, optimization of alloy composition, batch preparation and recovery of alloys, vacuum melting process and equipment, machining, cutting, welding and polishing, and many other material processing technologies. Zr-based amorphous alloys, as a new high specific strength material which can be formed by ultra-precision vacuum die-casting, have obvious advantages in forming process and cost in the application fields of partially replacing stainless steel and titanium alloys. Amorphous alloys are characterized by low solidification shrinkage, large elastic strain limit, high strength, high hardness and high corrosion resistance, which make them have excellent application prospects in the fields of high strength, precision thin-walled complex structures and elastic components. This paper focuses on the application and development of liquid metals in communication equipment, automotive parts and high-performance industrial fasteners, as well as the development of key processing technologies.

Key words: bulk amorphous alloys; industrialization; application; vacuum die-casting

先进热管理材料及技术的研究与应用进展

郭 宏

（有研科技集团有限公司，有研工程技术研究院有限公司，北京 100088）

摘 要：随着半导体和微电子技术的发展，无论是军用还是民用器件的功率密度越来越高，高功率密度电子器件的散热问题逐渐成为"卡脖子"问题。先进热管理材料及技术将为我国的高端军事装备、民用 5G 网络、计算机和 5G

智能手机的发展提供强有力的技术支撑。本报告分析了先进热管理材料及技术的需求背景，重点介绍新型碳复合材料、金刚石基散热材料、相变储热材料、热界面材料等新型散热材料的研究及应用进展；介绍热管、均热板、微槽道及热电制冷技术等主动散热技术及应用进展。探讨被动散热材料与主动散热技术的融合发展趋势。

Progress in Research and Application of Advanced Thermal Management Materials and Technologies

Guo Hong

(GRINM Group Corporation Limited, GRIMAT Engineering Institute Co., Ltd., Beijing 100088, China)

Abstract: With the development of semiconductor and microelectronics technology, power density is getting higher and higher, both in military and civilian devices. Heat dissipation of electronic devices with high-power density has gradually become the bottleneck. Advanced thermal management materials and the concerned technologies will provide strong technical support for the development of China's high-end military equipment and civilian 5G networks, computers and 5G smart phones. In this report the background of demand on advanced thermal management materials and technologies are analyzed, the research and application progress of new heat-dissipating materials, such as new carbon composite materials, diamond-based heat dissipation materials, phase change heat storage materials, and thermal interface materials are emphasized. Active heat dissipation technology and its application progress such as heat pipe, soaking plate, microchannel and thermoelectric refrigeration technology are also introduced. And the confusion development trend of passive heat dissipation materials and active heat dissipation technology are explored.

粉末冶金技术在航空发动机高温合金涡轮盘制备中的应用

田高峰

（北京航空材料研究院先进高温结构材料重点实验室，北京 100095）

摘　要：涡轮盘是航空发动机的最重要的核心热端部件之一，采用粉末冶金技术制备的镍基高温合金以其特有的优势，成为制备先进航空发动机涡轮盘的首选材料。本文阐述了粉末冶金高温合金在航空发动机部件上的应用和发展；对比分析了目前实际工程应用中最广泛采用的等离子旋转电极制粉+直接热等静压成形和氩气雾化制粉+热等静压（或热压实）+热挤压+等温锻造成形两种涡轮盘制备工艺的技术特点、合金体系和盘件性能；并简要介绍了粉末冶金高温合金材料研究的热点方向和粉末涡轮盘制备技术的发展趋势。

关键词：粉末冶金；航空发动机；高温合金；涡轮盘

Application of Powder Metallurgy Technology in Superalloy Turbine Disk of Aeroengine

Tian Gaofeng

(Science and Technology on Advanced High Temperature Structural Materials Laboratory,
Beijing Institute of Aeronautical Materials, Beijing 100095, China)

Abstract: Turbine disk is one of the most important hot section components of aeroengine, Nickel-based superalloy prepared by powder metallurgy technology has become the preferred material for the preparation of advanced aeroengine turbine disks due to its unique advantages. The application and development of powder metallurgy superalloy in aeroengine components are described in the paper. The technical characteristics, alloy system and disk properties of two kinds of turbine disk fabrication processes, i.e. plasma rotating electrode powder + hot isostatic pressing and argon atomization powder + hot isostatic pressing (or hot pressing) + hot extrusion + isothermal forging, which are widely used in practical engineering applications are compared and analyzed. The key issues of powder metallurgy superalloy and the recent technology development trend are also introduced.

Key words: powder metallurgy; aeroengine; superalloy; turbine disk

钢铁工业的资源消耗及废物减排的方向与途径

蔡九菊[1,2]

（1. 东北大学热能工程系；2. 中国金属学会能源与热工分会）

摘　要：基于工业系统与环境之间的物质交换，建立了钢铁工业消耗"四类资源"（矿物、能源、水和空气）排放"四类废物"（固体废物、废气、废水和废热）的概念，阐述了这四类资源的消耗量、废物产生量、污染物排放量三者间的相互联系及其对区域环境质量的影响。构建了各类资源效率的计算式，指出钢铁工业降低污染物排放量，必须从减少物耗、能耗、水耗尤其是空气消耗等天然资源做起，过量消耗空气是导致废气排放总量超出环境容量的主要原因之一。调研并统计了我国钢铁工业的资源消耗量和废物排放数据，比较高炉-转炉流程、全废钢-电炉流程的空气消耗量和废气排放量，强调减少空气消耗和废气产生量对降低污染物排放总量、改善区域大气环境质量的重要性。文中还详细地讨论了四类废物减排的方向与途径，指明：废钢、余热、废水和废气的循环再利用，可大幅度减少天然资源的消耗量和各种废物的产生量；生产流程是影响吨钢资源消耗量和废物产生量的关键因素，优化工艺流程和适时发展全废钢电炉是减少资源消耗和废物产生量，进而减少污染物排放量的有效途径。

关键词：钢铁工业；高炉转炉流程；全废钢电炉流程；资源消耗；废物排放；废物循环；资源效率；环境质量

Resources Consumption and Wastes Emission Reduction of the Iron and Steel Industry

Cai Jiuju[1,2]

(1. Department of Thermal Engineering, Northeastern University;
2. Energy and Thermal Engineering Branch, The Chinese Society for Metals)

Abstract: By investigating the material metabolism between industrial systems and the environment, four resources – ore, energy, water, and air – are consumed and four wastes – solid waste, waste gas, wastewater, and waste heat – are emitted, respectively, in the iron and steel industry. For each of the four resources, the relationship among resource consumption, waste generation, pollutant emissions and the ambient air quality were established. Methods for calculating the four resource efficiencies were proposed, which indicate that it is essential to reduce the resource consumption of ore, energy, water, and especially air to mitigate emissions from the iron and steel industry. Excessive air consumption is one of the major contributions that make the total emissions exceed the environmental capacity. Based on the data of resource consumption and waste emission in China, the air consumptions and waste gas emissions of BF-BOF route and EAF route were compared. It is emphasized that reducing the air consumption and waste gas emission are of great significance to reduce emissions and improve the air quality. Moreover, the direction and ways for reducing the four wastes were discussed in detail. The recovery and reuse of steel scrap, waste heat, wastewater, and waste gas significantly decrease the consumption of natural resources and the generation of every waste. The steelmaking route type is the critical factor influencing the resource intensity and environmental load. It is pointed out that the layout optimization of the steel manufacturing process and development of the EAF-route are effective to reduce resource consumption and the resulted emissions.

Key words: iron and steel industry; BF-BOF route; EAF route; resource consumption; waste emission; waste recovery; ambient air quality

钢铁冶金工程清洁高效燃烧技术现状与展望

张福明

（首钢集团有限公司，北京 100041）

摘 要： 钢铁工业是经济社会发展的重要基础产业，钢铁材料是国民经济发展中不可或缺的重要功能性材料和结构材料，在未来发展中仍将具有重要的主导地位。钢铁冶金过程的实质是铁素物质流在碳素能量流的驱动下，完成一系列复杂物理化学变化的过程，是典型的耗散结构。钢铁冶金过程的能量主要是通过碳素物质的燃烧而获得，工业燃烧是钢铁冶金过程不可或缺的重要单元过程。通过解析钢铁冶金过程燃料燃烧的过程，阐述了新一代钢铁制造流程理论及其主要特征，分析了炼铁高炉热风炉、焦炉、加热炉等钢铁冶金工程重要的工业燃烧技术发展现状和方向，探讨了清洁高效燃烧技术的发展理念、方法和目标。

关键词： 钢铁冶金；燃烧；物质流；能量流；节能减排；低碳绿色

真空冶金技术在资源综合利用领域的应用

徐宝强，杨 斌，戴卫平，刘大春，汤文通，李一夫，蒋文龙

（昆明理工大学，昆明鼎邦科技股份有限公司，云南昆明 650093）

摘 要： 真空冶金技术应用始于20世纪初期，开始用于炼钢的脱气，进入20世纪中后期以来，真空冶金技术的应用领域不断扩大，报告集中介绍了近年来我国真空冶金技术在锡、锌、铅、铜、铋资源综合利用以及镁、铝、钛冶金领域的新研究进展。实践表明，真空冶金技术在冶金过程资源综合利用以及冶金新技术的开发方面有着广阔的应用前景。

关键词： 真空冶金；资源综合利用；冶金新技术；研究动态

钢铁行业固废处理及资源化利用

王天义

（中国金属学会）

摘　要：报告首先介绍钢铁工业固体废弃物近几年来的新的法规、政策，以及相关的标准，然后分类讲解钢铁生产各个工艺阶段产生的固体废弃物和其处置利用途径，分析了含铁尘泥、转炉钢渣、脱硫灰、危险废弃物等的处理利用现状，最后展望了钢铁工业固废利用的发展趋势。

钢铁水污染全过程控制技术与应用

曹宏斌

（中科院过程工程研究所）

摘　要：钢铁是我国尤其是京津冀地区的支柱产业，水消耗量大且污水排放量大。此外，钢铁产业与城市高度融合，要实现钢铁行业的可持续发展，不仅要求满足环境质量要求而且生产过程必须产生合理的生产利润。科技创新引领企业低成本治污是支撑钢铁行业可持续发展的唯一途径。

在水专项等国家科技专项和广大钢铁企业的支持下，我国钢铁水污染控制技术已取得重大进展，初步形成了基于综合成本最小化和污染稳定达标的钢铁水污染全过程技术体系，并在绿色供水、过程节水与污染减排、废水处理与循环利用、多尺度水网络优化与智能管控等方面研发突破了绿色阻垢药剂、铁钢轧节水、干法除尘、焦化废水处理-安全焖渣、脱硫废液处理、高盐水直接制酸碱、综合废水复合污染低成本脱除与循环回用、酸性废水资源化、轧钢乳化液药剂强化破乳、钢铁工业园多尺度水网络优化与智能管控平台等核心关键技术。

焦化废水处理、脱硫废液处理、高盐废水制酸碱等关键技术已经实现应用，部分成果还实现行业推广。成套技术正在邯钢进行综合示范。预期成果有望为建设绿色钢铁工业园，实现超低排放和全局最优发挥重要作用，支撑我国钢铁污染控制由国际并跑向领跑转变。

环境承载力与钢铁行业绿色发展

杨晓东

（北京京诚嘉宇环境科技有限公司冶金清洁生产技术中心，北京　100053）

摘　要：十八大以来，我国环境保护政策确定了"以保护和改善环境质量"为核心的要求，陆续修订了《环保法》及配套颁布相关法规规章，出台了"气十条""水十条"等，开展"蓝天、碧水和净土"战役，推进钢铁行业"超低排放"。"十三五"以来，我国环境空气质量持续改善，但污染物排放总量仍处于高位，PM2.5 超标依然普遍，环境污染亦呈现新的变化。本文分析我国钢铁工业布局，结合区域环境空气污染特点和变化趋势，阐述行业发展的环境制约因素，提出环境承载力决定环境政策变化的观点，行业应以前瞻的做法，主动为之；探讨在生态文明建设和绿色制造的总体要求下，钢铁工业应通过制造流程工艺变革，跨行业跨领域产业创新，制造业间产业融合重构、服务城市建设等途径，构建资源高效利用、环境友好、综合效益优的钢铁企业持续稳定经营的绿色发展道路。

关键词：环境承载力；政策趋势；钢铁业；绿色发展

Environmental Carrying Capacity and Green Development of the Steel Industry

Yang Xiaodong

(Technical Center for Cleaner Production of Metallurgical Industry,
CERI Eco Technology Co., Ltd., Beijing 100053, China)

Abstract: Since the 18th National Congress of the Communist Party of China, the environmental protection policy has determined the priority requirements as "protecting and improving environmental quality". The "Environmental Protection Law" and related regulations were successively revised, followed by the promulgation of the "Action Plan for Prevention and Control of Water Pollution" "Action Plan for Prevention and Control of Air Pollution" and other relevant regulations. The "Blue sky、clean water and pure land" action was also carried out to promote the "ultra-low emissions" within the steel industry. China's ambient air quality has continued to improve since the introduction of the 13th Five-Year Plan. However, the total amount of pollutants remains at a high level and PM2.5 is still commonly exceeding the standard, new changes have presented in environment pollution now. This article analyzes the layout of China's steel industry, combines the characteristics and trends of regional environmental air pollution, expounds the environmental constraints of industry development, and puts forward the prospection that environmental carrying capacity determines environmental policy changes，and steel industry should take the initiative; Finally through manufacturing process evolution、cross-industry and cross-sector innovation、industrial integration、city service and other approach, steel industry can build a continuous and stable environmental friendly development path where resources can be used efficiently and the overall benefit can be optimized.

Key words: environmental carrying capacity; policy trends; steel industry; green development

钢铁烧结烟气全流程减排技术

魏进超[1,2,3]

（1. 烟气多污染物协同治理及资源化湖南省重点实验室，湖南长沙 410205；
2. 国家烧结球团装备系统工程技术研究中心，湖南长沙 410205；
3. 中冶长天国际工程有限责任公司，湖南长沙 410205）

摘 要：烧结工序的污染物负荷约占钢铁工业流程的 45%，是工业大气污染物控制的重点和难点，2019 年 4 月，生态环保部等五部门发布《关于推进实施钢铁行业超低排放的意见》，明确了烧结烟气有组织排放的超低排放指标。本文将烧结过程控制与末端治理，有组织和无组织排放控制相结合，开发基于烟气全流程减排的工艺及装备技术，实现低成本钢铁烧结烟气超低排放。

Flue Gas Full-process Emission Reduction Technology for Sintering of Iron and Steel

Wei Jinchao

钢铁企业超低排放技术路线与实践

朱晓华

（中冶节能环保有限责任公司）

摘　要：紧扣国家对钢铁行业超低排放改造要求，结合宝钢湛江钢铁、首钢、日照钢铁等大型钢铁企业的环保实践经验，重点介绍钢铁企业全流程粉尘控制、烧结烟气脱硫脱硝、高炉煤气精脱硫等方面的先进技术，为钢铁行业大气污染治理提供借鉴。

大型钢铁企业从内陆搬迁至沿海的分析与实践

吴礼云

（首钢京唐钢铁联合有限责任公司，河北唐山　063200）

摘　要：中国的钢铁工业布局正发生深刻变化，由于资源结构变化，环保压力及各项成本的上升，部分钢铁企业正在有序退出或异地搬迁，逐步由资源临近型的内陆布局转向市场临近型和物流临近型的沿海布局，产能布局与环境承载能力正逐步趋于协调。搬迁企业可充分利用原料资源优势，物流优势，市场优势，同时通过提升工艺技术水平，产品结构布局提高企业低成本运营及清洁生产水平，顺应循环经济和绿色环保对钢铁行业发展趋势要求。

关键词：钢铁企业；沿海布局；绿色搬迁

Analysis and Practice of Major Steel Industry Transfer from Inland to Coastal Areas

Wu Liyun

(Shougang Jingtang United Iron and Steel Co., Ltd., Tangshan 063200, China)

Abstract: The layout of China's steel industry is undergoing profound changes. Due to changes in resource structure, environmental pressures and rising costs, some steel companies are moving out of order or relocating from different places, gradually shifting from the inland layout of resources to market proximity and logistics.The adjacent coastal layout, capacity layout and environmental carrying capacity are gradually becoming more coordinated.Relocation enterprises can make full use of the advantages of raw material resources, logistics advantages, market advantages, and at the same time improve the low-cost operation and clean production level of enterprises by improving the technological level, product structure layout, and comply with the development trend of the steel industry in the circular economy and green environmental protection.

Key words: steel industry; coastal layout; relocation

激光诱导击穿光谱技术在钢铁在线分类中的初步应用

刘 佳[1]，沈学静[1]，史孝侠[2]，崔飞鹏[2]，
徐 鹏[2]，李晓鹏[2]

(1. 钢铁研究总院，北京 100081；2. 钢研纳克检测技术股份有限公司，北京 100081)

摘 要：废旧钢铁的高效回收利用既能降低成本，又可突破资源瓶颈制约，有利于钢铁工业可持续发展。废钢的形态各异、数量巨大，给分选带来了极大的挑战。因此，各种来源、成分的废钢几乎不经分类直接冶炼，不仅浪费资源降低效率，同时影响产品质量。本文基于激光光谱技术研制在线分拣系统，建立相关数学模型，实现了不同形态低碳钢、不锈钢以及高锰钢的自动化在线分类检测。结果表明，可自适应动态检测高度差异30mm样品，分拣准确率达到90%以上，分拣速度可达到90块/min。

关键词：在线；分拣；激光诱导击穿光谱；钢铁

Application of Laser Induced Breakdown Spectroscopy in Online Sorting of Steel

Liu Jia[1], Shen Xuejing[1], Shi Xiaoxia[2], Cui Feipeng[2],
Xu Peng[2], Li Xiaopeng[2]

(1. Central Iron and Steel Research Institute, Beijing 100081, China;
2. NCS Testing Technology Co., Ltd., Beijing 100081, China)

Abstract: The efficient recovery and utilization of waste steel can reduce costs and break through resource bottlenecks, which is conducive to the sustainable development of the steel industry. The shape of scrap is different and the number is huge, which brings great challenges to the sorting. Therefore, scrap steel of various compositions is directly smelted without classification, which not only wastes resources and reduces efficiency, but also affects product quality. In this paper, an online sorting system was developed based on LIBS and a mathematical model was established to realize automated online sorting of different forms of low carbon steel, stainless steel and high manganese steel. The results show that the sample with height difference of 30mm can be adaptively and dynamically detected, the sorting accuracy rate is over 90%, and the sorting speed can reach 90 pieces/min.

Key words: online; sorting; laser induced breakdown spectroscopy; steel

中国材料与试验团体标准体系（CSTM）

陈 鸣，杨植岗，李梦琪，王 矛，王 洋，王 蓬

（中关村材料试验技术联盟，北京 100081）

摘 要：2018 年 1 月 1 日《标准化法》的颁布和一系列深化标准化改革的政策文件出台，激发了团体标准的制修订热潮。本文系统全面的介绍了 CSTM 团体标准体系，包括其体系建设思路、创新点、创新目标，CSTM 标准的特点，已经建设的 CSTM 标准体系等，并对 CSTM 团体标准体系下的标准制修订实施情况进行了总结。源于创新的理念和体系架构，CSTM 标准体系得到了越来越多的社会认可，其标准也具有越来越大的社会影响力。期望本文能够对当前我国深化标准化工作改革、积极开展团体标准工作的其他社会团体起到一定的参考作用。

关键词：高质量发展；团体标准；领域；实物标准；材料属性；应用属性；通用技术

The Introduction of Chinese Society for Testing & Materials (CSTM)

Chen Ming, Yang Zhigang, Li Mengqi, Wang Mao, Wang Yang, Wang Peng

Abstract: The promulgation of the "Standardization Law" on January 1, 2018 and releasing a series of policy and documents to deepen the standardization reform is stimulating the upsurge of setting social group standards. This paper systematically and comprehensively introduces the group standard system of CSTM, including its thought of system construction, innovation, goals, the characteristics of the CSTM standard, the standard system of CSTM that has been built, and summarizes the implementation of the group standard system of CSTM. Due to innovative thought and architecture of CSTM system, CSTM has been more and more recognized by the society, and its standard has been increasingly influential. It is expected that this paper can play a reference role for other social groups that deepen standardization reform and actively develop social group standards in China.

Key words: high quality development; social group standards; fields; physical standards; material attributes; application attributes

无人天车与智能库管技术的研究与应用

王晓晨

（北京科技大学工程技术研究院信息技术研究所，北京 100083）

摘 要：报告内容为北京科技大学工程技术研究院针在钢铁企业智能工厂建设背景下，针对无人天车与智能库管方面的技术研究与实践成果。报告回顾无人天车与智能库管技术的发展历史，从系统架构、功能模块组成、工程建设任务、实施效果等方面进行了详细阐述，介绍北京科技大学在车体精准定位控制、库区智能调度、行驶路径规划、车辆物料识别、库区协同控制等技术的最新研发成果，以及针对板带、棒线企业的代表性应用案例，并总结了天车与库区运行效率提升的先进技术标准。

中冶京诚板带处理新技术新装备

王建兵

（中冶京诚轧钢所板带部）

摘　要：中冶京诚冷轧工程项目总承包业绩。为响应国家 2025 制造强国战略和清洁绿色环保生产，中冶京诚集各专业技术骨干，集数年之积累，研究开发了免酸洗板带清洗技术及装备，覆膜铁技术及装备，粉末彩涂技术及装备。实现了清洁化生产和无污染排放，节省了废弃物处理费用，实现了冶金国家队的社会责任。

冶金车间智能化建设研究与实践

赵振锐，甄景燕，尹国强

（河钢集团唐钢公司，河北唐山　063009）

摘　要：顺应实施制造业强国战略，推进"中国制造 2025"战略目标的实现，加快制造业整体的转型升级，从实际出发，探索冶金车间的智能化建设，对炼钢车间的生产进行全方面的管控，使信息流，物质流，能源流三流不落地，打造智能化冶金车间。

关键词：钢包管理；废钢管理；铁包计划；天车管理；炉次计划

Research and Practice on Intelligent Construction of Metallurgical Workshop

Zhao Zhenrui, Zhen Jingyan, Yin Guoqiang

(HBIS Group Tangsteel Company, Tangshan 063009, China)

Abstract: Complying with the implementation of the strategy of strengthening the manufacturing industry, promoting the realization of the strategic goal of "Made in China 2025", speeding up the transformation and upgrading of the manufacturing industry as a whole, starting from reality, exploring the intelligent construction of metallurgical workshops, and controlling all aspects of the actual construction plants, so as to keep the information flow, material flow and energy flow controlling and build an intelligent metallurgical workshop.

Key words: ladlemanagement; scrap management; ladle plan; crane management; plan scheduling

机器视觉技术及在钢铁生产中的应用

徐　科[1]，周　鹏[2]，贺　笛[1]，邓能辉[2]

（1. 北京科技大学钢铁共性技术协同创新中心，北京　100083；
2. 北京科技大学工程技术研究院，北京　100083）

摘　要：机器视觉技术就是通过光学成像方法获取图像或视频，并对采集的图像或视频进行分析处理，以替代人眼和大脑，实现目标的识别、跟踪和测量，是实现智能制造的基石。本报告主要介绍团队在机器视觉领域的理论研究工作，以及机器视觉技术在钢铁生产中的应用业绩与案例。通过光学成像、三维测量以及深度学习算法等关键技术的研究与开发，将机器视觉技术应用于产品质量在线检测、表面三维形貌以及三维尺寸测量、字符识别与物料追踪定位等，并在钢铁生产线上实现产业化应用。

Brief Introduction of Machine Vision and Its Application in Steel Production

轧钢过程异常工况诊断与容错控制

彭开香

（北京科技大学自动化学院，北京　100083）

摘　要：面向轧钢工业过程智能化制造的重大需求，围绕提升我国轧钢工业过程的可靠性和安全性为总体目标，针对轧钢工业过程对故障诊断、故障预测与容错控制理论与技术存在的巨大社会需求，及因现有方法存在不足而导致在使用安全性、可靠性、经济性方面显得不足等原因，从中提取出一些亟待解决且具有共性的理论与技术问题，以期建立一套轧钢工业过程异常工况智能诊断、预测与容错控制新方法：异常工况的在线监测与智能诊断、异常工况的智能预测与安全性评估、基于安全性评估的容错控制关键技术。拟重点解决一些关键科学问题：面向多指标的多级诊断系统融合的故障诊断与智能预测、基于安全性评估的故障诊断与容错控制协同设计。为以带钢热连轧为代表的复杂工业过程的安全监控、自主保障与智能运行提供理论支撑与技术保证。

关键词：轧钢工业过程；异常工况；故障诊断；容错控制

Fault Diagnosis and Fault Tolerant Control of Abnormal Conditions for Industrial Processes of Steel Rolling

Peng Kaixiang

(School of Automation and Electrical Engineering, University of Science and Technology Beijing, Beijing 100083, China)

Abstract: This presentation focuses on the great demand of intelligent manufacturing in complex industrial manufacturing process, aims to improve the reliability and safety of complex industrial manufacturing process in our country, for the huge social demand of the theory and technology of fault diagnosis, fault prediction and fault tolerant control for complex industrial manufacturing process, and for the shortage of existing methods, which lead to the lack of safety, reliability and economy, some common theoretical and technical issues which need to be solved have been extracted, the whole life cycle of hot strip mill is selected as the research object, a new method for intelligent diagnosis, prediction and fault tolerant control of abnormal working conditions in complex industrial manufacturing process is proposed: on-line monitoring and intelligent diagnosis of abnormal working conditions, intelligent prediction and safety evaluation of abnormal working conditions, key technologies of fault tolerant control based on security evaluation, experimental verification. Focus on solving some key scientific issues: fault diagnosis and intelligent prediction based on multi-index multi-level diagnosis system, cooperative design of fault diagnosis and fault tolerant control based on safety evaluation. It provides theoretical support and technical guarantee for the safety monitoring, independent guarantee and intelligent operation of the complex industrial manufacturing process represented by hot strip rolling.

Key words: industrial processes of steel rolling; abnormal conditions; fault diagnosis; fault tolerant control

长材生产线智能轧钢技术的研发与应用

房继虹

（北京金自天正智能控制股份有限公司，北京　100071）

摘　要：为了建设中国制造 2025 规划下的智慧工厂，北京金自天正智能控制股份有限公司在业内创造性提出多个轧线控制的关键技术，为其找到理论依据，同步针对性的推出切实可行的实现方案。通过把这些创新技术有机贯穿起来，融入到轧线控制的相关过程中去，形成轧钢智能化控制体系。运用该体系建设起来的控制系统，可以解决传统长材轧线的控制难题，满足轧材厂对于产品精度、生产产量、远程集中操控的需求，并可以实现智能轧钢。通过基础自动化系统功能细分和控制精度的大幅度提升，辅以生产大数据系统和前馈调节模型，使各个设备具备人工智能，使生产线本身具备自主调节和容错响应的能力，实现长材产线自主调节、全自动轧钢，并为轧材企业实现异地远程集控提供生产线端的支持和保障。该智能轧钢技术已在宝武集团韶关钢铁特轧厂高线工序上得以应用。

关键词：智能轧钢技术；长材生产线

Development and Application of Intelligent Technology of Rolling-steel for Long Product Line

Fang Jihong

(Beijing ARITIME Intelligent Control Co., Ltd., Beijing 100071, China)

Abstract: Beijing Aritime Intelligent Control CO., LTD has creatively proposed multiple key technologies of rolling lines in the metallurgical industry, aiming to build Smart Factory under the "Made in China 2025 Plan". Not only its principle and theoretical analysis but also feasible implementation approaches will be presented in detail. By integrating the innovative technologies into the control of the relevant rolling line processes, a hierarchy control system is developed. Based on the hierarchy structure, an intelligent rolling steel control system is designed to cope with the control challenges of the traditional long rolling-line automation so that the requirements of product quality, productivity and plant central control can be fulfilled successfully. By means of big-data based feed-forward systems model, every equipment is armed with artificial intelligence functions, which are achieved upon the functional segmentation of automation system and significant improvement of control accuracy. As a result, the product line is capable of self-regulation, fault-tolerant response, and full automatic steel rolling control. Furthermore, the guarantee and technical support at the product line level can be provided by the proposed intelligent technology of rolling-steel automation which has been successfully applied in the Shaoguan Rolling Steel Plant of Baowu Group.

Key words: intelligent technology of rolling-steel automation; long product line

既有工业建筑结构诊治与性能提升关键技术研究

常好诵

（中冶建筑研究总院有限公司）

摘 要：既有工业建筑结构诊治与性能提升关键技术研究旨在构建既有工业建筑全寿命结构诊治和寿命提升成套技术。在基于性能的结构诊治技术方面开展结构可靠度评定基础理论研究和钢结构疲劳、结构耐久性、结构振动及灾害等诊治和性能提升、绿色高效围护系统结构体系及节能评价关键技术研究；在基于功能转型的改造技术方面开展非工业化改造关键技术研究，形成既有工业建筑全寿命期结构诊治和性能提升成套技术，建立工业建筑大数据平台。

关键词：既有工业建筑；诊治；性能提升；关键技术

Key Technology Research on Existing Industrial Building Structure Diagnosis and Performance Improvement

Chang Haosong

(Central Research Institute of Building and Construction Co., Ltd., MCC)

Abstract: The project is closely linked to the green building and building industrialization guidelines, aiming to form a complete set of technology about industrial building whole life diagnosis and life promotion. Based on the investigation data of the existing industrial building types, environment and technique status, related research work has been carried out in the field of performance-based structural diagnosis and function transformation technology. The basic theory of structural reliability assessment, the diagnosis and treatment of steel structure fatigue, structural durability, structural vibration and disaster, the structural system of green and efficient envelope system and the key technology of energy-saving evaluation were studied in the performance-based structural diagnosis and treatment technology, and the transformation technology based on function was also studied. In the aspect of non-industrialized transformation, key technology research is carried out to form a complete set of technologies for diagnosis, treatment and performance improvement of existing industrial buildings, large data level of industrial buildings is established, and engineering application demonstration is carried out.

Key words: existing industrial building; diagnosis; performance improvement; key technology

锈损钢结构承载性能退化规律研究

徐善华

（西安建筑科技大学）

摘 要：许多长期处于工业、海洋大气等腐蚀环境下的钢结构工程，往往难以通过防护和构造措施避免锈蚀。钢结构锈蚀不仅造成资源巨大浪费，同时还对钢结构后续服役性能和安全性产生重要影响。本课题以锈损钢结构（热轧型钢、冷轧型钢）为研究对象，采用试验研究、理论分析、数值模拟相结合的方法，提出了钢结构耐久性检测新技术，建立了既有钢结构锈蚀程度表征参数体系与评定方法，明确了腐蚀环境下钢结构防腐涂层寿命终结标准与预测模型，提出了既有钢结构耐久性综合评价方法，解决了腐蚀环境既有钢结构耐久性检测与评价的技术难题；开展了腐蚀环境下"材料—构件—结构"多层次、"静力—地震—疲劳"多工况的性能评估理论研究，探明了钢结构锈蚀损伤对材料本构关系的影响规律，揭示了锈损钢构件承载性能退化规律与破坏机理，提出了锈蚀钢结构承载性能评估方法，建立了锈蚀钢结构疲劳寿命评估理论，解决了腐蚀环境下既有钢结构性能评估的关键技术难题。

关键词：锈损钢结构；承载性能；退化规律；评定方法

Research on Degradation Law of Bearing Capacity of Steel Structures with Corrosion Damage

Xu Shanhua

(Xi'an University of Architecture and Technology)

Abstract: Many steel structures exposed to the corrosive environment for a long time, including industry and ocean environments, were often difficult to avoid corrosion through protective and structural measures. Corrosion not only causes huge waste of resources, but also has an important impact on the service performance and safety of steel structure. In this paper, the corroded steel structure (hot-rolled steel and cold-formed steel) was taken as the research object, and the experimental research, theoretical analysis and numerical simulation were used. A new technology for durability testing of steel structures was proposed, and a characterization parameter system and evaluation method for corrosion degree of existing steel structures were established. The technical problems of durability testing and evaluation of existing steel structures in corrosive environment have been solved, and the performance evaluation theory of "material-component-structure" and "static-seismic-fatigue" under corrosive environment has been studied. The influence law of corrosion on material constitutive relation of steel structure was explored, and the degradation law and failure mechanism of bearing capacity of corroded steel members were revealed. The evaluation method of bearing capacity of corroded steel structure was proposed. The fatigue life evaluation theory of corroded steel structure was established, and the key technical problems of performance evaluation of existing steel structure under corrosive environment were solved.

Key words: steel structure with corrosion damage; bearing capacity; degradation law; evaluation method

1　炼铁与原料

大会特邀报告

分会场特邀报告

★ 炼铁与原料

炼钢与连铸

轧制与热处理

表面与涂镀

金属材料深加工

先进钢铁材料

粉末冶金

能源、环保与资源利用

钢铁材料表征与评价

冶金设备与工程技术

冶金自动化与智能化

建筑诊治

其他

焦炉炉门砖应用现状与发展趋势

武 吉，庞克亮，刘冬杰，王 超，朱庆庙，蔡秋野

（鞍钢股份技术中心，辽宁鞍山 114009）

摘 要：随着环保标准愈加严格和大高炉焦炭质量稳定需求，焦炉在炉砖材质、浇筑工艺、炉门结构优化方面取得新突破。对比分析了不同材质的单一、复合型炉门砖耐火材料性能，介绍了炉门浇筑工艺、炉门砖涂釉技术以及炉门砖结构优化新技术。从炉门砖耐火材料性能、结构、浇筑工艺展望了焦炉炉门砖的发展趋势，焦炉炉门砖将向着保温效果好、耐侵蚀、大型整体化、密封环保等方向发展。

关键词：焦炉；炉门砖；耐火性能；炉门结构

The Application State and Development Trend of Coke Oven Door Brick

Wu Ji, Pang Keliang, Liu Dongjie, Wang Chao, Zhu Qingmiao, Cai Qiuye

(Technology Center of Angang Steel Co., Ltd., Anshan 114009, China)

Abstract: With the increasing stringent environmental standards and the higher demands for the quality stability of the coke, new breakthroughs were achieved in the material performance, the casting process and the structure optimization of the door brick of coke oven. In this paper, the refractory performance of simple and complex which are made of different materials are analyzed and compared, and new techniques, such as casting, glazing, and structure optimization of the door brick are introduced. Then, the prospect on the development trend of the door brick is presented, the direction is being headed to the way of good insulation, anti-erosion, integral structure, good sealing, and better environmental protection effect.

Key words: coke oven; oven door brick; refractory performance; oven door structure

澳洲 FMG 粉矿烧结性能研究

王纪元，邱兰欣，阎丽娟

（宝钢股份有限公司，上海 201941）

摘 要：本文介绍了澳洲 FMG 粉矿基本物理化学质量及烧结性能，并进行不同比例 FMG 粉矿替代 BHP 烧结粉进行烧结杯试验研究，有利于烧结矿强度的提高及燃料消耗的降低，但烧结矿的 Al_2O_3 含量略有上升，相互替换对烧结生产的影响不大，能够满足生产要求。

关键词：FMG 粉矿；烧结；同化性能

Search on Sintering Properties of Australian FMG Powder Ore

Wang Jiyuan, Di Lanxin, Yan Lijuan

(Baosteel Co., Ltd., Shanghai 201941, China)

Abstract: This paper introduces the basic physicochemical quality and sintering performance of Australian FMG powder ore, and carries out the sintering cup test with different proportion of FMG powder ore instead of BHP sintering powder, which is conducive to the improvement of sinter strength and the reduction of fuel consumption, but the content of Al_2O_3 in sinter increases slightly, but the influence of mutual substitution on sintering production is little, it can meet the production requirements.

Key words: FMG powder; sintering; assimilation performance

无精粉的哈扬粉、国王混合粉的烧结性能比较研究

邸兰欣

（宝钢股份有限公司，上海　201941）

摘　要：褐铁矿的含铁量虽低于磁铁矿和赤铁矿，但因为它较疏松，易于冶炼，所以也是重要的铁矿石。本文分析使用无精粉条件下两种褐铁矿烧结杯的技术指标与过程参数的变化，并研究了其合理的替代比例。从化学性质、微观颗粒、工艺矿物学和烧结杯实验等维度，综合评价烧结性能。

关键词：烧结；铁矿石；褐铁矿

Technical Analysis on the Addition of HIY Fines and FMG Blend Fines at Sintering Process

Di Lanxin

(Baosteel Co., Ltd., Shanghai 201941, China)

Abstract: Although the iron content of limonite is lower than magnetite and hematite, it is also an important iron ore because it is loose and easy to smelt. In this paper, the changes of technical indexes and process parameters of two kinds of limonite sintering cups under the condition of using non-concentrate powder are analyzed. The reasonable substitution ratio was also studied. From the aspects of chemical properties, micro-particles, process mineralogy and sintering cup experiments, and the sintering performance is evaluated comprehensively.

Key words: sintering process; iron ore; the limonite

宝钢 1 号高炉稳定炉况生产实践

王 波，陈永明，宋文刚，王士彬

（宝钢股份炼铁厂，上海 200941）

摘 要：高炉的稳定顺行是高炉高产、优质、低耗、高效的基础。本文简要阐述了宝钢1号高炉炉况稳定性不强的表现及原因，重点介绍在"二对四"等不利条件下，从送风制度、装料制度、热制度、造渣制度、炉前作业等的调整方面，阐明了稳定炉况的具体措施。通过上述措施的实施，1号高炉炉况顺行良好，崩滑料明显减少，热负荷波动显著减小且较稳定，经济指标改善，炉缸温度稳定受控，铁水质量改善。

关键词：炉况；操作制度；原燃料；作业

Production Practice of Stable Furnace Conditions for Baosteel No.1 Blast Furnace

Wang Bo, Chen Yongming, Song Wengang, Wang Shibin

(Ironmaking Plant, Baoshan Iron & Steel Co., Ltd., Shanghai 200941, China)

Abstract: The stable and smooth operation of blast furnace is the basis of high yield, high quality, low consumption and high efficiency of blast furnace. This paper briefly expounds the manifestation and reasons of the instability of No. 1 BF in Baosteel, and emphatically introduces the specific measures for stabilizing the BF condition under the unfavorable conditions of "two to four", such as the adjustment of air supply system, charging system, heat system, slagging system and operation in front of the furnace. Through the implementation of the above measures, the blast furnace of No. 1 BF is running smoothly, the slump material is obviously reduced, the fluctuation of heat load is significantly reduced and more stable, the economic index is improved, the hearth temperature is stable and controlled, and the quality of iron is improved.

Key words: furnace condition;operation systeml;raw fuel;furnace operation

脱硫废液中硫代硫酸铵的再利用研究

贾楠楠[1]，芦 参[2]，张 展[1]，张大奎[3]，姚 君[1]，李冈亿[3]

（1. 鞍钢集团钢铁研究院，辽宁鞍山 114009；2. 鞍钢股份炼焦总厂，辽宁鞍山 114000；
3. 化学科技有限公司，辽宁鞍山 114000）

摘 要：本文介绍了脱硫废液的副盐含量对脱硫系统的影响；还对化学科技有限公司的脱硫废液提盐工艺进行了讲解；阐述了脱硫废液中硫代硫酸铵回收处理的意义；进行了硫代硫酸铵和过量浓硫酸的实验研究。得出主要结论如下：该反应能够生成硫酸铵，实现脱硫废液中硫代硫酸铵的再利用，降低了脱硫废液提盐后产生的硫代硫酸铵的处理成本，更有利于环保。

关键词：脱硫废液；硫代硫酸铵；硫酸铵；副盐

Reuse of Ammonium Thiosulfate in Desulfurization Waste Liquid

Jia Nannan[1], Lu Shen[2], Zhang Zhan[1], Zhang Dakui[3], Yao Jun[1], Li Gangyi[3]

(1. Ansteel Iron & Steel Research Institutes, Anshan 114009, China;
2. General Coking Plant of Angang Steel Co., Ltd., Anshan 114000, China;
3. Ansteel Chemical Technology Co., Ltd., Anshan 114000, China)

Abstract: This paper described the influence of the desulfurization waste liquid's by-salt content on the desulfurization system. It also explained the process of extracting salt from desulfurization waste liquid of Chemical Technology Co., Ltd. The significance of the recovery treatment of ammonium thiosulfate in desulfurization waste liquid were expounded. We studied on the experiment of ammonium thiosulfate and excess concentrated sulfuric acid. The main conclusions are as follows: The reaction is capable of producing ammonium sulfate, Reuse of ammonium thiosulfate in desulfurization waste liquid, Reduce the treatment cost of ammonium thiosulfate produced after desulfurization waste liquid extraction, More conducive to environmental protection.

Key words: desulfurization waste liquid; ammonium thiosulfate; ammonium sulfate; secondary salt

利用酸洗废水制备絮凝剂的工艺研究

张　展[1]，张大奎[2]，姚　君[1]，贾楠楠[1]，张馨予[1]，王晓楠[2]

（1. 鞍钢集团钢铁研究院，辽宁鞍山　114009；2. 鞍钢化学科技公司，辽宁鞍山　114021）

摘　要： 为了充分利用钢铁联合企业的酸洗废水和臭氧资源，本文从聚合氯化铁絮凝剂的氧化水解机理入手，对聚合氯化铁的聚合温度、臭氧进气速率及稳定剂选择进行研究。改变反应温度、臭氧进气速率和稳定剂种类，在不同工艺条件下制备聚合氯化铁絮凝剂。结果表明，氯化铁的聚合速率随着反应温度的升高先增长后下降,在45℃达到最大值；随着臭氧进气速率的增加逐渐增大，到达最大值（每分钟臭氧进气量与溶液体积比控制为13.33）之后不再提高；磷酸二氢盐是效果最好的稳定剂。

关键词： 氯化铁絮凝剂；臭氧；氧化水解

Research on Preparation of Polyferric Chloride by Pickling Wastewater

Zhang Zhan[1], Zhang Dakui[2], Yao Jun[1], Jia Nannan[1], Zhang Xinyu[1], Wang Xiaonan[2]

(1. Anshan Iron & Steel Group Research Institutes of Iron & Steel, Anshan 114009, China;
2. Ansteel Chemical Technology Co., Ltd., Anshan 114021, China)

Abstract: In order to make full use of the pickling wastewater and ozone resources of iron and steel complexes, this paper started with the oxidative hydrolysis mechanism of polyferric chloride flocculant, The polymerization temperature of the ferric chloride, the rate of ozone gas intake, and the choice of stabilizer were investigated. Polyferric chloride flocculant was prepared under different process conditions by changing the reaction temperature, the ozone intake rate and the type of the

梅钢四号烧结机余热锅炉改造实践

黄 辉，周康军，邱海雨

（上海梅山钢铁股份有限公司炼铁厂，江苏南京 210039）

摘 要：介绍了梅钢四号烧结机配套的余热回收系统，结合生产实践，分析了余热锅炉改造前所存在的主要问题，提出了相应的改进措施。通过设备改进优化，月平均吨矿产气量从 39.82kg/t 提高至 85.22kg/t，改造效果显著，为公司节能环保做出巨大贡献。

关键词：余热锅炉；蒸汽产量；节能环保

Retrofit Practice of Waste Heat Boiler of No. Fourth Sintering Machine in Mei Steel

Huang Hui, Zhou Kangjun, Qiu Haiyu

(Shanghai Meishan Iron and Steel Limited by Share Co., Ltd., Ironmaking Plant, Nanjing 210039, China)

Abstract: This paper introduces the waste heat recovery system of Mei Steel No. Fourth sintering machine, combines the production practice, analyzes the main problems existing before the transformation of waste heat boiler, and puts forward the corresponding improvement measures. Through equipment improvement and optimization, the average monthly steam production of one tonnage sinter is improved from 39.82kg/t to 85.22kg/t, the transformation effect is remarkable, for the company to make great contributions to energy conservation and environmental protection.

Key words: waste heat boiler; steam production; energy saving and environmental protection

提高炼焦加热速度对焦炭性能的影响

王 超[1]，庞克亮[1]，杜庆海[2]，朱庆庙[1]，武 吉[1]，蔡秋野[1]，刘福军[1]

（1. 鞍钢集团钢铁研究院，辽宁鞍山 114009；
2. 鞍钢股份有限公司炼焦总厂，辽宁鞍山 114021）

摘 要：为研究加热速度在炼焦过程中的影响，开展了炼焦煤黏结性试验及300kg炼焦试验，考察了加热速度变化对炼焦煤黏结性及焦炭性能的影响。研究结果表明：提高加热速度，有利于改善炼焦煤粘结性和提高焦炭质量，试验条件下以加热速度15℃/h，3h升至1065℃，降温速度10℃/h，降至1020℃，所得焦炭质量较好，同时提高加热速度可以提高弱黏性 1/3 焦煤用量 3%的同时实现焦炭质量有所提升，扩大炼焦煤资源。

关键词：炼焦煤；加热速度；塑性温度区间；焦炭质量

Influence of Increasing Coking Heating Rate on Coke Performance

Wang Chao[1], Pang Keliang[1], Du Qinghai[2],
Zhu Qingmiao[1], Wu Ji[1], Cai Qiuye[1], Liu Fujun[1]

(1. Ansteel Iron & Steel Research Institutes, Anshan 114009, China;
2. General Coking Plant of Angang Steel Co., Ltd., Anshan 114021, China)

Abstract: In order to study the influence of heating rate on the coking process, coking coal cohesiveness test and 300 kg coking test were carried out. The influence of heating rate on the cohesiveness of coking coal and coke properties was investigated. The results show that: increasing heating speed was beneficial to improving the cohesiveness of coking coal and coke quality. Under test conditions, the coke quality was better when the heating rate was 15℃/h, 3 hours to 1065℃, and the cooling rate was 10℃/h, falling to 1020℃. At the same time, increasing heating speed can increase the consumption of weak viscous 1/3 coking coal by 3%, while improving the coke quality and expanding coking coal resources.

Key words: coking coal; heating rate; plastic temperature range; coke quality

焦炉放散点火系统的改进与应用

朱庆庙[1]，庞克亮[1]，甘秀石[2]，王　超[1]，武　吉[1]

（1. 鞍钢集团钢铁研究院，辽宁鞍山　114009；2. 鞍钢股份炼焦总厂，辽宁鞍山　114031）

摘　要：针对焦炉生产过程中能够产生重大环境污染的环保装置进行了系统实践应用研究试验，提出了焦炉集气管自动放散点火装置的系统恢复及使用办法，并针对不同情况提出了现场生产岗位所需采取的不同的处理措施。同时针对焦炉自动放散点火装置日常的维护及保养提出了要求，以保证此套装置在应急条件下能够发挥作用，避免出现影响恶劣的大面积环境污染现象。

关键词：荒煤气；自动放散点火系统；应用与改进

Application and Improvement of Ignition Bleeding System on Coke Oven

Zhu Qingmiao[1], Pang Keliang[1], Gan Xiushi[2], Wang Chao[1], Wu Ji[1]

(1. Ansteel Iron & Steel Research Institutes, Anshan 114009, China;
2. Coking Plant of Angang Steel Co., Ltd., Anshan 114031, China)

Abstract: Based on coke oven production process to avoid major environmental pollution devices for practical application of the system of environmental protection and research experiments presented coke oven collector auto emission ignition systems restoration and the procedures to use and production posts is proposed for different circumstances required different treatment measures taken. While coke oven automatic relief requested ignition device for routine maintenance and repairs, to ensure that this unit can play a role in emergency conditions to avoid bad influence of large-area pollution.

Key words: crude oil gas; auto emission ignition system; application and improvement

4000m³高炉炉缸炉底耐材性能的试验分析

刘英才，张晓萍，李小静

（马鞍山钢铁股份有限公司技术中心，安徽马鞍山 243000）

摘　要： 为对炉缸炉底耐材的选用和生产监控提供支撑，在试验室考察了六种炉缸炉底耐材的理化性能指标以及高温和锌、碱作用的影响，并就锌、碱作用后的两种耐材进行了SEM形貌及能谱分析。从炉缸温度场要求考虑，采用具有阻热作用的陶瓷杯配合具有高导热性能炭砖的复合炉缸炉底结构有利于获得炉缸长寿。碱金属和锌的侵蚀对耐材的理化性能影响很大，尤其导热性能和耐压强度。耐材的性能指标是高炉炉缸炉底设计时的选材依据，但不是决定炉缸寿命的关键，减少有害元素对耐材的侵蚀有利于炉缸的监控和操作应对。

关键词： 高炉；炉缸炉底；耐材；性能指标

Experimental Analysis of Hearth Bottom Refractory Performance in 4000m³ BF

Liu Yingcai, Zhang Xiaoping, Li Xiaojing

(Technology Center, Maanshan Iron & Steel Co., Ltd., Maanshan 243000, China)

Abstract: In order to support the selection and production monitoring of refractory materials at the hearth bottom in BF, the physical and chemical properties of the refractory materials and the effects of Zinc or alkali at the bottom of 6 different hearths were investigated in the laboratory, and the SEM morphology and energy spectrum analysis of the two kinds of materials after Zinc or alkali treatments were carried out. Considering the temperature field requirements of the hearth, the ceramic cup with heat resistance and the composite hearth bottom structure with high thermal conductivity carbon bricks were beneficial to obtain long life of the hearth. The corrosion of alkali metals or Zinc has a great influence on the physical and chemical properties of refractory materials, especially thermal conductivity and compressive strength. The performance index of refractory materials is the basis for material selection in the design of hearth bottom, but it is not the key to determine the life of it, reducing the erosion of harmful elements on the refractory materials is beneficial to the monitoring and operation of the hearth.

Key words: blast furnace; furnace hearth bottom; refractory; performance

高炉返矿喷吹工艺探讨

汤楚雄，闫朝付

（中冶南方工程技术有限公司炼铁分公司，湖北武汉 430223）

摘　要：利用返矿具有良好的冶炼性能，从高炉风口直接喷吹入高炉。通过创造有利的冶炼条件，满足高炉接受返矿的操作要求，达到降低焦比的目的。从报道的粉矿喷吹的试验结果、喷吹过程热平衡计算、喷吹工艺形式等方面综合分析，喷吹返矿工艺是可行的；提出了高炉喷吹返矿初步目标：喷吹返矿量 75kg/t，降低焦比约 15kg/t。阐述了返矿粉碎、加热以及喷吹的工艺方案。尚需系统性地深入研究喷吹返矿对炉内操作的影响，结合工业试验，建立数学模型，确定完全消化烧结返矿的基本条件，并预测节能效果以及综合经济效益。

关键词：高炉；返矿；喷吹；节焦；讨论

Discussion on Injection of Under-screen Ore into Blast Furnaces

Tang Chuxiong, Yan Chaofu

(Ironmaking Business Division, WISDRI Engineering &
Research Incorporation Limited, Wuhan 430223, China)

Abstract: Injection process for under-screen ore powder to blast furnace is taking the advantage of its good comprehensive performance. Under-screen ore powder injection into the blast furnace, by creating favorable conditions to meet the operational requirements for blast to accept, may reduce coke ratio. From preliminary synthesis According to reports from the test results of ore powder blowing, the blowing process heat balance calculations for the blowing process, injection technology and other aspects, the process for under-screen ore powder to blast furnace is feasible; The basic objectives for the ore powder injection is the amount of 75kg/tHM, and the coke ratio reduce approximately 15kg/tHM. It is still necessary to systematically and deeply study the the impact on the furnace operation for the ore powder blowing. A mathematical model, combined with industrial test, would be used to determine the basic conditions for under-screen ore powder blowing and predict energy savings and economic benefit.

Key words: blast furnace; under-screen ore; injection; coke-saving; discussion

烧结喷加气体燃料均匀性的可视化技术研究

武　轶，李小静，张晓萍

（马钢技术中心，安徽马鞍山　243000）

摘　要：在烧结机料层表面喷入气体燃料可提高烧结矿质量，减少固体燃料的使用量，是一项烧结节能环保技术。此项技术的关键在于气体燃料的喷加方式，本研究自主开发了一套对各种气体喷加方式进行快速可视化筛选的技术。经试验验证，该可视化技术可以大大缩短气体燃料喷加方式及技术参数的筛选时间，能够直观地表现出不同喷加参数条件下气体燃料与空气混匀的效果，有助于发现喷加方式的不足之处，为后续改进提供技术依据。

关键词：气体燃料；烧结；均匀性；可视化

Research on Visualization Technology of Gas Fuel Injection Uniformity on Sintering Process

Wu Yi, Li Xiaojing, Zhang Xiaoping

(Technology Center, Maanshan Iron and Steel Co., Ltd., Maanshan 243000, China)

Abstract: Sinter quality is improved by injecting gas fuel from the top surface of the sintering machine. Solid fuel consumption is decreased by it. It is a saving energy and environment protection technology in sintering plants. The key of this technology is gas fuel injection method. This research develops independently a quick visualization screening technology of gas fuel injection method. It is demonstrated that this visualization technology can greatly shorten the screening time of gas fuel injection method and technical parameters. It can straightly show mixing effects of flammable gases and air with different injection parameter. It is helpful to find the deficiencies of the injection method. It supplies the technique foundation for the further modification.

Key words: gas fuel; sinter; uniformity; visualization

硫铵单元振动流化干燥床内件的改造

肖红春,赵 飞

(安徽马钢化工能源科技有限公司,安徽马鞍山 243021)

摘 要:本文主要分析了硫铵振动流化干燥床堵料、漏料以及筛板变形的原因,通过对干燥床筛板和布料机构进行改造,解决了筛板堆料、漏料等问题,降低清理频次,提高流化干燥床的设备性能。

关键词:振动流化干燥床;筛板;布料机构

Modification of Inner Parts in Vibration Fluidized Dryer of Ammonium Sulfate Unit

Xiao Hongchun, Zhao Fei

(Anhui Masteel Chemical Energy Technology Co., Ltd., Maanshan 243021, China)

Abstract: For plugging and leaking and deformation of sieve plate occurs in the vibration fluidized dryer of ammonium-sulfate unit, this paper analyses the reasons. By modifying the sieve plate and the distribution mechanism of the dryer, stacking and leaking that occurs in sieve plate are solved, the frequency of cleaning is reduced, and the performance of the dryer is improved.

Key words: vibration fluidized dryer; sieve plate; distributing mechanism

7m 焦炉边火道热工管理优化的实践

赵 华,周 鹏,马银华,党 平,代 成,彭 磊

(鞍钢股份有限公司鲅鱼圈钢铁分公司,辽宁营口 115007)

摘 要:焦炉边火道因热损失大、加热煤气量供给相对不足以及进入煤气热值相对低等原因导致炉头温度低这一情况,本文通过对炉门改型、增加辅助加热系统以及优化蓄热室封墙修补技术等方式,从而改善炉头温度,基本达到焦炉边炉热工管理要求。

关键词:炉头;热工

剩余氨水蒸氨工艺设备系统优化的生产实践

赵 华，赵恒波，周 鹏，武 斌，何亦光，代 成

(鞍钢股份有限公司鲅鱼圈钢铁分公司，辽宁营口 115007)

摘 要：围绕剩余氨水处理工艺影响因素进行分析，通过逐步对气浮焦油器、剩余氨水槽工艺设备改进提高剩余氨水脱除焦油能力，对蒸氨塔塔盘改进提高蒸氨塔处理剩余氨水的操作弹性，对蒸氨塔清扫装置方法的改进降低蒸氨系统焦油堵塞的影响，大幅提升单台蒸氨塔实际最大处理能力，降低了蒸汽使用量和废水处理难度。

关键词：剩余氨水；蒸氨；环保

三种熄焦车防碰撞技术在鞍钢鲅鱼圈的应用

赵 华，马银华，于庆泉，魏世明，何亦光，李 志

(鞍钢股份有限公司鲅鱼圈钢铁分公司，辽宁营口 115007)

摘 要：为了解决熄焦车撞车事故的发生，鲅鱼圈炼焦部先后自主研发了基于限位开关的防碰撞走行联锁控制技术、基于码牌识别定位的焦炉机车防碰撞技术以及引进一种基于超声波测距的焦炉车辆防碰撞技术，通过技术改造不断提升设备保障能力，可有效避免了熄焦车撞车事故的发生；同时通过对不同熄焦车防碰撞技术对比，为其他企业在选择应用上提供一些参考。

关键词：焦炉；防碰撞

高炉槽下称量系统误差补偿控制算法的优化

陈 明

(山信软件股份有限公司莱芜自动化分公司，山东济南 271104)

摘 要：本文主要介绍了高炉槽下称量系统 PLC 称量自动补偿算法，实现对称量过程自动闭环控制。该算法具有自动参数优化和误差补正的特点，能有效简化控制过程，提高控制精度。

关键词：称量误差；自动补偿；闭环控制；高炉

Optimization of Error Compensation Control Algorithms for Weighing System under Blast Furnace Groove

Chen Ming

(Shanxin Software Co., Ltd., Laiwu Automation Branch, Jinan 271104, China)

Abstract: This paper mainly introduces the PLC weighing automatic compensation algorithm of the weighing system under the blast furnace trough, which realizes the automatic closed-loop control of the symmetrical weighing process. The algorithm has the characteristics of automatic parameter optimization and error correction. It can effectively simplify the control process and improve the control accuracy.

Key words: weighing error; automatic compensation; closed-loop control; blast furnace

大型高炉炉缸长寿数值解析与操作对策

王训富，毛晓明

（宝山钢铁股份有限公司中央研究院，上海 200941）

摘　要： 在我国高炉大型化趋势背景下，炉缸长寿成为制约高炉长寿的关键，也越来越成为大高炉发挥优势的限制性环节。造成炉缸侵蚀的机理有很多，包括物理因素、化学因素等。然而铁水在炉缸内的流动是造成一切侵蚀的根本原因，文中以铁水流动的数值模拟为基础，分析了流动造成的炉缸内温度场、剪切应力的不均匀性分布，发现在铁口下方的侵蚀主要是由于流动造成对流换热系数高，导致 1150℃等温线位于耐材内而造成的。文中还分析了侧壁温度波动的原因以及不同操作制度对炉缸温度场分布的影响，提出了预防炉缸侧壁温度升高的对策措施。

关键词： 大型高炉；炉缸长寿；数值解析；操作对策

Numerical Analysis and Operational Countermeasures for Hearth Long Life Campaign of Large Blast Furnace

Wang Xunfu, Mao Xiaoming

(R&D Center of Baoshan Iron & Steel Co., Ltd., Shanghai 200941, China)

Abstract: Under the background of large-scale development of blast furnace in China, the hearth life of large blast furnace became the key to blast furnace longevity, and is becoming restrictive link in large scale blast furnace advantage. There are many mechanisms that cause hearth erosion, including physical and chemical factors. However, the flow of the molten iron in the hearth is the root cause, the paper on the basis of the numerical simulation of the hot metal flow, analyze the flow inside the hearth and the nonuniformity of temperature field, shear stress distribution, find at the bottom of the tap hole erosion is mainly caused by the highest flow convective heat transfer coefficient that lead to 1150℃ isotherm in refractory material. The causes of side wall temperature fluctuation and the influence of different operating systems on the temperature field distribution of hearth are analyzed in this paper.

Key words: large blast furnace; hearth long life campaign; numerical analysis; operational countermeasures

鞍钢利用镁基粘结剂生产镁质球团的工业试验

孙 彬，杨熙鹏，李建军，高宏庄，李林春，曾 宇，李晓春

（鞍钢股份有限公司炼铁总厂，辽宁鞍山 114021）

摘 要：利用镁基粘结剂代替膨润土进行球团生产工业试验结果表明：镁基粘结剂可以满足鞍钢带式球团焙烧机的生产需要。镁基粘结剂配比为 3.2%的情况下生产的，与配加膨润土 1.5%的普通酸性球团矿比较，生球质量未发生明显变化，镁质成品球团矿冶金性能有了明显改善，高炉使用透气性变好，稳定顺行，取得了日产提高 107t、综合焦比降低 4kg/tFe。

关键词：镁基粘结剂；镁质球团；工业试验

Industrial Test of Magnesium Pellets Produced by Binder base on Magnesium at Anshan Iron and Steel Company

Sun Bin, Yang Xipeng, Li Jianjun, Gao Hongzhuang,
Li Linchun, Zeng Yu, Li Xiaochun

(General Ironmaking Plant of Angang Steel Co., Ltd., Anshan 114021, China)

Abstract: The industrial test results of pellet production by binder base on magnesium instead of bentonite show that the binder base on magnesium can meet the production needs of belt roasting machine. Compared with ratio 3.2% Mg-based binder and ratio 1.5% bentonite, the quality of green pellets has not changed significantly, the metallurgical properties of finished Mg pellets have been improved obviously, the permeability of blast furnace is improved, the operation is stable and smooth, the daily output is increased by 107 t, and the comprehensive coke ratio is reduced by 4 kg/tFe.

Key words: binder base on magnesium ; magnesia pellet; industrial test

140t/h 干熄焦一次除尘入口膨胀节在线改为砌砖结构技术

张 雷，王君敏，洪耀明，陈艳伟，王 健

（鞍钢股份炼焦总厂，辽宁鞍山 114000）

摘 要：本文针对 140t/h 干熄焦装置设备运行中的薄弱部位：一次除尘两侧膨胀节容易出现烧损，造成膨胀节内部浇筑料脱落，外部波纹管高温变形损坏的问题。经过多次论证决定在线改型，将内部浇注料改为砌砖式结构，四年的生产运行未发现任何问题，延长了膨胀节的使用周期，并降低了年修成本。

关键词：一次除尘；膨胀节；砌砖式；措施

On-line Replacement of Expansion Joint of Primary Dust Removal Inlet for 140 t/h CDQ to Brick-laying Structure Technology

Zhang Lei, Wang Junmin, Hong Yaoming,
Chen Yanwei, Wang Jian

(Angang Steel Co., Ltd., Anshan 114000, China)

Abstract: Aiming at the weak parts of 140 t/h CDQ equipment in operation, the expansion joints on both sides of primary dust removal are prone to burn out, which causes the casting material falling off inside the expansion joints and the high temperature deformation and damage of external bellows. After many demonstrations, the online modification was decided, and the internal castable was replaced by brick structure. No problems were found in the four-year production and operation, which prolonged the service life of expansion joints and reduced the annual repair cost.

Key words: primary dedusting; expansion joints; brick construction; measure

提高烧结矿海砂配入量的实验研究

马 杰，程功金，薛向欣

（东北大学冶金学院资源与环境研究所，辽宁沈阳 110819）

摘 要： 本次实验通过微烧实验来验证添加适量的石灰石和硼镁铁矿能否实现提高烧结矿配料中新西兰海沙的配比，通过一系列实验得出：在石灰石配比为4%、硼镁铁矿配比在10%时，烧结矿配料中海沙的配比在15%达到最高，同时海沙配比在20%、25%、30%通过它们的烧结基础特性判断也在烧结矿配料合适区间内。

关键词： 新西兰海砂；硼镁铁矿；烧结基础特性；合适区间

Study on Improving the Ratio of Sea Sand in Sinter

Ma Jie, Cheng Gongjin, Xue Xiangxin

(School of Metallurgy, Northeastern University, Shenyang 110819, China)

Abstract: In this experiment, it was verified by micro-burning experiments whether the addition of proper amount of limestone and boromagnesium iron can improve the ratio of New Zealand sea sand in the sinter mix. Through a series of experiments, it is found that the limestone ratio is 4%, boron and magnesium. When the iron ore ratio is 10%, the ratio of sea sand in the sinter content reaches the highest at 15%, and the ratio of sea sand and sand at 20%, 25%, 30% is judged by the sintering base characteristics. Within the interval.

Key words: New Zealand sea sand; boron-magnesium iron ore; sintering basic characteristics; suitable interval

鞍钢 5 号高炉稳定顺行和提产攻关实践

赵东明[1]，张延辉[1]，刘振宇[1]，黄　勇[1]，朱建伟[2]，姜　喆[2]

（1. 鞍钢股份有限公司炼铁总厂，辽宁鞍山　114021；
2. 鞍钢股份有限公司技术中心，辽宁鞍山　114009）

摘　要：鞍山钢铁集团有限公司（鞍山钢铁）5 号高炉（2580m^3）大修开炉后一段时间，炉况出现了小幅波动。根据高炉用料特点及以往操作经验，并借助于高炉大修新安装的激光测料面、炉体炉缸智能检测等技术手段，对高炉布料制度和送风制度进行了一系列调整，最终达到了炉况长期稳定顺行，利用系数逐步提高的效果。

关键词：高炉；检测手段；稳定顺行；提产

Practice of Keeping Stable and Smooth Running and Raising Productivity on No. 5 Blast Furnace at Ansteel

Zhao Dongming[1], Zhang Yanhui[1], Liu Zhenyu[1],
Huang Yong[1], Zhu Jianwei[2], Jiang Zhe[2]

(1. General Ironmaking Plant of Angang Steel Co., Ltd., Anshan 114021, China;
2. Technology Center of Angang Steel Co., Ltd., Anshan 114009, China)

Abstract: After a large-scale renovation to No. 5 Blast Furnace (2580m^3) at Ansteel, a slight fluctuation on operation occurred. Based on experience and burden characteristics used in blast furnace, as well as making of the new-installed laser facility and body intelligent monitor system, the charging mode and blast system were adjusted accordingly. At last, the goal of long-term stable and smooth running and raising productivity have been achieved.

Key words: blast furnace; monitor system; stable and smooth running; raising productivity

高配比海砂矿制备氧化球团的试验研究

邢振兴，程功金，杨　合，薛向欣

（东北大学冶金学院，辽宁省冶金资源循环科学重点实验室，辽宁沈阳　110819）

摘　要：为提高海砂矿在钢铁企业中作为炼铁原料的使用量，本文依据优化配矿的原则，利用海砂矿制备了氧化球团，并对球团矿的冷态性能和还原性能进行了检测分析。研究表明：随着海砂矿配加量的增加，成球性能变差，生球落下强度以及抗压强度急剧下降，生球水分几乎不随海砂矿配加量的增加而发生变化，但海砂矿成品球团抗压强度不断降低。成品球团还原膨胀指数以及还原膨胀后球团的抗压强度逐渐降低，海砂矿球团的还原度以及还原后球团的抗压强度也逐渐下降。

关键词：海砂；氧化球团；抗压强度；还原性能

Experimental Research on Preparation of Oxidized Pellets with High Proportion Sea Sand Mine

Xing Zhenxing, Cheng Gongjin, Yang He, Xue Xiangxin

(School of Metallurgy, Northeastern University, Liaoning Key Laboratory of Recycling Science for Metallurgical Resources, Shenyang 110819, China)

Abstract: In order to improve the usage amount of sea sand mine as an ironmaking raw material in iron and steel enterprises, the paper based on the principle of optimized ore blending, oxidized pellets were prepared by using sea sand mine, and the cold properties and reduction properties of pellets were tested and analyzed. The research shows that with the increasing amount of sea sand mine, the ball-forming performance is deteriorated, the falling strength of green ball and crushing strength drop sharply, and the water content of green ball almost does not change with the increasing amount of sea sand mine, but crushing strength of finished pellets of sea sand mine is continuously reduced. The reduction swelling index of finished pellets and crushing strength of pellets after reduction swelling gradually decreased, and the degree of reduction of sea sand mine pellets and crushing strength of pellets after reduction also gradually decreased.

Key words: sea sand; oxidized pellets; crushing strength; reduction properties

球团直径对碳热还原过程传热传质的影响

刘程宏，丁学勇，代婧鑫，唐昭辉，董 越

（东北大学冶金学院，辽宁沈阳 110819）

摘 要：影响含碳复合球团的直接还原过程的因素众多，主要包括碳氧比、炉温、反应时间和球团直径。本文建立了含碳球团直接还原的非稳态数学模型，并通过实验数据验证了模型的准确性。运用模型研究了球团直径对碳热还原过程的影响规律。结果表明：球团直径的增加抑制了球团内的传热，球团内平均温度相应降低，减缓了球团内的还原速率，而对球团内还原气体的浓度和最终金属化率影响不大。

关键词：含碳球团；球团直径；数学模型；直接还原

Effect of Pellet Diameter on Heat and Mass Transfer in Carbothermal Reduction Process

Liu Chenghong, Ding Xueyong, Dai Jingxin, Tang Zhaohui, Dong Yue

(School of Metallurgy, Northeastern University, Shenyang 110819, China)

Abstract: There are many factors that affect the direct reduction process of carbon-containing composite pellets, including ratio of C/O, furnace temperature, reduction time and pellet diameter. In this paper, the unsteady mathematical model of direct reduction of carbon-containing pellets is reported, and then the accuracy of the model is verified by comparing experiments data and simulated results metallization rate and reduction degree. The influence of pellet diameter on carbothermal reduction process was studied by using the model. The results show that the increase of pellet diameter inhibits the heat transfer in the pellet, and the average temperature in the pellet decreases accordingly, which slows down the

reduction rate in the pellet. However, the change of pellet diameter has little effect on the molar concentration of reducing gases and the final metallization rate.
Key words: coal-based composite pellets; pellet diameter; mathematical model; direct reduction

高炉冶炼专家系统的研究现状与发展趋势

车玉满，郭天永，孙　鹏，姜　喆，姚　硕，费　静，刘炳南

（鞍钢集团钢铁研究院，辽宁鞍山　114009）

摘　要：从 20 世纪 90 年代开始，专家系统在国外高炉得到推广应用。同时期，我国采用多种模式开发与应用高炉专家系统，但由于操作理念、检测数据、维护等原因，国内高炉专家系统没有达到预期效果。在新形势下，结合信息化技术、物联网技术，专家系统应该向高炉集约化、可视化管理模式发展。

关键词：高炉；专家系统；集约化；可视化

Current Status and Development on Expert System of Blast Furnace

Che Yuman, Guo Tianyong, Sun Peng, Jiang Zhe, Yao Shuo, Fei Jing, Liu Bingnan

(Ansteel Iron & Steel Research Institutes, Anshan 114009, China)

Abstract: From 90s last century, the expert system began to put use in the blast furnace in overseas. At same time, the blast furnace expert system were developed by using three methods in Chinese, so far the, the blast furnace expert system did not realized their purpose, for some reasons of the operation belief difference, of fault data, of short of maintenance. At new situations, blast furnace expert system must combine with internet information technology and IT technology to lie on the blast furnace operated intensive and controlled visualization.
Key words: blast furnace; expert system; operated intensive; controlled visualization

高比例配用镜铁矿烧结技术研究

史先菊[1,2]，李光强[1]，李　军[2]，刘代飞[3]

（1. 武汉科技大学材料与冶金学院，湖北武汉　430081；2. 宝钢股份中央研究院武汉分院
（武汉钢铁有限公司技术中心），湖北武汉　430080；
3. 长沙理工大学能源与动力工程学院，湖南长沙　410114）

摘　要：本文对镜铁矿的制粒性能、成矿性能、烧结性能进行研究，研究表明：（1）镜铁矿亲水性差，自身难以成球，但其制粒选择性好，与亲水性好的细粒级铁矿粉混合制粒可强化镜铁矿制粒；（2）镜铁矿的高温反应性差，难以形成液相，且粒度细，容易在气流的作用下在液相中流动，最终聚集在液相边缘，阻碍液相的成片发展，从而使烧结矿的产质量降低。综合制粒性能、成矿性能、烧结性能的研究提出了镜铁矿配用比例 20%以上的较优配矿结构框架。

关键词：镜铁矿；制粒性能；成矿性能；配矿结构

Study on Sintering Technology of Specularite with High Ratio

Shi Xianju[1,2], Li Guangqiang[1], Li Jun[2], Liu Daifei[3]

(1. School of Material and Metallurgy, Wuhan University of Science and Technology, Wuhan 430081, China; 2.Wuhan Branch of Baosteel Central Research Instotute(R & D Center of Wuhan Iron & Steel Co., Ltd.), Wuhan 430080, China; 3. School of Energy and Power Engineering, Changsha University of Science and Technology, Changsha 410114, China)

Abstract: In this paper, the granulation, metallogenic and sintering properties of specularite are studied, and the following conclusions are drawn: (1) the hydrophilicity of specularite is poor, and it is difficult to form pellets by itself, but its granulation selectivity is good, and the granulation of specularite can be strengthened by mixing with fine-grained iron ore powder with good hydrophilicity; (2) the reactivity of specularite at high temperature is poor, it is difficult to form liquid phase, and its particle size is fine, so it easy to flow in liquid phase under the action of airflow, and eventually gather at the edge of the liquid phase, which obstruct the development of liquid phase, thus reducing the quality of sinter production. In view of the above shortcomings, the optimum ore blending structure of specularite with a ratio of more than 20% is proposed as follows: specularite + 30%~40% limonite + 10%~20% skeleton ore + other.

Key words: specularite; granulation performance; metallogenic performance; ore blending structure

生产大型高炉用焦炭配煤炼焦工艺选择

胡德生，孙维周

（宝山钢铁股份有限公司中央研究院，上海　201900）

摘　要：针对国内大型高炉数量增多，对焦炭质量要求提高，高质量焦炭产量提高，优质炼焦煤资源日趋紧张的局面，本文归纳总结近几十年来各国开发的配煤炼焦工艺技术，例如：大容积焦炉，捣固炼焦，煤预热，煤调湿，配型煤，DAPS，SCOPE21 等，分析比较各工艺特点、节能环保效果以及提高弱粘煤比例幅度，推荐适应生产大型高炉用焦炭配煤炼焦工艺路线。

关键词：炼焦；大容积焦炉；捣固炼焦；煤预热；煤调湿；配型煤；DAPS；SCOPE21

Selection of Coal Blending and Cokemaking Technology for Large Blast Furnace Used Coke

Hu Desheng, Sun Weizhou

(Centrel Research Institute, Baoshan Iron & Steel Co., Ltd., Shanghai 201900, China)

Abstract: For the situation of increasing amount of domestic large blast furnace, progressively quality requirements for coke and production, and increasingly scarce of high-quality coking coal resources, this paper summarizes the development of coal blending and cokemaking technology in recent decades, such as large volume of coke oven, Stamp-charge cokemaking, preheating coal, CMC, coal briquetting, DAPS and SCOPE21 etc, analyzing and comparing the process characteristics, energy-saving effects, environmental protection and the improvement of the weak caking coal blending ratio, eventually recommended the coal blending and cokemaking technology adapting to the production of large blast furnace.

Key words: coking; large volume coke oven; stamp-charge cokemaking; preheating coal; CMC; coal briquetting; DAPS; SCOPE21

湛江钢铁 1BF 稳定热负荷生产实践

王志宇，张永新

(宝钢湛江钢铁有限公司，广东湛江　524000)

摘　要： 本文对湛江钢铁 1 号高炉开炉以来稳定热负荷生产进行总结。1 号高炉热负荷变动主要集中于炉身下部及中部且脱落不均匀，渣皮循环脱落和粘结导致热负荷以及炉温波动。得出合理煤气流分布不是一成不变的，边缘及中心煤气流相对强弱与炉墙稳定与否并非有绝对的关系，边缘气流分布不均匀以及边缘过重更容易产生炉墙脱落严重及热负荷高且波动大等情况发生。在正常生产时，气流上保证中心，适当发展边缘，并且尽力保证边缘的均匀性以形成合理炉型，在炉墙脱落时及时查明原因并采取积极措施，保证在炉墙大幅度脱落时快速稳定热负荷及恢复炉况。

关键词： 热负荷；渣皮；煤气流

炉顶可视化技术在朝阳钢铁 2600m³ 高炉的应用

王光伟，胡德顺，刘喜亮，金　遥，关　爽，马秀生

(鞍钢集团朝阳钢铁有限公司，辽宁朝阳　122000)

摘　要： 系统介绍了炉顶可视化技术朝阳 2600m³ 高炉的应用情况，炉顶可视化技术实际应用效果进行积累、分析、总结，找出适合本高炉的料面形状；并通过炉顶可视化技术的数据，逐步优化高炉上料装料制度，指导高炉日常操作指标，找出最优的经济煤气流分布方案。

关键词： 可视化；料面形状；装料制度；煤气流

Application of Top Visualization Technology in Chaoyang Iron and Steel 2600m³ Blast Furnace

Wang Guangwei, Hu Deshun, Liu Xiliang, Jin Yao, Guan Shuang, Ma Xiusheng

(Chaoyang Iron and Steel Co., Ltd., Angang Group, Chaoyang 122000, China)

Abstract: The application of visualization technology of blast furnace top in Chaoyang 2600m³ is introduced systematically. The practical application effect of visualization technology of blast furnace top is accumulated, analyzed and summarized to find out the material surface shape suitable for the blast furnace. Through the data of visualization technology of blast furnace top, the charging system of blast furnace is optimized step by step, the daily operation index of blast furnace is guided, and the optimal economic gas flow distribution scheme is found.

Key words: visual; material surface shape; charging system; gas flow

朝阳钢铁 1#高炉使用修复风口实践

吕宝栋，李泽安，王振东，李 男，刘喜亮，王渐灵

（鞍钢集团朝阳钢铁有限公司，辽宁朝阳 122000）

摘 要：朝阳钢铁1#高炉2010年开炉生产，为应对钢铁寒冬，降低成本，从2013年开始部分使用修复风口，通过研究确立修复风口使用准则，稳定了炉况顺行，使修复风口与新风口使用寿命基本一致，并通过逐年增加修复风口使用比例，节省了大量成本。

关键词：风口；修复风口；成本

Practice of Repairing Tuyere in Chaoyang Iron and Steel No.1 BF

Lv Baodong, Li Zean, Wang Zhendong, Li Nan, Liu Xiliang, Wang Jianling

(Chaoyang Iron and Steel Co., Ltd., Angang Group, Chaoyang 122000, China)

Abstract: The No.1 Blast Furnace of Chaoyang Iron & Steel Corporation was blow-in in 2010. In order to cope with the cold winter of steel industry and reduce the costs, some repaired tuyeres were used since 2013. Through researching, the use criteria of the repaired tuyeres were established which makes the blast furnace running stably and smoothly, the service life of the repaired tuyeres were basically the same as that of new tuyeres. Thereby a large amount of cost was saved by increasing the use ratio of the repaired tuyeres year by year.

Key words: tuyere; repaired tuyere; cost

朝阳钢铁铁前系统科学管理高效生产实践

胡德顺，王光伟，刘喜亮

（鞍钢集团朝阳钢铁有限公司，辽宁朝阳 122000）

摘 要：系统总结了朝阳钢铁单高炉生产的铁前系统科学管理、高效生产实践的理念及实绩。认为"高炉长周期稳定顺行"就是铁前最大的效益，提出了"以控制重大事故为前提，以高炉长周期稳定顺行为中心"，降低成本；以"严控原燃料质量为突破口，以模板化管理为手段，以量化指标为依据，以趋势化管理为判断标准"提高铁前系统的管控能力。

关键词：科学管理；高效生产；长周期稳定顺行

High Efficiency Production Practice of Chaoyang Iron and Steel Iron Front System Scientific Management

Hu Deshun, Wang Guangwei, Liu Xiliang

(Anshan Iron and Steel Group Chaoyang Iron and Steel Co., Ltd., Chaoyang 122000, China)

Abstract: The concept and achievements of the scientific management and high efficiency production practice of the iron front system produced in Chaoyang iron and steel single blast furnace are systematically summarized. That long period of stable operation of blast furnace iron before "is the biggest benefit, put forward to control the accident as the premise, with long period smooth operation of blast furnace center, reduce costs; to strictly control the quality of raw materials and fuel as a breakthrough in the template management as a means to quantify the index according to the trend the management criterion" to improve the ability to control the system before iron.

Key words: scientific management; efficient production; long period stable operation

朝阳钢铁高炉炉缸管理实践

张洪宇，王光伟，胡德顺，刘喜亮

（鞍钢集团朝阳钢铁有限公司，辽宁朝阳　122000）

摘　要：介绍了朝阳钢铁高炉炉缸管理模式，通过建立炉芯温度和炉缸侧壁温度控制标准，对高炉炉芯温度和炉缸侧壁温度波动原因进行分析，并采取相应治理措施，确保了高炉生产稳定顺行，高炉生铁一级品率由63.6%显著提高至97.59%。

关键词：高炉；温度场；炉芯

Practice of the Management of the Hearth of Chaoyang Steel

Zhang Hongyu, Wang Guangwei, Hu Deshun, Liu Xiliang

(Anshan Iron and Steel Group Chaoyang Iron and Steel Co., Ltd., Chaoyang 122000, China)

Abstract: By investigating the change of the temperature field of the blast furnace hearth of Chaoyang iron and steel, and analyzing its reasons, it is considered that the temperature field of the hearth cylinder will fluctuate greatly when the temperature of the central temperature of the hearth is lower than the lower limit, which will affect the long life of the hearth, and according to the change of the working state of the hearth of Chaoyang iron and steel in recent years and the effect on the blast furnace. Formulate corresponding measures in time to ensure stable production of blast furnace.

Key words: blast furnace; the temperature field; the hearth

冶金石灰硅钙硫含量的测定-波长色散 X 射线荧光光谱法

胡 涛，崔 隽，沈 克，曾 莹，关 晖

（宝武股份武钢有限公司质检中心，湖北武汉　430080）

摘　要：采用一组石灰石、白云石标准样品经过灼烧后，覆盖冶金石灰中主量元素的含量范围，得到满足 X 荧光光谱法的校准曲线样品，通过熔融制样—波长色散 X 荧光测量待测元素特征波长的谱线强度，建立了波长色散 X 射线荧光光谱法（XRF）测定冶金石灰中硅、钙、硫的方法，各元素的检出限分别为 0.050%、1.30%、0.016%。试验结果表明：当硅、钙、硫质量分数分别为 4.47%、81.86%、0.107%时，五次测量结果的相对标准偏差（RSD）分别为 0.09%、0.04%和 0.49%。方法用于冶金石灰样品中硅、钙、硫的测定，与湿法测定值吻合较好，能满足常规分析要求。

关键词：熔融制样；冶金石灰；X 荧光光谱法

Metallurgical Lime Determination of Silicon Calcium and Sulfur Content Wavelength Dispersive X-ray Fluorescence Spectrometry

Hu Tao, Cui Juan, Shen Ke, Zeng Ying, Guan Hui

(Baowu WISCO Co., Ltd., Quality Inspection Center, Wuhan 430080, China)

Abstract: Using a set of linestone and dolomite standard samples after burning, Covering the content range of major elements in metallurgical lime, A calibration curve sample satisfying X-ray fluorescence spectroscopy was obtained, and the spectral line intensity of the characteristic wavelength of the element to be tested was measured by melting sample-wavelength X-ray fluorescence. Wavelength dispersive X-ray fluorescence spectrometry (XRF) was established to determine silicon and calcium in metallurgical lime. For the sulfur method,the detection limits of each element are 0.050%,1.30% and 0.016%, respectively.The test results show that when the mass fractions of silicon, calcium and sulfur are 4.47%, 81.86% and 0.107%, respectively,the relative standard deviations (RSD) of the five measuerments are 0.09%, 0.04% and 0.49%, respectively. The method is applied to the determination of silicon, calcium and sulfur in metallurgical lime samples, which is in good agreement with the wet method and can meet the requirements of routine analysis.

Key words: fusion sample preparation; metallurgical lime; X-ray fluorescence spectrometry

6m 焦炉配用高硫煤的生产实践

夏雷雷

（宝钢集团八钢公司炼铁分公司，新疆乌鲁木齐　830022）

摘 要：本文围绕优化化工系统过程控制指标，尝试通过配加部分高硫煤开展炼焦试验与生产实践工作，最终在保证焦炉煤气含硫达到要求前提下，焦炭质量及成本均得到改善。

关键词：高硫煤；炼焦；生产实践

Abstract: The paper focuses on optimizing process control index of chemical system, Coking test and production practice by adding part of high sulphur coal, finally, under the premise of ensuring the sulful content of coke oven gas to meet the requirements, the quality and cost of coke are improved.

Key words: high sulphur coal; coking; production practice

2500 高炉循环水系统冬季送水实践

付 涛

（宝武集团新疆八一钢铁公司炼铁厂，新疆乌鲁木齐 830022）

摘 要：由于生产经营需要，八钢 2500 高炉实施了冬季开炉工作，其间环境最低温度达-20℃左右，高炉开炉循环水系统送水工作需要做好专项保温、提温工作，主要是高炉软水、高低压净环水，其中炉体保温和泵房空冷器保温是关键。

关键词：循环水；冬季；–20℃；送水

A 高炉更换冷却壁操作实践

李刚义，黄银春

（新疆八一钢铁股份有限公司炼铁厂，新疆乌鲁木齐 830022）

摘 要：A 高炉利用中修对破损冷却壁进行更换，进行了停炉、检修、开炉等一系列工作，开炉后对高炉操作制度进行相应调整，保证了炉况稳定运行，经济指标进一步优化。

关键词：冷却壁；降料面；填料；开炉

浅谈 V 形锥流量计在测量八钢焦化煤气介质的应用

苗 钧，肖先锋

（宝钢集团八钢公司炼铁分公司，新疆乌鲁木齐 830022）

摘 要：介绍了宝钢集团新疆八一钢铁有限公司焦化分厂粗苯 v 形锥流量计加热系统的配置、控制原理和方式，并对实际使用中存在的问题进行了分析，提出改进办法。改造后的系统运行平稳。

关键词：流量计

八钢 2500m³ 高炉顶压控制及优化方案

李 杰

（新疆八一钢铁股份有限公司炼铁厂，新疆乌鲁木齐　830022）

摘　要：重点阐述了 TRT 控制高炉顶压系统的主要特点和控制的基本方法，通过对八钢高炉顶压系统运行现状的分析，指出了当前运行中存在的问题及原因分析，提出了优化改进方案，以保证高炉顶压控制的稳定可靠。

关键词：高炉煤气；余压透平发电装置（TRT）；高炉顶压调节控制

Abstract: The main characteristics of TRT control top pressure system of blast furnace weredescribed in detail.

Key words: blast furnace gas; top pressure recovery turbine (TRT); top pressure adjustment control of blast furnace

焦化除尘灰回配炼焦试验初步研究

夏雷雷，王雪超

（宝钢集团八钢公司炼铁分公司，新疆乌鲁木齐　830022）

摘　要：文章介绍了 60kg 试验焦炉与 6m 焦炉分别配加除尘灰的配煤试验，通过大量 60kg 试验焦炉与 6m 焦炉试验找到除尘灰在八钢现有配煤结构的合理配量，优化配煤结构，最终降低焦炭成本。

关键词：除尘灰；炼焦；试验研究

Abstract: The paper introduces the coal blending test of 60kg test coke oven and 6m coke oven with dust removal ash respectively, through a large number of 60kg test coke oven and 6m coke oven tests, the reasonable disposition of dust in the existing coal blengding structure of the Bayi iron and Steel corporation was found, optimizing coal blending structure and ultimately reducing coke cost.

八钢 2500 高炉经济指标改善途径思考

王宗乐

（新疆八一钢铁股份有限公司炼铁厂，新疆乌鲁木齐　830022）

摘　要：分析影响八钢 2500 高炉经济指标的因素，找出主因素，突破传统思路，分析论证后制定操作路线，为降成本找出新对策。

关键词：经济料；煤气富化

八钢 A 高炉长周期检修开炉达产实践

黄银春，马 锐

（宝钢集团八钢公司炼铁分公司，新疆乌鲁木齐 830022）

摘 要：对八钢 A 高炉长周期检修开炉快速达产进行了总结，通过科学精心的开炉准备、合理的制定开炉方案和在开炉过程中及时调节各项操作制度，实现了全焦开炉快速达产达效，大大地降低了开炉燃料消耗。

关键词：高炉；开炉；达产；实践

八钢干熄焦生产与优化操作实践

刘智江

（宝武集团八钢公司炼铁厂焦化分厂，新疆乌鲁木齐 830022）

摘 要：干熄焦可以显著改善炼焦生产环境，降低能耗，提高焦炭质量。本文结合新疆八钢焦化 140t/h 干熄焦生产实际，介绍了干熄焦生产过程中存在的问题，分析并制定了解决问题的优化操作措施，以实现干熄焦系统生产的稳定顺行。

关键词：干熄焦焦炭；循环风量；压力

Practice of Production and Optimization of Dry Coke Quenching in Bayi Iron & Steel Co.

Liu Zhijiang

(Coke Branch of Iron of Bayi Iron & Steel Co., Baosteel Group, Urumchi 830022, China)

Abstract: Dry quenching can significantly improve the coking production environment, and reduce energy consumption and increase coke production. The article introduces problems in the production of coke combined with coke production practiced in 140t/h dry coke quenching production line in Bayi Iron & Steel Co., analysis and formulate the optimal operation measures to solve the problem, in order to realize the stable operation of the dry coke quenching system.

Key words: dry coke quenching; circulating air volume; pressure

八钢 C 高炉钒钛冶炼理论与实践

刘永想，高小雷

（宝钢集团八钢公司，新疆乌鲁木齐 830022）

摘 要：介绍了八钢 C 高炉配加钒钛矿冶炼情况，分析了使用该矿后高炉冶炼规律及对高炉生产指标的影响。冶炼实践表明，钒钛矿冶炼关键是维持合适的炉温，大风量操作，吹透中心、适当疏松边缘、活跃炉缸，才能使高炉长期稳定顺行，以及人员思想观念上的转变。

关键词：高炉；钒钛矿冶炼；顺行；渣铁流动性

The Theory and Practice of the V-Ti Smelting in the C-Blast Furnace of the Eighth Steel

Liu Yongxiang, Gao Xiaolei

(Baosteel Group Xinjiang Bayi Iron and Steel Co., Ltd., Urumchi 830022, China)

Abstract: This paper introduces the smelting situation of vanadium and titanium ore in C blast furnace of Baotou Iron and Steel Co., Ltd. and analyzes the smelting law of blast furnace after using this mine and its influence on the production index of blast furnace. The smelting practice shows that the key to smelting vanadium-titanium ore is to maintain the appropriate furnace temperature, large air volume operation, blowing through the center, properly loose edge, active hearth, in order to make the blast furnace stable and smooth for a long time, as well as the change of personnel ideology.

Key words: blast furnace; V-Ti ore smelting; smooth; iron slag mobility

八钢高炉高锌负荷操作

张文庆，高小雷

（宝钢集团八钢公司，新疆乌鲁木齐 830022）

摘 要：对高炉入炉锌负荷及内部循环的情况、变化趋势进行监控，制定原燃料中锌带入量目标，同时同时通过调整高炉操作，制定高炉排锌措施，减少炉况波动。

关键词：锌富集；排锌；应用

八钢焦化 VOC 气体治理新技术的浅析

杜生强

（宝钢集团新疆八钢炼铁分公司，新疆乌鲁木齐 830022）

摘 要：本文充分结合钢铁联合企业的特性，在 VOC 化工尾气安全、可靠、有效的基础上，通过风机与管道将 VOC 气体输送至 180t 锅炉，进行高温燃烧处理。新技术安全、高效、投资省，为化工尾气治理技术打开了新的篇章。

关键词：VOC 气体；锅炉燃烧；新技术

Analysis on the New Technology of Coking VOC Gas Treatment in Bayi Iron and Steel Co.

Du Shengqiang

(Ironmaking Branch, Bayi Iron & Steel Co., Baosteel Group, Urumchi 830022, China)

Abstract: This paper fully combines the characteristics of steel joint ventures. On the basis of safe, reliable and effective collection of VOC chemical exhaust gas, VOC gas is transported to a 180t boiler for high-temperature combustion treatment through fans and pipelines. The new technology is safe, efficient and investment-saving, opening a new chapter for chemical tail gas treatment technology.

Key words: VOC gas; boiler burning; new technology

八钢欧冶炉配加高锌烧结矿生产实践

李维浩

（宝武集团八钢公司炼铁厂，新疆乌鲁木齐 830022）

摘 要： 通过对欧冶炉配加高锌烧结矿处置含锌废弃物的机理进行理论分析，锌元素在欧冶炉内具备循环富集条件，并主要以锌元素进入污泥方式排出，占比达到入炉锌含量的96%以上。八钢欧冶炉配加高锌烧结矿处置含锌废弃物，符合实现无害化、减量化和资源化的原则，能有效克服含锌废弃物堆储对环境产生的污染问题。通过欧冶炉配加高锌烧结矿生产性工业试验，其工艺流程能满足处置含锌废弃物生产组织和操作过程需要，不需要增加新的辅助处理设施，具有可处理冶金行业多种形态含锌废弃物的特点。

关键词： 欧冶炉；高新烧结矿；含锌废弃物；生产实践

八钢焦化干熄焦红焦装入系统技术改进

刘明钊

（宝武集团八钢公司炼铁分公司新区焦化焦炉作业区，新疆乌鲁木齐 830022）

摘 要： 本文介绍红焦装入系统在八钢新区焦化的应用情况，对红焦装入系统焦罐旋转不到位，电机车撞坏APS夹壁，电机车撞坏提升机吊钩，限位检测元件的干扰，电动缸限位存在的缺陷，料斗衬板结构和材质改进等情况进行了详细的分析，并结合实际情况采取有效的措施进行改进，经验证取得了良好的效果，大大增强了焦炉红焦装入焦罐系统运行过程中安全性能，提高了新区焦化的生产效率。

关键词： 旋转焦罐；载码体限位；程序控制图

非高炉炼铁技术的发展现状

田宝山

(宝钢集团新疆八一钢铁有限公司炼铁分公司，新疆乌鲁木齐 830022)

摘 要：高炉法炼铁对于世界经济与钢铁工业的发展以及人类的文明进步做出了极大的贡献，但高炉炼铁一定要使用焦炭，但焦煤资源非常有限；且随着高炉炉容的逐渐扩大，对于原料的指标要求也越来越高。在烧结、焦化厂生产的过程中产生的废水、废气含有酚氰、SO_2、NO_x、CO_2 等有害的物质，污染非常严重；高炉的炼铁工艺面临的巨大挑战促进了非高炉炼铁工艺的研发。为了克服高炉炼铁的种种缺点，人们研究开发了多种非高炉炼铁法。尽可能地不用或少用焦炭，使用块矿甚至粉矿，减少了较大污染源。为实现钢铁厂清洁生产、减少环境污染创造了条件。由于上述优势，近年来，非高炉炼铁工艺发展迅猛。

关键词：Corex；Finex；HIsmelt；欧冶炉

Development Status of Non-blast Furnace Ironmaking Technology

Tian Baoshan

(Baosteel Xinjiang Bayi Iron and Steel Co., Ltd., Urumchi 830022, China)

Abstract: Blast furnace iron making has made great contribution to the world economy, the development of iron and steel industry and the progress of human civilization. With the expansion of blast furnace capacity, the requirement of raw material is higher and higher. The waste water and waste gas produced in the process of sintering and coking plant contain phenolic cyanogen, SO_2, NO_x, CO_2 and other harmful substances, resulting in serious pollution. The great challenge of blast furnace ironmaking technology promotes the research and development of non-blast furnace ironmaking technology. In order to overcome the disadvantages of blast furnace iron-making, many non-blast furnace iron-making methods have been developed. Coke should be used as little or as little as possible, and lumps or even powders should be used to reduce major pollution sources. It creates conditions for realizing clean production and reducing environmental pollution in steel works. Due to the above advantages, in recent years, non-blast furnace ironmaking technology has developed rapidly.

Key words: Corex; Finex; HIsmelt; OY

八钢 B 高炉炉缸侧壁温度升高的原因分析及治理实践

舒 艺，刘晓勇

(宝武集团八钢股份公司炼铁厂，新疆乌鲁木齐 830022)

摘　要： 介绍了 B 高炉炉缸炉底结构的设计特点，结合炉缸实测数据，计算了炉缸炉底 1150℃ 等温线的分布情况，分析了炉缸侧壁温度异常升高的原因，并提供改进措施，取得了良好的效果。

关键词： 高炉；炉缸侧壁；侵蚀检测；控制措施

Abstract: The design features of hearth and hearth bottom structure of B BF are introduced. Based on the measured hearth data, the distribution of 1150℃ isotherm of hearth and hearth bottom is calculated. The causes of abnormal increase of hearth sidewall temperature are analyzed, and the improvement measures are provided, and good results are obtained.

Key words: blast furnace; hearth side wall; erosion detection; control measures

斜脱硫熔硫釜清液冷却器的设计与应用

戴　杰

（宝钢集团新疆八钢炼铁分公司，新疆乌鲁木齐　830022）

摘　要： 根据湿法脱硫中使用熔硫釜造成脱硫循环液温度高的原因进行冷却器的设计。针对熔硫釜产生的清液特点进行了冷却器的设计，并通过现场的应用提出了改进意见。

关键词： 湿法脱硫；熔硫釜；冷却器

Design and Application of Clean Liquid Cooler for Inclined Desulfurization Fusion Tank

Dai Jie

(Baosteel Group Xinjiang Bayi Iron Branch, Urumchi 830022, China)

Abstract: The cooler was designed according to the high temperature of desulfurization circulating liquid caused by the use of molten sulfur kettle in wet desulfurization. According to the characteristics of the clear liquid produced by the sulphur melting kettle, the cooler was designed, and some suggestions for improvement were put forward through field application.

Key words: wet desulfurization; sulfur melting kettle; cooler

基于"一带一路"沿线烧结烟气脱硫脱硝对策探讨

冶　飞，李维浩

（宝钢集团八一股份炼铁厂，新疆乌鲁木齐　830022）

摘　要： 基于"一带一路"沿线国家可能采用的烧结污染物排放标准，针对钢铁项目全球采购铁矿原料的构成状况，对铁矿原料含硫组分进行了具体分析，并对建设的烧结机头烟气污染物平均和最大浓度进行了对比分析，结合沿海和东南亚主要烧结厂烟气污染物浓度情况，及对目前主流烧结烟气脱硫脱硝工艺特点对比，归纳整理得出在"一带一路"沿线钢铁项目建设的烧结烟气污染物治理的可行性建议。

关键词："一带一路"；烧结机头烟气；污染物排放；烧结机烟气脱硫脱硝

Discussed for Countermeasures of Desulfurization and Denitration in Sintering Plant Based on One Belt and One Road Along the Steel Project Construction

Ye Fei, Li Weihao

(Iron of Bayi Iron & Steel Co., Baosteel Group, Urumchi 830022, China)

Abstract: Based on the possible emission standards of sintering pollutants along One Belt and One Road line, raw material composition of global iron ore procurement for steel projects, the sulfur content of sintering raw material was analyzed in detail, the average and maximum concentration of flue gas pollutants in the sintering head were compared and analyzed, combined with coastal and southeast Asia main sintering plant gas pollutant concentration, and compared with the current mainstream sintering flue gas desulfurization and denitrification process, it's put forward feasibility suggestion of controlling sintering gas pollution in One Belt and One Road steel project.

Key words: the Belt One Road; sintering head flue gas; pollutant discharge; desulfurization and denitration of sintering flue gas

炉前开口机旋转拖链优化设计

李海云

（宝钢集团新疆八钢股份公司炼铁厂烧结分厂，新疆乌鲁木齐　830022）

摘　要：文章以欧冶炉开口机为例，介绍旋转拖链在竖直方向布置的技术改造过程、重点对结构设计进行阐述。

关键词：旋转拖链；结构

欧冶炉绿色炼铁生产实践

陈若平

（宝钢集团八钢公司，新疆乌鲁木齐　830022）

摘　要：本文从八钢欧冶炉的工艺条件开展理论分析其拱顶处置废弃物的优势，并通过生产实践验证了欧冶炉在钢铁企业协同发展的作用，开辟出一条新的绿色炼铁之路。

关键词：欧冶炉；绿色炼铁

Production Practice of Green Ironmaking in the OY Smelter

Chen Ruoping

(Baosteel Group Xinjiang Bayi Iron and Steel Co., Urumchi 830022, China)

Abstract: In this paper, theoretically analyses the advantages of its vault for waste disposal based on the technological conditions of Bayi OY, and verifies the synergistic development of the smelting furnace in iron and steel enterprises through production practice, thus opening up a new green ironmaking road.

Key words: OY; green iron smelting

欧冶炉开炉降硅降焦比冶炼实践对比分析

田宝山，陈若平

（宝武集团八钢股份公司炼铁厂，新疆乌鲁木齐 830022）

摘 要：八钢欧冶炉3月7日11:58成功点火开炉，3月8日凌晨8:15第一炉顺利出铁。开炉快速降硅、降焦比，产能得以快速提升，各项指标较2018年开炉生产都有很大进步。

关键词：欧冶炉；金属化率；焦比；低硅

欧冶炉冷填洗涤器液位计设计及操作实践

季书民

（宝钢集团八钢公司炼铁分公司，新疆乌鲁木齐 830022）

摘 要：直观液位计，来观测水位。达到阻断煤气的作用。水位稳定，有效地封堵气化炉的煤气。

关键词：冷填洗涤器液位计装置；封堵煤气

Design and Operation Practice of Liquid Level Meter for Cold Fill Scrubber in Europe Smelter in the Bayisteel OY

Ji Shumin

(Baosteel Group Bayi Corporation Ironmaking Branch, Urumchi 830022, China)

Abstract: Intuitive level gauge to observe water level. To achieve the effect of blocking gas. The water level is stable, effectively blocking the gas in the gasifier.

Key words: the liquid level gauge device of the cold filling scrubber; blocks gas

浅析焦炉集气管压力的几种调节手段

占国保

(宝钢集团新疆八一钢铁有限公司炼铁分公司焦化分厂,新疆乌鲁木齐 830022)

摘 要：焦炉炼焦工艺中,集气管压力是一个非常重要的控制参数,其控制的稳定,对于改善环境,保证安全、节约能源、提高焦炭质量、延长焦炉寿命等方面具有非常重要的意义。本文主要对集气管压力的调节手段进行了归纳和简要分析。

关键词：集气管压力；调节控制；静态特性曲线

Analysis of Coke Oven Gas Pressure of Adjusting Means

Zhan Guobao

(Baosteel Group Xinjiang Bayi Iron and Steel Company Limited, Iron Smelting Branch, Urumchi 830022, China)

Abstract: Coking process, the collecting pipe pressure is a very important control parameters, its control stability, on the improvement of the environment、ensure safety、energy saving、improving the quality of coke、prolong the service life of coke oven etc. has very important significance. This paper focuses on the gas collector pressure adjustment method are summarized and analyzed briefly.

Key words: gas collector pressure; regulation and control; static characteristic curve

烧结机尾余热锅炉系统改造现状及趋势研究

马 猛

(宝钢集团八钢公司炼铁分公司,新疆乌鲁木齐 830022)

摘 要：分析了烧结工序中可回收利用的余热资源及其特性,在此基础上提出了烧结机尾烟气与冷却废气余热联合回收发电技术,并分析和研究了其技术优势和瓶颈,提出了烧结余热发电系统设计的一些建议。

关键词：烧结；烟气；冷却废气；余热联合回收；发电

烧结系统提高工业废水用量降低工序成本实践

谭哲湘,臧疆文,彭 丹

(宝武集团八钢股份炼铁厂烧结分厂,新疆乌鲁木齐 830022)

摘　要：分析钢铁工业废水的主要来源，探索在烧结系统提高工业废水用量的途径，达到了节约新水，回收铁元素，废水零排放，降低工序成本的效果。

关键词：烧结；废水用量；实践

Practice of Increasing Industrial Wastewater Consumption and Reducing Process Cost in Sintering System

Tan Zhexiang, Zang Jiangwen, Peng Dan

(Ironmaking Branch, Bayi Iron & Steel Co., Baowusteel Group, Urumchi 830022, China)

Abstract: Analysis of the main source of iron and steel industry waste water, in the use of large sintering machine and other processing methods were compared.eal late; sintering; advantages Recycling iron the effect of wastewater zero discharge.

Key words: sintering; wastewater consumption; advantage

一种新型振动式清扫器在生产实践中的运用

荆　斌，张忠德

（宝钢集团八钢公司炼铁厂，新疆乌鲁木齐　830022）

摘　要：本文对一种新型清扫器做了详细的说明，其清扫原理是通过摩擦及拍打皮带将皮带上粘的物料清扫下来，该清扫器结构简单，易制作，易安装，易维修不损皮带，清扫效果明显。

关键词：胶带机；振动清扫器

鞍钢炼铁总厂11号高炉3#热风炉长寿攻关

张荣军[1]，徐吉林[2]，王　志[1]

（1. 鞍钢股份炼铁总厂，辽宁鞍山　114000；
2. 鞍钢集团工程技术有限公司，辽宁鞍山　114000）

摘　要：鞍钢11号高炉3#热风炉为内燃式热风炉，其高温区扩张段局部（燃烧室侧）温度可达300℃以上，现场已在炉壳外部增设水冷箱控制，当前风温控制在1100~1150℃。燃烧器、拱顶等关键部位日常检修不可控因素多。在吸取国内外长寿热风炉成功经验的基础上，鞍钢股份炼铁总厂与鞍钢集团工程技术有限公司开展内燃式热风炉的长寿技术攻关，开发适合鞍钢生产操作高温长寿内燃式热风炉，以满足鞍钢未来的检修需要。

关键词：热风炉；国产；内燃式；长寿；设计

Long Life Tackling of No.3 Hot Blast Furnace of No.11 Blast Furnace in Angang Iron and Steel General Plant

Zhang Rongjun[1], Xu Jilin[2], Wang Zhi[1]

(1. General Iron Plant, Angang Co., Ltd., Anshan 114000, China;
2. Anshan Group Engineering Technology Co., Ltd., Anshan 114000, China)

Abstract: No. 3 hot blast stove of No. 11 BF in Angang is an internal combustion hot blast stove. The local temperature of expansion section in high temperature zone (combustion chamber side) can reach more than 300℃. Water-cooled tank has been added outside the furnace shell to control the temperature. At present, the air temperature is controlled at 1100~1150℃. There are many uncontrollable factors in daily maintenance of key parts such as burners and vaults. Based on the successful experience of longevity hot blast stoves at home and abroad, Angang Iron and Steel Co., Ltd. and Angang Group Engineering Technology Co., Ltd. have carried out technical research on longevity of internal combustion hot blast stoves, and developed high temperature and longevity internal combustion hot blast stoves suitable for Angang production and operation, in order to meet Angang's future maintenance needs.

Key words: hot blast stove; domestic; internal combustion; long life; design

高炉控料线停炉操作实践

王彩永，任艳军，何少松，冯凯民，袁兆锋，金亚丽

（河钢集团石钢公司炼铁厂，河北石家庄 050031）

摘 要： 文章对2018年11月30日河钢石钢0号高炉（580m³）控料线停炉工作进行了总结。通过停炉前制定周密工作计划，采用回收煤气、炉顶打水降温控料线将料面降至炉腹区域，在控料线过程中通过对各项参数的合理控制，减少了煤气外排时间，实现了安全、环保、快速降料面停炉。

关键词： 高炉；控料线；降料面；休风

Operation Practice of Cutting off the Charging Line of Blast Furnace

Wang Caiyong, Ren Yanjun, He Shaosong, Feng Kaimin, Yuan Zhaofeng, Jin Yali

(Ironmaking Plate of HBIS Group Shisteel Company, Shijiazhuang 050031, China)

Abstract: This paper summarizes the work of stopping the charging line of no. 0 bf (580m³) at Hegang Shigang on November 30, 2018. By making a careful work plan before stopping the furnace, the material surface is lowered to the belly of the furnace by adopting the recovery gas and the temperature drop control line of the top of the furnace.

Key words: blast furnace; control stockline; drop charge level; damping down

湘钢降低球团矿膨润土用量研究

汤 勇，刘敏媛，周云花，侯盼盼，严永红

（华菱湘钢技术质量部，湖南湘潭 411101）

摘 要：2017年以来以膨润土为黏结剂制备球团矿时，膨润土配比高达3.80%；膨润土质量较差，并且各厂家之间波动较大。为了降低膨润土用量，为此开展精矿和膨润土的造球和焙烧实验，通过实验得出：适宜的膨润土配比，和兴精为3.5%、巴西精为7.5%、澳洲精为2.0%；因和兴精成品球的强度比澳洲精高1000N以上，并且铁水成本较低，因此应采购一部分来降成本和膨润土配比；湘钢目前的膨润土供应商有6家，为保供应一般三家就可，因此至少应淘汰H和B；多数情况下膨润土的性能检测指标能反映膨润土的造球性能，但也有例外，如C膨润。因此，通过造球和焙烧实验来选择更好。

关键词：精矿；膨润土；造球；焙烧

Experimental Study on Minimizing Bentonite Mixture Proportion of Pelletizing in Xiangtan Iron & Steel Group

Tang Yong, Liu Minyuan, Zhou Yunhua, Hou Panpan, Yan Yonghong

(Technical & Quality Department of Valin Xiangtan Steel, Xiangtan 411101, China)

Abstract: Since the pelleting was done using bentonite as a binder in 2017, the bentonite proportion has been mixed as high as 3.80%; in this case, not only the bentonite quality is poor, but also its supply fluctuates very much. In order to minimize the bentonite proportion in final pellets, the test was done on the pelletizing and calcining of iron ore concentrates with bentonite mixtures. The test results showed that the proper bentonite proportion is 3.5% in Hexing concentrates, 7.5% in Brazil concentrates and 2.0% in Australia concentrates. Only part of Hexing concentrates were bought to minimize the pelting costs and bentonite mixture proportion because the strength of finished pellets of Hexing concentrates are 1000N higher than that of Australian concentrates and the costs of ironmaking are also lowered. There are 6 bentonite suppliers to XISC in total, and only three bentonite suppliers are demanded at most, therefore, H and B should be eliminated at least. In most cases, the test performance indexes of bentonite reflect the pelletizing performance of bentonite, but there are some exceptions, for instance, C bentonite. Therefore, it is better to choose the mixtures through experiments of pelletizing and calcining.

Key words: iron ore concentrates; bentonite; pelletizing; calcining

某非洲粉矿烧结性能研究

汤 勇，刘敏媛，周云花

（华菱湘钢技术质量部，湖南湘潭 411101）

摘　要：通过对某非洲粉矿（以下简称 LB 矿）烧结工艺及成品烧结矿冶化性能进行研究。单烧实验研究结果表明：LB 矿烧结矿质量指标次于南非粉，与 PB 粉相差不大，优于国王粉和混合粉。其烧结矿垂直速度为 23.06 mm/min，转鼓强度为 63.60%，燃耗为 66.16 kg/t，利用系数 1.832 t/($m^2 \cdot h$)。LB 矿单烧时的烧结矿产质量和成分都处于中等偏上。混烧实验研究结果表明：用 20%以内 LB 矿替代国王粉时烧结矿质量指标基本无变化；用 20%以内 LB 矿替代南非粉时烧结矿利用系数下降，但品位提高 0.31%，SiO_2 降低 0.11%。用 20%以内 LB 矿替代混合粉时可提高转鼓强度，品位提高 0.42%，SiO_2 降低 0.16%。LB 矿的低温还原粉化研究表明：其与南非粉和国王粉相当，较 PB 和混合粉好。因此，可用 LB 矿代国王粉和混合粉，并且能够替代其他铁矿粉用于烧结。

关键词：烧结；冶金性能；低温粉化；铝硅比

Study on Sintering Properties of Afican Iron Ore Fines

Tang Yong, Liu Minyuan, Zhou Yunhua

(Technology & Quality Department of Valin Xiangtan Steel, Xiangtan 411101, China)

Abstract: Both the sintering technology of Afican iron ore fines and the metallurgical chemical properties of their final sintered products are studied in this paper. The results of single ore fine-sintering experiment shows that the quality indexes of sintered LB iron ore fines are slightly lower than that of South Africa's iron ore fines and do not differ much from PB iron ore fines, but better than GUOWANG ore fines and mixed ore fines. The vertical rate of sintering is 23.06mm/min, the tumbler strength is 63.60%, the fuel consumption is 66.16kg/t, and the utilization coefficient is 1.832t/($m^2 \cdot h$). The quality for sintering and compositions of LB iron ore fines are in better or middle quality for single-ore fines sintering. The result of mixtures-sintering experiment shows that the sinter quality indexes keep basically stable when the LB ore fines below 20% substitute GUOWANG ore fines for sintering; When LB ore fines below 20% usage substitute the South African ore fines, the utilization coefficient of sintered ores decreases, but the grade increases by 0.31% and SiO_2 decreases by 0.11%. When LB ore fines below 20% substitute the mixed ores, the tumbler strength can be improved and sintering grade can be increased by 0.42% and SiO_2 can be reduced by 0.16%. The study on the low-temperature reduction pulverization of LB ore fines shows that the quality of such ore fines is comparable with South African ore fines and GUAOWANG ore fines and better than PB-type and mixed-type ore fines. Therefore, LB ore fines can be used as GUOWANG type ore fines and mixed ore fines for sintering, and can substitute other iron ore fines for sintering.

Key words: sintering; metallurgical properties; low temperature pulverization; aluminum-silicon ratio

大型高炉中心漏斗和边缘平台实验研究

赵华涛[1]，杜　屏[1]，卢　瑜[1]，张明星[1]，朱　华[2]

（1. 沙钢集团钢铁研究院铁前研究室，江苏张家港　215625；
2. 沙钢集团有限公司炼铁厂，江苏张家港　215625）

摘　要：利用冷布料实验模型研究了沙钢 5800 高炉不同时期典型布料制度的料层分布特点，结合实际高炉生产情况分析了布料平台对边缘气流稳定性的影响、矿焦比分布对高炉压差的影响，以及实际高炉炉缸工作情况对漏斗大小的影响规律。并利用冷布料实验设备对当前布料制度进行了优化实验，既保持较大的中心漏斗，维持高炉透气性，又防止边缘负荷过重，造成根部肥大。

关键词：中心漏斗；布料平台；矿焦比

Experimental Research of Burden Layer Distribution in Large Blast Furnace

Zhao Huatao[1], Du Ping[1], Lu Yu[1], Zhang Mingxing[1], Zhu Hua[2]

(1. Ironmaking Research Group, Institute of Research of Iron & Steel, Shasteel, Zhangjiagang 215625, China; 2. Ironmaking Plant of Shasteel, Zhangjiagang 215625, China)

Abstract: The burden layer distribution at different typical stages of 5800m^3 blast furnace in Shasteel was studied utilizing the cold charging model. The experimental results together with the actual operation results were utilized to analyze the effect of terrace length on the peripheral gas flow stability, radial O/C distribution of burden layer on the blast furnace overall pressure drop, and the hearth activity on the size of central funnel. And the current charging pattern was optimized by the cold charging model experiment, keeping a relative large funnel in the center and avoiding too big O/C which may result in big cohesive zone root.

Key words: burden layer; funnel terrace; O/C

输送带硫化工艺的优化及创新

陈小波

(河钢集团舞阳钢铁有限责任公司检修厂，河南舞钢　462500)

摘　要： 输送带在炼铁厂的物料传送中起到至关重要的作用，通过更换输送带热硫化接头的工艺特点和日常运行中输送带热硫化接头出现的问题，对硫化工艺进行了科学、合理的创新，并通过现场实践，探索出输送带设备实现高效、环保顺行的新模式。

关键词： 输送带；热硫化；工艺优化及创新；热硫化修补

水平管中高密相煤粉气力输送的模拟研究

代婧鑫[1]，丁学勇[1]，徐海法[2]，刘程宏[1]，唐昭辉[1]，董　越[1]

(1. 东北大学冶金学院，辽宁沈阳　110819；2. 宝钢中央研究院，上海　201900)

摘　要： 针对以氮气为输送介质的水平管高密相煤粉气力输送，基于颗粒动力学理论与摩擦应力模型，结合 Huilin 曳力模型构建了水平管高密相气力输送数值模型，并使用该模型进行模拟计算。模拟结果表明：随着表观气速的增加，气相压损和总压损均呈上升趋势，但气相压损在总压损中所占比例十分小；每条流动特性曲线上都会出现压降相对最低点，该点所对应的输送气速就是最经济速度；在设计输送时，要根据压降-气速流态图确定最经济速度，实际输送气速要稍大于最经济速度，以降低输送能耗，保证输送系统的稳定性。

关键词： 氮气；高密相；水平管；气力输送；数值模拟

Numerical Simulation on Pneumatic Conveying of Dense Phase Pulverized Coal in Horizontal Pipe

Dai Jingxin[1], Ding Xueyong[1], Xu Haifa[2], Liu Chenghong[1], Tang Zhaohui[1], Dong Yue[1]

(1. School of Metallurgy, Northeastern University, Shenyang 110819, China;
2. Baosteel Research Institute, Shanghai 201900, China)

Abstract: Aimed at pneumatic conveying of dense phase pulverized coal in horizontal, based on the kinetic theory of granular flows incorporating friction stress model for solids stress, Huilin drag model for the interaction between gas and particles. Therefore, a two-phase flow model was established for simulating on pneumatic conveying of dense phase pulverized coal in horizontal. The simulation results show that with the increase of superficial velocity, both gas phase pressure drop and total pressure drop show an upward trend, but gas phase pressure drop accounts for a small proportion of total pressure drop. The relative minimum point of pressure drop appears on each flow characteristic curve. The relative minimum point of pressure drop appears on each flow characteristic curve, and the superficial velocity corresponding to this point is the most economical speed. When designing, the most economical speed should be determined according to the pressure drop-gas velocity flow diagram. The actual superficial velocity should be slightly larger than the most economical speed to reduce the transportation energy consumption and ensure the stability of the conveying system.

Key words: nitrogen; dense phase; horizontal pipe; pneumatic conveying; numerical simulation

宝钢炼焦环保技术现状及探讨

余国普，张福行

（宝山钢铁股份有限公司炼铁厂，上海 200941）

摘 要： 炼焦是高消耗、高排放的生产单元，具有巨大的减排潜力。本文简要介绍了宝钢宝山基地炼焦现状，结合"城市钢厂"和国家超低排放要求，近年来宝山基地在焦炉环保治理上采用一系列技术措施，包括全密封装煤车、炭化室压力均压技术、焦炉机侧和焦侧烟尘治理、不同工艺的烟气净化和炉门、上升管等炉体耐材技术改进的应用，基本达到了有效抑制焦炉无组织排放和降低污染物排放浓度的目的。最后对其他排放控制进行了论述与展望。

关键词： 炼焦；焦炉；环保；装煤车；炭化室；均压

The Present State of Coking Environmental Protection Techniques and Discussion on Baosteel

Yu Guopu, Zhang Fuhang

(Iron-making Plant, Baoshan Iron & Steel Co., Ltd., Shanghai 200941, China)

Abstract: Coking is a high-consumption, high-emission production unit with huge potential for emission reduction. This paper briefly introduces the present situation of coking in baosteel, combined with the "Urban Steel Plant" and the national ultra-low emission requirements. In recent years, a series of technical measures have been adopted in the environmental protection of coke ovens in Baoshan base, including the pressure-tightening of fully sealed coal loading vehicles and carbonization chambers. And the application of new technologies such as monotonous technology, coke oven side and coke side soot treatment, flue gas purification of different processes and furnace refractory upper furnace door and riser tube have basically achieved effective suppression of coke oven unorganized emission and pollution reduction. The purpose of the concentration of the substance. Finally, other emission control is discussed and prospected.

Key words: coking; coke oven; environmental protection; coal loading vehicle; carbonization chamber; pressure equalization

烧结机机尾除尘超低排放关键技术及工程应用

臧疆文

（新疆八一钢铁股份有限公司炼铁厂，新疆乌鲁木齐 830022）

摘 要：本工程将烧结机机尾电除尘器改为电袋复合除尘器，针对机尾除尘系统的特点，通过对技术路线的选择、除尘器结构、净气室结构的选择、气流分布、过滤风速、滤袋选择、气流上升速度、烟气温度控制、清灰方式等关键技术的研究，实现了机尾除尘器的超低排放，改造1年多来，除尘器稳定运行，出口含尘浓度小于 $5mg/Nm^3$。

关键词：机尾除尘；超低排放；电袋复合除尘器；滤袋

Key Technology and Engineering Application of Ultra-low Emission of Sintering Machine Tail Dust Removal

Zang Jiangwen

(Ironmaking Plant of Xinjiang Bayi Iron and Steel Co., Ltd., Urumchi 830022, China)

Abstract: In this project, the electrostatic precipitator at the tail of sintering machine is replaced by the electrostatic bag compound precipitator. According to the characteristics of tail dust removal system. The key technology such as the selection of technical route, the structure of dust collector, the structure of air purification chamber, the distribution of air flow, the filtration air speed, the selection of filter bag, the rising speed of air flow, the control of flue gas temperature and the way of ash cleaning, etc was studied. The ultra-low emission of the tail dust collector was realized. It has been renovated for more than one year. Dust collectors run stably. The dust concentration at the outlet is less than $5mg/Nm^3$.

Key words: dedusting of sintering machine tail; ultra-low emission; electric bag composite precipitator; filter bag

烧结机柔性传动持续周期地摆动原因分析

马 云

（宝武集团新疆八一钢铁有限公司炼铁分公司烧结分厂，新疆乌鲁木齐 830022）

摘　要：通过对八钢 430 烧结机运行现状分析以及现场发生的事故原因分析，利用机械动力学理论研究方法，分析得出 430 烧结机传动链轮齿板磨损后，齿板与烧结机台车辊轮之间产生周期性负载力矩，使得柔性传动在周期变化的外负载作用下发生周期摆动现象，是 430 烧结机柔性传动装置持续周期摆动主要原因。

关键词：烧结机；柔性传动；周期摆动；链轮齿板；辊轮

某尾矿库膏体尾矿固结性研究

毛新福，刘　威，吴　迪，李　新

（中冶沈勘工程技术有限公司地质环境所，辽宁沈阳　110169）

摘　要：目前，膏体尾矿排放工艺在国内矿山企业应用较少，但作为一种节能、环保的生产工艺，必然是矿山企业未来发展的方向。某大型矿山是较早采用尾矿膏体排放的企业，目前正处于生产初期，对于膏体尾矿的固结性质亟待研究。本次研究主要是对膏体尾矿现有堆积体采用三种试验方法，获取大量参数，经过统计、计算、分析，最终总结出膏体尾矿固结性质的结论，为该矿山企业及设计单位解决了多项难题。

关键词：尾矿库；膏体尾矿；静力触探；固结系数；先期固结压力

Study on Consolidation of Paste Tailings in One Tailings Pond

Mao Xinfu, Liu Wei, Wu Di, Li Xin

(Shen Kan Engineering & Technology Corporation MCC, Shenyang 110169, China)

Abstract: At present, paste tailings discharge technology is seldom used in domestic mining enterprises, but as an energy-saving and environmental protection production technology, it is bound to be the future development direction of mining enterprises. A large mine is an enterprise that used tailings paste discharge earlier. It is in the early stage of production. The consolidation property of paste tailings needs to be studied urgently. This research mainly adopts three kinds of test methods for the existing accumulation of paste tailings, and obtains a large number of parameters. After statistics, calculation and analysis, the conclusion of consolidation properties of paste tailings is finally summarized, which solves many difficult problems for the mine enterprises and design units.

Key words: tailings pond; paste tailings; cone penetration test; consolidation coefficient; pre consolidation pressure

炼铁创新引领和高质量发展

刘文权[1]，宋文刚[2]，吴记全[3]

（1. 冶金工业规划研究院，北京　100013；2. 宝山钢铁股份有限公司，上海　201900；
3. 方大特钢科技股份有限公司，江西南昌　330012）

摘　要：本文从减量调整、创新发展、降本增效、标准引领、智慧制造、低碳和氢冶金等方面论述了炼铁转型升级和创新发展，实现炼铁高质量发展的途径，并指出了炼铁创新发展的方向:烧结烟气循环工艺、强力混合机在烧结机中的应用、烧结竖罐冷却和余热发电技术、高比例球团矿生产和应用、高炉冲渣水余热回收利用技术、煤气变压吸附技术等。

关键词：智慧制造；低碳；高质量

Ironmaking Innovation Leads to High Quality Development

Liu Wenquan[1], Song Wengang[2], Wu Jiquan[3]

(1. MPI, Beijing 100013, China; 2. Baosteel Co., Ltd., Shanghai 201900, China;
3. Fangda Special Steel Technology Co., Ltd., Nanchang 330012, China)

Abstract: Staring from the adjustment, innovation development, cost decreasing and benefit increasing, standard lead, Intelligent manufacturing, low carbon and hydrogen metallurgy etc aspects discusses the innovation and development, and the transformation and upgrading of ironmaking implementation way of ironmaking high quality development, and points out the developing direction of ironmaking innovation: sintering flue gas recirculation technology, strong mixing machine in the application of sintering machine, cooling and waste heat power generation technology, high percentage pellet production and application, blast furnace slag water waste heat recycling technology, the gas PSA(Pressure Swing Adsorption), etc.

Key words: intelligent manufacturing; low carbon; high quality

欧冶炉全氧风口喷吹顶煤气的工业实践

田 果

（宝钢集团八钢公司股份炼铁厂，新疆乌鲁木齐 830022）

摘 要：顶煤气回喷在炼铁工艺创新中是研究较多的项目。欧冶炉作为非高炉炼铁的代表工艺，进行全氧风口喷吹顶煤气的工业试验具有重大意义，首先实践C的减排工艺，一种新的煤气利用方式促进炼铁工艺技术的发展。

关键词：全氧风口；煤气喷吹；非高炉炼铁

Industrial Practice of Injecting Top Gas into Total Oxygen Tuyere of European Smelting Furnace

Tian Guo

(Baosteel Group Bayi Iorn Works Co., Ltd., Urumchi 830022, China)

Abstract: Top gas back injection is a research subject in ironmaking process innovation. As a representative process of nonblast furnace ironmaking, it is of great significance to carry out industrial experiments of injecting top gas into all-oxygen tuyeres. Firstly carbon emission reduction process practice is carried out, a new way of gas utilization promotes the development of ironmaking technology.

Key words: all-oxygen tuyere; gas injection; non-blast furnace ironmaking

电感耦合等离子体发射光谱法测定硅铁中的硼

王聪磊

（河钢集团石钢公司技术中心品质部，河北石家庄 050031）

摘　要： 应用电感耦合等离子体发射光谱方法测定了硅铁中硼，选择分析谱线和测量的最佳酸度条件，配制合适浓度标准溶液制作工作曲线，无需基体匹配，对共存元素的影响进行讨论，进行准确度、精密度、检出限测定，结果令人满意。

关键词： 电感耦合等离子体发射光谱法；硅铁中硼；工作曲线

鞍钢集团朝阳钢铁增建 4#热风炉工程设计实践

徐吉林[1]，杜春松[1]，王光伟[2]，李勇勤[2]，敖永和[2]，张式宝[2]

（1. 鞍钢集团工程技术有限公司，辽宁鞍山　114000；
2. 鞍钢集团朝阳钢铁分公司，辽宁朝阳　122000）

摘　要： 对鞍钢集团朝阳钢铁公司增建一座（4#）热风炉工程方案实施进行了总结。朝阳钢铁公司炼铁厂原配置3座国产内燃式热风炉（1#、2#、3#），经过多年运行，3座热风炉均出现了不同程度的损坏，其中2#热风炉尤为严重，由于朝阳钢铁公司仅一座高炉，此情况严重威胁了整个企业的生产运行。增建一座（4#）热风炉势在必行，待4#热风炉建成后，再逐一对原热风炉进行大修。4#热风炉已于2017年底投入运行，工程建设期间高炉正常生产，投运后风温1200℃，各部运行状态良好。2018年底完成了2#热风炉的在线大修工作，实现了增建4#热风炉的设计目标。

关键词： 内燃式热风炉；在线大修；设计目标

Engineering Design Practice of Additional (No.4) Hot Blast Stove Project in Chaoyang Iron and Steel Company of Ansteel

Xu Jilin[1], Du Chunsong[1], Wang Guangwei[2], Li Yongqin[2],
Ao Yonghe[2], Zhang Shibao[2]

(1. Ansteel Engineering Technology Co., Ltd., Anshan 114000, China;
2. Chaoyang Iron & Steel Co. of Ansteel, Chaoyang 122000, China)

Abstract: The implementation of an additional (No.4) hot blast stove project in Chaoyang Iron and Steel Company of Ansteel is summarized. Three domestic internal combustion hot stoves (No.1, No.2, No.3) were originally equipped in Chaoyang Iron and Steel Company's ironmaking plant. After years of operation, three hot stoves were damaged to varying degrees. Among them, the No.2 hot stove is especially serious. Because Chaoyang Iron and Steel Company only has only one blast furnace, this situation seriously threatens the production and operation of the whole enterprise. It is imperative to build an additional (No.4) hot blast stove. After the No.4 hot blast stove is completed, the original hot blast stoves will be overhauled one by one. No.4 Hot Blast Furnace has been put into operation since the end of 2017. During the construction period, the blast furnace is in normal production. The blast temperature is 1200℃ after its operation. By the end of 2018, the online overhaul of No. 2 hot blast stove was completed, and the design target of adding No. 4 hot blast stove was realized.

Key words: internal combustion stove; on-line overhaul; design goal

硅铁还原回收低品位钒钛磁铁矿有价元素研究

高子先[1]，程功金[1,2]，薛向欣[1,2]

（1. 东北大学冶金学院，辽宁沈阳　110819；
2. 辽宁省冶金资源循环科学重点实验室，辽宁沈阳　110819）

摘　要：由于该低品位钒钛矿具有钛含量高的特点，在高炉中比较难冶炼。本实验采用非高炉炼铁工艺，以硅铁为还原剂，研究了硅铁配比对低品位钒钛磁铁矿金属化率的影响。通过磁选分离工艺，研究了还原温度和时间对有价元素 Fe、V、Ti 和 Cr 回收率的影响。随着硅铁配比的增加，样品金属化率先增加后降低，配比为 1.2 时，金属化率达到最大。随着还原温度的增加，有价元素 Fe、V、Ti 和 Cr 的回收率逐渐升高。在还原产物中，元素 V 和 Ti 具有相同的分布规律，主要富集在黑钛石相中。随着还原时间的增加，Fe 回收率逐渐升高，V、Ti 和 Cr 的回收率先升高后降低。最佳工艺条件为还原温度为 1450℃、还原时间为 30 分钟，此时 Fe、V、Ti 和 Cr 的回收率分别为 88%，90%，90%和 92%。

关键词：低品位钒钛磁铁矿；硅铁；非高炉炼铁；磁选

Ferrosilicon as Reduction Reagent on Recoveries of Valuable Elements of Low-grade Vanadia-titania Magnetite

Gao Zixian[1], Cheng Gongjin[1,2], Xue Xiangxin[1,2]

(1. School of Metallurgy, Northeastern University, Shenyang 110819, China; 2. Liaoning Key Laboratory of Recycling Science for Metallurgical Resources, Shenyang 110819, China)

Abstract: Low-grade vanadia-titania magnetite has characteristics of high titanium content, so it is a bit more hard to be smelted in blast furnace process. Therefore, in this work, experiments were carried out to investigate effect of ferrosilicon ratio on degree of metallization and reduction temperature on recoveries of valuable elements Fe、V、Ti and Cr of low-grade vanadia-titania magnetite employing an reduction-magnetic separating process. With the increase of ferrosilicon, the degree of metallization increased first and then decreased. When ferrosilicon ratio was 1.2, the degree of metallization reached to maximum. The recoveries of Fe, V, Ti and Cr increased with the reduction temperature. V and Ti had the same distribution in reduction sample, mainly concentrated in anosovite. The recovery of Fe increased but the recoveries of V, Ti and Cr increased first and then decrease with time increasing. The best value of technological parameters were that the reduction temperature was 1450℃ and the reduction time was 30 min. At this point, the recoveries of Fe, V, Ti and Cr were 88%, 90%, 90% and 92%, respectively.

Key words: low-grade vanadia-titania magnetite; ferrosilicon; non-blast furnace ironmaking; magnetic separation

粘结剂对铁焦强度的影响研究

徐润生[1,2]，邓书良[1,2]，陈绍鹏[1,2]，王 炜[1,2]，黄晓明[1,2]

（1. 武汉科技大学省部共建耐火材料与冶金国家重点实验室，湖北武汉 430081；
2. 湖北省冶金二次资源工程技术研究中心，湖北武汉 430081）

摘 要：铁焦作为优化高炉炼铁工艺的一种新型炉料，面临着铁矿粉添加量较少，强度不足等困境。本文采用坩埚炼焦和 I 型转鼓研究了大比例铁矿粉添加量和配加粘结剂对铁焦强度的影响规律。结果表明，随着铁矿粉添加量的增加，铁焦的强度逐渐降低，当添加量为 30% 时，强度骤降，仅为 28.04%；在炼焦配煤中添加煤沥青可以显著增强铁焦的强度，当添加量为 10%，效果最佳。

关键词：气煤；鄂西矿；铁焦；转鼓强度；粘结剂

Effect of Binder on the Drum Strength of Ferro-coke

Xu Runsheng[1,2], Deng Shuliang[1,2], Chen Shaopeng[1,2],
Wang Wei[1,2], Huang Xiaoming[1,2]

(1. State Key Laboratory of Refractories and Metallurgy, Wuhan University of Science and Technology, Wuhan 430081, China; 2. Hubei Provincial Engineering Technology Research Center of Metallurgical Secondary Resources, Wuhan 430081, China)

Abstract: Ferro-coke as a new burden, it's can optimize the ironmaking process in blast furnace. However, ferro-coke is faced difficulties such as low iron ore powder addition and insufficient strength. In this paper, the effects of large proportion of iron ore powder addition, and added binder on the strength of iron coke were studied by crucible coking and I-type drum testing. The results show that, the drum strength of iron coke decreases with the increase of iron ore powder. When the addition amount of iron ore powder is 30%, the drum strength of ferro-coke is plummeted to 28.04%. The drum strength of ferro-coke can be significantly enhanced by adding coal-tar pitch in feedstock for coke making. The best strength is achieved when the coal-tar pitch addition amount is 10%.

Key words: gas coal; Western Hubei iron ore; ferro-coke; drum strength; binder

锌对焦炭冶金性能的影响研究

王 炜[1,2]，陈柏文[1,2]，徐润生[1,2]，昝日安[1,2]

（1. 湖北省冶金二次资源工程技术研究中心，湖北武汉 430081；
2. 武汉科技大学省部共建耐火材料与冶金国家重点实验室，湖北武汉 430081）

摘 要：本文对焦炭进行了不同锌量下的吸附，依据国标开展了焦炭反应性及反应后强度检测实验，探究了不同富锌量下焦炭冶金性能的变化规律。通过光学显微镜、扫描电子显微镜、X 射线衍射及 BET 比表面积检测，分析了负载锌的焦炭的微观形貌，微晶结构和气孔演变规律，最后运用第一性原理计算对锌的催化机理进行了解释。

关键词：焦炭；锌；气化反应；催化作用

Study on the Effect of Zinc on the Metallurgical Properties of the Coke

Wang Wei[1,2], Chen Bowen[1,2], Xu Runsheng[1,2], Zan Rian[1,2]

(1. Hubei Provincial Engineering Technology Research Center of Metallurgical Secondary Resources, Wuhan 430081, China; 2. State Key Laboratory of Refractories and Metallurgy, Wuhan University of Science and Technology, Wuhan 430081, China)

Abstract: In this paper, the coke samples were adsorbed at different zinc content. According to the national standard, the reaction and post-reaction strength tests were carried out to explore the changes of coke metallurgical properties under different zinc content. The macroscopic pore statistics under optical microscope, the microscopic morphology under electron microscope, the microcrystal structure detection under XRD and the integrated analysis of BET specific surface area and mesoporous structure were carried out. Finally, the catalytic mechanism of zinc is explained by first-principle calculation.

Key words: coke; zinc; gasification; catalysis

烧结矿还原对料层内焦炭消耗程度的影响

王　炜，王浩翔，徐润生，昝日安，杨代伟

（武汉科技大学钢铁冶金及资源利用省部共建教育部重点实验室，湖北武汉　430081）

摘　要：为了研究焦炭在高炉内的消耗行为，有必要研究矿焦耦合反应对焦炭消耗程度的影响。本文采用三维光学数码显微镜和扫描电镜对与三种不同还原性烧结矿耦合反应后的焦炭进行微观形貌检测。结果表明反应结束后下层焦炭的气化反应程度最高，中层焦炭与上层焦炭的气化反应程度依次降低。由此可见，在实际反应过程中，与铁矿石接触的焦炭的气化反应程度比未与铁矿石接触的焦炭的气化反应程度要高的多。

关键词：高炉；烧结矿；焦炭；耦合反应；微观形貌

Effect of Sinter Reduction on Coke Consumption in Material Layer

Wang Wei, Wang Haoxiang, Xu Runsheng, Zan Rian, Yang Daiwei

(Key Laboratory for Ferrous Metallurgy and Resources Utilization of Ministry of Education, Wuhan University of Science and Technology, Wuhan 430081, China)

Abstract: In order to study the consumption behavior of coke in blast furnace, it is necessary to study the effect of coke coupling reaction on coke consumption. In this paper, 3D optical digital microscope and SEM were used to detect the morphology of coke coupled with three different reductive sinter. The results showed that the gasification degree of coke in the lower layer is the highest after the end of the reaction. The degree of gasification reaction of the middle layer coke and the upper layer of coke decreased in turn. It can be seen that the gasification reaction of coke in contact with iron ore is much higher than that of coke not in contact with iron ore in the actual reaction process.

Key words: blast furnace; iron ore; coke; coupling reaction; micromorphology

煤胶质层指数测定影响因素分析

乌木提江，吴 琨，张启发

（宝武集团八钢公司制造管理部理化检验中心，新疆乌鲁木齐 830022）

摘 要： 煤的胶质层指数是判断煤结焦性的重要指标，也有企业作为煤结算的重要依据。煤的胶质层指数对炼焦厂炼焦炭有着重要作用。本文通过煤种类型、煤杯透气性、加热速度等方面研究影响煤胶质层指数的因素。

关键词： 煤；胶质层指数

Analysis of Factors Affecting the Measurement of Coal Colloid Layer Index

Wumu Tijiang, Wu Kun, Zhang Qifa

(Baowu Group Ba Steel Company Manufacturing Management Department Physical and Chemical Testing Center, Urumchi 830022, China)

Abstract: The index of colloid layer of coal is an important index to judge the coking property of coal. It is also an important basis for the settlement of the incoming coal by the eight-steel company. The index of colloid layer of coal plays an important role in the coking of coking plant. In this paper, the factors affecting the index of the coal colloid layer are studied by the types of coal, and the heating speed of the coal cup.

Key words: coal; colloidal layer index

添加废塑料对含碳压块热成型和焙烧还原的影响

孟庆民，龙红明，春铁军，魏汝飞，王 平，李家新

（安徽工业大学冶金工程学院，安徽马鞍山 243032）

摘 要： 在查明PS、SAN、PP三种热塑性废塑料热解和软熔特性基础上，考察了废塑料种类、模具及原料温度对含碳铁矿热压块强度的影响，确定了热压工艺参数。基于含碳铁矿热压块焙烧还原实验和还原后试样物相与微观结构分析，讨论了添加废塑料对含碳铁矿热压块还原的影响。结果表明：废塑料PS、SAN、PP低于热解开始温度下所具有良好的塑性使含碳铁矿热压块低温热成型成为可能。在C/O=1，废塑料添加量为10%时，适宜热成型条件为：模具温度120℃，含PS和SAN混合料温度为170~180℃，PP混合料温度为150℃。添加废塑料能够降低含碳铁矿热压块碳气化反应开始温度和剧烈反应温度，同时可以促进还原过程中铁的渗碳反应。

关键词： 含碳压块；废塑料；热成型；焙烧还原；微观结构

Waste Plastic Addition on the Thermoforming and Roasting Reduction of Carbon-bearing Iron Ore Briquette

Meng Qingmin, Long Hongming, Chun Tiejun, Wei Rufei, Wang Ping, Li Jiaxin

(School of Metallurgical Engineering, Anhui University of Technology, Maanshan 243032, China)

Abstract: On the basis of pyrolysis and softening characteristics of thermoplastic waste plastics including PS, SAN and PP, the effects on the strength of carbon-bearing iron ore briquette were investigated such as waste plastics type, mold temperature and raw material temperature, and the thermoforming parameters were determined. According to the roasting reduction experiment of carbon-bearing iron ore briquette, phase and microstructure analysis of samples after reduction, the effect of waste plastics addition on the reduction of carbon-bearing iron ore hot briquette was discussed. The results show that the waste plastic PS,SAN,PP have good plasticity when it is below the initial pyrolysis temperature, which makes it possible for thermoforming of the carbon-bearing iron ore briquette at low-temperature. When C/O=1 and the amount of waste plastic addition is 10%, the best thermoforming conditions are: mold temperature 120℃, mixture material temperature containing PS and SAN is 170~180℃, and mixture material temperature containing PP is 150℃. Waste plastic addition can reduce the initial temperature and intense reaction temperature of gasification reaction of carbon-bearing iron ore briquette, and promote the carburization reaction at the same time.

Key words: carbon-bearing iron ore briquette; waste plastic; thermoforming; roasting reduction; microstructure

高炉铜冷却壁长寿技术分析

张 勇[1]，龚卫民[2]，贾国利[2]，赵满祥[2]，焦克新[3]，佘雪峰[3]

（1. 首钢集团有限公司技术研究院，北京 100043；2. 首钢股份公司，
河北迁安 064404；3. 北京科技大学，北京 100083）

摘 要： 本文对行业内众多铜冷却壁的破损案例进行了统计分析，结果认为：破损的主要原因是边缘煤气流控制不合理、冷却制度管控不精细，炉体维护不及时和缺乏渣皮监控手段等，并最终提出铜冷却壁长寿的关键是要建立"以渣皮控制为核心"的理念，坚持"一稳、二均，三监控"的操作。

关键词： 高炉；铜冷却壁；长寿技术

汉钢 2#高炉炉缸侧壁温度升高护炉实践

毛洁成，王纪民

（陕西汉中钢铁有限责任公司炼铁厂，陕西汉中 712400）

摘 要： 对汉钢 2#高炉炉缸侧壁温度升高的护炉实践进行了总结。2018 年针对 2#高炉炉缸侧壁温度有 4 个点升高异常，已威胁到高炉的安全生产，结合生产实际情况，通过采取钛矿护炉、调整风口进风状态、调整装料制度、炉

体灌浆、强化冷却和加强炉前铁口维护等技术措施，使炉缸侧壁温度得到有效控制，为高炉安全运行提供了保证。

关键词：高炉；炉缸；侧壁温度；护炉

欧冶炉氧气风口喷吹煤气的结构变化

田 果

（宝钢集团八钢股份公司炼铁厂，新疆乌鲁木齐 830022）

摘 要：欧冶炉是非高炉炼铁的一种形式。风口综合喷吹鼓风一直是制约风口长寿的关键。欧冶炉氧气风口历经了风口氧气氮气综合喷吹、风口喷煤试验、风口氧气压缩空气综合喷吹、风口雾化水氧气综合喷吹，直至现在欧冶炉进行的风口煤气喷吹，欧冶炉氧气风口的结构形式出现了较多的变化。

关键词：氧气风口；非高炉炼铁；结构形式

Industrial Practice of Injecting Top Gas into Total Oxygen Tuyere of European Smelting Furnace

Tian Guo

(Baosteel Group Bayi Iron Works Co., Ltd., Urumchi 830022, China)

Abstract: Ouyelu is a form of blast furnace tuyere combined blowing blast tuyere is restricted by the key to longevity Ouyelu after oxygen tuyere tuyere oxygen nitrogen injection tuyere tuyere oxygen coal injection test comprehensive comprehensive compressed air blowing tuyere oxygen atomizing water injection, comprehensive until now Ouyelu gas injection of tuyere tuyere Ouyelu oxygen structural form the more changes.

Key words: all-oxygen tuyere; non-blast furnace ironmaking; structural style

巴西南部高硅粉矿在湛江钢铁的试验研究和工业使用实践

章苇玲，牛长胜

（宝钢湛江钢铁有限公司制造管理部，广东湛江 524000）

摘 要：本文通过对巴西南部高硅粉矿的试验研究，了解该矿的基础性能，在此基础上进行了烧结杯试验和工业生产试验。试验表明：南部高硅粉除了 SiO_2 含量较高外，其他杂质含量低，粒度偏细，具有较好的烧结基础特性；在保证烧结矿目标成分指标的前提下，使用一定比例的南部高硅粉替代其他粉矿，有利于明显改善烧结矿成品率、转鼓强度以及低温还原粉化率等质量指标，但会降低烧结生产率。

关键词：南部高硅粉；烧结；基础性能；配矿

Experimental Research and Industrial Use Practice of Southern Brazil's High Silica Iron Ore

Zhang Weiling, Niu Changsheng

(Technique & Products Manufacturing Department of Baosteel Zhanjiang Iron and Steel Co., Ltd., Zhanjiang 524000, China)

Abstract: The basic properties of the Southern Brazil's high silica powder ore in southern Brazil were studied by experimental study. On this basis, sinter pot experiments and industrial-scale experiments were carried out. The experiments results showed that the high silica iron ore had low impurity content except the high content of SiO_2, and had good sintering basic characteristics. On the premise of guaranteeing the target composition index of sinter, using a certain proportion of the high silica powder ore could obviously improve the sinter yield, drum strength and low temperature reduction pulverization rate, but the sintering production rate was reduced.

Key words: southern brazil's high silica iron ore; sintering; basic characteristics; ore matching

风口流场与温度场分布的数值模拟研究

许海法[1]，孙成烽[2]，廖哲晗[2]，寇明银[3]，徐 健[2]

（1. 宝山钢铁股份有限公司中央研究院，上海 201900；2. 重庆大学材料科学与工程学院，重庆 400044；3. 北京科技大学冶金与生态工程学院，北京 100083）

摘 要：风口是炼铁过程中的重要部件，它的使用寿命是能否获得高产的关键。风口的工作环境是一个密闭高温的工作环境，无法直接对其观测研究。本文以计算流体力学方法来模拟风口的流场和温度场，探讨入水口流速对水室内的流场与温度场的影响关系，通过优化风口结构来达到增强冷却效果的目的。结果表明，增加入水口流速，冷却水在水室内的最高流速也随之明显增加；入水口流速越高，冷却效果越好，但是冷却不均匀，风口上部冷却效果优于下部；通过优化风口结构，冷却结果得到明显改善。

关键词：风口；数值模拟；流场；温度场；结构优化

Numerical Simulation of Flow and Temperature Distributions inside the Furnace Tuyere

Xu Haifa[1], Sun Chengfeng[2], Liao Zhehan[2], Kou Mingyin[3], Xu Jian[2]

(1. Baosteel Research Institute, Shanghai 201900, China; 2. College of Materials Science and Engineering, Chongqing University, Chongqing 400044, China; 3. School of Metallurgical and Ecological Engineering, University of Science and Technology Beijing, Beijing 100083, China)

Abstract: Tuyere is an important part in ironmaking process, and its service life is the key to obtain high production. The working condition of the tuyere is sealed and high temperature, which can not be directly observed and studied. In this paper, computational fluid dynamics (CFD) method is used to simulate the flow field and temperature field inside the tuyere, the influence of inlet velocity on the flow field and temperature field in water chamber is discussed, and the cooling effect is enhanced by optimizing the structure of tuyere. The results show that the maximum velocity of cooling water in the water chamber increases with the increase of the inlet velocity, the higher the inlet velocity, the better the cooling effect, but the cooling effect is uneven, and the cooling effect of the upper part of the tuyere is better than that of the lower part, by optimizing the structure of the tuyere, the cooling result is obviously improved.

Key words: tuyere; numerical simulation; flow field; temperature field; structure optimization

高炉风口焦炭质量特性研究

张文成[1,2], 张小勇[3], 郑明东[1]

（1. 安徽工业大学冶金学院，安徽马鞍山 243002；2. 宝钢研究院梅钢技术中心，
江苏南京 210039；3. 安徽工业大学化学与化工学院，安徽马鞍山 243002）

摘 要：通过对高炉风口焦炭的质量分析，探讨了风口不同深度的风口焦炭的粒度、化学成分及焦炭热性能的变化趋势。研究结果表明，焦炭粒度随着风口深度而变化，小于3mm的焦炭呈现先增后减再增加的分布，在0.5~1m的位置有一个小峰值。从渣铁比例来看，则随着风口距离增加呈现先减后增的趋势。从工业分析来看，各粒度等级及各深度范围的风口焦的灰分均有所增加，风口焦全部挥发分都有所降低。从风口焦炭灰成分分析来看，钾、钠和锌元素在0.5m的范围内大块风口焦和小块风口焦均高于入炉焦，在风口位置产生富集；风口焦炭中的硅和铝含量远低于入炉焦，硅在炉内被还原；风口焦炭中的钙和镁含量远高于入炉焦，受到高炉内灰渣影响明显。从焦炭催化指数看，风口焦炭催化指数MCI远大于入炉焦，在高炉内在焦炭外部粘连的渣铁其催化性远大于焦炭本身所带灰分的催化，风口焦较入炉焦的反应性大幅度提高，反应后强度大幅度下降。

关键词：焦炭；风口；高炉；化学成分

Study on Quality Characteristics of Coke in Blast Furnace Tuyere

Zhang Wencheng[1,2], Zhang Xiaoyong[3], Zheng Mingdong[1]

(1. Institute of Metallurgy, Anhui University of Technology, Maanshan 243002, China; 2. Meishan Technology Center, Baosteel Research Institute, Nanjing 210039, China; 3. Institute of Chemical Industry, Anhui University of Technology, Maanshan 243002, China)

Abstract: By analyzing the quality of coke in blast furnace tuyere, the change trend of particle size, chemical composition and thermal properties of coke in tuyere of different depth was discussed. The results show that the coke particle size varies with the tuyere depth, and the coke size less than 3mm shows a distribution of first increasing, then decreasing and then increasing, with a small peak value at the position of 0.5~1m. According to the proportion of slag and iron, it decreases first and then increases with the increase of tuyere distance. From the perspective of industrial analysis, the ash content of tuyere coke in each particle size and depth range increased, and all volatiles of tuyere coke decreased. From the composition analysis of tuyere coke ash, potassium, sodium and zinc elements in the 0.5m range of large tuyere coke and small tuyere coke are higher than the furnace coke, and the concentration occurs in the tuyere location. The content of silicon and aluminum in tuyere coke is much lower than that in coke. The content of calcium and magnesium in tuyere coke is much higher than that in coke. From the perspective of coke catalytic index, tuyere coke catalytic index MCI is much higher than that of coke, and the catalytic property of slag iron sticking to coke outside the blast furnace is much higher than that of coke with ash. The reactivity of tuyere coke is much higher than that of coke, and the intensity of the reaction is greatly reduced after the reaction.

Key words: coke; tuyere; blast furnace; chemical composition

在线激光料面探测系统在鞍钢高炉的应用

高征铠[1]，戴建华[1]，赵东铭[2]，张洪宇[3]，李建军[2]，胡德顺[3]，高　泰[4]

(1. 北京科技大学冶金与生态工程学院，北京　100083；2. 鞍山钢铁集团有限公司炼铁厂，辽宁鞍山　114021；3. 鞍钢集团朝阳钢铁有限公司，辽宁朝阳　122000；4. 北京神网创新科技有限公司，北京　100083)

摘　要： 2018年6月鞍钢集团朝阳钢铁有限公司1号高炉($2600m^3$)安装了一套在线激光料面探测系统，取得了很好的使用效果。鉴于该系统可直观地对高炉布料操作提供重要帮助，鞍钢决定在本部8座高炉陆续应用该料面探测系统。本文主要介绍了鞍钢朝阳1号高炉和本部5号高炉在生产中使用激光探测系统实时扫描料面，并根据扫描结果指导布料操作调整的过程。朝阳1号高炉通过料制调整使料面形状和煤气流分布得到改善，高炉稳顺，燃料比降低，产量增加。鞍钢本部5号高炉通过实际操作料线校正和调整获得了较为稳定的操作料面，高炉稳定、顺行状况得到较大改善。

关键词： 高炉；可视化；激光；料面探测

Application of On-line Laser Burden Surface Detection System on Blast Furnaces of Ansteel Group

Gao Zhengkai[1], Dai Jianhua[1], Zhao Dongming[2], Zhang Hongyu[3], Li Jianjun[2], Hu Deshun[3], Gao Tai[4]

(1. University of Science & Technology Beijing Metallurgical and Ecological Engineering School, Beijing 100083, China; 2. Ansteel Group Co., Ltd., Iron-making Plant, Anshan 114021, China; 3. Ansteel Group Chaoyang Iron & Steel Co., Ltd., Chaoyang 122000, China; 4. Shenwang Pioneer Tech. Corporation Beijing, Beijing 100083, China)

Abstract: In June 2018, a set of on-line laser burden surface detection system was installed on No.1 blast furnace ($2600m^3$) of Ansteel group Chaoyang Iron & Steel Co., Ltd., which achieved good results. Considering that the system can directly provide important help to the charging operation of blast furnaces, Ansteel Group decided to apply the system to the eight blast furnaces at the headquarters. In this paper, the application of on-line laser burden surface detection systems on Chaoyang No.1 BF and No.5 BF of Ansteel Group is introduced. The profile of burden surface and distribution of gas flow are improved by adjusting the charging process of Chaoyang No.1 BF, and the stability and smooth operation of the blast furnace have been greatly improved of No.5 BF at Ansteel Group.

Key words: blast furnace; visualization; laser; burden surface detection

铁精矿粉配比对球团制备和冶金性能的影响

杨子超,储满生,柳政根,唐 珏,高立华,葛 杨

(东北大学冶金学院,辽宁沈阳 110819)

摘 要:通过对不同铁精矿粉配比的球团进行氧化焙烧,得到成品球团矿,系统地研究了某铁精矿粉配比对球团冶金性能的影响机理。结果表明,随铁精矿 A 含量增加,球团矿的抗压强度先趋平缓而后显著上升,铁精矿 A 对抗压强度影响较大,还原膨胀率及低温还原粉化率呈降低的趋势。研究得出最佳制备参数:膨润土 0.7%、焙烧温度 1270℃、焙烧时间 30min、矿粉粒度及膨润土粒度均为–200 目,最佳处理参数:铁精矿 A:铁精矿 B 为 6:4,球团抗压强度达 4372 N,还原膨胀率最低为 9.24%,低温还原粉化程度 $RDI_{+3.15}$ 为 93.91%,从冶金性能上看,其具备良好的还原膨胀和还原粉化性能,满足高炉冶炼需求。

关键词:铁矿粉;优化配矿;氧化球团;冶金性能

Effect of Iron Concentrate Mixing Ratio on Pellet Preparation and Metallurgical Properties

Yang Zichao, Chu Mansheng, Liu Zhenggen, Tang Jue, Gao Lihua, Ge Yang

(School of Metallurgy, Northeastern University, Shenyang 110819, China)

Abstract: The granules of different iron concentrate powders were oxidized and calcined to obtain the finished pellets. The influence mechanism of the ratio of iron concentrates on the metallurgical properties of the pellets was systematically studied. The results showed that with the increase in the content of iron ore A, the compressive strength of the pellets tends to gentle then increased significantly, iron ore A great influence on the compressive strength, and reduction expansion coefficient RDI rate tended to decrease. The best preparation parameters were obtained: bentonite 0.7%, calcination temperature 1270℃, calcination time 30 min, mineral powder particle size and bentonite particle size –200 mesh, optimal treatment parameters: iron concentrate A:iron concentrate B was 6:4, pellet compressive strength of 4372 N, the lowest reduction expansion coefficient was 9.24%, the degree of $RDI_{+3.15}$ 93.91% from a metallurgical performance point of view, which have a good expansion and reduction differentiation reducing performance to meet blast furnace requirements.

Key words: iron ore fines; optimized ore blending; oxide pellets; metallurgical properties

武钢一号高炉提高煤气利用率生产实践

龚 臣,唐少波

(武钢有限一高炉,湖北武汉 430081)

摘 要:本文对武钢 1 号高炉由于配矿变化,Zn 负荷升高,炉前渣铁排放等因数导致的炉墙平凡结厚,煤气利用率大幅下降的现象,通过改善装配料制度和炉前渣铁排放管理,调整上下部制度及强制排碱等措施。达到了很好的

效果，高炉炉型恢复正常，煤气利用率得到改善，燃料比大幅降低，大大的降低了生铁冶炼成本，达到了预期效果。

关键词：1 号高炉；配矿；Zn；煤气利用率；制度

针对近年国内高炉大型化的思考

覃开伟[1]，王中华[1]，王 涛[2]

（1. 河钢集团宣钢公司技术中心，河北张家口 075100；
2. 河钢集团宣钢公司炼铁厂，河北张家口 075100）

摘 要：2005 年以后，国内钢铁企业走上高炉大型化快车道，高炉平均炉容大幅上升，超大型高炉数量快速增加。但从实际管理经营效果来看，高炉超大型化，并未能带来理想中的效益和竞争优势。分析认为，资源因素、技术及管理因素、市场因素，是导致这种状况的主要原因。企业推进高炉大型化时应当充分考量区位因素、市场定位等现实条件，不宜盲目追求大型化。

关键词：高炉；资源；技术；市场

Consideration on Recent Trend of Large-Scale Blast Furnace

Qin Kaiwei[1], Wang Zhonghua[1], Wang Tao[2]

(1. Xuan Steel Technical Centre of HBIS, Zhangjiakou 075100, China;
2. Xuan Steel Iron-making Plant of HBIS, Zhangjiakou 075100, China)

Abstract: Since 2005, the trend of large-scale blast furnace has spread rapidly in China, the average scale of blast furnaces increased, the amount of super large-scale BF increased definitely. But when we consider the result of this trend, we can see that large –scale BF did not bring back expected benefit and advantage in competition. From the assessment we find that, resource factors、technical & operation factors、market factors are the main causes of this situation. So, we suggest that steel –making corps should take the market influence and location influence into consideration when process the large-scale BF strategy, avoiding the blindly following of the trend.

Key words: blast furnace; resource; technology; market

钛、钒、铬对氧化球团三维显微矿相结构的影响

唐 珏，储满生，欧 聪，柳政根

（东北大学冶金学院，辽宁沈阳 110819）

摘 要：基准球团内部赤铁矿分布紧密，多数呈大颗粒团块状，孔洞分布均匀，以小型孔洞为主，硅酸盐渣相含量较少，以均匀分布的细小粒状为主；含钛球团内部纯赤铁矿含量较基准球团少，分布稀散，多数呈小晶粒互联状，仅有少部分为大颗粒团矿状，孔洞尺寸偏大，数量较多，硅酸盐渣相含量相对基准球团明显增加，并以大颗粒形式

存在；含钒球团内部赤铁矿结构疏松，多数为发育不完全的小晶粒互联状，孔洞尺寸偏大，数量较多，偏析分布，渣相以大颗粒和小颗粒耦合嵌布；含铬球团内部赤铁矿含量虽与基准球团相近，但多数为小晶粒互联状，孔洞呈大孔洞和小孔隙交接形态，阻碍赤铁矿大范围联晶，渣相含量较多，且部分形成大颗粒偏析。

关键词：有价组元；氧化球团；三维结果；赤铁矿

Effect of Titanium, Vanadium and Chromium on Three-dimensional Microscopic Phase Structure of Oxidized Pellet

Tang Jue, Chu Mansheng, Ou Cong, Liu Zhenggen

(School of Metallurgy, Northeastern University, Shenyang 110819, China)

Abstract: The hematite in the reference pellets was distributed closely. Most of them were large particles with uniform distribution of pores which mainly was small pores. The content of silicate slag phase was small and mainly distributed in fine particles. The content of pure hematite in titanium-bearing pellets was less than that in the benchmark pellets, and the distribution is sparse. Only a few of them were in the form of large particle agglomerations, with large size and large number of holes. The content of silicate slag increased obviously compared with the reference pellets and existed in the form of large particles. The hematite structure inside vanadium bearing pellets was loose and most of them were small interconnecting grains with incomplete development. The holes in vanadium bearing pellets were distribution of segregation with large size and quantity. The slag phase was intercalated with large and small particles. Although the hematite content in the chrome-containing pellets was similar to that of the reference pellets, most of them were interlinked with small grains. The pore morphology of large pore and small pore in chrome-containing pellets prevented hematite from joining crystals in a large scale. The content of slag phase was large and some of them form large particle segregation.

Key words: valuable component; oxidized pellet; three-dimensional results; hematite

宣钢 1#高炉炉役后期指标提升实践

路 鹏，裴生谦，王洪余，吕志敏，褚润林，王 斌

(河北钢铁集团宣钢公司炼铁厂，河北宣化 075100)

摘 要：对宣钢 1#高炉炉役后期操作实践进行了总结。通过下部缩小风口面积、使用长风口，提高鼓风动能；上部优化布料矩阵，采用大矿批，稳定煤气流分布；维持合理的理论燃烧温度；阶段性钛矿护炉、临时堵风口等措施，克服了炉役后期不利条件的影响，在连续保持炉况 14 个月稳定顺行的基础上，技术经济指标不断提升，2019 年 2 月份日产合格生铁完成 6185t/d，入炉大焦比 337kg/t,煤比 170kg/t，为企业生产稳定及成本降低创造了良好条件。

关键词：风口；鼓风动能；布料矩阵；理论燃烧温度；钛矿护炉

The Practice of Improving the Indices at Post Campaign in Xuansteel's No. 1 BF

Lu Peng, Pei Shengqian, Wang Hongyu, Lv Zhimin, Chu Runlin, Wang Bin

(Xuansteel Iron Works, Hebei Steel Group, Xuanhua 075100, China)

Abstract: This article summarized the production practice of reducing cooling stave damage in Xuansteel's No. 1 BF at its mid-late campaign. A series of measures were taken, such as using long tuyere to improve the blast kinetic energy, optimizing burden distribution to stabilize gas flow distribution, improving the quality of raw material, improving pressure of the blast furnace roof, injecting for the wall and furnace shell, and so on, contained the trend of cooling stave damage at its mid-late campaign, realized long-term smooth operation and improvement of economic indexes.

Key words: long tuyere; blast kinetic energy; burden matrix; heat load; crude fuel

烧结烟气 CO、NO$_x$ 及二噁英的协同处理技术

申明强

（河钢邯钢炼铁部，河北邯郸 056015）

摘 要：烧结机头烟气中含有大量的 CO，目前尚无治理措施。CO 的无组织排放不仅造成了环境的污染，更造成了能源的浪费。据统计，烧结烟气中 CO 携带能量约占烧结消耗总能量的 15%~20%，如果采取措施使其氧化成 CO$_2$，将其携带的能量释放到烧结烟气中正好满足 SCR 及二噁英的脱除温度要求，从而实现烧结烟气污染物的同步脱除。把脱除 CO、NO$_x$ 及二噁英后的烟气热量进行回收，将产生显著的经济效益，切实实现节能减排的目标。

关键词：烧结烟气；CO；NO$_x$；二噁英；脱除

Co-treatment of CO, NO$_x$ and Dioxins in Sintering Flue Gas

Shen Mingqiang

(Iron-making Department of HBIS Group Hansteel Company, Handan 056015, China)

Abstract: There is a large amount of CO in the flue gas of sintering machine head, and no control measure at present. The non-organized emission of CO not only pollutes the environment, but also wastes energy. According to statistics, the energy carried by CO in sintering flue gas accounts for about 15%~20% of the total energy consumed by sintering. If measures are taken to oxidize it into CO$_2$, the energy carried by it will be released into sintering flue gas to meet the requirements of SCR and dioxin removal temperature, thus achieving the simultaneous removal of pollutants in sintering flue gas. Recovery of flue gas heat after removal of CO, NO$_x$ and dioxins will produce significant economic benefits and achieve the goal of energy saving and emission reduction.

Key words: sintering flue gas; CO; NO$_x$; dioxin; removal

7.63m 焦炉湿熄焦焦炭水分的控制措施

赵松靖，张爱平

（江苏沙钢集团有限公司焦化厂，江苏苏州 215000）

摘 要：沙钢 7.63m 焦炉在投产之初，采用湿法熄焦工艺，焦炭水分波动大，且较高，未能达到 3%~4%的设计水

平，对5800m³大容积高炉正常生产带来了一定不利影响。本文分析了焦炭水分过高的原因以及针对这些问题进行处理的一些措施，为降低大型焦炉的焦炭水分起到了很好的借鉴作用。

关键词：7.63m焦炉；湿熄焦；红焦；焦炭水分；原因分析；处理措施

通过荷载效应分析高炉炉壳上涨及解决实践

卢维强[1]，张俊杰[1]，段国建[2]

(1. 中天钢铁集团有限公司,江苏常州　213011；2. 中冶京诚工程技术有限公司,北京　100176)

摘　要：高炉的长寿、经济、环保和安全运行对钢铁企业生产具有重要意义，但随着高炉的大型化，操作的高压和高强度化，特别随着高炉薄壁炉衬技术的发展，高炉的长寿和安全运行成为当今钢铁行业需跟进和研究的课题，其中高炉炉壳上涨问题就是急需解决的课题，本文通过荷载效应分析高炉炉壳上涨原因，并针对不同的高炉采取不同的措施，对大修的高炉研发了一种抑制炉壳上涨的双炉底板结构解决炉壳上涨问题，对还在生产的高炉采用结构加固的方法解决炉壳上涨问题。

关键词：高炉；薄壁炉衬；炉壳上涨

新常态下的焦炭质量预测模型的研究与应用

宋林娜，张　磊

(河钢集团邯钢公司生产制造部，河北邯郸　056015)

摘　要：介绍了环保限产新常态下，邯钢不同炉型长结焦时间下配煤结构的研究与应用，对邯钢适应限产新常态长结焦时间下的配煤操作积累了经验。同时，降低了配煤成本，创造了经济效益。

关键词：环保限产；配煤结构；预测模型

Research and Application of Coke Quality Prediction Model in the New Normal State

Song Linna, Zhang Lei

(HBIS Hansteel Produce and Manufacture Department, Handan 056015, China)

Abstract: This paper introduces the research and application of coal distribution structure under the new normal of environmental protection limited production, and accumulates the experience of coal distribution operation under the new normal long coking time. At the same time, the cost of coal distribution has been reduced and economic benefits have been created.

Key words: environmental production restrictions; eoal distribution structure; prediction model

炉料性能及锌富集对高炉波动影响的分析

肖志新，王齐武，陈令坤，郑华伟，刘栋梁

（武汉钢铁有限公司，湖北武汉　430083）

摘　要：对2018年10月~2019年2月武钢有限1号（2200m³）和8号高炉(4117m³)受原燃料质量波动和Zn富集造成炉况不同程度的变化进行了分析。结果表明：1号高炉在炉料来源、质量和结构频繁波动的情况下，再叠加Zn负荷升高，高炉出现了反复结厚的问题。与前一年同期相比燃料消耗（494.1kg/t Fe）增加18kg/t Fe，煤气利用率下降（50.03%）4.3%；8号高炉在以稳定原料质量为方针的操作条件下运行稳定，与前一年同期相比燃料消耗(501.8kg/t Fe)降低3.1kg/t Fe，利用系数增加（2.31t/m³·d）0.1t/m³·d；同时，形成了处理1号高炉反复结厚的技术措施。

关键词：高炉波动；原料性能；Zn富集；结厚处理

Analysis for the Effect of Charge Performance and Zinc Enrichment on Blast Furnace

Xiao Zhixin, Wang Qiwu, Chen Lingkun, Zheng Huawei, Liu Dongliang

(Wuhan Iron and Steel Co., Ltd., Wuhan 430083, China)

Abstract: The effect of charge property and Zinc enrichment on the performance of No.1 and No.8 blast furnace was analyzed form October 2018 to February 2019 in the Wuhan Iron and Steel Co., Ltd. The result showed that a repeatedly accretion of the No.1 blast furnace took place because of frequent fluctuation on the source, quality and composition of charges and a unfavorable superposition of zinc enrichment; The fuel consumption of the No.1 blast were increasing by 18kg/t Fe and the gas utilization ratio of it was decreasing by 4.3% respectively, compared with the data of 494.1kg/t Fe and 50.03% in the same period a year ago; The No.8 blast furnace ran smoothly for the operation principle of stable charge quality , its fuel consumption was decreasing by 3.1 kg/t Fe and using coefficient of capacity was increasing by 0.1t/m³·d, compared with the data of 501.8kg/t Fe and 2.31t/m³·d in the same period a year ago; moreover, A technical measure was formed to deal with the repeatedly accretion of the No.1blast furnace.

Key words: fluctuation of blast furnace; charge performance; zinc enrichment; removal of thickness

鞍钢超厚料层双层预烧结新工艺研究与工业试验

周明顺[1]，王义栋[2]，赵东明[3]，朱建伟[1]，姜　涛[4]，钟　强[4]

（1. 鞍钢集团钢铁研究院，辽宁鞍山　114009；2. 鞍山钢铁集团公司，
辽宁鞍山　114021；3. 鞍钢股份有限公司炼铁总厂，辽宁鞍山　114021；
4. 中南大学资源加工与生物工程学院，湖南长沙　410083）

摘 要：厚料层烧结可提高烧结矿成品率和转鼓强度，降低烧结固体燃耗和氮氧化物和碳氧化物排放，代表铁矿烧结的发展方向。但对于烧结铁料粒度较细且配加比例较高的原料条件，一般地，当料层厚度超过720~750mm时，由于料层透气性变差，烧结矿产率会明显下降。针对厚料层烧结存在的提产瓶颈问题，鞍钢集团钢铁研究院开发了双层预烧结新工艺，新工艺可显著改善大比例细粒铁精矿料层透气性，烧结提产效果显著，解决超厚料层烧结产率低的问题。但新工艺烧结矿的冷强度有一定程度下降，其主要原因是下层烧结过程缺氧导致。目前，通过适当延长预烧时间，可一定程度缓解下层烧结缺氧，取得明显改善烧结矿冷强度的效果。在鞍钢炼铁总厂二烧车间 360m² 烧结机和4号、5号高炉同时开展的近7个月的工业试验结果表明，双层预烧结新工艺在生产实践上是完全可行的，取得了 16.11%的烧结增产效果，高炉使用双层预烧结新工艺的烧结矿顺行情况良好，高炉利用系数、燃料比及风量与基准期基本一致。该新工艺的实施在国内是烧结领域一项颠覆性的技术创新，但新工艺在某些工艺参数上尚需继续完善，以期取得最佳的烧结提产、降低烧结固体燃耗和氮氧化物及碳氧化物排放的效果。

关键词：铁矿烧结；超厚料层；双层预烧结；预烧时间

Research and Industrial –scale Pilot on Double-layer Pre-sintering New Process for Ultra-thick Bed Layer

Zhou Mingshun[1], Wang Yidong[2], Zhao Dongming[3],
Zhu Jianwei[1], Jiang Tao[4], Zhong Qiang[4]

(1. Iron and Steel Research Institute, Ansteel Group Corporation, Anshan 114009, China; 2. Ansteel Group Corporation, Anshan 114021, China; 3. Iron-making Plant, Angang Steel Co., Ltd., Anshan 114021, China; 4. School of Minerals Processing and Bioengineering, Central South University, Changsha 410083, China)

Abstract: Ultra-deep bed Layer sintering can improve sinter yield and drum strength, reduce solid fuel consumption and emission of nitrogen oxides and carbon oxides, representing the development direction of iron ore sintering. However, for ore iron sinter with finer particle size and higher proportion of raw materials, generally, when the thickness of the bed layer exceeds 720~750 mm, the sinter yield will decrease significantly due to the poor permeability of the sinter mix bed layer. Aiming at solveing the problem of increasing production in sintering with thick bed layer, AISC Iron and Steel Research Institute has developed a new double-layer pre-sintering process. The new process can significantly improve the permeability of large proportion fine iron concentrate layer, the effect of sintering production is remarkable, and solve the problem of low sintering yield in ultra-deepbed layer. However, the cold strength of sinter in the new process has decreased to a certain extent, which is mainly caused by the lack of oxygen in the lower Layer sintering process. At present, by properly prolonging the pre-sintering time, the oxygen deficiency in the lower sintering layer can be alleviated to a certain extent, and the cold strength of sinter can be significantly improved. The industrial test results of 360 m² sinter machine in No. 2 sintering workshop and No. 4 and No. 5 blast furnace of Angang Iron-making Plant for nearly 7 months show that the new double-layer pre-sintering process is completely feasible in production practice, and 16.11% sintering yield increase has been achieved. The sinter of the new double-layer pre-sintering process used in blast furnace goes smoothly and the blast furnace utilization is good. Coefficient, fuel ratio and air volume are basically consistent with the reference period. The implementation of this new process is a subversive technological innovation in sintering field in China. However, some technological parameters of the new process need to be further improved in order to achieve the best sintering output, reduce solid fuel consumption and emission of carbon oxides.

Key words: iron ore sintering; ultra-deep bed; double pre-sintering; pre-sintering time

八钢烧结配加除尘灰工艺影响及分析

王雪超，秦 斌

(宝钢集团八钢公司，新疆乌鲁木齐 830055)

摘 要：结合八钢炼铁自身特点，在烧结生产过程中对配加除尘灰进行生产实践，通过配加不同比例的各类除尘灰，分析除尘灰配加入烧结系统的可行性，以及对烧结过程及质量的影响，进而确定八钢烧结配加除尘灰合适的配比方案。

关键词：烧结；除尘灰；转股

斗轮减速机润滑装置的改进

王明江

(武钢有限公司炼铁厂，湖北武汉 430081)

摘 要：斗轮取料机由于斗轮减速机在结构和润滑装置的设计上存在的问题，导致减速机发生断齿和轴承烧蚀的设备故障，针对存在的问题，对斗轮减速机的结构和润滑装置进行了改进。该文设计思路清晰，对斗轮减速机的故障制定了切实可行的改造方案。通过改造实施，效果明显。

关键词：斗轮减速机；润滑装置；改进

斗轮机斗轮轴承防尘密封装置的改进

王明江，邬显敏

(武钢炼铁厂，湖北武汉 430081)

摘 要：本文分析了 QLK1200·51 型斗轮取料机的斗轮轴轴承防尘密封装置在生产过程中出现的主要缺陷及产生这些缺陷的原因，改进了斗轮轴轴承密封存在的问题，保证了斗轮取料机的正常运行。

关键词：轴承；密封装置；改进

富氧干馏技术制备富氢还原气体试验研究

张小蕊，邹 冲，赵俊学，马 成，李 宝，吴 浩

(西安建筑科技大学冶金工程学院，陕西西安 710055)

摘　要：结合高炉炼铁中气基还原的优势及低阶煤低温富氧干馏技术的优势，通过调节干馏条件来控制煤气成分以便得到富氢煤气供高炉还原铁矿石而用。本研究进行了不同含量的 H_2、CH_4 和 CO 气氛下煤低温干馏实验，考察了煤气成分、焦油收率和组成的变化。结果表明：H_2 配比的提高，有助于煤气中 CH_4、CO 及 C_nH_m 的提升；CH_4 配比的增加，有效地提高了煤气中 H_2、CH_4 及 C_nH_m 的含量；CO 气氛有助于煤干馏煤气中 H_2、CH_4 及 C_nH_m 的生成。同时，H_2、CH_4 和 CO 气氛可提高煤气产率与焦油收率，且可使焦油中轻质焦油的份额增加。

关键词：煤低温干馏；富氧干馏；富氢煤气；铁矿石还原

Experimental Study on Preparation of Hydrogen-rich Reducing Gas by Oxygen-enriched Dry Distillation

Zhang Xiaorui, Zou Chong, Zhao Junxue, Ma Cheng, Li Bao, Wu Hao

(School of Metallurgical Engineering, Xi'an University of Architecture and Technology, Xi'an 710055, China)

Abstract: Based on the advantages of gas-based reduction in blast furnace ironmaking and the advantages of low-rank coal low-temperature oxygen-enriched dry distillation technology, the gas component is controlled by adjusting the dry distillation conditions to obtain a hydrogen-rich gas for use in blast furnace reduction iron ore. In this study, low temperature dry distillation experiments of coal under different H_2, CH_4 and CO atmospheres were carried out, and the changes of gas composition, tar yield and composition were investigated. The results show that the increase of H_2 ratio contributes to the increase of CH_4, CO and C_nH_m in gas; the increase of CH_4 ratio effectively increases the content of H_2, CH_4 and C_nH_m in gas; CO atmosphere contributes to coal dry distillation gas Generation of H_2, CH_4 and C_nH_m. At the same time, the H_2, CH_4 and CO atmospheres increase the gas yield and tar yield and increase the fraction of light tar in the tar.

Key words: low-temperature carbonization of coal; oxygen-rich dry distillation; hydrogen-rich gas; iron ore reduction

大型钢铁企业焦炉煤气净化技术应用前景探讨

吴江伟

（宝钢湛江钢铁有限公司炼铁厂，广东湛江　524072）

摘　要：近年，产能结构优化、绿色环保、低成本成为了我国钢铁行业的主旋律，高炉、焦炉的发展也逐渐趋向于大型化，而配套的焦炉煤气净化系统规模普遍较小，成为制约企业降低成本的瓶颈之一。本文讨论了大型钢铁联合企业焦炉煤气净化系统的技术应用前景，即大型化以及深度脱硫两个方向，介绍了大型化的优势以及在湛江钢铁的应用情况、焦炉煤气深度脱硫技术应用现状。

关键词：焦炉煤气；大型化；深度脱硫

Discussion on the Application Prospect of COG Purification System in Iron and Steel Complex

Wu Jiangwei

(Baosteel (Zhanjiang) Iron & Steel Co., Ltd., Zhanjiang 524072, China)

Abstract: In recent years, optimization of capacity structure, environmental protection and low cost have become the main theme of China's iron and steel industry. The development of blast furnaces and coke ovens has gradually tended to be large-scale. The scale of the supporting coke oven gas purification system is generally small, which has become one of the bottlenecks restricting the cost reduction of enterprises. This paper discusses the application prospects of coke oven gas purification system in large-scale iron and steel complex, i.e. large-scale and deep desulfurization, and introduces the advantages of large-scale and its application in Zhanjiang Iron and Steel Co., Ltd. and the application status of deep desulfurization technology of coke oven gas.

Key words: coke oven gas; large scale; deep desulfurization

铝酸盐型高炉渣的物理化学性质

庞正德[1,2]，严志明[1,2]，吕学伟[1,2]，蒋宇阳[1,2]，凌家伟[1,2]

（1. 重庆大学材料科学与工程学院，重庆 400044；
2. 重庆钒钛冶金及先进材料重点实验室，重庆 400044）

摘　要： 随着高品位铁矿石资源的消耗，利用低品位、难处理的铁矿石将成为缓解铁矿石资源枯竭的可选途径。针对大量使用高铝铁矿资源必然造成高炉渣高 Al_2O_3 的特点，使得高炉渣系由硅酸盐渣系向铝酸盐渣系转变。本研究针对高 Al_2O_3 含量的 $CaO-MgO-Al_2O_3-SiO_2-TiO_2$ 渣系，重点讨论了 Al_2O_3，Al_2O_3/SiO_2 对炉渣高温物理化学性质（黏度、硫容量、密度和表面张力）的影响。此外，本文采用合适的理论模型，预测了不同组分下炉渣的物理化学性质。本研究旨在为高铝铁矿资源在高炉工艺中的大规模使用提供科学依据和理论指导。

关键词： 铝酸盐系炉渣；黏度；硫容量；密度；表面张力

Physicochemical Properties of Alumina Based Slag for Blast Furnace

Pang Zhengde[1,2], Yan Zhiming[1,2], Lv Xuewei[1,2], Jiang Yuyang[1,2], Ling Jiawei[1,2]

(1. College of Materials Science and Engineering, Chongqing University, Chongqing 400044, China;
2. Chongqing Key Laboratory of Vanadium-Titanium Metallurgy and Advanced Materials, Chongqing 400044, China)

Abstract: With the consumption of high-grade iron ore resources, the use of low-grade and difficult-to-handle iron ore will become an alternative way to alleviate the depletion of iron ore resources. High alumina iron ore will become an important resource for future steel production, and the utilization of high alumina iron ore in the blast furnace will inevitably increase the Al_2O_3 content in the slag, resulting in the slag system translation from silicate based to aluminate based. In this case, the $CaO-MgO-Al_2O_3-SiO_2-TiO_2$ slag system with high alumina content was studied in this work. The effects of Al_2O_3 on the high temperature physicochemical properties (viscosity, sulfide capacity, density and surface tension,) were studied experimentally. In this study, the physicochemical properties of the slag system in a wide range of composition were predicted by using suitable models. This work will provide scientific basis and theoretical guidance for the large-scale utilization of high-alumina iron ore in the blast furnace process.

Key words: alumina based slag; viscosity; sulfide capacity; density; surface tension

安钢 1#烧结机热风烧结技术的研制与应用

王军锋，周东锋，李发展

（安阳钢铁股份公司炼铁厂，河南安阳 455004）

摘 要：烧结烟气内循环工艺作为整个烧结生产系统的一部分，与主系统有机地融合在一起，通过自动控制，调整新系统与原有系统的温度、压力、风量、含氧量四个方面的平衡，既能回收利用一部分烟气显热，降低燃料消耗，又能减少排入大气的烟气量总量，降低主抽风机、主电除尘器和脱硫装置负荷，为治理烧结烟气污染创造有利条件，是传统烧结工艺实现节能减排、绿色发展的思路创新和有效探索。

关键词：热风烧结；烟气循环；节能减排；协同治理

Development and Application of Hot Air Sintering Technology for 1# Sintering Machine at Angang

Wang Junfeng, Zhou Dongfeng, Li Fazhan

(Anyang Iron and Steel Group Co., Ltd., Anyang 455004, China)

Abstract: Sintering flue gas in the circulation process as part of the whole sintering production system, with the main system organically fuses in together, through the automatic control, adjustment of the new system and old system of temperature, pressure, air volume, four aspects of oxygen balance, can utilize the sensible heat of flue gas recycling, reduce fuel consumption, and can reduce the amount of smoke gas discharged into the atmosphere, reduce the main exhauster, main electric dust remover and desulfurization unit load, create favorable conditions for control of sintering flue gas pollution, is a traditional sintering idea of realizing the energy conservation and emissions reduction, development of green technology innovation and the effective exploration.

Key words: hot air sintering; flue gas circulation; energy conservation and emissions reduction; collaborative governance

碱金属碳酸盐对焦炭气化反应的催化动力学研究

程可扬，方云鹏，吴 亮，祝宇涵，朱荣锦，胡义卿，张生富

(重庆大学材料科学与工程学院，重庆 400044)

摘 要：为了研究碱金属碳酸盐对焦炭气化反应的催化作用，采用热重分析法对不同碱金属碳酸盐（Na_2CO_3 与 K_2CO_3）添加量的焦炭气化反应动力学进行研究。结果表明，碱金属碳酸盐对焦炭气化反应的正催化作用明显。随着碱金属碳酸盐的含量增加，焦炭气化反应最大反应速率增加，起始反应温度与激烈反应温度向低温区移动，且碳酸钾的作用效果明显大于碳酸钠。碱金属碳酸盐的催化机理主要为降低焦炭气化反应过程的活化能，在低含量时随着其含量的增加催化效果愈发明显，但到达一定值时，降低效果基本不再增加，主要原因为在此过程中化学反应是

焦炭气化反应的限制性环节,当碱金属碳酸盐含量较低时主要受一级反应控制,含量较高时受二级反应控制。

关键词:焦炭;气化反应;碱金属碳酸盐;气化反应动力学

Kinetics of Coke Gasification Reaction under the Catalysis of Alkali Carbonates

Cheng Keyang, Fang Yunpeng, Wu Liang, Zhu Yuhan,
Zhu Rongjin, Hu Yiqing, Zhang Shengfu

(College of Materials Science and Engineering, Chongqing University, Chongqing 400044, China)

Abstract: In order to study the catalytic effect of alkali carbonates on coke gasification reaction, the kinetics of coke gasification reaction with different alkali carbonates (Na_2CO_3 and K_2CO_3) was studied by thermogravimetric analysis. The results show that alkali carbonate has a significant positive catalytic effect on coke gasification reaction. With the increase of alkali carbonates content, the maximum reaction rate of coke gasification reaction increases, the temperatures both the initial reaction and the intense reaction move to the low temperature zone, and the effect of potassium carbonate is significantly greater than that of sodium carbonate. The catalytic mechanism of alkali carbonates is mainly to reduce the activation energy of coke gasification reaction process. The effect is more obvious with the increase of their content at low content, but when the content reaches a certain value, the reduction effect will not increase basically. The main reason is that the reaction is the restrictive link in coke gasification reaction, and the main reason is that it is controlled by the first-order reaction when the content of alkali metal carbonate is low and by the second-order reaction when the content is higher.

Key words: coke; gasification reaction; alkali carbonates; kinetics of gasification reaction

电磁作用对白云鄂博铌精矿渣金分离的影响

文 明,赵增武,李保卫

(内蒙古科技大学大学白云鄂博矿多金属资源综合利用国家重点实验室,内蒙古包头 014010)

摘 要:包头白云鄂博矿是以铁、铌、稀土为主的大型多金属矿,矿石性质非常复杂,该矿产资源拥有巨大的利用价值,探索其有价组元的高效利用具有重要意义。本文通过对高温状态下渣铁电磁分离的实验研究,在未施加磁场和施加磁场的实验条件下,探究了电磁作用对渣铁分离的影响。实验结果表明,在还原温度 1300℃、保温 60min 时,铌矿物的还原次序是 $FeNb_2O_6 \rightarrow NbO_2 \rightarrow NbC$;渣金分离温度为 1550℃、保温 20min 时,施加电磁作用可以使渣中铁的含量快速降低,碳化铌大部分保留在铁中。

关键词:白云鄂博矿;铌精矿;电磁作用;物相转变;NbC

Effect of Electromagnetic Action on Separation of Metal from Slag of Bayan Obo Niobium Concentrate

Wen Ming, Zhao Zengwu, Li Baowei

(Key Laboratory of Integrated Exploitation of Bayan Obo Muti-Metal Resources, Inner Mongolia University of Science and Technology, Baotou 014010, China)

Abstract: Baotou Bayan Obo Mine is a large-scale polymetallic ore mainly composed of Fe,Nband Re,the ore is very complicated has great value to be utilized. Exploring the high utilization of its valuable components is of great importance. In this paper, the experimental study on the electromagnetic separation of slag iron under high temperature conditions is carried out. Under the experimental conditions without applying magnetic field and applying magnetic field, the influence of electromagnetic action on slag-iron separation is explored. The results showed that when the reduction temperature is 1300℃ and the reduction time is 60min, the phase transition process of niobium minerals is $FeNb_2O_6 \rightarrow NbO_2 \rightarrow NbC$. The separation of metal from slag temperature is 1550℃ and the reduction time is 20min, the application of electromagnetic action can rapidly reduce the content of iron in the slag, and most of the NbC remains in the iron.

Key words: bayan obo mine; niobium minerals; electromagnetic effect; phase transfer; NbC

高炉炉缸结构与炉衬选型

卢正东[1,3]，陈令坤[1]，张正东[2]，向武国[2]，顾华志[3]，李承志[3]

(1. 宝钢股份中央研究院武汉分院炼铁所，湖北武汉 430080；
2. 宝钢股份武汉钢铁有限公司炼铁厂，湖北武汉 430073；
3. 武汉科技大学省部共建耐火材料与冶金国家重点实验室，湖北武汉 430081)

摘　要： 针对高炉炉缸结构和炉衬选型，对铸铁冷却壁、铜冷却壁和大块炭砖、复合炭砖不同组合形式的炉缸开展了温度场分析。结果表明，采用相同炭砖类型时，不同冷却壁选型对炉缸温度场影响较小，采用铸铁冷却壁即可满足长寿要求。未来炉缸优化方向在于提高炉衬炭砖导热系数，强化炉缸冷却效果，促进渣铁保护层的形成。采用铸铁冷却壁+大块炭砖（导热系数 20 W/m·K）时，炉缸热面 1150℃凝固线形成时对应炉衬残余厚度为 710mm，采用铸铁冷却壁+复合炭砖（小块炭砖导热系数 140W/m·K，大块炭砖导热系数 20W/m·K）时，1150℃凝固线形成时对应炉衬残余厚度为 960mm，上述两种结构炉缸均可满足长寿服役要求。

关键词： 高炉长寿；炉缸结构；冷却壁；炭砖；温度场

Hearth Structure and Lining Selection of Blast Furnace

Lu Zhengdong[1,3], Chen Lingkun[1], Zhang Zhengdong[2],
Xiang Wuguo[2], Gu Huazhi[3], Li Chengzhi[3]

(1. Wuhan Branch of Baosteel Central Research Institute, Wuhan 430080, China; 2. Ironmaking Plant of Baosteel Wuhan Iron and Steel Co., Ltd., Wuhan 430073, China; 3. The State Key Laboratory of Refractories and Metallurgy, Wuhan University of Science and Technology, Wuhan 430081, China)

Abstract: According to the hearth structure and lining selection of blast furnace, the temperature field of hearth with different combinations of cast iron cooling stave, copper cooling stave, large carbon brick and composite carbon brick was analyzed. The results show that when the same type of carbon block is used, the different types of cooling stave have little influence on the temperature field of hearth, and the cast iron cooling stave can meet the requirement of longevity. In the future, the optimization direction of hearth is to improve the thermal conductivity of carbon brick lining, strengthen the cooling effect of hearth and promote the formation of slag-iron protective layer. For cast iron cooling stave + large carbon brick (thermal conductivity 20 W/m·K), the residual thickness of furnace lining is 710 mm at 1150℃. For cast iron cooling stave + composite carbon brick (small carbon brick thermal conductivity 140 W/m·K, large carbon brick thermal

焦粉粒度对烧结减排的影响

刘敏媛，严永红，汤 勇，刘守文

（华菱湘钢，湖南湘潭 411100）

摘　要：氮氧化物、烟尘和二氧化硫并称为 3 种主要的大气污染物，而烧结是重要的污染源。本项目主要考察焦粉粒度对烧结矿质量及减排的影响。通过实验得出：无论从烧结矿质量还是从减排来看，都应提高焦粉中 1~3mm 的比例，此时烧结矿转鼓强度、利用系数高，冶金性能好，减排少。烟气成分中无论是 CO、NO_x、SO_2 分成三类，中间粒级 1~3mm 一类，粗粒级 3~5mm 和>5mm 一类、细粒级<0.5mm 和 0.5~1mm 一类；CO 从高到低为细粒级、中间粒级、粗粒级；NO_x 从高到低为细粒级、粗粒级、中间粒级；SO_2 从高到低为细粒级、中间粒级、粗粒级；从减排来看，焦粉粒度 1~3mm 较适宜。

关键词：焦粉；粒度；质量；减排

Abstract: Nitrogen oxides, fume dust and sulfur dioxides are three major air pollutants, and the sintering is a major pollution source. This paper investigates mainly the influence of coke sizes on sinter quality and emission reduction. It is concluded from experiments that the coke size proportions between 1 and 3mm in coke powder should be increased in terms of sinter quality and emission reduction. In this case, we can get higher sinter drum strength, higher utilization coefficient, better metallurgical performance and little emission. No matter CO, NO_x and SO_2 are classified as three categories in flue gas compositions, the intermediate particle sizes between 1 and 3mm are categorized , the coarse particle sizes between 3 and 5mm and the coarse particle sizes more than 5mm are categorized, and the fine particle sizes less than 0.5mm and 0.5~1mm are   categorized. CO is graded as a fine grain, intermediate grain and coarse grain from high to low; NO_x is graded as a fine grain, coarse grain, intermediate grain from high to low; SO_2 is graded as a fine grain, intermediate grain and coursed grain from high to low. From the point of view of emission reduction, coke particle sizes between 1 and 3mm are more appropriate.

Key words: coke powder; particle size; quality; emission reduction

新型碳铁复合炉料冷压制备工艺的实验研究

鲍继伟，储满生，韩　冬，曹来更，柳政根，唐　珏

（东北大学冶金学院，辽宁沈阳 110819）

摘　要：碳铁复合炉料作为低碳炼铁新技术，具有提高高炉生产效率、降低焦比，减少 CO_2 排放的优点。本研究提出了利用粘结剂进行对辊成型和炭化处理的碳铁复合炉料冷压制备工艺，主要考察了各种工艺参数对碳铁复合炉料抗压强度的影响并进行了优化。结果表明，优化的碳铁复合炉料制备工艺参数为 30%铁矿、45%烟煤 1、20%烟煤 2、5%无烟煤，5%粘结剂 B 或者粘结剂 C，成型压力 3t/cm，炭化温度 1000℃，炭化时间 4h。此工艺条件下，添加粘结剂 B 时，碳铁复合炉料炭化前后抗压强度分别达 618N 和 3053N，金属化率达 86.34%；添加粘结剂 C 时，碳铁复合炉料炭化前后抗压强度分别达 1094N 和 3575N。所制备碳铁复合炉料具有较高的机械强度，满足高炉生产要求。

关键词：低碳炼铁；碳铁复合炉料；抗压强度；粘结剂；成型压力；炭化温度；炭化时间

Experimental Study on the Cold Pressing Preparation Process of New Ironmaking Burden-iron Carbon Agglomerates

Bao Jiwei, Chu Mansheng, Han Dong, Cao Laigeng,
Liu Zhenggen, Tang Jue

(School of Metallurgy, Northeastern University, Shengyang 110819, China)

Abstract: As a new low carbon ironmaking technology, iron carbon agglomerates (ICA) has the advantages of improving blast furnace production efficiency, reducing coke ratio and reducing CO_2 emission. In this study, the cold pressing preparation process of ICA with binders through roll molding and carbonization treatment was proposed. The effects of various process parameters on the compressive strength of ICA was investigated and optimized. The results showed that the optimized preparation process parameters of ICA are 30% iron ore, 45% bituminous coal 1, 20% bituminous coal 2, 5% anthracite and adding 5% binder B or binder C, molding pressure of 3t/cm, carbonization temperature of 1000℃, and carbonization time of 4h. Under the above preparation process parameters, when binder B is added, the compressive strengths of ICA before and after carbonization reach 618N and 3053N respectively, and the metallization rate reaches 86.34%. When binder C is added, the compressive strengths of ICA before and after carbonization reach 1094N and 3575N, respectively. ICA has a high mechanical strength, which can meet the requirements of blast furnace production.

Key words: low carbon ironmaking; iron carbon agglomerates; compressive strength; binder; molding pressure; carbonization temperature; carbonization time

新疆低质煤热解过程中煤的拉曼结构特征研究

姚 利，朱子宗，武 强，孙 灿，宗 楷

（重庆大学冶金工程系，重庆 400044）

摘 要： 拉曼光谱是一种有效的煤结构研究方法。通过对在不同温度下热解的煤进行拉曼光谱检测，可以发现煤微晶结构及其演化与拉曼光谱特征参数间存在密切联系。新疆低质煤的拉曼光谱实验结果表明在不同温度下热解所得半焦的拉曼光谱一阶模都存在两个明显的特征峰，分别为相对宽缓的 D 峰和更为尖锐的 G 峰。随热解温度的升高，煤焦中峰面积比 A_D/A_A 逐渐增加，A_G/A_A 不断降低。在热解过程中，改性剂 ZBS 的作用主要体现在低温段，同时会推迟焦炭石墨化进程，有利于降低焦炭的反应性。

关键词： 新疆低质煤；热解过程；拉曼光谱；结构演化

Study on the Evolution Characteristics of Coal's Roman Structure in Pyrolysis Process of Low-quality Coal in Xinjiang

Yao Li, Zhu Zizong, Wu Qiang, Sun Can, Zong Kai

(Chongqing University, Chongqing 400044, China)

Abstract: Raman spectroscopy is an effective method for studying coal structure. The Raman spectra of pyrolyzed coal at different temperatures show that there is a close relationship between the micro-crystalline structure and evolution of coal and the characteristics of Raman spectra. The experimental results of Raman spectra of Xinjiang low-quality coal show that there are two distinct characteristic peaks in the first-order mode of Raman spectra of semi-coke obtained by pyrolysis at different temperatures, which are relatively gentle D peak and sharper G peak. With the increase of pyrolysis temperature, the peak area ratio of coal char to A_D/A_A increases gradually, while A_G/A_A decreases continuously. In the pyrolysis process, the effect of modifier ZBS is mainly reflected in the low temperature section, and it will delay the graphitization process of coke, which is conducive to reducing the reactivity of coke.

Key words: Xinjiang low-quality coal; pyrolysis process; Raman spectrum; structural evolution

高炉煤气高效、高值和创新利用技术

刘文权[1]，吴记全[2]

（1. 冶金工业规划研究院，北京　100013；2. 方大特钢科技股份有限公司，江西南昌　330012）

摘　要：本文从高炉煤气基本特性出发，论述了高炉煤气高效、高值利用的多种途径，指出了钢化联产有利于平衡钢铁市场由于价格波动带来的市场风险，实现钢铁尾气的超值利用，多次高效增值。有利于延长产业链条，优化产业结构，走钢化联产、低碳减排的可持续发展道路。

关键词：高炉煤气；产业链；钢化联产

High Efficiency, High Value and Innovative Utilization Technology of Blast Furnace Gas

Liu Wenquan[1], Wu Jiquan[2]

(1. MPI, Beijing 100013, China; 2. Fangda Special Steel Technology Co., Ltd., Nanchang 330012, China)

Abstract: Based on the basic characteristics of blast furnace gas, this paper discusses various ways of high efficiency, high value and innovative utilization of blast furnace gas, and points out that the combined production of steel is beneficial to balance the market risk caused by price fluctuation in the steel market, realize the over-value utilization of steel exhaust, and increase the value of multiple high efficiency. It is beneficial to extend the industrial chain, optimize the industrial structure, and take the road of sustainable development of Steel chemical combined production and low carbon emission reduction.

Key words: blast furnace gas; the industrial chain; steel chemical combined

根据实用性分类配煤提高 M40 指标的方法

周俊兰，张明星，杜　屏

（江苏（沙钢）钢铁研究院炼铁研究室，江苏张家港　215600）

摘 要：沙钢在 2017 年上半年 1~6 号焦炉 M40 指标低于 84%的现象增多，焦炭质量下滑。为了提高 M40 指标，稳定焦炭质量，通过对沙钢来煤进行工业分析、粘结性评价、镜质组反射率、是否混煤等全方位检测和评价，参照国标分类标准对 29 种炼焦煤种根据实用性重新分类，结合现场配煤单，参考基氏流动度和单种炼焦煤镜质组反射率进行煤种替代配煤研究。使用新的配煤方案后，焦肥煤比例下降，M40 指标逐渐上升。

关键词：实用性分类；配煤；基氏流动度；M40

Method to Promote M40 Index According to Coal Blending Based on Applied Classification

Zhou Junlan, Zhang Mingxing, Du Ping

(Ironmaking Group, Iron and Steel Research Institute, Shasteel/Jiangsu, Zhangjiagang 215600, China)

Abstract: The M40 index in Shasteel's 1~6# blast furnace was lower than 84% during first half year of 2017, coke quality was declining. In order to promote M40 index and steady coke quality, coals arrived in Shasteel were tested and analyzed with industrial analysis, caking property, vitrinite reflectance, mixed coal degree. Then based on application and national standard the 29 coking coals were reclassified. Combining coal blending sheet on site and gieseler fluidity and vitrinite reflectance, coals replacement in coal blending sheet was carried out. After using new coal blending method, the ratio of coking coal and fat coal was declined and M40 index was increased

Key words: applied classification; coal blending; gieseler fluidity; M40

韶钢 7 号高炉护炉操作实践

陈生利，匡洪锋，陈国忠

（宝武集团广东韶关钢铁有限公司炼铁厂，广东韶关 512123）

摘 要：针对高炉本体长高，炉身冷却壁大量烧坏，炉缸区域炉墙侵蚀，炉缸侧壁温度偏高现象，总结出炉龄后期的护炉操作技术，为今后高炉长寿管理提供了技术依据。

关键词：炉龄后期；炉体长高；护炉

Operation Practice of Furnace Maintenance of BF #7 at SGIS

Chen Shengli, Kuang Hongfeng, Chen Guozhong

(Baowu Group Guangdong Shaoguan Iron and Steel Co., Ltd., Shaoguan 512123, China)

Abstract: In view of the phenomenon that the blast furnace body is high, the cooling walls of the blast furnace body are burnt out, the furnace walls in the hearth area are eroded, and the temperature of the side walls of the blast furnace is high, the operation technology of protecting the blast furnace in the late stage of the campaign is summarized, which provides the technical basis for the management of blast furnace longevity in the future.

Key words: late stage of campaign; furnace body growing taller; furnace maintenance

韶钢 2200m³ 高炉无计划长期停炉恢复炉况实践

齐万兵，匡洪锋，曹旭博

（宝武集团广东韶关钢铁有限公司炼铁厂，广东韶关 512123）

摘　要：2018 年 2 月初韶钢 7 号高炉因喷渣事故无计划长期停炉达 54 天。为顺利恢复炉况，制定了详细全面的安全复产方案，并对设备进行了全方位的升级改造。经专家组审核通过后，7 号高炉按照"炉缸冻结处理"模式以两个风口送风复产。复产过程中通过埋煤氧枪加热炉缸、合理控制操作参数、强化炉前出渣铁等措施，在 25 天内恢复正常炉况，两天后利用系数稳定在 2.73t/m³ 以上，此次复产，积累了丰富的操作经验，也提高了事故的处理能力。

关键词：高炉；无计划停炉；复产；恢复炉况

Practice of Recovering Furnace Conditions after Long Term Unscheduled Shutdown for a 2200m³ BF at SGIS

Qi Wanbing, Kuang Hongfeng, Cao Xubo

(Baowu Group Guangdong Shaoguan Iron and Steel Co., Ltd., Shaoguan 512123, China)

Abstract: At the beginning of February 2018, the No. 7 blast furnace of SGIS was shutdown without schedule for 54 days due to the slag spraying accident. In order to restore the furnace conditions smoothly, a detailed and comprehensive safety recovery plan was worked out, and the equipment had been upgraded in all directions. After passing the expert group's examination and approval, the No. 7 BF was recovered for production in accordance with the "hearth freezing treatment" mode with two outlets. In the process of reproducing, the normal furnace conditions was restored within 25 days by heating the hearth with a burial oxygen lance, controlling the operating parameters reasonably, and strengthening the slag and iron in the casthouse. After two days, the utilization coefficient was stable above 2.73t/m. This reproduction accumulated rich experience and improved the handling capacity of the accident.

Key words: blast furnace; unplanned shutdown; reproduction; recovery of BF conditions

韶钢六号高炉高生矿比生产实践

李国权，王　振，柏德春

（宝武集团广东韶关钢铁有限公司炼铁厂，广东韶关 512123）

摘　要：韶钢六号高炉 2015 年 4 月 19 日投产，有效炉容 1050m³，为降低炼铁成本六号高炉通过以块代球生产实践，根据高生矿比炉料特点形成一套稳定的操作模型，入炉生矿比例由 20%逐步增加至 30%以上，实现了经济型生产炉料结构。

关键词：高炉；高生矿比；实践

Production Practice of High Rawore Ratio for BF #6 at SGIS

Li Guoquan, Wang Zhen, Bai Dechun

(Baowu Group Guangdong Shaoguan Iron and Steel Co., Ltd., Shaoguan 512123, China)

Abstract: BF #6 of SGIS was put into operation on April 19, 2015, with an effective capacity of 1050m. A stable operation model was formed according to the characteristics of high rawore ratio for charging BF #6 in order to reduce the cost of ironmaking. The proportion of rawore charging is increased gradually from 20% to more than 30%, thus realizing the economical production burden structure.

Key words: BF; high rawore ratio; practice

巨型高炉炉缸死料柱焦炭研究

牛 群[1]，陈艳波[2]，郑朋超[2]，程树森[1]，
徐文轩[1]，于兆斌[3]，马元明[3]

（1. 北京科技大学冶金与生态工程学院，北京 100083；2. 首钢京唐钢铁联合有限责任公司炼铁作业部，河北唐山 063210；3. 方大炭素新材料科技有限公司技术研发部，甘肃兰州 730084）

摘 要：巨型高炉炉缸焦炭行为的研究对深入了解炉缸工作状态及高炉长寿有着重要的意义。本文在巨型高炉停炉时对炉缸不同高度中心焦炭进行取样，详细对其成分、微观结构、物相和无机矿物质含量进行了研究。此外，还分析了死料柱焦炭变化对高炉炉缸长寿的影响。研究发现，高炉渣能够大量渗入焦炭内部；铁口中心线以下区域，只要有死料柱存在的地方就会有高炉渣的存在，高炉渣的存在为炭砖热面形成保护层提供了物质基础。

关键词：高炉；死料柱；焦炭；微观结构；保护层

Study on Deadman Coke of a Huge Industrial Blast Furnace

Niu Qun[1], Chen Yanbo[2], Zheng Pengchao[2], Cheng Shusen[1],
Xu Wenxuan[1], Yu Zhaobin[3], Ma Yuanming[3]

(1. School of Metallurgical and Ecological Engineering, University of Science and Technology Beijing, Beijing 100083, China; 2. Ironmaking Department, Shougang Jingtang United Iron and Steel Co., Ltd., Tangshan 063210, China; 3. Technology Research and Development Department, Fangda Carbon New Material Co., Ltd., Lanzhou 730084, China)

Abstract: The study on the coke behavior of a huge blast furnace hearth is of great significance for understanding the working condition of the hearth and the longevity of the blast furnace. In this paper, the central coke samples from different heights of the hearth was obtained after blow out and cool down of the blast furnace and the coke composition, microstructure, mineral phase and inorganic mineral content were studied in detail. The influence of the changes of deadman coke on the longevity of blast furnace hearth was also analyzed. The results show that the blast furnace slag can penetrate into the interior of the deadman coke; Below the center line of the taphole, as long as there is a deadman, there will be blast furnace slag. The presence of blast furnace slag provides a material basis for the formation of a protective layer on the hot surface of the carbon brick.

Key words: blast furnace; deadman; coke; microstructure; protective layer

热风炉焊接残余应力周向分布与消除

陈 辉[1,2]，孙 健[2]，王 伟[2]，梁海龙[2]，熊 军[3]，
王长水[3]，王 建[3]，要志超[3]

（1. 首钢集团有限公司技术研究院，绿色可循环钢铁流程北京市重点实验室，北京 100043；
2. 首钢集团有限公司技术研究院，钢铁技术研究所，北京 100043；3. 首钢京唐
联合钢铁有限公司炼铁作业部，河北唐山 063200）

摘 要：残余应力集中并在运行过程中不断释放是影响承压装备开裂、失效的原因之一。首次尝试对正在施工的热风炉炉壳焊接前后应力分布情况，提出了检测方案，并获得了周向上应力分布不均，且最大应力值超过许用应力的事实。退火消除可以消除近70%以上的残余应力，但炉壳开孔附近因保温效果差，影响应力消除效果。大型工艺装备残余应力检测与消除是必要的，建议推广。

关键词：热风炉；残余应力；退火；应力消除

Circumferential Distribution and Elimination of Welding Residual Stress for Hot Blast Stove

Chen Hui[1,2], Sun Jian[2], Wang Wei[2], Liang Hailong[2], Xiong Jun[3],
Wang Changshui[3], Wang Jian[3], Yao Zhichao[3]

(1. Shougang Group, Shougang Research Institute of Technology, Beijing Key Laboratory
of Green Recyclable Process for Iron & Steel Production Technology, Beijing 100043, China;
2. Shougang Group, Shougang Research Institute of Technology, Beijing 100043, China;
3. Shougang Jingtang Iron & Steel Co., Ltd., Tangshan 063200, China)

Abstract: The concentration of residual stress and its continuous release during operation were one of the reasons affecting the cracking and failure of pressure equipment. For the first time, the stress distribution before and after welding of the hot blast stove shell under construction was attempted, and the detection scheme was put forward. The fact that the circumferential stress distribution was uneven and the maximum stress value exceeds the allowable stress was obtained. Removal of residual stress by annealing can eliminate more than 70%, but the effect of stress relief was affected by poor thermal insulation near the opening of furnace shell. It was necessary to detect and eliminate the residual stress of large-scale process equipment, and it was suggested that it should be popularized.

Key words: hot blast stove; residual stress; annealing process; stress relief

并罐高炉无钟炉顶设备结构对布料偏析的影响

徐文轩[1]，高永会[2]，谢建军[2]，程树森[1]，牛　群[1]，梅亚光[1]

(1. 北京科技大学冶金与生态工程学院，北京　100083；
2. 河钢集团邯钢公司邯宝炼铁厂，河北邯郸　056009)

摘　要：本文以国内大型并罐高炉为研究对象，利用离散单元法分析了并罐高炉无钟炉顶设备结构包括上料主皮带方位角、料罐出口位置、料罐出口倾角、中心喉管直径和旋转溜槽结构对料面炉料偏析分布的影响。研究结果指出，上料主皮带中心线与两并罐对称面不重合会导致左右料罐布料不对称，由于布料偏析具有遗传性，因此在高炉设计时应避免上料主皮带中心线与两并罐对称面不重合。料罐出口位置位于料罐中心或靠近高炉中心线以及料罐出口倾角为70°时有利于将大颗粒布入高炉中心，小颗粒布入高炉边缘。在中心喉管不堵料的前提下，缩小中心喉管直径有利于减小料面炉料流量偏析。最后，使用料磨料结构的方溜槽可以有效减小料面炉料流量偏析。

关键词：并罐；高炉；设备结构；布料；偏析；离散单元法

Influence of Bell-less Top Equipment Structure on the Segregation of Burden Distribution in Blast Furnace with Parallel Hopper

Xu Wenxuan[1], Gao Yonghui[2], Xie Jianjun[2], Cheng Shusen[1],
Niu Qun[1], Mei Yaguang[1]

(1. School of Metallurgical and Ecological Engineering, University of Science and Technology Beijing, Beijing 100083; 2. HBIS Group Hansteel Company Hanbao Ironmaking, Handan 056009 China)

Abstract: In this study, the domestic large blast furnace with parallel hopper was research object, the influence of bell-less top equipment structure including the azimuth angle of main feeding belt, the outlet position of hopper, the outlet tilting angle of hopper, the diameter of center throat tube and the rotating chute structure on the segregation of burden distribution on the stock surface was analyzed based on Discrete Element Method. The research result shows that the burden distribution is asymmetric during the left and right hopper discharging process due to the main feeding belt centerline not coincidence with the symmetry plane of two parallel hoppers, the main feeding belt centerline should coincidence with the symmetry plane of two parallel hoppers during the blast furnace design process due to the genetic characteristic of burden distribution segregation. It is beneficial to distribute large particles into the center of blast furnace and small particles into the edge of blast furnace when the outlet position of hopper locates in the center of hopper or near the centerline of blast furnace and the hopper outlet tilting angle is 70°. Decreasing the diameter of center throat tube is good to reduce the flux segregation of burden distribution in the stock surface when the clogging phenomenon not occur in the center throat tube. Finally, using the rectangular rotating chute and adding the rock box on the rotating chute can reduce the flux segregation of burden distribution in the stock surface effectively.

Key words: parallel hopper; blast furnace; equipment structure; burden distribution; segregation; discrete element method

真空碳酸钾法脱硫工艺优化和改进

杨志军，杨 涛

（宝武集团广东韶关钢铁有限公司炼铁厂，广东韶关 512123）

摘 要：介绍了焦炉煤气真空碳酸钾脱硫工艺，结合生产实际情况，分析了生产过程中脱硫效率偏低、脱硫液变质、脱硫废液处理等故障问题产生的原因，并从原料控制、生产操作及工艺等方面，提出了相应的改进措施。改进后，脱硫效率提升，提高系统生产稳定性。

关键词：真空碳酸钾；脱硫效率；故障分析；改造

Optimization and Improvement of Vacuum Potassium Carbonate Desulfurization Process

Yang Zhijun, Yang Tao

(Baowu Group Guangdong Shaoguan Iron and Steel Co., Ltd., Shaoguan 512123, China)

Abstract: The desulfurization process with vacuum potassium carbonate for coke oven gas is introduced. Combining with the actual production situation, the causes of the low desulfurization efficiency, the deterioration of desulfurization liquid and the treatment of desulfurization waste liquid in the process are analyzed, and the corresponding improvement measures are put forward from the aspects of raw material control, production operation and process. After improvement, the desulfurization efficiency and the production stability of the system have been improved.

Key words: vacuum potassium carbonate; desulfurization efficiency; fault analysis; renovation

装炉煤细度对焦化生产的影响研究

王永亮，郭 飞，何 波，陈 康，嵇长发

（青岛特殊钢铁有限公司焦化厂，山东青岛 266400）

摘 要：把单种煤、配合煤的细度作为唯一变量进行设计，利用小焦炉试验得出结果，并进行了数据全面分析。分析表明，无论是单种煤还是配合煤中＜0.5mm煤粉比例增加，均会造成焦炭CSR和M25的下降、M10上升，对焦炭的粒级分布也会造成影响。同时，结合实际生产情况，阐述了配合煤的细度过高对焦化生产带来的负面影响。

关键词：细度；小焦炉实验；单种煤；配合煤；CSR；M25；煤气净化

Study on the Influence of Fineness of Charging Coal on Coking Production

Wang Yongliang, Guo Fei, He Bo, Chen Kang, Ji Changfa

(Qingdao Special Iron and Steel Co., Ltd., Qingdao 266400, China)

Abstract: The fineness of a single coal or blended coal is taken as the only variable to design the experiment, and the results are obtained by the small coke oven test, and the results are analyzed. The analysis shows that the increase of pulverized coal ratio below 0.5mm will result in the decrease of CSR and M25 and the increase of M10, which will also affect the particle size distribution of coke. At the same time, combined with the actual production situation, the influence of excessive fineness of blended coal on coking production was expounded.

Key words: fineness; small coke oven experiment; single coal; blended coal; CSR; M25; gas purification

宝钢湛江钢铁高炉长寿技术设计与应用

周 琦，贾海宁，苏 威，沙华玮

（宝钢湛江钢铁有限公司，广东湛江 524072）

摘 要： 宝钢湛江钢铁基地建有 2 座 $5050m^3$ 高炉，年产铁水 823 万 t，设计无中修情况下一代炉龄 22 年。1 号高炉于 2015 年 9 月 25 日点火投产，2 号高炉于 2016 年 7 月 15 日点火投产。本文从长寿的角度，简述了湛钢高炉长寿技术的设计与应用。

关键词： 高炉；长寿技术；设计

Design & Practice of Long Service-life of Blast Furnaces in Baosteel Zhanjiang Iron & Steel Works

Zhou Qi, Jia Haining, Su Wei, Sha Huawei

(Baosteel Zhanjiang Iron & Steel Co.,Ltd., Zhanjiang 524072, China)

Abstract: There are two $5050m^3$ blast furnaces in Baosteel Zhanjiang Iron & Steel base, designed annual iron production of 8.23 million tons, and 22 year-service life. No.1 blast furnace (BF) was put into operation on September 25, 2015, and No.2 blast furnace (BF) was put into operation on July 15, 2016. This paper briefly introduced the design and technology employed for prolong service life of furnace.

Key words: blast furnace(BF); service life; design

湛钢烧结节能环保生产实践

李 明

(宝钢湛江钢铁有限公司炼铁厂,广东湛江 524000)

摘 要:本文介绍了湛江钢铁烧结单元自设计至投产以来,为烧结过程节能环保所开展的系列举措,并在生产实践中取得了良好效果,烧结工序能耗逐年降低,各项环保指标排放稳定,创造了巨大的经济和社会效益。

关键词:烧结;节能;环保

Pratice of Energysaving and Environmental of Zhansteel Sintering Process

Li Ming

(The Ironmaking Plant of Zhanjiang Iron & Steel Co., Ltd., Zhanjiang 524000, China)

Abstract: From the design to the production, series of measures has been taken by Zhanjiang Iron and Steel Sintering Unit in orden to ensure that the sintering process is energy-saving and environmental. And it has achieved good results in the production practice. The energy consumption of the sintering process is decreasing year by year, and the emission of various environmental indicators is stable. It has created enormous economic and social benefits.

Key words: sintering; energysaving; environmental

韶钢烧结适宜焦粉粒度组成的技术研究

毛爱香,蓝伯洋

(宝武集团广东韶关钢铁有限公司制造管理部,广东韶关 512123)

摘 要:通过对韶钢现用配矿方案进行烧结适宜焦粉粒度组成的技术研究,创造性地提出了韶钢烧结历年来未明确的焦粉适宜粒度组成,为韶钢烧结用焦粉破碎工艺提供了科学依据;该研究成果应用于韶钢实际生产中后,年降低烧结固体燃料成本 4400 万元以上。

关键词:烧结;配矿方案;焦粉粒度组成;固体燃耗

Study on Sintering Technology of Suitable Particle Size Composition of Coke Powder at SGIS

Mao Aixiang, Lan Boyang

(Baowu Group Guangdong Shaoguan Iron and Steel Co., Ltd., Shaoguan 512123, China)

Abstract: Through the research on the technology of suitable size composition of coke powder for sintering in the existing ore blending scheme at SGIS, the suitable size composition of coke powder for sintering, which has not been specified over the years, has been creatively put forward, which provides a scientific basis for the crushing process of coke powder for sintering. After the research results are applied to the actual production at SGIS, the cost of solid fuels for sintering has been reduced over RMB 44 million yuan annually.

Key words: sintering; blending scheme; coke powder size composition; solid fuel

BLOOM LAKE 精粉烧结试验研究和生产实践

欧阳希，毛爱香，余 骏，黄承芳

（宝武集团广东韶关钢铁有限公司制造管理部，广东韶关 512123）

摘 要： BLOOM LAKE 精粉是一种很纯净的矿粉，是属以赤铁矿为主的粗精粉，具有品位高、Al_2O_3 含量低的特点。通过烧结试验研究，在韶钢当前原料条件下，确定 BLOOM LAKE 精粉的合适使用配矿结构。烧结使用 BLOOM LAKE 精粉的配矿结构后，烧结矿质量稳定，烧结矿产量提升，缓解了铁烧平衡。高炉使用 BLOOM LAKE 精粉配矿结构后，炉况保持较好的顺行状态，产量基本保持在 8000t/d 以上，燃料消耗逐步降低。通过分析，BLOOM LAKE 精粉配矿结构可满足高炉使用要求。

关键词： BLOOM LAKE 精粉；配矿结构；试验研究；高炉使用

Research and Production Practice of BLOOM LAKE Fine Powder Sintering Trial

Ouyang Xi, Mao Aixiang, Yu Jun, Huang Chengfang

(Baowu Group Guangdong Shaoguan Iron and Steel Co.,Ltd., Shaoguan 512123, China)

Abstract: The fine powder of BLOOM LAKE iron concentrate is a kind of very pure mineral powder, namely a kind of crude refined powder mainly composed of hematite, with high grade and low content of Al_2O_3. The appropriate ratio of using it has been determined through the sintering experiment research in accordance with the current material conditions at SGIS. After using the ratio for blending for sintering, the quality of the sinter was stable and the output of the sinter was increased, which alleviated the balance of iron-making and sintering. When the sinter blended according to the ratio is charged into BF, the furnace conditions can keep in good state, the output is basically maintained above 8000t/d, and the fuel consumption is gradually reduced. The blending can meet the use requirements of blast furnace through the analysis.

Key words: BLOOM LAKE iron ore concentrate; ore blending structure; experiment research; application in BF

要努力提高高炉炼铁喷煤比

王维兴

（中国金属学会，北京 100081）

摘　要： 提高高炉喷煤比是我国炼铁技术发展方向，可以炼焦短缺、低污染物排放、降低生产成本、降低投资、允许使用高风温；近年来，我国喷煤比在下降，不利于提高企业效能，要尽快扭转；提高喷煤比的技术措施：保持炉缸热量充沛、提高煤粉燃烧率、改善炉料质量、提高炉料透气性、使用高质量煤粉等。

关键词： 努力提高；高炉；喷煤比

Strive to Increase the Coal Injection Ratio of Blast Furnace

Wang Weixing

(The Chinese Society for Metals, Beijing 100081, China)

Abstract: Increasing the coal injection ratio of BF is the development direction of Chinese ironmaking technology, which can solve the shortage of coke and achieve low pollutant emissions, thereby reducing production costs and investment. In recent years, the coal injection ratio of BF in China has been continuously decreasing, which is not conducive to improving the efficiency of enterprises, so it is necessary to reverse this situation as soon as possible. The measures to increase the coal injection ratio are as follows. Maintaining the furnace hearth with sufficient heat, increasing the combustion rate of the coal powder, improving the quality of the furnace charge, improving the gas permeability of the furnace charge, and using high-quality coal powder.

Key words: striveincrease; blast furnace; injecting coal rate

降低高炉燃料比的作用

王维兴

（中国金属学会，北京　100081）

摘　要： 多年来，我国高炉平均燃料比在 545kg/t 左右，比国际先进水平高 40kg/t，约有 20 几座高炉燃料比在 500kg/t 左右，大多数高炉燃料比偏高。降低燃料比可以降低炼铁生产成本，减少污染物排放，降低炼铁工序能耗，实现炼铁绿色生产。不再追求高冶炼强度、高产，可以降低燃料比；提高高炉入炉矿石铁品位、热风温度，改善炉料质量（特别是焦炭质量），提高高炉操作水平等技术措施均可以降低燃料比。

关键词： 降低；高炉；燃料比

Effect on Reducing the Fuel Ratio of Blast Furnace

Wang Weixing

(The Chinese Society for Metals, Beijing 100081, China)

Abstract: Over the years, the average fuel ratio of BF in China is around 545kg/t, which is 40kg/t higher than the international advanced level. There are more than 20 BF in China with a fuel ratio of about 500kg/t and most BFs have a higher fuel ratio. Reducing the fuel ratio can reduce the production cost and energy consumption of ironmaking as well as reduce the emission of pollutants, and realize the green production of ironmaking. Nowadays, high smelting strength and high yield are no longer pursued. The fuel ratio is reduced by improving the BF operation level, improving the iron grade of the ore, increasing the hot blast temperature, and improving the quality of the furnace charge (especially the quality of coke).

Key words: reduction; blast furnace; fuel ratio

基于烧结基础特性的烧结优化配矿模型

潘禹竹，柳政根，储满生，高立华，唐珏

（东北大学冶金学院，辽宁沈阳 110819）

摘　要： 通过烧结基础特性实验以及烧结杯实验，在控制其余变量条件下，得到某企业主要烧结原料——赤铁矿 A、赤铁矿 B、磁铁矿 A 的单因素烧结基础特性以及相应烧结矿冶金性能。积累前期实验数据并预处理后，通过 SPSS 软件进行线性拟合建模，将烧结基础特性与烧结矿冶金性能进行联系，可得出精度较高的目标函数，用于现场烧结矿的性能预测。

关键词： 烧结基础特性；烧结矿；冶金性能；线性建模；SPSS

Ore-matching Optimization Model based on the Basic Characteristics of Sintering

Pan Yuzhu, Liu Zhenggen, Chu Mansheng, Gao Lihua, Tang Jue

(School of Metallurgy, Northeastern University, Shenyang 110819, China)

Abstract: According to the basic sinter characteristic experiments and the sinter pot tests, controlling other relevant variables, the basic sinter characteristics and metallurgical properties, varying with the content of hematite A, hematite B and magnetite A from a steel enterprise were obtained. After the preliminary experimental data are accumulated and pre-processed, linear fitting modeling was carried out by SPSS software, and the basic characteristics of sintering were connected with metallurgical properties. The objective function with high precision would be obtained for the prediction of metallurgical properties in sintering.

Key words: basic sintering characteristics; sinter; metallurgical properties; linear method; SPSS

超高品位铁精矿气基直接还原的可行性研究

黄欧瑾，应自伟，吴宇航，林兴良，李靖宇

（东北大学冶金学院，辽宁沈阳 110819）

摘　要： 本文针对之前提出的超高品位铁精矿直接还原-熔分制备高纯铁新工艺中的直接还原进行可行性研究。研究结果表明：超高品位铁精矿氧化球团在 900℃、$H_2/CO=2/5$ 与 1050℃、$H_2/CO=5/2$ 两种还原条件下的膨胀率分别为 243%与 51%，不能应用气基竖炉直接还原工艺；添加 4%CaO 试剂后还原膨胀率分别降低至 3.2%和 8.8%，可以满足气基竖炉生产要求。

关键词： 高纯铁；超高品位铁精矿；直接还原；熔分

Discussion on Feasibility of Production of Direct Reduction Iron by Usingultra-high Grade Iron Concentrate

Huang Oujin, Ying Ziwei, Wu Yuhang, Lin Xingliang, Li Jingyu

(School of Metallurgy, Northeastern University, Shenyang 110819, China)

Abstract: The feasibility of production of direct reduction iron by using ultra-high grade iron concentrate is discussed in this paper, which is included in new process of preparing high purity iron by direct reduction and melting separation of ultra-high grade iron concentrate. The results show the ultra-high grade iron concentrate oxidized pellet have the problem of high reduction swelling index (243% in the system of reduction temperature 900℃ and $H_2/CO=2/5$ and 51.20% in the system of reduction temperature 1050°C and $H_2/CO=5/2$), and it can hardly meet the requirements of gas based shaft furnace direct reduction process; the reduction swelling index can be decreased to 3.2% and 8.8% after adding 4% CaO reagent in raw materials, and it can meet the requirents of gas based shaft furnace direct reduction process.

Key words: high purity iron; ultra-high grade iron concentrate; direct reduction; melting separation

高炉冶炼高钒高铬型钒钛磁铁矿资源综合利用关键技术

薛向欣[1,2]，程功金[1,2]，杨 合[1,2]，段培宁[1,2]

（1. 东北大学冶金学院，辽宁沈阳 110819；
2. 辽宁省冶金资源循环科学重点实验室，辽宁沈阳 110819）

摘 要： 高钒高铬型钒钛磁铁矿作为一类特殊的多金属共伴生矿，矿物组成复杂，综合利用难度极大，在本课题研究之前国内外无工业化生产实践。本文分别从该矿球团矿制备与性能、烧结矿制备与性能以及高炉炉料结构与操作等方面对高炉冶炼该矿资源综合利用关键技术进行了研究。球团方面，涉及该矿磨矿前后的增量化配加、配矿影响及有价组元 V_2O_5、TiO_2、Cr_2O_3、B_2O_3、CaO、MgO 对球团冶金性能的影响。烧结方面，涉及该矿烧结基础特性、增量化配加、配矿及配碳量、碱度、硼铁矿及 MgO 等对烧结矿冶金性能的影响。炉料结构与高炉操作方面，涉及工业化试验到工业化生产的合理炉料结构、球团利用系数、烧结利用系数、高炉利用系数、综合焦比等及铁水含钒等。针对该矿的系列研究实现了从实验室研究到工业化试验与应用的跨越，为高钒高铬型钒钛磁铁矿的大规模利用起到了指导和借鉴作用。

关键词： 高钒高铬型钒钛磁铁矿；球团矿；烧结矿；炉料结构；高炉操作

Key Technology of Resource Comprehensive Utilization for High-Chromium and High-Vanadium Vanadium-Titanium Magnetite Smelting in the Blast Furnace

Xue Xiangxin[1,2], Cheng Gongjin[1,2], Yang He[1,2], Duan Peining[1,2]

(1. School of Metallurgy, Northeastern University, Shenyang 110819, China; 2. Liaoning Key

Laboratory of Recycling Science for Metallurgical Resources, Shenyang 110819, China)

Abstract: High-chromium and high-vanadium vanadium-titanium magnetite (HCHVVTM), as one kind of particular multi-metal co-associated ore, has complex mineral compositions and fairly difficult comprehensive utilization technologies, and the industrial production had been elusive before this study carried out by our team. Pellet preparation and properties, sinter preparation and properties, blast furnace burden structure and operation, etc, were investigated for the key technology of resource comprehensive utilization for high-chromium and high-vanadium vanadium-titanium magnetite smelting in the blast furnace in this paper. In the pelletizing study, systematic investigations on incremental addition and ore proportioning of HCHVVTM and effect of valuable components including V_2O_5, TiO_2, Cr_2O_3, B_2O_3, CaO and MgO were carried out. In the sintering study, basic characteristics of sintering, incremental addition and ore proportioning of HCHVVTM and effect of carbon addition, basicity, paigeite and MgO were carried out. In the furnace burden and operation study, reasonable furnace burden, pelletizing utilization coefficient, sintering utilization coefficient, blast furnace coefficient, comprehensive coke ratio, [v]% in the melted iron, etc were carried out. The technological leapfrogging from laboratory investigations to industrial tests and applications was realized, providing the guidance and reference for the utilization of high-chromium and high-vanadium vanadium-titanium magnetite on a large scale.

Key words: high-chromium and high-vanadium vanadium-titanium magnetite; pellet; sinter; furnace burden structure; blast furnace operation

辽西高钒高钛型磁铁矿资源综合利用关键技术

程功金[1,2]，薛向欣[1,2]，高子先[1]，杨合[1,2]，
邢振兴[1]，马杰[1]，刘学志[1]

（1. 东北大学冶金学院，辽宁沈阳　110819；
2. 辽宁省冶金资源循环科学重点实验室，辽宁沈阳　110819）

摘　要： 辽西高钒高钛型磁铁矿矿物组成复杂，综合利用难度极大，针对该矿的系统性研究较为匮乏。本文分别从辽西高钒高钛型磁铁矿造块、直接还原、直接提钒等方面对该矿资源综合利用关键技术进行了研究，丰富了辽西高钒高钛型磁铁矿的研究体系，为该矿未来的高效利用起到了指导和借鉴作用。

关键词： 辽西高钒高钛型磁铁矿；造块；直接还原；直接提钒

Key Technology of Resource Comprehensive Utilization for Liaoxi High-Vanadium and High-Titanium Magnetite

Cheng Gongjin[1,2], Xue Xiangxin[1,2], Gao Zixian[1], Yang He[1,2],
Xing Zhenxing[1], Ma Jie[1], Liu Xuezhi[1]

(1. School of Metallurgy, Northeastern University, Shenyang 110819, China; 2. Liaoning Key Laboratory of Recycling Science for Metallurgical Resources, Shenyang 110819, China)

Abstract: Liaoxi high-vanadium and high-titanium magnetite (HVHTM) has complex mineral compositions and fairly difficult comprehensive utilization technologies, and systematic studies have still been scarce. Agglomeration, direct reduction and direct vanadium-extracting, etc, were investigated for the key technology of resource comprehensive utilization in this paper, enriching the researching system and providing the guidance and reference for the efficient utilization for HVHTM.

Key words: high-vanadium and high-titanium magnetite; agglomeration; direct reduction; direct vanadium-extracting

SCOPE21 工艺现状及应用推广存在的问题

黄世平[1]，Alex Wong[2]，刘守显[1]，牛　虎[2]

(1. 河钢工业技术服务有限公司，河北石家庄　050023；
2. 济南戴耳塔干燥设备有限公司，山东济南　250101)

摘　要： 从工艺技术角度出发，分析新日铁中断推广已成熟的 SCOPE21 工艺的原因，针对所存在的问题提出解决方案以开发出更加先进的焦煤预处理工艺。

关键词： 炼焦煤；DAPS；SCOPE21；高温流化床；快速预热；弱粘煤

Current Status of SCOPE21 Process and its Problems in Application

Huang Shiping[1], Alex Wong[2], Liu Shouxian[1], Niu Hu[2]

(1. HBIS Industrial Technology Service Co., Ltd., Shijiazhuang 050023, China;
2. Jinan Delta Drying Equipment Co., Ltd., Jinan 250101, China)

Abstract: Analysis were conducted on the reasons why Nippon Steel has discontinued the application of the mature SCOPE21 process from the perspective of technology, and solutions to the existing issues were proposed to develop a more advanced coking coal pretreatment process.

Key words: coking coal; DAPS; SCOPE21; high temperature fluidized bed; rapid heating; slightly-caking coal

无气隙热阻球墨铸铁冷却壁的研制

宗燕兵[1]，张学栋[1]，周传禄[2]，辛虹霓[2]，万奇林[1]

(1. 北京科技大学冶金与生态工程院，北京　100083；
2. 天铭冶金设备有限公司，山东济南　271100)

摘　要： 球墨铸铁冷却壁是当前高炉使用最为普遍、使用量最大的冷却设备之一。本研究设计并建造了一座 1:1 全尺寸冷却壁热态试验炉及其配套试验系统。以莱芜天铭冶金设备有限公司研制的新型球墨铸铁冷却壁（齐鲁壁）为主要研究对象，与该厂的传统铸铁冷却壁（天铭壁）以及铸钢冷却壁进行了热态对比试验。试验结果表明，在相同冷却水流速和炉膛温度条件下，齐鲁壁冷却水带走热流密度最大，导热效果优于其他类型冷却壁；齐鲁壁内部蓄热能力更强，从炉膛内吸收的热量更多；在综合导热系数的比较上，齐鲁壁要优于天铭壁与铸钢冷却壁。

关键词： 球墨铸铁；冷却壁；热态试验；温度场

Development of Nodular Cast Iron Cooling Stave without Air Gap Thermal Resistance

Zong Yanbing[1], Zhang Xuedong[1], Zhou Chuanlu[2], Xin Hongni[2], Wan Qilin[1]

(1. School of Metallurgy and Ecological Engineering, University of Science and Technology Beijing, Beijing 100083, China;
2. Tianming Metallurgical Equipment Co., Jinan 271100, China)

Abstract: Nodular cast iron cooling stave is one of the most widely used cooling equipment in blast furnace. In this study, a 1:1 full-size cooling stave thermal test furnace and its supporting test system were designed and constructed. The new nodular cast iron stave (Qilu stave) developed by Laiwu Tianming Metallurgical Equipment Co., Ltd. is the main research object. The hot state experiment was used to compare it with a conventional cast iron stave (Tianming stave) and a cast steel stave. The experimental results show that, under the same cooling water velocity and chamber temperature, qilu stave has the largest heat flux carried away by cooling water, and the heat conduction effect is better than other cooling staves. Qilu stave internal heat storage capacity is stronger, more heat absorption from the furnace; In the comparison of comprehensive thermal conductivity, qilu stave is better than tianming stave and cast steel stave.

Key words: nodular cast iron; cooling stave; thermal test; the temperature field

鞍钢 4#高炉热风围管吊挂设计总结

赵晓峰，张 帅，张维巍

（鞍钢集团工程技术有限公司铁烧室，辽宁鞍山 114000）

摘 要： 本文从讨论新 4#高炉热风围管自身结构、重量、热风输送等物理特性入手，分析热风围管吊挂的结构形式及安全状况，总结了新 4#高炉热风围管吊挂的设计过程，并对吊挂的合理结构设计、材料的选用提出了建议。

关键词： 热风围管；结构；吊挂

Summary of Hanging Design of Hot Blast Pipe in No. 4 BF at Ansteel

Zhao Xiaofeng, Zhang Shuai, Zhang Weiwei

(Anshan Iron & Steel Group Engineering Technology Co., Ltd., Anshan 114000, China)

Abstract: This article from discusses new 4# blast furnace hot blast physical properties and so on bustle pipe own structure, weight, hot blast transportation to obtain, the analysis hot blast bustle pipe hangs the structural style and the safe condition, summarized the design process which the new 4# blast furnace hot blast bustle pipe hangs, and to the reasonable structural design which, the material hung selects put forward the proposal.

Key words: Hot blast bustle pipe; structure; hangs

2 炼钢与连铸

大会特邀报告

分会场特邀报告

炼铁与原料

★ 炼钢与连铸

轧制与热处理

表面与涂镀

金属材料深加工

先进钢铁材料

粉末冶金

能源、环保与资源利用

钢铁材料表征与评价

冶金设备与工程技术

冶金自动化与智能化

建筑诊治

其他

2.1 炼　　钢

迁钢 210t 转炉经济炉龄探索

孙　亮，刘风刚，刘珍童，毕泽阳，赵艳宇

（北京首钢股份有限公司炼钢作业部，河北唐山　064404）

摘　要：文章介绍了迁钢 210t 顶底复吹转炉经济炉龄探索情况，得出了现情况下迁钢最佳经济炉龄为 5200 炉，及相应的补护炉和复吹孔维护方案。结果表明：耐材成本增加 0.95 元/t，转炉后吹率由 4.3%降低到 2.2%、吹炼时间缩短 18s、转炉平均终点氧活度降低 29×10^{-6}、碳氧积降低 0.00012、终渣 TFe 降低 0.5%、转炉吹损降低 0.75%、金属料损失降低 1.35 元/t，转炉平均脱磷率提高 1.1%，溅渣料成本降低 0.21 元/t、溅渣氮气降低 $0.6m^3$/t、溅渣时间缩短 21s，铝金属成本降低 0.65 元/t，检修费用增加 0.38 元/t，综合降低炼钢成本 0.88 元/t。

关键词：转炉；炉龄；成本；质量

Exploration on Economic Age of 210t Converter in Qian Steel

Sun Liang, Liu Fenggang, Liu Zhentong, Bi Zeyang, Zhao Yanyu

(Beijing Shougang Limited by Share Ltd., Steelmaking Operation Department,
Tangshan 064404, China)

Abstract: The paper introduces the exploration of economic furnace life of 210t top-bottom combined blown converter in Qian Steel. It is concluded that the optimum economic furnace life for Qian Steel is 5200, and the corresponding maintenance schemes for repairing furnace and combined blown hole are obtained. The results show that the cost of refractory increases by 0.95 yuan/t, the blowing rate after converter decreases from 4.3% to 2.2%, the blowing time decreases by 18 seconds, the average end-point oxygen activity of converter decreases by 29×10^{-6}, the carbon and oxygen volume decreases by 0.00012, the final slag TFe decreases by 0.5%, the blowing loss of converter decreases by 0.75%, the loss of metal material decreases by 1.35 yuan/t, the average dephosphorization rate of converter increases by 1.1%, the cost of slag splashing decreases by 0.21 yuan/t, and the slag nitrogen gas decreases by the slag splashing time is shortened by 21 seconds, the cost of aluminium metal is reduced by 0.65 yuan/t, the overhaul cost is increased by 0.38 yuan/t, and the steelmaking cost is reduced by 0.88 yuan/t.

Key words: converter; furnace life; cost; quality

RH 无铬耐火材料的应用与改进

吕志勇，邢维义，于海岐，苏小利

（鞍钢股份有限公司鲅鱼圈钢铁分公司，辽宁营口　115007）

摘　要：介绍了 RH 无铬耐火材料在鞍钢股份有限公司鲅鱼圈钢铁分公司使用情况，通过分析各部位耐火材料的侵蚀状况和损毁形式，开展技术创新，插入管的使用寿命提高到接近含铬耐火材料的水平，真空室异常下线的次数显著降低，保证了精炼生产的稳定运行。

关键词：无铬耐火材料；插入管；改进

炼钢生产模式比较及应用

董金刚

（宝山钢铁股份有限公司制造管理部，上海　201900）

摘　要：根据转炉生产能力和连铸生产能力之间的关系，炼钢厂的生产分为以转炉为中心的生产模式和以连铸为中心的生产模式。以转炉为中心生产模式追求高效率生产，以连铸为中心生产模式追求高连续性生产。在实际生产中，根据不同阶段的生产特点，这两种生产模式可以相互转换。对于以产定销或者小品种、大批量生产的炼钢厂，最优的生产模式是以连铸为中心；对于按合同组织生产或者生产品种和规格多、批量小的炼钢厂，最优的生产模式是以转炉为中心。为提高钢水纯净度以及炼钢高效率生产，当 A 钢铁厂未来只有三个高炉生产时，可采取以转炉中心的生产模式，即两座转炉改为脱磷炉，四座转炉作为脱碳炉；B 炼钢厂，在三个高炉和四座转炉情况下，四台连铸机时炼钢厂是以连铸为中心的生产模式，五台连铸机时炼钢厂是以转炉为中心的生产模式，后者比前者在产能规模、生产效率、物流平衡方面都有明显的优势。

关键词：转炉；连铸；生产模式；脱磷；脱碳

Comparison and Application of Steelmaking Production Patterns

Dong Jingang

(Baoshan Iron & Steel Co. Manufacturing Management Department, Shanghai 201900, China)

Abstract: Based on the relationship between converter production capacity and continuous casting production capacity, there are production mode centered on converter and centered on continuous casting in steelmaking plant. The production mode centered on converter is the pursuit of efficient production, the production mode centered on continuous casting is aim to high continuous production. According to the characteristics of different stages of production, these two production modes can be converted to each other in actual production. For steel plant that produce first and then sell or produce large quantities of small varieties, the production model centered on continuous casting is the best. For steel mills that produce under contract or produce multi-variety specifications and small batch, the production model centered on converter is the best. To improve the purity of steel and high efficiency production, A steelmaking plant can use the production model centered on converter when there are only three blast furnaces in future, that is, two converters will be changed to dephosphorization furnaces, and four converters will be used as decarbon furnaces. In B steelmaking plant with three blast furnaces and four converters, the production mode centered on continuous casting is used when there are four continuous casting machines, the production mode centered on converter is used when there are five continuous casting machines, the latter has obvious advantages over the former in terms of productivity scale production efficiency and logistics balance.

Key words: converter; continuous casting; production model; dephosphorization; decarbon

超低碳钢中的硫化物析出行为

张 峰

(宝山钢铁股份有限公司中央研究院，上海 201900)

摘 要：基于特定的硫含量变化，采用非水溶液电解提取+扫描电镜/透射电镜观察相结合的方法，研究了超低碳钢中硫化物夹杂物的组成和存在形式，及其相应的形貌、尺寸、数量、尺寸分布变化，以及对成品试样的显微组织的影响。结果表明，在实验条件下，钢中的硫化物夹杂物尺寸，绝大部分集中在 0.005~0.2μm 和 0.2~0.5μm 范围。钢中硫含量从 0.0068%增加至 0.0353%时，两者所占比例分别从 80.83%降低至 67.29%，和从 11.39%增加至 24.92%，但两者对应的硫化物夹杂物的总量能够保持稳定，且波动范围在同一数量级之内；随着钢中硫含量的增加，钢中的硫化物夹杂物逐渐由尺寸较大的胶囊形、长条形 MnS，演变成以类球形、椭球形 MnS 为核心的 MnS、Cu_2S 复合析出，以及部分单个析出的、尺寸细小的 Cu_2S 夹杂物，且以两者的复合析出为主，平均尺寸稳定在 100nm 左右；随着钢中硫含量的增加，成品试样的平均晶粒尺寸先是有所增加，然后快速降低。钢中的硫含量为 0.0102%时，成品试样的平均晶粒尺寸达到 32.4μm，均高于硫含量为 0.0068%、0.0255%和 0.0353%时，成品试样的平均晶粒尺寸 22.9μm、20.1μm 和 18.6μm。

关键词：硫含量；超低碳钢；非水溶液电解；硫化物夹杂物；析出行为；显微组织；晶粒尺寸

Precipitation Behavior of Sulfide Inclusions in Ultra-low Carbon Concentration Steel Sheets

Zhang Feng

(Central Research Institute, Boashan Iron & Steel Co., Ltd., Shanghai 201900, China)

Abstract: Based on the change of given sulfur concentration, the type and composition, the size and number, and the size distribution and the change of finished sample micro-structure in ultra-low carbon steel sheet was discussed by SEM/TEM and methods after the finished steel sample was extracted from steel matrix by electrolysis method in a non-aqueous solution. The results show that for present work, the size of most sulfide inclusions is focused on the range of 0.005 to 0.2μm and the range of 0.2 to 0.5μm, respectively. As the increase of sulfur from 0.0068% to 0.0353%, the proportion of which will decrease from 80.83% to 67.29%, and increase from 11.39% to 24.92%, respectively, and the total number of sulfide inclusions will keep stabilized and focus on the same quantity class. On the other hand, as the increase of sulfur in steel, the sulfide inclusions will be changed from the bigger capsule form and elongated form MnS, evolves to the near spheroid form and ellipsoidal form combination of MnS and Cu_2S, some of the smaller single Cu_2S inclusions. Most of the sulfide inclusions is the combination of MnS and Cu_2S, and the average size of which is about 100nm or so. The size of sulfide inclusions above will affect the grain size of finished sample in high sulfur steel. As the sulfur concentration is 0.0102%, the corresponding average grain size of the finished sample is 32.4μm, and is higher than that of 22.9μm, 20.1μm and 18.6μm, which the corresponding sulfur concentration is 0.0068%, 0.0255% and 0.0353%, respectively.

Key words: sulfur concentration; ultra-low carbon steel; non-aqueous solution electrolysis; sulfide inclusion; precipitation behavior; micro-structure; grain size

迁钢 210 吨转炉自动出钢技术的开发与应用

江腾飞，朱 良，成天兵

（首钢股份公司迁安钢铁公司炼钢作业部，河北迁安 064404）

摘 要：介绍了迁钢 4#转炉自动出钢控制技术的设计理念，包括安全控制逻辑的建立，出钢模式的设计，以及相关设备的自动控制方式。针对于炉口下渣检测，钢包净空识别，分别说明了图像分析技术在自动出钢技术中的应用。迁钢自动出钢控制技术已经应用于实际生产，投入率为100%，成功率在95%以上，降低了岗位劳动强度，提升了过程控制的稳定性。

关键词：自动出钢；出钢模式；出钢曲线；炉口下渣

Development and Application of Automatic Tapping Technology for 210t Converter at Qiangang

Jiang Tengfei, Zhu Liang, Cheng Tianbing

(Shougang Qian'an Iron and Steel Co., Ltd., Qian'an 064404, China)

Abstract: The design concept of automatic tapping control technology for No. 4 converter in Qiangang is introduced. Including the establishment of safety control logic, the design of tapping mode, and the automatic control mode of related equipment. Aiming at slag detection and ladle clearance recognition, the application of image analysis technology in automatic tapping technology is explained respectively. The automatic tapping control technology of Qiangang has been applied in actual production. The input rate is 100% and the success rate is more than 95%. It reduces the labor intensity of the post and improves the stability of process control.

Key words: automatic tapping; tapping mode; tapping curve; slag overflow

鞍钢提高转炉废钢比的生产实践

李伟东，何海龙，李 冰，李 泊

（鞍钢股份有限公司炼钢总厂，辽宁鞍山 114021）

摘 要：为提高转炉废钢比，鞍钢采取了一系列措施。通过提高铁水入炉温度，增加了转炉热量来源；通过采取措施降低各工序温度损失及中间包过热度，降低了转炉出钢温度。实施后，转炉废钢比逐年提高。最高月份达到了 182.77kg/t 钢。废钢比调节能力得到显著提升。

关键词：转炉；废钢比；钢水温度

Production Practice of Improving Scrap Ratio of Converter of Anshan Iron and Steel Group Coporation

Li Weidong, He Hailong, Li Bing, Li Bo

(Steel Plant in Angang Co., Ltd., Anshan 114021, China)

Abstract: To improve the scrap ratio, Anshan Iron and Steel Group has taken a series of measures.The heat source of converter is increased by increasing the temperature of molten iron entering converter; By taking measures to reduce the temperature loss in each process, the temperature of the output of converter is reduced. After implementation, converter scrap ratio is improved year by year. The highest monthe was 182.77 Kilogram per ton steel. The regulation capacity of scrap ratio has been improved significantly.

Key words: converter; scrap ratio; the molten steel temperature

镁质熔剂在转炉炼钢中的应用研究

李伟东，何海龙，孙振宇，李　冰，李　泊

（鞍钢股份有限公司炼钢总厂，辽宁鞍山　114021）

摘　要：本文针对转炉冶炼用镁质熔剂进行研究。分析了各镁质熔剂的造渣机理、冷却效应以及对转炉脱磷、转炉渣量、转炉溅渣护炉的影响。总结出各镁质熔剂的使用原则。

关键词：白云石；菱镁石；轻烧白云石；轻烧镁球

Application of Magnesium Flux in Converter Steelmaking

Li Weidong, He Hailong, Sun Zhenyu, Li Bing, Li Bo

(Steel Plant in Angang Co., Ltd., Anshan 114021, China)

Abstract: Magnesium flux for converter smelting is studied in this paper. The slagging mechanism and cooling effect of magnesium fluxes on dephosphorization,slag content and slag splash protection of converter are analyzed.The selection of magnesium fluxes was summarized.

Key words: dolomite; magnesite; light burning magnesium ball; light burning dolomite

90t顶吹转炉冶炼超低磷钢工艺开发

尚世震[1]，臧绍双[1]，李　泊[1]，陶功捷[1]，王　爽[2]，高立超[1]，潘瑞宝[1]

（1. 鞍钢股份有限公司炼钢总厂，辽宁鞍山　114021；

2. 鞍钢集团钢铁研究院，辽宁鞍山　114009）

摘　要：介绍了鞍钢股份有限公司炼钢总厂 90t 氧气顶吹转炉采取双渣冶炼、LF 炉脱 P 生产超低磷钢水的新工艺。采用该工艺替代原双联工艺生产的超低磷钢水能将成品磷含量稳定降低 0.0019%，硫、氮、全氧含量均达到较低水平，完全能够满足品种质量的要求。

关键词：顶吹转炉；超低磷钢；磷含量

Process Development of Smelting Molten Ultra-low Phosphor Steel by 90t Oxygen Top-Blown Converter

Shang Shizhen[1], Zang Shaoshuang[1], Li Bo[1], Tao Gongjie[1],
Wang Shuang[2], Gao Lichao[1], Pan Ruibao[1]

(1. General Steelmaking Plant of Angang Co., Ltd., Anshan 114021, China;
2. Ansteel Iron & Steel Research Institutes, Anshan 114009, China)

Abstract: The new process for smelting molten ultra-low phosphor steel by double slag process in 90t oxygen top-blown converter and dephosphorizing process in LF in General Steelmaking Plant of Angang Steel Co., Ltd. was introduced. The phosphor content in finish product could be stably reduced by 0.0019% after adopting the new process in stead of the former duplex process. The content of these elements such as sulphur, nitrogen and total oxygen in the molten steel reached to the floor level, which could completely meet the requirements asked by both steel grades and quality.

Key words: top-blownconverter; ultra-lowphosphorsteel; phosphorcontent

鞍钢 KR 法脱硫浸入深度研究与应用

宋吉锁，何海龙，刘鹏飞，孙　群，李伟东，乔冠男

（鞍钢股份有限公司炼钢总厂，鞍山辽宁　114021）

摘　要：针对鞍钢 KR 法脱硫效果不好的问题开展了工业试验，得出了搅拌桨的最佳浸入深度是 1.5m，据此建立了 KR 脱硫插入深度模型。应用新模型后，铁水预处理脱硫效率提到 94.2%，脱硫粉剂消耗降至 6.4kg/t 铁。

关键词：铁水预处理；KR 法脱硫；浸入深度

Development of Model with Deep Immersion for Desulphurization by KR Method in Ansteel and Application of the Model

Song Jisuo, He Hailong, Liu Pengfei, Sun Qun, Li Weidong, Qiao Guannan

(General Steelmaking Plant of Angang Steel Co., Ltd., Anshan 114021, China)

Abstract: With regard to the problem of poor effect of desulphurization by KR method in Ansteel, the industry experiment was carried and thus it was concluded that the best deep immersion for the stirring paddle was 1.5 meter and the model for the stirring paddle with deep immersion for KR method was then established. After the model was used, the desulphurization ratio in pretreatment of hot metal was improved to 94.2% and the consumption of desulphurizer was reduced to 6.4 kg per tone hot metal.

Key words: pretreatment of hot metal; desulphurization by KR method; deep immersion

260t 转炉轻烧镁球冶炼脱磷生产实践

李玉德，李叶忠，朱国强，查松妍，齐志宇

(鞍钢股份有限公司炼钢总厂，辽宁鞍山　114000)

摘　要： 本文以 260t 复吹转炉为研究对象，分析冶炼过程终点温度、终点氧值、炉渣碱度和返干时机等对 260t 复吹转炉轻烧镁球冶炼的脱磷效果及影响特点，从而指导现场操作。研究结果表明：采用轻烧镁球进行转炉冶炼，转炉出钢氧值应大于 0.04%；炉渣碱度在 2.5~3.3；转炉出钢温度小于 1680℃；返干时间小于 2min，转炉脱磷效率较高，可实现轻烧镁球冶炼终点磷含量的稳定控制。

关键词： 转炉冶炼；轻烧镁球；脱磷；炉渣碱度；出钢温度

Production Practice of Light Burning Magnesium Ball Smelting Dephosphorization in 260t Converter

Li Yude, Li Yezhong, Zhu Guoqiang, Zha Songyan, Qi Zhiyu

(Angang Steel Co., Ltd., Steel Plant, Anshan 114000, China)

Abstract: This paper takes 260t re-blowing converter as the research object, analyzes the dephosphorization effect and the influence characteristic of 260t re-drying converter light burning magnesium ball smelting, such as tapping temperature, tapping oxygen, slag basicity and back drying, in order to guide the operation on the spot. The results show that the oxygen value of the converter should be greater than 0.04%. The slag basicity is 2.5~3.3; The converter tapping temperature is less than 1680℃; The drying time is less than 2min, and the efficiency of dephosphorization in converter is higher. Realization of stable control of the phosphorus content at the end of light burning magnesium sphere smelting.

Key words: converter smelting; light burning magnesium ball; dephosphorization; slag basicity; tapping temperature

鞍钢 260t 转炉 IF 钢顶渣改质工艺研究与应用

李　冰，李　泊，朱国强，齐志宇，何文英，孙振宇，高立超

(鞍钢股份有限公司炼钢总厂，辽宁鞍山　114021)

摘　要： 对 260t 转炉钢包顶渣改质工艺进行了研究，分析了不同条件下改质剂对改质效果的影响，优化了转炉前、后挡渣工艺，改进了转炉出钢过程钢渣改质工艺，确定了小粒白灰与改质剂的最佳配比及加入方式，同时改进钢包

底吹氩搅拌工艺提高改质效果。结果表明，采取是上述措施后，钢包顶渣氧化性显著降低，FeO 由 12.52%降低至 9.54%，中间包全氧由 28.75ppm 降低至 22.37ppm，对减少及控制 Al_2O_3 夹杂起到了重要作用，提高了连铸坯的洁净度。

关键词：转炉；顶渣改质；氧化铁；吹氩

Research and Application of IF Steel Ladle Top Slag Improving Process on Ansteel 260t Conventer

Li Bing, Li Bo, Zhu Guoqiang, Qi Zhiyu, He Wenying, Sun Zhenyu, Gao Lichao

(Steelmaking Plant, Angang Steel Co., Ltd., Anshan 114021, China)

Abstract: This article undertakes a study on IF steel ladle top slag improving process for Ansteel 260t converter, which analysis affects of improving effect from reforming agents under different conditions, in order to ensure process improving direction. Emphasis of process improvement: one is optimize slag stopping process before and after converter which reduce slag amount, decrease oxidizing of ladle slag, the other is improve slag reforming process during converter tapping, the optimum proportion of granule lime and reforming agents as well as adding way are confirmed, meanwhile improve ladle bottom argon injection process in order to enhance reforming effect. The results show that after taking measures, oxidizing of ladle top slag is significantly decreased which plays an important role in reduce，FeO reduced from 12.52% to 9.54%, total oxygen content in tundish reduced from 28.75ppm to 22.37ppm and control Al_2O_3 inclusion, improve cleanliness of continuous cast slab.

Key words: conventer; top slag improvement; FeO; Argon injection

转炉单渣留渣高效冶炼技术的研究与应用

刘忠建，王忠刚，高志滨，赵立峰，公　斌，张　丽

（莱芜钢铁集团银山型钢有限公司炼钢厂，山东济南　271104）

摘　要：通过单渣留渣工艺的研究，分析了单渣留渣工艺的脱磷原理。掌握了该工艺的规律和特点，建立了单渣留渣工艺自动炼钢模型并成功开发了留渣高效开吹打火技术、快速精准留渣技术、低碱度终渣循环冶炼技术以及低热值铁水高效低能耗冶炼技术等。应用单渣留渣工艺后，石灰消耗降低 15.67kg/t，白云石消耗降低 1.60kg/t，烧结矿消耗降低 10.07kg/t，钢铁料消耗减少 18kg/t，渣量降低 25.16kg/t。

关键词：单渣留渣工艺；高效；脱磷；消耗

Research and Application of High Efficiency Smelting Technology with Single Slag Retaining in Converter

Liu Zhongjian, Wang Zhonggang, Gao Zhibin, Zhao Lifeng, Gong Bin, Zhang Li

(Steelmaking Plant of Laiwu Iron and Steel Group Yinshan Section Steel Co., Ltd., Jinan 271104, China)

Abstract: The dephosphorization principle of single slag residue technology was analyzed through the study of single slag residue technology. The rules and characteristics of the process are mastered, the automatic steelmaking model of single slag retention process is established, and the technologies of slag retention, high efficiency open blowing and ignition, rapid and accurate slag retention, low basicity final slag recycling and low calorific value hot metal high efficiency and low energy consumption are developed. After single slag retention process, the consumption of lime, dolomite, sinter, iron and steel materials and slag decreased by 15.67kg/t, 1.60kg/t, 10.07kg/t, 18kg/t and 25.16kg/t respectively.

Key words: single slag residue retention process; high efficiency; dephosphorization; consume

莱钢钢包全程智能吹氩工艺开发与应用

薛 志，王忠刚，高志滨，丁洪周

（莱钢集团银山型钢炼钢厂，山东济南 271104）

摘 要：本文从设备升级改造、自动控制系统及智能控制模型建立以及精炼节点控制方面，介绍了莱钢钢包吹氩工艺的优化与创新。通过开发钢包全程智能吹氩工艺，实现了转炉出钢过程全程智能吹氩，精炼进站钢水成分、温度、渣况提升明显，优化精炼节点控制模型后，大幅降低了精炼工序消耗，缩短了精炼时间，提高了精炼生产效率。

关键词：高效低成本；转炉；底吹

Development and Application of Intelligent Argon Blowing Process for Ladle in Laigang

Xue Zhi, Wang Zhonggang, Gao Zhibin, Ding Hongzhou

(Laiwu Steel Group Yinshan Section Steel Co., Ltd., Jinan 271104, China)

Abstract: In this paper, the optimization and innovation of ladle argon blowing process in Laigang are introduced from the aspects of equipment upgrading, automatic control system, intelligent control model establishment and refinery node control. Through the development of intelligent argon blowing technology for ladle, the whole process of converter tapping is realized. The composition, temperature and slag condition of molten steel in refining station are improved obviously. After optimizing the control model of refining node, the consumption of refining process is reduced greatly, the refining time is shortened, and the refining production efficiency is improved.

Key words: high efficiency and low cost; converter; bottom-blown

脱磷剂在转炉深脱磷中的作用研究

尚 游

（莱芜钢铁集团银山型钢炼钢厂，山东济南 271104）

摘 要：由于目前铁水磷高达 0.150%~0.210%，而钢水磷要求较低，转炉为了达到深脱磷目的往往采用双渣甚至双联操作，这就造成了转炉渣量上升，金属回收率的降低，增加了转炉冶炼成本。为了实现转炉冶炼过程渣料消耗的降低，控制转炉消耗成本，莱芜钢铁集团型钢炼钢厂根据实情况调整入炉结构，研究将氧化铁皮、石灰粉直接加入炉内代替部分造渣料。通过实践相关数据分析，达到了提高转炉脱磷率的目的。

关键词：氧化铁皮；石灰粉；深脱磷；降低成本

Study on the Role of Dephosphorizing Agent in Deep Dephosphorization of Converter

Shang You

(Laiwu Iron and Steel Group Steelmaking Plant, Jinan 271104, China)

Abstract: Because of the high phosphorus content in hot metal and the low requirement of phosphorus in molten steel, the converter usually adopts double slag or even double operation in order to achieve the purpose of deep dephosphorization, which results in the increase of converter slag, the decrease of metal recovery and the increase of converter smelting cost. In order to reduce slag consumption in converter smelting process and control the cost of converter consumption, the steel-making plant of Laiwu Iron and Steel Group adjusts the feeding structure according to the actual situation, and studies the direct addition of iron oxide scale and lime powder to the furnace to replace part of the slag-making material. Through the analysis of practical data, the purpose of improving the dephosphorization rate of converter is achieved.

Key words: iron oxide scale; lime powder; deep dephosphorization; cost reduction

KR 法铁水预脱硅工艺研究与应用

王 强，王忠刚，高志滨，薛 志，尚 游

（山东钢铁股份有限公司莱芜分公司，山东莱芜 271104）

摘 要：开发 KR 法铁水预脱硅技术在现有工艺设备的基础上，通过研究铁水脱硅理论，包括脱硅原理、KR 法脱硅可行性研究、炉渣发泡研究，开发新型氧化铁类脱硅剂，建立铁水脱硅工艺模型，相关配套设备改造等措施，铁水硅含量平均降低 0.14%，调整了入炉铁水硅含量，降低了转炉渣料消耗，同时达到了减少转炉喷溅的目的。工艺路线科学合理，使用效果良好，具有良好的社会推广价值，对钢铁企业具有积极的借鉴意义。

关键词：铁水脱硅；脱硅剂；KR 法

Research and Application of Hot Metal Desilication with KR Technology

Wang Qiang, Wang Zhonggang, Gao Zhibin, Xue Zhi, Shang You

(Shandong Laiwu Iron and Steel Group, Laiwu 271104, China)

Abstract: The hot metal desilication with KR technology on the basis of existing equipment, to study on desilication theory of hot metal, including desilication principle, feasibility study on desilication by KR method, study on foaming of slag. To develop new desilication agent, build desiliconization process model, modification of equipment, trim the silicon content, the average decrease of silicon content in hot metal is 0.14%, reduce slag consumption, reducing splash of converter. The technological route is scientific and reasonable, with a good result in application. It has good social popularization value and positive reference significance for iron and steel enterprises.

Key words: hot metal desilication; desiliconization; KR technology

转炉底吹供气元件选择与维护讨论

王 东

（中冶京诚工程技术有限公司转炉冶炼室，北京 100176）

摘 要： 转炉底吹供气元件的选择与维护是转炉炼钢的重要课题，也是困扰设计人员和生产一线技术人员的技术难题。本文结合实际生产情况，通过对底吹元件的选型、数量、布置进行了论述，为转炉底吹系统的选择和优化提供借鉴和参考。同时，为实现转炉底吹长寿命及良好的复吹效果，在保证转炉"底吹眼"裸露可见的情况下，论述了转炉底吹系统进行日常管理和维护。

关键词： 转炉；底吹；供气元件；维护

Discussion on Selection and Maintenance of Bottom Blowing Components in Converter

Wang Dong

(Capital Engineering & Research Incorporation Ltd., Beijing 100176, China)

Abstract: The selection and maintenance of bottom blowing gas supply components is an important subject in converter steelmaking. And it is also a technical problem that puzzles designers and front-line technicians of steel plant. According to the actual production situation, the selection, quantity and arrangement of bottom blowing elements is discussed in this paper. It can provide reference for the selection and optimization of converter bottom blowing system. At the same time, in order to achieve long service life of bottom blowing and good combined blowing effect, the daily management and maintenance of converter bottom blowing system are discussed. The purpose is to make the bottom blowing hole of converter exposed and visible.

Key words: converter; bottom blow; gas supply component; maintenance

转炉自动化炼钢控制技术研究

万雪峰[1,2]，曹 东[1,2]，王丽娟[1,2]，马 勇[3]，赵 亮[1,2]，高学中[3]

（1. 海洋装备用金属材料及其应用国家重点实验室，辽宁鞍山 114009；
2. 鞍钢集团钢铁研究院，辽宁鞍山 114009；
3. 鞍钢股份有限公司炼钢总厂，辽宁鞍山 114021）

摘　要：在 100t 顶吹转炉上研究了炉气变化与吹炼末期熔池之间的关系，并依据规律分别建立二级过程控制数学模型，拓展了自行开发的 AOA 系统功能，使之同时具备高、低碳钢自动控制能力。并通过工业试验，应用该系统对转炉吹炼过程及终点进行控制，取得高碳钢一次性拉碳直接出钢率 92.3%、低碳钢 98.6%的实绩，并在系统中增设终点磷含量预测功能，一拉命中率达到 83.1%。

关键词：AOA；炉气分析；顶吹转炉；动态控制

Research of Converter Automation Steelmaking Control Technology

Wan Xuefeng[1,2], Cao Dong[1,2], Wang Lijuan[1,2], Ma Yong[3],
Zhao Liang[1,2], Gao Xuezhong[3]

(1. State Key Laboratory of Marine Equipment Made of Metal Material and Application, Anshan 114009, China; 2. Iron and Steel Academe of Ansteel Group, Anshan 114009, China; 3. General Steelmaking Plant of Ansteel Co., Ltd., Anshan 114021, China)

Abstract: Research of relationship between off-gas compositions variation and bath state in 100t top blowing converter were carried on, and the secondary process control model was also established, and the fuction of self-developed AOA system with both high carbon and low carbon steel grade automation controlling ability was added. Based on the industrial test, blowing process and end could be controlled applying this system and direct tapping rate arrived at 92.3% for high carbon steelgrade and 98.6% for low carbon steel grade. Forecasting function to phosphorus content was installed in AOA system, the single-time hit rate of phosphorus content w[P] at the blowing end could reach 83.1%.

Key words: AOA (Ansteel off-gas analysis technology); off-gas analysis; top blowing converter; kinetic control

高碳铬轴承钢氮化钛的控制

李广帮，魏崇一，廖相巍

（鞍钢集团钢铁研究院，辽宁鞍山　114009）

摘　要：对轴承钢中形成氮化钛夹杂的条件进行热力学分析，由于选分结晶及氮和钛的平衡浓度积下降，氮化钛在凝固过程中析出、长大，为了抑制氮化钛析出，在某钢厂分别采取冶炼各工序控制钛含量和氮含量的措施。控制铁水、废钢及铁合金，优化钢中酸溶铝及精炼渣中氧化钛的含量，在采用高碱度和低碱度渣操作时，与之前高碱度渣操作相比可以将钢液的钛含量由 0.0045%分别降至 0.0028%和 0.0013%的水平。通过优化转炉及精炼操作，加强连铸保护浇铸，可以将钢液的氮含量由 0.0050%降低到 0.0038%的水平。

关键词：转炉；轴承钢；氮化钛夹杂；氮含量；钛含量

Control of Titanium Nitride Inclusion in High Carbon Chromium Bearing Steel

Li Guangbang, Wei Chongyi, Liao Xiangwei

(Ansteel Iron & Steel Research Institute, Anshan 114009, China)

Abstract: Thermodynamics conditions of titanium nitride formation in High Carbon Chromium Bearing steel have been investigated. With the solidification and the decrease of equilibrium concentration product of N and Ti, TiN separates out and grows up. In order to prevent TiN precipitation, measures of controlling Ti and N contents have been adopted in each process in a steel company. With the controlling of the qualities of molten iron, steel scrap and steel alloy, optimization of contents of acid-melting aluminum in steel and TiO_2 in slag, Ti content in molten steel can be decreased from 0.0045% to 0.0028% with high basicity slag and 0.0013% with low basicity slag, respectively. With the optimization of technics in BOF and refining processes and enhance the protection casting level in continuous casting, N content in the molten steel can be decreased from 0.0050% to 0.0038%.

Key words: converter; bearing steel; TiN; nitrogen content; titanium content

钢包底吹智能控制技术研究与应用

李卫东

（上海梅山钢铁公司设备部，江苏南京 210039）

摘　要：本文围绕钢包底吹闭环控制，采用视觉识别技术，通过钢包内钢水液面视频图像，判断钢包底吹搅拌强度，计算出底吹气体阀门设定值，调节底吹气体流量，实现闭环控制，直至符合工艺控制目标要求。评价同一钢包不同阶段底吹搅拌命中目标的优劣，选出该钢包的最佳控制系数，优化控制下一次该钢包底吹搅拌，加快命中目标速度和出钢阶段的"盲吹"控制。 实施投用以来，运行稳定，不仅实现从出钢开始到吹氩结束全程全自动底吹闭环控制，降低了劳动强度，而且节约合金消耗，提高钢水质量，实现了智能控制。

关键词：视觉识别；炉外精炼；底吹；闭环控制；智能控制

Research and Application of Intelligent Control Technology for Ladle Bottom Blowing

Li Weidong

(Shanghai Meishan Iron and Steel Company Equipment Department, Nanjing 210039, China)

Abstract: This paper focuses on the closed-loop control of the bottom of the ladle, adopts the visual recognition technology, judges the bottom blowing strength of the ladle through the video image of the molten steel in the ladle, calculates the set value of the bottom blowing gas valve, adjusts the bottom blowing gas flow, and realizes the closed-loop control until Meet the requirements of process control objectives. Evaluate the advantages and disadvantages of the bottom blow mixing target in different stages of the same ladle, select the best control coefficient of the ladle, optimize the control of the bottom of the ladle, and accelerate the "blind blowing" control of the target speed and the tapping stage. Since the implementation of the application, the operation is stable, not only realizes the full-automatic bottom-blowing closed-loop control from the start of tapping to the end of argon blowing, which reduces the labor intensity, saves alloy consumption, improves the quality of molten steel, and realizes intelligent control.

Key words: visual recognition; furnace refining; bottom blowing; closed-loop control; intelligent control

重轨钢质量控制实践

王 宁[1]，常宏伟[1]，金纪勇[2]，张 锐[1]，孙振宇[3]，李 旭[3]

(1. 鞍钢股份有限公司产品发展部，辽宁鞍山 114021；2. 鞍钢股份有限公司技术中心，辽宁鞍山 114009；3. 鞍钢股份有限公司炼钢总厂，辽宁鞍山 114021)

摘 要：本文介绍了鞍钢重轨钢的生产特点，通过研究重轨钢的生产工艺和影响因素，采用提高转炉终点碳含量、保证 VD 真空脱气效果、优化合金加入工艺、改善 LF 化渣条件和 LF 成渣方式等措施，降低了钢水中的氢、氧、氮和夹杂物含量，这些控制措施的实施使钢轨内在质量满足了现行钢轨标准的要求，并形成了鞍钢重轨钢的纯净化生产工艺。

关键词：真空脱气；重轨钢；夹杂物；纯洁化

Quality Control Practice of Heavy Rail Steel

Wang Ning[1], Chang Hongwei[1], Jin Jiyong[2], Zhang Rui[1], Sun Zhenyu[3], Li Xu[3]

(1. Product Development Department of Ansteel, Anshan 114021, China; 2. Technology Center of Ansteel, Anshan 114009, China; 3. Steelmaking Plant of Ansteel, Anshan 114021, China)

Abstract: This paper introduces the production characteristics of heavy rail steel made in Ansteel, especially researching of the heavy rail steel production process and influences. Through improving final carbon content in converter, ensuring the ability of vacuum degassing, optimizing the alloy add process and improving the condition of melting slag and slag formation in ladle furnace to cutting down the hydrogen, oxygen, nitrogen and nonmetallic inclusions in molten steel. All of smelting control methods above ensure the internal quality of rail steel which is able to meet the requirements of TB/2344—2012. The new production process of clean heavy rail steel in Ansteel has formed.

Key words: vacuum degassing; heavy rail steel; nonmetallic inclusions; clean

基于图像分析的钢包底吹搅拌在线监视与自动控制

张才贵[1]，郭海滨[1]，孙玉军[1]，徐建雄[2]

(1. 宝钢上海梅山钢铁股份有限公司炼钢厂，江苏南京 210039；
2. 镭目科技有限责任公司，湖南衡阳 421000)

摘 要：钢包底吹搅拌是炼钢二次精炼广泛采用的精炼方式之一，搅拌强度对二次精炼钢水质量是很重要的。由于吹气管路泄漏或是堵塞原因，目前普遍采用的固定钢包底吹搅拌压力或流量不能满足钢水处理质量要求。本文介绍了宝钢上海梅山钢铁股份有限公司炼钢厂基于图像分析的钢包底氩搅拌在线监视与自动控制装置。该装置通过图像计算钢包钢水裸露面积和钢渣面积比例，并与标准面积比例进行比较，判断强度大小，反馈给流量调节阀，自动调节流量调节阀开度，从而获得合适的搅拌强度,满足钢水质量要求。

关键词：钢包；搅拌；图像；监视；自动控制

Online Monitoring and Automatic Control of Ladle Bottom Stirring by Using Image Analysis

Zhang Caigui[1], Guo Haibin[1], Sun Yujun[1], Xu Jianxiong[2]

(1. Steel-Making Plant of Baosteel Shanghai Meishan Iron & Steel Corporation, Nanjing 210039, China;
2. Ramon Technology Co., Ltd., Hengyang 421000, China)

Abstract: Ladle bottom stirring is one method used widely of secondary metallurgy in steel-making process.Stirring intensity is important to the quality of liquid steel during secondary metallurgy.Permanent pressure or flow of ladle bottom stirring now often used can not fufill the requirement of the quality of liquid steel because of leakage or block of the gas supplies. A device by which the intensity of ladle bottom stirring can be monitored online and controlled automatically based on image analysis in Baosteel Meishan Iron & Steel Corporation was introduced in the article.The working principle of the device was described as below: calculates the area proportion of bared steel and slag in the ladle; judges the intensity of bottom stirring by compared with the standardized area proportion;gives instructions to the flow valve which will adjust automatically soon after. As a result,suitable intensity of bottom stirring was acquired which meeted the quality requirement of the liquid steel.

Key words: ladle; stirring; image; monitoring; automatic control

低氧含硫易切削钢的工艺优化

刘　峰，马玉强，莫　丹，司焕庆

（石家庄钢铁有限责任公司，河北石家庄　050031）

摘　要：针对低氧含硫易切削钢（0.015%≤[S]≤0.035%，[Al]≥0.020%）生产中易出现的非金属夹杂物超标，氧含量偏高，可浇性差等问题，通过优化转炉工艺、精炼造渣、脱氧工艺和连铸工艺解决了钢水可浇性问题，实现了单中包连浇炉次≥8炉，钢材全氧含量稳定控制在10×10^{-6}以下，非金属夹杂物的控制满足了高端客户需求。

关键词：低氧含硫易切削钢；工艺优化；氧含量；可浇性；非金属夹杂物

Process Optimization of Reducing Oxygen Content in Free-cutting Steel

Liu Feng, Ma Yuqiang, Mo Dan, Si Huanqing

(Shijiazhuang Iron & Steel Co., Ltd., Shijiazhuang 050031, China)

Abstract: Low-oxygen and sulfur-containing free-cutting steel (0.015%≤[S]≤0.035%, [Al]≥0.020%) exceeds the standard for non-metallic inclusions, high oxygen content, By optimizing the converter process, The molten steel flow problem is solved by optimizing the converter process, refining slag system, deoxidation process and continuous casting process, Achieved the number of continuous casting furnaces ≥ 8 furnaces, Steel's total oxygen content is stable under 10×10^{-6}, The control of non-metallic inclusions meets the needs of high-end customers.

Key words: low-oxygen and sulfur-containing free-cutting steel; process optimization; oxygen content; fluidity; non-metallic inclusion

分钢种精炼窄成分渣系工艺技术开发与应用

王 键，刘文凭，郭庆军，王玉春

（山钢股份有限公司莱芜分公司炼钢厂，山东济南 271126）

摘 要： 针对 LF 精炼存在的缺少不同品种钢科学合理的目标渣系，精炼过程成渣速度慢、渣系组分波动大等问题，分钢种确定精炼窄成分渣系；在转炉终点实施动态脱氧和顶渣处理工艺，稳定进 LF 精炼钢水氧化性；采用硅平衡法稳定控制炉渣组分。该工艺技术的应用使得上下炉次精炼渣碱度波动在 0.5 以内，夹杂物总级别≤2.0 合格率达到 95%以上。

关键词： 精炼渣系；窄成分控制；碱度；夹杂物；工艺控制

Development and Application of Process Technology for Refining Narrow Composition Slag System of Steel

Wang Jian, Liu Wenping, Guo Qingjun, Wang Yuchun

(Shandong Iron and Steel Co., Ltd., Laiwu Branch Steel Mill, Jinan 271126, China)

Abstract: In view of the lack of scientific and reasonable target slag system in LF refining, the slow rate of slag formation and the wide fluctuation of slag system components, the narrow slag system was determined by steel separation. Dynamic deoxidation and top slag treatment process are carried out at the end of the converter, and the oxidization of steel into LF refining is stabilized. Silicone equilibrium method was used to stabilize the slag composition. The application of this technology has caused the alkaline fluctuation of the refined slag in the upper and lower furnace to be within 0.5, and the total level of inclusions ≤2.0 has a pass rate of more than 95 %.

Key words: refining slag system; narrow component control; alkali; folders; process control

基于 SOM-RELM 的 LF 炉钢水温度预测

信自成[1]，张江山[1]，张军国[2]，路博勋[2]，
李军明[2]，郑 瑾[2]，刘 青[1]

（1. 北京科技大学钢铁冶金新技术国家重点实验室，北京 100083；
2. 河钢股份有限公司唐山分公司，河北唐山 063000）

摘　要：为实现 LF 精炼炉钢水终点温度的精准控制，以唐钢生产数据为依托，利用分类模型(SOM)结合正则化极限学习机模型(RELM)的方法对 LF 精炼钢水终点温度进行了预测。首先，利用现场生产规则、SPSS 软件及 Matlab 软件，提取了影响终点温度的主要因素，并对数据进行了预处理；然后，构建 SOM 神经网络分类模型，并对数据进行分类；最后，利用分类后的数据构建对应类别的 RELM 终点温度预测模型，进而建立分类别的钢水终点温度预测模型。结果表明，SOM-RELM 模型通过对特征分散的数据进行先分类处理后再预测的方法，提高了模型的预测精度；当钢水终点温度容差为±5℃时，采用单一的 BP 神经网络和 RELM 模型的命中率分别为 86.5%和 91.0%，SOM-RELM 模型的命中率达到了 93.3%，可以更有效地提高模型的可靠性和指导实际生产。

关键词：LF 精炼；温度预测；SOM 神经网络；RELM 正则化极限学习机

Temperature Prediction of Molten Steel in a Ladle Furnace Based on SOM-RELM Model

Xin Zicheng[1], Zhang Jiangshan[1], Zhang Junguo[2], Lu Boxun[2],
Li Junming[2], Zheng Jin[2], Liu Qing[1]

(1. State Key Laboratory of Advanced Metallurgy, University of
Science & Technology Beijing, Beijing, 100083, China;
2. Hesteel Group Tangsteel Company, Tangshan 063000, China)

Abstract: A SOM-RELM network model based on the principal of self-organizing map model and regularized extreme learning machine model was developed to predict the end point temperature of molten steel in a ladle furnace using the actual data of Tangsteel. Specifically, the main influencing factors were firstly identified by following the refining rules and using the data treatment by SPSS and Matlab software. Second, a SOM prediction model was established to categorize the obtained data. At last, a RELM model was introduced into the SOM model to carry out the endpoint temperature prediction. The results show that the SOM-RELM model improves the prediction accuracy of the model by first classifying and then forecasting the data with scattered features. When the tolerance is ±5℃, the hit rate of the single BP neural network and RELM model is 86.5% and 91.0% respectively, and the hit rate of the SOM-RELM model is 93.3%. It can effectively improve the reliability of the model and guide the actual production.

Key words: ladle furnace; temperature prediction; self-organizing map model; regularized extreme learning machine model

莱钢转炉炉体维护技术研究与实践

王　健，张昭平，任科社，郭庆军，杨普庆

（山东钢铁股份有限公司莱芜分公司炼钢厂，山东济南　271126）

摘　要：针对转炉炉体在运行过程中侵蚀速度快，维护方式单一，效果差，炉体安全稳定性差等问题，对转炉炉体侵蚀原理进行研究，通过优化溅渣工艺，实现炉体溅渣层精准控制，通过创新炉体维护方式，提高炉体薄弱部位维护能力，在减缓炉体侵蚀速度同时实现炉体各部位均衡，为提升转炉冶金效果提供保障。

关键词：炉体；溅渣工艺；均衡

Research and Practice on the Maintenance Technology of Furnace Body of Laigang Converter

Wang Jian, Zhang Zhaoping, Ren Keshe, Guo Qingjun, Yang Puqing

(Shandong Iron and Steel Co., Ltd., Laiwu Branch Steel Plant, Jinan 271126, China)

Abstract: Aiming at the problems such as fast erosion speed, single maintenance mode, poor effect and poor safety and stability of converter body during operation, the erosion principle of converter body was studied, and the precise control of slag sputtering layer was realized by optimizing slag sputtering technology. By innovating the way of furnace body maintenance, improving the capacity of weak parts of furnace body, and at the same time reducing the erosion speed of furnace body, all parts of furnace body are balanced, so as to provide guarantee for improving the metallurgical effect of converter.

Key words: sprinkle; slag process; equilibrium

转炉动态精准控制技术研究与应用

郭伟达，王　键，张昭平，任科社，杨普庆

（山东钢铁股份有限公司莱芜分公司炼钢厂，山东济南　271126）

摘　要： 针对转炉烟气分析智能控制中存在的操作模型适应性差，过程变量多，转炉终点参数无有效检测装置及终点控制参数精度低等问题，对转炉冶炼过程烟气各成分变化和炉内反应速率进行分析，查找其关联性，通过建立转炉过程反应的动态控制模型，实现吹炼过程炉内参数实时在线预报，并通过炉气分析对动力学模型进行全程的动态校正，实现转炉动态精准控制，提高转炉终点控制精度。

关键词： 烟气分析；动态控制；在线预报；动态校正

Research and Application of Converter Dynamic Precise Control Technology

Guo Weida, Wang Jian, Zhang Zhaoping, Ren Keshe, Yang Puqing

(Shandong Iron and Steel Co., Ltd., Laiwu Branch Steelmaking Plant, Jinan 271126, China)

Abstract: Aiming at the problems of poor adaptability, many process variables, no effective detection device and low precision of endpoint control parameters in the intelligent control of converter flue gas analysis, this paper analyzes the changes of flue gas components in converter smelting process and the reaction rate in furnace, and finds its correlation, By establishing the dynamic control model of the converter process reaction, the real-time on-line prediction of the parameters in the blowing process is realized, and the dynamic correction of the dynamic model is carried out through the furnace gas analysis, the dynamic precision control of the converter is realized, and the control precision of the converter endpoint is improved.

Key words: flue gas analysis; dynamic control; on-line forecast; dynamic correction

供异型钢高裂纹敏感铸坯质量控制研究

郭 达，刘文凭，张昭平，杨普庆，郭庆军

（山钢股份有限公司莱芜分公司炼钢厂，山东济南 271126）

摘 要：为提升异型钢铸坯质量，莱钢炼钢厂通过转炉过程炉渣氧化性精准控制、转炉终点钢水预脱氧、成分均质化控制、无缺陷铸坯控制、全流程低过热控制等技术研究与创新，开发出供异型钢高裂纹敏感铸坯质量控制研究工艺，实现异型钢铸坯质量本质化稳定，而且生产成本和工序能耗得到有效降低。

关键词：异型钢；高裂纹敏感；本质化稳定

For Special-shaped Steel High Crack Sensitive Cast Blank Quality Control Research

Guo Da, Liu Wenping, Zhang Zhaoping, Yang Puqing, Guo Qingjun

(Shandong Iron and Steel Co., Ltd., Laiwu Branch Steel Mill, Jinan 271126, China)

Abstract: In order to improve the quality of the special steel cast blanks, Lai steel steel mill through the transfer process furnace slag oxidation precision control, transfer furnace end steel water pre-oxygen, into technical research and innovation, such as homogenization control, defect-free casting blanks, and the production cost and process energy consumption were effectively reduced.

Key words: special steel; high crack sensitivity; essential stability

转炉高效低成本冶炼集成技术研究与应用

任科社，郭庆军，杨普庆，张昭平

（山东钢铁股份有限公司莱芜分公司炼钢厂，山东济南 271126）

摘 要：通过系统分析研究转炉整个工艺流程，找出制约生产的关键环节，从炉气分析静态控制技术开发、非对称大流量底吹冶金工艺技术研究、异夹角顶枪优化等几个方面开展工作，形成了一套转炉高效低成本冶炼集成技术。

关键词：转炉；炉气分析；非对称大流量底吹模型；异夹角氧枪

Research and Application of Converter Integrated Technology with High Efficiency and Low Cost

Ren Keshe, Guo Qingjun, Yang Puqing, Zhang Zhaoping

(Shandong Iron and Steel Co., Ltd., Laiwu Branch Steel Mill, Jinan 271126, China)

Abstract: By systematically analyzing and studying the whole process flow of converter, the key links restricting production are found. A set of integrated technology of high efficiency and low cost smelting for converter is formed, which includes the development of static control technology for furnace gas analysis, the research of asymmetric large flow bottom blowing metallurgical technology, and the optimization of different angle top lance.

Key words: converter; furnace gas analysis; asymmetric bottom blowing model with large flow; different angle oxygen gun

钛微合金化钢精炼工艺研究与应用

王　键，刘文凭，任科社

（山钢股份有限公司莱芜分公司炼钢厂，山东济南　271126）

摘　要： 针对钛微合金元素控制范围波动大，回收率不稳定，钢水洁净度不够的问题，通过研究钛微合金机理，分析影响钛元素回收率的因素主要是钢中氧含量、精炼过程操作、微合金化时机，通过优化转炉终点动态脱氧制度、完善精炼操作模式，根据脱氧方式调整钛微合金化时机，稳定钛元素的回收率，减少碳氮化物的析出，提高钛微合金化钢的洁净度，稳定铸坯和轧材性能。

关键词： 钛微合金化；脱氧制度；回收率

Research and Application of Titanium Microalloying Steel Refining Process

Wang Jian, Liu Wenping, Ren Keshe

(Shandong Iron and Steel Co., Ltd., Laiwu Branch Steel Mill, Jinan 271126, China)

Abstract: In allusion to the problems of large control range of titanium microalloy elements, unstable recovery rate and insufficient purity of molten steel are pointed out, by studying the mechanism of micro titanium alloy, analysis the factors affecting the recovery of titanium mainly are the oxygen content in steel, refining process operation, micro alloying time, by optimizing the dynamic deoxidation system at the converter end, perfecting the refining operation mode, adjust the timing of titanium microalloying according to deoxidation mode stability of titanium recovery, reduce the precipitation of carbon and nitrogen compounds, improve the cleanliness of ti microalloyed steel, stable billet and rolling properties.

Key words: titanium microalloying; deoxygenation system; recovery rate

耐火材料对切割丝用钢洁净度的影响

李　阳，杜高锋，杜沣庭，武锦涛，王　硕

（东北大学冶金学院，辽宁沈阳　110819）

摘　要： 选用 SiO_2、Al_2O_3、$MgO-Al_2O_3$ 和 $MgO-CaO$ 四种不同材质的耐火材料冶炼 95Cr 切割丝用钢，考察这四种不同材质的耐火材料对于切割丝用钢洁净度的影响，并在此基础上，进一步分析了各种材质耐火材料与钢液的相互作用。实验结果表明，不同材质耐火材料对钢中各元素含量均有较大的影响，不同材质耐火材料对钢中 T.O 及 S 有显著影响，其中以 MgO-CaO 材料最佳，可将全氧含量控制在 5×10^{-6} 以下，其夹杂物数量较少，尺寸较小，主要为镁类及镁类复合夹杂，因此冶炼 95Cr 等切割丝用洁净钢时，MgO-CaO 材料是最优质的耐火材料选择。

关键词：切割丝钢；耐火材料；夹杂物；洁净钢

Influence of Refractory on Cleanliness of Steel for Cutting Wire

Li Yang, Du Gaofeng, Du Fengting, Wu Jintao, Wang Shuo

(School of Metallurgy, Northeastern University, Shenyang 110819, China)

Abstract: Four different refractory materials, SiO$_2$, Al$_2$O$_3$, MgO- Al$_2$O$_3$ and MgO-CaO, were selected to smelt 95Cr cutting wire steel. The influence of these four different refractory materials on the cleanliness of cutting wire steel was investigated. On this basis, the interaction between various refractory materials and molten steel was further analyzed. The experimental results show that refractory materials of different materials have great influence on the content of various elements in steel, and refractory materials of different materials have significant influence on T.O and S in steel, of which MgO-CaO material is the best, which can control the total oxygen content below 5×10^{-6}, and its inclusions are small in number and size, mainly magnesium and magnesium compound inclusions, so MgO-CaO material is the best choice for smelting clean steel for cutting wires such as 95Cr.

Key words: steel for cutting wire; refractory matter; occluded foreign substance; clean steel

复吹转炉底枪布置对熔池搅拌混匀的影响

孙建月[1,3]，杨晓江[2]，杜文斌[1,3]，周泉林[2]，张 全[2]，钟良才[1,3]

（1. 东北大学冶金学院，辽宁沈阳 110004；2. 河钢乐亭钢铁有限公司，河北唐山 063016；
3. 东北大学多金属共生矿生态化冶金教育部重点实验室，辽宁沈阳 110004）

摘 要：通过物理模拟和数学模拟研究了不同底枪布置方案对于 200t 复吹转炉熔池搅拌混匀的影响。结果表明，底枪非对称布置能够使熔池内的流体产生非对称的流动，在竖直截面形成较大的回流区，在水平截面形成大小不同且无对称轴的环流区，"死区"体积与底枪对称布置相比减小约 50%，获得较短的熔池混匀时间。对称底枪布置使熔池内的流体分别在竖直截面和在水平截面形成有对称轴且相对独立的环流区，在小的底吹气量时，熔池的混匀时间长。

关键词：复吹转炉；熔池；搅拌混匀；底枪布置；物理模拟；数学模拟

Influence of Bottom Tuyere Configurations on Bath Stirring and Mixing in a Combined Blown Converter

Sun Jianyue[1,3], Yang Xiaojiang[2], Du Wenbin[1,3], Zhou Quanlin[2],
Zhang Quan[2], Zhong Liangcai[1,3]

(1. School of Metallurgy, Northeastern University, Shenyang 110004, China;

2. HBIS Laoting Iron & Steel Co., Ltd., Tangshan 063016, China;
3. Key Laboratory of Ecological Metallurgy of Polymetallic Ore of Ministry of Education, Northeastern University, Shenyang 110004, China)

Abstract: Influence of bottom tuyere configurations on bath stirring and mixing in a combined blown converter was investigated with physical modeling and mathematical simulation. It was shown from the investigation that asymmetric bottom tuyere configurations can obtain asymmetric fluid flowing in converter bath. Larger recirculation zones form on vertical sections. Circulation zones with different size and no symmetric axes can be found on horizontal sections. Volume of "dead zone" in the converter bath is reduced by about 50% in comparison with symmetric bottom tuyere configurations and shorter bath mixing time can be obtained. Symmetric bottom tuyere configurations form and relatively independent circulation regions with symmetric axes on vertical and horizontal sections, respectively and have longer complete mixing time at small bottom gas flow rate.

Key words: top and bottom combination blown converter; bath; stirring and mixing; bottom tuyeres configuration; physical modeling; mathematical simulation

电感耦合等离子体发射光谱法测定锰铁中 P 含量的测量不确定度评定

李玉洁，陈高莉，崔　隽，陈晓燕，刘丽荣

（武钢有限质检中心，湖北武汉　430080）

摘　要： 本文根据中国金属学会分析测试方法 CSM 04 02 94 01－2001《锰铁　铝、磷含量的测定　电感耦合等离子体发射光谱法》中锰铁磷含量的检测方法，对涉及测量过程中可能影响测量结果的称量、溶液体积、标准溶液浓度、测量重复性等方面进行了分析，评定出一个样品中 P 元素的测定结果的扩展不确定度。

关键词： 不确定度；磷；ICP；锰铁

Evaluation of Uncertainty for Determination of P in Manganese-iron Alloy by ICP Method

Li Yujie, Chen Gaoli, Cui Jun, Chen Xiaoyan, Liu Lirong

(WISCO Limited Quality Inspection Center, Wuhan 430080, China)

Abstract: Based on CSM 04 02 94 01—2001 "Determination of Al and P content in manganese-iron alloy by ICP method", This paper analyzes the Weighing process, solution volume, the concentration of standard solution and measurement repeatability that may affect the measurement results, and evaluates the extended uncertainty of the determination results of P in a sample.

Key words: uncertainty; P; ICP; manganese-iron alloy

烟气分析智能系统在转炉冶炼中的应用

郭伟达，王　键，杨普庆，任科社，张昭平

（山东钢铁股份有限公司莱芜分公司炼钢厂，山东济南　271126）

摘　要：本文主要介绍山钢股份莱芜分公司炼钢厂4号60t转炉，烟气分析智能系统的应用情况，经过与钢铁研究总院的共同努力，建立了国内首家小吨位转炉智能炼钢系统，进一步提高了品种钢炼成率、提高了终点碳温双命中率、缩短了冶炼周期、降低了出钢温度和降低了耐材消耗等，取得了较好的经济价值。

关键词：炼钢；烟气分析；小吨位转炉；经济价值

Application of Smoke Analysis Intelligent System in Converter Smelting

Guo Weida, Wang Jian, Yang Puqing, Ren Keshe, Zhang Zhaoping

(Shandong Iron and Steel Co., Ltd., Laiwu Branch Steel Mill, Jinan 271126, China)

Abstract: This paper mainly introduces the application of Smart System for Smoke Analysis in No.4 60-ton Converter of Laiwu Branch of Shangang. Through joint efforts with the General Institute of Iron and Steel Research, the first Smart Smelting System for Small-tonnage Converter in China has been established, which further improves the yield of variety steel, increases the double hit rate of end carbon temperature, shortens the smelting cycle, reduces the tapping temperature and reduces the tapping temperature. The consumption of refractories has achieved good economic value.

Key words: steel-making; flue gas analysis; small tonnage converter; economic value

气体[N]产生废次降的控制实践

陈代明，王永胜

（中冶京诚工程技术有限公司冶金工程事业部炼钢所，北京　100176）

摘　要：受钢铁料原材料影响，介绍了某炼钢厂采取应对措施，降低铁耗，提高废钢比例后，给转炉冶炼控氮带来较大影响。目前，根据多次冶炼工艺优化方案，某炼钢厂掌握了氮超炉次多的根本原因和控氮的难点所在。针对以上现象，通过有效手段解决存在的问题。最终，某炼钢厂制定了操作规程，保证了钢水去氮效果。

关键词：提高废钢比例控氮有效手段；去氮效果

Control Practice of Waste Inferior Degrade Production due to Gas[N]

Chen Daiming, Wang Yongsheng

(Steel Making Division, Metallurgical Industrials Business Unit, Capital Engineering &

Research Incorporation Limited, Beijing 100176, China)

Abstract: Influenced by raw materials of hot metal and steel scrap, the measures taken by one steelmaking plant to reduce hot metal consumption and increase scrap ratio, which have a great impact on [N] control of variety steels. At present, according to the optimization scheme of smelting process, the steelmaking plant master the root cause of more times of nitrogen overheating and the difficulty of nitrogen control. In view of the above phenomena, the existing problems can be solved by effective means. Finally, the steelmaking plant formulated the operation rules, which ensured the effect of nitrogen removal of molten steel.

Key words: nitrogen removal effect by effective means of increasing proportional; nitrogen control of scrap steel

我国转炉氧枪及喷头的发展现状

冯 超

（北京科技大学冶金与生态学院，北京 100083）

摘 要：我国转炉近年来趋于大型化发展，氧枪及喷头是实现安全生产和提高冶炼效果的关键装置，本文对我国目前氧枪和喷头的发展现状进行了系统性的总结，并分析各类氧枪和喷头的特点及使用现状，希望对从事转炉冶炼的工作者有所帮助。

关键词：转炉；氧枪；喷头；特点

Development Status of Converter Oxygen Lance and Nozzle in China

Feng Chao

(School of Metallurgical and Ecological Engineering, University of Science and Technology Beijing, Beijing 100083, China)

Abstract: Oxygen lance and nozzle are the key devices to realize safe production and improve smelting effect. This paper systematically summarizes the current development status of oxygen lance and nozzle in China, and analyses the characteristics and application status of various oxygen lances and nozzles, hoping to be helpful to the workers engaged in converter smelting.

Key words: converter; oxygen lance; nozzle; characteristics

ER70S-G 焊丝钢冶炼技术探讨

赵晓锋，张志强，王建忠

（河钢集团宣钢公司，河北宣化 075100）

摘 要：宣钢在开发生产 ER70S-G 焊丝钢过程中发现，由于钢中钛含量较高极易与钢中氧、氮反应生成高熔点的 TiN 和 Ti_2O_3 等夹杂物，导致钢中钛含量控制不稳定，在浇注过程中引起浸入式水口絮流，影响生产顺行，产品质量不稳定。通过控制钢中氮含量，优化精炼渣系，降低渣中（FeO）、钢中[O]含量等手段保证了生产顺行，产品质量稳定。

关键词：ER70S-G；焊丝；钛含量；水口絮流

Discussion on Welding Wire Steel Smelting Technology of ER70S-G

Zhao Xiaofeng, Zhang Zhiqiang, Wang Jianzhong

(Hesteel Group Xuansteel Company, Xuanhua 075100, China)

Abstract: During the development and production of ER70S-G wire steel, Xuanhua Iron and Steel Co., Ltd. found that the high content of titanium in steel easily reacted with oxygen and nitrogen in steel to form high-melting TiN and Ti_2O_3 inclusions, resulting in unstable control of titanium content in steel. In the pouring process, the immersion nozzle flocculation is caused, which affects the smooth production and the product quality is unstable. By controlling the nitrogen content in the steel, optimizing the refining slag system, reducing the content of [O] in the slag (FeO) and steel, the production is guaranteed to be smooth and the product quality is stable.

Key words: ER70S-G; welding wire; titanium content; clogging at submerged entry nozzle

宣钢 120t 转炉高效化生产实践

张利江，张明海

（河钢集团宣钢公司，河北宣化　075100）

摘　要：介绍了宣钢 120t 转炉通过提高一倒出钢率，优化氧枪及出钢口参数、应用钢包加盖技术，提高了转炉生产效率。实施优化后，转炉冶炼周期由 40～43min 缩短至 32～34min，日产炉数由原来的 29～31 炉提高到 37～39 炉，适应了新生产模式下的铁钢平衡，实现了高效化生产。

关键词：转炉；高效化；优化

Practice of High Efficiency Production of 120t Converter in Xuanhua Steel

Zhang Lijiang, Zhang Minghai

(HBIS Group Xuansteel, Xuanhua 075100, China)

Abstract: The production efficiency of the 120t converter has been improved by increasing the first inverted tapping rate, optimizing the parameters of oxygen lance and tapping hole, and applying ladle capping technology. After optimization, the smelting cycle of converter is shortened from 40～43 minutes to 32～34 minutes, and the daily production of converter is increased from 29～31 to 37～39 stoves, which adapts to the iron-steel balance under the new production mode and realizes high efficiency production.

Key words: converter; high efficiency; optimization

干式除尘转炉自动炼钢高拉碳工艺研究

张立君，王金龙，王宏斌，张明海

(河钢宣化钢铁集团有限责任公司，河北宣化 075100)

摘 要：宣钢在炼钢工艺流程及主要工艺装备采用成熟、先进、可靠的国产化干法除尘设备，在炼钢工艺技术上积极创新，研发具有自主知识产权的创新技术，干式除尘转炉自动炼钢高拉碳工艺就是其中的一种。干式除尘转炉冶炼终点高拉碳的技术难题曾是宣钢 150t 转炉工程开产之初技术系统一直担心的问题，由于干法除尘的限制，150t 转炉要想实现终点高碳、低磷就要双渣操作，双渣操作又容易引起电除尘设备发生泄爆。但是随着干式除尘转炉自动炼钢高拉碳工艺实施以后，通过对早期制定的泄爆预防"七步法"进行改进，很好的解决了转炉终点高拉碳的难题，同时又避免了电除尘器发生泄爆。通过实施后的对比，取得了很好的效果，终点碳较以前有了很大的提高，提高了钢水的纯净度，改善了铸坯质量。

关键词：干法除尘；双渣；高碳；低磷；泄爆

The Study on the Technology of High Carbon Steel Automatic Dry Dust Converter

Zhang Lijun, Wang Jinlong, Wang Hongbin, Zhang Minghai

(Steel Making Plant of Xuanhua Steel, Xuanhua 075100, China)

Abstract: Xuanhua Steel Company adopts domestic dry dedusting equipment is mature, advanced and reliable in steelmaking process and main equipment, positive innovation in steelmaking technology, developed with independent intellectual property rights of technology innovation, dry dust converter automatic steelmaking high carbon technology is one of them. Technical problems of dry dedusting of converter smelting high carbon steel end point was 150 tons of converter production opened at the beginning of the engineering technology system has been worried about the problem, because the dry dust limit of 150 tons of converter to achieve high end point carbon, low phosphorus will double slag operation, double slag operation and easy to cause the electric dust removal equipment occur venting. But with the dry dust converter automatic steelmaking process by high carbon, explosion prevention of "seven steps" to improve the formulation of the early release, it solved the problem of high end point carbon, while avoiding the occurrence of explosion of electric precipitator. Through the comparison, good results have been achieved. The end carbon has been greatly improved compared with before, which improves the purity of molten steel and improves the quality of billets.

Key words: dry dust removal; double slag; high carbon; low phosphorus; explosion venting

邯钢 260t 转炉顶底复吹过程数值模拟研究及应用

范 佳[1]，李太全[2]，高福彬[1]，靖振权[1]

(1. 河钢集团邯钢公司技术中心，河北邯郸 056015；

2. 河钢集团邯钢公司邯宝炼钢厂，河北邯郸　056015）

摘　要：本文基于有限单元法，通过建立多组不同底吹砖数量、安装位置以及底吹气体流量方案，对邯钢邯宝炼钢厂260t转炉的顶底复吹过程进行数值模拟研究。通过计算结果对比，最终确定将转炉底吹布局由原来的双环–12块底吹砖布局优化为单环–8块底吹砖布局。通过对生产现场底吹布局优化，底吹砖一次使用寿命从不超过3000炉延长至5000炉，同时有效改善了熔池搅拌效果，碳氧积从0.0026降低到0.0023以下，终点氧含量平均降低90ppm，钢水质量得到了有效提升。

关键词：有限单元法；260t转炉；顶底复吹；数值模拟；熔池搅拌；碳氧积

Research and Application of Numerical Simulation on Top-bottom Combined Blowing Process of 260 ton Converter in Han-Steel

Fan Jia[1], Li Taiquan[2], Gao Fubin[1], Jing Zhenquan[1]

(1. HBIS Group Han-Steel Company Technology Center, Handan 056015, China;
2. HBIS Group Han-Steel Company Hanbao Steel-making Plant, Handan 056015, China)

Abstract: In this paper, based on the finite element method, the top-bottom combined blowing process of 260 ton converter in Hanbao Steel-making Plant of Han-Steel. was numerically simulated by establishing a number of different groups of bottom-blown bricks, installation positions and bottom-blown gas flow schemes. By comparing the calculation results, it is determined that the bottom blowing layout of converter should be adjusted from the original double-ring-12 bottom blowing bricks to the single-ring-8 bottom blowing bricks. By optimizing the layout of bottom blowing in production site, the service life of bottom blowing brick is extended from 3000 to 5000 furnaces, and the stirring effect of molten pool is effectively improved. The carbon and oxygen volume are reduced from 0.0026 to less than 0.0023, the oxygen content of end point is reduced by 90 ppm on average, and the quality of steel is effectively improved.

Key words: finite element method; 260 ton converter; top-bottom combined blow; numerical simulation; bath stirring; carbon oxygen product

攀钢半钢冶炼转炉脱磷工艺技术研究

陈　均[1]，曾建华[1]，梁新腾[1]，喻　林[2]

（1. 攀钢集团研究院有限公司，钒钛资源综合利用国家重点实验室，四川攀枝花　617000；
2. 攀钢集团西昌钢钒有限公司，四川西昌　615000）

摘　要：本文针对攀钢半钢冶炼时辅料消耗大、终渣氧化性高且脱磷效果不佳的问题，通过对转炉脱磷理论以及冶炼过程脱磷规律的研究，确定了前期脱磷率偏低、中期返干是影响脱磷率的主要因素；通过对终渣岩相的进一步分析，确定了磷在渣中的主要富集相，解释了后期依靠提高炉渣氧化性和增加辅料消耗不能显著提高脱磷率的原因。通过采用"留渣加料"、含铝复合造渣剂造渣以及降低冶炼后期枪位等技术措施后，转炉成渣时间由4.3min缩短到3.2min，返干比例由56%显著降低到18%，在总渣料消耗平均减少5.69kg/t$_{钢}$的情况下转炉冶炼全程脱磷率由79.4%提高到84.1%，终渣TFe含量平均降低2.49个百分点。通过对脱磷工艺参数的优化，在提高转炉脱磷效率的同时，降低了转炉冶炼成本。

关键词：半钢冶炼；转炉脱磷；岩相分析；留渣加料

Research on Dephosphorization Technology by Semi-steel Steelmaking Converter of Pangang

Chen Jun[1], Zeng Jianhua[1], Liang Xinteng[1], Yu Lin[2]

(1. State Key Laboratory of Vanadium and Titanium Resources Comprehensive Utilization, Pangang Group Research Institute Co., Ltd., Panzhihua 617000, China; 2. Pangang Group Xichang Vanadium & Steel Co., Ltd., Xichang 615000, China)

Abstract: To solve the problem of high slagging materials consumption, high oxidablitiy of slag and low dephosphorization effect, theory of dephosphorization and dephosphorization rules were researched and factors affected dephosphorization rate were confirmed. And according to the lithofacies of slag, enriched phase of phosphorous in slag was found out, reasons of why dephosphorization rate couldn't be improved remarkable by high slagging materials consumption and high oxidablitiy were explained. By adopting the measures of "slag-remaining –charging material", aluminiferous-slag former and lower the lance position, slag-forming time was reduced from 4.3min to 3.2min, ration of re-dry was reduced remarkably from 56% to 18%, total dephosphorization rate was improved from 79.4% to 84.1 under the condition that total material consumption was reduced by 5.69kg/t$_钢$, in addition, TFe content of slag is reduced by 2.49 percent point on average. According to optimization of dephosphorization parameters, dephosphorization rate was improved and cost of BOF refining was reduced as well.

Key words: semi-steel steelmaking; dephosphorization; lithofacies analysis; slag-remaiing charging material

老厂房 150t 转炉铁水一罐到底工艺研究与应用

田云生，郭永谦，孙　拓，张远强，李志广

（安阳钢铁股份有限公司，河南安阳　455004）

摘　要： 介绍安钢二炼轧厂通过对现有厂房、设备等参数的实地测量及考证，制订合理的实施方案，最大限度地利用现有的生产设备，通过关键设备的改造（如铁水罐、过跨车等），实现了150t转炉铁水"一罐到底"的工艺改造过程。从整个试车流程来看，过程中出现的问题较少，绝大部分工艺均在预期的效果之内。安钢150t转炉铁水"一罐到底"系统工艺具有简化生产流程、降低能源消耗、稳定转炉操作、节约增效等优点，在环保要求严格的今天，有着较好的应用前景。

关键词： 转炉炼钢；一罐到底；铁水罐；铁水包；过跨车

Research and Application of Hot Metal Get the Ladle Done at One Go for 150 t Converter in Old Factory Building

Tian Yunsheng, Guo Yongqian, Sun Tuo, Zhang Yuanqiang, Li Zhiguang

(Anyang Iron and Steel Co., Ltd., Anyang 455004, China)

Abstract: Introduce the Angang No. 2 refining and rolling mill to formulate the project technical transformation implementation plan through the on-site measurement and research on the existing plant, equipment and other parameters, to maximize the use of existing production equipment, through the transformation of key equipment (such as iron water tanks), over-span, etc.), the realization of 150t converter molten iron "one can the end" process transformation. From the perspective of the entire test run, there are fewer problems in the process, and most of the processes are within the expected results. Angang's 150t converter molten iron "one can end" system process has the advantages of simplifying production process, reducing energy consumption, stabilizing converter operation, energy saving and efficiency, and conforming to the green, high efficiency and environmental protection concept of today's factory development.

Key words: converter steelmaking; one can end; iron water tank; ladle; over-street

一种低碳高磷系列钢造渣冶炼方法

佟 迎

（宝武集团广东韶关钢铁有限公司炼钢厂，广东韶关　512123）

摘　要： 低碳高磷系列钢对炉渣碱度、枪位控制、化渣及脱磷效果的要求都不高，针对这一特点，使用生白云石作为主造渣料，取代石灰和轻烧这两种传统的主造渣料，同时采取低枪位、大氧压、强底吹的冶炼控制模式。大幅度降低该系列钢种的造渣料消耗成本，同时也降低了脱磷率，节约磷铁合金消耗成本；低枪位吹炼加快生产节奏，提高转炉生产效率。

关键词： 渣料；生白云石；脱磷率；低枪位；低成本

A Slag Smelting Method for Low Carbon and High Phosphorus Steel Series

Tong Ying

(Shaoguan Iron & Steel Co., Ltd., Baosteel Group, Shaoguan 512123, China)

Abstract: Low carbon and high phosphorus steel series have low requirements for slag basicity, lance level control, slagging and dephosphorization effect. In view of this characteristic, raw dolomite is used as main slag-making material instead of lime and light burning two traditional slag-making materials, and the smelting control mode of low lance level, high oxygen pressure and strong bottom blowing is adopted. The slagging material consumption cost of this series of steels has been greatly reduced, the dephosphorization rate has been reduced, and the consumption cost of ferrophosphorus alloy has been saved. The production rhythm has been accelerated and the converter production efficiency has been improved by low lance blowing.

Key words: slag-making material; raw dolomite; dephosphorization rate; low lance level; low cost

韶钢120t铁水包加废钢提升废钢比生产实践

佟 迎

（宝武集团广东韶关钢铁有限公司炼钢厂，广东韶关　512123）

摘　要：在不需要辅助热源，不影响转炉兑铁，不影响铁包寿命及正常周转的情况下，向铁包加入一定量的废钢，利用铁水包空包物理热来加热废钢。此举，开拓了提升转炉废钢加入量的新路径，达到进一步提高转炉废钢比，降低铁水消耗的目的；同时，充分回收利用一部分自然散失的热能。

关键词：铁水包；废钢；加热；废钢比

Production Practice of Raising Scrap Ratio by Adding Scrap Steel to 120tn Ladle at Shaoguan Iron and Steel Co.

Tong Ying

(Shaoguan Iron & Steel Co., Ltd., Baosteel Group, Shaoguan 512123, China)

Abstract: In the case of no need of auxiliary heat source, no influence on converter iron mixing, no influence on ladle life and normal turnover, a certain amount of scrap steel is added to the ladle, and the physical heat of empty ladle is used to heat the scrap steel. This opens up a new way to increase the scrap content of converter, to further improve the scrap ratio of converter and reduce the consumption of molten iron, while fully recovering and utilizing a part of the natural heat dissipated.

Key words: Ladle; scrap steel; heat; scrap ratio

冷轧基料冶炼工艺优化与实践

吕圣会，李海洋，王克忠

（泰山钢铁集团有限公司，山东莱芜　271100）

摘　要：通过优化冷轧基料冶炼工艺路线，实施钢包顶渣改质、改善炉后吹氩质量、控制钢中的[Al]s 含量损失、连铸全程保护性浇注等措施，铸坯夹杂物控制水平、连铸连浇炉数及板坯质量得到了提高，实现了冷轧基料不走精炼，吨钢成本降低 100 元。

关键词：冷轧基料；工艺优化；成本降低；夹杂物控制

Optimization Practice of Smelting Process of Base Material for Cold Rolling

Lv Shenghui, Li Haiyang, Wang Kezhong

(Taishan Iron and Steel Group Co., Ltd., Laiwu 271100, China)

Abstract: By optimizing the base material for cold rolling smelting process, the implementation of the ladle slag modification and improvement of furnace after argon blowing quality, control of [al] s content in steel, continuous casting full protection casting and other measures, the inclusion control level continuous casting and casting heats and slab quality improved, the base material for cold rolling away refining, the cost per ton of steel reduced 100 Yuan.

Key words: base material for cold rolling; process optimization; cost reduction; inclusion control

铝基脱磷剂对高磷锰硅合金还原脱磷的影响

孙 灿，朱子宗，宗 楷，焦万宜

（重庆大学材料科学与工程学院，重庆 400044）

摘 要：为研究脱磷剂对锰硅合金还原脱磷的影响，在1400℃，采用铝基脱磷剂，以$CaO-SiO_2-Al_2O_3$为覆盖渣，对$w(P)$为1%的高磷锰硅合金进行还原脱磷热力学实验。重点研究了铝基脱磷剂对锰硅合金脱磷效果、脱磷产物及合金组织的影响。研究结果表明，铝基脱磷剂和熔体中的P发生反应生成AlP，并上浮到渣中排出，当脱磷剂用量为合金质量的8%时，脱磷率可达60%；通过对脱磷后合金微观结构的分析发现，过量的铝基脱磷剂会与合金中以及冶炼容器带入的C反应生成Al_4C_3，Al_4C_3可与由合金中夹杂物形成的微裂纹进入的水分子发生水解反应，生成的气体与微裂纹相互作用导致合金粉化。

关键词：锰硅合金；铝基脱磷剂；脱磷效果；脱磷产物；合金组织

Effect of Al-based Dephosphorization Agent on Reduction and Dephosphorization for High Phosphorus MnSi Alloys

Sun Can, Zhu Zizong, Zong Kai, Jiao Wanyi

(College of Material Science and Engineering, Chongqing University, Chongqing 400044, China)

Abstract: In order to study the effect of Al-based dephosphorization agent on reduction and dephosphorization for high phosphorus MnSi alloys, reduction and dephosphorization thermodynmics of $w(P)$ MnSi alloys(1%) at 1400℃ was investigated by using Al-based dephosphorization agent and $CaO-SiO_2-Al_2O_3$ slags. The present study focused on the effect of Al-based dephosphorization agent on dephosphorization result, dephosphorization product and alloy structure. The results show that the Al-based dephosphorization agent reacts with P in the melt to form AlP, and it floats up to the slag, when the amount of dephosphorization agent is 8% of the alloy mass, the dephosphorization rate can reach 60%. Through the analysis of the microstructure of the alloy after dephosphorization, it is found that the excess Al-based dephosphorization agent reacts with the C in the alloy and the smelting container to form Al_4C_3, and the Al_4C_3 can react with water molecules to form a gas. The interaction between gas and microcracks leads to alloy powdering.

Key words: MnSi alloys; Al-based dephosphorization agent; dephosphorization result; dephosphorization product; alloy structure

X80管线钢非金属夹杂物控制研究

李战军[1,2,3]，初仁生[1]，刘金刚[1]，郝 宁[1]

（1. 首钢技术研究院宽厚板所，北京 100043；2. 绿色可循环钢铁流程北京市重点实验室，北京 100043；3. 北京市能源用钢工程技术研究中心，北京 100043）

摘　要：本文研究了"铁水脱硫预处理—转炉—LF 精炼—RH 真空精炼—板坯连铸"工艺 X80 管线钢非金属夹杂物的控制，采用此工艺生产的 X80 管线钢 T[O]≤10ppm，实现了高洁净度控制；并对冶炼过程中的夹杂物转变机理进行研究，使钢坯中的夹杂物得到有效的控制：从形态上看，以球形夹杂为主；从成分上看，以高熔点的钙铝酸盐和 CaS 的复合夹杂物为主；从尺寸上看，尺寸控制在≤8μm。通过板坯轧制后，采用金相评级的方法对轧材中的非金属夹杂物进行评级，非金属夹杂物的合格率稳定控制在 99.6%以上，其中 A 类非金属夹杂物评级为 0，B 类非金属夹杂物评级≤1.5。

关键词：X80 管线钢；冶炼工艺；洁净度；非金属夹杂物

Study on Control of Non-metallic Inclusions in X80 Pipeline Steel

Li Zhanjun[1,2,3], Chu Rensheng[1], Liu Jingang[1], Hao Ning[1]

(1. Shougang Research Institute of Technical Department of Heavy and Medium Plate, Beijing 100043, China; 2. Beijing Key Laboratory of Green Recyclable Process for Iron & steel Production Technology, Beijing 100043, China; 3. Beijing Engineering Research Center of Energy Steel, Beijing 100043, China)

Abstract: In this paper, the smelting process of "Hot metal desulphurization pretreatment-Basic oxygen furnace-LF refining-RH refining-continuous casting slab" is studied to control of non-metallic inclusions in X80 pipeline steel. The X80 pipeline steel T[O]≤10ppm produced by this process realizes high cleanliness control. The transformation mechanism of inclusions in the process of smelting was studied, so that the inclusions in the billet could be effectively controlled: spherical inclusions are dominant in morphology; high melting point calcium aluminate and CaS composite inclusions are dominant; the size is controlled ≤8μm. After slab rolling, the metallographic grading method was used to grade the non-metallic inclusions in the rolled steel. The qualified rate of non-metallic inclusions is stably controlled over 99.6%, of which the grade of A-type non-metallic inclusions is 0, and that of B-type non-metallic inclusions is less than 1.5.

Key words: X80 pipeline steel; smelting process; cleanliness; non-metallic inclusions

应用极值分析标准评估轴承钢中大尺寸夹杂物分布特征

郭洛方，高永彬，陈殿清，赵东记，张　杰，雷　富

（青岛特殊钢铁有限公司中特研究院青钢分院（技术中心），山东青岛　266409）

摘　要：本文基于金相检测的大颗粒夹杂物的尺寸，应用 ASTM E2283 中极值分析的标准化方法（最大似然估计法）对轴承钢不同冶炼阶段所取试样中夹杂物尺寸分布进行了统计分析，并采用 SEM/XPS 对大尺寸夹杂物进行了成分检测。分析结果表明：在逆程周期为 1000 的检测条件下，通过极值分析方法可计算出试样中相对应 99.90%概率的最大夹杂物尺寸，客观描述了轴承钢中夹杂物尺寸分布特征；某一视场或不同视场内发现两类的大尺寸夹杂物：$CaO-Al_2O_3-(MgO)$ 类和 Al_2O_3 类，两类夹杂物具有不同的尺寸分布规律；中间包浇注 25min 时 Al_2O_3 类夹杂物出现大尺寸概率明显小于 $CaO-Al_2O_3-(MgO)$ 类夹杂物，说明大尺寸夹杂物主要为 $CaO-Al_2O_3-(MgO)$ 类夹杂物。

关键词：极值分析；轴承钢；夹杂物；尺寸分布

Size Distribution Characteristics of Large Size Inclusions in Bearing Steel by Extreme Value Analysis Standard

Guo Luofang, Gao Yongbin, Chen Dianqing, Zhao Dongji, Zhang Jie, Lei Fu

(Qingdao Special Steel Co., Ltd., Research Institute of Wire Materials, Qinggang Rranch (Technology Center), Qingdao 266409, China)

Abstract: In this paper, based on the metallographic examination of large inclusion size, inclusion size distributions of samples taked at different bearing steel production process stages were statistically analyzed by using extreme value analysis standard method of ASTM E2283(the maximum likelihood method), and the large size inclusions compositions were detected by SEM/XPS testing method. The results of analysis indicate that: in the condition of return period equal to 1000, the size of an inclusion corresponding to the 99.90% probability value could be calculated, thus the size distribution features of inclusions in bearing steel can be objectively described; two types of large size inclusions which are $CaO-Al_2O_3-(MgO)$ type and Al_2O_3 type were found in one same or different metallographic microscope fields of view, these two different type inclusions have different size distribution regulation; When tundish casting 25min, the Al_2O_3 type inclusions are apparently less likely to be found than $CaO-Al_2O_3-(MgO)$ type inclusions, it illustrates that the larger size inclusions are mainly $CaO-Al_2O_3-(MgO)$ type.

Key words: extreme value analysis; bearing steel; inclusion; size distribution

热送热装在青岛特钢的实践与应用

张军卫，刘　澄，刘政鹏，钟　浩

（中信特钢集团青岛特钢有限公司工艺所，山东青岛　266409）

摘　要： 青岛特钢公司充分利用前期设计的热送热装装置，通过无缺陷铸坯生产、设备整修、自动化控制系统完善，通过试样取送分析与结果传递的信息化改进，通过科学化的生产组织管理，使得铸坯的热装比例达到50%以上、高温直送铸坯温度达到620℃以上，每年为公司减低燃耗4.16×10^5GJ。热送热装的实施，每年能为公司带来2000多万元的经济效益。

关键词： 热送热装；热装比；热送温度；燃耗；经济效益

Practice and Application of Hot Delivery and Hot Loading in Qingdao Special Steel

Zhang Junwei, Liu Cheng, Liu Zhengpeng, Zhong Hao

(Citic Special Steel Group Qingdao Special Steel Co., Ltd., Qingdao 266409, China)

Abstract: Qingdao special steel company make full use of the preliminary design of hot delivery and hot installation charging device, with no defects casting billet production, equipment repairs, automation control system is perfect, and through the analysis of the sample take to send the transfer information to improve, through scientific production organization management, make the slab hot charging ratio above 50%, high temperature sent straight slab temperature of

620 ℃ or above, more than a year for the company to reduce fuel consumption of 4.16 x 10⁵ GJ. The implementation of hot delivery and hot installation can bring more than 20 million yuan of economic benefits to the company every year.

Key words: hot delivery and hot installation; hot charging ratio; heat transfer temperature; fuel consumption; economic benefits

钢包底吹过程物理模拟研究

支保宁[1,2]，赵定国[1,2]，张印棠[1,2]，王书桓[1,2]，张福君[1,2]

（1. 华北理工大学冶金与能源学院，河北唐山 063210；
2. 唐山特种冶金及材料制备重点实验室，河北唐山 063210）

摘　要： 针对某厂 90t 钢包进行底吹氩气水模拟实验，根据原厂尺寸按照 1∶3 的比例制作有机玻璃钢包模型，研究了不同气体流量下对卷渣行为以及钢包混匀时间的影响，阐述了钢包卷渣形成和混匀机理，研究表明，气体流量越大，卷渣现象越严重，随着底吹流量增加，液面凸起直径和钢液裸露面积逐渐增大，最大直径达到 16cm，钢液在空气中二次氧化越严重，造成钢液的污染，增加底吹气体显著加快了钢包混匀时间。

关键词： 钢包精炼；底吹氩；卷渣；混匀时间

Physical Simulation of Ladle Bottom Blowing Process

Zhi Baoning[1,2], Zhao Dingguo[1,2], Zhang Yintang[1,2],
Wang Shuhuan[1,2], Zhang Fujun[1,2]

(1. College of Metallurgy and Energy, North China University of Science
and Technology, Tangshan 063210, China; 2. Tangshan Key Laboratory of Special
Metallurgy and Materials Manufactory, Tangshan 063210, China)

Abstract: The bottom argon blowing gas-water simulation experiment was carried out for a 90t ladle in a factory. The PMLA ladle model was made according to the original size of the factory according to the ratio of 1∶3. The effects of different gas flow rates on slag entrainment behavior and mixing time of the ladle were studied. The mechanism of slag entrainment formation and mixing was expounded. The results show that the larger the gas flow rate, the more slag entrainment occurs. With the increase of bottom blowing flow rate, the diameter of liquid surface bulge and the exposed area of molten steel gradually increase. The maximum diameter of molten steel reaches 16 cm. The more serious the secondary oxidation of molten steel in air is, the more contaminated the molten steel is, and the increase of bottom blowing gas remarkably speeds up the mixing time of ladle.

Key words: ladle refining; blowing argon treatment; slag entrapment; mixing time

韶钢铁水一罐制改造的探索与实践

黄纯旭，陈　贝，邓长付

（宝武集团广东韶关钢铁有限公司炼钢厂，广东韶关　512123）

摘　要：本文通过对韶钢生产现场铁水一罐制工艺的改造实施及运行情况进行分析总结，阐明了铁水一罐制工艺特点和优势，为老厂实施改造提供了切实可行的经验措施。

关键词：钢铁冶金；老厂；铁水一罐制；改造

Exploration and Practice of One-pot System Reform of SGIS

Huang Chunxu, Chen Bei, Deng Changfu

(Steel-making Plant, Baowu Group, Shaoguan 512123, China)

Abstract: This paper analyzes and summarizes the transformation and implementation of the hot metal one-pot process at the production site of SGIS, and clarifies the characteristics and advantages of the hot metal one-pot process, and provides practical and feasible measures for the transformation of the old factory.

Key words: ferrous metallurgy; old factory; hot metal one-pot process; transformation

550A 磨球钢圆钢断裂原因分析

叶德新，邓湘斌，曾令宇，冯杰斌

（宝武集团广东韶关钢铁有限公司炼钢厂，广东韶关　512123）

摘　要：针对550A磨球钢圆钢断裂的问题，取样进行宏观形貌、成分、金相组织、扫描电镜等理化性能检测分析，同时对生产过程进行详细调查。结果表明，生产过程中圆钢弯曲大，冷却速度快，后续未经去应力处理直接矫直，应力过大导致圆钢产生裂纹甚至断裂。通过原因分析，采取相应措施，有效的避免了圆钢断裂。

关键词：550A；磨球钢；断裂；应力

Analysis of Fractures on Grinding Ball Steel 550A

Ye Dexin, Deng Xiangbin, Zeng Lingyu, Feng Jiebin

(Baowu Group Guangdong Shaoguan Iron and Steel Co., Ltd.,
Shaoguan 512123, China)

Abstract: Spectroscopic analyzer, microscope, scanning electron microscope, etc are used to test and analyze the physical and chemical properties of the samples for the fractures on grinding ball steel 550A, and the production process is investigated in detail. The results show that the round bars have been bent to much and cooled at fast speed during the process. The subsequent straightening is conducted without stress relief which directly results in cracking or even fracture of the round bar due to the excessive stress. The fractures on round steel bars have been effectively avoided by means of cause analysis and adopting corresponding measures.

Key words: 550A; grinding ball steel; fracture; stress

20 钢连铸圆管坯的研发和生产

李祥才,张 虎,柯家祥

(青岛特殊钢铁有限公司棒材研究所,山东青岛 266400)

摘 要:根据客户需要及青钢设备状况,青岛特钢采用 LD 转炉→LF 精炼炉→CC 连铸工艺开发生产了 ϕ250mm 规格 20 钢连铸圆管坯。产品实物质量表明,其化学成分满足客户协议要求,同时连铸圆管坯中心疏松、缩孔、中心裂纹、中间裂纹、皮下裂纹、皮下气泡低倍指标均满足客户协议要求,用户反馈使用良好,青岛特钢已经具备了生产 ϕ250mm 规格 20 钢连铸圆管坯的能力。

关键词:20 钢;连铸圆管坯;中心疏松;缩孔;中心裂纹

Research and Production of 20 Steel for Continuous Casting Round Tube Blank

Li Xiangcai, Zhang Hu, Ke Jiaxiang

(Round Bar Reseach Institute, Qingdao Special Iron and Steel Co., Ltd., Qingdao 266400, China)

Abstract: According to customer needs and equipment conditions of Qinggang, 20 steel of ϕ250mm size for continuous casting round tube blank was developed and produced adopting the process route of LD→LF→CC by Qingdao Special Iron and Steel Co., Ltd., The product quality showed the chemical composition of 20 steel satisfied the requirements of customer agreement, at the same time the center porosity、shrink cavity、center crack、medium crack、sub-surface crack、sub-surface blister of macro-indexes all met the requirements of customer agreement, the quality of 20 steel was fully qualified and the customer feedback the use of 20 steel for continuous casting round tube blank was good. Qingdao Special Iron and Steel Co., Ltd. has possessed the capability to produce the 20 steel of ϕ250mm size for continuous casting round tube blank.
Key words: 20 steel; continuous casting round tube blank; center porosity; shrink cavity; center crack

承钢 150t 转炉分钢种高效脱磷方法与低成本控制

韩德文,仇 军,胡凤伟,郭盈伟,张文彪,高建国

(河钢股份有限公司承德分公司板带事业部,河北承德 067002)

摘 要:通过对承钢 150t 转炉"双联"工艺条件下,铁水中 Si、Ti、部分 Mn 和部分 Cr 等元素在提钒工序经过氧化进入钒渣,造成半钢炼钢渣系单一[1],化渣困难,脱磷困难。结合承钢 150t 产品成分要求的特点,成功开发出一系列高效低成本精益控制方法,有效降低了合金、熔剂及钢铁料成本。

关键词:脱磷;方法;双联工艺;成本;半钢炼钢

High Efficiency Dephosphorization Method and Low Cost Control for 150t Converter of Bearing Steel

Han Dewen, Qiu Jun, Hu Fengwei, Guo Yingwei, Zhang Wenbiao, Gao Jianguo

(Hegang Co., Ltd., Chengde Branch Board Division, Chengde 067002, China)

Abstract: Under the "double connection" process condition of the 150-ton converter of the steel, elements such as Si, Ti, some Mn, and some Cr in the iron water were oxidized into vanadium slag during the vanadium process, resulting in a single system of semi-steel steelmaking slag[1] The slag is difficult and the phosphorus is difficult. Combined with the characteristics of 150-ton product composition requirements of Chenggang, a series of high-efficiency and low-cost lean control methods were successfully developed, which effectively reduced the cost of alloys, fluxes and steel materials.

Key words: dephosphorization; methodology; dual process; cost; half steel steelmaking

转炉冶炼低硫钢的控硫方法

赵 科, 邓长付

(宝武集团广东韶关钢铁有限公司炼钢厂, 广东韶关 512123)

摘 要： 分析了转炉冶炼低硫钢回硫原因, 脱硫捞渣后残余渣量和废钢带入硫含量是造成回硫的主要因素。结合生产实践, 通过工艺改进, 转炉终点硫含量能稳定的控制在0.006%以内, 具备低硫钢开发生产的能力。

关键词： 转炉；低硫钢；回硫

Method of Sulphur Control for Converter Smelting of Low Sulphur Steel

Zhao Ke, Deng Changfu

(Baowu Group Guangdong Shaoguan Iron and Steel Co., Ltd., Shaoguan, 512123, China)

Abstract: The reason of resulfurization in converter smelting of low sulphur steel is analyzed. The main factors of resulfurization are residual slag after desulphurization and deslagging, and sulphur content in scrap. Combining production practice, and through process improvement, the sulfur content at the final end of smelting in converter can be controlled to within 0.010%, which means it is able to develop and produce low sulphur steel.

Key words: BOF; low sulphur steel; resulfurization

钢中复合夹杂物/钢基体的电势差与电偶腐蚀的关系

侯延辉[1,2], 刘林利[1,2], 李光强[1,2], 李腾飞[1,2], 刘 昱[1,2]

(1. 耐火材料与冶金省部共建国家重点实验室, 武汉科技大学, 湖北武汉 430081;

2. 钢铁冶金及资源利用省部共建教育部重点实验室，武汉科技大学，湖北武汉 430081）

摘 要：为了揭示高强度管线钢中典型复合夹杂物诱发点蚀的机理，本文以 Al-Ti-Mg 脱氧钢为例，采用第一性原理计算，浸泡试验，扫描电镜等研究了钢中复合夹杂物/钢基体的电势差与电偶腐蚀的关系。结果表明，在 3.5% NaCl 腐蚀环境中，MnS 夹杂物起阳极作用，优先腐蚀和溶解；$MgAl_2O_4$ 和 Al_2O_3 起阴极作用，导致铁基体的腐蚀；$MgTiO_3$ 和 $MgTi_2O_4$ 的不同端面同时起阳极和阴极的作用，因此对点蚀的影响不明显。

关键词：夹杂物；点蚀；电偶腐蚀；第一性原理；电势差

The Correlation between Potential Difference and Galvanic Corrosion of Composite Inclusions/steel Matrix in Steel

Hou Yanhui[1,2], Liu Linli[1,2], Li Guangqiang[1,2], Li Tengfei[1,2], Liu Yu[1,2]

(1. State Key Laboratory of Refractories and Metallurgy, Wuhan University
of Science and Technology, Wuhan 430081, China;
2. Key Laboratory for Ferrous Metallurgy and Resources Utilization of Ministry
of Education, Wuhan University of Science and Technology, Wuhan 430081, China)

Abstract: In order to reveal the mechanism of galvanic pitting corrosion initiation induced by typical composite inclusions, first-principles calculations, combining with immersion tests, scanning electron microscopy was used to study the correlation between electronic work function and galvanic corrosion of Al-Ti-Mg killed steel. The results show that MnS inclusions act as anodes in the electrochemical environment, preferentially corroding and dissolving; $MgAl_2O_4$ and Al_2O_3 inclusions act as cathodes, leading to the corrosion of Fe matrix; different end planes of $MgTiO_3$ and $MgTi_2O_4$ act as both anodes and cathodes, so they have little effect on pitting corrosion.

Key words: inclusions; pitting corrosion; galvanic corrosion; first-principles; potential difference

CaF_2 和 Na_2O 对钢渣含磷相析出过程的影响

王达志，包燕平，王 敏

（北京科技大学钢铁冶金新技术国家重点实验室，北京 100083）

摘 要：为了研究 CaF_2 和 Na_2O 对钢渣含磷相析出过程的影响，以 $CaO-SiO_2-Fe_2O_3-P_2O_5$ 炉渣为研究对象，分别添加 5%的 CaF_2 和 Na_2O，通过热态实验探讨了 CaF_2 和 Na_2O 的加入对钢渣中含磷相的成分组成的影响。通过 FactSage7.2 热力学软件计算了钢渣冷却过程中含磷相的析出过程，结果表明：$CaO-SiO_2-Fe_2O_3-P_2O_5$ 渣系在冷却过程后期析出的含磷相为 $Ca_5P_2SiO_{12}$。F 元素主要与 P 元素结合进入含磷相，在冷却过程中析出的含磷相为 $Ca_5(PO_4)_3F$。Na 元素同样与 P 元素结合进入含磷相，析出 $Na_2Ca_2P_2O_8$ 相和少量 Na_3PO_4 相。Na_3PO_4 相的含量随着 Na_2O 的增加呈现增加的趋势。

关键词：CaF_2；Na_2O；钢渣；含磷相；析出过程

Effects of CaF₂ and Na₂O on the Precipitation Process of Phosphorus-containing Phase in Steel Slag

Wang Dazhi, Bao Yanping, Wang Min

(State Key Laboratory of Advanced Metallurgy, University of
Science and Technology Beijing, Beijing 100083, China)

Abstract: In order to study the effect of CaF₂ and Na₂O on the precipitation process of phosphorus-containing phase in steel slag, 5% CaF₂ and Na₂O were added to CaO-SiO₂-Fe₂O₃-P₂O₅ slag respectively. The effects of CaF₂ and Na₂O on the composition of phosphorus-containing phase in steel slag were investigated by thermal experiments. The precipitation process of phosphorus-containing phase in steel slag during cooling process was calculated by FactSage 7.2 thermodynamic software. The results show that the phosphorus-containing phase of CaO-SiO₂-Fe₂O₃-P₂O₅ is Ca₅P₂SiO₁₂ at the later stage of cooling process. F element is mainly combined with P element into phosphorus-containing phase, and the phosphorus-ontaining phase precipitated during cooling process is Ca₅(PO₄)₃F. Na also binds with P into phosphorus-containing phase and precipitates Na₂Ca₂P₂O₈ and a small amount of Na₃O₄. The content of Na₃PO₄ increases with the increase of the content of Na₂O.

Key words: CaF₂; Na₂O; steel slag; phosphorus-containing phase; precipitation process

煤粉中氮含量的检验

崔隽，张巧燕，沈克，胡涛，李玉洁，陈高莉

（宝钢股份武钢有限公司质检中心，湖北武汉 430080）

摘 要： 煤粉燃烧生成大量的氮氧化合物，是大气中氮氧化物的主要人为来源，在全球环境变化中起着重要作用。煤在燃烧时生成氨等气体，会腐蚀燃煤设备及管道，但这些气体又可以被回收利用，用来生产化工产品。本项目对煤粉的试样量、助溶剂以及工作曲线进行了选择，研究了脉冲加热样品熔融-热导法测定煤粉中的氮含量。试验结果表明：该方法操作简单，准确度高，重复性好，试样检测时间控制在 30 分钟以内。

关键词： 煤粉；氮；脉冲加热样品熔融-热导法

Inspection of Nitrogen Content in Pulverized Coal

Cui Jun, Zhang Qiaoyan, Shen Ke, Hu Tao, Li Yujie, Chen Gaoli

(Quality Inspection Center of Baosteel Co., Ltd., WISCO Limited, Wuhan 430080, China)

Abstract: The combustion of pulverized coal produces a large number of nitrogen oxides, which are the main anthropogenic sources of nitrogen oxides in the atmosphere and play an important role in global environmental change. When coal burns, it produces ammonia and other gases, which will corrode coal-fired equipment and pipelines, but these gases can be recycled for the production of chemical products. In this project, the sample size, cosolvent and working curve of pulverized coal were selected, and the determination of nitrogen content in pulverized coal by pulse heating sample melting-thermal conductivity method was studied. The test results show that the method is simple, accurate and reproducible. The test time of the sample is controlled within 30 minutes.

Key words: pulverized coal; nitrogen; pulse heating sample melting-thermal conductivity method

硅脱氧弹簧钢 55SiCr 炼钢过程氧化物夹杂变化研究

孟耀青[1,2]，赵铮铮[2]，赵昊乾[1,2]，吕海瑶[1,2]，赵 垒[2]

（1. 邢台钢铁有限责任公司技术中心，河北邢台 054027；
2. 河北省线材工程技术研究中心，河北邢台 054027）

摘 要： 利用 FEI Explorer 4 自动扫描电镜对硅脱氧弹簧钢 55SiCr 炼钢过程样中氧化物夹杂的成分、尺寸、数量、形貌进行检测，统计分析过程变化情况。结果表明：从 LF 到连铸中间包的过程中，13μm 以上氧化物夹杂由 SiO_2 类转变为 $CaO-SiO_2-Al_2O_3-MgO-MnO-(Na_2O)$。随着 LF 精炼及软吹处理，氧化物夹杂中 Al_2O_3 和 MgO 占比逐步升高，软吹处理后，13μm 以上的夹杂物的数量显著减少。铸坯中 20μm 以上的氧化物夹杂易发生析晶现象，部分夹杂物的结晶相为 $2MgO·SiO_2$。

关键词： 弹簧钢；硅脱氧；氧化物夹杂；夹杂物结晶

Study on the Change of Oxide Inclusions during Si-killed Spring Steel 55SiCr Steelmaking Process

Meng Yaoqing[1,2], Zhao Zhengzheng[2], Zhao Haoqian[1,2], Lv Haiyao[1,2], Zhao Lei[2]

(1. Technology Center of Xingtai Iron & Steel Co., Ltd., Xingtai 054027, China;
2. Hebei Engineering Research Center for Wire Rod, Xingtai 054027, China)

Abstract: The change of the composition, size, quantity and morphology of oxide inclusions in samples of Si-killed spring steel 55SiCr during steelmaking process were examined using an FEI Explorer 4 automated scanning electron microscope. The results show that the oxide inclusions above 13 micron change from SiO_2 type to $CaO-SiO_2-Al_2O_3-MgO-MnO-(Na_2O)$ type in the process from LF to tundish. The proportion of Al_2O_3 and MgO in oxide inclusions increases gradually along with LF refining and soft bubbling. After soft bubbling, the number of inclusions above 13 micron decreases significantly. Most of the oxide inclusions above 20 micron in bloom are prone to crystallization, and the crystalline phase of some inclusions is $2MgO·SiO_2$.

Key words: spring steel; Si-killed; oxide inclusions; inclusions crystallization

转炉留渣双渣工艺前期脱磷热力学及实践

孙学刚

（新疆八一钢铁股份有限公司第一炼钢厂，新疆乌鲁木齐 830022）

摘 要： 为实现转炉留渣+双渣工艺吹炼前期一次倒渣的高效脱磷，应用正规离子溶液模型对脱磷反应热力学规律进行了计算，分析了影响转炉的渣－金间磷分配比 L_P 的主要因素；同时，对热力学计算和现场试验结果进行了对比分析，转炉吹炼前期脱磷较佳的工艺控制条件是：炉渣碱度 R 控制在1.5左右，一次倒渣温度控制在1330～1360℃，渣中(FeO)控制在16%～17%。在冶炼过程中，铁水的成分和温度的稳定性对留渣+双渣工艺过程操作顺利控制影响较大。

关键词：留渣+双渣；脱磷；正规离子溶液模型

Thermodynamics and Practice of SGRS Process during BOF

Sun Xuegang

(No. 1 Steel-making Plant, Xinjiang Bayi Iron & Steel Co., Ltd., Urumchi 830022, China)

Abstract: In order to dephosphorized efficiently at the first deslagging with SGRS (slag generation reduced steelmaking) method in converter, the rules that the thermodynamic law of dephosphorization reaction is calculated with regular ionic solution model, and the main factors affecting the phosphorus partion ratio (L_P) of converter was analyzed; Meanwhile comparative analysis are made between the thermodynamic calculation and the site test results. The optimal process condition for dephosphorization in the early stage of converter is that the basicity of salg is controlled around 1.5, the first deslagging temperature is controlled at 1330-1360℃ and the mass fraction of (FeO) of slag is 16%-17%. During melting, hot metal composition and temperature stability have great influence upon the successful control of SGRS process.

Key words: SGRS; dephosphorization; regular ionic solution model

八钢汽车大梁钢（B510L）钢中夹杂物探讨

李立民

（宝武集团新疆八一钢铁股份有限公司，新疆乌鲁木齐　830022）

摘　要：通过汽车大梁钢生产工艺优化，系统性分析了LF精炼、连铸浇注过程钢水中氧氮变化、夹杂物的组成变化、夹杂物的数密度和面密度变化及不同夹杂物百分数变化，得出八钢汽车大梁钢洁净化水平为钢中氧含量≤16$\times10^{-6}$%，氮含量≤70$\times10^{-6}$%，夹杂物数量≤78个/mm^2，影响钢水洁净化主要问题是钙处理过程和连铸保护浇注。

关键词：夹杂物；钙处理；保护浇注

Discussion on Inclusions in Eight Steel Big Beam Steel(B510L)

Li Limin

(Baowu Xinjiang Bayi Iron & Steel Stock Co., Ltd., Urumchi 830022, China)

Abstract: By studying the automobile big beam steel production process, the systematic analysis of the LF refining, continuous casting molten steel casting process of oxygen and nitrogen changes, the composition of inclusions, the inclusions of the change in the number density and surface density and percent changes different inclusions, it is concluded that the eight steel car big joist steel clean level of oxygen content in steel 16×10^{-6}% or less, the nitrogen content in 70 by

10^{-6}%, or less inclusions number 78 or less/was, affect the main problem is steel clean calcium process and casting mould casting.

Key words: inclusion; calcium treatment; protection of pouring

炉底厚度维护对转炉炉衬寿命的影响

韩东亚

（新疆八一钢铁股份有限公司炼钢厂，新疆乌鲁木齐 830022）

摘 要：文章介绍了八钢第二炼钢厂 120t 转炉在多年的生产实践过程中，结合其自身的炉衬砌筑、冶炼钢种、操作技能等特点，逐步深入探索炉底厚度的控制，采用溅渣护炉、渣补、铺生铁等手段将炉底厚度控制在 800～1000mm，节约耐材成本和提高转炉作业率的同时，炉龄可安全达到 10000 炉的水平。

关键词：炉衬砌筑；炉底厚度；炉底维护

Effect of Bottom Thickness Maintenance on Converter Lining Life

Han Dongya

(Xinjiang Bayi Iron & Steel Co., Ltd., Urumchi 830022, China)

Abstract: This paper introduces that the 120t converter of Bagang in the years of production practice, combined with its own characteristics of lining masonry, steel smelting, operation skills and other characteristics, gradually in-depth exploration of the furnace, slag splashing, laying pig iron and other means to control thickness of the furnace bottom at 800~1000mm, saving the cost of refractory, improve the converter operation rate, the furnace age can reach more than 10000.

Key words: lining masonry; furnace bottom; bottom maintenance

钛强化低合金高强钢夹杂物的控制研究

吾 塔，李立民，吴 军，卜志胜，丁 寅，刘军威

（新疆八一钢铁股份有限公司，新疆乌鲁木齐 830022）

摘 要：通过对钛强化 Q355D 的成分优化设计。针对钛元素容易氧化及与钢中氮结合能力强的特点，炼钢过程严格控制氮含量，钢水经 LF 还原后加入钛铁，钢中的 TO≤0.0030%, N≤0.0055%。采用扫描电镜分析钛强化钢材非金属夹杂类型。分析显示，大颗粒夹杂基本消除，粒度小于 15μm 的夹杂重要组成为 Al_2O_3、CaO、MnS，另一类夹杂为 TiN 和 Ti(C,N)。

关键词：炼钢；夹杂物；氮

Study on Inclusion Control of Titanium Strengthened Low Alloy High Strength Steel

Wu Ta, Li Limin, Wu Jun, Bu Zhisheng, Ding Yin, Liu Junwei

(Xinjiang Bayi Iron & Steel Stock Co., Ltd., Urumchi 830022, China)

Abstract: The composition of titanium reinforced Q355D was optimized. In view of the fact that titanium element is easy to oxidize and has a strong ability to combine with nitrogen in steel, the nitrogen content is strictly controlled in the steelmaking process, and after the reduction of molten steel by LF, titanium and iron are added, TO≤0.0032% and N≤0.0055% in steel. The types of nonmetallic inclusions in titanium strengthened steel were analyzed by SEM. The results show that the large particle inclusions are almost eliminated, and the important inclusions with particle size less than 15μm are Al_2O_3, CaO, MnSand the other inclusions are TiN and Ti (C, N).

Key words: steelmaking; inclusions; nitrogen

2.2 连　　铸

薄板坯 9SiCr 合金工具钢高温热塑性研究

李具中[2]，朱万军[1]，王春锋[2]，齐江华[1]，孙宜强[1]，蔡　珍[1]，邱　晨[2]

（1. 宝钢股份中央研究院武汉分院，湖北武汉　430080；
2. 武汉钢铁有限公司条材厂，湖北武汉　430083）

摘　要： 为评估薄板坯连铸连轧工艺(CSP 工艺)生产合金工具钢 9SiCr 的可行性，利用 Gleeb-1500 热模拟机研究了合金工具钢 9SiCr 的高温力学性能，并采用扫描电镜观察了不同温度区间下 9SiCr 的断口形貌，分析了不同温度区域的断裂机理。研究结果表明，9SiCr 钢在 1200~600℃存在两个脆性温度区，第Ⅰ脆性区温度＞1170℃，表现为熔融断口；第Ⅲ脆性区为 820~600℃，表现为脆性解理断口；存在一个塑性良好区为 1170~820℃，表现为韧窝断口。利用高温热塑性结果，并通过工艺优化，在 CSP 产线实现了合金工具钢 9SiCr 的稳定生产，最薄规格 1.5mm。

关键词： 薄板坯；合金工具钢 9SiCr；高温热塑性；断裂机理

High-temperature Thermoplastic Study of 9SiCr of CSP

Li Juzhong[2], Zhu Wanjun[1], Wang Chunfeng[2], Qi Jianghua[1], Sun Yiqiang[1], Cai Zhen[1], Qiu Chen[2]

(1. Wuhan Branch of Baosteel Central Research Institute, Wuhan 430080, China;
2. Wuhan Steel Processing Co., Ltd., CSP, Wuhan 430083, China)

Abstract: To research the production availability of the 9SiCr-alloy-tool steel by CSP process, the mechanical properties under high-temperature of 9SiCr steel were studied by the thermal simulator, and the appearances of fractures at different temperatures were observed by the scanning electron microscope. The result of research showed that the 9SiCr steel had two brittle temperature ranges, the first one of which was the temperature range between 600 to 820℃ with cleavage brittle fracture, the second one between 820 to 1170℃ with dimple fracture, and the last one higher than 1170℃ with molten fracture. Through the process optimization, the 9SiCr steel can be produced steadily by the CSP process with the thinnest specification of 1.5mm.

Key words: CSP; 9SiCr steel; high-temperature thermoplastic; fracture mechanism

连铸生产模式比较及应用

董金刚

（宝山钢铁股份有限公司制造管理部，上海　201900）

摘　要：连铸有恒拉速生产模式和恒通钢量生产模式，恒拉速生产在浇铸过程中保持恒定拉速或很小的拉速波动，但浇铸周期受浇铸宽度的影响大；恒通钢量生产在所有炉次保持恒定或较小波动的浇铸周期。恒拉速生产模式和恒通钢量生产模式在铸流宽度设计、无合同板坯数量、组浇次策略、生产稳定性和生产节奏上都有差异。对于生产节奏快、制造周期波动小、品种相对较少的铸机或炼钢厂更适于采用恒通钢量的生产模式，以实现高效率、高稳定性生产；对于生产节奏慢、制造周期波动大、品种多的铸机或炼钢厂更适于采用恒拉速的生产模式，以实现相对稳定的连续性生产。A 钢厂 0 号铸机主要生产汽车钢种，更适于恒通钢量生产模式，为此，需要采取措施降低冶炼周期、稳定不同钢种的 RH 精炼周期，形成 $t_{冶炼}:t_{精炼}:t_{浇铸}=38:34:40$min 的高效率的工序时间匹配关系，满足产能提升的需要。

关键词：连铸；恒拉速；恒通钢量；工序

Comparison and Application of Continuous Casting Production Patterns

Dong Jingang

(Baoshan Iron & Steel Co. Manufacturing Management Department, Shanghai 201900, China)

Abstract: There are constant casting speed production mode and constant pass steel quantity production mode in continuous casting. The constant casting speed production maintain constant speed or very small speed fluctuations during casting, but the casting cycle is greatly affected by the casting width. The constant pass steel quantity production maintain constant casting cycle or very small casting cycle fluctuations. There are different in casting width design、group cast strategy、production stability and production rhythm, between constant casting speed production mode and constant pass steel quantity production mode. Those continuous casting machines or steel mills with fast production tempo, small production cycles fluctuations and fewer varieties are more suitable for the use of constant pass steel quantity production mode for efficient and stable production. Those continuous casting machines or steel mills with slow production pace, wide production cycles fluctuations and more varieties are more suitable for the use of constant casting speed production mode for continuous production. Because of mainly producing automobile steels in A steelmaking mill, No. 0 continuous casting machine is more suitable for constant pass steel quantity production mode, so measures need to be taken to reduce smelting cycles and to stabilize the RH refining cycle of different steels, forming efficient process time matching relationship:

smelting cycle 38 minutes, refining cycle 34 minutes, casting cycle 40 minutes to meet the need for capacity enhancement.
Key words: continuous casting; constant casting speed; constant steel quantity; procedure

260t 钢包水口尺寸对卷渣高度的水模拟研究

彭春霖，张晓光，郭庆涛，贾吉祥，柴明亮，杨 骥

(鞍钢集团钢铁研究院炼钢技术研究所，辽宁鞍山 114009)

摘 要：依据相似原理，通过水模拟实验，对 260t 钢包浇注过程中钢水旋涡产生及卷渣过程进行研究，考察不同水口直径对产生旋涡临界高度的影响。结果表明，产生旋涡的临界高度随水口直径增大而有所增加。当水口尺寸从 75mm 增至 80mm 时，产生旋涡的临界高度增加幅度较小。

关键词：钢包；浇注过程；旋涡；临界高度；水力学模拟

Water Simulation Study on Critical Height of Vortex from the Different Nozzle Diameter of the 260t Steel Ladle

Peng Chunlin, Zhang Xiaoguang, Guo Qingtao,
Jia Jixiang, Chai Mingliang, Yang Ji

(Steelmaking Technology Research Institute, Ansteel Iron and
Steel Research Institute, Anshan 114009, China)

Abstract: Based on analogy principle, vortex and slag entrapment in a 260t steel ladle were studied during by water model experiments. The effects of different nozzle diameter on critical height of vortex were investigated. The results show that the critical height of vortex will rise with increasing of nozzle diameter. When the nozzle diameter increased from 75mm to 80mm, the increase amplitude of critical height of vortex was small.

Key words: ladle; casting process; vortex; critical height; hydraulics simulation

高铝双相钢连铸 T 坯质量控制及改进研究

胡署名[1]，常文杰[1]，允占英[2]，王迎春[3]，杨启宇[2]，吕宪雨[1]

(1. 宝山钢铁股份有限公司炼钢厂，上海 201900；2. 宝山钢铁股份有限公司制造部，
上海 201900；3. 宝山钢铁股份有限公司研究院，上海 201900)

摘 要：针对宝钢高铝双相钢连铸 T 坯质量缺陷导致的产品冲压分层问题，对高铝双相钢连铸 T 坯进行了检测分析，通过铸坯浇铸末期的凝固过程研究，探讨了 T 坯疏松和缩孔缺陷形成的机理，找出了导致产品分层的主要原因。并且结合现场生产工况，对高铝双相钢浇铸末期的工艺参数进行了分析，优化和改进了相关工艺参数，从而有效减轻了 T 坯的疏松和缩孔缺陷，成功地解决了宝钢高铝双相钢连铸 T 坯质量缺陷导致的产品分层问题。

关键词：双相钢；T坯；缩孔；中心疏松；质量控制

Study on Quality Control and Improvement of Continuous Casting T Slab of High Aluminum Dual-phase Steel

Hu Shuming[1], Chang Wenjie[1], Kang Zhanying[2],
Wang Yingchun[3], Yang Qiyu[2], Lv Xianyu[1]

(1. Steelmaking Plant, Baoshan Iron & Steel Co., Ltd., Shanghai 201900, China;
2. Manufacturing Department, Baoshan Iron & Steel Co., Ltd., Shanghai 201900, China;
3. Research Institute, Baoshan Iron & Steel Co., Ltd., Shanghai 201900, China)

Abstract: Based on the quality defect of continuous casting T slab and the product lamination problem in baosteel, the continuous casting T slabs of high aluminum dual-phase steel were detected and analyzed in this study. Through the study of solidification process at the end of casting, the mechanism of shrinkage cavity and porosity defects of T slab was discussed, and the main reasons of product lamination were found out. In addition, combining with the processing conditions, the process parameters at the end of the casting of high aluminum dual-phase steel were analyzed, and the related process parameters were optimized and improved. As a result, the shrinkage cavity and porosity defects of T slabs were reduced effectively. The quality defect and the product lamination problem of continuous casting T slab of high aluminum dual-phase steel at baosteel were solved successfully.

Key words: dual-phase steel; T slab; shrinkage cavity; center porosity; quality control

外加电场对浸入式水口内壁结瘤行为的影响研究

田 晨[1]，翟晓毅[2]，肖国华[2]，徐 斌[2]，孙江波[2]，
李海斌[2]，袁 磊[1]，于景坤[1]

（1. 东北大学冶金学院，辽宁沈阳 110819；2. 河钢集团邯钢公司，河北邯郸 056015）

摘 要：基于夹杂物瞬间摩擦荷电理论，本文通过对浸入式水口内壁施加直流电场来抑制水口结瘤行为。结果表明：利用外加电场的方法可以抑制夹杂物在水口内壁的粘附行为，进而防止夹杂物与水口间的进一步烧结粘附和反应行为，新生成的结瘤物的致密度更高，更稳定，从而达到保护水口的目的。同时由于施加外加电场后夹杂物间也会形成一定排斥力，进而使得钢中夹杂物无法进一步团聚长大。使得钢中夹杂物尺寸同样得到细化。

关键词：铝镇静钢；浸入式水口；水口结瘤；直流电场

Effect of Clogging Behavior on the Submerged Entry Nozzle by External Electric Field

Tian Chen[1], Zhai Xiaoyi[2], Xiao Guohua[2], Xu Bin[2], Sun Jiangbo[2],
Li Haibin[2], Yuan Lei[1], Yu Jingkun[1]

(1. School of Metallurgy, Northeastern University, Shenyang 110819, China;

2. HBIS Group Hansteel Company, Handan 056015, China)

Abstract: Based on the theory of instantaneous friction charge of inclusions, the clogging behavior of the submerged entry nozzle is prevented by applying DC electric field in this paper. The results show that the adhesion behavior of inclusions on the nozzle can be inhibited by applying electric field. The sintering adhesion and reaction behavior between inclusions and nozzle can also be prevented. The newly clogging is more dense and stable, which achieving the purpose of protecting nozzle. At the same time, due to the external electric field, the inclusions will form a certain repulsive force, which makes the inclusions cannot further agglomerate and grow in the molten steel. Finally, the inclusions size in the steel are also refined.

Key words: aluminum-killed steel; the submerged entry nozzle; clogging; DC electric field

异型坯单点浇注条件下结晶器控制模型研究与应用

公斌，王忠刚，卢波，赵立峰，张丽，刘忠建

（莱芜钢铁集团银山型钢有限公司炼钢厂，山东济南　271104）

摘　要： 在异型坯非平衡单点浇注条件下，异型坯结晶器内的流及温度场出现非对称变化，原有的结晶器—冷水的冷却条件已经不能完全适应铸坯坯壳的均匀生长条件。通过研究含铝钢异型坯塞棒中间包非平衡单点浇注技术条件下温度场流场模拟研究，从而达到优化结晶器水缝结构、减少非平衡布流单点浇注时结晶器钢水流场及温度场分布不对称影响，促进铸坯坯壳生长的均匀性，减少或避免铸坯表面裂纹质量问题的发生。

关键词： 非平衡单点浇注；非对称温度场及流场；保护渣；水缝结构；坯壳

Research and Application of Mold Control Model under Single Point Casting of Blank

Gong Bin, Wang Zhonggang, Lu Bo, Zhao Lifeng, Zhang Li, Liu Zhongjian

(Laiwu Iron and Steel Group Yinshan Profile Steel Co., Ltd., Steelmaking Plant, Jinan 271104, China)

Abstract: Under the condition of non-equilibrium single-point casting, the flow and temperature field in the mould of special-shaped billet show asymmetric changes. The original cooling condition of the mould-cold water can not fully adapt to the uniform growth condition of the billet shell. By studying the simulation of temperature field and flow field under the condition of non-equilibrium single-point casting technology for aluminium-containing steel blank plug tundish, the structure of water gap in the mould can be optimized, the asymmetry of distribution of molten steel flow field and temperature field in the mould can be reduced, the uniformity of slab shell growth can be promoted, and the quality problem of surface cracks on the slab can be reduced or avoided.

Key words: non-equilibrium; single-pointcasting asymmetric temperature field and flow field; mold powder; Watercrevice structure; shell

异型坯 C345 工程用钢的研究与开发

张　丽，王忠钢，赵立锋，公　斌，刘忠建

（莱芜钢铁集团银山型钢有限公司炼钢厂，山东济南　271104）

摘　要：通过分析异型坯含铝钢生产过程中存在晶器钢水流场、温度场分布不对称、不均匀等问题，开发新的生产工艺，比如采用结晶器非对称冷却技术，增加浇注点侧翼缘部位冷却强度，使结晶器内铸坯坯壳能够均匀生长；同时保护渣采用非对称布料技术，非浇注点部位与浇注点控制不同渣层厚度，可解决非浇注点侧渣面火焰大和浇注点保护渣结壳问题；并且通过采用非平衡布流单点浇注技术等技术，实现了异型坯铝脱氧钢的批量、稳定生产，提高了产品质量，满足了客户需求。

关键词：非对称冷却；单点浇注；非对称布料

Research and Development of C345 Engineering Steel for H-beam

Zhang Li, Wang Zhonggang, Zhao Lifeng, Gong Bin, Liu Zhongjian

(Laiwu Iron and Steel Group Yinshan Profile Steel Co., Ltd., Steelmaking Plant, Jinan 271104, China)

Abstract: By analyzing the problems of asymmetric flow field, asymmetric distribution of temperature field and non-uniformity in the production process of aluminium-containing steel with special-shaped billet, a new production process was developed, such as using asymmetric cooling technology of mould to increase the cooling intensity of the flange part of the casting point, so that the billet shell in the mould could grow uniformly; at the same time, the asymmetric distribution technology of powder and the non-casting point part were adopted. Controlling different slag layer thickness with pouring point can solve the problems of large flames on slag surface at non-pouring point side and mold slag crust at pouring point, and realize batch and stable production of profiled aluminium deoxidized steel by adopting non-balanced flow distribution single point pouring technology, improve product quality and meet customer needs.

Key words: asymmetric cooling; single point casting; asymmetric cloth

莱钢宽厚板线轧材折叠黑线缺陷原因分析及对策

赵立峰，王忠刚，卢　波，张　丽，公　斌，刘忠建

（莱芜钢铁集团银山型钢有限公司，山东济南　271104）

摘　要：莱钢宽厚板线轧材折叠黑线缺陷是目前制约热送热装率的主要因素，本文通过金相组织分析方法分析了宽厚板线轧材折叠黑线缺陷形成机理，进一步明确了该缺陷是在轧制过程形成的非裂纹缺陷。通过数值模拟验证了倒角结晶器的可行性及合理的倒角角度，实践证明倒角结晶器+角部钝化方案可解决宽厚板轧材折叠黑线缺陷，对提

高铸坯热送率，减少精整量，提高金属收得率、增加铸坯显热利用率、提产节能增效作用显著。

关键词：宽厚板；折叠黑线；倒角结晶器；角部钝化

Analysis and Countermeasure of Folding Black Line Defect in Wide and Heavy Plate Rolling in Laigang

Zhao Lifeng, Wang Zhonggang, Lu Bo, Zhang Li, Gong Bin, Liu Zhongjian

(Laiwu Iron and Steel Group Yinshan Profile Steel Co., Ltd., Steelmaking Plant, Jinan 271104, China)

Abstract: The folding black line defect of wide and heavy plate and wire rolling in Laiwu Iron and Steel Co. is the main factor restricting the hot delivery and hot charging rate at present. In this paper, the forming mechanism of folding black line defect of wide and heavy plate and wire rolling is analyzed by means of metallographic structure analysis method, which further clarifies that the defect is a non-crack defect formed in rolling process. The feasibility and reasonable chamfering angle of chamfering mould are verified by numerical simulation. Practice proves that chamfering mould + corner passivation scheme can solve the folding black line defect of wide and heavy plate rolling. It has remarkable effect on improving hot delivery rate of slab, reducing finishing quantity, increasing metal yield, increasing sensible heat utilization rate of slab and increasing energy saving and efficiency of production.

Key words: wide and thick plate; folded black line; chamfered mould; corner passivation

鞍钢电磁冶金技术应用及展望

郭庆涛[1,2]，贾吉祥[1,2]，唐雪峰[2]，彭春霖[1,2]，康 磊[1,2]，廖相巍[1,2]

（1. 海洋装备用金属材料及其应用国家重点实验室，辽宁鞍山 114009；
2. 鞍钢集团钢铁研究院，辽宁鞍山 114009）

摘 要：本文介绍了鞍钢电、磁冶金技术研究进展，重点介绍了外加电场钢液净化技术、传统电磁搅拌、螺旋电磁搅拌技术冶金基本原理及应用实例，指出了电、磁技术在冶金领域的技术特点和优势，并对其今后的研究方向进行了展望。

关键词：电磁净化；电磁搅拌；螺旋搅拌

Application and Prospect of the Electromagnetic Metallurgy in Ansteel

Guo Qingtao[1,2], Jia Jixiang[1,2], Tang Xuefeng[2], Peng Chunlin[1,2], Kang Lei[1,2], Liao Xiangwei[1,2]

(1. State Key Laboratory of Metal Materials for Marine Equipment and Application, Anshan 114009, China;
2. Iron & Steel Research Institutes of Ansteel Group Corporation, Anshan 114009, China)

Abstract: This paper introduces the research progress of electric and magnetic metallurgy technology in Ansteel, with emphasis on the application of electric field on the molten steel purification process, traditional electromagnetic stirring and helical electromagnetic stirring technology, points out the technical characteristics and advantages of electric and magnetic technology in the field of metallurgy, and looks forward to its research direction in the future.

Key words: electromagnetic purification; EMS; helical EMS

静态凝固末端强冷在八钢连铸机上的应用

秦 军，卜志胜，陈晓山

（宝武集团八钢公司制造管理部，新疆乌鲁木齐 830022）

摘 要：本文描述了板坯连铸机凝固末端强冷模式的原理，通过在八钢 4 号连铸机 250 断面上进行的静态凝固末端水量强冷试验表明，该模式需建立在稳态浇注的前提上，此时板坯内部的中心偏析级别有所改善，内部中心线裂纹封锁率降低，而板坯的表面纵裂及角裂率略有升高，总体效果理想。

关键词：板坯连铸机；静态凝固末段强冷；中心偏析；评级

Effect of Static Solidification Ends Strong Cooling on Bayi Steel Slab Caster

Qin Jun, Bu Zhisheng, Chen Xiaoshan

(Manufacturing Management Department, Bayi Iron & Steel Co., Urumchi 830022, China)

Abstract: This paper describes the principles and efficacy of static solidification ends strong cooling technology for the slab caster. Through the static solidificetion ends strong cooling experiment on section 250 of No.4 continuous caster of Bayi Steel, the practice shows that this mode should be buint on the premise of steady pouring and at this time the central segregation level inside the slab is improved, the internal centerline crack blocking rate is reduced, howerver the surface longitudinal crack and corner carck blocking rate of the slab slightly increased. The overall effect is fine.

Key words: slab caster; static solidification ends strong cooling; center segregation; rating

45#钢圆钢表面纵裂纹成因分析及改进措施

宋德山

（华菱湘钢炼钢厂，湖南湘潭 411101）

摘 要：本文通过对 45#钢圆钢表面纵裂纹缺陷进行分析，表明 45#钢圆钢表面纵裂纹是因钢中 N、H 气体含量高所致。通过实施降低 45#钢中气体含量措施，45#钢圆钢表面纵裂纹率由 1.28%降低到 0.30%以下。

关键词：45#钢；气孔；表面裂纹

Cause Analysis and Improvement Measures for Longitudinal Cracks on Surface of 45# Roung Steel Bars

Song Deshan

(Steelmaking Division of Valin Xiangtan Steel, Xiangtan 411101, China)

Abstract: In this paper, the longitudinal crack defects were analyzed on the surface of 45# round steels.The result shows that the longitudinal crack on the surface is caused by the higher contents of N and H gas in steels.The longitudinal cracking rates of 45# round steels bar surface have decreased from 1.28% to 0.30% by taking actions to minimize the gas contents in steel.

Key words: 45# carbon round steel bar; porosity; surface crack

矩形连铸坯夹杂物控制技术的研究与应用

王 键,王玉春,郭 达,刘文凭,谭学样

(山东钢铁股份有限公司莱芜分公司炼钢厂,山东济南 271126)

摘 要: 通过研究液态低熔点夹杂物控制、矩形坯连铸机无氧化浇注、稳态无氧化浇铸系统洁净度集成等控制技术、弥散型气泡幕墙技术,提高钢水洁净度,降低钢种氧含量,改善铸坯表面质量,有效解决夹杂物含量高和铸坯表面缺陷造成的轧材缺陷问题。

关键词: 矩形铸坯;钢水洁净度;夹杂物控制;表面质量

Research and Application of Control Technology for Rectangular Continuous Casting Inclusions

Wang Jian, Wang Yuchun, Guo Da, Liu Wenping, Tan Xueyang

(Shandong Iron and Steel Co., Ltd., Laiwu Branch Steel Mill, Jinan 271126, China)

Abstract: Through research on the control of liquid low melting point inclusions, the non-oxidation casting of rectangular billet continuous casting machine, the cleanliness integration of steady-state non-oxidation casting system, and the control technology, diffuse bubble curtain wall technology, to improve the purity of molten steel, reduce the oxygen content of steel, improve the surface quality of slab, and solve the problem of rolled defects caused by high inclusion content and surface defects of slab effectively.

Key words: rectangular slab; purity of molten steel; inclusion control; surface quality

IF 钢 Al-Ca-O 系细微夹杂物控制研究与实践

李俊伟,赵 元,袁少江,杨成威,代碧波

(武汉钢铁有限公司制造管理部,湖北武汉 430000)

摘　要：本文通过分析细微夹杂缺陷成分、缺陷坯位置、炼钢工艺过程，弄清了细微 Al-Ca-O 系夹杂物的重要来源为钢包氧化性渣，采取了 RH 精炼结束顶渣改质工艺、中间包挡墙工艺、线圈式下渣检测设备三项措施后，基本杜绝了汽车外观零件和内板可视零件细微夹杂缺陷。

关键词：IF 钢；细微夹杂物；顶渣改质；中间包挡墙

Control and Practice on Al-Ca-O Tiny Inclusion in Interstitial Free Steel

Li Junwei, Zhao Yuan, Yuan Shaojiang, Yang Chengwei, Dai Bibo

(Wuhan Iron and Steel Co., Ltd., Wuhan 430000, China)

Abstract: SEM analysis shows that the tiny inclusions in IF steel are Al-Ca inclusions. An important source of Mg-Al-Ca tiny inclusions through the analysis of the position of defect blank, technological process and production data of oxidation is ladle oxidizing slag, and the adoption of RH refining to top slag modification, tundish retaining wall technology and coil type slag detection equipment can effectively reduce the occurrence of Al-Ca tiny inclusions.

Key words: interstitial free steel; tiny inclusion; top slag modification; tundish retaining wall

铌微合金化小方坯生产实践

郭伟达，王　键，郭　达，任科社，谭学样

（山东钢铁股份有限公司莱芜分公司炼钢厂，山东济南　271126）

摘　要：针对莱钢铌微合金化小方坯脱方、角裂质量状况，对铌合金化的强化作用、铸坯脱方的成因进行分析，查找连铸机工艺、设备的存在问题，通过采用高精度结晶器，优化二次冷却制度，降低钢水过热度，保障关键设备功能精度，铸坯冷却均匀性得到显著改善，有效解决了铌微合金化钢种小方坯脱方、角裂。

关键词：小方坯；铌微合金化；脱方；成因；措施

The Production Practice of Small Square Blanks in Micro-alloying

Guo Weida, Wang Jian, Guo Da, Ren Keshe, Tan Xueyang

(Shandong Iron and Steel Co., Ltd., Laiwu Branch Steel Mill, Jinan 271126, China)

Abstract: In view of the quality status of squaring and corner cracking of Nb-microalloyed billet in Laiwu Iron and Steel Co., the strengthening effect of Nb alloying and the causes of slab squaring are analyzed, and the problems existing in the process and equipment of continuous caster are found out. By adopting high precision crystallizer, the secondary cooling system is optimized, the superheat of molten steel is reduced, the function accuracy of key equipment is ensured, and the cooling uniformity of billet is significantly improved. The squaring and corner cracking of Nb-microalloyed billet have been solved.

Key words: billet; Nb-microalloying; rhombic deformation; causes; measure

浸入式水口穿孔原因分析及对策

汪 雷，龚志翔，李 新，王 洛，王俊北，樊明宇

（马鞍山钢铁股份有限公司，安徽马鞍山 243000）

摘 要：为了探究浸入式水口穿孔的原因，对用后的连铸浸入式水口开展了在线测量及残砖的理化性能检测，分析了浸入式水口穿孔的机理。结果表明，水口抗氧化性能差、复合部位材料热膨胀不匹配等是造成穿孔的主要原因。采取提高原料质量和稳定性、改进结合部位材料复合水平、加大探伤与质检等措施降低穿孔事故的发生率。

关键词：浸入式水口；穿孔；分析；对策

Analysis on Origin of Perforation of Submerged Nozzle and Counter Measure

Wang Lei, Gong Zhixiang, Li Xin, Wang Luo, Wang Junbei, Fan Mingyu

(Maanshan Iron and Steel Co., Ltd., Maanshan 243000, China)

Abstract: In order to explore the reasons for perforation of submerged nozzle, the physical and chemical properties of submerged nozzle were tested and the mechanism of perforation was analyzed. The results show that the main reasons for perforation are poor oxidation resistance of the nozzle and mismatch of thermal expansion of composite materials. Measures such as improving the quality and stability of raw materials, improving the composite level of materials at bonding sites, and increasing flaw detection and quality inspection are taken to reduce the incidence of perforation accidents.

Key words: submerged nozzle; perforation; analysis; counter measure

连铸塞棒中间包冶金集成技术研究与实践

王 键，谭学样，郭 达，王玉春

（山东钢铁股份有限公司莱芜分公司炼钢厂，山东济南 271126）

摘 要：介绍了通过研究中间包动态精准吹氩系统、大包喇叭式水口无二次氧化开浇技术、中间包环式氩封保护技术、带氩封环浸入式水口保护技术、中间包双挡墙挡坝技术、"双高"缓冷型保护渣连铸塞棒中间包冶金集成技术，改善连铸塞棒中间包冶金效果，降低夹杂物含量，提升铸坯质量，有效解决了轧制孔洞、起皮、裂边等质量问题。

关键词：塞棒中间包；保护浇注；夹杂物；优化

Continuous Casting Stopper Integrated Package Metallurgy Research and Practice

Wang Jian, Tan Xueyang, Guo Da, Wang Yuchun

(Shandong Iron and Steel Co., Ltd., Laiwu Branch Steel Mill, Jinan 271126, China)

Abstract: This paper introduces the research on the dynamic precision argon blowing system of the tundish, the secondary oxidation open pouring technology of the big bag La nozzle, the argon sealing protection technology of the tundish ring, the immersion nozzle protection technology with argon sealing ring, and the long life refractory material of the tundish. Technology, tundish double retaining wall dam technology, "double high" slow cooling type slag continuous casting plug tundish metallurgical integration technology, improve the metallurgical effect of continuous casting plug tundish, reduce inclusion content, improve slab quality, effectively solve the quality problems of rolling holes, peeling, cracking and so on.

Key words: stopper tundish; protective casting; inclusions; optimization

高裂纹敏感性钢铸坯表面质量的研究与控制

王键，王玉春，郭达，刘文凭，谭学样

（山东钢铁股份有限公司莱芜分公司炼钢厂，山东济南 271126）

摘 要：通过研究不同元素微合金化工艺、LF精炼过程微合金元素精准控制技术、分区间动态控制冷却技术、关键设备功能精度保障技术、中间包保护浇注集成技术，提高微合金元素回收率，改善铸坯凝固组织、传热、冷却效果，提升铸坯表面质量，降低铸坯表面缺陷率，有效解决铸坯横裂纹和角部裂纹造成的判废和轧制裂边等质量问题。

关键词：高裂纹敏感性钢坯；微合金化；横裂纹和角部裂纹；裂边

Research and Control on the Surface Quality of High Crack Sensitivity Steel Cast Blanks

Wang Jian, Wang Yuchun, Guo Da, Liu Wenping, Tan Xueyang

(Shandong Iron and Steel Co., Ltd., Laiwu Branch Steel Mill, Jinan 271126, China)

Abstract: Through research on different microalloying elements process, precision control technology of microalloying elements in LF refining process, dynamic control cooling technology between zones, functional safety guarantee technology of key equipment, and tundish protection and pouring integration technology, increase the recovery rate of microalloying elements, improve the solidification structure, heat transfer and cooling effect of the slab, improve the surface quality of the slab, reduce the surface defect rate of the slab, and solve quality problem of defects and rolling cracks caused by the transverse cracks and corner cracks of the slab effectively.

Key words: high crack sensitive billet; microalloying; transverse crack and corner crack; cracked edge

六流方坯连铸中间包内型结构优化

夏振东，胡 增，胡 睿，岳 强

（安徽工业大学冶金工程学院，安徽马鞍山 243032）

摘 要：基于某钢厂六流小方坯连铸中间包，通过数学物理模拟的方式研究了不同控流装置下，六流小方坯连铸中间包内流体的停留时间分布曲线、流场和温度场，对中间包挡墙结构和参数进行了优化。结果表明在现有工艺和中间包内腔结构条件下：挡墙开孔设计为圆形，孔径为239 mm，开孔轴线方向和挡墙垂直方向夹角为15°。挡墙结构优化后有效增加了中间包钢水平均停留时间，死区体积相对减少，各流的流动一致性得到提升，同时有利于温度场的均匀化。

关键词：中间包；水模；RTD 曲线；数值模拟；优化

Optimization of Six-Strand Tundish Configuration for Billet Continuous Casting

Xia Zhendong, Hu Zeng, Hu Rui, Yue Qiang

(School of Metallurgical Engineering, Anhui University of Technology, Maanshan 243032, China)

Abstract: The residence time distribution curve, flow and temperature field of the fluid in the six-strand billet continuous casting tundish under different flow control devices were studied by means of the mathematical and physics simulation based on the six-strand billet continuous casting tundish of a steel mill. The tundish structure and parameters of the tundish are optimized. The results show that the hole of the retaining wall is circular, the diameter is 239 mm, and the angle between the axis of the opening and the vertical direction of the retaining wall is 15° under the existing process and tundish inner structure. The average residence time of the tundish steel is effectively increased, the dead volume is relatively reduced, the flow consistency of each flow is improved, and the temperature field is uniformed after the optimization of the retaining wall structure.

Key words: tundish; water modeling; RTD curve; numerical simulation; optimization

板坯连铸过程结晶器内钢液面异常波动的机理

苏志坚[1,2]，江中块[1,2,3]，陈 进[1,2]，徐承乾[1,2]，范 围[1,2]

(1. 东北大学材料电磁过程研究教育部重点实验室，辽宁沈阳 110004；2. 东北大学冶金学院，辽宁沈阳 110004；3. 上海梅山钢铁股份有限公司，江苏南京 210039)

摘 要：结晶器液面异常波动现象是当前连铸过程的关键共性问题，严重影响生产顺行和钢坯质量的提升。利用液面检测和 FFT 分析法对某厂的海量液面波动数据（不同钢种，拉速参数，铸坯宽度下）进行频谱分析。结果表明：在包晶钢板坯连铸过程中，结晶器内液面异常波动主频率随着碳含量的增加先增加后减小；在生产不同钢种时，增加拉坯速度时液面异常波动主频率相应增大,增加铸坯宽度时液面异常波动主频率几乎不变。通过以上规律推论,液面异常波动主要是由铸坯不稳定鼓肚产生,其频率与结晶器内钢液固有频率相近时会产生共振导致液面大振幅波动。

关键词:板坯连铸;结晶器液面异常波动;共振频率;包晶钢

Study on Mechanism of Large Amplitude Mold Level Fluctuation in Slab Continuous Casting Process

Su Zhijian[1,2], Jiang Zhongkuai[1,2,3], Chen Jin[1,2], Xu Chengqian[1,2], Fan Wei[1,2]

(1. Key Laboratory of Electromagnetic Processing of Materials (Ministry of Education), Northeastern University, Shenyang 110004, China; 2. Northeastern University School of Metallurgy, Shenyang 110004, China; 3. Shanghai Meishan Iron and steel Co., Ltd., Nanjing 210039, China)

Abstract: Large amplitude mold level fluctuation is frequently observed in continuous casting specially for peritectic steels, which is proved to deteriorate product quality severely. In this paper, both liquid level detection and FFT analysis method are used to analyze the frequency spectrum of mass mold level fluctuation data in mold (different steel grades, casting speed and slab width), the result shows: In the slab continuous casting process of peritectic steels, with the increase of carbon content, the main frequency and equivalent amplitude of large amplitude mold level fluctuations is first increasing and then decreasing. In the production of various grades of steel, increase casting speed, the main frequency of large amplitude mold level fluctuation increase, increase casting width the main frequency of large amplitude mold level fluctuation change little. Based on these results, a new mechanism can be obtained: large amplitude mold level fluctuation mainly caused by the bulgings that passing through the rollers, and when this frequency approaches the natural frequency of the molten steel in mold, a resonance occurs and resulting in large amplitude (large amplitude) fluctuation of mold level.

Key words: slab continuous casting; large amplitude mold level fluctuation; resonance frequency; peritectic steel

IF 钢在双辊薄带连铸过程中微观结构演变规律的研究

张华龙,张同生,王万林,吕培生

(中南大学冶金与环境学院,湖南长沙 410083)

摘　要:薄带连铸过程中,结晶辊内钢水的凝固行为属于亚快速凝固范畴。该凝固过程会提升杂质和合金元素在钢中的固溶度,同时细化晶粒。这会导致薄带钢在后续热处理过程中微观组织转变和沉淀相的析出规律不同于传统铸坯热处理过程。目前,有关钢水亚快速凝固过程中微观结构演变规律方面的研究还较少,亟需建立一种合适的方法对其展开分析研究,从而提高产品性能和生产的稳定性。本研究选取 IF 钢作为实验材料,采用熔滴凝固技术模拟 IF 钢的亚快速凝固过程,研究了钢带的卷曲温度和冷却速率对 IF 钢微观结构的影响。研究表明:提升卷曲温度和降低冷却速率均有利于第二相的析出;卷曲温度对晶粒尺寸的影响不明显,提升冷却速率可以起到细化晶粒的作用。

关键词:薄带连铸;亚快速凝固;微观结构;晶粒;IF 钢

Study on the Evolution of Microstructure of IF Steel during Twin-roll Strip Continuous Casting

Zhang Hualong, Zhang Tongsheng, Wang Wanlin, Lv Peisheng

(School of Metallurgy and Environment, Central South University, Changsha 410083, China)

Abstract: During the strip casting process, the solidification behavior of the molten steel in the crystallization rolls belong to the sub-rapid solidification category. This solidification process enhances the solid solubility of impurities and alloying elements in the steel, while also refining the grains. This will lead to the microstructural transformation and precipitation phase of the thin strip steel during the subsequent heat treatment process is different from the traditional slab heat treatment process. In addition, there are few studies on the microstructure evolution of the sub-automatic solidification process of molten steel. Therefore, it is necessary to establish a suitable method to carry out analysis and research to improve product performance and production stability. In this study, IF steel was selected as the experimental material, and the sub-rapid solidification process of IF steel was simulated by the droplet solidification technique. The effects of the crimping temperature and cooling rate of the steel strip on the microstructure of IF steel were studied. Studies have shown that: lifting and lowering the cooling rate coiling temperature conducive to the precipitation and coarsening of the second phase; Influence of coiling temperature on the grain size is not obvious, it may act to enhance the cooling rate of grain refinement.

Key words: strip casting; sub-fast solidification; micro structure; grain; interstitial-free steel

钎具钢轧后角部裂纹形成机理研究

张建康，王万林，周乐君，陈俊宇，薛利文

（中南大学冶金与环境学院，湖南长沙 410083）

摘 要： 针对某钢厂生产的钎具钢，在进行一次初轧后发现角部出现了几处裂纹，切割之后采用扫描电镜、金相显微镜等仪器对裂纹的特征及微观组织进行观察分析。结果表明，裂纹附近组织粗大，附近存在脱碳现象，裂纹为铸态裂纹，并且存在较多钙铝酸盐、钙硅酸盐和氧化物。因此，在实际生产中，应对炼钢工艺高温过程进行改进控制，减少类似的缺陷发生。

关键词： 钎具钢；角部裂纹；铸坯缺陷；连铸坯

Study on the Formation Mechanism of Drill Steel Corner Crack after Rolling

Zhang Jiankang, Wang Wanlin, Zhou Lejun, Chen Junyu, Xue Liwen

(School of Metallurgy and Environment, Central South University, Changsha 410083, China)

Abstract: Several cracks were found in the corner of drill steel produced by a steelworks after a blooming rolling. After cutting, the characteristics and microstructure of the cracks were observed and analyzed by means of scanning electron microscopy and metallographic microscope. The results show that the structure near the crack is thick and decarbonization exists. The crack is casting crack, and there are more calcium aluminates, calcium silicates and oxides. Therefore, in actual production, the high temperature process of steelmaking process should be improved and controlled to reduce the occurrence of similar defects.

Key words: drill steel; corner crack; defects in the original slab; continuous casting bloom

QP980钢亚快速凝固研究

徐 慧，王万林，曾 杰，路 程，朱晨阳，吕培生

（中南大学冶金与环境学院，湖南长沙 410083）

摘 要：QP980是第三代超高强钢的代表性钢种之一，其生产方式主要有传统连铸和CSP连铸生产工艺。由于QP980钢中的Si、Mn含量很高，包晶反应剧烈，采用CSP及传统连铸方式生产的铸坯存在较严重的表面质量缺陷。本研究采用新型融滴凝固技术生产QP980钢，通过金相显微镜、场发射电子扫描显微镜观察，对比使用薄带连铸方式和CSP连铸方式生产出的QP980钢的微观组织差异。研究了冷却速度对QP980钢的晶粒度、二次枝晶间距的影响，并通过金相、能谱和硬度实验进一步分析了在快速冷却条件下QP980钢的相态变化。研究结果表明：与CSP工艺相比，采用薄带连铸技术生产的QP980钢的凝固组织晶粒更加均匀细小。CSP连铸工艺的冷速为2.4~5.6℃/s，薄带连铸技术的冷速为247~439℃/s。随冷速的大幅度增加，其微观组织得到了明显的细化，二次枝晶间距由86μm减小到13μm，硬度也有所增加。

关键词：薄带连铸；融滴凝固技术；QP980钢；微观组织；二次枝晶间距

Effect of Cooling Rate on Microstructure and Properties of QP980 Steel by using Rroplet Solidification Technique

Xu Hui, Wang Wanlin, Zeng Jie, Lu Cheng, Zhu Chenyang, Lv Peisheng

(School of Metallurgy and Environment, Central South University, Changsha 410083, China)

Abstract: QP980 is one of the representative steels of the third generation ultra-high strength steel. Its production methods mainly include traditional continuous casting and CSP continuous casting process. Because of the high content of Si and Mn in QP980 steel and the intense peritectic reaction, the slab produced by CSP and traditional continuous casting has serious surface quality defects. In this study, QP980 steel was produced by melt-drop solidification technology. The microstructures of QP980 steel produced by thin strip continuous casting and CSP continuous casting were compared by metallographic microscope and field emission electron scanning microscope. The effect of cooling rate on grain size and secondary dendrite spacing of QP980 steel was studied. The phase change of QP980 steel under rapid cooling was further analyzed by metallographic, energy spectrum and hardness experiments. The results show that, compared with CSP process, the solidified grains of QP980 steel produced by droplet solidification technique are more uniform and fine. The cooling rate of CSP continuous casting process is 2.4~5.6℃/s, and that of thin strip continuous casting process is 247~439℃/s. With the increase of cooling rate, the microstructure was refined obviously, the secondary dendrite spacing decreased from 86μm to 13μm, and the hardness also increased.

Key words: strip casting; droplet solidification technique; QP980 steel; microstructure; secondary dendrite spacing

板坯连铸过程流动、传热及夹杂物分布的数值模拟研究

张立峰，陈 威

（北京科技大学冶金与生态工程学院，北京 100083）

摘 要：连铸过程结晶器内多相流分布、传热和凝固直接影响了流场形态、液位波动及氩气泡和夹杂物被凝固坯壳捕获等现象。本研究通过建立三维数学模型研究了连铸结晶器内的多相流动、传热、凝固以及夹杂物的运动和捕获等现象。结果表明随着吹氩流量的增大，流场形态从双环流逐渐转变为复杂流和单环流，且复杂流和单环流下液位波动较大。电磁制动能够提高夹杂物在弯月面处的去除率，并且降低铸坯表层一定厚度内的夹杂物数量。夹杂物分布预测结果表明铸坯中夹杂物分布存在铸坯中心及距内弧1/4宽度处的条状聚集区。

关键词：多相流；流场形态；电磁制动；夹杂物分布；板坯连铸

Mathematical Modeling on the Fluid Flow, Heat Transfer, and Inclusion Distribution during the Continuous Casting Slab Strand

Zhang Lifeng, Chen Wei

(School of Metallurgical and Ecological Engineering, University of
Science and Technology Beijing, Beijing 100083, China)

Abstract: The multiphase flow, heat transfer and solidification during the continuous casting process directly affect the flow pattern, surface level fluctuations and the entrapment of argon bubbles and inclusions by the solidified shell. In the current study, a three-dimensional mathematical model was established to investigate the multiphase flow, heat transfer, solidification and the transport of inclusions in the continuous casting mold. The results show that with the increase of the argon flow rate, the flow pattern gradually changed from double roll flow to complex flow and single roll flow. The surface level fluctuation under complex flow and single roll flow was larger compared with the double roll flow. The inclusion removal fraction was increased after the application of the FC-Mold. The number of inclusions in a certain thickness below the slab surface was also reduced accordingly. The inclusion distribution results indicate that two accumulation peaks of inclusions along the thickness of the slab were existed, including the centerline of the slab thickness and the 1/4 thickness from the loose side.

Key words: multiphase flow; flow patterns; FC-mold; inclusion distribution; slab continuous casting

含硼钢板坯角横裂缺陷的成因与工艺控制

江中块[1,2]，夏兆东[1]，蔡兆镇[2]

（1. 上海梅山钢铁股份有限公司，江苏南京 210039；2. 东北大学冶金学院，辽宁沈阳 110004）

摘 要： 含硼钢连铸板坯在中国生产量非常大，但含硼钢板坯角横裂缺陷发生率非常高，严重影响生产顺行和板坯质量的提升。本文利用铸坯高温热模拟 Gleeble 检测、缺陷组织晶相分析、BN 热力学与动力学行为分析等手段，揭示了硼钢板坯角横裂缺陷的成因，并制订了现场控制方法。研究结果表明：合理地控制含硼钢钢水氮与硼元素含量，必要时采用 Ti 元素固氮，采用合理的二冷强度，可有效抑制 BN 在晶界上的析出量。通过现场采取相应的控制手段，含硼钢板坯角横裂缺陷发生率达到了较低值，生产得以顺行。

关键词： 板坯连铸；含硼钢；角横裂；工艺

Cause of Formation Transverse Corner Crack and Process in Boron Steel CC Slab

Jiang Zhongkuai[1,2], Xia Zhaodong[1], Cai Zhaozhen[2]

(1. Shanghai Meishan Iron and steel Co., Ltd., Nanjing 210039, China;
2. Northeastern University School of Metallurgy, Shenyang 110004, China)

Abstract: Quantity of boron steel CC slab is very large in China. However, transverse corner crack of boron steel CC slab is severe. It prevents from increasing slab quality and produce smoothly. In this paper, with checking the high temperature plasticity by Gleeble, observing the metallographic structure morphologies, and thermodynamics analysis, causes of formation transverse corner crack in Boron steel CC slab are discovered, and the process is formation. The result shows: In the slab continuous casting process of boron steels, with the adjust the quantity of B element and N element, adding Ti element if necessary, and modifying secondary intensity, the quantity of BN in crystal boundary get lower. Based on these results, the ratio of transverse corner crack of boron steel CC slab is decreased lowly and produce get smooth.

Key words: slab continuous casting; boron steel; transverse corner crack; process

电磁搅拌下板坯连铸结晶器内的瞬态三相流动

安 文，刘中秋，李向龙，吴存友，李宝宽

（东北大学冶金学院，辽宁沈阳 110819）

摘 要： 采用大涡模拟（LES）和 VOF 模型对电磁搅拌下板坯连铸结晶器内的瞬态三相流动特征进行了数值模拟研究。为验证模型的准确性，将预测的磁感应强度与工厂实际测量数据进行了对比，结果吻合较好。模拟结果表明，在电磁搅拌作用下，结晶器内钢液流速分布更加均匀，沿拉坯方向的流动趋于平缓，渣-金界面会产生较大波动，在水口两侧会出现两个对称的涡旋状波动形貌，弯月面处的渣-金界面波动幅度较大。

关键词： 大涡模拟；VOF；板坯连铸；电磁搅拌

Three-phase Flow in Slab Mould under Electromagnetic Stirring

An Wen, Liu Zhongqiu, Li Xianglong, Wu Cunyou, Li Baokuan

(School of Metallurgy, Northeastern University, Shenyang 110819, China)

Abstract: Numerical simulation of multiphase flow in slab continuous casting mould under electromagnetic stirring was carried out by using large eddy simulation(LES) and VOF model. In order to verify the accuracy of the model, the magnetic induction intensity is compared with the actual measured data in the factory, and the data agree well. The simulation results show that under the electromagnetic stirring, the velocity distribution of molten steel in the mold is more uniform, the flow along the casting flow direction is faster and tends to be gentle. There are two symmetrical vortices on both sides of the nozzle, and the fluctuation amplitude of the slag-metal interface at the meniscus is larger. The results of the study have certain guiding significance for practical production.

Key words: large eddy simulation; vof; slab continuous casting; electromagnetic stirring

CSP 薄板坯表面纵裂缺陷原因分析与控制

巩彦坤，张志克，杨学雨，王　鹏

（河钢集团邯钢公司，河北邯郸　056015）

摘　要：针对 CSP 薄板坯连铸机生产过程中的铸坯表面纵裂缺陷问题，借助宏观形貌、金相组织、扫描电镜能谱分析，结合现场生产实践，对纵裂产生原因进行了分析研究。通过钢水化学成分优化、结晶器工艺参数控制、保护渣性能优化、冷却水工艺优化和浸入水口对中精度等设备方面采取综合的优化措施，有效控制了铸坯纵裂缺陷发生，纵裂缺陷率稳定控制在 0.038%以下，有效改善了热卷板纵裂缺陷发生。

关键词：CSP；薄板坯；纵裂缺陷；分析与控制

Analysis and Control of Longitudinal Crack on CSP Thin Slab Surface

Gong Yankun, Zhang Zhike, Yang Xueyu, Wang Peng

(HBIS Group Hansteel Company, Handan 056015, China)

Abstract: Aiming at the problem of longitudinal crack on slab surface in the production process of CSP thin slab continuous caster, the causes of longitudinal crack were analyzed and studied by means of macro-morphology, metallographic structure, scanning electron microscopy and energy spectrum analysis combined with field production practice. The longitudinal crack defect of slab was effectively controlled by adopting comprehensive optimization measures in the aspects of chemical composition optimization of molten steel, process parameter control of mould, powder performance optimization, cooling water process optimization and immersion nozzle alignment accuracy. The longitudinal crack defect rate was stably controlled below 0.038%, and the longitudinal crack of hot coil plate was effectively improved. Defects occur.

Key words: CSP; thin slab; longitudinal crack defect; analysis and control

新一代连铸结晶器电液直驱伺服缸智能振动

刘　玉，蔡春扬，王永猛，彭晓华，龙　灏，李新有

（中冶赛迪技术研究中心有限公司，重庆　401122）

摘　要：本文对新一代连铸结晶器电液直驱伺服缸智能振动（CEDD 振动）的基本原理、系统组成、技术特点、工程应用情况进行了介绍，为冶金连铸机结晶器振动装备的升级改造提供了一种全新的更优的解决方案。

关键词：连铸结晶器振动；电液直驱伺服系统；智能控制

New Generation Continuous Casting Mould Oscillation by Electro-hydraulic Direct Drive Servo Cylinder

Liu Yu, Cai Chunyang, Wang Yongmeng, Peng Xiaohua, Long Hao, Li Xinyou

(Cisdi R & D Co., Ltd., Chongqing 401122, China)

Abstract: The basic principle, system composition, technical characteristics and engineering application of a new generation continuous casting mould oscillation by electro-hydraulic direct drive servo cylinder are introduced in this paper, which provides a new and better solution for upgrading and renovating the metallurgical continuous casting mould oscillation.

Key words: continuous casting mould oscillation; electro-hydraulic direct drive servo system; intelligent control

基于高频超声检测的铸坯内部缺陷三维重构可视化

丁　恒[1]，李　雪[1]，王　柱[2]，卫广运[3]，刘宏强[3]，黎　敏[1]

（1. 北京科技大学钢铁共性技术协同创新中心，北京　100083；2. 北京科技大学机械工程学院机电系，北京　100083；3. 河钢集团钢研总院，河北石家庄　050023）

摘　要：铸坯作为钢铁生产的中间产品，其内部质量对下游产品质量有重要的影响。传统铸坯质量检测大多只能对表面缺陷进行检测，而无法检测内部缺陷。针对工业生产中对铸坯内部质量控制的需求，提出了基于高频超声显微技术的铸坯内部缺陷三维重构可视化方法，通过该方法可以得到材料内部缺陷的尺寸、数量和空间分布特征。以 42CrMo 铸坯为检测对象，对铸坯实施不同压下量的处理，检测结果发现：当压下量为 16mm 时，对铸坯内部疏松有明显地改善作用，但继续增大压下量，则会诱发铸坯内部裂纹的产生。为验证这一结果，对经过不同压下量处理后的铸坯，继续进行轧制处理，检测结果表明：压下量为 24mm 的铸坯，在轧制后的材料内部出现宏观裂纹，进一步证实了过大的压下量会导致铸坯内部出现裂纹，进而影响下游产品的内部质量。该方法可以为优化连铸工艺参数提供一种快速、准确、三维可视化的检测和评价手段。

关键词：铸坯；内部缺陷；高频超声显微技术；三维重构可视化

3D Reconstruction Visualization for Internal Defects of Slab based on High Frequency Ultrasonic Testing

Ding Heng[1], Li Xue[1], Wang Zhu[2], Wei Guangyun[3], Liu Hongqiang[3], Li Min[1]

(1. Collaborative Innovation Center of Steel Technology, Beijing University of Science and Technology, Beijing 100083, China; 2. School of Mechanical Engineering, Beijing University

of Science and Technology, Beijing 100083, China; 3. HBIS Group Technology Research Institute, Shijiazhuang 050023, China)

Abstract: As a middle product of steel production, the slab has an important influence on the quality of downstream products. Traditional slab quality inspection can only detect surface defects and cannot detect internal defects. Aiming at the demand for internal quality control of slab in industrial production, a three-dimensional reconstruction visualization method for internal defects of slab based on high-frequency ultrasonic microscopy is proposed. The size, quantity and spatial distribution characteristics of internal defects of materials can be obtained by this method. The 42CrMo slab was used as the test object, and the slab was treated with different reductions. The results show that when the reduction is 16mm, the internal looseness of the slab is obviously improved, but the reduction continued to increase which will induce the occurrence of internal cracks. In order to verify this result, the slab after the different reductions were continuously rolled, and the results show that the slab with a reduction of 24 mm appear macroscopic cracks inside the rolled material, further confirming the quantity of the reduction will lead to crack of slab inside, which in turn affects the internal quality of downstream products. This method can provide a fast, accurate and 3d visualization method for detecting and evaluating the parameters of continuous casting process.

Key words: slab; internal defects; high frequency ultrasound microscopy; 3D reconstruction visualization

大圆坯连铸机轻压下仿真计算

曹学欠，陈 杰

（中冶京诚工程技术有限公司炼钢工程技术所，北京 100176）

摘 要： 基于我公司设计的铸机条件，建立了大圆坯连铸轻压下的有限元模型，得到了不同压下量下铸坯变形规律。结果表明：Ø800mm 在固相率 0.35 时，单辊压下量需大于 10mm，压下效果才能传递至铸坯芯部。根据连铸机的设计标准，当铸坯表面应变值超过 0.015 时，铸坯表面极容易产生裂纹，即单辊的最大压下量不应超过 8mm。所以圆坯不适合于实施轻压下工艺。

关键词： 连铸；大圆坯；轻压下；有限元

Finite Element Analysis of Process of Soft Reduction for Round Bloom Continuous Casting

Cao Xueqian, Chen Jie

(Steelmaking and Continuous Casting Engineering Division, Capital Engineering & Research Inc. Ltd., Beijing 100176, China)

Abstract: Based on the round bloom continuous casting machine that designed by CERI, the finite element model of soft reduction was established, and got the change for transformation of bloom on the various reduction amount. The results show that: Ø800mm in the solid fraction of 0.35, the single roll reduction should be greater than 10mm, the effect can be passed to the bloom core. According to the design standard of the continuous casting machine, when the surface strain value of the slab exceeds 0.015, the surface of the slab is very easy to crack, that is, the maximum reduction of single roll should not exceed 8mm. Therefore, round bloom is not suitable for the implementation of the soft reduction process.

Key words: continuous casting; round bloom; soft reduction; finite element method

连铸坯热芯大压下技术轧制规程实验研究

李睿昊[1]，李海军[1]，李天祥[1]，李双江[2]，郭子强[2]

（1. 东北大学轧制技术及连轧自动化国家重点实验室，辽宁沈阳　110819；
2. 河钢集团钢研总院，河北石家庄　050000）

摘　要：随着连铸坯断面尺寸增加，芯部质量问题日益凸显，一系列针对连铸坯质量改善的重压下技术被相继提出，针对重压下易出现内裂纹的缺点，连铸坯热芯高温大压下技术顺势而生。在工艺应用过程中，应当延续重压下多辊协作变形的特点，还是利用传统轧制的思路实行单道次大压下，成为热芯大压下研究的重要内容。本文通过对连铸小方坯直接进行轧制实验，使用数值模拟还原变形时的温度场分布和等效应力状态，并将单道次和多道次两种大压下中间坯轧制成最终成品板材，通过检测中间坯和成品的组织、性能，最终确定单道次热芯大压下工艺在改善宏观缩孔和微观晶粒度方面具有一定优势。

关键词：热芯大压下轧制技术；连铸坯；数值模拟；芯部性能

Experimental Research on Rolling Procedure of Hot-Core Heavy Reduction Rolling Tech for Continuous Casting Billet

Li Ruihao[1], Li Haijun[1], Li Tianxiang[1], Li Shuangjiang[2], Guo Ziqiang[2]

(1. State Key Laboratory of Rolling and Automation, Northeastern University, Shenyang 110819, China;
2. Technology Research Institute, HBIS Group, Shijiazhuang 050000, China)

Abstract: With the increase of size of continuous casting billet, the quality problems of the center of billet become more and more important. A series of improvement technologies for continuous casting billets have been developed. However, Heavy reduction is prone to internal cracks because of the liquid core in the deformation position. In order to solve this problem, the Hot-Core Heavy Reduction Rolling (HHR^2) Technology has begun to develop. In the process of application of HHR^2 tech, whether it is to continue the characteristics of multi-roll collaborative deformation under heavy reduction tech or to use the traditional rolling idea to carry out single reduction has become an important part of HHR^2 research. In this paper, through direct rolling experiment of continuous casting billet, temperature field distribution and effective strain state during reduction deformation are simulated by numerical simulation, and the final product is rolled from mono-pass and mul-pass HHR^2 intermediate billet. Finally, through testing the structure and properties of intermediate billet, the mono-pass HHR^2 technology has certain advantages in improving macroscopic shrinkage and micro-grain size.

Key words: hot-core heavy reduction rolling technology; continuous casting billet; numerical simulation; central property

板坯连铸结晶器窄面锥度对铸坯传热行为的影响

牛振宇[1]，蔡兆镇[1]，刘志远[2]，王重君[2]，陈长芳[2]，朱苗勇[1]

（1. 东北大学冶金学院，辽宁沈阳　110819；2. 唐山中厚板材有限公司，河北唐山　063600）

摘 要：板坯连铸过程中结晶器锥度对铸坯表面质量、生产效率有重要影响。为优化板坯连铸结晶器锥度，本文通过建立结晶器-钢系统瞬时热/机耦合有限元模型，模拟分析了不同锥度下钢凝固传热及收缩行为、气隙及保护渣分布。结果表明，锥度为 1.05%时，结晶器中部处铸坯角部与宽、窄面铜板间存在较大间隙，导致此处渣膜变厚、气隙生成。同时，铸坯角部降温速率下降。当锥度增至 1.2%时，宽、窄面角部及偏离角处渣膜有所减薄，使得铸坯角部降温速率有所升高

Thermal Behavior of Slab under Different Mold Tapers during Continuous Casting

Niu Zhenyu[1], Cai Zhaozhen[1], Liu Zhiyuan[2], Wang Chongjun[2], Chen Changfang[2], Zhu Miaoyong[1]

(1. School of Metallurgy, Northeastern University, Shenyang 110819, China;
2. Tangshan Medium and Thick Plate Co., Ltd., Tangshan 063600, China)

Abstract: During slab continuous casting, the mold taper has a significant influence on the surface quality of slab and production efficiency. To optimize the slab mold taper, a transient thermo-mechanical coupled finite element model of mold-steel system was developed in the present work, based on which the thermal distortion and temperature evolution of solidified shell as well as the distributions of air gap and mold flux were simulated. The results show that a large gap between slab corner and copper plates at both wide and narrow faces appears at the middle part of mold under the taper of 1.05%, which leads to the thick mold flux film and formation of air gap. Meanwhile, the cooling rate of slab corner reduces. While, increasing the mold taper to 1.2% could thin the corner flux film of wide and narrow face. Consequently, the cooling rate of slab corner increases.

巨能特钢 2 号连铸机的技术特点和应用

樊伟亮，陈 杰，曹学欠

（中冶京诚工程技术有限公司，北京 100176）

摘 要：本文主要介绍了中冶京诚（CERI）为巨能特钢改造的 2 号方坯连铸机的重要技术参数和改造设备（如结晶器足辊、液压振动、扇形段、拉矫机等）的特点，介绍了该铸机配置的二冷模型、电磁搅拌及轻重压下模型等先进技术，以及取得的实际效果。

关键词：方坯连铸；中冶京诚；技术特点；轻重压下技术

The Technological Characteristic and Application of No.2 Bloom Caster in Shandong Juneng Special Steel Co., Ltd.

Fan Weiliang, Chen Jie, Cao Xueqian

(Capital Engineering & Research Incorporation Limited, Beijing 100176, China)

Abstract: This paper mainly introduces the important technical parameters of 2# bloom caster in Juneng Special Steel Co.,Ltd. and the characteristics of reforming equipment (such as mould and foot roll, hydraulic vibration, sector section, stretch straightener, etc.). It also introduces the advanced technology of secondary cooling model, electromagnetic stirring and soft and hard reduction model of the caster, as well as the actual results have been achieved.

Key words: bloom caster; CERI; technological characteristic; soft and hard reduction

基于 CA 模型的多元合金微观组织模拟

高晓晗[1]，孟祥宁[1,2]，崔 磊[1]，朱苗勇[1,2]

（1. 东北大学冶金学院，辽宁沈阳 110819；
2. 多金属矿生态冶金重点实验室，辽宁沈阳 110819）

摘 要： 本文基于热传输和溶质传输建立了一个二维四元合金元胞自动机(CA)模型，同时考虑了热过冷，溶质过冷和曲率过冷对枝晶生长的影响，并引入偏心正方形算法模拟了任意角度的等轴晶生长。数值模拟结果表明，模型能够合理描述任意角度的晶粒生长过程，且能准确描述柱状晶生长过程中生成的次级枝晶。建立的 CA 模型与 LGK 解析模型进行了对比，验证了模型准确性。

关键词： 多元合金；CA 模型；数值模拟；枝晶

Numerical Simulation of Microstructure Evolution based on Multiple Alloy CA Model

Gao Xiaohan[1], Meng Xiangning[1,2], Cui Lei[1], Zhu Miaoyong[1,2]

(1. School of Metallurgy, Northeastern University, Shenyang 110819, China;
2. Key Laboratory for Ecological Metallurgy of Multi-metallic Ores, Shenyang 110819, China)

Abstract: In this paper, a two-dimensional cellular automaton model for Quaternary alloys is established based on heat and solute transport. The effects of thermal, solute and curvature undercooling on dendrite growth are considered. Decentered square algorithm is introduced to simulate equiaxed crystal growth at any angle. The numerical simulation results show that the model can reasonably describe the grain growth process at any angle and accurately describe the secondary dendrites in the columnar crystal growth process. The established CA model is compared with the LGK analytical model to verify the accuracy of the model.

Key words: multicomponent alloy; CA model; numerical simulation; dendrite

低碳钢角部纵裂缺陷分析及控制实践

秦 伟，刘红军

（华菱涟源钢铁公司 210 转炉厂，湖南娄底 417000）

摘 要： 针对低碳钢生产过程中出现角部纵裂纹的情况进行了分析，提出了优化结晶器锥度、控制铜板表面质量、优化提速制度、优化结晶器冷却强度等措施，有效地控制了低碳钢角纵裂的发生。

关键词：连铸；角纵裂；低碳钢；结晶器

Analysis and Control Practice of Longitudinal Corner Cracks in Low Carbon Steel

QinWei, Liu Hongjun

(210 BOF Plant, Valin Lianyuan Steel Co., Loudi 417000, China)

Abstract: This paper analyzes the occurrence of longitudinal corner cracks in the production of low carbon steel. It proposes measures such as optimizing the mold taper, controlling the surface quality of the copper plate, optimizing the casting speed-up system, optimizing the cooling intensity of the mold etc. which have effectively controlled the occurrence of longitudinal corner cracks in low carbon steel.

Key words: continuous casting; longitudinal corner cracks; low carbon steel; casting mold

高效自清洗喷嘴在板坯连铸机上的应用

吕宪雨，张威东，赵晓波

（宝山钢铁股份有限公司，上海 200941）

摘　要：本文结合宝钢4号板坯连铸机加装自清洗喷嘴的实践，介绍了在线自清洗喷嘴在板坯连铸机上的实际应用。研究了连铸机生产过程中喷嘴堵塞的在线自清洗处理工艺，加装自清洗系统后，通过监测连铸机各个冷却回路的压力与流量的对应异常变化，从而判断回路喷嘴的堵塞情况，及时将堵塞物通过特制的自清洗喷嘴及系统将喷嘴堵塞物清除，喷嘴水气通道重新畅通，喷嘴的雾化及其他功能恢复，使连铸坯的均匀冷却得以保证，板坯的内部裂纹得到控制，铸坯的质量得到明显提高，也避免了因喷嘴堵塞严重造成的被迫停机，连铸机的作业率得到了提高。

关键词：板坯连铸机；自清洗喷嘴；内部质量；作业率

Application of Efficient Self-cleaning Nozzle in Slab Caster

Lv Xianyu, Zhang Weidong, Zhao Xiaobo

(Baoshan Iron and Steel Co., Ltd., Shanghai 200941, China)

Abstract: This paper introduces the practical application of on-line self-cleaning nozzle on slab continuous caster, based on the practice of installing self-cleaning nozzle in No. 4 slab casters of Baosteel.

　　The on-line self-cleaning treatment process of nozzle clogging in the production process of continuous caster was studied. After installing self-cleaning system, it is timely that to judge the blockage of nozzles in the loop and to remove the blockage through specific self-cleaning nozzle and system, by monitoring the abnormal changes of pressure and flow rate in each cooling circuit of the continuous caster. Nozzle water passage resumes unimpeded, and the atomization and other functionsof the nozzles are restored, which would ensure the uniform cooling of the continuous casting slab and control the internal crack of the slab. In this way, the quality of the slab will get a remarkable improvement, the forced shutdown caused

by the nozzle clogging will be avoided, and the operation rate of the continuous casterwill be enhanced.
Key words: slab caster; self-cleaning nozzle; internal quality; operation rate

含 B_2O_3 无氟保护渣的微观结构和黏度特性研究

吴 婷[1,2]，钟 磊[1]，Shama Sadaf[1]，廖直友[1,2]，王海川[1,2]，王万林[3]

（1. 安徽工业大学冶金工程学院，安徽马鞍山 243032；2. 冶金减排与资源综合利用教育部重点实验室，安徽马鞍山 243002；3. 中南大学冶金与环境学院，湖南长沙 410083）

摘 要：为响应绿色、环保发展的需求，无氟连铸保护渣成为了国内外研究热点，其中，以硼硅酸钙（$Ca_{11}Si_4B_2O_{22}$）为主要析晶相的含 B_2O_3 无氟保护渣具有较好的应用前景。本文结合分子动力学模拟、旋转黏度计测试、红外光谱实验以及 FactSage 热力学计算，研究 B_2O_3 对无氟保护渣的微观结构和黏度特性的影响机制。结果表明，$CaO-SiO_2-B_2O_3$ 三元渣系形成了稳定的$[SiO_4]^{4-}$四面体、$[BO_3]^{3-}$三面体和$[BO_4]^{4-}$四面体结构单元，B_2O_3 使 Si-O 网络结构一定程度上聚合，而对 B-O 网络结构聚合度的影响因其含量而异，红外光谱所得实验结果与分子动力学模拟结果很吻合；$w(B_2O_3)=4\%\sim12\%$范围内增加，无氟保护渣 1300℃黏度和液相线温度降低，而熔渣网络结构聚合度基本不变，表明 B_2O_3 使无氟保护渣 1300℃黏度降低的主要原因是熔渣过热度的增加；由于析晶矿相和固溶体比例的变化，$w(B_2O_3)=4\%\sim6\%$范围内无氟保护渣黏度-温度曲线呈现碱性渣特征，$w(B_2O_3)=8\%\sim12\%$范围内无氟保护渣黏度-温度曲线呈现出酸性渣特征。

关键词：无氟保护渣；B_2O_3；微观结构；黏度特性

Study on Microstructure and Viscosity of B_2O_3-Containing and Fluoride-free Mold Fluxes

Wu Ting[1,2], Zhong Lei[1], Shama Sadaf[1], Liao Zhiyou[1,2], Wang Haichuan[1,2], Wang Wanlin[3]

(1. School of Metallurgical Engineering, Anhui University of Technology, Maanshan 243032, China; 2. Key Laboratory of Metallurgical Emission Reduction & Resource Recycling (Ministry of Education), Anhui University of Technology, Maanshan 243002, China; 3. School of Metallurgy and Environment, Central South University, Changsha 410083, China)

Abstract: In response to the needs of green and environmental protection development, fluoride-free mold fluxes has become a research hotspot at home and abroad. Among them, the B_2O_3 containing and fluoride-free slag with calcium boron silicate ($Ca_{11}Si_4B_2O_{22}$) as the main crystallization phase has a good application prospect. Based on molecular dynamics simulation, rotating viscometer test, infrared spectrum experiment and FactSage thermodynamic calculation, this paper studies the influence mechanism of B_2O_3 on the microstructure and viscosity characteristics of fluoride-free slag. The results showed that the ternary residue system of $CaO-SiO_2-B_2O_3$ formed stable structural units of $[SiO_4]^{4-}$ tetrahedron, $[BO_3]^{3-}$ trihedron and $[BO_4]^{4-}$ tetrahedron. B_2O_3 polymerized Si-O network structure to some extent, and the influence on the polymerization degree of B-O network structure was different due to its content. The infrared spectrum experimental results are in good agreement with molecular dynamics simulation results. In the range of $w(B_2O_3)=4\%\sim12\%$, the viscosity at 1300℃ and liquidus temperature of fluoride-free mold fluxes decrease, while the polymerization degree of slag network structure remains basically unchanged, indicating that the main reason of B_2O_3 reduces the viscosity of fluoride-free slag at

1300℃ is the increase of superheat of slag. Due to the changes in crystallization phase and solid solution ratio, the viscosity-temperature curve of fluoride-free slag in the range $w(B_2O_3)$=4%~6% shows the characteristics of alkaline slag, while the viscosity-temperature curve of fluoride-free mold fluxes in the range $w(B_2O_3)$=8%~12% shows the characteristics of acidic slag.

Key words: fluoride-free mold fluxes; B_2O_3; microstructure; viscosity

基于热丝法的保护渣溶解 Al_2O_3 动力学研究新方法

陈富杭，唐 萍，文光华，谷少鹏

（重庆大学材料科学与工程学院冶金系，重庆 400044）

摘 要：旋转柱体法和共焦激光扫描显微镜法常被用于 Al_2O_3 在保护渣中的溶解动力学研究，但前者无法实现原位观察，后者存在操作复杂等缺陷。热丝法可弥补前述两种方法的不足。本文探究了利用热丝法研究 Al_2O_3 在保护渣中溶解动力学的方法，验证了实验结果的正确性，并对实验重现性的控制方法进行了研究。结果表明，利用热丝法进行实验得到的碱度对 Al_2O_3 溶解速率的影响规律与旋转柱体法的结果相似；将 Al_2O_3 颗粒的粒径控制小于 350μm，实验的重现性好；实验所采用的 Al_2O_3 颗粒密度应略大于熔渣的密度。

关键词：热丝法；原位；Al_2O_3；溶解动力学

A Novel Method of Research on Dissolution Kinetics of Alumina in Mold Flux based on Hot Thermocouple Technique

Chen Fuhang, Tang Ping, Wen Guanghua, Gu Shaopeng

(Department of Materials Science and Engineering Chongqing University, Chongqing 400044, China)

Abstract: Rotating cylinder method and confocal laser scanning microscope are used to be applied to the research on dissolution kinetics of alumina in mold flux. But the former can't observe in situ, the latter has defects such as complex operation. Hot thermocouple technique can make up the deficiencies of the preceding methods. This paper explores the approach of researching dissolution kinetics of alumina in mold flux by hot thermocouple technique. It verifies the correctness of experiments and explores the controlling means of the reproducibility of experiments. The results show that the effect of basicity on dissolution rate of alumina by hot thermocouple technique is similar to the consequence of rotating cylinder method. The reproducibility of experiments is good when the diameters of alumina particles are less than 350μm. The density of alumina particles using in experiments should be slightly bigger than the density of slag.

Key words: hot thermocouple technique; in situ; alumina; dissolution kinetics

异型坯连铸技术专利分析

高 仲，王 颖，陈卫强，陈 杰

（中冶京诚工程技术有限公司炼钢工程技术所，北京 100176）

摘　要：基于中国知识产权网数据信息，本文针对在国内申请的异型坯连铸相关专利技术，从申请人角度对国内外企业，以及从异型坯连铸机工艺和设备角度对中间包、结晶器、二冷区、保护渣和品种钢开发等，进行分析。总结出异型坯连铸技术的专利现状和发展方向：国内企业越来越重视专利申请，在我国申请的异型坯连铸相关专利数量远多于国外企业。设计中间包不但需要合适的流场，而且还需要具备浸入式水口和中间包快换功能。结晶器铜板需要被高效且均匀的冷却、同时还需要考虑结晶器铜板的易加工性。二冷区喷嘴布置，既要满足纵向的冷却合理性，还要兼顾横向的冷却合理性。

关键词：异型坯；专利；连铸；设备

Analysis of the Patents of Beam Blank Continuous Casting Technology

Gao Zhong, Wang Ying, Chen Weiqiang, Chen Jie

(Capital Engineering & Research Incorporation Limited(CERI) Steelmaking and Continuous Casting Engineering Division (S&C), Beijing 100176, China)

Abstract: Based on the data and information of China intellectual property right net, this paper analyzes the patents related to beam blank continuous casting technology applied in China from the perspective of the applicant, and the process and equipment of beam blank caster. It is concluded that the patents status and development direction of the beam blank continuous casting technology: Domestic enterprises pay more and more attention to patents application, in our country the number of patent application by domestic enterprises is far more than foreign enterprise. The design of tundish not only needs proper flow field, but also needs SEN and tundish quick change function. The mold copper plate needs to be cooled efficiently and evenly, and the workability of the mold copper plate needs to be considered. The nozzle layout of secondary cooling zone should not only satisfy the longitudinal cooling rationality, but also take into account the transverse cooling rationality.

Key words: beam blank; patent; continuous casting; equipment

重轨钢凝固末端机械压下对溶质偏析行为的影响

祭　程[1]，关　锐[1]，朱苗勇[1]，吴国荣[1,2]，陈天明[2]，李红光[2]

（1. 东北大学冶金学院，辽宁沈阳　110819；
2. 攀钢集团研究院有限公司钒钛钢研究所，四川攀枝花　617022）

摘　要：钢在连铸过程中，其富集溶质元素的长距离传输导致宏观偏析。该质量缺陷难以在后续加热及轧制过程有效消除，严重影响最终轧材的综合力学性能。本文针对重轨钢大方坯连铸过程，在准确描述其坯壳形貌基础上，建立了耦合凝固末端机械压下过程的多相凝固模型，揭示了机械压下对重轨钢大方坯宏观偏析改善机制。研究结果表明：凝固终点之前实施压下方可有效调控大方坯内部富集溶质液相的流动，进而改善宏观偏析缺陷。随着压下量由0mm 增大至 12mm，重轨钢大方坯心部富集溶质的液相流动速度不断减小并发生反向流动，碳中心偏析度由 1.23 降低至 1.08。

关键词：重轨钢；连铸；多相凝固模型；机械压下；溶质偏析

Effect of Mechanical Reduction Technology in the Heavy Rail Bloom Continuous Casting on the Solute Segregation

Ji Cheng[1], Guan Rui[1], Zhu Miaoyong[1], Wu Guorong[1,2], Chen Tianming[2], Li Hongguang[2]

(1. School of Metallurgy, Northeastern University, Shenyang 110819, China; 2. V and Ti Steels Institute, Pangang Group Research Institute Co., Ltd., Panzhihua 617022, China)

Abstract: During the continuous casting process of steel, the long-distance transmission of segregated solute results in the macro-segregation of continuous casting steel. This internal quality defect is hardly to be eliminated the subsequent heat treatment or rolling process, which seriously deteriorates the mechanical properties of the final rolling materials. In the present work, multi-phase solidification model for continuous casting process of heavy rail steel coupling the mechanical reduction was developed based on the accurately-predicted morphology of the solidified shell. Mechanism of mechanical reduction on reducing the bloom macro-segregation of heavy rail steel was investigated. Results indicate that a greater reduction should be implemented before the solidification end for effectively promoting the reverse flow and thus improving the macro-segregation. With the reduction amount continuously increased from 0 mm to 12 mm, flow of the solute enriched liquid steel constantly decreases, and the reverse flow occurs. The degree of carbon macrosegregation decreases from 1.23 to 1.08.

Key words: heavy rail bloom; continuous casting; multi-phase solidification model; mechanical reduction; solute segregation

大圆坯凝固末端电磁搅拌流动行为数值模拟

郭壮群，罗森，张文杰，王卫领，朱苗勇

（东北大学冶金学院，辽宁沈阳 110819）

摘 要：连铸过程中合理的末端电磁搅拌有利于混匀铸坯中心的高温钢液与低温钢液，促进柱状晶向等轴晶转变，同时降低糊状区的过热度，促进溶质的重新分布。本文通过 Ansys fluent 软件建立了大圆坯连铸过程 3D-2D 相结合的多物理场数学模型，研究不同末搅参数对钢液流动行为的影响。研究表明，随着末端电磁搅拌电流的增加，搅拌速度明显增大，当搅拌电流从 120A 增加到 400A，切向速度由 0.0030m/s 增大到 0.029m/s。这使得钢液在水平截面的旋转运动冲刷清洗凝固界面，促进了等轴晶的行程与溶质的重新分布，减小中心偏析。同样地，当拉速从 0.23m/min 提高到 0.28m/min 时，切向速度从 0.00306m/s 增大到 0.01162m/s。因此，也可以通过改变拉速的方式来达到改善末搅流场的效果。

关键词：连铸；圆坯；末端电磁搅拌；数值模拟

Numerical Simulation of Fluid Flow in Liquid Core of Strand during Round Bloom Continuous Casting with Final Electromagnetic Stirring

Guo Zhuangqun, Luo Sen, Zhang Wenjie, Wang Weiling, Zhu Miaoyong

(School of Materials and Metallurgy, Northeastern University, Shengyang 110819, China)

Abstract: Reasonable final electromagnetic stirring(F-EMS) during casting is conducive to the mixing of high temperature molten steel and low temperature molten steel, interruption of columnar dendrites and enlargement of the center equiaxed zone, reducing the superheat of the mushy zone and promoting the redistribution of solute. In this paper, the mathematical model of 3D-2D in the process of large round billet continuous casting (CC) was established by Ansys fluent software. Studies have shown that as the electric current increases, the agitation speed increases significantly. When the electric current increases from 120A to 400A, the tangential velocity increases from 0.0030m/s to 0.029m/s. This makes the molten steel scouring and cleaning the solidification interface in the horizontal section, which not only promotes the equiaxed crystal stroke but also reduces the central segregation and eliminates center segregation. Similarly, when the pulling speed is increased from 0.23 m/min to 0.28 m/min, the tangential speed is increased from 0.00306 m/s to 0.01162 m/s. Therefore, the effect of improving the final agitating flow field can also be achieved by changing the pulling speed.

Key words: continuous casting; round bloom; final electromagnetic stirring(F-EMS); numerical simulation

旋流情况下晶粒移动行为研究

王 鹏，罗 森，刘光光，王卫领，朱苗勇

（东北大学冶金学院，辽宁沈阳 110819）

摘 要：为了研究枝晶在凝固过程中的生长和移动规律，本文提出了相场法（PF）和格子玻尔兹曼方法（LBM）耦合模型（PF-LBM）用于模拟在流动熔体中的枝晶生长和移动。在本模型中，相场法（PF）用于模拟枝晶生长过程中的形貌变化和溶质传输，格子玻尔兹曼方法（LBM）用于模拟凝固过程中的熔体流动。枝晶的移动由牛顿第二定律确定，并通过追踪笛卡尔网格中的拉格朗日点实现。本模型用于模拟旋流情况下的枝晶生长和移动，模拟结果显示，在旋流情况下，枝晶在自身旋转的同时会绕计算域中心旋转，且旋转半径越来越大。枝晶在移动过程中周围溶质分布均匀，形貌与等轴晶相近。

关键词：旋流；PF-LBM 模型；枝晶生长；溶质分布；移动

Study on the Motion of Dendrite under Swirling

Wang Peng, Luo Sen, Liu Guangguang, Wang Weiling, Zhu Miaoyong

(School of Materials and Metallurgy, Northeastern University, Shenyang 110819, China)

Abstract: In order to study the growth and motion of dendrite during solidification, a phase field method(PF) and lattice Boltzmann method(LBM) coupling model was proposed to simulate dendrite growth and motion in flowing melt. In this model, the phase field method (PF) is used to simulate the dendritic morphology and solute transport during dendritic growth. The lattice Boltzmann method (LBM) is used to simulate the melt flow during solidification. The movement of the dendrites is determined by Newton's second law and is achieved by tracking the Lagrangian points in the Cartesian grid. This model is used to simulate the growth and motion of dendrite in a rotational fluid flow. The simulation results show that the dendrites rotate around the center of the domain under swirling, and the radius of rotation is larger and larger. During the movement of dendrites, the distribution of solute is uniform and the morphology is close to equiaxed crystal.

Key words: swirling; PF-LBM model; dendritic growth; motion; solute concentration

连铸结晶器电磁制动下导电壁面对湍流运动的影响

吴颖东,刘中秋,李宝宽

(东北大学冶金学院,辽宁沈阳 110819)

摘 要:采用大涡模拟研究全幅一段式电磁制动下连铸结晶器内瞬态湍流运动行为,分析了绝缘壁面和导电壁面两种边界条件对湍流的影响。模拟得到水平速度的瞬时值和时均值与超声多普勒测速仪测量值吻合良好。利用 Q 准则数等值面对结晶器内的典型三维湍流涡结构进行了可视化分析,结果表明,壁面绝缘时,流场流动不稳定,且存在低频率的振荡,振荡时间间隔不固定。壁面导电时,射流的低频振荡得到了很好的抑制,流场呈现稳定的双辊流结构。

关键词:连铸结晶器;电磁制动;湍流;导电壁面

Effect of an Electrically-Conducting Wall on Turbulent Flow in a Continuous-Casting Mold with an Electromagnetic Brake

Wu Yingdong, Liu Zhongqiu, Li Baokuan

(School of Metallurgy, Northeastern University, Shenyang 110819, China)

Abstract: Transient turbulent flow in continuous casting mould under single-ruler electromagnetic braking was studied by large eddy simulation (LES). The influence of two boundary conditions, insulative walls and conductive walls, on turbulence in mold was analyzed. The instantaneous and time-averaged horizontal velocity obtained by simulation are in good agreement with the measurement of ultrasonic Doppler velocimetry (UDV). The isosurface of Q criterion was used to analyze the typical 3d turbulent eddy structure in mold. The results show that as walls are insulated, the flow field is unstable with a low frequency oscillation, but the time interval of changeover is flexible. Under the electrically-conducting wall condition, the low-frequency oscillation of jet is well suppressed, and the flow field presents a stable double-roll flow pattern.

Key words: continuous-casting mold; electromagnetic brake; turbulent flow; electrically-conducting walls

微合金钢高温板坯全连续淬火新技术开发及应用

蔡兆镇[1],刘志远[1,2],王重君[2],赵佳伟[1],朱苗勇[1]

(1. 东北大学冶金学院,辽宁沈阳 110819;2. 唐山中厚板材有限公司,河北唐山 063610)

摘 要:研究分析了含 Nb 微合金钢高温淬火过程不同淬火温度及冷却速度下的组织结构转变及析出特点,确定了实现微合金钢板坯表层组织高塑化的最佳淬火控制起始温度为950℃、最佳冷却速度≥5℃/s。基此结合连铸板坯温度场的计算,确定了最佳淬火位置为扇形段 12 段末,并设计开发了连铸机在线全连续淬火装备,现场实施应用表明本研发的工艺与装备满足了微合金钢连铸坯热送要求。

关键词：微合金钢；板坯；热送；表面淬火

Development and Application of New Continuously Quenching Technology for High Temperature Micro-alloyed Steel Slab

Cai Zhaozhen[1], Liu Zhiyuan[1,2], Wang Chongjun[2], Zhao Jiawei[1], Zhu Miaoyong[1]

(1. School of Metallurgy, Northeastern University, Shenyang 110819, China;
2. Tangshan Heavy Plate Co., Ltd., Tangshan 063610, China)

Abstract: The evolution of the microstructure of the continuous casting slab of Nb containing alloyed steel under different quenching temperatures and cooling rate were analyzed, and the optimum quenching temperature 950℃ and the optimum cooling rate ≥5℃/s for having higher ductility of slab surface microstructure were obtained. Moreover, the optimum position of slab surface quenching, where is just after the end of No.12 segment, was determined by the slab heat transfer simulation, and the quenching equipment was developed. The application results show that the present process and equipment can meet the requirement of hot charging for micro-alloyed slab well.

Key words: micro-alloyed steel; continuous casting slab; surface quenching; hot charging

IF 钢 Al-Ca-O 系细微夹杂物控制研究与实践

李俊伟，赵　元，袁少江，杨成威，代碧波

（武汉钢铁有限公司制造管理部，湖北武汉　430000）

摘　要：本文通过分析细微夹杂缺陷成分、缺陷坯位置、炼钢工艺过程，弄清了细微 Al-Ca-O 系夹杂物的重要来源为钢包氧化性渣，采取了 RH 精炼结束顶渣改质工艺、中间包挡墙工艺、线圈式下渣检测设备三项措施后，基本杜绝了汽车外观零件和内板可视零件细微夹杂缺陷。

关键词：IF 钢；细微夹杂物；顶渣改质；中间包挡墙

Control and Practice on Al-Ca-O Tiny Inclusion in Interstitial Free Steel

Li Junwei, Zhao Yuan, Yuan Shaojiang, Yang Chengwei, Dai Bibo

(Wuhan Iron and Steel Co., Ltd., Wuhan 430000, China)

Abstract: SEM analysis shows that the tiny inclusions in IF steel are Al-Ca inclusions. An important source of Mg-Al-Ca tiny inclusions through the analysis of the position of defect blank, technological process and production data of oxidation is ladle oxidizing slag, and the adoption of RH refining to top slag modification, tundish retaining wall technology and coil type slag detection equipment can effectively reduce the occurrence of Al-Ca tiny inclusions.

Key words: interstitial free steel; tiny inclusion; top slag modification; tundish retaining wall

连铸结晶器非正弦振动装置试验研究

张兴中[1]，周　超[1,2]，任素波[1]

(1. 燕山大学国家冷轧板带装备及工艺工程技术研究中心，河北秦皇岛　066004；
2. 河北农业大学海洋学院，河北秦皇岛　066003)

摘　要：针对双液压缸驱动的结晶器非正弦振动装置系统复杂、投资及维护费用高等问题，本文提出一种双伺服电机驱动的结晶器非正弦振动装置，设计并制造试验样机。首先阐述其工作原理；其次给出实现德马克、复合函数非正弦振动波形伺服电机的转动规律；最后进行了实验室试验验证。该装置能够较好地实现结晶器非正弦振动，可降低设备投资、提高铸坯质量、有助于提高拉坯速度。

关键词：连铸；结晶器；非正弦振动；振动装置；试验

Experimental Investigation on Non-sinusoidal Oscillator of Continuous Casting Mold

Zhang Xingzhong[1], Zhou Chao[1,2], Ren Subo[1]

(1. National Engineering Research Center for Equipment and Technology of Cold Strip Rolling, Yanshan University, Qinhuangdao 066004, China;
2. Ocean College of Hebei Agricultural University, Qinhuangdao 066003, China)

Abstract: Due to the disadvantages of the complex system, vast investment and high maintenance for mold non-sinusoidal oscillator driven by the hydraulic cylinder, a new oscillator synchronously driven by double servomotors was proposed in this paper, which experimental prototype was designed and manufactured. Firstly, the working principle of the oscillator was described. Secondly, the rotating rules of servomotor to realize non-sinusoidal oscillation waveform for DEMAG and composite functions were given. Finally, the laboratory experiments were made. The oscillator can realize non-sinusoidal oscillation well, which provide a feasible scheme for reducing oscillator investment, increasing casting speed and enhancing slab quality.

Key words: continuous casting; mold; non-sinusoidal oscillation; oscillator; experiment

连铸坯凸型辊压下工艺技术发展及其应用

逯志方[1,2]，赵昊乾[1,2]，田新中[1,2]，孟耀青[1,2]，王晓英[1,2]

(1. 邢台钢铁有限公司技术中心，河北邢台　054027；
2. 河北省线材工程技术研究中心，河北邢台　054027)

摘　要：随着连铸坯断面的增加及产品质量要求的不断提升，连铸坯中心偏析、疏松等内部组织问题日益突出。凸型辊压下工艺在改善连铸坯内部质量方面效果显著，在国内外连铸机上广泛应用。本文对日本新日铁、韩国浦项以及国内台湾中钢、大连特钢等企业的凸型辊压下工艺技术的发展及应用情况进行了详细的调研，结果显示，近几年在以东北大学为代表的科研院所与特钢企业的共同努力下，我国已成功掌握了连铸坯凝固末端凸辊压下工艺关键技术，连铸坯实物质量得到了大幅提升。

关键词：连铸坯；凸型辊；中心偏析；中心疏松；压下

The Development and Application of Convex Roll Reduction Technology in Continuous Casting Bloom

Lu Zhifang[1,2], Zhao Haoqian[1,2], Tian Xinzhong[1,2],
Meng Yaoqing[1,2], Wang Xiaoying[1,2]

(1. Technology Center, Xingtai Iron and Steel Co., Ltd., Xingtai 054027, China;
2. Wire Engineering Technology Research Center, Xingtai 054027, China)

Abstract: With the increasing size of continuous casting bloom and improvement of the product quality, the macrostructure like center segregation and porosity has become increasingly prominent. The technology of convex roll reduction is widely applied in conticaster and makes a remarkable result on the macrostructure. This paper is researched in detailed on the development and application of convex roll reduction technology in Continuous Casting Bloom of the special steel enterprise including Nippon steel, POSCP, China Steel Corp, and so on. The research shows that china has mastered the key technology of convex roll reduction, with the joint efforts of scientific research institutions represented by Northeastern University and special steel enterprise, and the quality of continuous casting bloom is achieved substantial improvement.

Key words: continuous casting bloom; convex roll; center segregation; center porosity; reduction

3 轧制与热处理

大会特邀报告

分会场特邀报告

炼铁与原料

炼钢与连铸

★ 轧制与热处理

表面与涂镀

金属材料深加工

先进钢铁材料

粉末冶金

能源、环保与资源利用

钢铁材料表征与评价

冶金设备与工程技术

冶金自动化与智能化

建筑诊治

其他

高强抗震螺纹钢连铸连轧短流程轧机研究与应用

戴江波[1]，王保元[1]，刘宏[2]，张业华[2]

（1. 中冶南方武汉钢铁设计研究院有限公司轧钢事业部，湖北武汉 430080；
2. 宝武集团襄樊钢铁重材有限公司，湖北襄阳 441100）

摘 要：宝武集团襄樊钢铁长材有限公司为了降低螺纹钢生产成本，采用了电炉冶炼钢坯直接轧制的短流程生产工艺，本文结合生产实际，利用全球第一家非线性大型有限元 MARC 软件，基于旋转轧辊刚性面接触摩擦引领钢坯运动、钢坯热轧塑性变形的热力耦合有限元法，建立了在 825℃ 至 1050℃ 的钢坯粗轧有限元仿真模型，为短流程无加热炉加热、无槽连续轧制提供了理论计算的手段，提出了襄钢螺纹钢短流程低温小钢坯生产关键在粗轧机轧制力的确定，揭示了其生产线产品质量不稳定的原因，并依据研究分析结果完善了螺纹钢轧制工艺制度。

关键词：短流程；高强螺纹钢；节能减排；无槽；有限元

Application and Research of Short Flow Rolling Mill for the High Strength Seismic Thread Steel Continuous Casting and Rolling

Dai Jiangbo[1], Wang Baoyuan[1], Liu Hong[2], Zhang Yehua[2]

(1. WISDRI Wuhan Iron and Steel Design Institute Co., Ltd., Wuhan 430080, China;
2. Baowu Group Xiangyang Steel Co., Ltd., Xiangyang 441100, China)

Abstract: In order to reduce the production cost of threaded steel, Baowu Group Xiangfan Iron and Steel Long Bar Co., Ltd. adopted the short-process production technology of direct rolling of billet by electric furnace smelting. In this paper, combined with the actual production, the first large-scale non-linear finite element MARC software in the world was used to establish the thermo-mechanical coupled finite element method based on the contact friction of rigid surface of rotating roll leading billet movement and hot-rolling plastic deformation of billet. The finite element simulation model of rough rolling of billet at 825~1050℃ provides a theoretical calculation method for short-process without reheating furnace and continuous rolling without groove. The determination of rolling force of rough rolling mill is the key to the production of low-temperature billet with short-process of thread steel in Xianggang, and the reasons for the unstable quality of products in the production line are revealed. Based on the analysis results, the rolling process of thread steel is improved.

Key words: short process; high strength thread steel; conserve energy and reduce emissions; slotless rolling; finite element

热处理工艺对 9%Ni 钢低温韧性影响的研究

杜 林，朱莹光，侯家平，张宏亮

（鞍钢集团钢铁研究院，辽宁鞍山 114009）

摘 要：研究了不同轧制温度下热处理工艺对 9%Ni 钢低温韧性的影响，利用 X 射线衍射仪、扫描电镜和透射电镜，研究热处理工艺对 9%Ni 钢显微组织及力学性能的影响。基于内耗仪模拟了碳原子配分的过程。结果表明：经 QLT 工艺得到的显微组织以回火索氏体为主，QT 工艺得到的显微组织以回火屈氏体为主，QLT 的显微组织比 QT 的明显细化，Ni 和 Mn 的元素富集区在回火过程生成的逆转变奥氏体，净化基体的三者的共同作用是提高 9%Ni 钢韧性的原因。

关键词：9%Ni 钢；低温韧性；逆转变奥氏体；内耗

Study on the Influence of Heat Treatment Process on Cryogenic Toughness of 9%Ni Steel

Du Lin, Zhu Yingguang, Hou Jiaping, Zhang Hongliang

(Ansteel Group Iron and Steel Research Institute, Anshan, 114009, China)

Abstract: The influence of heat treatment process on the cryogenic toughness of 9%Ni steel at different finishing rolling temperatures was studied. The influence of heat treatment process on the microstructure and mechanical properties of 9%Ni steel was studied by using X-ray diffraction, scanning electron microscopy and transmission electron microscopy. The process of carbon atom partition is simulated based on multi-function internal friction test equipment. Results show that the QLT process of microstructure is given priority to with tempered sorbite and the QT process of microstructure is given priority to with tempered troostite. The microstructure of QLT process is smaller than that of QT process. Reversed austenite is produced in the tempering process of Ni and Mn element enrichment region. The ferrite matrix was purified by QLT process. Those are main reasons to improve cryogenic toughness of 9%Ni steel.

Key words: 9%Ni steel; cryogenic toughness; reversed austenite; internal friction

先进压水堆核电站用 SA-738Gr.B 特厚钢板的开发

胡海洋，孙殿东，胡昕明，王　爽，颜秉宇，欧阳鑫

（鞍钢股份有限公司，辽宁鞍山　114009）

摘　要：通过采用微合金化设计以及合理的生产工艺，成功开发出先进压水堆核电站用 SA-738Gr.B 特厚钢板。性能检测和显微组织分析结果表明：钢板具有良好的室温、高温拉伸性能，弯曲性能和低温冲击性能优异，而且具有较高的抗脆断性能，1/4 厚度处的显微组织为贝氏体回火组织，组织均匀，奥氏体晶粒度为 8.5 级，各项性能指标均满足相关技术要求。

关键词：压水堆核电站；SA-738Gr.B 钢板；特厚板；高强韧钢

Development of SA-738Gr.B Extra-heavy Plate for Advanced PWR Nuclear Power Station

Hu Haiyang, Sun Diandong, Hu Xinming, Wang Shuang, Yan Bingyu, Ouyang Xin

(Angang Steel Company Limited, Anshan 114009, China)

Abstract: Through microalloying design and reasonable production process, the extra-heavy plate of SA-738Gr.B for advanced PWR nuclear power station has been successfully developed.The results of performance test and microstructural analysis show that the steel plate has good tensile properties at room temperature and high temperature, excellent bending properties and low temperature impact properties.It also has high brittle fracture resistance.The microstructure of the studied steel at 1/4 thickness is tempered bainite with homogeneous structure and its austenite grain size is 8.5 grade.All performance indicators meet the relevant technical requirements.

Key words: PWR nuclear power station; SA-738Gr.B steel plate; extra-heavy plate; high strength and toughness steel

20NCD14-7钢最佳热处理工艺研究

王 爽，孙殿东，胡海洋，颜秉宇

（鞍钢集团钢铁研究院，辽宁鞍山 114009）

摘 要：为了研究核电蒸汽发生器支承用20NCD14-7钢热处理工艺同性能之间的关系，对20NCD14-7钢的淬火加回火热处理工艺进行了正交试验设计并优化，采用三水平四因素的试验方案进行了不同组合方式的热处理工艺试验。正交试验的结果表明：影响性能因素的主次顺序为：回火温度>回火时间>淬火时间>淬火温度，并确定了20NCD14-7钢的最佳热处理工艺。

关键词：20NCD14-7钢；淬火；回火；正交试验

Research on Heat-treatment Process for 20NCD14-7 Steel

Wang Shuang, Sun Diandong, Hu Haiyang, Yan Bingyu

(Ansteel Iron & Steel Research Institute, Anshan 114009, China)

Abstract: Orthogonal experiment design of three factors, four levels and nine groups was used during heat-treatment experiment of 20NCD14-7 steel in order to attain the relation between heat-treatment process and properties of 20NCD14-7 steel for SG support on nuclear power plant. The calculation and analysis of variance show that the sequence of influential significance is as follows: temper temperature, temper time, quenching time, quenching temperature. The optimum heat-treatment system was obtained according to orthogonal experiment.

Key words: 20NCD14-7 steel; quenching; tempering; orthogonal experiment

热连轧板"唇印结疤"成因分析与对策

郭 斌

（宝钢股份中央研究院武汉分院，湖北武汉 430080）

摘 要：对热连轧板"唇印结疤"缺陷的形貌、分布以及形成原因进行了系统的分析，并提出了消除措施。"唇印结疤"宏观呈云状疤块，类似唇印，呈网状分布，发生在钢板的上表面，沿长度方向间断、无规律，有时连续分布，在宽度方向时左、时右、时中，在板宽1/4～1/3处居多，随着厚度增加，疤块越明显、严重；微观分析表明钢板表面"唇印结疤"是一种网状裂纹，其形成与连铸坯上的缺陷或Cu、Ni、As在晶界富集有关；系统调查表明连铸坯、加热坯、粗轧后中间坯表面均普遍存在由氧化铁皮下Cu、Ni、As富集产生的奥氏体晶界网状裂纹；钢坯除鳞时一次氧化铁皮未除净或二次氧化铁皮清除不及时，将出现裂纹中Ni、Cu、As富集严重或裂纹表面明显氧化，导致裂纹不能轧合；此外，除鳞喷嘴流量、重叠量大，使钢板表面冷却不均匀，局部过冷部位的裂纹也不能轧合，上述两种情况均导致钢板表面形成网状裂纹——"唇印结疤"；采取更换除鳞冷却水集管和水嘴，整改定宽机异常漏水点，严格执行加热工艺，控制Cu、As等残余元素含量等措施彻底消除了"唇印结疤"。

关键词：热连轧板；结疤；Cu、Ni、As；表面裂纹；除鳞

Formation Analyze and Countermeasure of Lip Print Scab Defect on Hot Rolled Strip

Guo Bin

(Wuhan Branch of Baosteel Central Research Institute, Wuhan 430080, China)

Abstract: The reasons of formation had scientifically researched for lip print scab defect on hot rolled strip, and the measure of solve scab defect had been put forward. The macroscopical features of lip print scab defect are cloudy scars, network-like distribution, similar lipstick, on the upper surface of the strip, irregularly and disjunctive distribution along the length, sometimes continuous distribution along the length, left time right and sometimes middle along width, majority on 1/4-1/3 of width, more obvious and serious as the thickness increases. Through microscope analysis, the results show that the lip print scab defect is surface network cracks, the cause of formation is related to the surface defect of slab or the concentration Cu, Ni, As at grain boundary. Systematic investigation shows that the austenitic reticulate crack produced by Cu, Ni, As enrichment is common below surface oxidizing iron of continuously cast slab, heated slab, roughed slab. When enrichment of Cu, Ni, As in the cracks or the oxidation of cracks is serious because one iron oxide skin has not been eliminated or two iron oxide skin is out of time by descaling, the slab surface network cracks cannot be rolled. In addition, the local overcooling cracks cannot be rolled because uneven cooling of plate surface caused by large flow rate and large overlap of water descaling sprayer. The above two aspect lead to the formation of cracks- lip print scab defect. The lip print scab defect had been thoroughly eliminated by replacing the descaling collection tube and the nozzle, by rectification of the abnormal leakage point of the fixed width machine, by strictly enforcing the heating process, by controlling the content of residual elements such as Cu, As, and so on.

Key words: hot rolled strip; scab; Cu、Ni、As; surface crack; descaling

EDC线材表面黄锈的分析和解决

尹 一，郭大勇，刘 祥，徐 曦，安绘竹

（鞍钢集团钢铁研究院，辽宁鞍山 114001）

摘 要：EDC线材表面产生黄色锈层，通过检验分析确定了锈层的成分，分析了锈层产生的原因和影响因素，提出解决办法，经过系统地改进有关工艺设备条件，消除了EDC线材表面黄锈。

关键词：EDC线材；表面黄锈；影响因素；改进措施

Analysis and Solution of Surface Yellow Rust of EDC Wire Rod

Yin Yi, Guo Dayong, Liu Xiang, Xu Xi, An Huizhu

(Iron and Steel Research Institute of Angang Group, Anshan 114001, China)

Abstract: Yellow rust occurred on the surface of EDC wire rods, the composition of rust layer was determined by test and analysis, the reason of rust layer occurring and affecting factors were studied, and solution method was presented. Through improving relative technique and equipment conditions systemically, the yellow rust on the surface of EDC wire rod was eliminated.

Key words: EDC wire rod; surface yellow rust; affecting factors; improving measurements

厚板轧机关于水梁印问题的分析及优化

陈国锋

(宝山钢铁股份有限公司厚板部，上海 201900)

摘 要： 水梁印是板坯加热的一种常见问题。水梁印问题严重时，对轧制力控制精度影响大，最终造成钢板尺寸和板形不良等问题。一般的厚板轧线控制系统并无专门的对水梁印的温度和轧制力计算，导致水梁印位置的相关计算偏差过大。本文结合厚板轧制过程控制的特点，围绕水梁印对轧制力计算的影响进行了分析，并对水梁印的温度以及轧制力计算进行了优化，提高了轧制力计算精度。

关键词： 水梁印；厚板轧制；轧制力

Analysis and Improvement for Water Beam Mark during Heavy Plate Rolling

Chen Guofeng

(Heavy Plate Department of Baosteel Iron and Steel Co., Ltd., Shanghai 201900, China)

Abstract: Water beam mark is an usual problem during slab reheating. When the water beam mark problem is serious too much, the rolling force calculation accuracy will be influenced, and finally leads to the plate dimension and unflattens problem. Usually there is not any temperature and rolling force calculation specially for water beam mark in heavy rolling line control system, so the calculation error is too much. Basing on the heavy plate rolling process characteristic, the temperature and rolling force calculation for beam mark were analyzed and optimized, and then the rolling force calculation accuracy were improved.

Key words: water beam mark; heavy plate rolling; rolling force

湛江厚板 MULPIC 系统头尾遮挡技术的应用及优化

陈国锋

(宝山钢铁股份有限公司厚板部，上海 201900)

摘　要：TMCP 钢板自然空冷一段时间后容易出现返浪现象。其中钢板头尾的各种长度方向、宽度方向上的翘曲、混合类型的翘曲等最难控制，而且这种浪形问题严重时会对后工序造成严重影响。MULPIC 轧后冷却系统的头尾遮挡技术是解决 TMCP 钢板头尾返浪的重要技术手段。本文对生产现场常见的头尾板形不良问题进行了归纳分析，并基于 MULPIC 头尾遮挡技术提出了多种解决措施。而且针对 MULPIC 头尾遮挡技术，提出了重要功能优化，提高了 MULPIC 系统的头尾遮挡的位置跟踪精度、动态流量精度以及板形控制能力。

关键词：MULPIC；头尾遮挡；板形

The Application and Optimization for Head Tail Masking of Zhanjiang Heavy Plate Mill MULPIC Cooling System

Chen Guofeng

(Heavy Plate Department of Baosteel Iron and Steel Co., Ltd., Shanghai 201900, China)

Abstract: The TMCP plate un-flatness comes again easily after air cooling down. And the plate head and tail un-flatness along length direction and width direction and combination of then is very difficult to solve. This will influence the later production processes if the un-flatness is too much serious. Head tail masking of MULPIC is an important technology method to solve plate head and tail un-flatness problem. The usual unflatness problem was analyzed and concluded in this paper, and various relative solutions were proposed based on the MULPIC head tail masking technology as well. And some function optimizations were proposed for MULPIC head tail masking, and will be good for its tracking position and dynamic flow accuracy and plate flatness control.

Key words: MULPIC; head tail masking; flatness

极限宽厚比淬火钢板研发与应用

韩　钧，付天亮，王昭东，王国栋

(东北大学轧制技术及连轧自动化国家重点实验室，辽宁沈阳 110004)

摘　要：东北大学在喷嘴射流角度、喷嘴宽向流量均匀性、不同淬火方式进行了研究，成功应用到涟钢超薄板淬火机、最小生产厚度 4mm、湘钢超宽板淬火机、最大生产宽度 5m、舞钢特厚板淬火机、最大生产厚度 300mm，使极限宽厚比淬火板的研发与应用取得了较大进步，为我国热处理产品的升级换代做出了贡献，为国家重大建设工程与国防建设所需的热处理产品研发生产奠定了坚实基础。

关键词：超薄；超宽；特厚；淬火机；宽厚比

Development and Application of Quenching Steel Plate with Ultimate Width-thickness Ratio

Han Jun, Fu Tianliang, Wang Zhaodong, Wang Guodong

(State Key Laboratory of Rolling Technology and Automation,
Northeastern University, Shenyang 110004, China)

Abstract: Northeast University has studied the nozzle jet angle, nozzle wide flow uniformity and different quenching methods. It has been successfully applied to Lianyuan Steel ultra-thin plate quenching machine, minimum production thickness 4mm, Xiangtan Steel ultra-wide plate quenching machine, maximum production width 5m, Wuyang Steel ultra-thick plate quenching machine and maximum production thickness 300mm. It has made great progress in the development and application of quenched plate with extreme width-thickness ratio, contributed to the upgrading of heat treatment products in China, and laid a solid foundation for the research and development of heat treatment products needed for major national construction projects and national defense construction.

Key words: ultra-thin; ultra-wide; ultra-thick; quenching machine; width-thickness ratio

基于余弦定理的钢种相似度评估应用研究

王金涛，张志超，杨 军，单旭沂

（宝山钢铁股份有限公司热轧厂，上海 201900）

摘 要：为了更稳定的进行新钢种的拓展试生产，经常根据新试产钢种的成分寻找历史生产实绩中成分最接近的钢种，进行相关工艺的比对，确定合适的工艺参数。论文借鉴基于余弦定理的评估方法，对钢种成分进行定量分析与计算评判，量化评估了钢种成分间的相似度。结果表明，该种方法可有效解决新钢种比对选择的问题，以更合理选择诸如精轧轧制力学习系数的问题，为类似工作的开展提供了拓展思路。

关键词：余弦定理；相似度；钢种；热轧

An Comparability Evaluation Method about Steel Grade based on Cosine Theorem

Wang Jintao, Zhang Zhichao, Yang Jun, Shan Xuyi

(Hot Rolling Plant, Baoshan Iron & Steel Co., Ltd., Shanghai 201900, China)

Abstract: In order to carry out the trial production of new steel grades more steadily, the similar steel grade picked up to analyze in the historical production database play an important role according to the components comparability. And the more similar steel grade is selected, the more relative process can be compared, the more appropriate technological parameters can be determined. In this paper, the evaluation method based on cosine theorem is used to quantitatively analyze and evaluate the composition of the steel grade, and the similarity between the steel grades are evaluated quantitatively. The results show that this method can effectively solve the problem of comparable steel grade selection, for instance, determining more reasonable learning coefficients of finishing rolling force, which provides a new idea for similar work.

Key words: cosine theorem; comparability; steel grade; hot rolling

退火温度对170MPa级IF钢组织和性能影响

王占业，杨　平，李　进

（马鞍山钢铁股份有限公司技术中心，安徽马鞍山　243000）

摘　要： 采用Multipas连续退火模拟机模拟了不同退火温度的IF钢退火实验，研究了退火温度对170MPa级IF钢的组织和力学性能的影响。实验结果表明，退火温度750℃时，试验钢为晶粒未充分再结晶，随着退火温度的升高，试验钢都为完全再结晶组织，铁素体晶粒为8.0级，铁素体晶粒形状为完全等轴状晶粒。随着退火温度的升高，屈服强度和抗拉强度呈降低趋势，A_{80}延伸率、n值和r值均呈上升趋势。退火温度高于780℃时，IF钢具有较好的综合力学性能。

关键词： 退火温度；IF钢；组织性能

Effect of Annealing Temperature on Microstructure and Properties of 170MPa IF Steel

Wang Zhanye, Yang Ping, Li Jin

(Technology Center, Maanshan Iron & Steel Co., Ltd. , Maanshan 243000, China)

Abstract: The annealing experiments of IF steel at different annealing temperatures were simulated by Multipas continuous annealing simulator. The effect of annealing temperature on the structure and mechanical properties of 170MPa grade IF steel was studied. The experimental results show that the grain of the test steel is not fully recrystallized at 750℃. With the increase of annealing temperature, the test steel is fully recrystallized, the ferrite grain is 8.0 grade, and the ferrite grain shape is completely equiaxed grain. With the increase of annealing temperature, the yield strength and tensile strength decreased, while the elongation, n value and R value increased. IF steel has better comprehensive mechanical properties when annealing temperature is higher than 780℃.

Key words: annealing temperature; interstitial free steel; microstructure and properties

冷连轧机组生产薄规格带钢划伤问题的分析与控制

王　静[1]，孙荣生[2]，辛利峰[1]

（1. 鞍钢股份冷轧厂，辽宁鞍山　114003；2. 鞍钢集团钢铁研究院，辽宁鞍山　114003）

摘　要： 本文主要针对大压缩比薄规格带钢在冷连轧机组高速生产中经常出现的表面划伤缺陷进行了介绍，并详细的分析了产生表面划伤的影响因素，如负荷分配、乳化液润滑、轧辊辊面及测张辊辊面磨削管理等，最后提出了控制薄带钢表面划伤的措施。

关键词：冷连轧机组；薄带钢；划伤

The Analysis and Control of the Scratch Defect of the Thin Strip Produced by TCM

Wang Jing[1], Sun Rongsheng[2], Xin Lifeng[1]

(1. The Cold-Rolling Plant of Ansteel Corporation, Anshan 114003, China;
2. Ansteel Group Iron and Steel Research Institute, Anshan 114003, China)

Abstract: This paper mainly introduced the surface scratch of the thin strip with high compression that produced by the tandem cold mill with high speed. Then it analyzed the influenced factor detailed, such as load distribution, emulsion lubricating, roll surface and tension roll surface management. At last it brought out the control measurements for the surface scratch defect.

Key words: the tandem cold rolling mill; thin strip; scratches

耐候桥梁钢板 Q345qNH/Q370qNH 的试制开发

向 华[1]，王敬忠[2]，秦 军[1]，陈晓山[1]，赵 虎[1]

（1. 新疆八一钢铁股份有限公司，新疆乌鲁木齐 830022；
2. 西安建筑科技大学，陕西西安 710055）

摘 要：随着桥梁建筑行业对桥梁结构用钢材的耐腐蚀性能要求越来越高[1]，为了提高钢结构桥梁使用寿命，耐候桥梁钢得到了越来越广泛的应用。通过与西安建筑科技大学合作并结合八钢公司多年来在 Q355NH 耐候钢与 Q345qE 桥梁钢的生产经验基础上，八钢公司成功开发了 Q345qNH/Q370qNH 耐候桥梁钢，产品质量满足国家标准要求并具备充足的力学性能富余量。

关键词：Q345qNH 钢板；控制轧制；低温韧性；耐候性

Development of Weather-resistant Bridge Steel Q345qNH/Q370qNH

Xiang Hua[1], Wang Jingzhong[2], Qin Jun[1], Chen Xiaoshan[1], Zhao Hu[1]

(1. Xinjiang Bayi Iron and Steel Co., Ltd., Urumchi 830022, China;
2. Xi'an University of Architecture and Technology, Xi'an 710055, China)

Abstract: As the bridge construction industry has higher requirements for the corrosion resistance of steel for bridge structures, in order to improve the service life of the bridge, weather-resistant bridge steel has been more and more widely used. Based on the cooperation with Xi'an University of Architecture and Technology and the production of Q355NH weathering steel and Q345qE bridge steel by Bayi Steel Co., Ltd., Bayi Steel successfully developed Q345qNH/Q370qNH weathering bridge steel. The product quality meets the national standard requirements and has sufficient performance the amount.

Key words: Q345qNH steel plate; controlled rolling; low temperature toughness; weather resistance

12MnNiVR 钢板拉伸试样分层开裂原因分析

欧阳鑫，胡昕明，王　储，胡海洋，孙殿东，李广龙

（鞍钢集团钢铁研究院，辽宁鞍山　114009）

摘　要：利用金相观察，扫描电镜，能谱分析等手段，研究了 12MnNiVR 钢轧后拉伸试样中的夹杂物和显微组织，分析产生拉伸试样分层现象的原因。结果表明，12MnNiVR 钢在连铸坯凝固的过程中钢坯中心出现以合金元素 C、Mn 为主的偏析，造成轧后钢板中存在大量的长条状 MnS 夹杂物和条带状回火索氏体组织，组织的不均匀致使试样在拉伸过程中产生局部的应力集中，进而出现开裂分层现象，导致力学性能下降。

关键词：12MnNiVR 钢；夹杂物；显微组织；分层现象

Analysis in Delamination and Cracking of 12MnNiVR Steel Tensile Sample

Ouyang Xin, Hu Xinming, Wang Chu, Hu Haiyang, Sun Diandong, Li Guanglong

(Anshan Iron and Steel Group Steel Research Institute, Anshan 114009, China)

Abstract: The inclusion and microstructure in rolled tensile specimen of 12MnNiVR steel were studied with optical metallograph, SEM, energy dispersive spectrum analysis on the element distribution and so on to explore the cause for laminating in tensile specimen. The results show that the segregation of main microalloying elements such as C and Mn in the continuous cast billet of 12MnNiVR steel will result in a great number of MnS streaky inclusion and ribbon tempered sorbite existing in the rolled specimen, and as a resuit, the stress of the sample is concentrated in local area during the tensile process, consequently the crack and lamination will be induced along crystal plane where the defects concentrate.

Key words: 12MnNiVR steel; inclusion; microstructure; lamination

IF 系列钢再结晶退火时选择性氧化的研究

卢秉仲，周宏伟

（本钢板材技术研究院钢轧工艺所，辽宁本溪　117000）

摘　要：热浸镀锌板的镀锌层表面质量好坏完全取决其表面因素，例如选择性氧化、晶界的密度。时至今日，带钢表面微观结构的研究体系也日趋成熟，利用微观结构知识体系相关理论，完全有能力控制好镀锌层的微观结构。IF TiNb 钢在镀锌前，考虑带钢表面物理因素的同时，研究并分析退火炉的露点对带钢表面选择性氧化的影响。依照带钢反应的情况和带钢表面状态得出相关结论。通过在退火后对带钢各种化学元素的检验，如 B，Si 和 Mn 等。研究分析了带钢表面以及包括带钢深度方向上，化学组分元素的影响。

关键词：IF 钢；露点温度；选择性氧化；化学成分

Research of Selective Oxidation in IF-Series Steel during Annealing

Lu Bingzhong, Zhou Hongwei

(BX STEEL Co., Ltd., Benxi 117000, China)

Abstract: Galvanized coating quality is highly dependent of steel surface parameters such as selective oxidation and the density of grains boundaries. Today, the knowledge of the starting microstructure of the steel surface is established, and the prediction ability is adequate to control the galvanized coatings microstructure. The aim of this paper is to give the keys of this understanding. In the first part of this paper we analyse the effect of annealing dew-point on selective oxidation prior to galvanizing in the case of IF TiNb steel. The physical parameters of the steel surface are also taken into account. The results are discussed in terms of steel reactivity. In the second part, we study the effect of the chemical composition on the steel surface and also in steel thickness after annealing by examining steels with various composition in B, Si and Mn.

Key words: IF steel; DP temperature; selective oxidation; chemical component

鞍钢冷轧连退低碳钢性能影响因素研究

郭洪宇，王　越，付　薇，刘英明

（鞍钢股份有限公司产品发展部，辽宁鞍山　114021）

摘　要： 鞍钢冷轧 3#线、4#线建设连续退火生产线后，在低碳钢性能方面，与罩式炉存在较大差异，在用户使用过程中出现了一些问题。通过对连退工艺特点的分析及查询相关论文，以及到先进企业德国蒂森、韩国现代走访学习，找到了影响连退低碳钢性能的原因。低碳钢的性能与化学成分、热轧工艺、冷轧、连退工序均紧密关联，通过调整化学成分及热轧工艺，是最有效的调整性能的手段，其次是冷轧压缩比，而以往认为的影响较大的连退温度，实际对低碳钢性能的影响较小，本文从带钢特性、化学成分、热轧工艺制度、冷轧压缩比、退火温度等角度，充分分析了各影响因素对连退低碳钢性能的造成的影响，为连退低碳钢性能调整做出了指导。

关键词： 低碳钢；连续退火；冷轧板；性能；工艺控制

Q235B 焊管开裂原因分析

张爱梅

（宝钢集团八钢公司制造管理部，新疆乌鲁木齐　830022）

摘　要： 通过对断口宏观检查、断口微观分析、化学成分分析、金相组织检验、非金属夹杂分析等手段、探讨分析碳素结构钢 Q235B 在制作焊管过程中出现裂纹原因。结果表明：焊接过程中形成焊接方向应力集中，是焊管断裂的主要原因；焊接过热区出现的异常粗大魏氏组织，导致焊缝脆化，促进了焊管的断裂。

关键词： Q235B 钢；焊管；断裂分析

Failure Analysis of Q235B Steel Welded Pipe

Zhang Aimei

(Manufacturing Management Department, Bayi Iron & Steel Co., Baosteel Group, Urumchi 830022, China)

Abstract: Fracture of straight welded pipe was investigated by macroscopical fracture observation, micro-fracture analysis, material compositions analysis, metallographic examination, non-metallic inclusion analysis and finite element analysis. The results show that the fracture mode is brittle. The fracture is mainly caused by stress concentration which resulted from initial crack generating during weld procedure and the toughness decreases due to the Widmanstatten structure existing in weld overheated zone which promotes fracture in some extent

Key words: Q235B steel; welded pipe; failure analysis

平整工作辊崩边原因分析及对策

曹七华，方　俊，刘文杰，唐　华

（宝钢股份武钢有限冷轧厂，湖北武汉　430080）

摘　要： 轧辊是平整机的重要部件，本文主要研究某平整机工作辊在轧制时的崩边问题，找到真正原因，采取措施进行控制，确保生产顺行。

关键词： 连退平整机 SPM；工作辊崩边；倒角修复；着色探伤

Analysis and Solution of Work Roll Side Damage for SPM

Cao Qihua, Fang Jun, Liu Wenjie, Tang Hua

(Baosteel Wuhan Iron and Steel Co., Ltd., Cold Rolled Sheet Plant, Wuhan 430080, China)

Abstract: The rolls are very important parts of SPM, in this paper, the edge breakage of a work roll of SPM during rolling is studided. Find the real reason, take measures to control, ensure production goes smoothly.

Key words: SPM; work roll side damage; chamfering repair; dye penetrant inspection

热轧标签打印焊挂系统

窦　刚，蔡　炜

（中冶南方工程技术有限公司，湖北武汉　430081）

摘　要：热轧标签打印焊挂系统以六轴工业机器人为基础，设计开发了焊钉整理供料机构、标签打印供料机构、焊钉标签抓取焊挂机构，能够将散乱的焊钉整理整齐后单个输出、打印输出标签、自动抓取标签及焊钉并将标签定位焊挂在成捆棒材的端面。系统采用 3D 视觉传感器获取待焊挂部位的三维数据，通过视觉检测算法判断选择合适的焊挂点，可对棒材、型钢、高线等多种热轧产品进行标签焊挂。整个系统技术先进，运行可靠，维护方便。

关键词：机器人；标签；焊挂；视觉

17CrNiMo6 钢的等温正火工艺研究

相　楠

（河钢集团石家庄钢铁有限责任公司技术中心，河北石家庄　050031）

摘　要：本文对 17CrNiMo6 渗碳钢进行了不同热处理工艺试验，分析奥氏体化后不同冷却方式及等温时间对硬度及组织的影响。采用金相法检测边部及心部组织，采用硬度分析检测不同等温正火工艺后硬度。试验结果表明：采用 930℃保温 2h 风冷 700℃后 680℃保温 4h，边部和心部金相组织为铁素体和球状珠光体及片状珠光体，硬度为 166HBW。

关键词：高速齿轮；等温正火；渗碳钢；组织

Study on Isothermal Normalizing of 17CrNiMo6

Xiang Nan

(Shisteel Company of Hesteel Group, Shijiazhuang 050031, China)

Abstract: In this paper, different heat treatment experiments were carried out on 17CrNiMo6 carburized steel, and the effect of different cooling methods and isothermal time after austenitization on hardness and structure were analyzed. Metallographic method was used to inspect structure of the edge and core, and hardness was detected after different isothermal normalizing processes. The results show that the metallographic structure of the edge and core is Ferrite+spherical Pearlite+flake Pearlite, and the hardness is 166 HBW.

Key words: high speed gear; isothermal normalizing; carburized steel; metallographic structure

一种冷连轧机轧制油的选择方法

贾生晖，赵利明

（天津市新宇彩板有限公司，天津　300382）

摘　要：冷连轧机选择轧制油要综合考虑工艺装备、产品大纲、工艺要求、产品要求，轧制油不仅要满足工艺技术要求，还要符合环保，具有经济适用的特点。

关键词：轧制油；工艺；装备；产品；环保；经济

Choice Method of Rolling Oil for Tandem Cold Rolling Mill

Jia Shenghui, Zhao Liming

(Tianjin Xinyu Color Coated Board Co., Ltd., Tianjin 300382, China)

Abstract: The selection of rolling oil for tandem cold rolling mill should consider comprehensively the process equipment, product outline, process requirements and product requirements. The rolling oil should not only meet the process technical requirements, but also conform to the environmental protection and has the characteristics of economy and applicability.

Key words: rolling oil; technology; equipment; products; environmental protection; economics

低淬火开裂的锚具钢开发

陈定乾

（湘潭钢铁有限公司，湖南湘潭　411101）

摘　要： 从锚具的淬火开裂试样分析，不能准确判断开裂原因，但是从裂纹宏观和微观形态来看，明显的淬火裂纹，淬火开裂均发生在应力比较集中的位置。由于不能确定裂纹源，无法分析其具体产生的原因。从锚具生产企业的反馈，开发之前的40CrM开裂比例约为2%~3%，比例较大，严重影响生产效率，需要开发低淬火开裂的锚具。本文通过从冶炼、轧制方面控制入手，降低硫化物、氮、氢、氧含量和棒材细化晶粒、消除应力。热顶锻后试样上不得有裂口和裂缝。这种低淬火开裂锚具钢棒材加工成锚具后，淬火开裂的比例为0.01%以下。

关键词： 低淬火开裂；锚具钢；40CrM

The Development of Anchor Steel with Low Quenching Cracking

Chen Dingqian

(Xiangtan Iron and Steel (XISC) Co., Ltd., Xiangtan 411101, China)

Abstract: Through the analysis of quenching cracking of anchorage specimens, the cause of cracking cannot be accurately judged. From the perspective of the macro and micro morphology of cracks, obvious quenching cracks and cracking occur at stress concentration locations. Since the source of cracking cannot be determined, it is impossible to analyze the specific cause. The feedbacks from the anchor manufacturers show that the carcking ratio of 40CrM before development is about 2%~3%. The large proportions of carcking ratio seriously affect the production efficiency. Therefore, it is necessary to develop anchors with low quenching cracking. This paper proceeds with control from smelting and rolling, reduces the content of sulfide, nitrogen, hydrogen and oxygen, refines grain fineness and relieves stress. No cracks are allowed on the specimen after hot forging. When this anchor steel bar with low quenching cracking is processed into the anchor, the proportion of quenching cracking fall below 0.01%.

Key words: low quenching cracking; anchor steel; 40CrM

B、Ti 元素对 SPHC 酸洗板屈服平台的影响研究

李 霞[1]，岳重祥[1]，李 冉[2]，李化龙[1]

（1. 江苏省（沙钢）钢铁研究院，江苏张家港 215625；
2. 江苏沙钢集团有限公司总工办，江苏张家港 215625）

摘 要：采用工业试验方法，研究了 B、Ti 对 SPHC 热卷板组织性能的影响。在常规 SPHC 钢基础上添加微量 Ti，可以析出大量富 N 的 Ti(N, C)，减少间隙原子 C、N 的数量，减弱屈服平台现象。但晶粒细化，不利于屈服平台的消除。添加微量 Ti 不能避免热卷板 SPHC 在推拉式酸洗线开卷时产生横折印缺陷。在常规 SPHC 钢基础上同时添加微量 Ti 和微量 B，可以在析出富 N 的 Ti(N, C)后进一步形成 BN，进一步减少间隙原子 C、N 的数量，减弱屈服平台现象；同时，钢板的组织和强度与常规 SPHC 钢基本一致。

关键词：SPHC；屈服平台；横折印；夹杂物

Effects of B, Ti Elements on the YPE of SPHC P.O. Plate

Li Xia[1], Yue Chongxiang[1], Li Ran[2], Li Hualong[1]

(1. Institute of Research of Iron and Steel, Shasteel, Zhangjiagang 215625, China;
2. Chief Engineer Office, Jiangsu Shasteel Group, Zhangjiagang 215625, China)

Abstract: Industrial tests were performed to study the influence of B and Ti on the microstructure and properties of SPHC pickled and oiled plate in this paper. After adding trace element Ti to the conventional SPHC steel, Ti(N, C) precipitates enriched with N are formed to reduce the amount of interstitial atoms C and N, and the YPE (Yield Point Elongation) weakens. However, grains are also refined that is harmful to the elimination of YPE. The addition of trace element Ti into SPHC hot rolled plate can not avoid the occurrence of cross break defects during decoiling process in the push-pull pickling line. After adding trace elements Ti and B to the conventional SPHC steel, in addition to Ti(N, C) precipitates, precipitation of BN further reduces the amount of interstitial atoms C and N, and the YPE weakens. The microstructure and strength of steel plate are rather the same when compared with the conventional SPHC steel.

Key words: SPHC; yield piont elongation; cross break defect; inclusion

加热炉自动燃烧控制模型的研发与应用

李小新，向永光，亓鲁刚

（河钢唐钢信息自动化部，河北唐山 063000）

摘 要：加热炉燃烧控制技术，就是在各种燃烧工况条件下，找到合理的最佳空燃比，以提高炉温控制精度和加热速度。本模型主要是研究炉内热交换机理包括有关辐射、对流和传导的关系，建立各段温度控制模型；研究压力、温度的滞后效应，燃烧热值的波动、轧线轧制节奏的变化对加热炉温度控制的影响。结合模糊控制、预测控制等先

进控制算法，建立加热炉控制模型，开发加热炉自动燃烧系统，实现温度自动控制，减少人为失误；提高控制精度，优化炉温曲线，提高加热质量与稳定性；进一步优化空燃比，提高加热炉热效率，实现节能降耗。

关键词：加热炉；空燃比；温度控制模型；模糊控制

Research and Application of Automatic Combustion Control Model for Heating Furnace

Li Xiaoxin, Xiang Yongguang, Qi Lugang

(Information and Automation Department of Tangshan Iron and Steel Co., Ltd., Tangshan, 063000, China)

Abstract: The combustion control technology of heating furnace is to find a reasonable optimal air-fuel ratio under various combustion conditions in order to improve the accuracy and heating speed of furnace temperature control. The main part of this model is to study the heat exchange mechanism of the furnace, including the relation between radiation, convection and conduction, and establish the temperature control model of each section; the effect of pressure, the hysteresis effect of the temperature, the fluctuation of the heat value of the combustion and the change of the rolling rhythm of the rolling line on the temperature of the heating furnace are studied. Combined with the advanced control algorithms such as fuzzy control and predictive control, the control model of the heating furnace is established, the automatic combustion system of the heating furnace is developed, the automatic temperature control is realized, human error is reduced, the control precision is improved, the furnace temperature curve is optimized, and the heating quality and the stability are improved; the air-fuel ratio is further optimized, the heat efficiency of the heating furnace is improved, and the energy-saving and consumption.

Key words: heating furnace; air-fuel ratio; temperature control model; fuzzy control

GCr15 线材网状碳化物控冷工艺研究

熊钟铃

（华菱湘钢销售部，湖南湘潭 411100）

摘 要：通过实验对比，在不同的吐丝温度及风机和辊道速度控制条件下，选取合适的工艺参数。避免 GCr15 线材轧后网状碳化物的大量析出，进而控制碳化物析出的级别，使 GCr15 线材网状碳化物能满足客户要求，提高产品质量。

关键词：轴承钢线材；吐丝温度；风机风量；网状碳化物

冷轧板表面斑迹控制技术研究

张建波，刘 坤，张 涛，吴首民

（浙江联鑫板材科技有限公司，浙江杭州 310014）

摘　要：冷轧板表面缺陷与乳化液的性能指标有关，本文通过分析造成板表面缺陷的发生原因，通过对乳化液系统的改造，加强机组的管理措施，使得乳化液的性能指标回归到正常的水平，从而有效控制板表面缺陷的发生率，提升了板表面的清洁程度。

关键词：冷轧；乳化液；表面缺陷；清洁度

Surface Cleanliness Control Technology for Cold Rolling Steel

Zhang Jianbo, Liu Kun, Zhang Tao, Wu Shoumin

(Zhejiang Lianxin Steel Plate Technology Co., Ltd., Hangzhou 310014, China)

Abstract: The surface defect of cold rolling steel is related to the performance of emulsion. In the paper, based on the analysis of the cause of the surface defect, the performance index of emulsion has been returned to the normal level by strengthens the management and reforming the emulsion system. Thus effectively control the incidence of surface defects of the steel, improving the cleanliness of the steel surface.

Key words: cold rolling; emulsion; surface defects; cleanliness

38MnSiVS 非调质钢的组织性能预测模型

卢世康，陈雨来，余　伟

（北京科技大学工程技术研究院，北京　100083）

摘　要：本文以 38MnSiVS 中碳非调质钢为研究对象，采用光学显微镜（OM）、扫描电子显微镜（SEM）、透射电镜（TEM）和定量金相方法，确定各样品中铁素体的铁素体体积分数 f_a，铁素体晶粒尺寸 d_a 以及珠光体片层间距 S_0。热力学计算第二相 V(C,N) 体积分数约为 0.002，TEM 下观察了析出物粒子的形貌，第二相 V(C,N) 粒子尺寸大多在 4~8nm 范围内。通过数值回归方法，建立材料组织-性能预测模型的物理冶金模型。其中对屈服强度预测模型误差控制在±3.0%以内，对抗拉强度预测模型误差控制在±3.5%以内，室温下冲击功预测模型误差在±5.0%以内。

关键词：非调质钢；第二相；沉淀强化；组织-性能预测模型

Prediction Model of Microstructure and Properties of 38MnSiVS Non-quenched and Tempered Steel

Lu Shikang, Chen Yulai, Yu Wei

(Engineering Technology Institute, University of Science and Technology Beijing, Beijing 100083, China)

Abstract: In this paper, the 38MnSiVS medium carbon non-quenched and tempered steel was taken as the research object, the ferrite volume fraction f_a, ferrite grain size d_a and pearlite lamellar spacing S_0 were determined by Optical Microscopy (OM), Scanning Electron Microscopy (SEM), Transmission Electron Microscopy (TEM) and quantitative metallography. The volume fraction of V (C, N) is about 0.002 by thermodynamics. The morphology of precipitates is observed by TEM.

The size of V (C, N) is mostly in the range of 4~8 nm. The physical metallurgical model of material microstructure-property prediction model was established by numerical regression method. The error of yield strength prediction model is within ±3.0%, tensile strength prediction model is within ±3.5% and absorbed energy prediction model is within ±5.0% at room temperature.

Key words: non-quenched and tempered steel; second phase; precipitation strengthening; structure-property prediction model

高碳钢盘条 72A 表面氧化铁皮控制研究

王海宾[1]，于 聪[2]，贾建平[1]，曹光明[2]，沈俊杰[1]，刘振宇[2]

（1. 河钢集团宣钢公司，河北宣化 075100；2. 东北大学轧制技术及连轧自动化国家重点实验室，辽宁沈阳 110819）

摘 要： 本文通过氧化增重实验对高碳钢盘条（72A）高温氧化行为进行了系统研究，分析了不同温度条件下的氧化动力学行为，并采用热模拟实验机研究了不同吐丝温度下氧化铁皮的演变规律，同时采用场发射电子探针表征氧化铁皮厚度及断面微观形貌。实验结果表明，72A 在低温区 1050～1150℃内氧化增重曲线遵循直线规律，在高温区 1200～1250℃内遵循抛物线规律；吐丝温度的升高，增加了氧化铁皮总厚度，促进了 FeO/Fe_3O_4 界面上 Fe_3O_4 的还原反应，使 FeO 层所占比例增加，原始 Fe_3O_4 含量减小；根据实验室基础理论研究，制定了最优吐丝温度，提高盘条表面氧化铁皮的除鳞性能，满足了下游用户的使用需求。

关键词： 高碳钢；氧化铁皮；氧化动力学；吐丝温度

Study on Controlling of Surface Oxide Scale of High Carbon Steel Wire Rod 72A

Wang Haibin[1], Yu Cong[2], Jia Jianping[1], Cao Guangming[2], Shen Junjie[1], Liu Zhenyu[2]

(1. HBIS Group Xuangang Company, Xuanhua 075100, China; 2. The State Key Laboratory of Rolling Automation, Northeastern University, Shenyang 110819, China)

Abstract: In this paper, the high-temperature oxidation behavior of high carbon steel wire rod (72A) was systematically studied by oxide weight gain experiment. The oxidation kinetics behavior at different temperature conditions was analyzed, and the thermom-mechanical simulator was used to study the evolution of iron oxide scale at different laying temperatures, while using electron probe microanalysis (EPMA) to characterize the thickness of the scale and cross-sectional microscopic morphology. The experimental results indicated that the oxidation weight gain curve of 72A follows the linear law in the low temperature range of 1050~1150℃, and follows the parabolic law in the high temperature range of 1200~1250℃. The total thickness of the scale increases with the laying temperature, promotes the reduction of Fe_3O_4 on the FeO/Fe_3O_4 interface, increases the proportion of the FeO layer, and reduces the content of the original Fe_3O_4. According to the basic theoretical research of the laboratory, the optimal laying temperature was established and improves the descaling performance of the scale on the surface of the wire rod, which satisfies the requirement of customers.

Key words: high carbon steel; oxide scale; oxidation kinetics; laying temperature

16mm 厚度 Q690CFD 高强煤机钢弯曲不合原因分析

薛如锋

(宝钢湛江钢铁有限公司厚板厂，广东湛江 524000)

摘 要：针对 16mm 厚度的 Q690CFD 高强煤机钢在弯曲试验过程中出现的开裂现象，对缺陷部位材料进行了扫描电镜、金相和化学成分进行分析。结果表明：在弯曲试验过程中，裂纹首先从近外表面存在夹杂物的位置开裂，然后向试样的内部延伸，而形成开口裂缝；材料抗拉强度高，比标准要求高出约 357MPa，在同样弯曲变形量情况下，使近外表面夹杂物受到拉应力大而容易形成开裂源。

关键词：Q690CFD；开裂；夹杂物；强度

Analysis of the Causes of Bending Misalignment of Q690CFD High Strength Coal Machine Steel with 16mm Thickness

Xue Rufeng

(Heavy Plate of Baosteel Zhanjiang Iron and Steel Co., Ltd., Zhanjiang 524000, China)

Abstract: For the cracking phenomenon of the Q690CFD high-strength coal steel with a thickness of 16mm during the bending test, the material of the defect was analyzed by scanning electron microscopy, metallurgy and chemical composition. The results show that during the bending test, the crack first cracks from the position where the inclusions exist on the near surface, and then extends to the inside of the sample to form an open crack; The tensile strength of the material is nearly 357 MPa higher than the standard requirement. In the same bending deformation, the near-surface inclusions are subjected to tensile stress and are more likely to form a cracking source.

Key words: Q690CFD; cracking; inclusions; strength

连退在线平整轧制力异常原因分析及对策

孙超凡[1,2]，王雅晴[1,2]，方圆[1,2]，张宝来[3]，胡小明[3]

(1. 首钢集团有限公司技术研究院，北京 100043；2. 绿色可循环钢铁流程北京市重点实验室，北京 100043；3. 首钢京唐钢铁联合有限责任公司镀锡板事业部，河北唐山 063200)

摘 要：通过对带钢显微组织和力学性能进行检测，对产线运行带速、均热温度等工艺执行情况进行分析，明确了某厂连退产线在线平整轧制力异常偏大的原因。结果表明，连退炉区带速和温度波动大，带钢长度方向部分区域退火再结晶不充分，冷轧遗传的冷硬态纤维组织导致带钢强度偏大，进而造成平整轧制力异常偏大。通过控制炉区带速平稳，匹配合理的温度制度，可以有效解决该类问题，保障生产顺行。

关键词：平整；轧制力异常；原因；控制策略

Cause Analysis and Control Strategy of Online Temper Rolling Force Abnormity in Continuous Annealing Line

Sun Chaofan[1,2], Wang Yaqing[1,2], Fang Yuan[1,2], Zhang Baolai[3], Hu Xiaoming[3]

(1. Shougang Group Co., Ltd., Research Institute of Technology, Beijing 100043, China; 2. Beijing key Laboratory of Green Recyclable Process for Iron & Steel Production Technology, Beijing 100043, China; 3. The Tinplate Department of Shougang Jingtang Iron and Steel Co., Ltd., Tangshan 063200, China)

Abstract: In order to confirm the cause of the on line temper rolling force abnormity in a continuous annealing line, the microstructure and mechanical property of the strip steel were detected, and the running speed and soaking temperature of the strip steel during the annealing process were also analysed. The result showed that due to the large fluctuations of the running speed and soaking temperature of the strip steel, parts of the the strip steel along the rolling direction recrystallized incompletely, thus the fibrous microstructure formed in the previous cold rolling were preserved during the subsequent annealing process, causing the strength of the strip enhanced and the on line temper rolling force increased markedly. Through controlling the smoothness of the running speed of the strip and matching with proper soaking temperature, the problem above can be solved well and industrialized production can be guaranteed.

Key words: on line temper rolling; rolling force abnormity; cause; control strategy

连续退火炉炉辊结瘤类缺陷原因分析及控制措施

李 军，李 源，杨麒冰，周诗正

（宝钢股份武钢有限冷轧厂，湖北武汉 430080）

摘 要： 本文介绍了连续退火炉内炉辊结瘤类缺陷（麻点、鱼鳞纹等）的形貌和产生原因，通过分析炉内气氛恶化是产生炉辊结瘤类缺陷的主要原因，并提出了控制结瘤类缺陷的相关措施。

关键词： 连续退火；炉辊结瘤；缺陷分析；气氛

Cause Analysis and Control Measures of Nodulation Defects in Rollers of Continuous Annealing Furnace

Li Jun, Li Yuan, Yang Qibing, Zhou Shizheng

(Cold Mill of Baosteel Wuhan Iron and Steel Ltd., Wuhan 430080, China)

Abstract: In this paper, the shape and causes of nodulation defects (pitting, fish scale, etc.) of furnace rolls in continuous annealing furnace are introduced. By analyzing that the deterioration of atmosphere in the furnace is the main cause of nodule defects in the roll, the relevant measures to control nodule defects are put forward.

Key words: continuous annealing; nodulation of roller; defect analysis; atmosphere

JMatPro 在核电 SA738 Gr.B 钢热处理工艺设计中的应用

张舒展[1]，李艳梅[1]，杨梦奇[1]，姜在伟[2]，叶其斌[1]

（1. 东北大学轧制技术及连轧自动化国家重点实验室，辽宁沈阳 110819；
2. 南京钢铁股份有限公司，江苏南京 210035）

摘 要：应用 JMatPro 软件对核电 SA738 Gr.B 钢进行了材料特性参数模拟计算和热处理参数的制定。结果表明，核电 SA738 Gr.B 钢的 Ac_1=695℃，Ac_3=807℃，Ms=399℃，淬透性能良好，淬火组织主要为贝氏体。在 A_3 温度以上难溶碳化物数量较少，仅有 MC 型碳化物稳定存在；在 670℃回火时会沿晶界析出大尺寸的 $M_{23}C_6$ 型碳化物，严重影响钢的韧性，故回火加热温度应控制在 670℃以下。

关键词：JMatPro 软件；SA738 Gr.B；模拟计算；热处理；碳化物

Application of JMatPro in Heat Treatment Process Design of SA738 Gr. B Steel in Nuclear Power Plant

Zhang Shuzhan[1], Li Yanmei[1], Yang Mengqi[1], Jiang Zaiwei[2], Ye Qibin[1]

(1. State Key Laboratory of Rolling and Automation, Northeastern University, Shenyang 110819, China;
2. Nanjing Iron and Steel Co., Ltd., Nanjing 210035, China)

Abstract: JMatPro software was used to simulate the material characteristic parameters and determine the heat treatment parameters of SA738 Gr.B steel. The results show that Ac_1=695℃, Ac_3=807℃ and Ms=399℃ of the SA738 Gr.B steel which have good hardenability, the quenching microstructure is mainly bainite. There are few insoluble carbides above A_3 temperature. Only MC carbides are stable. When tempering at 670℃, large-sized $M_{23}C_6$ carbides will precipitate out along grain boundary, which seriously affects the toughness of steel. Therefore, tempering temperature should be controlled below 670℃.

Key words: JMatPro software; SA738 Gr.B; simulated calculation; heat treatment; carbide

U 型钢板桩轧制翘曲的有限元分析

杨洋，刘凯，苏磊，吴功军，刘杨

（河钢唐钢型钢厂，河北唐山 063000）

摘 要：以 PU400×170 热轧 U 型钢板桩为对象，采用 ABAQUS 有限元软件对不同孔型压力配比和压下率配置条件下精轧孔型的轧制过程进行仿真计算，分析了 U 型钢板桩轧制翘曲产生的原因，得到了合适孔型压力和压下率配比，解决了 U 型钢板桩轧制翘曲问题。

关键词：U 型钢板桩；有限元；轧制翘曲；压下率配比

Finite Element Analysis of U-sheet Pile Rolling Warpage

Yang Yang, Liu Kai, Su Lei, Wu Gongjun, Liu Yang

(HBIS Section Steel Plant, Tangshan 063000, China)

Abstract: Aimed at hot rolled U-sheet pile of PU 400×170, simulated the rolling process of finish rolling pass under different pass pressure ratio and reduction ratio with finite element software ABAQUS, analyzed the causes of U-sheet pile rolling warpage, obtained appropriate ratio of pass pressure and reduction ratio, solved the problem of U-sheet pile rolling warpage.

Key words: U-sheet pile; finite element analysis; rolling warpage; reduction ratio

轧钢车间降尘工艺及设备综述

徐言东[1]，韩　爽[1]，王占坡[2]，程知松[1]

（1. 北京科技大学工程技术研究院，北京　100083；2. 冶金自动化研究设计院，北京　100071）

摘　要： 轧钢车间在钢铁生产过程中是实现成材的最后一环，棒材、板带生产线进行轧制时，钢材进、出轧机时温度较高，遇到空气就会发生氧化反应，产生大量的氧化粉尘及烟气，若处理不当，便会漂浮在空气中，会造成一方面降低产品的表面质量和性能；另一方面对操作人员和设备造成损害，污染环境。随着国家对环境保护的重视，国内外相关机构和学者对轧钢车间降尘工艺进行了深入研究，探索最新的技术和工艺。本文综合以往的研究和生产实践，指出目前轧钢车间降尘工艺的发展方向是斯普瑞喷雾降尘系统、BEC 环保抑尘处理工艺设备的应用。

关键词： 轧钢车间；降尘(抑尘)；设备；氧化粉尘

Summary of Dust Suppression Process and Equipment of the Steeling Rolling Plant

Xu Yandong[1], Han Shuang[1], Wang Zhanpo[2], Cheng Zhisong[1]

(1. Engineering and Technology Research Institute, University of Science and Technology Beijing, 100083, China; 2. Automation Research and Design Institute of Metallurgical Industry, Beijing 100071, China)

Abstract: The rolling plant is the last link in the steel production process. When the bar and plate production line is working, the temperature of the steel entering and leaving the rolling mill is high, an oxidation reaction occurs as it encounters the air, and a large amount of oxidized dust and gas are generated. If it is not handled properly, it will float in the air, which will reduce the surface quality and performance of the product on the one hand, also causes damage to operators and equipment and pollutes the environment on the other hand. With the country's emphasis on environmental protection, Related institutions and scholars at home and abroad have conducted in-depth research on dust suppression process in rolling plant, explore the latest technologies and processes. This paper combines previous research and production practices to point out that the current development direction of the dust suppression process in rolling mills is the application of Spree spray dust suppression system and BEC dust environmental protection treatment process equipment.

Key words: rolling plant; dust suppression; equipment; oxidized dust

冷轧处理线卷取机带头定位精度改进

黄海生，胡剑斌

（新余钢铁集团有限责任公司冷轧厂，江西新余 338001）

摘 要：采用带头与卷取机芯轴橡胶套筒定位区动态接触的动态自动定位技术，因不同规格、钢种钢带工艺参数不同，钢带穿带传输过程中在传送辊上的打滑量不同，及传输通道上的下垂量不同，带头到达卷取机芯轴定位区，计算带头位置与实际带头位置存在不稳定的偏差，带头定位精度不稳定。采用改进后的静态带头自动定位技术，穿带过程中预留了足够的钢带打滑及下垂富余量，且带头与定位区中心线在助卷皮带咬入口静止状态下接触，消除了影响带头定位精度的主要因素，带头定位精度能稳定在±15mm 的误差范围内。

关键词：卷取机；带头印；带头自动定位；改进

以高端钢板桩产品为定位的钢板桩系列产品研发

王君珂，杨 洋，苏 磊，吴功军，刘 杨

（河钢唐钢型钢厂，河北唐山 063000）

摘 要：钢板桩是一种广泛应用于建筑基础施工，特别是地基项目水利工程的建筑型材，具有有高强度、轻型、止水性好、耐久性强、施工效率高、占地少等独特的优点，具有广阔的市场前景。河钢唐钢型钢厂对国内外先进钢板桩企业产品特点进行对比，分析了钢板桩型钢的关键技术控制点，并结合现有工艺设备特点，运用有限元分析等方法自主研发 3#、4#钢板桩系列型钢生产孔型系统及工艺。

Research and Development of Steel Sheet Pile Series Products Oriented by High-end Steel Sheet Pile Products

Wang Junke, Yang Yang, Su Lei, Wu Gongjun, Liu Yang

(HBIS Section Steel Plant, Tangshan 063000, China)

宽厚板 3.5m 线镰刀弯刮框的原因分析及预防措施

谢富强，邹星禄，何明涛，乔 坤

（江苏沙钢集团有限公司钢板总厂，江苏张家港 215600）

摘 要：钢铁是建筑的基础，现市场对中厚板的需求日趋上升，而中厚板生产技术也日趋成熟，为保障生产稳定性，各厂对事故的分析和预防都作为重点。本文主要讲解了宽厚板 3.5m 线影响镰刀弯的几个可能原因，同时针对一次镰刀弯过大造成的一次刮框事故进行了系统的剖析，通过分析做出相应的预防措施及解决方法，提升 3.5m 生产指标，减少生产故障。

关键词：3.5m 中厚板；镰刀弯刮框；预防措施

Reasons Analysis and Preventive Measures of 3.5m Wire Sickle Bending Scraper for Wide Plate

Xie Fuqiang, Zou Xinglu, He Mingtao, Qiao Kun

(Jiangsu Shagang Group Co., Ltd., Steel Plate Factory, Zhangjiagang 215600, China)

Abstract: Iron and steel is the foundation of building. The market demand for medium and heavy plate is increasing day by day, and the production technology of medium and heavy plate is becoming more and more mature. In order to ensure the stability of production, every factory focuses on accident analysis and prevention. This paper mainly explains several possible reasons for the influence of 3.5m line on sickle bending of wide and thick plate. At the same time, a scraping accident caused by too large sickle bending is systematically analyzed. Through analysis, corresponding preventive measures and solutions are made to improve 3.5m production index and reduce production failure.

Key words: 3.5m middle thick plate; sickle bending scratch frame; preventive measures

浅谈 3.5m 超快速冷却设备与板形控制

谢富强，何明涛，邹星禄，乔 坤

（江苏沙钢集团有限公司钢板总厂，江苏张家港 215600）

摘 要：本文简单介绍了宽厚板 3.5m 线车间的超快速冷却设备，3.5m 线超快冷设备应用东北大学的二级控制系统，拥有先进的二级模型及自学习能力。本文分析了控冷过程中，由于温度不均匀造成的热应力的变化，因此而产生的板形缺陷问题。研究了冷却过程中为保持板形良好，如何修改超快冷的各项参数设置及各个功能的合理使用，以达到改善板形的目的，提高产品合格率。

关键词：3.5m 中厚板；超快冷；板形控制

Discussion on 3.5m Ultra-fast Cooling Equipment and Plate Control

Xie Fuqiang, He Mingtao, Zou Xinglu, Qiao Kun

(Jiangsu Shagang Group Co., Ltd. Steel Plate Factory, Zhangjiagang 215600, China)

Abstract: This article briefly introduces the ultra fast cooling equipment in the 3.5m line workshop of my wide thick plate, and the 3.5m line ultra fast cooling equipment is applied to the secondary control system of Northeastern University. It has advanced secondary model and self-learning ability. At the same time, the problem of plate defect caused by the change of thermal stress caused by uneven temperature during the cooling control process is analyzed. In order to keep the shape of the plate in good condition, how to modify the parameters of ultra-fast cooling was studied in order to improve the shape and improve the yield of the product.

Key words: 3.5m medium thick plate; ultra-fast cold; plate control

型钢轧辊磨损规律分析与研究

郭 平，张明海，谢海深

（河钢集团宣钢公司，河北宣化 075100）

摘 要：型钢孔型的磨损情况直接关系到轧线生产效率和产品质量，同时对轧机料型的调整、负差的控制等都有很大的影响。所以通过了解和掌握轧制过程中各架次孔型的磨损规律，可以及时对料型进行调整和轧槽更换，并采取相应的措施以达到稳定生产提高轧槽吨位和产品质量的目的。结果表明：在轧制角钢过程中通过采取改善轧辊材质、修改成品孔型、优化成品前轧槽侧壁斜度、提高轧辊加工精度、制定合理的轧槽吨位和压下制度等措施对于降低轧槽的磨损程度和不均匀性起到了关键作用，从而达到提高轧制吨位和成品质量的目的。

关键词：轧辊；角钢；磨损；规律；吨位

Research on Wear of Roll for Section Steel

Guo Ping, Zhang Minghai, Xie Haishen

(HBIS Group Xuansteel, Xuanhua 075100, China)

Abstract: The wear of the type steel pass is directly related to the production. Efficiency and product quality of the rolling line, and it has a great influence on the adjustment of the material type and the contronl of the negative difference. Therefore, by understanding and grasping the wear rules of all pass in the rolling process. We can adjust the material type and replace the grave in time, and take corresponding measures to achive stable production and raise the tonnage and production quality of the groove. The results showed that in the rolling process by improving steel roll material, modify the finished product pass,fininshed before the optimization of groove angle, roll to improve machining precision, making groove tonnage and the pressure of the system and reasonable measures to reduce the wear degree of groove and inhomeogeneity plays a key role,so as to improve the rolling tonnage and produce quality.

Key words: roll; angel; wear; regular; tonnage

柔性支撑技术在矫直机换辊中的应用

栗增杰，李艳辉，李光伟，白世军

（河钢集团邯钢公司中板厂，河北邯郸 056000）

摘 要：针对矫直机换辊装配入位困难的问题，提出柔性自适应装配，将柔性支撑技术应用于矫直机接轴抱紧装置中，设计可调柔性支撑单元，可调柔性支撑单元阈值可调，可根据其受力情况改变自身刚性，降低接轴抱紧装置刚性，增加装配自适应性及精度补偿功能；利用有限元分析软件 ABAQUS，对可调柔性支撑单元进行有限元仿真，仿真结果与设计原理相符，从而验证其理论可行性。

关键词：柔性自适应装配；柔性支撑；调姿补偿；阈值

The Application of Type of Flexible Support in Roll Change of PPL

Li Zengjie, Li Yanhui, Li Guangwei, Bai Shijun

(Hebei Iron and Steel Group Co., Ltd., Hangang Plate Plant, Handan 056000, China)

Abstract: In view of the difficulty of roll changing of leveller, the adaptive principle of flexible assembly was presented. The flexible supporting technology is applied to the shaft clasping device of the straightening machine. The flexible support unit was designed. The threshold value of flexible support unit can be adjusted. According to the force, the flexible support platform can change the rigidity. By reducing the rigidity of the shaft holding device, the adaptability and precision compensation function of the assembly is increased. The finite element simulation of the adjustable flexible support unit is carried out by using the finite element analysis software ABAQUS. The simulation results are in good agreement with the design principle, so as to verify the theoretical feasibility.

Key words: flexible adaptive assembly; flexible support; pose adjustment compensation; threshold value

钢管控制冷却工艺的试验研究

洪 汛，邸 军，谷大伟

（鞍钢股份有限公司无缝钢管厂，辽宁鞍山 114021）

摘 要：为了降低成本，鞍钢无缝厂进行了 80 级钢管控冷工艺的试验研究。结果表明，在合理控制冷却参数后，可提高钢筋的屈服强度 35～70MPa，抗拉强度 40～80MPa，减少了合金成分的添加，而且产品表面质量有所改善。在产品质量满足标准的情况下，试验的产品吨钢成本降低。

关键词：控制冷却；强度；流量；温度；冷却速度

Experimental Study on Controlled Cooling Process of Steel Tubes

Hong Xun, Di Jun, Gu Dawei

(Production Cooperate Center of Anshan Iron and Steel Group Corporation, Anshan 114021, China)

Abstract: In order to reduce costs, the Angang Seamless Factory conducted a pilot study on the 80-stage steel tube controlled cooling process. The results show that after reasonable control of the cooling parameters, the yield strength of the steel bars can be increased by 35～70MPa, the tensile strength is 40～80MPa, the addition of alloy components is reduced, and the surface quality of the products is improved. In the case where the product quality meets the standard, the cost per ton of steel tested is reduced.

Key words: controlled cooling; intensity; flow; temperature; cooling rate

单机架可逆冷轧机的厚控系统升级与优化技术

陈跃华，吴有生，王志军

（中冶南方工程技术有限公司，湖北武汉　430223）

摘　要：为满足大量存量单机架可逆冷轧机厚控系统的优化改造需求，中冶南方自主开发了一套厚度控制升级优化系统。通过对原有轧机一二级系统的升级改造，结合了一级二级系统各自的优势，由二级系统中带自学习功能的高精度工艺模型和大数据分析优化程序为一级系统提供高精度和优化的设定值、效率系数、修正系数等参数，一级系统依托二级的支持，运用模糊自适应技术、加减速预控补偿、厚控张力解耦控制等技术，大幅提高了原有 AGC 系统的控制精度和适应性。

关键词：单机架冷轧机；厚度控制；升级优化

Upgrade and Optimization Technology of RCM AGC System

Chen Yuehua, Wu Yousheng, Wang Zhijun

(WISDRI Engineering Co., Ltd., Wuhan 430223, China)

Abstract: In order to satisfy the need for upgrade and optimization of a large number of RCM AGC systems, WISDRI developed the AGC upgrade and optimization system. The method introduced by this article has take the advantage of basic automation and process control system of RCM. Base on the high precise technology model with self-study function and big data analysis program, the process control system can provide high precise and optimized set-points, efficiency coefficients and correction parameters to basic automation system. With the support from process control system, basic automation system utilize fuzzy self-adaption technology, compensation method in acceleration and deceleration, decouple of AGC and tension control system, etc. With the successful application in a RCM mill, the control accuracy and adaptability of the original AGC system have been greatly improved.

Key words: RCM; AGC; upgrade and optimize

基于图像识别技术的加热炉自动上料系统

蔡　炜[1]，叶理德[1]，吉　青[2]，祝兵权[2]

（1. 中冶南方工程技术有限公司技术研究院，湖北武汉　430081；
2. 中冶南方工程技术有限公司电气自动化设计所，湖北武汉　430081）

摘　要：为了实现加热炉上料自动化，本文针对上料台架的特点和工艺流程设计一套无人化解决方案，在不改动上料台架机械装置的情况下，仅通过增加摄像机等设备，并采用图像识别技术来实现加热炉上料工序的智能化改造。

关键词：加热炉上料；图像识别；自动控制

Ti-IF 钢铁素体轧制实践及工艺探讨

蔡 珍，梁 文，汪水泽，何龙义

（宝钢股份中央研究院武汉分院（武钢有限技术中心），湖北武汉　430080）

摘　要：本文研究了 Ti-IF 钢铁素体轧制热轧产品的组织、力学性能、氧化铁皮、析出物和织构特征，并对铁素体轧制工艺要点进行探讨，结果表明：铁素体轧制具有晶粒粗大、强度低、塑性好、氧化铁皮薄、γ 织构强、析出物弥散细小等特点，但铁素体轧制容易产生组织、取向不均匀的问题，对钢卷下线后的冷却和润滑轧制提出更高的要求。

关键词：铁素体轧制；热轧产品；织构；润滑轧制；组织性能均匀性

Practice and Discussion of Ferrite Rolling in Ti-IF Steel

Cai Zhen, Liang Wen, Wang Shuize, He Longyi

(Wuhan Branch of Baosteel Central Research Institute (R&D Center of Wuhan Iron & Steel Co., Ltd.), Wuhan 430080, China)

Abstract: Characteristics of ferrite rolling were studied in Ti-IF steel, including microstructure, mechanical properties, oxide scales, precipitates and textures, and the key process technologies were discussed. The results revealed that ferrite rolling have the characteristics of coarser ferrite grain size, lower strength, better plasticity, stronger γ texture and finer precipitates. However, problems of uneven microstructure in length and width directions of coils and uneven microstructure, texture in thickness direction could easily occur in ferrite rolling, thus cooling after coiling and rolling lubrication should be paid more attention.

Key words: ferrite rolling; hot-rolled products; texture; rolling lubrication; homogeneity of microstructure and mechanical properties

无缝钢管某 PQF 连轧机组扩大产品尺寸范围初步探讨

李琳琳，郭海明，张 尧，张卫东，杨永海

（鞍钢股份有限公司无缝钢管厂，辽宁鞍山　114200）

摘　要：为扩大产品规格范围，在 177PQF 连轧管机组原有孔型的基础上，以生产 121×27 DZ65 产品无缝钢管为例，介绍如何充分利用现有工具，不增加新工具准备成本的基础上拓宽既有的产品组距，且在壁厚范围拓展的难点、改进方案方面做了充分的分析和预估，以脱管机-微涨减孔型为改进重点，且成功试生产，为该机组的产品规格组距进一步拓展做了有益的尝试。

关键词：PQF 连轧管机；脱管机；壁厚范围；孔型设计

Preliminary Discussion on the Range of Product Size of a PQF Continuous Rolling Mill for Seamless Steel Tubes

Li Linlin, Guo Haiming, Zhang Yao, Zhang Weidong, Yang Yonghai

(Ansteel Co., Ltd., Seamless Steel Tube Plant, Anshan 114200, China)

Abstract: In order to expand the product specification range, based on the original hole type of the 177PQF continuous rolling mill unit, the production of 121×27 DZ65 seamless steel pipe is taken as an example to introduce how to make full use of the existing tools without adding new tool preparation costs. Widening the existing product group distance, and making full analysis and estimation in the difficulty and improvement of the wall thickness range, taking the de-pipe machine-micro-increase and reduction type as the improvement focus, and successfully trial production, the product specification group of the unit has been further expanded to make a useful attempt.

Key words: PQF continuous rolling mill; pipe stripping machine; wall thickness range; hole design

优化 177PQF 脱管机孔型

郭海明，张　尧，李琳琳，张卫东，杨永海

（鞍钢无缝钢管厂，辽宁鞍山　114001）

摘　要：孔型设计是 PQF 限动连轧管机的核心技术，孔型参数的选择将直接影响到轧制过程是否顺利和产品的几何尺寸精度、表面质量。本文介绍了鞍钢无缝厂177PQF 限动连轧管机组因品种开发及成本管控，造成薄壁管增多脱管机频繁出现青线困扰，通过优化脱管机孔型，减少了脱管机青线的产生，提高产品的表面质量，使生产顺利进行的成功经验，为存在同类问题的其他机组提供一些参考经验。

关键词：青线；脱管机；孔型

Optimization of Extractor Caliber in 177 PQF Seamless Tube Unit

Guo Haiming, Zhang Yao, Li Linlin, Zhang Weidong, Yang Yonghai

(Ansteel Co., Ltd., Seamless Steel Tube Plant, Anshan 114001, China)

Abstract: Pass design is the core technology of PQF restricted continuous rolling mill. The selection of pass parameters will directly affect the smooth rolling process and the geometric accuracy and surface quality of products. This paper introduces the successful experience of 177PQF limited continuous rolling mill in seamless plant of Anshan Iron and Steel Co., Ltd. in the production of thin-walled pipes, because of variety development and cost control, the green line troubles frequently occur in the increasing stripper of thin-walled pipes. By optimizing the pass of stripper, the production of green line of stripper is reduced, the surface quality of products is improved, and the production is carried out smoothly. Other units of the problem provide some reference experience.

Key words: blue line; pipe removal machine; caliber

京唐公司热轧检查线生产效率提升的研究与实践

伊成志,曹艳生,肖 楠,崔惠民

(首钢京唐钢铁联合有限责任公司,河北唐山 063200)

摘 要:本文通过对影响京唐公司2250mm检查线工作效率因素的详细分析,开发了开卷器运行距离自动定位功能和钢板取样长度的自动控制功能,增加了采样剪剪切过程自动化连锁,实现一键式剪切,优化了取样小车制动模式。系列措施的实施取得了良好效果,检查线工作效率得到了大幅提升。

关键词:热轧;检查线;生产效率;自动化改造

Research and Practice on Working Efficiency Improvement of Hot Rolling Inspection Line of Jingtang Company

Yi Chengzhi, Cao Yansheng, Xiao Nan, Cui Huimin

(Shougang Jingtang United Iron & Steel Co., Ltd., Tangshan 063200, China)

Abstract: Based on the detailed analysis of the factors affecting the working efficiency of the 2250mm inspecting line of Jingtang company, this paper developed the auto-position function of the running distance of the uncoiling device and the auto-control function of the sampling length of the steel plate, increased the automatic interlocking of the shearing process of the sampling shear, realized the one-key shearing, and optimized the braking mode of the sampling trolley. The implementation of a series of measures has achieved good results, inspecting line working efficiency has been greatly improved.

Key words: hot rolling; inspection line; working efficiency; automated transformation

攀钢热连轧普碳钢低温轧制技术研究与应用

肖 利[1],陈 永[1],方淑芳[2],张中平[2],刘 勇[2],文永才[1]

(1. 攀钢集团研究院有限公司,四川攀枝花 610031;
2. 攀钢集团攀枝花钢钒有限公司,四川攀枝花 610031)

摘 要:介绍了攀钢热连轧实现普碳钢低温轧制的技术路线并根据该技术路线的指引对实现普碳钢低温轧制的设备能力校核、粗轧温度优化控制、精轧温度优化控制的关键过程进行了分析。这种低温轧制技术的应用,普碳钢晶粒直径减小0.8μm,屈服强度提高了25MPa,抗拉强度提高了18MPa,延伸率降低1.2%,力学性能满足标准要求。普碳钢低温轧制技术的应用,取得了节能、提高成材率、降低生产成本、提高产品质量的良好效果并为后续冷轧料、品种钢低温轧制技术开发奠定了科学基础。

关键词:热轧带钢;低温轧制;温度优化控制;温降

Pan Steel Hot Strip Plain Carbon Steel Low Temperature Rolling Technology Research and Application

Xiao Li[1], Chen Yong[1], Fang Shufang[2], Zhang Zhongping[2], Liu Yong[2], Wen Yongcai[1]

(1. Pangang Group Research Institute Co., Ltd., Panzhihua 610031, China;
2. Panzhihua Steel Group Panzhihua Steel Vanadium Co., Ltd., Panzhihua 610031, China)

Abstract: Realization of plain carbon steel low temperature rolling technology route and according to the technical route of the guidelines for plain carbon steel cold rolling plant capacity checking, roughing rolling temperature optimization control, finish rolling temperature optimization control of the key process are analyzed Pangang hot strip mill is introduced. With the application of the low temperature rolling technology, the grain diameter of plain carbon steel is 0.8mm, the yield strength is increased by 25MPa, the tensile strength is increased by 18MPa, the elongation is decreased by 1.2%, and the mechanical properties meet the standard requirements. Application of plain carbon steel of low temperature rolling technology to save energy, improve yield and reduce production cost and improve the good quality of the products did not follow cold-rolled material, varieties of steel, low temperature rolling and laid a scientific foundation.

Key words: hot rolling strip steel; low temperature rolling; temperature optimal control; temperature drop

低温环境服役大壁厚管件 X80 钢板的研发

张志军[1]，邓建军[1]，龙 杰[1]，高 雅[1]，祝 鹏[2]，程海林[1]

（1. 河钢集团舞钢公司，河南平顶山 462500；
2. 中油管道机械制造有限公司，河北廊坊 065000）

摘 要： 为响应"一带一路"国家发展战略，结合中俄东线天然气管道工程站场低温环境（-45℃）用 D1422 大口径 X80 热挤压三通需求，河钢舞钢联合中油管道机械有限公司成功研制出适合低温环境服役用 X80 管件钢板，中油管道机械公司压制成三通，经中国石油天然气集团公司管材研究所可靠性试验检验，各项力学指标完全符合相关标准，钢板成功应用于中俄东线天然气管道工程。

关键词： 中俄东线；低温环境；管件钢；大口径；高韧；X80

Research and Development of X80 Steel Plate for Large Wall Thickness Pipe Fitting in Low Temperature Service

Zhang Zhijun[1], Deng Jianjun[1], Long Jie[1], Gao Ya[1], Zhu Peng[2], Cheng Hailin[1]

(1. HBIS Group Wusteel Company, Pingdingshan, 462500, China;
2. China Petroleum Pipeline Machinery Manufacturing Co., Ltd., Langfang 065000, China)

Abstract: In response to the national development strategy of One Belt and One Road and in combination with the demand of hot pressed tee fitting of D1422 large diameter X80 grade steel pipe for China-Russia East Natural Gas Pipeline Project in -45℃ low temperature service, HBIS Wusteel and China Petroleum Pipeline Machinery Manufacturing Co., Ltd. successfully develope X80 steel plate for pipe fitting in low temperature service, which is pressed into tee fitting by China

Petroleum Pipeline Machinery Manufacturing Co., Ltd. and tested by CNPC Tubular Goods Research Institute. All mechanical properties fully comply with relevant standards. The steel plate is successfully applied in the China-Russia East Natural Gas Pipeline Project.

Key words: China-Russia east natural gas pipeline project; low temperature environment; pipe fitting steel; large diameter; high toughness; X80

U 型钢孔型设计

安华荣，王 伟，左 岩，卜俊男

（鞍钢股份有限公司大型厂，辽宁鞍山 114021）

摘 要： 文章介绍了 U 型钢的孔型特点，并给出 U 型钢孔型系统设计的方法和基本参数。

关键词： U 型钢；孔型设计

Design Method of U-shaped Steel for Coal Roadway Support

An Huarong, Wang Wei, Zuo Yan, Bu Junnan

(Heavy Section Mill of Angang New Steel Co., Ltd., Anshan 114021, China)

Abstract: In this paper, it is introduced the characteristics of pass designU-shaped Steel for Coal Roadway Support, giving the method of pass design U-shaped Steel for Coal Roadway Support and parameters design.

Key words: U-shaped steel for coal roadway; supportpass design

热连轧带钢温度对轧后板形影响的研究

杨立庆，饶 静，张文兴

（安阳钢铁股份有限公司第二炼轧厂，河南安阳 455004）

摘 要： 针对热连轧带钢轧后板形问题，提出带钢横向温度不均匀分布是造成板形恶化的主要原因，并得出轧后板形总是向边浪发展这一趋势。通过温度凸度的提出，分析了轧后温度变化符合高次方曲线特征。采用板坯直热装、降低冷却速率、合理的堆放方式及控制钢卷出库温度等措施，有效改善轧后板形缺陷。

关键词： 热连轧带钢；轧后板形；温度凸度；内应力

The Research of the Influence of Strip Temperature on Strip Shape after Hot Rolling

Yang Liqing, Rao Jing, Zhang Wenxing

(Anyang Iron and Steel Co., Ltd., Anyang 455004, China)

Abstract: Aim at the problem of strip shape after hot rolling, the main reason of strip shape deterioration caused by uneven distribution of transverse temperature is presented. After rolling, the strip shape always develops to the side wave. The concept of temperature crown is proposed, the temperature change after rolling conforms to the characteristics of high power curve. Some measures such as direct heating, reducing cooling rate, reasonable stacking and controlling the outbound temperature of strip can effectively improve the strip shape after rolling.

Key words: hot rolling strip; strip shape after rolling; temperature crown; internal stress

低碳钢工艺设计对冷板组织的影响研究

李 雯

（梅山钢铁公司技术中心，江苏南京 210039）

摘 要：本文研究了 C 含量为 0.02%及 0.04%两种低碳钢在 1200℃和 1135℃两种出炉温度轧制后热轧卷的金相组织及机械性能的差异，并从 C 含量及二相粒子析出的角度对该种差异进行了分析。出炉温度对 AlN、BN 的析出影响虽未在热卷中直接观察到，但从冷轧板后续退火时碳化物析出形态的明显差异，可间接证明低温出炉促进了 AlN、BN 的析出，而 AlN、BN 的析出又为退火时碳化物的析出提供了更有利的形核质点，从而促进了碳化物在其上的聚集析出。由此，通过对冷轧低碳钢的碳含量、热轧出炉温度及退火温度对冷板金相组织及碳化物析出的分析，可为不同用途的冷轧板选择不同的工艺设计提供借鉴。

关键词：冷轧低碳钢；出炉温度；碳化物；氮化铝

Effect of Low Carbon Steel Procession on Microstructure of Cold Rolling Steel

Li Wen

(Technology Center of Meishan Iron & Steel Co., Nanjing 210039, China)

Abstract: The difference of microstructure and mechanical property of two kinds of hot rolling steel sheets, the former carbon content was 0.02%, the other carbon content was 0.04%, were investigated after respective 1200℃ and 1135℃ tapping temperature and rolling, then the effect of carbon content and the precipitation of second phrase particles on the difference of hot rolling steel sheets were discussed. The effect of tapping temperature on aluminium nitride and boron nitride precipitation weren't observed directly, but which can be proved indirectly from different carbide morphology of corresponding cold rolling annealing steel that lower tapping temperature promoted more aluminium nitride and boron nitride precipitation because aluminium nitride and boron nitride could supply more nucleation sites which promoted carbide precipitation further when cold rolling steel annealed. Therefore, above analysis of carbon content, tapping temperature and annealing temperature on microstructure and carbide precipitation of cold rolling steel could provide some reference for different procession according to different production function.

Key words: low-carbon cold rolling steel; tapping temperature; carbide; aluminium nitride

推拉式酸洗机组工艺段技术改造

陈普，侯元新，吕圣才，刘军燕

（山东泰山钢铁集团有限公司冷轧部，山东莱芜 271100）

摘 要：介绍了泰钢冷轧部推拉式酸洗机组技术特点。分析了造成酸洗钢带板面发黄的主要原因是漂洗槽漂洗效果不良、热风干燥器烘干及钢带边部吹扫效果差，通过对950推拉式酸洗机组漂洗槽的喷淋管、喷嘴进行改造，在热风干燥器风箱内设置上下喷吹装置，将吹扫装置改造为上吹风管可自动升降的结构，解决了酸洗钢带板面发黄的问题，提高了酸洗钢带表面质量。

关键词：酸洗机组；发黄；漂洗槽；热风干燥器；吹扫装置

Technical Reform of Process Section of Push-pull Pickling Unit

Chen Pu, Hou Yuanxin, Lv Shengcai, Liu Junyan

(Shandong Taishan Iron and Steel Group Co., Ltd., Laiwu 271100, China)

Abstract: The technical characteristics of Push-pull Pickling Line in cold rolling section of Taigang are introduced. The main reasons for yellowing of pickling steel strip plate surface are analyzed as follows: poor rinsing effect of rinsing tank, poor drying effect of hot air dryer and poor cleaning effect of steel strip edge. Through reforming spraying pipe and nozzle of rinsing tank of 950 Push-pull Pickling unit, setting up and down spraying device in air box of hot air dryer, the blowing device is transformed into blowing device. The automatic lifting structure of the upper blowing duct solves the problem of yellowing on the surface of the pickling steel strip and improves the surface quality of the pickling steel strip.

Key words: pickling machine; yellowing; rinsing tank; hot air dryer; blowing device

韶钢中棒线轧钢工艺技术创新及优化改造

周小兵，李学保

（中国宝武集团广东韶关钢铁有限公司，广东韶关 512123）

摘 要：韶钢中棒线由于工程设计存在缺陷，自投产以来存在4低：生产效率低，质量控制水平低、成材率低和尺寸精度低。为了改善现状，近年来进行了孔型系统改造、导卫系统改造、剪切工艺优化等一系列创新手段，有效地提高了中棒质量、生产稳定性，大幅度释放特钢产能。

关键词：特钢；孔型改造；导卫改造；剪切工艺

Technological Innovation and Optimum Reform of Medium Bar Rolling in Shaoguan Iron and Steel Co., Ltd.

Zhou Xiaobing, Li Xuebao

(Baosteel Group Shaoguan Iron & Steel Co., Ltd., Shaoguan 512123, China)

Abstract: Defects in engineering design of middle bar line at Shaoguan Iron and Steel Co., Ltd.. Since its commissioning, there have been four low levels: the production efficiency is low, the quality control level is low, low yield and low dimensional accuracy. To improve the status quo, in recent years, a series of innovative means such as roll pass system reform has been carried out, reform of Guidance System, shearing process optimization. The quality and the stability of production of medium rod are improved effectively. Large-scale release of special steel production capacity.

Key words: special steel; pass reformation; guide transformation; shearing process

针对特殊钢棒材控轧控冷工艺的温度场优化

张华鑫，程知松，余 伟，徐言东

（北京科技大学工程技术研究院，北京 100083）

摘 要： 通过对特殊钢棒材控轧控冷工艺过程各个环节传热关系以及边界条件的分析，利用有限差分法建立特殊钢棒材控轧控冷过程的温度场计算模型。并结合生产实践，利用该模型对整个控轧控冷过程进行了温度模拟，得出轧件在同一截面位置上心部、中部和外表面上从出炉到出水箱共18道次的温降曲线，计算结果和实测值吻合较好。

关键词： 特殊钢棒材；控轧控冷；有限差分法；温度场；温降

Temperature Field Optimization for Controlled Rolling and Controlled Cooling Process of Special Steel Bars

Zhang Huaxin, Cheng Zhisong, Yu Wei, Xu Yandong

(Institute of Engineering Technology, USTB, Beijing 100083, China)

Abstract: By analyzing the heat transfer relationships and boundary conditions of special steel bars throughout the controlled rolling and controlled cooling process, and by using the finite difference method to establish the temperature field calculation model before taking it to simulate the temperatures of the whole controlled rolling and cooling process, it is concluded that, on the basis of production practice, there are altogether 18 temperature drop curves of the rolled piece from the furnace to the outlet tank on the core, middle and outer surfaces at the same section position, which accords with the measured values.

Key words: special steel bars; controlled rolling and controlled cooling; finite difference method; temperature filed; falling of temperature

帘线钢珠光体片间距的测定

李珊珊，王 涛，赵 磊

(中信泰富特钢研究院青钢分院，山东青岛 266000)

摘 要：本文采用扫描电镜对某高碳热轧钢棒的片状珠光体进行测量，并依据其不同珠光体区域的碳含量。分为最小含碳量区域间距法、平均含碳量区域片间距法和最大含碳量片间距法三种测量方法对片间距进行测量，并对测量结果进行分析讨论，检测结果表明：最大碳含量—小片间距法的测量方法可操作性强，最大碳含量区域合金元素含量较高，更能反映出该钢的理化属性，可以用来表征珠光体片间距。

关键词：帘线钢；珠光体；片间距；含碳量

Determination of Pearlitic Plate Spacing in Steel Cord

Li Shanshan, Wang Tao, Zhao Lei

(Qinggang Branch of CITIC Pacific Special Steel Institute, Qingdao 266000, China)

Abstract: In this paper, the lamellar pearlite of a high carbon hot rolled steel rod was measured by scanning electron microscope and the carbon content in different pearlite regions was determined. There are three measuring methods: the minimum carbon content area spacing method, the average carbon content area spacing method and the maximum carbon content area spacing method. The results show that the measurement method of the maximum carbon content—small chip spacing method should be operated. The alloy element content in the maximum carbon content area is higher, which can better reflect the physical and chemical properties of the steel, and can be used to characterize the pearlite chip spacing.

Key words: cord steel; pearlite; the spacing; carbon content

时效处理对大形变珠光体钢丝扭转性能的影响

王 猛，黄 波，胡乃志

(中信特钢集团青岛特殊钢铁有限公司，山东青岛 266043)

摘 要：通过扫描电镜、X射线、室温拉伸和扭转试验，研究了大变形量桥梁缆索镀锌钢丝用热轧盘条SWRS82B拉拔变形过程中的组织演变过程和力学性能分析，结果显示：盘条拉拔过程中，随着真应变达到1.8时，钢丝的强度从1265MPa提高到1875MPa；组织中的珠光体沿着拉拔方向呈纤维状，珠光体中盘层间距缩小到50nm。同时模拟镀锌过程进行了盐浴等温时效处理，随着时效时间的延长，拉拔变形的纤维状珠光体组织中的渗碳体相发生熔断和球化；同时随着时效时间延长，钢丝抗拉强度增长到2035MPa，然后又减小到1829MPa，扭转性能从21次降低到6次。

关键词：桥梁缆索；冷拉钢丝；珠光体；时效处理；盐浴；扭转；渗碳体球化

Effect of Annealing Treatment on Torsion Property of Heavily Cold Drawn Pearlitic Steel Wire

Wang Meng, Huang Bo, Hu Naizhi

(Qingdao Special Steel Co., Ltd. of Citic Special Steel Group, Qingdao 266043, China)

Abstract: The evolutions of the microstructure and mechanical properties of the SWRS82B steel wire during cold drawing and subsequent aging treatment at different temperatures was investigated using scanning electron microscope, X-ray diffraction, room-temperature tensile test and torsion test. Experimental results showed that the tensile strength of pearlitic wire was increased from 1265 MPa to 1875 MPa with the increasing drawing strain of 1.8. As the drawing strain was up to 1.8, almost all pearlite lamellae were elongated along the axial direction and the interlamellar spacing was about 50 nm. As the annealing temperature increased, the tensile strength of wires would be increased from 1875MPa to 2035MPa, and then decreased to about 1829 MPa. The twist angle of wire would be decreased from about 21 cycles to only about 6 cycles. The amorphous cementite produced in cold drawing would be transformed into crystalline state. And the crystal size of cementite should be increased as the increase of annealing temperature.

Key words: bridge cable; cold drawn wire; pearlitic; annealing treatment; salt bath; torsional property; spheroidization of cementite

高速线材智能夹送辊控制系统的研究应用

谭秋生，王正洁

（中信特钢集团青岛特殊钢铁有限公司高线厂，山东青岛 266409）

摘 要： 随着冶金行业产品市场竞争激烈，用户对特种线材的性能和质量要求不断提高，高速线材生产线对成品区的质量控制愈加严格，尤其在夹送辊的控制线材的尾部圈型以及成品后堆钢的方面，传统的控制不仅不稳定，而且在出现问题时的可追溯性方面也存在较大的困难。

本文主要阐述了高速线材生产线吐丝机前夹送辊对尾部圈型控制的工艺过程以及控制原理，通过研究发现：控制过程中尾部圈型乱产生的原因主要是减定径机组、夹送辊、吐丝机三者线速度不匹配造成，结合日常生产过程中出现的问题，利用 ABB ACS800 变频器的 DTC 技术和 SIEMENS S120 伺服控制器工艺包软件，对控制程序不断地优化和改进，夹送辊的控制功能趋于完善，使产品的成材率提高的同时，也带来一定的经济效益。

关键词： 夹送辊；伺服控制；PLC；优化；研究

Study on the Automatic Control System of Intelligent Pinch Roll for High Speed Wire Mill

Tan Qiusheng, Wang Zhengjie

(CITIC Qingdao Special Steel High Speed Wire Mill, Qingdao 266409, China)

Abstract: Along with the development of metallurgical industry product market competition, performance and quality as well as end users of special wire requirements continue to increase, the high speed wire rod production line quality of finished area is more and more strict control, especially in the tail ring clamp control wire roll in the heap and finished steel, traditional control not only is not stable, there is great difficulty in dealing with problems traceability.

This paper mainly expounds the domestic and imported high speed wire production line feeding machine front clip production process feed roller and control principle, through the study found: control causes chaos tail ring type process is mainly reducing sizing mill, pinch roll, spinning machine of three line speed does not match. According to the problems in the daily production process, using DTC ABB ACS800 inverter direct torque control technology, and SIEMENS S120 servo controller software package program, the control program is continuously optimized and improved, and the control function of the pinch roll is improved. The yield of the product increase at the same time, also will bring certain economic benefits.

Key words: prich roll; servo control; PLC; optimize; research

基于 800xA 平台的 ABB 线材减定径生产线

王廷雷，谭秋生，王正洁

（中信特钢集团青岛特殊钢铁有限公司高线厂电气作业区，山东青岛　266409）

摘　要： 随着现代化工业技术以及生产工序自动化的迅速发展，对钢材的品种规格、尺寸精度以及性能等都提出了更高的要求。衡量一个国家钢铁生产发展水平的标志就是钢材的轧制技术水平。基于 800xA 平台的线材减定径机线具有传统系统控制无法比拟的优点，同时减定径为热轧生产的核心部分，因此研究热轧机的减定径机组热轧过程具有重要意义。本文介绍了高速线材生产线的主要设备及自动化系统设置，对关键设备的控制原理及其软件实现也作了详细分析。

关键词： 减定径；自动控制；热轧机

ABB Wire Reducing and Sizing Production Line based on the Platform of 800xA

Wang Tinglei, Tan Qiusheng, Wang Zhengjie

(CITIC Qingdao Special Steel High Speed Wire Mill, Qingdao 266409, China)

Abstract: With the rapid development of modern industrial technology and production process automation, it puts forward higher requirements for the steel specifications, size precision and performance. Symbol to measure a country's level of iron and steel development and production is the steel rolling technology level. Wire reducing and sizing mill line based on the 800xA platform has incomparable advantages compared with traditional control system, at the same time reducing and sizing mill is a core part of the hot rolling production, so it is important to research the hot rolling process of hot rolling mill reducing sizing mill. This paper introduces the setting of main equipment and automation system of high speed wire production line, and analyzes the control principle of the key equipment and its software realization in detail.

Key words: reducing and sizing mill; automatic control; hot rolling mill

斯太尔摩风冷线工艺优化对焊接用钢 ER50-6 组织的影响

郝文权，钟 浩

（青岛特殊钢铁有限公司研发中心，山东青岛 266409）

摘 要：通过调整保温罩开启等手段进一步降低盘条出保温罩的温度，以降低盘条在保温罩外发生相变的概率，降低淬火组织等级。

关键词：焊接用钢；淬火组织；搭接点

Effect of Technological Optimization of Stelmore Air Cooling Line on Microstructure of Welding Steel ER50-6

Hao Wenquan, Zhong Hao

(Qingdao Special Steel Co., Ltd., Research Institute of Wire Materials, Qinggang Branch (Technology Center), Qingdao 266409, China)

Abstract: By adjusting the opening of the heat preservation cover and other means, the temperature of the wire rod exiting the heat preservation cover is further reduced, so as to reduce the probability of phase transformation of the wire rod outside the heat preservation cover and reduce the quenching structure grade.

Key words: welding steel; quenched structure; lap joint

水冷式加热炉节能措施的探讨分析

陈海明，胡学民，查安鸿

（宝武集团广东韶关钢铁有限公司特轧厂，广东韶关 512123）

摘 要：韶钢棒一生产线加热炉为水冷式加热炉，加热炉是轧钢生产工序的能源消耗大户，占整个轧钢工序能耗的70%，以韶钢棒一生产线加热炉的为研究对象，分析研究了影响加热炉产品单耗的主要影响因素，并提出相应的节能措施，并有针对性地在钢坯入炉物理热量上提出了相应的解决办法，为加热炉的节能降耗提供了现实有效措施。

关键词：加热炉；煤气单耗；钢坯入炉物理热量；节能措施

Discussion and Analysis of Energy-saving Measures for Water-cooling Reheating Furnace

Chen Haiming, Hu Xuemin, Zha Anhong

(Baowu Group Guangdong Shaoguan Iron and Steel Co., Ltd., Shaoguan, 512123, China)

Abstract: The reheating furnace of SGIS is a water-cooling reheating furnace. The reheating furnace is a major energy consumer in steel rolling process, accounting for 70% of the energy consumption in the whole rolling process. Taking the reheating furnace in the production area of SGIS as the research object, the main factors affecting the unit consumption of reheating furnace products are studied and analyzed, and the corresponding energy-saving measures are put forward. In terms of physical heat, the corresponding solutions are put forward, which provide practical and effective measures for energy saving and consumption reduction of reheating furnace.

Key words: reheating furnace; gas consumption per unit; physical heat of billet charged into furnace; energy saving measures

大棒 1#加热炉二级优化控制系统研究与开发

蒋国强，张宝华，陈建洲

（宝武集团广东韶关钢铁有限公司特轧厂，广东韶关　512123）

摘　要：本文针对宝钢特钢韶关有限公司大棒 1#加热炉二级优化控制系统开展研究工作，开发了钢坯加热过程数学模型，实现了大棒 1#加热炉物料及温度跟踪、工艺查询及修改、历史数据查询和钢坯详细温度记录等功能模块，所开发的二级优化控制系统可对钢坯加热过程进行有效控制，为大棒 1#加热炉钢坯加热质量提高及节能降耗做出贡献。

关键词：大棒 1#加热炉；二级优化控制系统；钢坯加热过程

Research and Development of L2 Optimizing Control System for Reheating Furnace #1 of Large-size Bar Rolling Line

Jiang Guoqiang, Zhang Baohua, Chen Jianzhou

(Baowu Group Guangdong Shaoguan Iron and Steel Co., Ltd., Shaoguan 512123, China)

Abstract: The research work on the L2 optimizing control system of the reheating furnace #1 of the large-size bar rolling line of Baosteel Special Steel Shaoguan Co., Ltd. has been carried out, and the mathematical model of the bloom reheating process has been developed. The function modules of material and temperature tracking, process inquiry and modification, historical data inquiry and detailed temperature record of blooms of the reheating furnace #1 have been realized. The L2 optimizing control system can effectively control the bloom reheating process, and contribute to the improvement of bloom reheating quality and energy saving and consumption reduction of the reheating furnace.

Key words: reheating furnace#1 for large-size bar rolling line; L2 optimizing control system; bloom reheating process

低碳高含硫钢易切削钢轧制技术应用

李学保，周小兵，潘泽林

（宝武集团广东韶关钢铁有限公司特轧厂，广东韶关　512123）

摘　要：韶钢特钢每年生产低碳高含硫钢如 1215（美标）、1215CW、1215YG、1214Bi 等合计约 2 万吨，此类钢种生产难点为：轧制温度过高容易打滑，轧制温度过低头部容易开裂造成堆钢，成材率低生产成本高，是目前生产难度较高的钢种之一。国内市场上硫系易切削钢占总易切削钢量 90% 以上，主要用于汽车、仪器、机床、五金及标准件等领域，市场发展前景极大。韶钢特钢为了抢占市场先机，从加热工艺、轧制工艺和设备功能精度三方面攻关，解决低碳高含硫钢易切削钢轧制难题。

关键词：1215；易切削钢；均热温度；轻压下工艺

Application of Rolling Technology for Low Carbon and High Sulphur Free Cutting Steel

Li Xuebao, Zhou Xiaobing, Pan Zelin

(Baowu Group Guangdong Shaoguan Iron and Steel Co., Ltd., Shaoguan 512123, China)

Abstract: Special Steel of Shaoguan Iron and Steel Co., Ltd. produces low carbon and high sulphur steels, such as 1215 (US Standard), 1215CW, 1215YG, 1214Bi and so on, with total annual output about 20,000.00 ton. The production difficulties of this kind of steels are as follows: it is easy to slip if the rolling temperature is too high and crack at the head which will cause cobbling when the rolling temperature is too low. It results in low yield and high production cost, it is one of the most difficult steel grades to be produced. Chalcogenide free cutting steel accounts for more than 90% of the total free cutting steel in the domestic market, mainly used in automobiles, instruments, machine tools, hardware and standard parts and other fields, the market has great prospects for development. In order to seize the market opportunity and solve the rolling problem of low carbon and high sulfur free cutting steels, the key problems of reheating process, rolling process and functional accuracy of equipment are tackled.

Key words: 1215; free-cutting steel; soaking temperature; soft reduction process

17CrNiMo6 圆钢表面裂纹成因分析与对策

吴学兴，岳　峰，钟芳华

（宝武集团广东韶关钢铁有限公司制造管理部，广东韶关　512123）

摘　要：采用 Gleeble3800 热模拟试验机、金相显微镜等手段，分析了减速机用 17CrNiMo6 齿轮钢的高温塑性和膨胀曲线，得出 17CrNiMo6 齿轮钢高温脆性区域及相变转化温度。基于上述结果，结合生产实践表明，通过调整连铸二冷冷却制度、提高连铸坯缓冷温度、保证缓冷效果、控制轧材加热升温速率等措施，能有效避免圆钢表面应力裂纹的产生。

关键词：高温塑性；缓冷温度；应力裂纹

Analysis and Countermeasure on the Formation of Surface Stress Cracks on Steel Rods 17CrNiMo6

Wu Xuexing, Yue Feng, Zhong Fanghua

(Baowu Group Guangdong Shaoguan Iron and Steel Co.,Ltd., Shaoguan 512123, China)

Abstract: The high-temperature plasticity and expansion curves of gear steel 17CrNiMo6 used in reducer were analyzed by adopting the Gleeble3800 thermal simulation testing machine and metallographic microscope. Subsequently, the high-temperature brittle zone and phase transformation temperature of the gear steel were obtained. Based on the above results and combined with the production practice in SGIS, the surface stress cracks can be effectively avoided by adjusting the secondary cooling system of continuous casting, increasing the temperature of the slow cooling process for billets, ensuring the slow cooling effect, and controlling the reheating rate for steel rods.

Key words: high temperature plasticity; slow cooling process temperature; stress cracking

超厚特种钢板连续辊式淬火装备技术与应用

付天亮，王昭东，邓想涛，王国栋

（东北大学轧制技术及连轧自动化国家重点实验室，辽宁沈阳 110819）

摘 要：传统浸入式淬火冷速低、均匀性差、板形差，制约了100mm以上超厚特种钢板的研发与应用。基于高温壁面液固有序传热、断面大温度梯度导热、尺寸效应引起的组织差异等机理研究，阐明壁面高效传热与内部梯度导热动态平衡原理，开发出多束阵列射流淬火、非对称淬火、高低压连续淬火等新技术，研制成功国际首套超厚特种钢板连续辊式淬火装备，以及配套分级、往复、间歇等多路径淬火工艺，实现100~300mm厚、单重最大50t钢板连续辊式淬火生产。结果表明，300mm厚钢板淬火心部冷速≥0.52℃/s，淬火后整板温度均匀性≤±8℃，淬火后整板平直度≤3mm/m，达到国际最好水平。板形合格率由不足53%升至98%，性能合格率由不足70%升至99.6%，基本实现一次淬火代替两次淬火。

关键词：特厚钢板；辊式淬火；传热机理；装备技术

Technology and Application of Continuous Roller Quenching Equipment for Ultra Heavy Special Steel Plate

Fu Tianliang, Wang Zhaodong, Deng Xiangtao, Wang Guodong

(State Key Laboratory of Rolling and Automation, Northeastern University, Shenyang 110819, China)

Abstract: The traditional immersion quenching has low cooling rate, poor uniformity and poor shape, which restricts the development and application of ultra heavy steel plate over 100mm. Based on the mechanism research of high temperature fluid-wall sequence heat transfer, large section great gradient heat conduction, and the microstructure difference caused by size effect, the dynamic balance principle of high efficiency wall heat transfer and internal gradient heat conduction is expounded, and then, new technologies such as multi array jet quenching, asymmetric quenching and continuous quenching under high and low pressure are developed, and the international first ultra heavy plate continuous roller quenching equipment is developed successfully, which realized 100~300mm thickness, maximum weight of 50 tons ultra heavy plate continuous quenching production. The results show that the core cooling rate of 300mm thick steel plate is more than 0.52℃/s, the temperature uniformity of the whole plate after quenching is less than ±8℃, and the flatness of the whole plate after quenching is less than 3mm/m, reaching the best level in the world. The qualified rate of flatness increased from less than 53% to 98%, and the qualified rate of performance increased from less than 70% to 99.6%, one quenching instead of two quenching is basically realized.

Key words: ultra heavy plate; roller quenching; heat transfer mechanism; equipment and technology

鞍钢 5500mm 线异型断面钢板生产技术研究

王若钢，李靖年，丛津功，李新玲，姚 震，韩 旭

（鞍钢股份有限公司鲅鱼圈钢铁分公司厚板部，辽宁营口 115007）

摘 要： 异型断面钢板是一种通过控制轧制过程速度和辊缝，实现板材在厚度在沿钢板长度方向上发生变化，满足桥梁、造船业及汽车制造业等关键部件使用要求，以达到资源高效利用的目的。国内对异型断面钢板的研究尚在起步阶段，主要在于钢板在异型断面段轧制过程中，金属沿长度与宽度的延伸量无法精确计算，导致前期坯料设计精确度不够、轧制过程中变厚度阶段尺寸规格误差较大以及没有稳定轧制工艺等诸多原因导致异型断面钢板研究进展缓慢。鞍钢 5500mm 线通过发挥自身装备优势，根据异型断面钢板轧制过程中金属流动规律，制定出合理的坯料设计、加热工艺以及轧制工艺，实现 2000 吨芜湖至合肥国家高速公路安徽省林头至陇西段改扩建工程用 Q345qDNH-LP 桥梁钢板异型断面钢板供货，受到用户一致好评。

关键词： 异型断面桥梁用钢；坯料设计；轧制技术；金属流动；形状控制

Study on Production Technology of 5500mm Profile Steel Plate in Angang

Wang Ruogang, Li Jingnian, Cong Jingong, Li Xinling, Yao Zhen, Han Xu

(Bayuquan Iron & Steel Subsidiary Company of Angang Steel Co., Ltd., Yingkou 115007, China)

Abstract: Profiled section steel plate is a kind of steel plate which can change its thickness along the length of steel plate by controlling the rolling speed and roll gap. It can meet the use requirements of key components such as bridge, shipbuilding industry and automobile manufacturing industry, so as to achieve the purpose of efficient utilization of resources. Domestic research on profiled steel sheet is still in its infancy. The main reason is that the elongation of metal along length and width can not be calculated accurately during the rolling process of profiled steel sheet, which leads to inadequate accuracy of billet design in the early stage, large size error in the variable thickness stage of rolling process, and lack of stable rolling process, and many other reasons lead to the slow progress of research on profiled steel sheet. By giving full play to its own equipment advantages and according to the metal flow law during the rolling process of profiled section steel plate, Anshan

测厚仪精度影响因素浅析

常福刚，赵立军，丛津功，王若钢，翟忠军，安宇恒

（鞍钢股份鲅鱼圈钢铁分公司，辽宁营口 115007）

摘　要：钢板厚度测量是中厚板生产过程中的一个重要参数，随着钢板轧制厚度的自动控制的广泛应用，厚度的自动检测变得更为重要，γ射线测厚仪反应速度快，测量精度高，可连续测量并易与计算机联网，实现钢板厚度自动控制。

关键词：厚度；精度；温度；材质

Elementary Analysis on the Influencing Factors of Thickness Meter Accuracy

Chang Fugang, Zhao Lijun, Cong Jingong, Wang Ruogang, Zhai Zhongjun, An Yuheng

(Angang Bayuquan Iron and Steel Branch, Yingkou 115007, China)

Abstract: Plate thickness measurement is a important parameter in production process of heavy plate.With widely use of thickness automatic control in heavy plate rolling ,thickness automatic measurement become more and more important . γray thickness measurement gauge has a fast response,it can continuously measure the tinckness of plate.It can be easly connected with computer in order to realize the automatic control of plate tinckness.

Key words: thickness; precision; temperature; materiral

超宽 X80M 管线钢板冷矫板形的研究

周　强，王亮亮，姚　震，王若钢，丛津功，李新玲

（鞍钢股份公司鲅鱼圈钢铁分公司，辽宁营口 115007）

摘　要：针对超宽 X80M 管线钢板轧制后不平度超差的问题，进行了原因分析，通过提高冷矫直机标定精度和优化冷矫工艺技术参数等措施，实现超宽 X80M 管线钢板不平直度由 30mm/2m 降低至 3mm/2m 以内。

关键词：X80M 管线钢板；冷矫；平直度；工艺参数

Study on Cold Correction of Ultra Wide X80M Pipeline Steel Plate

Zhou Qiang, Wang Liangliang, Yao Zhen, Wang Ruogang,
Cong Jingong, Li Xinling

(Angang Bayuquan Iron and Steel Branch, Yingkou 115007, China)

Abstract: Contraposed unflatness of X80M pipeline steel exceed standard, by analyzing the causes, formulate corresponding measures, such as improve calibration accuracy and optimize process parameters, get obvious results, unflatness reduced from 30mm/2m to 3mm/2m.

Key words: X80M pipeline steel; cold leveling; unflatness; parameter

基于 AMEsim 软件研究轧机液压辊缝控制系统震荡问题

张秀超，赵立军，肖争光，朱丽娜，丛津功，王若钢

（鞍钢股份鲅鱼圈钢铁分公司厚板部，辽宁营口 115000）

摘 要：本文基于 AMEsim 软件建立液压辊缝控制系统 AMEsim 模型，模拟真实轧制过程液压辊缝控制系统震荡冲击，通过分析 AMEsim 软件仿真的结果，研究系统震荡产生的原因，提出系统优化方案，并应用于现场实际，解决系统震荡。

关键词：液压辊缝控制系统；AMEsim 模型；震荡

Based on AMEsim Software Research the Shaking Problem of Rolling Mill Hydraulic Gap Control System

Zhang Xiuchao, Zhao Lijun, Xiao Zhengguang, Zhu Lina,
Cong Jingong, Wang Ruogang

(The Heavy Plate Department of Ansteel Bayuquan Iron & Steel Subsidiary, Yingkou 115000, China)

Abstract: This paper building the model of hydraulic gap control system based on AMEsim software, to simulate the shaking problem of rolling mill hydraulic gap control system when plate thread in mill and plate thread out mill during real rolling process, through analysis the AMEsim software simulation results, research the reason of system shaking and propose the system optimization scheme and apply, solve the shaking problem.

Key words: hydraulic gap control system; AMEsim-model; shaking

汽包钢蓄热式室式炉生产轧制裂纹分析

韩 旭，刘长江，王若钢，丛津功，段江涛，罗 军，李新玲

（鞍钢股份有限公司鲅鱼圈钢铁分公司厚板部，辽宁营口 115007）

摘 要：本文就厚板部 5500 产线利用室式炉生产的大钢锭产品汽包钢 13MnNiMoR 表面质裂纹进行深入研究，从加热及轧制工艺上提出了改善建议。

关键词：汽包钢；裂纹

Analysis of Rolling Cracks in Regenerative Chamber Furnace of Drum Steel

Han Xu, Liu Changjiang, Wang Ruogang, Cong Jingong, Duan Jiangtao, Luo Jun, Li Xinling

(Heavy Plate Department, Bayuquan Iron and Steel Branch, Angang Co., Ltd., Yingkou 115007, China)

Abstract: In this paper, the surface quality cracks of drum steel 13MnNiMoR produced by chamber furnace in 5500 heavy plate production line are studied in depth, and suggestions for improvement in heating and rolling process are put forward.

Key words: drum steel; crack

延长炉底辊挂腊周期的实践

王亮亮，周 强，李新玲，王若钢，丛津功，姚 震

（鞍钢股份公司鲅鱼圈钢铁分公司，辽宁营口 115007）

摘 要：针对 5500 产线 1#热处理炉炉底辊频繁结瘤的问题，通过分析结瘤的原因，制定了相应的改进措施，有效改善了炉底辊结瘤的问题。其次，对于初期的结瘤，提出了一种无需停炉和升降温的在线处理技术。最后，提出了一种有效修磨炉底辊的方法。

关键词：炉底辊；抛丸；结瘤；修磨

Practice of Prolonging Wax Hanging Period of Bottom Roller

Wang Liangliang, Zhou Qiang, Li Xinling, Wang Ruogang, Cong Jingong, Yao Zhen

(Angang Bayuquan Iron and Steel Branch, Yingkou 115007, China)

Abstract: Contraposed 1# heat treatment furnance of 5500 production line about scale tumor, by analyzing the causes, formulate corresponding measures, get obvious results. Second contraposed early scale tumor, proposed a online method that not need to shut down furnace and change temperature. Last proposed a method coping roll.

Key words: roll; blasting; scale tumor; coping roll

变角度刀梁式静电涂油机在高速薄带材的使用

任予昌[1]，彭　强[1]，张东方[1]，龚　艺[1]，黄秋华[2]

（1. 宝武钢铁武汉有限公司冷轧厂，湖北武汉　430081；
2. 武汉华伟奇机械成套设备有限公司，湖北武汉　430081）

摘　要：某冷轧连续退火生产线以非半成品生产为主，由于市场需要开发直接出货的新产品，为防止表面锈蚀的产生，在生产线上增设静电涂油机，由于新增涂油机为防锈油，在切换其他产品时，防锈油易污染其他辊面，尽量放置在靠近卷曲机部位，由于空间原因，选择了一个带钢有15℃倾角的位置。通过现场的安装调试，达到了设计最初的要求，并申请了专利技术。同时对现场使用的涂油机的原理、油品选择、涂油机构造等进行了简单的介绍。

关键词：冷轧板；变角度；静电涂油机；专利技术

Variable Angle Statical Electricity Coating Oil Machine Used in High Speed Line

Ren Yuchang[1], Peng Qiang[1], Zhang Dongfang[1], Gong Yi[1], Huang Qiuhua[2]

(1. The Cold-Rolling Mill of Wuhan Iron and Steel Co., Ltd., Wuhan 430081, China;
2. Wuhan Huaweiqi Machinery Complete Equipment Co., Ltd., Wuhan 430081, China)

Abstract: A cold-rolled continuous annealing process line is mainly produced to develop new products that are shipped directly, in order to prevent surface rust from occurring, the statical electricity coating oil machine is added to the process line. Rust-resistant oil is easy to pollute other roller surfaces, and the statical electricity coating oil machine is placed as close to the tension reel as possible. Through on-site installation and commissioning, the new design requirements and patent technology was applied. At the same time, the main structure of the statical electricity coating oil machine was introduced.

Key words: cold rol led sheet; variable angle; statical electricity coating oil machine; patent technology

SKP振动痕分析研讨

罗明辉，朱壁学，李子武

（广州JFE钢板有限公司，广东广州　511464）

摘　要：在汽车钢板生产工艺中，光整机(SKP)是常用的一道工序，经光整后的汽车钢板性能更佳。汽车钢板生产线例如CGL、CAL，还有PL-TCM产线，在生产的过程中，均容易出现钢板振动痕，这种品质缺陷影响到钢板成材率，给公司生产经营等均带来损失。本文结合现公司产线实际产生的振动痕品质缺陷案例，以及实际有效的解决方案进行了分类汇总，进而分析研讨。

关键词：SKP；振动痕

冷轧酸轧机组激光焊机焊缝断带分析

吉学军，李保卫，张核新

（首钢京唐钢铁联合有限责任公司，河北唐山 063200）

摘　要：本文通过酸轧机组激光焊机焊缝断带的原因分析，认为激光束位置偏斜是影响焊缝质量的主要因素。提出及实施了解决措施并取得良好的效果。

关键词：激光焊机；断带；焊缝

Analysis on Weld Strip Breakage of Laser Welder for Pickling and Cold Rolling Continuous Production Line

Ji Xuejun, Li Baowei, Zhang Hexin

(Shougang Jingtang Iron & Steel United Co., Ltd., Tangshan 063200, China)

Abstract: Based on the analysis of the causes about strip breakage for laser welder of Pickling and Cold Rolling Continuous Production Line, it is considered that the Laser beam position deviation is the main factor affecting the quality of the weld seam. Put forward and implement some improvement measures, good results have been achieved.

Key words: laser welder; strip breakage; welding seam

连退镀锡基板斑迹类缺陷研究

吉学军，胡小明，张文亮

（首钢京唐钢铁联合有限责任公司，河北唐山 063200）

摘　要：本文通过连退线镀锡基板斑迹类缺陷原因分析，指出平整机吹扫系统、结露滴落残留及支撑辊轴承漏油为斑迹类缺陷产生的主要因素。提出及实施了相应的改进措施并取得良好的效果。

关键词：连退；镀锡基板；斑迹类缺陷

Study on Residual Patches Defects about Continuous Annealing Raw Tin Coil

Ji Xuejun, Hu Xiaoming, Zhang Wenliang

(Shougang Jingtang Iron & Steel United Co., Ltd., Tangshan 063200, China)

Abstract: In this paper, the reasons of the residual patches defects about continuous annealing raw tin coil are analyzed, it is pointed out that the cleaning system of temper mill, the condensation dripping residual and oil leakage of the backup roll bearings are the important causes of the residual patches defects. Put forward and implement some improvement measures, good results have been achieved.

Key words: continuous annealing; raw tin coil; residual patches defects

冷轧带钢连续退火炉绿色制造探索与实践

何建锋，王　鲁，李庆胜

（宝钢日铁汽车板有限公司，上海　201900）

摘　要： 连续退火炉是连退、热镀锌机组的主要工艺设备，是产品性能、质量的保证，也是冷轧工序的能源环保重点。为适应城市钢厂要求，退火炉的绿色制造是冷轧单元面临的重要课题。本文总结了宝钢冷轧1800单元在连续退火炉稳定生产、带钢温度控制、质量管理、节能减排以及退火炉的维护评价等方面进行了系统实践，以便与大家共同交流，共同学习。

关键词： 连续退火；绿色制造；退火炉

Green Manufacturing Exploration and Practice of Cold Rolling Strip Continuous Annealing Furnace

He Jianfeng, Wang Lu, Li Qingsheng

(Baosteel /NSC Automotive Steel Sheets Co., Ltd., Shanghai 201900, China)

Abstract: Continuous annealing furnace is the main process equipment for continuous annealing line and hot-dip galvanizing line, which is the guarantee of product performance and quality. It also plays an important role in energy reduction and environmental protection of cold rolling process. In order to meet the requirements of the city steel plant, the green manufacturing of the annealing furnace becomes an important issue of the cold rolling unit. This paper summarizes the exploration and practice of continuous annealing furnace of Baosteel cold rolling 1800 unit in stable production, strip temperature control, quality management, energy saving and emission reduction, maintenance evaluation of annealing furnace, etc. So as to share with you all and learn together.

Key words: continuous annealing; green manufacturing; annealing furnace

基于Python的变形抗力预测系统的实现与应用

彭　诚

（本钢板材股份有限公司冷轧厂，辽宁本溪　117000）

摘　要： 在冷轧生产中优化数学模型往往涉及大量的相关联的数据的收集、整理、分析等问题。计算机程序是解决这类问题的重要工具，可以通过这种方式对大量的数据进行拟合和可视化处理，最终得到期望的结果。可视化系由 Python 语言和 SQL 数据库开发，将最小二乘的思想应用在冷轧轧制力优化中，取得了很好的应用效果。

关键词：冷轧；Python；最小二乘法；大数据；可视化

Implementation and Application of Python based on Deformation Resistance Prediction System

Peng Cheng

(Cold Rolling Mill of Bengang Steel Plates Co., Ltd., Benxi 117000, China)

Abstract: In the cold rolling process, Mathematical model optimization often involves a large number of related data collection, sorting, analysis and other issues. Computer programs are an important tool for solving such problems. In this way, a large number of data can be fitted and visualized to achieve the desired results. The visualization system is developed by Python language and SQL database, and the idea of least square is applied to the optimization of cold rolling force. It has obtained the very good applying effects in the practical application.

Key words: cold rolling; python; least square method; big data; visualization

双机架四辊 CVC 平整机辊形配置优化与应用

文　杰[1]，于　孟[1]，莫志英[2]，王永强[1]，张宝来[2]，刘学良[2]，李永新[2]

(1. 首钢集团有限公司技术研究院京唐技术中心，北京　100041；
2. 首钢京唐钢铁联合有限公司镀锡板事业部，河北唐山　063200)

摘　要：针对双机架四辊 CVC 平整机生产镀锡基板时存在支撑辊不均匀磨损与板形控制问题，提出了双机架四辊平整机生产镀锡基板时的辊形配置优化设计策略——支撑辊辊形设计首要目标是具有较好的自保持性，而工作辊辊形应在保证平整机板形控制能力的基础上尽可能做到简单、易磨削。本文在考虑了支撑辊磨损和热膨胀对辊形影响的基础上，采用了 VCL 辊形技术对四辊 CVC 平整机的支撑辊辊形进行了优化设计，并在工作辊上采用了平辊辊形。生产实践表明，该辊形配置在板形控制、提高辊形自保持性、降低大轧制力工况下弯辊力水平等方面取得了良好效果。

关键词：镀锡板；双机架平整机；CVC；VCL；板形

Optimization and Application of Roll Contour Configuration on Double Stands 4-h CVC Temper Mill

Wen Jie[1], Yu Meng[1], Mo Zhiying[2], Wang Yongqiang[1], Zhang Baolai[2], Liu Xueliang[2], Li Yongxin[2]

(1. Jingtang Technology Center of Research Institute of Technology, Shougang Group Corporation, Beijing 100041, China; 2. Tinplate Department, Shougang

Jingtang United Iron and Steel Co., Ltd., Tangshan 063200, China)

Abstract: In order to solve the problems of uneven wear of back-up roll and flatness control on production of TMBP (tin mill black plate) with double stands 4-h CVC temper mill, the optimal design strategy of roll contour configuration is put forward. The primary goal of back-up roll contour design is the good self-preservation. The work roll contour should be simple and easy to grind on the basis of ensuring the flatness control ability. In this paper, the influence of wear and thermal expansion on roll contour is considered. The VCL technology is used to optimize the back-up roll contour, and the flat roll contour is used on work roll. The practice shows that the roll contour configuration has achieved good results in flatness control, improving self-preservation of roll contour and reducing work roll bending force under the condition of large rolling force.

Key words: tinplate; double stands temper mill; CVC; VCL; flatness

基于焊缝3D检测系统应用的超高强钢焊接工艺优化研究

朱健华，何建锋

（宝钢日铁汽车板有限公司，上海 201900）

摘 要： 焊接工序作为热镀锌及合金产品生产的头道工序，在确保产业流水线的顺利运转中显得尤为重要。通过在窄搭接焊机上搭建三维激光扫描仪和涡流检测设备，利用C#开发在线分析软件处理采集到的数据，从而实现对焊缝缺陷的实时监测。在焊缝检测结果的基础上根据现场生产、调试情况，优化生产工艺参数，摸索超高强钢的作业方式。实现了超高强钢焊接工艺优化，形成了一整套全面可行的解决方案。

关键词： 焊缝3D检测；超高强钢；焊接工艺

Research on Welding Process Optimization of Ultra-high Strength Steel based on the Application of Welding Seam 3D Detection System

Zhu Jianhua, He Jianfeng

(Baosteel-Nippon Steel Automotive Steel Sheets Co., Ltd., Shanghai 201900, China)

Abstract: Narrow lap welding is the first important process to ensure continuous production in hot dip galvanized steel production lines. We equip 3D laser line scan camera and eddy current testing system to improve reliability and monitoring timeliness in welding process. The collected data is analyzed through software developed by C#, thereby realizing real-time monitoring of weld defects. Based on the results of weld inspection, the production process parameters are optimized according to the on-site production and commissioning conditions, and the operation mode of ultra-high-strength steel is explored. The welding process optimization of ultra-high strength steel is realized and formed a comprehensive set of feasible solutions.

Key words: welding seam 3D detection; ultrahigh-strength steel; welding process

钢种成分和工艺对连退 DC01 产品力学性能的影响

王业科，辜蕾钢，刘显军，杨 薇

（中冶赛迪工程技术股份有限公司，重庆 401122）

摘　要：DC01 是连续退火机组生产的常见品种，但相对来说，连退对原料钢卷钢种成分和工艺要求比罩式炉更苛刻。本文结合实验数据和实际生产数据，系统论述了钢种化学成分、热轧工艺、冷轧压下率，以及连续退火机组生产过程中的退火温度、均热时间、过时效温度和时间以及平整和拉矫延伸率等工艺控制工序对 DC01 产品力学性能的影响。为了获得良好的力学性能，在实际生产过程中必须严格控制钢种成分和工艺参数。

关键词：连续退火；化学成分；工艺；力学性能

Effect of Steel Composition and Process on Mechanical Properties of DC01 Product in CAL

Wang Yeke, Gu Leigang, Liu Xianjun, Yang Wei

(CISDI Engineering Co., Ltd., Chongqing 401122, China)

Abstract: DC01 is a common product of CAL, but relatively speaking, CAL has more stringent requirements on the composition and process of raw material than bell furnace. Based on the experimental data and actual production data, the effects of chemical composition of steel grades, hot rolling process, cold rolling reduction rate, annealing temperature, soaking time, over-aging temperature and time as well as SPM and tension leveller elongation on mechanical properties of DC01 products are systematically discussed in this paper. In order to obtain good mechanical properties, steel composition and process parameters must be strictly controlled in the actual production process.

Key words: CAL; chemical composition; process; mechanical properties

剪切断面质量对焊接的影响及控制

张　志，黄爱军，夏大斌

（浙江龙盛薄板有限公司，浙江绍兴 312369）

摘　要：因焊接工艺对于剪切板材的端面质量要求较高，本文从实践中总结，确定出高强度板材分条剪切侧向间隙，控制剪切端面质量，达到焊接工艺要求。

关键词：纵剪；圆盘剪；断面质量；侧向间隙

Influence of Longitudinal Section Quality on Welding and Its Control

Zhang Zhi, Huang Aijun, Xia Dabin

(Zhejiang Lonsen Sheet Co., Ltd., Shaoxing 312369, China)

Abstract: As the welding process has high requirements on the quality of the end face of the shearing plate, this paper summarizes from the practice, determines the lateral gap of the high-strength plate strip shearing, controls the quality of the shearing end face, and meets the welding process requirements.

Key words: longitudinal shear; disc scissors; section quality; lateral clearance

首钢智新电工钢连退机组入口和出口张力辊装置方案的设计与分析

张 伟[1,2], 何云飞[1,2], 张乐峰[1,2], 孟祥军[1,2]

(1. 北京首钢国际工程技术有限公司,北京 100043;
2. 北京市冶金三维仿真设计工程技术研究中心,北京 100043)

摘 要: 以电工钢连续退火机组入、出口张力辊为研究对象,介绍了入、出口张力辊组的设计原理、结构和设计方法,对提供初张力时的张力辊压辊电机功率进行了计算,并就入口张力辊在两种不同工况下,出口张力辊在三种不同的工况下,张力辊电机功率进行了详细的论述计算。

关键词: 连续退火机组;张力;入口张力辊;出口张力辊;电机功率计算

The Design and Analysis of the Entry and Exit Bridle Roll in the Silicon Steel Continuous Annealing Line

Zhang Wei[1,2], He Yunfei[1,2], Zhang Lefeng[1,2], Meng Xiangjun[1,2]

(1. Beijing Shougang International Engineering Technology Co., Ltd., Beijing 100043, China;
2. Metallurgical Engineering 3-D Simulation Design Engineering Technology Research Center of Beijing, Beijing 100043, China)

Abstract: This article introduces the design principle, structure and design method of the bridle rolls at the entry and exit of the silicon steel continuous annealing line. The motor power of the hold down roller is calculated when the initial tension is provided, and the motor power of the bridle roller at the entry side under two different working conditions and at the exit side under three different working conditions is discussed and calculated in detail.

Key words: continue annealing line; strip tention; entry bridle roll; exit bridle roll; motor power caculation

酸洗工艺参数对热轧板酸洗效果影响的研究

张 鹏[1]，孙 力[1]，王 峰[2]，任振远[3]，刘连喜[2]

（1. 河钢集团钢研总院，河北石家庄 050023；2. 河钢集团衡板公司，河北衡水 053000；3. 唐山不锈钢有限责任公司，河北唐山 063100）

摘 要：使用盐酸浓度分别为 18%、14% 与 10% 的溶液对 SPHC 产品进行酸洗，酸洗的温度为 75℃、80℃、85℃，酸洗时间为 42s、54s 和 66s，构建了酸碱减重率模型，对酸洗工艺参数对酸洗效果的影响进行分析。结果表明：在酸液浓度一定的条件下，酸洗温度比时间的影响更大，为现场制定合理的酸洗工艺制度提供了依据。

关键词：氧化铁皮；酸洗温度；酸洗时间；酸液浓度

Effect of Pickling Process Parameters on Pickling Effect of Hot Rolled Sheet

Zhang Peng[1], Sun Li[1], Wang Feng[2], Ren Zhenyuan[3], Liu Lianxi[2]

(1. Hesteel Technology Research Institute, Shijiazhuang 050023, China; 2. Hengshui Sheet Co., Ltd., Hesteel Group, Hengshui 053000, China; 3. Tangshan Stainless Steel Co., Ltd., Tangshan 063100, China)

Abstract: SPHC products were pickled with solutions containing 18%, 14% and 10% hydrochloric acid respectively. The pickling temperature was 75℃, 80℃, 85℃, and the pickling time was 42 seconds, 54 seconds and 66 seconds. The acid-base weight loss rate model was constructed to analyze the effect of pickling process parameters on pickling effect. The results show that under the condition of certain acid concentration, the pickling temperature has more influence than time, which provides a basis for the establishment of a reasonable pickling process system.

Key words: iron oxide scale; pickling temperature; pickling time; acid concentration

浅谈冷轧薄宽规格产品炉内擦伤的控制与研究

李国栋

（本钢板材股份有限公司冷轧厂，辽宁本溪 117000）

摘 要：本钢冷轧产品薄宽规格主要供应汽车用户，在生产过程中出现炉内擦伤缺陷。本钢通过内部原因分析，改善酸轧板形，调整连退工艺，有效控制了炉内擦伤缺陷的发生。

关键词：冷轧；薄宽规格；板形；炉内擦伤

Control and Research on Furnace Scratch in Cold Rolled Thin and Wide Specification Products

Li Guodong

(Bengang Steel Plates Co., Ltd., The Cold Rolling Mill, Benxi 117000, China)

Abstract: The thin and wide specifications of cold rolled products of Benxi Iron and Steel Co. are mainly supplied to automobile users, and there are scratch defects in furnace during production. Benxi Iron and Steel Co. has effectively controlled the occurrence of scratch defects in furnace by analyzing the internal causes, improving the shape of acid rolling plate and adjusting the continuous and retreating process.

Key words: cold rolling; thin and wide specifications; flatness; scratch in furnace

冷连轧过程中宽度变化时的头尾板形优化策略

任延庆

(本钢板材股份有限公司冷轧厂，辽宁本溪 117000)

摘　要： 针对冷连轧机组变宽度变规格时带头带尾板形不良的问题，从板形前馈控制和弯辊力设定计算两方面进行了理论分析。基于影响函数法，建立了冷轧板形设定计算模型。模拟计算了不同宽度带钢的工作辊与中间辊弯辊力设定值。根据理论分析和模拟计算结果，优化了变宽度规格时带头尾板形控制策略，通过跟踪采集实际生产数据表明，带头尾板形控制精度显著提高。

关键词： 冷连轧；板形；弯辊力

Head-to-tail Flatness Optimization Strategy for Width Variation during Cold Rolling

Ren Yanqing

(Bengang Steel Plates Co., Ltd., The Cold Rolling Mill, Benxi 117000, China)

Abstract: In order to solve the problem of the bad flatness of the head and tail in the tandem cold rolling mills the theoretical analysis is made from the aspects of the flatness feedforward control and the setting values of the roll bending force. The setup model of flatness is established based on the influence function method. The setting values of work roll bending force and intermediate roll bending force are simulated with different strip width. The flatness control strategy of the strip head and tail is optimized according to the theoretical analysis and simulation results. Through tracking and collecting the actual production data, it is shown that the precision of the flatness control of the strip head and tail is significantly improved.

Key words: the cold rolling; flatness; bending force

酸洗轧机联合机组清洗效果对汽车板表面质量影响

张彦雨，李 宁

(本钢板材股份有限公司冷轧厂，辽宁本溪 117000)

摘 要：对于每一条酸洗轧机联合机组来说，工艺段对产品表面质量起到决定性作用。随着汽车板的表面质量逐渐提高，为满足汽车厂对汽车面板的高质量要求，我们通过调整酸洗速度温度、pH 值、喷嘴喷梁、挤干辊管理和添加缓蚀剂钝化剂，加强酸洗漂洗效果，最终达到保证汽车面板表面质量的目的。

关键词：酸洗速度温度；pH 值；挤干辊；汽车板质量

The Effect of Picking Mill Combination on the Surface of Automobile Plate

Zhang Yanyu, Li Ning

(Bengang Steel Plates Co., Ltd., The Cold Rolling Mill, Benxi 117000, China)

Abstract: For each picking mill combined unit, cleaning effect is the most basic guarantee of product surface quality. At the same time, with the gradual improvement of the surface quality of the car boart, but also to meet the automobile factory surface quality requirements of the car panel, we adjust the pickling speed and pickling temperature、pH value、nozzle spray beam、drain roll management and adding corrosion inhibitor passivator, strengthen acid rinsing effect, finally achieve the goal of guarantee the quality of automobile panel surface.

Key words: picking speed temperature; pH value; squeeze roll; automotive board quality

本钢冷轧 1#酸轧机组厚度不合控制优化

张 勇

(本钢板材股份有限公司冷轧厂，辽宁本溪 117000)

摘 要：为了控制冷轧普碳产品厚度不合，通过技术和管理手段，达到提升产品厚度精度的目的。根据本钢冷轧 1#酸轧机组生产运行现状，扩大轧机第 3~4 机架 AGC 调节范围、对 3#~4#测厚仪增设防护装置及参数校正、优化酸洗工艺段停车排酸用时，使第 3~4 机架 AGC 调节呈现为不饱和状态、测厚仪零点漂移得到有效遏制、酸洗工艺段排酸时间延长 3min 而粗糙度满足 $Ra2$~$3\mu m$ 要求，实现了厚度波动范围被准确控制。不断满足高端用户要求的厚度公差、降低厚度不合市场异议率、提高 1#酸轧机组成材率、进一步提升冷轧普碳产品的市场竞争力。

关键词：厚度不合；AGC 调节；快速排酸；X 射线测厚仪

The Thickness Deviation Control Optimization of Benxi Steel Cold Rolling of 1 # Pickling Line Tandem Cold Rolling Mill

Zhang Yong

(Bengang Steel Plates Co., Ltd., The Cold Rolling Mill, Benxi 117000, China)

Abstract: In order to control the thickness deviation of cold rolled, carbon products, through technology and management, achieve the goal of improve product thickness accuracy. According to the current production run of Benxi steel cold rolling 1# pickling line tandem cold rolling mill, to 3#~4# thickness gauge add protection device and the parameter calibration, extend the AGC adjustment range of mill frame 3~4, optimize the pickling process section of parking acid removal unavailable, 3~4 rack AGC regulation to unsaturated state, thickness gauge to effectively curb zero drift, the pickling process section acid removal prolonged 3 minutes and roughness meet the requirement in Ra2~3 microns, to achieve the thickness range is accurate control. Constantly to meet the requirements of high-end users' thickness tolerance, to reduce the thickness of rolled out market objection rate and improve the 1 # pickling line tandem cold rolling mill, further enhance the market competitiveness of the cold-rolled, carbon products.

Key words: thickness deviation; AGC regulation; rapid acid removal; X-ray thickness gauge

卡罗塞尔卷取机挫伤缺陷分析及控制方案

田永强，周三保，陈 俊，刘 挺，张 波

（攀钢集团西昌钢钒有限公司，四川西昌 615000）

摘 要：卡罗塞尔双卷筒式卷取机作为最先进的冶金卷取设备之一，其结构复杂，控制精度高；同时分卷速度快，带钢失张段变短，头尾超差段减少，有益于提高成材率和生产效率，在冷连轧机组上得到广泛的应用，也容易产生卷取缺陷，特别是挫伤缺陷。本文针对因卷取过程产生的挫伤缺陷进行了分析并提出了改进方案，取得好的效果。

关键词：卡罗塞尔卷取机；松卷；挫伤

Abstract: As one of the most advanced metallurgical coiling equipment, Carrassel double drum reel has complex structure and high control precision, the fast speed coil separation the strip tension loss section is shorter, the head and tail out-of-tolerance section is reduced, which is beneficial to improve the yield and production efficiency, and it is widely used in cold continuous rolling mill, it is also easy to produce coiling defects.Especially the contusion defects. In this paper, analysis the contusion defects of Carrassel reel and puts forward the improvement scheme, get good results.

N含量超标对冷轧低合金高强钢280VK力学性能影响浅析

狄彦军，赵小龙，罗晓阳，赵占彪，王 强

（酒钢宏兴股份有限公司碳钢薄板厂，甘肃嘉峪关 735100）

摘 要：针对 CSP 流程生产的冷轧低合金高强钢 280VK 冶炼过程中 N 含量控制超标异常钢卷，对 N 含量正常及异常工艺钢带经正常退火工艺生产后产品力学性能检测分析，N 含量超标钢带最终成品强度偏低，不满足标准要求。通过 SEM 对比分析了正常（A）试样和异常（B）试样显微形貌及金相组织。结果表明，B 试样中 TiN 析出量较大，晶界和晶内都有分布，析出物尺寸较大，大部分尺寸在 1.0~1.8μm 左右，组织中铁素体晶粒尺寸较大，析出强化和细晶强化效果削弱；A 试样组织中 TiN 析出量相对较少，主要分布在晶界处，钉扎位错明显。析出物尺寸较小，大部分尺寸在 70nm 左右，抑制铁素体晶粒长大，组织中铁素体晶粒尺寸较小，析出强化和细晶强化作用明显。今后在工业化生产冷轧低合金高强钢 280VK 过程中，针对出现 N 含量超标钢卷，通过降低罩式炉退火温度 20~40℃，可保证产品强度达标。

关键词：低合金高强钢；280VK；细晶强化；析出强化；再结晶；退火温度

Analysis on the Effect of N Content Exceeding the Standard on the Mechanical Properties of Cold Rolled High Strength Low Alloy Steel 280VK

Di Yanjun, Zhao Xiaolong, Luo Xiaoyang, Zhao Zhanbiao, Wang Qiang

(Carbon Steel Strip Plant of JISCO Hongxing Iron & Steel Co., Ltd., Jiayuguan 735100, China)

Abstract: In view of the abnormal steel coil whose N content is controlled to exceed the standard in the 280VK smelting process of cold rolled low alloy high strength steel produced by CSP process, the mechanical properties of the product after normal annealing process are tested and analyzed, and the final finished product strength of the steel strip with N content exceeding the standard is low and does not meet the standard requirements. The microstructure and microstructure of normal (A) samples and abnormal (B) samples were compared and analyzed by SEM. The results show that the precipitation of TiN in B sample is large, the grain boundary and crystal are distributed, the size of precipitates is large, most of the sizes are about 1.0~1.8μm, the grain size of Ferrite in microstructure is larger, the effect of precipitation strengthening and fine grain strengthening is weakened, and the precipitation amount of TiN in sample A is relatively small, mainly distributed at grain boundary, and the pinning dislocation is obvious. The size of precipitates is small, and most of them are about 70nm, which can inhibit the grain growth of ferrite. The grain size of ferrite in microstructure is small, and the effect of precipitation strengthening and fine grain strengthening is obvious. In the process of industrial production of cold rolled low alloy high strength steel 280VK in the future, in view of the occurrence of N content exceeding the standard steel coil, by reducing the annealing temperature of cover furnace by 20~40℃, the strength of the product can be guaranteed to reach the standard.

Key words: high strength low alloy steel; 280VK; refined crystalline strengthening; precipitation strength; recrystallization; annealing temperature

基于 CSP 流程的镀铝锌锌花不均缺陷分析

孙朝勇

（酒钢集团宏兴股份公司碳钢薄板厂，甘肃嘉峪关 735100）

摘 要：酒钢热镀铝锌机组使用 CSP 原料，自 2017 年开始锌花不均缺陷成为镀铝锌生产的头号难题。本文介绍了镀铝锌锌花不均缺陷的表现形式，讨论了各类锌花不均缺陷产生的主要原因，提出了锌花不均缺陷的预防措施。

关键词：镀铝锌；锌花不均；预防措施

Analysis of Spangles Inhomogeneous Defect for Galvalume in CSP Process

Sun Chaoyong

(Carbon steel thin plate plant of Hongxing Iron & Steel Co., Ltd., Jiuquan Iron and Steel Group Corporation, Jiayuguan 735100, China)

Abstract: The galvalume plating unit of JISCO uses CSP materials.Since 2017, Spangles inhomogeneous defect become a challenge in galvalume production. In this paper, we introduced the defect appearance features，and discussed the cause of the defect.It also raised some pre-measures to uneven the defect.

Key words: galvalume; inhomogeneous spangle; pre-measures

酒钢CSP线600MPa级热轧双相钢生产工艺研究

杨 华[1]，昝理平[1]，王云平[1]，成洋洋[2]，刘 靖[2]

（1. 甘肃酒钢集团宏兴钢铁股份有限公司钢铁研究院，甘肃嘉峪关 735000；
2. 北京科技大学材料科学与工程学院，北京 100083）

摘 要： 针对酒钢CSP线，设计了600MPa级热轧双相钢的化学成分，测定了其动态CCT曲线。根据动态CCT，分别采用三段式与两段式冷却模式进行了现场试制，对试制热轧板进行了显微组织观察与力学性能检测。结果表明：通过化学成分的合理设计及关键工艺参数的合理控制，热轧板的显微组织为铁素体+马氏体，屈服强度均达到325MPa以上，抗拉强度均达到600MPa以上，屈强比为0.55~0.58，伸长率均达到27.0%以上，并且其应力-应变曲线为连续屈服。因此，该成分的钢可在酒钢CSP短流程线上生产出DP600热轧双相钢。

关键词： CSP短流程生产线；热轧双相钢；成分与工艺设计；动态CCT曲线；组织性能

Process Study of 600MPa Grade Hot-rolled Dual Phase Steel in CSP Production Line in JISCO

Yang Hua[1], Lin Liping[1], Wang Yunping[1], Cheng Yangyang[2], Liu Jing[2]

(1. Iron and Steel Institute, JISCO, Jiayuguan 735000, China; 2. School of Materials Science and Engineering, University of Science and Technology Beijing, Beijing 100083, China)

Abstract: For the JISCO's CSP line, the chemical composition of hot-rolled dual phase steel of 600 MPa grade was designed. The CCT curve of the test steel was measured. According to the dynamic CCT, the three-stage and two-stage cooling modes were used for on-site trial production, and the microstructure and mechanical properties of the trial-produced hot-rolled sheets were examined. The results show that the microstructure of the hot-rolled sheets are ferrite and martensite by reasonable control of the chemical composition and key process parameters，and the yield strength, tensile strength and ductility of DP600 are more than 325MPa, 600MPa, and 27.0%, respectively. Yield ratio is 0.55~0.58. The stress-strain curve is continuous and has no yield point. Therefore, DP600 hot-rolled dual phase steel can be produced on CSP short process line of JISCO.

Key words: compact thin slab production line; hot-rolled dual phase steel; composition and process design; dynamic CCT curve; microstructure and properties

酒钢冷连轧机二级模型的工艺优化应用

周文宾

(酒钢集团宏兴钢铁股份有限公司，甘肃嘉峪关　735100)

摘　要： 本文介绍了酒钢冷轧 UCM 连轧机二级模型的工艺优化应用，重点阐述通过 UCM 轧机二级模型系统在现场应用中存在的问题，通过对二级模型预设定、负荷分配和模型参数的优化，实现冷连轧机组各类差异化产品的稳定轧制。酒钢冷连轧机二级模型的工艺优化，使冷轧 UCM 轧机在成品粗糙度、卷取张力等工艺控制上更加灵活，而差异化的工艺参数控制为机组连续高质生产提供保障。

关键词： UCM；二级模型；工艺控制

Process Optimization Application of Two - level Model of Cold Tandem Mill at JISCO

Zhou Wenbin

(Research Institute, JISCO Hongxing Iron & Steel Co., Ltd., Jiayuguan 735100, China)

Abstract: This paper introduces the process optimization application of the two-level model of UCM tandem cold rolling mill at Jisco, and focuses on the problems existing in the field application of the two-level model system of UCM rolling mill. The process optimization of cold tandem mill model makes cold rolling UCM mill more flexible in the process control of finished product roughness and coiling tension, and the differential process parameters control provides guarantee for the continuous high quality production of the unit.

Key words: UCM; two-level mode; process control

冷轧带钢乳化液斑迹研究及控制措施

翁　星

(酒钢集团宏兴钢铁股份有限公司，甘肃嘉峪关　735100)

摘　要： 介绍了影响冷轧带钢退火后产生乳化液斑迹缺陷的主要因素和控制措施。以酒钢碳钢薄板厂 2018 年 1 月份发生的 SPCC 钢种尾部 700m 左右出现轻微黑色乳化液斑迹质量事件为攻关对象，通过大量实测数据对冷轧带钢表面清洁性与各个影响因素的关系进行研究。 通过对酸轧生产过程中可能造成下线带钢表面残留物质各个影响因素分析，为冷轧带钢表面乳化液斑迹缺陷控制提供了指导；通过影响因素制定出改善冷轧带钢乳化液斑迹缺陷的控制手段。

关键词：冷轧带钢；乳化液斑迹；控制措施

Study on the Emulsion Stain Trace of the Cold-rolled Strip and Its Control Measures

Weng Xing

(Research Institute, JISCO Hongxing Iron & Steel Co., Ltd., Jiayuguan 735100, China)

Abstract: This paper is about the main causes of the emulsion stain trace of cold-rolled strip after annealing and its control measures. Based on a large amount of measured data, it's a study of the relationship between the cleanliness of the cold-rolled strip and the various influencing factors in order to solve the problem that a slight black emulsion spot quality event occurred on 700 meters to the tail of the SPCC steel in the Carbon Steel Sheet Factory of Jiuquan Steel Company in January 2018. Through the analysis of the factors of the residual material on the surface of the strip during the acid rolling production process, it provides guidance to control the emulsion stain trace of the cold-rolled strip and works out the methods to improve the defects of the emulsion stain trace of the cold-rolled strip.

Key words: cold-rolled strip; emulsion stain; control measures

浅谈"一贯质量管理"

赵占彪

（酒钢宏兴股份有限公司碳钢薄板厂，甘肃嘉峪关　735100）

摘　要：随着社会的发展，全国各企业升级改造过程中需要的高性能钢材越来越多，相应的对钢材质量的要求也随之不断提升，故保证按时稳定供货的条件下，钢材质量的好坏，决定着企业能否盈利，能否生存的关键，一贯质量管理方式是保证产品质量的有效途径，实施一贯制质量管理是保证产品质量的有力保障。

关键词：一贯质量管理；产品研发；升级；技能；职责

Discussion on "Consistent Quality Management"

Zhao Zhanbiao

(Research Institute, JISCO Hongxing Iron and Steel Co., Ltd., Jiayuguan 735100, China)

Abstract: With the development of society, more and more high performance steel products are needed in the process of upgrading and transformation of enterprises throughout the country, and the corresponding requirements for steel quality are also continuously raised. Therefore, the quality of steel products is good or bad under the condition of timely and stable supply. The consistent quality management is an effective way to guarantee product quality, and the implementation of consistent quality management is the effective guarantee of product quality.

Key words: consistent quality management; product development; upgrading; skills; responsibilities

热镀铝锌硅板面边厚缺陷攻关

刘海军，马维杰

（酒钢集团宏兴股份公司碳钢薄板厂，甘肃嘉峪关　735100）

摘　要：为了研究主要针对酒钢热镀铝锌硅产品升级过程中，暴露出产品板面边厚缺陷，进行研究攻关，设计制作新型气刀边部挡板，解决边厚和错边卷取的问题。

关键词：热镀铝锌硅；气刀边部挡板；边厚

Study on the Defects of the Side-thickness of the Hot-dip 55%Al-Zn-Si Line

Liu Haijun, Ma Weijie

(JISCO Hongxing Iron & Steel Co., Ltd., Jiayuguan 735100, China)

Abstract: In order to study the defects of plate surface thickness in the upgrading process of hot-dip 55%Al-Zn-Si products in Jiuquan Iron and Steel Co., Ltd., the key problems were studied, and a new type of air knife edge baffle was designed and made to solve the problems of edge thickness and irregular edge coiling.

Key words: hot-dip 55%Al-Zn-Si line; air knife edge baffle; edge thickness

轧机液压 AGC 系统故障分析

毛成海

（酒钢集团宏兴钢铁股份有限公司，甘肃嘉峪关　735100）

摘　要：轧机液压 AGC 系统是轧机控制系统的核心，液压 AGC 系统的故障划分为以下几类：稳定型故障、精度型故障、响应速度型故障、元件失灵或失效型故障，这四类故障的机理分析是监测与诊断的基础，通过分析这四类故障的原因和总结一些发生过的典型案例，我们可以较全面地掌握液压 AGC 系统的故障机理，在发生故障后迅速准确地找出事故原因，提高事故处理效率。

关键词：轧机；液压 AGC；故障

Fault Analysis of Mill Hydraulic AGC System

Mao Chenghai

(JISCO Hongxing Iron and Steel Co., Ltd., Jiayuguan 735100, China)

Abstract: The AGC system of rolling mill is the core of the rolling mill control system. The faults of the AGC system can be divided into the following categories: stable faults, precision faults, response speed faults, component faults or failure faults. The mechanism analysis of these four types of faults is the basis of monitoring and diagnosis. By summarizing some typical cases, we can grasp the failure mechanism of AGC system more comprehensively, find out the cause of the accident quickly and accurately after the failure, and improve the efficiency of accident treatment.

Key words: mill; AGC; fault

罩式退火炉黑斑缺陷分析及防范措施

王生东

（酒钢集团宏兴钢铁股份有限公司，甘肃嘉峪关 735100）

摘 要： 针对罩式退火炉产生批量黑斑缺陷，本文对产生黑斑的机理进行了详细的分析，并提出了改进的措施。同时对采取措施前后黑斑缺陷率进行了对比，验证了措施的有效性。

关键词： 退火；乳化液；氢气吹扫；黑斑

Analysis and Preventive Measures of Black Spot Defect of Hood Type Annealing Furnace

Wang Shengdong

(Research Institute, JISCO Hongxing Iron & Steel Co., Ltd., Jiayuguan 735100, China)

Abstract: Aiming at the defects of batch blackspot produced by the blank-type annealing furnace, the mechanism of blackspot produced is analyzed in detail, and the improvement measures are put forward. At the same time, the effectiveness of the measures was verified by comparing the defect rates of black spots before and after the measures were taken.

Key words: annealing; emulsion; hydrogen purge; black spot

冷轧 Ti 系超低碳烘烤硬化钢的耐时效性能研究

周三保[1]，唐梦霞[1]，张功庭[2]

（1. 攀钢集团西昌钢钒有限公司，四川凉山 615000；
2. 攀钢集团研究院有限公司，四川攀枝花 617000）

摘 要： 通过自然时效、有预变形的加速时效和无预变形的加速时效，研究了 Ti 系超低碳烘烤硬化钢的耐时效性能。结果表明：通过三种不同时效方式后，试验钢的屈服强度和屈服平台长度均增加。自然时效 7 个月后，试验钢的屈服强度增加值和屈服平台长度增加值已经接近具有较好耐时效性能下限。相比于有预变形的加速时效后的试验钢，无预变形的加速时效后的试验钢的屈服强度增加值和屈服平台长度增加值更接近自然时效 7 个月后的试验钢屈服强度增加值和屈服平台长度增加值。因此试验钢的最佳使用期限为不超过 7 个月,无预变形的加速时效条件更能反应试验钢的耐常温时效性能。

关键词:超低碳烘烤硬化钢;自然时效;加速时效;预变形

Research on Aging Resistance of Cold-Rolling Ti-killed Ultra-Low Carbon Bake-Hardening Steel

Zhou Sanbao[1], Tang Mengxia[1], Zhang Gongting[2]

(1. Xichang Steel & Vanadium Limited Company of Pangang Group, Liangshan, 615000, China;
2. Research Institute of Pangang Group, Panzhihua, 617000, China)

Abstract: Researched on the aging resistance of cold-rolling Ti-killed ultra-low carbon bake-hardening steel by natural aging, accelerated aging with pre-strained and accelerated aging without pre-strained. The results show that yield strength and yield platform length of the test steel have increased by the three different aging ways. The added value of test steel's yield strength and yield platform length have closed to the bottom of good aging resistance after natural aging 7 months. Compared to steel after accelerated aging with pre-strained, the added value of test steel's yield strength and yield platform length of steel after accelerated aging without pre-strained are nearer the natural aging steel's. The test steel's optimum service life is not more than 7 months. Accelerated aging without pre-strained is more applicable to evaluate the steel's aging resistance.

Key words: cold-rolling Ti-killed ultra-low carbon bake-hardening steel; natural aging; accelerated aging; pre-strained

冷轧平整机组异常高速甩尾问题探讨

王 勇,冯志新,文纪刚,邢顺治,刘立恒

(鞍钢股份有限公司冷轧厂,辽宁鞍山 114021)

摘 要:冷轧平整机组生产过程中,带尾自动减速,需要进行精确控制。但在实际生产过程中,偶尔会发生自动减速失控问题,导致高速甩尾,造成机架间堆钢,甚至损坏一系列设备的严重后果。为此,本文在分析带尾自动减速原理的基础上探讨了一种简单有效的预防措施。

关键词:平整机组;带尾自动减速;高速甩尾

Discuss on the Abnormal High-speed Tail out of Temper Rolling Mill

Wang Yong, Feng Zhixin, Wen Jigang, Xing Shunzhi, Liu Liheng

(Cold Mill of Angang Steel Co., Ltd., Anshan 114021, China)

Abstract: During the production process of temper rolling mill, strip tail automatic decelerating need to be controlled. But during the actual production process, occasional automatic deceleration runaway cause high-speed tail out, even piling-up at the stand of temper rolling mill and equipment damage occurs. This study discusses preventive actions on the basis of analyzing the principles of strip tail automatic deceleration.

Key words: temper rolling mill; trip tail automatic deceleration; high-speed tail out

冷轧汽车板点状锈蚀产生机理及控制措施

辛利峰，阮国庆，孙荣生

（鞍钢股份有限公司冷轧厂，辽宁鞍山 114021）

摘 要： 点状锈蚀缺陷在冷轧钢板出厂时仅存在锈蚀源，并未形成明显的锈蚀斑迹，质检不易发现，是冷轧汽车板在应用过程中的主要缺陷，本文利用电镜扫描获得点状锈蚀缺陷成分，结合钢厂生产制造过程，对点状锈蚀缺陷进行分类，分析产生原因，采取有针对性的措施，避免问题的重复发生。

关键词： 冷轧钢板；点状锈蚀；机理；控制

The Mechanism and Control Measures about Spot Rust on the Cold Rolling Automobile Strip

Xin Lifeng, Ruan Guoqing, Sun Rongsheng

(Cold Mill of Angang Steel Co., Ltd., Anshan 114021, China)

Abstract: The spot rust on the cold rolling automobile strip is mainly defect during customers using process, it's not easy to find before using, because it's only rust source on the strip in cold rolling producing process before shipping and not clearly spot rust on the strip. The chemical composition of spot rust defect was analyzed by SEM in this paper, and considering cold rolling strip producing process, the classification and causes analysis was done about spot rust, and the effective measures to avoid spot rust appearing again was brought at last.

Key words: cool rolling strip; spot rust; mechanism; control

连退立式活套双塔改单塔控制的实现

赵天鑫，张国强，王 弢，昌 亮，张国立

（鞍山钢铁集团公司冷轧厂，辽宁鞍山 114021）

摘 要： 冷轧厂连退生产线设计有入口和出口两个立式双塔活套，目的是在焊机焊接和光整换辊过程中保证炉区带钢运行稳定，避免出现问题时直接导致炉区急降速甚至停车，产生废品。每个活套均由两个卷扬构成，速度、张力协调控制，但当其中一个卷扬出现问题时，只能停机处理，且时间较长，本文提出了一种单塔控制的预案，避免重大设备事故的发生。

关键词： 活套；张力控制；双塔改单塔

Realization of Double Towers Converted to Single Tower Control in Continuous Annealing Line

Zhao Tianxin, Zhang Guoqiang, Wang Tao, Chang Liang, Zhang Guoli

(Anshan Steel Corporation Cold Rolling Plant, Anshan 114021, China)

Abstract: Two vertical double-tower loopers are designed for the continuous annealing line of cold rolling mill. The purpose is to ensure the stable operation of the strip steel in the furnace area during welding and roll changing of the welding machine, and to avoid the rapid reduction or even stopping of the furnace area when problems occur, and resulting in waste products. Each looper is composed of two hoists, speed and tension coordinated control, but when one of the hoists has problems, it can only be stopped for a long time. This paper proposes a single tower control plan to avoid major equipment accidents.

Key words: looper; tension control; double tower to single tower

轮盘卷取卷筒轴承拆装方法的改进

王桂玉，褚国嵩，张国强

（鞍钢股份有限公司冷轧厂，辽宁鞍山 114021）

摘 要： 本文详细论述了冷轧联合机组轮盘卷取机卷筒轴轴承拆装方法的改进，大幅缩短作业时间，为同类型机组检修作业做出了探索和实践，可以推广和借鉴。

关键词： 轮盘卷取；卷筒轴；轴承；拆装

酸轧联合机组生产高强钢工艺的探讨

王 静[1]，孙荣生[2]，辛利峰[1]

（1. 鞍钢股份冷轧厂，辽宁鞍山 114003；2. 鞍钢集团钢铁研究院，辽宁鞍山 114003）

摘 要： 本文主要对高强钢在酸轧联合机组的生产情况进行了分析，详细的介绍了高强钢在焊接、酸洗及轧制过程中出现的问题及解决的措施，最后提出了保证高强钢在酸轧联合机组稳定生产及产品质量的关键控制点的工艺参数。

关键词： 高强钢；酸轧；生产工艺

The Discuss of the Technology of the High Strength Steel on the Continuous Pickling and Rolling Line

Wang Jing[1], Sun Rongsheng[2], Xin Lifeng[1]

(1. The Cold-Rolling Plant of Ansteel Corporation, Anshan 114003, China;
2. Ansteel Group Iron and Steel Research Institute, Anshan 114003, China)

Abstract: The production of the high strength steel on the continuous pickling and rolling line was analyzed in this paper, the problem and measurement of the high strength steel on the welding, pickling and rolling was introduced detailed, the technology parameter of the key point that ensure the production and quality of the high strength steel on the continuous pickling and rolling line was brought at last.

Key words: high strength steel; pickling and rolling; production technology

轧机轧制线的校核与调整

姜大鹏，王云良，张国强，王 荣

（鞍山钢铁集团公司冷轧厂，辽宁鞍山 114021）

摘 要：基于联合机组轧机的斜楔与阶梯垫复合式轧制线调整装置的结构和传统的计算方法，提出利用弯缸实际伸出量与弯缸理论计算伸出量比较，快速校验和标定轧机轧制线的方法，在实际得到广泛地应用，并取得了显著的效果。为同类设备中轧制线标高校验和调整提供了参考。

关键词：轧制线；标高；斜楔；阶梯垫；弯缸

Check and Adjustment of Mill Passline

Jiang Dapeng, Wang Yunliang, Zhang Guoqiang, Wang Rong

(Angang Iron & Steel Group Cold Strip Works, Anshan 114021, China)

Abstract: Based on the structure of Passline adjustment device and traditional calculating method about the tapered-step wedge of TCM rolling mill, according to compare between the actual extension elongation and theoretical calculating elongation of bending cylinder, introducing a new method for rapid check and calibration of the rolling mill passline, using widely in practice and gaining outstanding effect. A reference is provided about check and adjustment of passline level for other analogical cold-rolled mills.

Key words: passline; elevation; tapered wedge; step wedge; bending cylinder

轴承座润滑油含水原因分析及维护解决方案

李 军

（本钢板材冷轧厂，辽宁本溪 117000）

摘 要：五机架六辊轧机的支撑轴承采用强力稀油润滑方式，通过油的循环带走热量，保持轴承在不高的环境下工作。但是轴承座中润滑油含水量过大已成为行业的共性问题，它不仅破坏了润滑条件，也缩短了轴承使用寿命。本文对润滑油含水的原因进行了初步分析，并提出了轴承座润滑油中的水不是轧制乳化液侵蚀造成的，而是冷凝水集聚形成的观点。几年来在轴承维护中努力减少润滑油中的含水量，采取了一系列有效控制措施，使轧机调试期间受损的 48 组轴承劣化趋势得到遏制，延长了轴承使用寿命，取得较为可观的经济效益。

关键词：轴承；冷凝水；稀油润滑；维护

Cause Analysis and Maintenance Solution of Bearing Chock Lubricating Oil Moisture

Li Jun

(Bengang Steel Plates Co., Ltd., Cold Rolling Mill, Benxi 117000, China)

Abstract: The support bearing of the five-stand and six-high rolling mill is lubricated with strong thin oil, which takes away the heat through the oil circulation and keeps the bearing working in a low environment. However, excessive water content of lubricating oil in the bearing seat has become a common problem in the industry, which not only destroys the lubrication conditions, but also shortens the service life of the bearing. This paper makes a preliminary analysis on the causes of water content in lubricating oil, and puts forward that the water in bearing base lubricating oil is not caused by rolling emulsion erosion, but condensed water agglomeration. In recent years, a series of effective control measures have been taken to reduce the water content in lubricating oil during bearing maintenance, which has curbed the deterioration trend of 48 sets of bearings damaged during rolling mill commissioning, extended the service life of bearings, and achieved considerable economic benefits.

Key words: bearing; condensed water; thin oil lubrication; maintenance

光整机轴承座装配轧辊轴颈工艺改进

李 军

（本钢板材冷轧厂，辽宁本溪 117000）

摘 要：轧辊轴承是光整机的一个重要部件，其装配、调整、维护质量对轴承寿命、镀锌机组作业率有直接影响。轧辊轴承在使用过程中暴露出的某些设计缺陷，通过小改革进行了改进，为企业带来显著效益。

关键词：光整机；轴承；维护；改进

Modification of Roll-neck in Roll Bearing Chocks Assembly of Skin-Pass Mill

Li Jun

(Bengang Steel Plates Co., Ltd., Cold Rolling Mill, Benxi 117000, China)

Abstract: The Roll bearing is an important part of skin-pass mill, whose assembly, adjustment and maintenance quality have direct influence on bearing life and operating rate of CGL. Some design defects of roll bearings exposed in the process of rolling have been improved through small reforms, and bringing remarkable benefits to enterprises.

Key words: skin-pass mill; bearing; maintenance; improvement

锰含量对于 SWRCH22A 调质后组织和性能的影响

郭晓培[1]，阮士朋[1,2]

（1. 邢台钢铁有限责任公司技术中心，河北邢台 054027；
2. 河北省线材工程技术研究中心，河北邢台 054027）

摘 要： 研究了 Mn 含量从 0.80%~1.0% 降低到 0.70%~0.80% 以后，SWRCH22A 的组织和性能变化。结果表明：通过降低 Mn 含量，其淬回火后芯部硬度降低，淬透性能变化不大。

关键词： 锰含量；硬度；淬回火；淬透性

Effect of Manganese Content on Microstructure and Properties of SWRCH22A Quenched Tempering

Guo Xiaopei[1], Ruan Shipeng[1,2]

(1. Technology Center, Xingtai Iron and Steel Crop.,Ltd., Xingtai 054027, China;
2. Hebei Engineering Research Center for Wire Rod, Xingtai 054027, China)

Abstract: The microstructure and properties of SWRCH22A were studied after Mn content decreased from 0.80%~1.0% to 0.70%~0.80%.The results show that the hardness of the core decreases and the hardenability does not change much after tempering by reducing Mn content.

Key words: manganese content; hardness; quenching and tempering; hardenability

Q690 钢新型差温轧制工艺的有限元模拟与验证

江 坤[1]，蔡庆伍[1]，余 伟[2]

（1. 北京科技大学钢铁共性技术协同创新中心，北京 100083；

2. 北京科技大学工程技术研究院，北京　100083）

摘　要： 针对厚板轧制后心部变形小，中心晶粒尺寸粗大的问题，采用模拟与实验相结合的方法，建立了厚板轧制的有限元仿真模型，研究了道次间差温轧制对钢板表面、心部、1/4 处变形量及晶粒尺寸的影响，并与传统均温轧制进行了对比，在国家高效轧制中心进行了轧制实验验证。结果表明，道次间差温轧制相比均温轧制在心部和 1/4 处有更大的变形量，从而促进静态再结晶的发生并细化晶粒尺寸，实验值相比均温轧制分别细化了 11μm、10.5μm。

关键词： 差温轧制；静态再结晶；有限元模拟；奥氏体晶粒尺寸

Finite Element Simulation and Verification of New Temperature Gradient Rolling Process for Q690 Steel

Jiang Kun[1], Cai Qingwu[1], Yu Wei[2]

(1. Collaborative Innovation Center of Steel Technology, USTB, Beijing 100083, China;
2. Institute of Engineering Technology, USTB, Beijing 100083, China)

Abstract: Aiming at the problem of small core deformation and coarse grain size after rolling of thick plate, a finite element simulation model of thick plate rolling was established by the combination of simulation and experiment. The effect of inter-pass temperature gradient rolling on deformation and grain size at the surface, core and 1/4 of the plate was studied, and compared with the traditional uniform temperature rolling. The rolling test was carried out in the national high-efficiency rolling center. The results show that the inter-pass temperature gradient rolling has a larger deformation at the core and 1/4 than the uniform temperature rolling, which promotes the occurrence of static recrystallization and refines the grain size. Compared with the uniform temperature rolling, the experimental values are refined by 10.5μm and 11μm, respectively.

Key words: temperature gradient rolling; static recrystallization; finite element simulation; austenite grain size

铁素体区轧制 IF 钢组织性能

常文杲[1]，余　伟[1,2]

（1. 北京科技大学工程技术研究院，北京　100083；
2. 高效轧制国家工程研究中心，北京　100083）

摘　要： IF 钢铁素体区轧制在能耗、带钢表面质量上存在优势，但对钢板性能能否达到奥氏体区终轧的效果一直存有争议。以 Ti 微合金化-IF 钢为研究对象，在实验室物理模拟了铁素体区热轧-卷取、冷轧、罩式退火、连续退火工艺过程，通过拉伸试验、EBSD、TEM、SEM、XRD 等分析方法，研究了铁素体区热轧 IF 钢的组织和性能。结果表明：经过（800℃终轧）铁素体区热轧和高温卷取，IF 钢组织为均匀的再结晶铁素体晶粒和粗大的析出物，形成了弱的 α 织构和 γ 织构。冷轧后 α 和 γ 织构加强。退火后得到了均匀细小的晶粒，α 织构减少，γ 织构增强，γ 织构成为主要织构。840℃连续退火后，IF 钢的性能达到屈服强度 106MPa、抗拉强度 297 MPa、延伸率 52%、n 值 0.26、r 值 2.3。连续退火后的成形性能优于罩式退火。

关键词： IF 钢；铁素体区轧制；组织；织构

Microstructure and Properties of IF Steel Rolled in Ferrite Zone

Chang Wengao[1], Yu Wei[1,2]

(1. Engineering Technology Institute, University of Science and Technology Beijing, Beijing 100083, China; 2. National Engineering Research Center for Advanced Rolling Technology, Beijing 100083, China)

Abstract: IF steel rolling in ferrite zone has advantages in energy consumption and surface quality of strip steels. but whether the properties of steel plate can reach the effect of end rolling in austenite zone is still controversial. Taking Ti microalloyed IF steel as the research object, the process of hot rolling-coiling, cold rolling, bell annealing and continuous annealing in ferrite region was simulated in laboratory. The microstructure and properties of hot-rolled IF steel in ferrite region were studied by means of tensile test, EBSD, TEM, SEM and XRD. The results show that the microstructure of IF steel is uniform recrystallized ferrite grain and coarse precipitate after hot rolling in ferrite region (finishing at 800℃), and weak α-fiber and γ-fiber are formed. After cold rolling, the α-fiber and γ-fiber textures are strengthened. After annealing, uniform and fine grains were obtained, the α-fiber is transformed into γ-fiber. The γ-fiber becomes the main texture. After continuous annealing at 840℃, the yield strength, tensile strength, elongation, n value and r value of IF steel are 106 MPa, 297 MPa, 52%, 0.26 and 2.3. The formability of continuous annealing is better than that of batch annealing.

Key words: IF steel; ferritic hot rolling; microstructure; texture

激光熔覆技术在冷轧支撑辊辊颈修复应用实践

林上辉，潘勋平

（宝钢日铁汽车板有限公司，上海 200941）

摘 要： 本文对某冷轧机支撑辊辊颈修复再造进行研究分析，通过小样进行激光熔覆试验并分析其相关性能。制定了适合该支撑辊辊颈激光熔覆修复的工艺参数和修复流程。应用该修复工艺对支撑辊辊颈磨损缺陷进行激光熔覆修复处理，并从微观组织、力学性能等方面对熔覆层进行了测试。测试表面激光熔覆修复冶金结合过渡层性能满足力学性能要求，激光熔覆修复表面处理后表面硬度、截面微观组织满足使用要求。采用该激光熔覆工艺对支撑辊辊颈进行修复并上机使用。

关键词： 激光熔覆；支撑辊；修复

The Application of Laser Cladding Technology in Repairing the Neck of Backup Roll in Cold Rolling

Lin Shanghui, Pan Xunping

(Baosteel-Nippon Steel Automotive Steel Sheet Co., Ltd., Shanghai 200941, China)

Abstract: In the paper, the repairing process of backup roll neck of a rolling mill was analyzed. Laser cladding tests were carried out on the sample and their related properties were analyzed. The laser cladding technological parameters and repairing process suitable for repairing the backup roll neck was developed. The optimised laser cladding process was applied to repair the wear defect of roll neck from a backup roll. The properties of the composite surface such as hardness and microstructure after surface treatment by laser cladding can meet the requirements of backup roll during using. The backup roll neck was repaired and put into use finally by the laser cladding technology.

Key words: laser cladding; backup roll neck; repair

4　表面与涂镀

大会特邀报告

分会场特邀报告

炼铁与原料

炼钢与连铸

轧制与热处理

★ 表面与涂镀

金属材料深加工

先进钢铁材料

粉末冶金

能源、环保与资源利用

钢铁材料表征与评价

冶金设备与工程技术

冶金自动化与智能化

建筑诊治

其他

热浸镀锌铝镁镀层微观组织试验研究

吕家舜,徐闻慧,杨洪刚,李 锋,苏皓璐

(鞍钢集团钢铁研究院,辽宁鞍山 114009)

摘 要:研究了热浸镀工艺对于镀层的微观结构的影响,利用扫描电镜(SEM)及能谱分析(EDX)观察了Zn-Al-Mg镀层表面以及截面的微观结构、合金层的形貌、镀层中各相的成分组成,利用辉光放电发射光谱仪(GDS)分析了镀层中各元素沿深度方向的分布,利用电子探针(EPMA)分析了镀层中各元素的分布,利用X射线光电子能谱(XPS)分析了镀层表面元素,利用X射线衍射(XRD)分析了镀层的相组成。结果表明,镀层中各元素沿镀层的深度方向的分布并不均匀,Mg在镀层表面富集,镀层组织呈现多相混合结构,以Zn晶粒、$MgZn_2$与Zn组成的共晶为主,同时存在块状富铝相以及一层较薄的合金层。

关键词:热浸镀;Zn-Al-Mg镀层;X射线光电子能谱(XPS);镀层微观组织

Experiments Investigation on Microstructure of Galvanized Zn–Al–Mg Coated Steel Sheet

Lv Jiashun, Xu Wenhui, Yang Honggang, Li Feng, Su Haolu

(Ansteel Iron & Steel Research Institute, Anshan 114009, China)

Abstract: The facts of galvanizing parameters on microstructure of hot-dipped Zn-Al-Mg coating steel strip were investigated. The surface and cross sectional microstructure of Zn-Al-Mg coating and the chemical competition and the morphology of alloy layer were analyzed by SEM and EDX. The element distribution in the depth direction of coating was analyzed by GDS. The element distribution in coating was analyzed by EPMA. The phase composition of coating were analyzed XRD. The elements' content in coating was explored by XPS. It is proved that the composition elements distribution is asymmetry in the depth direction in coating, and zinc and magnesium is enriched in coating surface. Coating structure takes on multiphase commix, $MgZn_2$ and pure Zinc binary eutectic are the main phase and at the same time there are some zinc-rich nubby phase and a filmy alloy layer.

Key words: hot dip galvanizing; Zn-Al-Mg alloy coating; XPS; coating microstructure

马钢高导电热浸镀锌产品的研究与开发

杨兴亮,王 滕,杨 平,刘 劼

(马鞍山钢铁股份有限公司技术中心,安徽马鞍山 243000)

摘 要:基于马钢的热浸镀生产工艺,本文介绍了马钢高导电热浸镀锌钢板的研究与开发情况。通过导电性测试、中性盐雾实验、涂装性能测试等实验手段,得到优化的系统工艺。实现了马钢高导电热浸镀锌钢板的成功开发,产品综合工艺性能指标优异,均超过目标客户的要求。马钢高导电热镀锌产品已实现工业化的批量生产,并在伺服器、存储器等制造领域得到应用。

关键词:高导电;热浸镀锌;伺服器

Research and Development of High Conductivity Hot Dip Galvanizing Products in Masteel

Yang Xingliang, Wang Teng, Yang Ping, Liu Jie

(Technology Center, Maanshan Iron & Steel Co., Ltd., Maanshan 243000, China)

Abstract: Based on the hot-dip galvanizing process in Masteel, the research and development of high conductivity hot dip galvanized steel sheet in Masteel are introduced in this paper. Through conductivity test, neutral salt spray test and painting performance test, the optimized system process was obtained. The successful development of high conductivity hot dip galvanized steel sheet in Masteel has been realized, the comprehensive process performance of the product is excellent, which exceeds the requirements of the target customers. Masteel's high conductivity hot-dip galvanizing products have been industrialized in batches, and have been applied in the fields of servo and memory manufacturing.

Key words: high conductivity; hot dip galvanizing; servo

电视机背板用彩涂板涂层技术研究

施国兰,蔡 明,谷 曦,钱婷婷

(马鞍山钢铁股份有限公司,安徽马鞍山 243000)

摘 要:通过对涂层润滑性能、柔韧性能、耐磨性能等技术研究,设计了满足电视机背板加工成型需求的彩涂板涂层技术,应用效果良好。

关键词:电视机背板;彩涂板;涂层;技术

Study on Coating Technology of Color Coated Plates for TV Backplane

Shi Guolan, Cai Ming, Gu Xi, Qian Tingting

(Maanshan Iron & Steel Co., Ltd., Maanshan 243000, China)

Abstract: This article studied the lubrication performance、flexibility、wear resistance of coating, developed pre-coated sheet products for TV Backplane usage and have been obtained good application results.

Key words: TV backplane; color coated plates; coating; technology

烧结电除尘环境下碳钢防护涂层失效行为

高　鹏[1,2]，陈义庆[1,2]，武裕民[1,2]，王佳骥[1,2]，艾芳芳[1,2]，
李　琳[1,2]，钟　彬[1,2]，肖　宇[1,2]，伞宏宇[1,2]，苏显栋[1,2]

（1. 海洋装备用金属材料及其应用国家重点实验室，辽宁鞍山　114009；
2. 鞍钢集团钢铁研究院焊接与腐蚀研究所，辽宁鞍山　114009）

摘　要：采用水性无机富锌涂料在 St12 钢板表面制备了防护涂层，在烧结机头电除尘器中进行了挂片试验。研究结果表明，水性无机富锌涂层的失效过程为点状损伤萌生后逐步扩大，进而演变为溃疡状损伤；在烧结机头电除尘器环境下服役一年后，防护涂层表现出良好的防护能力，阳极板材料未发生显著腐蚀。

关键词：水性无机富锌涂料；电除尘；阳极板；腐蚀；防护涂层

Failure Behavior of Protective Coating on Carbon Steel under Sintering Electric Remove Condition

Gao Peng[1,2], Chen Yiqing[1,2], Wu Yumin[1,2], Wang Jiaji[1,2], Ai Fangfang[1,2],
Li Lin[1,2], Zhong Bin[1,2], Xiao Yu[1,2], San Hongyu[1,2], Su Xiandong[1,2]

(1. State Key Laboratory of Metal Material for Marine Equipment and Application, Anshan 114009, China;
2. Iron & Steel Research Institute of Angang Group, Anshan 114009, China)

Abstract: Waterborne inorganic zinc-rich coatings were coated on St12 steel sheets.Coupon test was performed in the electrostatic precipitator at the head of sintering machine. The results show that the deterioration processes of waterborne inorganic zinc-rich coatings includes the pitting initiation, corrosion pits expand, formation of ulcer-like corrosion.The coatings serviced for 1year under sintering electric eemove condition, no obvious corrosion occurred on positive plate, coatings had good protection properties.

Key words: waterborne inorganic zinc-rich coatings; electrostatic precipitation; positive plate; corrosion; protective coating

生产运营期钢结构防腐的探讨与实践

殷　栋，苏会德，吴新辉

（山东钢铁集团日照有限公司工程部，山东日照　276800）

摘　要：冶金工程投入生产运营后，由于钢结构大部分处于高空，生产运营期二次防腐时，由于除锈难度大，采用常规油漆防腐，难以达到理想效果，因此钢结构的二次防腐成为了冶金企业质量控制的难点。通过社会实践调研和实际应用，采用水性金属带锈防腐漆可以达到不错的防腐效果。

关键词：冶金工程；生产运营；钢结构；水性金属带锈防腐漆

不同纳米添加剂对微弧氧化陶瓷膜层耐腐蚀性能的影响

张 宇[1,2]，刘 鑫[1]，曾 丽[1]，李红莉[1]，宋仁国[3]

（1. 浙江工业职业技术学院机械工程学院，浙江绍兴 312000；
2. 浙江工业大学特种装备制造与先进加工技术教育部/浙江重点实验室，浙江杭州 310014；
3. 常州大学材料科学与工程学院，江苏常州 213164）

摘 要：本文通过在微弧氧化制备过程中添加 TiO_2 和 Al_2O_3 纳米粉末，采用 SEM、XRD、XPS、电化学工作站等，研究不同纳米添加剂对微弧氧化陶瓷膜层耐腐蚀性能的影响。结果表明，纳米添加剂在微弧氧化过程中都能够有效地参与到陶瓷膜层的制备过程中，使得陶瓷膜层耐腐蚀性能等有很大的提高，且含有纳米 Al_2O_3 添加剂的膜层性能比含有纳米 TiO_2 添加剂的膜层性能要好。

关键词：微弧氧化；纳米添加剂；耐腐蚀

Effects of Different Nano-additives on Corrosion Resistance of Micro-arc Oxidation Films

Zhang Yu[1,2], Liu Xin[1], Zeng Li[1], Li Hongli[1], Song Renguo[3]

(1. School of Mechanical Engineering, Zhejiang Industry Polytechnic College, Shaoxing 312000, China; 2. Key Lab of E&M, Ministry of Education & Zhejiang Province, Zhejiang University of Technology, Hangzhou 310014, China; 3. School of Materials Science and Engineering, Changzhou University, Changzhou 213164, China)

Abstract: In this paper, the effects of different Nano-Additives on the corrosion resistance of micro-arc oxidation ceramic coatings were studied by adding nano-powders of titanium dioxide and alumina during the preparation of micro-arc oxidation, by means of SEM, XRD, XPS and electrochemical workstation. The results show that Nano-Additives can effectively participate in the preparation of ceramic coatings during micro-arc oxidation, which greatly improves the corrosion resistance of ceramic coatings, and the performance of coatings containing nano-Al_2O_3 additive is better than that of coatings containing nano-TiO_2 additive.

Key words: micro-arc oxidation; nano-additives; corrosion resistance

防氧化涂料降低弹簧脱碳层深度的生产实践

彭 超[1]，王旭冀[2]，朱建成[2]

（1. 华菱湘钢高线厂，湖南湘潭 411100；2. 华菱湘钢技术质量部，湖南湘潭 411100）

摘 要：为研究不同涂料对弹簧钢盘条脱碳层深度的影响，研究了其氧化与脱碳规律的基础上在汽车用弹簧钢

55SiCrA 二火钢坯进行了两种防氧化涂料试验。使用喷涂法将涂料涂敷于试样表面，经高线加热炉高温加热后轧制成规格 ϕ16mm 的盘条，通过金相和电子显微镜分析了涂层在高温下的保护效果，得出结论：弹簧钢坯脱碳层深度涂料 A 防脱碳效果明显，脱碳层深度在 0.04~0.06mm，涂料 B 脱碳层深度在 0.07~0.09mm，未涂盘条脱碳层深度在 0.09~0.12mm；盘条表面的微观缺陷未有明显的变化，电镜下未发现涂料残留。

关键词：弹簧钢；涂料；脱碳层

入锌锅温度对连续热镀锌 Fe-Al 抑制层的影响

李婷婷，岳重祥，李化龙

（江苏省(沙钢)钢铁研究院，江苏张家港　215625）

摘　要：在带钢连续热镀锌工业化生产过程中，向锌液中添加少量 Al，采用不同带钢入锌锅温度进行连续热镀锌。利用扫描电镜（SEM）和能谱仪（EDS）对热镀锌镀层的形貌、相结构和成分进行分析，研究带钢连续热镀锌过程中带钢入锌锅温度对镀层的影响。结果表明：带钢连续热镀锌过程中，向锌液中添加少量 Al 可使钢基板和锌镀层间形成 Fe-Al 中间层，抑制 Fe-Zn 合金层的形成，提高镀层的粘附性；而带钢入锌锅温度显著影响 Fe-Al 抑制层的结构，进而影响 Fe-Zn 合金层厚度，并最终影响镀层性能。

关键词：热镀锌；入锌锅温度；Fe-Al 抑制层；热镀锌层

Effect of Strip-entry Temperature on the Fe-Al Inhibition Layer during Continuous Hot-dip Galvanizing Process

Li Tingting, Yue Chongxiang, Li Hualong

(Institute of Research of Iron and Steel, Shasteel, Zhangjiagang 215625, China)

Abstract: During industrial production of hot-dip galvanization, a small amount of Al was added into zinc bath and different strip-entry temperatures were tested. By scanning electronic microscope (SEM) and energy spectrometer (EDS), the morphology, phase structure and composition of the coating layer were characterized to study the effect of strip- entry temperature on the coating. The results showed that, during continuous hot-dip galvanizing process, the added Al formed a Fe-Al intermediate layer between the substrate and the zinc coating to inhibit the formation of Fe-Zn alloy layer and improved the adhesion of the coating; the strip-entry temperature significantly affected the structure of Fe-Al inhibition layer, thereby the thickness of Fe-Zn alloy layer, and ultimately coating performance.

Key words: hot-dip galvanizing; strip-entry temperature; Fe-Al inhibition layer; hot dip galvanizing coating

两种锌铝镁镀层板黑点缺陷分析

杜　江，许秀飞

（中冶赛迪工程技术股份有限公司板带事业部，重庆　401122）

摘 要：本文对来自国内 A 钢厂生产的中铝锌铝镁小黑点缺陷和 B 钢厂生产的低铝锌铝镁大黑点缺陷进行了分析，认为都是由于结晶组织异常造成的；小黑点缺陷是由于结晶时成分偏析或者冷却速度太低造成的，在局部区域出现了大晶粒、大面积的共晶组织，在仓储时出现了氧化现象，造成光线的散射，看起来就呈现出黑色小点，建议提高镀浴成分的均匀性并提高镀后冷却速度，或者添加 Ti 或 B 等合金元素；大黑点缺陷是由于镀层内有杂质，在冷却凝固过程中，杂质点附近大量形核、优先凝固，形成了大量细小的晶粒，反射光线较强的富锌相树枝晶很瘦小，而对光线反射较弱共晶相面积较大，所以看起来这一区域就是一个大黑点，建议改用内加热法生产的高纯度的合金锭。

关键词：连续热浸镀；锌铝镁；镀层组织；黑点缺陷；合金熔炼

Research of Two Kinds of Black Point Defects in Zinc-Aluminum-Magnesium Coated Sheet

Du Jiang, Xu Xiufei

(Iron & Steel Business Division, CISDI Engineering Co., Ltd., Chongqing 401122, China)

Abstract: In this article, we studied small size black point defect in Zinc-Aluminum-Magnesium coated sheet with medium content zinc and magnesium from company A, and large size black point defect in Zinc-Aluminum-Magnesium coated sheet with low content zinc and magnesium from company B. As results, it is thought that both defects are caused by abnormal crystalline structure. Segregation of components or a too low cooling rate during crystallization leads to small size black point defect. Large size grain and large area eutectic structure appear in partial area, oxidized during storage, causing light scattering, which makes it look like small black points. It is advised to avoid this defect by improving the uniformity of the composition of the plating bath and the cooling rate, or adding alloy element like titanium or boron. For the large black point, nearby the impurities in the coating, numerous crystal nucleuses form and crystallize preferentially, producing numerous small size grains. Zinc rich dendritic phase with strong reflection ability is tiny and area of eutectic phase with weak reflection ability is large, which makes it look like large black points. It is advised to avoid this defect by changing high purity alloy ingot produced by internal heating method.

Key words: continuous hot dip coating; zinc-aluminum-magnesium; coating structure; black point defect; alloy melting

中国钢制品及型材热浸镀锌线设计选型建议

徐言东[1]，马树森[2]，顾 洋[3]

（1. 北京科技大学工程技术研究院，北京 100083；2. 唐山市开平鑫德热镀锌技术有限公司，河北唐山 063000；3. 北京科技大学钢铁技术协同创新中心，北京 100083）

摘 要：笔者选择钢制品及型材热浸镀锌生产发达地区（如河北、天津、江苏、浙江、山东等地区）考察调研，对当前钢制品及型材热浸镀锌行业的工艺、技术、装备水平、设计形式、生产现状等进行了详细的考察，并做了解、分析和对比。通过组织这次中国钢制品及型材热浸镀锌领域生产技术考察活动，汇集了各方经验，解答了诸多疑问，解决了一些实际问题，提出了一些生产线设计选型建议。

关键词：热浸镀锌；生产；装备；现状

Suggestions on Design Selection of Hot Dip Galvanizing Line for Steel Products and Profiles in China

Xu Yandong[1], Ma Shusen[2], Gu Yang[3]

(1. Engineering and Technology Research Institute, University of Science and Technology Beijing, Beijing 100083, China; 2. Tangshan Kaiping Xinde Hot Dip Galvanizing Technology Co., Ltd., Tangshan 063000, China; 3. Collaborative Innovation Center of Steel Technology, University of Science and Technology Beijing, Beijing 100083, China)

Abstract: Author selected steel products and profiles in hot dip galvanizing to produce research and development in developed regions (such as Hebei, Tianjin, Jiangsu, Zhejiang, Shandong, etc.). The process, technology, equipment level, design form and production status of the current hot-dip galvanizing industry of steel products and profiles were examined in detail, and understood, analyzed and compared. Through the organization of the production technology inspection activities of Chinese steel products and profiles in hot dip galvanizing, the experience of various parties was gathered, many questions were answered, some practical problems were solved, and some design suggestions for production line design were proposed.

Key words: hot dip galvanizing; production; equipment; status

激光熔覆技术现状及发展

周 丰，贵永亮，胡宾生，扈理想

（华北理工大学冶金与能源学院，河北唐山　063210）

摘　要：激光熔覆技术是现代表面工程技术极具发展的技术之一，它的优势在于加工精度高、易于数字化控制、有效地减小热影响区、加工效率更高以及设备维护更方便，主要应用于对材料的表面改性和对产品表面的修复。本文综述了激光熔覆技术的原理及优点；介绍了激光熔覆工艺方式；阐述了激光熔覆的材料体系及工艺参数；分析了激光熔覆技术在工业中的应用；指出了激光熔覆技术存在的主要问题和今后的研究方向。

关键词：激光熔覆；工艺方式；耐磨；耐蚀

Present Situation and Development of Laser Cladding Technology

Zhou Feng, Gui Yongliang, Hu Binsheng, Hu Lixiang

(College of Metallurgy and Energy, North China University of Science and Technology, Tangshan 063210, China)

Abstract: Laser cladding technology is one of the most developed technologies in modern surface engineering technology. Its advantages are high processing precision, easy digital control, effective heat reduction zone reduction, higher processing efficiency and more convenient equipment maintenance. Surface modification of the material and repair of the surface of the product. This paper reviews the principle and advantages of laser cladding technology; introduces the laser cladding process; describes the material system and process parameters of laser cladding; analyzes the application of laser cladding

technology in industry; points out the laser cladding technology. The main problems and future research directions.
Key words: laser cladding; process mode; wear resistance; corrosion resistance

Zn-Al-Mg 合金镀层的微观组织及相组成分析

蔡 宁，黎 敏，赵晓非，其其格，曹建平，杨建炜

（绿色可循环钢铁流程北京市重点实验室，首钢集团有限公司技术研究院，北京 100043）

摘 要： Zn-Al-Mg 镀层钢板具有优异的切口保护性能和耐大气腐蚀性能，该镀层在 20 世纪 80 年代开始逐渐商业化，但 Zn-Al-Mg 镀层的相组成至今尚存在争议。为了更好地控制镀层结构、镀层钢板的成形性能、耐腐蚀性能，本论文详细分析了 Zn-Al-Mg 镀层钢板的微观组织及相结构。通过扫描电镜、XRD、辉光光谱、透射电镜等分析手段，确定了 Zn-Al-Mg 镀层由 Zn、Al、$MgZn_2$ 三相组成。透射电镜分析发现镀层与钢板的界面存在 10nm 厚的 Al-Fe 非晶态合金层，然后依次形成 $MgZn_2$ 合金层、颗粒状 $MgZn_2$ 与 Zn 的共晶组织层、纯 Zn 层或者片状 $MgZn_2$ 与 Zn 的共晶组织层。

关键词： Zn-Al-Mg 合金；镀层；微观组织；相组成

Microstructure and Phase Analysis on Zn-Al-Mg Alloy Coatings

Cai Ning, Li Min, Zhao Xiaofei, Qi Qige, Cao Jianping, Yang Jianwei

(Beijing Key Laboratory of Green Recyclable Process for Iron & Steel Production Technology, Research Institute of Technology of Shougang Group Co., Ltd., Beijing 100043, China)

Abstract: Zn-Al-Mg coated steels have excellent cut edges protection and atmospheric corrosion resistance. It has been commercialized since 1980s, but the phase composition of Zn-Al-Mg coating is still controversial. In order to better control the coating structure, the formability and corrosion resistance of the coated steel sheet, the microstructure and phase structure of the Zn-Al-Mg coated steel sheet were analyzed in detail in this paper. After analyzing with scanning electron microscopy, X-ray diffraction (XRD), glow discharge spectrum(GDS) and transmission electron microscopy (TEM), it was determined that the Zn-Al-Mg coatings were consisted with three phases, they were Zn, Al and $MgZn_2$. TEM analysis showed that there was a 10nm thick amorphous Al-Fe alloy layer at the interface between the coating and the steel plate, and then $MgZn_2$ alloy layer, eutectic structure layer of granular $MgZn_2$ and Zn, pure Zn layer, eutectic structure layer of lamellar $MgZn_2$ and Zn.
Key words: Zn-Al-Mg alloys; coatings; microstructures; phase composition

纯锌镀层和锌铝镁合金镀层在循环盐雾中的腐蚀行为

黎 敏，郝玉林，龙 袁，姚士聪

（首钢集团有限公司技术研究院，北京 100041）

摘　要：研究了纯锌镀层（GI）和锌铝镁合金镀层（ZM）在 PV1210 循环腐蚀 30 周期内的腐蚀行为，利用扫描电镜（SEM）、电化学工作站和 X 射线衍射（XRD）分析了腐蚀产物的组成。结果表明：ZM 具有更好的耐蚀性，腐蚀过程中，共晶相中的 Mg 优先溶解，Mg 离子可以缓冲 pH 值升高，抑制镀层表面碱化，抑制了 ZnO 的形成，降低了 ZM 镀层的阴极反应速率和腐蚀速率。

关键词：锌铝镁镀层；热镀锌板；石击；循环腐蚀

The Corrosion Behaviour of Zn and Zn–Al–Mg Coated Steel in the Cyclic Salt Spray Test

Li Min, Hao Yulin, Long Yuan, Yao Shicong

(Shougang Research Institute of Technology, Beijing 100041, China)

Abstract: The corrosion behaviour of pure zinc and zinc–magnesium–aluminium alloy has been studied during 30cy of exposure in PV1210 cyclic corrosion. The composition of corrosion products is analysed using sweep electron microscope (SEM), electrochemical workstation and X-ray diffraction (XRD). An improved corrosion resistance of ZM is observed. Magnesium dissolved preferentially. Magnesium buffered the pH at cathodic sites, which inhibit the ZnO formation, reduce the cathodic reaction and corrosion rate on ZM.

Key words: Zn–Al–Mg coated steel; galvanized sheets; stone chipping; cyclic corrosion

高强汽车用酸洗钢酸洗表面发黑研究与控制

肖厚念

（宝钢股份武钢有限冷轧厂，湖北武汉　430080）

摘　要：高强汽车用钢以热代冷逐渐成为趋势，热轧酸洗钢替代冷轧退火板具有高强化、低成本优势。但同时也对表面质量提出了更高要求，酸洗后表面具备色泽均匀光洁，无明显色差。本文以典型汽车酸洗钢典型发黑问题，从热轧工艺氧化铁皮改善、酸洗工艺改进等方面分析与研究，结合生产工艺管控点提出有效的改善措施。

关键词：汽车酸洗钢；氧化铁皮；酸洗；发黑；控制

Research and Control of Surface Blackening of Pickling Steel for High Strength Automobile

Xiao Hounian

(Cold Mill of Baosteel Wuhan Iron and Steel Ltd., Wuhan 430080, China)

Abstract: It is a trend to replace cold rolled plate with hot rolled plate for high strength automobile, Hot-rolled pickling steel has the advantages of high strengthening and low cost instead of cold-rolled annealed plate. But at the same time, it also puts forward higher requirements for surface quality. After pickling, the surface is uniform and smooth without obvious color difference. In this paper, typical pickling steels for automobiles are analyzed and studied from the aspects of iron oxide scale improvement and pickling process improvement in hot rolling process. Combining with the production process control points, effective improvement measures are put forward.

Key words: pickling steel for automobiles; iron oxide scale; pickling; blackening; control

Al-Si 镀层微观结构及变形能力研究

郭　健，魏焕君，耿志宇，潘文娜，杨丽娜，李　勃

(唐山钢铁集团有限责任公司技术中心，河北唐山　063016)

摘　要：本文利用扫描电镜观察了铝硅镀层相结构，利用热模拟机模拟了铝硅镀层在不同保温时间和变形速率下镀层厚度的变化规律及变形能力，利用辉光光谱仪测定了铁元素向合金层中扩散的规律。实验结果表明：铝硅镀层由富铝相、富硅相、富铁相组成，其中富铝相比例最高，约占 80%，富硅相约占 10%，富铁相为少量的块状结构；随着保温时间的延长，镀层中合金层的厚度从 4.03μm 长到 29.72μm，铁元素由基体向镀层表面扩散，浓度达到 50%；铝硅镀层在高温下有良好的变形能力。

关键词：Al-Si 镀层；镀层微观结构；镀层厚度；镀层变形能力

Study on Microstructure and Deformation Ability of Al-Si Coating

Guo Jian, Wei Huanjun, Geng Zhiyu, Pan Wenna, Yang Lina, Li Bo

(Tangshan Iron and Steel Group Technology Center, Tangshan 063016, China)

Abstract: In this paper, the phase structure of Al-Si coating was observed by scanning electron microscopy. The variation of coating thickness and deformation capacity of Al-Si coating under different holding time and deformation rate were simulated by thermal simulator. The iron element was measured by glow spectrometer. The law of diffusion. The experimental results show that the aluminum-silicon coating consists of aluminum-rich phase, silicon-rich phase and iron-rich phase. The aluminum-rich phase is the highest, accounting for 80%, the silicon-rich phase is about 10%, and the iron-rich phase is a small block structure. With the increase of holding time, the thickness of the alloy layer in the coating is from 4.03μm to 29.72μm, and the iron element diffuses from the substrate to the surface of the coating to a concentration of 50%; the aluminum-silicon coating has good deformability at high temperature.

Key words: Al-Si coating; coating microstructure; coating thickness; coating deformation ability

基板表面形貌对镀锡层覆盖能力影响的研究

谢志刚，李顺祥，张青树

(上海梅山钢铁股份有限公司，江苏南京　210039)

摘　要：镀锡板的耐蚀性与镀锡层的覆盖率有关，本文通过试验研究了镀锡层覆盖率与镀锡基板表面形貌之间的关系，研究结果表明：同一表面粗糙度的镀锡基板，镀锡层越厚则覆盖率越高；在镀锡量较低时，较低的表面粗糙度更有利于获得覆盖率较好的镀层；对于低锡量镀锡板（镀锡量小于 1.1g/m^2）的基板，为保证镀锡层的覆盖率，表面

粗糙度应不大于0.26μm。

关键词：基板形貌；镀锡；覆盖率

Study on the Influence of Substrate Surface Morphology on the Coating Coverage Rate of Tin Plating

Xie Zhigang, Li Shunxiang, Zhang Qingshu

(Shanghai Meishan Iron & Steel Co., Ltd., Nanjing 210039, China)

Abstract: The corrosion resistance of tin plating plates is related to the coverage of tin plating layers. In this paper, the relationship between tin plating coverage and the surface morphology of the substrate is studied by experiments. As the amount of tin plating thickness is low, the lower surface roughness is more favorable to obtain better coating. For low tin plating plates(less than 1.1 g/m^2), the surface roughness should be less than 0.26μm to ensure the coating coverage.

Key words: surface morphology; tin plating; coverage rate

宝钢无机润滑 GA 热镀锌汽车板产品的生产与应用

张 军[1]，温乃盟[2]，卢海峰[3]

(1. 宝山钢铁股份有限公司冷轧厂，上海 200941；2. 宝山钢铁股份有限公司研究院，上海 200941；3. 宝山钢铁股份有限公司湛江钢铁冷轧厂，广东湛江 524000)

摘 要：无机润滑GA汽车板产品是一种改善汽车用热镀锌板冲压成形性能的新型产品，本文概述了该产品的生产工艺，对该产品的主要理化性能、汽车厂实际冲压性能和耐蚀性能等进行对比分析，结果表明，无机润滑GA汽车板产品具有良好的表面摩擦特性和成形性，同时该产品满足汽车行业对冲压成形性能、磷化性能和涂装后的耐腐蚀性能等常规涂装工艺的生产要求。

关键词：合金化热镀锌；无机润滑膜；汽车板

Production and Application of Inorganic Solid Lubricant Coated Galvannealed Automobile Steel Sheet

Zhang Jun[1], Wen Naimeng[2], Lu Haifeng[3]

(1. Cold Rolling Mill, Baoshan Iron & Steel Co., Ltd., Shanghai 200941, China; 2. Research Institute, Baoshan Iron & Steel Co., Ltd., Shanghai 200941, China; 3. Cold Rolling Mill, Zhanjiang Iron & Steel Co., Ltd., Zhanjiang, 524000, China)

Abstract: The inorganic solid lubricant galvannealed automobile steel sheet was a new kind production for the purpose of improving the press formability of automobile body panels. In this paper, the production process, the physical and chemical properties, and the press formability and the corrosion resistance of the new kind production used in automobile plant were discussed. The results showed that the inorganic solid lubricant galvannealed automobile steel sheet has good surface friction characteristics and press formability. Meanwhile, the new kind production meets the requirements of press formability, phosphate properties and corrosion resistance after coating in automobile industry.

Key words: galvannealing; inorganic solid lubricant film; automobile steel sheet

液中放电沉积参数对工件性能的影响

顾晓辉[1]，何 星[2]，张晨昀[2]

（1. 上海江南轧辊有限公司，上海 201919；2. 上海理工大学，上海 200093）

摘 要： 电火花加工技术在低刚度零件加工等方面具有深远的实际应用意义。本研究中选用 TiC 粉末制备工具电极，在煤油中进行液中放电沉积，研究了峰值电流对沉积层硬度、厚度、表面粗糙度等的影响。结果表明随着峰值电流的增加，电极的消耗率以及其粗糙度会随之增加，而工件的单位面积损耗率却逐渐降低，改性层的硬度和厚度都会在峰值电流达到某一值时得到其最大值。

关键词： 电火花加工；液中放电沉积

Effect of Liquid Discharge Deposition Parameters on Workpiece Performance

Gu Xiaohui[1], He Xing[2], Zhang Chenyun[2]

(1. Shanghai Jiangnan Roll Co., Ltd., Shanghai 201919, China;
2. Shanghai University of Technology, Shanghai 200093, China)

Abstract: EDM technology has a practical application in the processing of low stiffness parts. In this study, TiC powder was used to prepare tool electrodes, which were deposited in kerosene by liquid discharge. The effects of peak current on hardness, thickness and surface roughness of the deposited layer were studied. The results show that with the increase of peak current, the consumption rate and roughness of the electrode will increase, while the unit area loss rate of the workpiece will gradually decrease, and the hardness and thickness of the modified layer will get its maximum value when the peak current reaches a certain value.

Key words: EDM technology; liquid discharge deposition

无铬钝化热镀锌钢板摩擦条纹的分析与改善

杨 芃，张 云

（宝山钢铁股份有限公司武汉钢铁有限公司冷轧厂，湖北武汉 430083）

摘 要： 本文对涂油无铬钝化热镀锌钢板进行分条加工出现的加工摩擦条纹缺陷的成分进行了分析，通过是或否涂油、新油或循环油、无铬钝化或三价铬钝化等各影响因素的实验室模拟实验对该缺陷的产生机理进行了分析，发现该缺陷是由于毛毡使无铬钝化膜减薄，使锌尖峰被毛毡磨损下来的锌粉与油的混合物，油在其中起显色和富集作用。

提出采用不涂油无铬钝化热镀锌钢板，或采用三价铬钝化热镀锌钢板的解决方案。

关键词：热镀锌钢板；无铬钝化；加工摩擦条纹；黑条纹

Analysis and Improvement of Friction Striation Defect on the Oiled Cr-free Passivation Hot-Dip Galvanizing Sheet

Yang Peng, Zhang Yun

(Cold Rolling Plant, Wuhan Iron and Steel Company Limited of Baosteel, Wuhan 430083, China)

Abstract: This paper analyses the components of processing friction striation defect produced during the oiled Cr-free passivation hot-dip galvanizing sheet is slitted. By lab simulation tests of several effects, such as yes or no oiling, new oil or cycle oil, Cr-free passivation or Cr^{3+} passivation, the mechanism of the defect is found, and it is that the felt reduces the thickness of Cr-free passivation film, wears the zinc peaks, and the mix of zinc powder and oil is formed, and the role of oil is black effects and enrichment. So, the solutions are suggested that using oil-free Cr-free passivation hot-dip galvanizing sheet or using Cr^{3+} passivation hot-dip galvanizing sheet.

Key words: hot-dip galvanizing sheet; Cr-free passivation; processing friction striation; black striation

直读光谱法测定锌合金中的铝含量

黄碧芬，王君祥，李艳

(宝钢湛江钢铁有限公司制造管理部，广东湛江 524072)

摘 要：铝是热镀锌中非常关键的金属元素，锌锅中铝含量的设定和稳定是确保热镀锌品镀层性能和质量的重要前提。本文采用直读光谱仪分析方法测定锌合金中的铝含量，对该方法的准确度和精度进行全面探讨。该分析方法操作简单，可快速检测，并且具有良好的准确度和精度，可满足镀锌机组工艺上的检测要求。

关键词：直读光谱仪；铝含量；准确度；精度；快速

Direct Reading Spectrometric Determination of Aluminum in Zinc Alloys

Huang Bifen, Wang Junxiang, Li Yan

(Products &Technique Management Department, Baosteel Zhanjiang Iron and Steel Co., Ltd., Zhanjiang 524072, China)

Abstract: Aluminum is a very important metal element in hot-dip galvanizing, and the setting and stability of aluminium in zinc pot is an important precondition to ensure the properties and quality of hot-dip galvanization.In this paper,the content of aluminum in zinc alloy is determined by direct-reading spectrum analysis method, and the accuracy and precision of this method are discussed.The analysis method is simple to operate, can be detected quickly, and has good accuracy and precision, which can satisfy the testing requirements of the galvanized unit process.

Key words: direct-reading spectrometer; aluminum content; accuracy; precision; quick

软熔处理对锡铁合金层耐蚀性能的影响研究

彭 强，任予昌，赵 柱，张东方，刘晓峰

（武汉钢铁有限公司，湖北武汉 430000）

摘 要：利用电化学测试和表面分析手段研究了软熔处理温度和时间对镀锡钢板中合金层及其在3.5% NaCl 溶液中腐蚀电化学行为的影响，结果显示：软熔处理温度和时间增加，镀锡钢板的合金层数量增多，腐蚀电位正移，对应腐蚀速率下降；基板与合金层界面更易受到腐蚀。

关键词：软熔处理；镀锡板；合金层

Study on the Effect of Soft Soluble Treatment on the Erosion Resistance of Tin Iron Alloy Layer

Peng Qiang, Ren Yuchang, Zhao Zhu, Zhang Dongfang, Liu Xiaofeng

(The Cold-Rolling Mill of Wuhan Iron and Steel Co., Ltd., Wuhan 430000, China)

Abstract: By means of electrochemical tests and surface analysis, the effects of the temperature and time of soft solution on the alloying layer in tin-plated steel plate and its corrosion electrochemistry behavior in 3.5 % NaCl solution were studied. The results showed that the temperature and time of soft treatment increased, and the number of alloy layers of tin-plated steel plate increased. Corrosion potential is positive shift, corresponding to the decrease of corrosion rate; The parts adjacent to the bare substrate and the alloy layer are more susceptible to corrosion.

Key words: flexible treatment; tin plating; alloy layer

控制热镀锌汽车外板锌渣缺陷的工艺创新方法

吴价宝，张雨泉

（宝武集团武汉钢铁有限公司冷轧薄板总厂，湖北武汉 430083）

摘 要：热镀锌汽车板外板对冲压和涂漆的使用要求高，锌渣缺陷对汽车板表面质量和冲压的影响巨大，控制热镀锌汽车外板的锌渣缺陷是汽车板产品质量控制的最关键点。本文在综合国内外热镀锌理论研究成果的基础上，对热镀锌"锌渣"形成机理进行了分析。结合国内外先进热镀锌生产线汽车外板生产准备和生产过程中，锌锅中铁的去除，锌渣的控制方法，提出了一套去除锌锅中铁的新方法，并在国内先进汽车板生产线得以实践，通过反复实践得出了一套控制锌渣和去除锌锅内铁的创新方法。通过在检修前、检修过程、检修开机后、生产汽车外板过程中采用相应的工艺方法，实现了有效控制汽车外板锌渣缺陷，极好地控制了表面质量。

关键词：热镀锌汽车外板；锌渣；去除铁；工艺方法

冷轧电镀锡板生产之关键技术的探讨与分析

何云飞[1,2]，周玉林[1,2]，刘宏文[1,2]，侯俊达[1,2]

（1. 北京首钢国际工程技术有限公司，北京　100043；
2. 北京市冶金三维仿真设计工程技术研究中心，北京　100043）

摘　要：本文介绍了镀锡板产品的生产工艺及产品用途与存在的问题，并着重探讨了与镀锡板生产相关的各冷轧工序的关键技术与特点及其发展趋势，有利于冷轧电镀锡板生产技术的推广，对新建镀锡板厂的设备选型、先进技术采用与产品质量的提高具有指导意义。

关键词：电镀锡板；冷轧；退火；平整

Analysis and Discussion on the Key Technology for Cold Rolled Electrolytic Tin-plate Production

He Yunfei[1,2], Zhou Yulin[1,2], Liu Hongwen[1,2], Hou Junda[1,2]

(1. Beijing Shougang International Engineering Technology Limited Corporation, Beijing 100043, China; 2. Metallurgical Engineering 3-D Simulation Design Engineering Technology Research Center of Beijing, Beijing 100043, China)

Abstract: Technological process and application of tin-plate product were introduced in this article, and it also discussed the importance and weaknesses about the key technologies, together with the characteristics, and development tendency of the related cold-rolling procedure during tin-plate production.

This article has contributed to the popularization of the producing technology of cold-rolled electrolytic tin-plate, and has a research significance for the equipment designation, advanced technology adoption, and product quality enhancement of newly built tin-plate plant.

Key words: electrolytic tin-plate; cold-rolling; annealing; skin-pass mill

一种热镀纯锌产品黑点缺陷成因及解决办法

李鸿友，张晶晶，马　峰，艾厚波

（本钢板材股份有限公司冷轧厂，辽宁本溪　117000）

摘　要：通过采用高倍显微镜分析了纯锌带钢酸洗去掉锌层后的表面状态，发现在黑点缺陷部位存在轻微高度差。通过调节炉内张力发现对此缺陷有明显影响，张力越大缺陷越严重，反之缺陷减轻。结合机组实际情况进行调整后缺陷得到明显改善。

关键词：热镀锌；黑点；汽车板

Causes and Solutions of Black Spots Defect in Hot－Dip Galvanized Zinc Products

Li Hongyou, Zhang Jingjing, Ma Feng, Ai Houbo

(Bengang Steel Plates Co., Ltd., The Cold Rolling Mill, Benxi 117000, China)

Abstract: The surface state of pure zinc strip after pickling and removing zinc layer was analyzed by using SEM, it is found that there is a slight height difference in the black spot defect area. By adjusting the tension in the furnace, it is found that the defect is obviousiy affected. The larger the tension, the more serious the defect is and vice versa, the defect is reduced. According to the actual situation of the line, the defects have been significantly improved after adjustment.

Key words: hot dip galvanizing; black spot; automobile plate

电镀锌磷化生产工艺的优化

孙晨航，曹　洋，车晓宇，曹志强

（本钢板材股份有限公司冷轧厂，辽宁本溪　117000）

摘　要： 电镀锌机组是一条以生产高质量磷化汽车板为目标进行设计的电镀锌产线，但在磷化生产开发过程中，发现由于磷化不均、磷化斑点、边部枝晶等问题，导致产品合格率较低，而且无法满足磷化产品的持续稳定生产，需要对磷化的生产工艺进行优化改造。

关键词： 磷化浓度；磷化温度；电镀锌机组

Optimization of Production Process of Electroplating Zinc Phosphate

Sun Chenhang, Cao Yang, Che Xiaoyu, Cao Zhiqiang

(Bengang Steel Plates Co., Ltd., The Cold Rolling Mill, Benxi 117000, China)

Abstract: The galvanizing unit of this steel is an electrogalvanizing production line designed to produce high-quality phosphating automobile plates. However, during the development of phosphating production, it was found that due to uneven phosphating, phosphating spots, and edge dendrites, etc. Problems, resulting in low product acceptance rate. Moreover, it can not satisfy the continuous and stable production of phosphating products, and it is necessary to optimize the production process of phosphating.

Key words: phosphate concentration; phosphate temperature; galvanized unit

本钢浦项热镀锌锌锅铝含量成分稳定控制

王子昂，赵兴时

（本钢板材股份有限公司冷轧厂，辽宁本溪　117000）

摘　要：本钢浦项镀锌机组隶属于本钢板材冷轧厂，目前以生产高档汽车内板、外板及高档家电板为主要目标。近年来随着社会和科技的进步，钢铁市场对高档钢板的需求日益增大，客户对我厂生产的产品质量水平越发严格，这就对镀锌板表面提出了更高的要求，为满足客户需求，快速占领高端市场的份额，提高镀锌工艺生产过程中的稳定性保证产品质量变得尤为重要。锌锅成分是镀锌过程中关键控制参数，本文针对锌锅成分的稳定控制，从锌液、锌层、锌渣等方面通过理论分析锌锅铝的消耗途径并结合生产现场实际如何保证铝含量的稳定控制。

关键词：热镀锌；锌液；锌层；锌渣；铝含量

Stable Control of Aluminum Composition in the Hot Dip Strip Coating

Wang Ziang, Zhao Xingshi

(Bengang Steel Plates Co., Ltd., The Cold Rolling Mill, Benxi 117000, China)

Abstract: Benxi steel posco galvanized units belonging to Benxi steel plate, abrasion is to produce high-grade car plate and high-grade home appliance plate as the main target in recent years, with the progress of society and science and technology, increasing demand for high-grade steel plate, steel market customers for our products quality more strictly, it has put forward higher requirements on the surface of a galvanized sheet, in order to meet customer needs, quickly occupied the high-end market share, improve the stability is particularly important in the process of producing galvanized zinc pot composition is a key control parameters in the process of galvanized, aiming at the stability control of zinc pot ingredients, from zinc liquid zinc layer zinc slag and other aspects through the theoretical analysis of zinc pot aluminum consumption and combined with the production site how to ensure the stability of aluminum content control.

Key words: hot dip galvanized; zinc liquid; the zinc layer; zinc slag; aluminum content

基于电镀锌生产线数据采集系统的分析与实现

张　勇

（本钢板材股份有限公司冷轧厂，辽宁本溪　117000）

摘　要：对基于电镀锌生产线数据采集系统的作用进行了简介，详细的分析了数据采集系统实现的关键点，并阐述了如何实现整个系统。通过数据采集系统实现电镀锌生产线工艺数据参数、设定值以及工艺设备优化和管理。

关键词：电镀锌；数据采集系统；数据库；工艺数据

Analysis and Realization of Data Collection System based on Electrolytic Galvanizing Line

Zhang Yong

(Bengang Steel Plates Co., Ltd., The Cold Rolling Mill, Benxi 117000, China)

Abstract: Introduces function that data collection system based on electrolytic galvanizing line. Detail analysis data collection system; describe how to realize the whole system. Process parameter、set values and process equipment can be optimized and managed by data collection system in electrolytic galvanizing line.

Key words: electrolytic galvanizing line; data collection system; database; process data

高强 IF 钢合金化热镀锌镀层相结构和抗粉化性能影响因素分析

周诗正[1]，李　军[1]，张伟浩[1]，杜小峰[2]

（1. 宝武集团武钢有限冷轧厂，湖北武汉　430083；
2. 宝钢股份中央研究院武汉分院（武钢有限技术中心），湖北武汉　430083）

摘　要：采用扫描电镜（SEM）、ICP、及弯曲试验研究了生产线产出的三种合金化镀锌板（GA）镀层微观组织、相结构及抗粉化性能差异。结果表明，随着钢基 P、Si 成分含量的升高，在相同合金化保温时间内，完成 Zn-Fe 扩散所需的温度升高；抗粉化性能对比结果表明，在表面只存在 δ_1 相的情况下，镀层 Fe 含量越低，对应的镀层粉化性能更好。

关键词：合金化镀锌板；镀层；相结构；粉化

Analysis of Factors Affecting Phase Microstructure and Powdering-Resistance of High Strength IF Galvannealed Steel Sheets

Zhou Shizheng[1], Li Jun[1], Zhang Weihao[1], Du Xiaofeng[2]

(1. Cold Rolling Mill of WISCO, Wuhan 430083, China;
2. Technology Center of WISCO, Wuhan 430083, China)

Abstract: The microstructure, phase morphology and powdering performance of three kinds of industrially produced hot-dip galvannealed (GA) coatings were investigated by means of SEM, ICP, and 60 degree bending test. The results show that, the diffusion temperature of Zn/Fe increases with the increase of P and Si in strip steel. Enhancement the the powdering resistance behavior is better when then coating is cosisting of less Fe for all δ_1 phase on coating surface.

Key words: hot-dip galvannealed steel sheet; coating; phase constitution; powdering

浅谈冷轧锌铝镁产品发展现状与创新实践

王海东[1]，严江生[1]，陈浩杰[2]，王骏飞[2]

（1. 宝钢工程技术集团有限公司冷轧事业部，上海　201900；
2. 宝钢湛江钢铁有限公司冷轧厂，广东湛江　524033）

摘　要：本文介绍了锌铝镁镀层产品国内外相关技术发展现状与趋势，并论述了宝钢锌铝镁开发历程及创新实践。

宝钢锌铝镁项目的实施将很好地填补宝钢锌铝镁产品空白，对于建设或改造同类产线，具有很好的借鉴价值，也将对我国锌铝镁冶金技术进步产生深远影响。

关键词：冷轧；锌铝镁；创新；实践

Development & Innovation Practice of ZnAlMg Coated Product

Wang Haidong[1], Yan Jiangsheng[1], Chen Haojie[2], Wang Junfei[2]

(1. Cold Rolling Department of Baosteel Engineering & Technology Group Co., Ltd., Shanghai 201900, China; 2. Cold Rolling Mill Baosteel Zhanjiang Iron & Steel Co., Ltd., Zhanjiang 524033, China)

Abstract: The present developments of ZnAlMg coated product is introduced. The development & innovation practice of Baosteel cold rolling ZnAlMg coated product is described. Based on filling in gaps of ZnAlMg coated product, Baosteel Zhanjiang ZnAlMg project will provide the reference value for the similar production line.

Key words: cold rolling; ZnAlMg; innovation; practice

5 金属材料深加工

大会特邀报告

分会场特邀报告

炼铁与原料

炼钢与连铸

轧制与热处理

表面与涂镀

★ 金属材料深加工

先进钢铁材料

粉末冶金

能源、环保与资源利用

钢铁材料表征与评价

冶金设备与工程技术

冶金自动化与智能化

建筑诊治

其他

1006 线材制作 T 铁时表面鼓泡原因分析

王宁涛[1,2]，阮士朋[1,2]，张 鹏[1,2]，王利军[1,2]

（1. 河北省线材工程技术研究中心，河北邢台 054027；
2. 邢台钢铁有限责任公司，河北邢台 054027）

摘 要：1006 线材制成的 T 铁，在镀锌后发生鼓泡。通过对试样的组织、成分、硬度的检测，发现鼓泡部位的变形程度是整个截面中最严重的部位，且试样中氢的含量较高。在排除了其他影响因素外，最可能的是氢聚集引起的氢致鼓泡。

关键词：1006；T 铁；鼓泡；氢

Analysis of the Causes of Surface Bubbling on the T-iron Made of 1006 Wire

Wang Ningtao[1,2], Ruan Shipeng[1,2], Zhang Peng[1,2], Wang Lijun[1,2]

(1. Hebei Engineering Research Center for Wire Rod, Xingtai 054027, China;
2. Xingtai Iron & Steel Co., Ltd., Xingtai 054027, China)

Abstract: The bubble occurs on the T-iron made of 1006 after galvanizing.By testing the hardness ,structure and composition of the specimen's,it is found that the deformation degree of the bubbling part is the most serious part in the whole section,and the hydrogen content in the sample is relatively high. After the exclusion of other factors, The possible reason is the hydrogen induced bubbling caused by hydrogen accumulation.

Key words: 1006; T-iron; bubbling; hydrogen

加热温度对 B 柱热冲压成形性影响的研究

郭 晶，陈 宇，陈虹宇

（本钢板材股份有限公司技术研究院，辽宁本溪 117000）

摘 要：采用数值模拟方法,研究了加热温度对 B 柱热冲压成形性能的影响。通过设计模具型面，添加压料板，合理优化热冲压工艺参数，建立了 B 柱热冲压成形的有限元模型。通过数值分析，得到了加热温度对 B 柱热冲压性能的影响规律。分析结果表明：在其他工艺参数相同的条件下，随着加热温度 840~960℃变化，B 柱的成形后最低温度和最高温度逐渐升高，马氏体含量逐渐增加到 100%。抗拉强度随着温度增加到 930℃呈增加趋势，当温度继续升高，抗拉强度反而下降。同样 B 柱硬度和回弹量都随着温度的增加也呈现先增大后减小的变化规律。最后，通过对比确定了合理的热冲压成形加热温度，为板料热冲压成形的设计提供可靠依据。

关键词：加热温度；B 柱；热冲压；抗拉强度；回弹

Research on Hot Stamping Formability of B-PILLAR by Heating Temperature

Guo Jing, Chen Yu, Chen Hongyu

(Technology Research Institute of Benxi Steel Plate Co., Ltd., Benxi 117000, China)

Abstract: In order to study the influence of heating temperature on the hot stamping performance of B-PILLAR, numerical simulation method was used. The finite element model of B-PILLAR hot stamping was established by designing die surface, adding press plate and optimizing process parameters. Through numerical analysis, the effect of heating temperature on the hot stamping performance of B-PILLAR was obtained. Under the same conditions, The analysis results showed that as the heating temperature of 840~960℃, B-PILLAR after forming the lowest temperature and highest temperature gradually raised, and martensite content gradually increased to 100%. Tensile strength increased as the temperature increased to 930℃, while with temperatures continue to rise, tensile strength reduced. Similarly, the hardness and springback of B-PILLAR also increase and then decrease with the increase of temperature. Finally, the reasonable heating temperature of hot stamping is determined by comparison, which provides a reliable basis for the design of sheet metal hot stamping.

Key words: heating temperature; B-PILLAR; hot stamping; tensile strength; springback

热处理工艺对 Q125 套管力学性能影响的试验

解德刚，吴 红，赵 波，袁 琴，王善宝

（鞍钢集团钢铁研究院，辽宁鞍山 114009）

摘 要：本文对一种 Q125 套管进行了热处理工艺参数对力学性能影响的试验研究。研究结果表明，回火温度对 Q125 套管力学性能影响最为显著，并对屈服强度和抗拉强度的影响呈线性关系。根据试验钢尺寸可将淬火加热温度、淬火保温时间及回火保温时间固定，通过调整回火温度即可获得满足 Q125 套管钢要求的综合力学性能。通过对不同回火温度下析出相的检测与分析可知，回火碳化物的尺寸与形态的控制对钢管强度、硬度均匀性及优异综合力学性能的获取至关重要。

关键词：套管；Q125；热处理；力学性能；碳化物

Effect of Heat Treating on Mechanical Properties of Q125 Casing

Xie Degang, Wu Hong, Zhao Bo, Yuan Qin, Wang Shanbao

(Technology Center of Angang Steel Co., Ltd., Anshan 114009, China)

Abstract: Experimental investigation was taken for improving mechanical properties of the Q125 Casing. The results showed that tempering is the most significant factor on mechanical properties. The yield strength and tensile strength vary linearly with tempering temperature. According to the size of steel, the quenching heating temperature, quenching holding time and tempering holding time can be fixed. By adjusting the tempering temperature, the comprehensive mechanical properties can be obtained which meeting to API 5CT. Through the detection and analysis of precipitates at different tempering temperatures, it can be seen that the control of the size and morphology of tempered carbides is very important to

obtain the strength, hardness uniformity and excellent comprehensive mechanical properties of the Q125 casing.
Key words: casing; Q125; heat treatment; mechanical properties; carbide

82A 高碳钢热变形 Arrhenius 本构模型研究

张彭磊[1]，苏　岚[1]，胡　磊[2]，沈　奎[2]，麻　晗[2]，米振莉[1]

（1. 北京科技大学工程技术研究院，北京　100083；
2. 江苏省（沙钢）钢铁研究院，江苏张家港　215625）

摘　要： 采用 Gleeble-3500 热模拟试验机对 82A 高碳钢进行不同变形温度及不同应变速率下的单道次等温热压缩试验，变形温度范围为 800～1000℃，应变速率范围为 0.1～10s^{-1}，最大真应变为 0.6，测得不同工艺条件下的应力-应变曲线。利用试验获得的应力应变数据建立了应变补偿型 Arrhenius 本构模型。结果表明，随着应变速率的增大和变形温度的降低，82A 高碳钢的流变应力值显著增加；通过建立的应变补偿型 Arrhenius 本构方程预测的流变应力值与试验值相吻合，具有较高的预测精度。
关键词： 高碳钢；热变形；Arrhenius 模型；本构模型

Study on Hot Deformation Arrhenius Constitutive Model of 82A High Carbon Steel

Zhang Penglei[1], Su Lan[1], Hu Lei[2], Shen Kui[2], Ma Han[2], Mi Zhenli[1]

(1. Institute of Engineering Technology, University of Science and Technology Beijing,
Beijing 100083, China; 2. Institute of Research of Iron and Steel,
Shasteel/Jiangsu Province, Zhangjiagang 215625, China)

Abstract: The single-pass isothermal compression test of 82A high carbon steel was carried out on Gleeble-3500 thermal simulator at different deformation temperatures and different strain rates, The deformation temperature ranged from 800℃ to 1000℃, the strain rate ranged from 0.1s^{-1} to 10s^{-1}, the maximum true strain is 0.6, and the stress-strain curve under different deformation conditions is obtained. The strain-compensated Arrhenius constitutive model was established by using the stress-strain data obtained from the experiment. The results show that the flow stress value of 82A high carbon steel increases significantly with the increase of strain rate and the decrease of deformation temperature. The experimental values of flow stress are consistent with values calculated by the strain-compensated Arrhenius constitutive equation, it proves that the model has good prediction accuracy.
Key words: high carbon steel; hot deformation; Arrhenius model; constitutive model

镀锌家电板白痕缺陷分析与改进

赵　刚[1]，毛一标[2]，曹　垒[1]

（1. 张家港扬子江冷轧板有限公司，江苏张家港　215625；

2. 江苏沙钢集团有限公司总工程师办公室，江苏张家港 215625）

摘　要：沙钢冷轧镀锌家电板白痕缺陷从 2017 年 8 月份开始，困扰冷轧连续稳定生产，影响镀锌产品质量，每月白痕判次率超 3%，2017 年 12 月份白痕缺陷判次达到 8.7%。针对镀锌家电板白痕缺陷，酸轧及镀锌采取的大量攻关工作，改善效果甚微。通过对热轧工艺优化后，白痕缺陷得到很大改善，白痕判次比例大幅下降，2018 年 5 月份判次比例下降到 1.6%，6 月份白痕判次比例更是控制在 1% 以下。本文重点从白痕缺陷形貌、镀前镀后特点以及白痕与卷取温度对应关系等方面分析，并制定有效措施予以改进。

关键词：镀锌家电板；白痕缺陷；分析与改进

Analysis and Improvement of White Mark Defect of Galvanized Electric Appliance Plate

Zhao Gang[1], Mao Yibiao[2], Cao Lei[1]

(1. Zhangjiagang Yangtze River Cold Rolled Plate Co., Ltd., Zhangjiagang 215625, China;
2. Chief Engineer Office, Jiangsu Shasteel Group Co., Ltd., Zhangjiagang 215625, China)

Abstract: The white mark defect of galvanized electric appliance plate began in August of 2017. It has plagued the continuous and stable production of cold rolling, which affects the quality of galvanized products. The monthly rate of white mark is over 3%, and the number of white mark defects in December of 2017 is 8.7%. In view of the white scratch defect of galvanized household appliance plate, acid rolling and galvanizing through a large number of tackling work, the improvement effect is very slight. After optimizing hot rolling process, the white mark defect has been greatly improved, the proportion of white marks has fallen sharply, the proportion of judgment in May was reduced to 1.6% in May, and the white mark in June was judged. The secondary proportion is less than 1%. In this paper, the reasons are analyzed from the appearance of white mark defects, the rule of production, the characteristics of the defects before plating and the relationship between the white mark defect and the coiling temperature, and the effective measures are made to improve.

Key words: galvanized household appliances board; white mark defect; analysis and improven

高强钢汽车板冷轧点状缺陷分析与改善

赵　刚[1]，毛一标[2]，曹　垒[1]

（1. 张家港扬子江冷轧板有限公司，江苏张家港 215625；
2. 江苏沙钢集团有限公司总工程师办公室，江苏张家港 215625）

摘　要：沙钢冷轧生产的汽车钢主要有 HC340LA、HC380LA 牌号为代表的高强钢汽车板，以及以 DC03、DC04 及 SLD 牌号为代表的深冲汽车板（IF 钢）。沙钢冷轧生产高强钢汽车板曾经出现下表面点状缺陷。分厂成立技术攻关组，从热轧、冷轧及普冷机组全面排查跟踪，最终锁定为酸轧胶辊上粘附的异物，攻关解决胶辊表面异物问题，制定有效措施，并形成长效管理机制，避免此类缺陷的批量发生。

关键词：高强钢汽车板；点状缺陷；分析与改善

Analysis and Improvement of Point Defects in High Strength Automobile Plate Cold Rolling

Zhao Gang[1], Mao Yibiao[2], Cao Lei[1]

(1. Zhangjiagang Yangtze River Cold Rolled Plate Co., Ltd., Zhangjiagang 215625, China;
2. Chief Engineer Office, Jiangsu Shasteel Group Co., Ltd., Zhangjiagang 215625, China)

Abstract: The automobile steel produced by cold rolling of Shagang steel is mainly with HC340LA and HC380LA brand as the high strength automobile board, as well as the deep drawing board (IF steel) represented by DC03, DC04 and SLD brand. The spot defect on the lower surface of the high strength cold rolling automobile plate are detected, The branch set up a technical team include the hot rolling and cold rolling, finally locked as the acid rolling rubber roll attached to the foreign objects, tackling the problem of the surface of the rubber roll to solve the problem, and establish effective measures, and form a long-term management mechanism to avoid such defects.

Key words: high strength auto plate; point defect; analysis and improvement

冷轧普冷卷板形问题分析及改善

赵 刚[1]，毛一标[2]，曹 垒[1]

（1. 张家港扬子江冷轧板有限公司，江苏张家港 215625；
2. 江苏沙钢集团有限公司总工程师办公室，江苏张家港 215625）

摘 要： 冷轧普冷卷板形问题有多种，如中浪、边浪、1/4浪等，产生的原因也有多种情况，比如热轧终轧温度高低、酸轧轧辊磨削精度优良与否、精密喷射系统是否正常、板形自动控制是否投用、轧制参数是否合理等方面，冷轧卷板形问题直接造成后道工序二次缺陷，对于薄板甚至出现断带问题，如中浪和1/4浪在炉内外，带钢表面温度急剧上升和下降，有浪位置会产生起筋、褶皱最终断带，另外，浪高超过一定范围，极易触碰到普冷产线的风冷喷嘴，产生划伤缺陷，尽管划伤经过普冷平整机后，划伤基本无手感，未判次；再如边浪，尤其是单边浪问题，带钢在运行过程会出现跑偏、刮边断带问题。为此，本文重点从冷轧普冷卷板形问题进行分析，主要从提高热轧终轧温度和改善冷轧酸轧5#机架工作辊辊形磨削精度入口，并制定有效的改善对策，冷轧普冷卷板形改善明显，减少跑偏断带及划伤缺陷，大大地提高了后道工序的满意度。

关键词： 冷轧普冷卷；板形问题；分析及改善

Analysis and Improvement of Cold Rolled Ordinary Cold Rolled Sheet

Zhao Gang[1], Mao Yibiao[2], Cao Lei[1]

(1. Zhangjiagang Yangtze River Cold Rolled Plate Co., Ltd., Zhangjiagang 215625, China;
2. Chief Engineer Office, Jiangsu Shasteel Group Co., Ltd., Zhangjiagang 215625, China)

Abstract: There are many problems in cold rolling of cold rolling plate, such as medium wave, edge wave, 1/4 wave and so on. There are many reasons, such as hot rolling finish rolling temperature, fine grinding precision of acid rolling roll, normal precision injection system, automatic control of plate shape, reasonable rolling parameters and so on. The problem directly causes two defects in the post process. For thin plates and even broken bands, such as medium wave and 1/4 wave in and outside the furnace, the surface temperature of the strip rises sharply and drops, the position of the wave will produce the stiffened and wrinkled final broken band. In addition, the wave height exceeds a certain range, and it is very easy to touch the nozzles of the cold production line and produce scratch and lack. In spite of the scratch, the scratch is basically no hand, not a judgement, and the wave, especially the problem of single side waves, will occur in the course of running and the problem of cutting and breaking. Therefore, this paper focuses on the analysis of cold rolling cold rolling plate shape problem, focusing on improving the final rolling temperature of hot rolling and improving the precision entrance of the working roll shape grinding of the 5# rack of acid rolling mill, and formulating effective improvement measures, the improvement of cold rolling plate shape is obvious, reduction of running off band and scratch defect, and the satisfaction of the post process is greatly improved.

Key words: cold rolled plain coil; flatness problem; analysis and improvement

冷轧普冷卷异物压入问题分析及改善

赵 刚[1]，毛一标[2]，曹 垒[1]

（1. 张家港扬子江冷轧板有限公司，江苏张家港 215625；
2. 江苏沙钢集团有限公司总工程师办公室，江苏张家港 215625）

摘 要： 冷轧带钢异物压入缺陷判次量一直高居不下，异物压入主要有：一是线状异物压入，酸轧入口矫直辊开口度，加之热卷头尾板形不良，穿带过程中产生的擦划伤，主要发生在头尾，呈线状无规律分布；二是雨点状及柳叶状异物压入，硅钢卷氧化铁较多，酸洗后带钢中部仍然有黑色氧化铁，在轧机入口 8-2 张力辊，长年累月，会粘附到张力辊胶辊表面，又会转印到其他钢种，主要集中在中部区域，呈雨点状及柳叶状，断续通长分布；三是水滴状异物压入，机组长时间生产窄料，窄料边部钢屑残留轧机入口测张辊，若未及时清理，生产宽料时，残留物粘附在带钢边部，一般在距两侧边部 30cm 以内，呈水滴状断续通长分布。据统计，异物压入判次量每月高居不下，从 2017 年 6 月份，每月多则判次超 600t，少则 88t，判次比例最高月份达到 1.68%，最少的月份也有 0.21%。针对异物压入问题，分厂成立技术攻关组，技术人员通过采取积极有效的措施，如加强工装件点巡检、强化下线钢卷检查，发现问题及时处理等措施，形成冷轧带钢异物压入缺陷控制管理规定，异物压入问题得到了有效控制，实现每月判次量在 150t 以内，判次比例控制在 0.3%左右。

关键词： 冷轧带钢；异物压入；分析及改善

Analysis and Improvement of the Defect of Impurity Pressed in Cold Rolled Sheet

Zhao Gang[1], Mao Yibiao[2], Cao Lei[1]

(1. Zhangjiagang Yangtze River Cold Rolled Plate Co., Ltd., Zhangjiagang 215625, China;
2. Chief Engineer Office, Jiangsu Shasteel Group Co., Ltd., Zhangjiagang 215625, China)

Abstract: The grade of the defect of impurity pressed in cold rolled sheet is always high. The impurity pressed mainly includes linear impurity pressed, opening of leveling roll at the entrance of acid rolling, bad shape of hot coil head and tail, scratches in the process of strip piercing mainly occur at the head and tail, showing linear and irregular distribution; and raindrop and willow leaf impurity pressed. Press in, silicon steel coil iron oxide is more, after pickling, there is still black iron oxide in the middle of strip steel. 8-2 tension roll at the entrance of rolling mill will adhere to the surface of tension roll cots for many years, and will be transferred to other kinds of steel, mainly concentrated in the middle area, raindrop and willow leaf shape, intermittent and long distribution. Input, the unit produces narrow material for a long time, and the residual steel scrap at the edge of narrow material is left at the entrance of the tension measuring roll. If it is not cleaned up in time, the residue adheres to the edge of strip steel in the production of wide material, generally within 30 cm from the edges of both sides, showing a drop-like intermittent long distribution. According to statistics, the number of impurity pressed judgments is still high every month. From June 2017, the number of impurity pressed judgments is more than 600 tons per month and less than 88 tons per month. The proportion of impurity pressed judgments reaches 1.68% in the highest month and 0.21% in the least month. Aiming at the problem of impurity pressed, the branch plant set up a technical research team. Through taking active and effective measures, such as strengthening the inspection of tooling points, strengthening the inspection of off-line coils, finding problems and timely handling, the technical personnel formed the control and management regulations for impurity pressed defects of cold rolled strip, and the problem of impurity pressed was effectively controlled. The monthly judgment quantity is less than 150 tons, and the judgment proportion is controlled at about 0.3%.

Key words: cold-rolled strip; impurity pressed; analysis and improvement

TRIP 钢的各向异性性能研究

杨登翠，阚鑫锋，李义强，孙蓟泉

（北京科技大学钢铁共性技术协同创新中心，北京 100083）

摘 要： 由于高强 TRIP 钢在轧制过程中形成了较强的各向异性，本文对比了具有各向异性的 TRIP600、TRIP800 两种实验钢变形后的性能变化。从三个不同方向取样（与轧制方向成 0°、45°及 90°）的两种 TRIP 钢通过单向拉伸试验测定其基本力学性能参数，得到工程应力应变及真实应力应变曲线，并通过真实应力、真实应变计算出应变硬化指数 n 和厚向异性指数 r。结果表明：两种 TRIP 钢的延伸率及 n 值均在轧制方向取最高值；TRIP600 的强度和 n 值较 TRIP800 略低，但延伸率和 r 值较高。为得到最佳成形效果，强度与塑性、n 值与 r 值的匹配是钢材选择的重要考虑因素。

关键词： TRIP 钢；成形性能；单向拉伸；应变硬化指数；厚向异性指数

Study on Anisotropic Properties of TRIP Steel

Yang Dengcui, Kan Xinfeng, Li Yiqiang, Sun Jiquan

(Collaborative Innovation Center of Steel Technology, University of
Science and Technology Beijing, Beijing 100083, China)

Abstract: Because of the strong anisotropy of high-strength TRIP steel during rolling, this paper compares the performance changes of anisotropic TRIP600 and TRIP800 after deformation. The basic mechanical properties of the two TRIP steels sampled from three different directions (0°, 45° and 90° with rolling direction) were measured by uniaxial tensile test, and the engineering stress-strain and true stress-strain curves were obtained. The strain hardening index n and the thick anisotropy index r were calculated from the true stress and the true strain. The results show that the elongation and n value of the two TRIP steels are the highest in the rolling direction; the strength and n value of TRIP600 are slightly lower than TRIP800, but the elongation and r value are higher. In order to obtain the best forming effect, the matching of strength and plasticity, n and r values is an important consideration for steel selection.

Key words: TRIP steel; formability; biaxial tensile test; strain hardening index; thick anisotropy index

MPM 连轧限动芯棒失效形式及延长寿命的方式

谷大伟，洪 汛

（鞍钢股份有限公司无缝钢管厂，辽宁鞍山 114021）

摘 要： 芯棒是钢管变形的重要工具，芯棒表面质量的好坏，直接影响到钢管产品的壁厚精度和内表面质量，文章针对连轧限动芯棒失效的缺陷形式，进行分类研究，重点探讨了减少芯棒表面缺陷产生的时间，延长芯棒使用寿命的方式。

关键词： 芯棒；失效；使用寿命；温度；润滑；轧制力

The Failure for Mandrel Retaining MPM and the Measures of Improving the Life-span

Gu Dawei, Hong Xun

(Production Cooperate Center of Anshan Iron and Steel Group Corporation, Anshan 114021, China)

Abstract: Mandrel is an important tool for steel tube deformation, and the surface quality directly affects the wall thickness accuracy and surface quality. Due to the failure for mandrel retaining MPM, this article mainly studies the way to prolong the service life by reducing the time of surface of mandrel.

Key words: mandrel; failure; service life; temperature; lubricate; rolling force

微张力对大方坯初轧过程的影响

张 迪[1]，米振莉[1]，苏 岚[2]

（1. 北京科技大学工程技术研究院，北京 100083；
2. 北京科技大学钢铁共性技术协同创新中心，北京 100083）

摘 要： 在大方坯初轧过程中微张力是非常重要的要素而且必须严格控制。采用有限元显式动力学分析软件 ANSYS/LS-DYNA 建立三维实体模型，分别对大方坯轧制过程施加前张力和后张力载荷。通过分析轧制过程中轧制力变化、大方坯宽展、整体应力应变水平、角部应力应变分布影响的变化，得出结论：前后张力的变化可以显著改变轧制过程中的轧制力、轧件宽展以及角部的应力应变，后张力对应力应变的影响幅度要大于前张力，为企业生产实践提供理论依据。

关键词： 微张力；大方坯轧制；轧件角部；数值模拟

Finite Element Analysis of the Micro-tension Influence on Process of Square Billet

Zhang Di[1], Mi Zhenli[1], Su Lan[2]

(1. Institute of Engineering Technology, University of Science and Technology Beijing, Beijing 100083, China; 2. Collaborative Innovation Center of Steel Technology, University of Science and Technology Beijing, Beijing 100083, China)

Abstract: Micro-tension are very important and must be strictly controlled in bloom rolling process. By finite element analysis software ANSYS/LS-DYNA, we established the three-dimensional model and imitated the rolling processes through changing micro-tension. The conclusion is drawn by analyzing the changes of rolling force, wide spread of square billet, overall stress and strain level, and stress and strain distribution of corner in rolling process: The change of tension can obviously change rolling force, wide spread of square billet and the stress and strain on the corner of rolled bar. The influence of backward tension on the stress and strain is greater than that of forward tension, which provide a theoretical basis for the production practice of enterprises.

Key words: micro-tension; bloom rolling; rolled piece corner; finite element analysis

NJT600 高强钢的焊接金相组织分析

罗扬[1]，马成[1]，孙力[1]，白丽娟[1]，
谷秀锐[1]，孙晓冉[1]，周士伟[2]，刘晓龙[2]

（1. 河钢集团钢研总院，河北石家庄 050023；2. 河钢集团销售总公司，河北石家庄 050023）

摘　要：为评价 NJT600 高强钢板带的焊接性能，分别与不同材料进行了 CO_2 气体保护焊实验，并对焊接部位的微观组织进行了考察。结果表明：在整个焊接部位，熔合区是魏氏组织最为严重的位置，其评级主要取决于焊丝材料和焊接工艺；对于 NJT600 与 NJT600、NJT600 与 Q235、NJT600 与 Q345 焊接件，魏氏组织最高均为 2 级，能够满足农机产品对于焊接质量的要求。

关键词：NJT600；高强钢；微观组织；焊接质量

Microstructure Analysis of the Welded Position of NJT600 High Strength Steel

Luo Yang[1], Ma Cheng[1], Sun Li[1], Bai Lijuan[1], Gu Xiurui[1],
Sun Xiaoran[1], Zhou Shiwei[2], Liu Xiaolong[2]

(1. Technology Research Institute, HBIS Group, Shijiazhuang 050023, China;
2. Sales Company, HBIS Group, Shijiazhuang 050023, China)

Abstract: In order to evaluate the welding quality of NJT600 high strength steel sheets, CO_2 gas shielded arc welding experiments of different materials were carried out separately, and microstructures of the welded positions were investigated. The results indicate that the Widmanstatten structure in the fusion zone is the most serious in the whole welding area.

Widmanstatten structure rating mainly depends on welding wire material and welding process. For NJT600 and NJT600, NJT600 and Q235, NJT600 and Q345 welding parts, the highest Widmanstatten structure rating is level 2, which can meet the requirements of agricultural machinery products for welding quality.
Key words: NJT600; high strength steel; microstructure; welding quality

高速线材斯太尔摩冷却线仿真计算模型开发

李会健[1]，米振莉[1]，苏 岚[2]

（1. 北京科技大学工程技术研究院，北京 100083；
2. 北京科技大学钢铁共性技术协同创新中心，北京 100083）

摘 要：由于高速线材在斯太尔摩冷却线上不同位置搭接密度不同，导致盘条不同位置冷却不同，造成不均匀性，影响盘条的力学性能。本文采用有限差分法对高速线材搭接点和非搭接点分别建立了斯太尔摩风冷线温度场及相变场耦合仿真计算模型，考虑了风机风速、辊道速度以及盘条规格这些因素对高速线材不同位置冷却过程的影响。以钢厂稳定生产的 ϕ5.5mm 82A 帘线钢现场实测数据对模型进行可靠性验证，模型仿真计算温度与实测温度误差在±15℃以内。分析了佳灵装置对线材不同位置冷却速度的影响。
关键词：高速线材；搭接点；非搭接点；数学模型

Development of Simulation Calculation Model for Stelmor Cooling Line of High Speed Wire

Li Huijian[1], Mi Zhenli[1], Su Lan[2]

(1. Collaborative Innovation Center of Steel Technology, University of Science and Technology Beijing, Beijing 100083, China; 2. Institute of Engineering Technology, University of Science and Technology Beijing, Beijing 100083, China)

Abstract: Due to the different lapping density of high-speed wire at different positions of Stelmor cooling line, the cooling of wire rod at different positions is different, resulting in non-uniformity and affecting the mechanical properties of wire rod. In this paper, the finite difference method is used to establish the temperature field and phase transition field coupling simulation model of high-speed wire lap and non-lap, and the influence of wind speed, roller speed and wire size on the cooling process of high-speed wire at different positions is considered. In steel production ϕ5.5mm 82A cord steel field measured data verifies the reliability of the model, model simulation and the measured temperature error within ±15℃. The influence of Jialing unit on the cooling speed of wire at different positions was analyzed.
Key words: high speed wire; lap joint; non-lap joint; mathematical model

钢丝电解磷化工艺探讨及应用

周斌斌，段建华，廖 建，彭 凯

（新余新钢金属制品有限公司，江西新余 338004）

摘 要：介绍了钢丝传统磷化和电解磷化发展历史、反应机制、工艺流程及过程控制要点，并通过不同磷化和拉拔试验对电解磷化、浸泡式磷化、连续在线磷化工艺进行比较，试验结果表明，电解磷化工艺相对于传统磷化工艺，其磷化膜质量更好，磷化层覆盖完全，磷化层结合力更强且牢固。且相对传统高温磷化工艺，电解磷化技术具有节省能源、消耗低、无磷渣产生等优点，符合当前绿色环保、低碳经济的发展要求。

关键词：电解磷化；传统磷化；磷化膜；拉拔；环保

Research and Application of Wire Electrolytic Phosphating Process

Zhou Binbin, Duan Jianhua, Liao Jian, Peng Kai

(Xinyu Xinsteel Metal Products Co., Ltd., Xinyu 338004, China)

Abstract: The electrolytic phosphating principle, process and parameters are introduced in this article. Two drawing experiments had been arranged to find the difference of electrolytic and conventional phosphating, which show that electrolytic phosphating technology have the advantages of better surface, better firmly and coincident phosphating film, and environmental protection.

Key words: electrolytic phosphating; conventional phosphating; phosphating film; wire drawing; environmental protection

高强钢回弹控制技术

马闻宇[1,2,3]，杨建炜[1,2,3]，郑学斌[1,2,3]，张永强[1,2,3]，王宝川[1,2,3]，李春光[1,2,3]

（1. 首钢集团有限公司技术研究院，北京 100043；2. 绿色可循环钢铁流程北京市重点试验室，北京 100043；3. 北京能源用钢工程研究中心，北京 100043）

摘 要：为提高数值模拟软件对回弹的预测精度，材料的包辛格效应得到越来越多的关注。材料的包辛格效应主要通过 Yoshida-Uemori (Y-U) 随动强化模型来描述。在本文中，根据 DP590 的拉压力学曲线，对 P590 材料的 Y-U 随动强化模型的模型参数进行了确定。进而，模型参数输入到仿真软件中进行 U 形件的仿真模拟。进而比较了仿真值和试验值之间的相近度，以及压边力对回弹的影响。压边力主要有 50 kN、100 kN 和 200 kN。由试验结果可知，仿真值与回弹值非常接近。随着压边力的增加，回弹量下降。仿真预测误差在 0.02% 之内。通过对 Y-U 模型和 Hill 模型进行比较发现，Y-U 模型要远远好于 Hill 模型。所以，在高强钢仿真模拟中有必要考虑材料的包辛格效应。

关键词：高强钢；回弹；包辛格效应；仿真模拟；材料模型

The Springback Control Technology for High Strength Steel

Ma Wenyu[1,2,3], Yang Jianwei[1,2,3], Zheng Xuebin[1,2,3], Zhang Yongqiang[1,2,3], Wang Baochuan[1,2,3], Li Chunguang[1,2,3]

(1. Research Institute of Technology of Shougang Group Co., Ltd., Beijing 100043, China; 2. Beijing Key Laboratory of Green Recycling Process for Iron & Steel Production Technology, Beijing 100043, China;

3. Beijing Engineering Research Center of Energy Steel, Beijing 100043, China)

Abstract: To improve the springback in finite element (FE) model, the Bauschinger effect of the high strength steel gained much attention. And the Yoshida-Uemori (Y-U) hardening material model is studied for the Bauschinger effect. In this study, the Yoshida-Uemori (Y-U) hardening material model is described and the parameters for this material model were calculated for DP590 according to the cycle tensile-compress tests. The material parameters for Y-U model of DP590 were put into the FE software and the simulations were conducted for U-stamping. The blank holder force was changed, including 50 kN, 100 kN and 200 kN. The corresponding experimental tests were conducted. The results were measured. It can be seen from the experiments that the springback value of U part decreases with the increase of the blank holder force. The application of Y-U model in the FE software can improve the predicting accuracy of FE model significantly, and the deviation can reach the minimum value of 0.02%, which is much better than the Hill material model. So it is necessary to consider the Bauschinger effect of the material in FE simulation.

Key words: high strength steel; springback; Bauschinger effect; finite element simulation; material model

微合金化高碳盘条生产钢帘线研究

周志嵩，孙　忍，姚海东，殷建创，苗为钢

（江苏兴达钢帘线股份有限公司，江苏泰州　225721）

摘　要：本文以 C72D2、C72D2Cr 和 C82D2 高碳钢盘条作为钢帘线生产的试验材料，其中 Cr 元素含量为 0.351%，研究 Cr 元素的添加对钢丝组织及性能的影响。结果表明：Cr 元素添加可以细化珠光体的片层间距，增加钢丝的抗拉强度和断面收缩率。与 C72D2 和 C82D2 相比，C72D2Cr 具有较好的加工硬化率，当拉拔真应变为 3.93 时，单丝的抗拉强度为 3705MPa，扭转值为 88.4。C72D2Cr 生产 3+9+15×0.175 帘线时捻股断丝率为 2.7 次/吨，破断力为 2176N，微合金化盘条生产的 3+9+15×0.175 钢帘线可以达到超高强度帘线的破断力水平。

关键词：微合金化；显微组织；抗拉强度；扭转

Study on Steel Cords Produced by Microalloying High-carbon Wire Rods

Zhou Zhisong, Sun Ren, Yao Haidong, Yin Jianchuang, Miao Weigang

(Jiangsu Xingda Steel Tyre Cord Co., Ltd., Taizhou 225721, China)

Abstract: C72D2, C72D2Cr and C82D2 wire rods were used as test materials for steel cords production and the content of Cr in C72D2Cr wire rod was 0.351%. The effect of chromium microalloying on microstructure and mechanical properties of the steel wires was investigated. Experimental results show that the Cr additioninto steel wires refined the interlamellar spacing of pearlite and increased the tensile strength and rate of reduction in area of the wires. C72D2Cr wire rod presents better work hardening rate than C72D2 and C82D2 wire rods. When the drawing strain of Cr-containing filamentwas 3.93, the tensile strength was 3705MPa and the torsion circle was 88.4. The 3+9+15×0.175 steel cord produced from C72D2Cr wire rod presented that the wire breaking rate of twisting was 2.7 times/ton and the breaking force was 2176N. The 3+9+15×0.175 steel cord produced from microalloying wire rods can reach the breaking force of super-high strength cord.

Key words: microalloying; microstructure; tensile strength; torsion circle

二十辊 SUNDWIG 轧机支承辊轴承外圈爆裂原因分析及对策

袁海永，李 明，周 军

（宁波宝新不锈钢有限公司，浙江宁波 315800）

摘 要：结合二十辊 SUNDWIG 轧机支承辊轴承的结构特点和工作条件，论文分析了轴承在使用过程中出现外圈爆裂的机理及原因，提出了相应的改进措施，并进行了实际使用验证。

关键词：SUNDWIG 轧机；支承辊轴承；爆裂

Cause Analysis and Countermeasure of Outer Ring Cracking of Backup Roll Bearing of 20-high SUNDWIG Mill

Yuan Haiyong, Li Ming, Zhou Jun

(Ningbo Baoxin Stainless Steel Co., Ltd., Ningbo 315800, China)

Abstract: Combining with the structure characteristics and working conditions of the support roll bearing of 20-high SUNDWIG mill, the mechanism and causes of the outer ring burst of the bearing in use are analyzed, and the corresponding improvement measures are put forward, and the actual application is verified.

Key words: SUNDWIG mill; backup roll bearing; crack

钢绞线用钢 SWRH82B 拉拔影响因素研究

贾元海[1]，田 鹏[2]，张俊粉[1]，李吉伟[1]，王 林[1]

（1. 河钢承钢钒钛工程技术研究中心，河北承德 067102；
2. 河钢承钢板带事业部，河北承德 067102）

摘 要：在对线材应用现状进行分析的基础上，根据断口的宏观形貌将钢绞线常见的断裂形式分为平直状、杯锥状、菊花状、斜茬状和杯凸状。其中杯锥状断裂比例最高，其主要与连铸坯心部质量有关。通过对影响钢绞线拉拔性能的因素进行了探讨，提出改善连铸坯质量、适当的时效处理、优化轧制工艺，强化包装和优化拉拔工艺有效降低钢绞线拉拔过程的断丝率。

关键词：钢绞线；拉拔性能；连铸坯

Study on Influencing Factors of Drawing Steel SWRH82B for Strand

Jia Yuanhai[1], Tian Peng[2], Zhang Junfen[1], Li Jiwei[1], Wang Lin[1]

(1. V-Ti Research Center for Engineering and Technology, HBIS Chengsteel, Chengde 067102, China;
2. Plate and Strip Business Department, HBIS Chengsteel, Chengde 067102, China)

Abstract: Based on the analysis of the wire applied status, the steel strand fracture modes were divided into flat-shaped, cup-convex, cup-cone, chrysanthemum shaped and inclined shaped according to the macroscopic fracture morphology. Cup-cone shaped fracture was the highest proportion in these fracture behaviors and primarily related to the internal quality of continuous casting billet. Through discussed the factors influencing performance of steel strand wire, measures to reduce the broken rate of steel strand wire included improve the billet quality, suitable aging treatment, optimize the rolling process, ensure the quality of packaging and optimize wire drawing process.

Key words: steel strand wire; drawing performance; continuous casting billet

浅谈辊道对棒材超声全截面校验信号的影响因素与克服措施

张进科，汪建军，张 浩

（宝钢特钢韶关有限公司，广东韶关 512123）

摘 要： 棒材自动超声旋转旋转头检测系统校验全截面探伤信号，信号幅度控制非常困难，难以实现连续走出稳定的信号，校验调节时间长短，影响校验效率。本文主要从辊道对超声校验全截面信号的影响因素方面介绍棒材超声校验信号的不确定性因素，为优化棒材超声全截面校验信号效率提供论证依据，提升作业效率同时合理把控质量过程控制。

关键词： 辊道；样棒；同心度；导套

The Influencing Factors and Overcoming Measures of the Roller Table on the Ultrasonic Check Signal of the Whole Section of the Bar are Briefly Discussed

Zhang Jinke, Wang Jianjun, Zhang Hao

(Baosteel Special Steel Shaoguan Co., Ltd., Shaoguan 512123, China)

Abstract: It is very difficult to control the signal amplitude, and it is difficult to realize the continuous stepping out of the stable signal. The length of calibration adjustment time affects the calibration efficiency. This paper mainly introduces the uncertainty factors of the ultrasonic calibration signal of the bar from the influence factors of the roller table on the ultrasonic calibration signal of the whole section, so as to provide the demonstration basis for optimizing the efficiency of the ultrasonic calibration signal of the whole section of the bar, improving the operating efficiency and reasonably

0Cr17Ni12Mo2 不锈钢弹簧断裂原因分析

郭燕飞，赵星明，李居强，孙克强，邢献强

（中钢集团郑州金属制品研究院有限公司，河南郑州 450001）

摘　要：某批次 0Cr17Ni12Mo2 奥氏体不锈弹簧钢丝绕制的压簧在使用过程中出现个别弹簧异常断裂，通过生产质控过程调查、化学成分分析、金相检验、体视显微镜和扫描电镜断口观察等方法对弹簧断裂的原因进行了分析。结果表明，弹簧断裂起源于表面缺陷处，弹簧在交变载荷应力作用下在表面缺陷处形成应力集中，裂纹在表面缺陷处萌生、扩展并造成弹簧失稳断裂。弹簧表面缺陷的形成与原料表面缺陷、拉拔钢丝润滑等因素有关，为避免弹簧使用过程中提前断裂，应严格控制弹簧表面质量。

关键词：不锈弹簧钢丝；断裂；表面缺陷；金相检验；断口分析

Fracture Analysis of 0Cr17Ni12Mo2 Stainless Steel Spring

Guo Yanfei, Zhao Xingming, Li Juqiang, Sun Keqiang, Xing Xianqiang

(Sinosteel Zhengzhou Research Institute of Steel Wire & Steel
Wire Products Co., Ltd., Zhengzhou 450001, China)

Abstract: A batch of pressure springs made of 0Cr17Ni12Mo2 austenitic stainless spring steel wire appeared abnormal breakage in use. The causes of spring breakage were analyzed by means of quality control process investigation, chemical composition analysis, metallographic examination, stereo microscope and scanning electron microscope fracture observation. The results show that the spring fracture originates from the surface defect. Under the action of alternating load stress, the spring forms stress concentration at the surface defect, and the crack sprouts and propagates at the surface defect, resulting in the unstable fracture of the spring. The formation of surface defect of spring is related to material surface defect and lubrication of drawing steel wire. In order to avoid premature rupture of spring in service, the surface quality of spring should be strictly controlled.

Key words: stainless spring steel wire; fracture; surface defect; metallographic examination; fracture analysis

新能源客车超高强钢方矩管链模成形产品质量仿真研究

熊自柳[1]，齐建军[1]，孙　力[1]，董伊康[1]，王　健[1]，
罗　扬[1]，丁士超[2]，孙　勇[2]

（1. 河钢集团钢研总院，河北石家庄　050000；2. 昆士兰大学，澳大利亚昆士兰　4072）

摘 要：本文采用 Abaqus 软件模拟了超高强钢在各向同性和混合硬化材料模型条件下新能源客车方矩管的链模成形工艺，对链模成形超高强钢回弹规律、残余应力分布，方矩管边波、箭形尺寸等产品质量指标等进行了研究。结果表明链模成形生产超高强钢方矩管时非变形区残余应力远小于材料屈服极限，边波控制在–0.15%~0.25%之间，腹板箭形尺寸低于 0.30mm；材料应力应变模型、屈服极限、模具尺寸均影响材料回弹值和模具设计；采用混合硬化模型时超高强钢回弹值略低于各向同性模型；超高强钢屈服极限增加时腹板和法兰边的回弹值增加，但成形角回弹减少；随着法兰边模具长度增加，方矩管回弹值增加；随着成形角度增加，回弹在某一角度存在最小值。

关键词：链模成形；新能源客车；方矩管；超高强钢；模拟仿真

Research on Simulation of Chain-die Forming of New Energy Bus AHSS Rectangle Tube Product Quality

Xiong Ziliu[1], Qi Jianjun[1], Sun Li[1], Dong Yikang[1], Wang Jian[1],
Luo Yang[1], Ding Shichao[2], Sun Yong[2]

(1. HBIS Group Technology Research Institute, Shijiazhuang 050000, China;
2. The University of Queensland, Queensland 4072, Australia)

Abstract: Chain-die forming process of new energy bus rectangle tube with Ural high strength steels under isotropic and combined hardening models were simulated using Abaqus software, and spring back rule, residual stress distribution, sidewave and bow dimension of rectangle tube were researched. The results show that residual stress much less than yield stress, edge wave is between –0.15%~0.2%, bow of web dimension is under 0.30mm. The spring back value and model design are affected by material hardening model, yield strength, model dimension. AHSS spring back value using combined hardening model is lower than using isotropic model. The web and flange spring back value increase and forming angle decrease when yield strength of AHSS increasing. Spring back value of rectangle increase with flange length increasing, exist minimum value with forming angle increasing.

Key words: chain-die forming; new energy bus; rectangle tube; advanced high strength steel; simulation

1500MPa 级超高强钢的成形特性研究

纪登鹏[1,2]，连昌伟[1,2]，韩 非[1,2]

（1. 宝钢股份有限公司中央研究院，上海 201900；
2. 汽车用钢开发与应用技术国家重点实验室（宝钢），上海 201900）

摘 要：为了研究宝钢新试验 1500MPa 级超高强钢的成形特性，本文以该新试验钢种及多种超高强钢作为研究对象，通过单向拉伸试验、成形极限试验及扩孔试验表征材料的成形性能。结果显示，新试验超高强钢断后延伸率达到了 12.9%，延伸率高于同强度级别的 MS1500，但低于 QP980。新试验超高强钢在拉伸过程中瞬时 n 值下降后在均匀应变的后期逐渐上升。新试验超高强钢的成形极限低于 QP980。新试验超高强钢的扩孔率为 31.3%，略微高于 QP980。数据分析对比表明，在类似强度级别的超高强钢中，新试验超高强钢具有优异的整体成形性能。新试验超高强钢中残余奥氏体含量影响材料的瞬时 n 值走势。材料的局部成形性随着组织中残余奥氏体含量的降低得到略微改善。

关键词：超高强钢；成形性；残余奥氏体

Research on Forming Characteristics of Ultra High Strength Steel of 1500MPa Grade

Ji Dengpeng[1,2], Lian Changwei[1,2], Han Fei[1,2]

(1. Research Institute of Baosteel, Shanghai 201900, China; 2. State Key Laboratory of Development and Application Technology of Automotive Steels (Baosteel), Shanghai 201900, China)

Abstract: In order to investigate the forming characteristics of Baosteel new trial ultra high strength steel of 1500MPa grade, uniaxial tensile test, forming limit test and hole expansion test were conducted on this steel and several other kinds of ultra high strength steels to characterize material forming properties. Results show that the new trial ultra high strength steel has an elongation after fracture of 12.9% which is higher than that of MS1500 with the same strength grade, but lower than that of QP980. During tension, instantaneous n value of the new trial ultra high strength steel decreases and then increases gradually at the end of uniform deformation. Forming limit of the new trial ultra high strength steel is lower than that of QP980. Hole expansion ratio of the new trial ultra high strength steel is 31.3% which is slightly higher than that of QP980. Comparison of experimental data reveals that the new trial ultra high strength steel has excellent global formability within steels with similar strength grade. The amount of retained austenite in the new trial ultra high strength steel affects variation of instantaneous n value. Local formability of material is improved slightly with decreasing the amount of retained austenite.

Key words: ultra high strength steel; formability; retained austenite

超高强钢在汽车座椅骨架轻量化中的应用

徐栋恺[1,2]，胡 晓[1,2]，陈自凯[1,2]

（1. 宝山钢铁股份有限公司研究院，上海 201900；
2. 汽车用钢开发与应用技术国家重点实验室(宝钢)，上海 201900）

摘 要： 迫于市场对轻量化水平要求的日益严苛，汽车座椅骨架系统对用材和加工工艺提出了新的要求。本文针对前、后排座椅骨架关键零部件用材进行升级，验证使用 1000MPa 及以上的超高强度钢的可行性。同时，结合零件特点，应用先进成形工艺提升超高强钢的可制造性。结果表明：在结构性能满足要求的前提下，与现有用材水平相比骨架关键零部件的减重幅度可达 25%~50%，应用超高强钢是实现轻量化的有效途径。
关键词： 汽车座椅骨架；超高强钢；先进成形工艺；轻量化

Application of Ultra High Strength Steels on Seat Frame Lightweight

Xu Dongkai[1,2], Hu Xiao[1,2], Chen Zikai[1,2]

(1. Research Institute, Baoshan Iron & Steel Co., Ltd., Shanghai 201900, China; 2. State Key Laboratory of Development and Application Technology of Automotive Steels(Baosteel), Shanghai 201900, China)

Abstract: Automotive seat frame products need to be upgraded with support of materials and processing technologies due to the increasingly strict requirement of lightweight, In this paper, ultra high strength steels (UHSS) with tensile strength above 1000MPa are utilized to achieve lightweight for key components. Also, advanced forming technologies are applied to increase manufacturability of UHSS. As a result, key components are capable of reducing weight by about 25% to 50% with the constraint of structure performance. It indicates that the application of UHSS is a feasible way to achieve lightweight for seat frame.

Key words: seat frame; ultra high strength steel; advanced forming technology; lightweight

1800MPa 级别超高强钢管气胀成形特性研究

程 超[1,2]，韩 非[1,2]，石 磊[1,2]

（1. 宝山钢铁股份有限公司研究院，上海 201900；
2. 汽车用钢开发与应用技术国家重点实验室（宝钢），上海 201900）

摘 要：文章研究了 B1800HS 超高强钢的气胀成形特性。采用实验和仿真分析结合，获得了 B1800HS 热态拉伸性能、径向压扁规律和极限自由胀形量。结果表明：温度上升时，管坯流动应力明显降低，且随着温度的升高，流动应力不断下降；随着应变速率的增加，流动应力不断增大。加热状态下，焊缝位置对径向压扁回弹影响较小，且随着温度上升，管材的回弹明显减低。自由胀形的胀破压力随增压速率的增大而升高，极限胀形率随增压速率的增大而增大；胀破压力随温度的升高而降低，极限胀形率随温度升高而略有升高。结果可以指导采用 B1800HS 高温下设计和加工复杂截面管状零件。

关键词：B1800HS；气胀成形；成形特性；极限胀形率

Experiments and Simulation on Hot Gas Formability of B1800HS

Cheng Chao[1,2], Han Fei[1,2], Shi Lei[1,2]

(1. Research Institute, Baoshan Iron & Steel Co., Ltd., Shanghai 201900, China; 2. State Key Laboratory of Development and Application Technology of Automotive Steels (Baosteel), Shanghai 201900, China)

Abstract: The hot gas formability of B1800HS UHSS was investigated with experiments and simulation. And the uniaxial tensile properties, radial flattening law and limit bulging rate were obtained. The results show that, the higher temperature resulted in the lower stress, and the higher strain rate resulted in higher stress. There was no crack under radial flatting tests, and under heated condition, the springback was few affected by the position of welding beam and decreased remarkably with higher temperature. The bursting pressure increased with increasing pressurization rate and deceased with increasing temperature, and the limit bulging rate increased with both increasing pressurization rate and temperature. The results could be used to design and hot gas forming complex cross-section tubular parts at high temperatures using the B1800HS.

Key words: B1800HS; hot gas forming; formability; limit bulging rate

热辊弯工艺在汽车前保险杠总成上的轻量化应用

张骥超[1,2]，石 磊[1,2]，杨智辉[1,2]

（1. 宝山钢铁股份有限公司研究院，上海 201900；
2. 汽车用钢开发与应用技术国家重点实验室（宝钢），上海 201900）

摘　要：汽车保险杠是轻量化材料与轻量化结构应用的最典型总成之一。本文提出了一种基于热辊弯工艺的保险杠总成轻量化设计方案，防撞梁零件采用热辊弯工艺将高刚性日字形截面与超高强度热冲压材料进行结合，在相近的结构性能前提下，防撞梁零件重量增加8%，总成重量增加14%，轻量化水平接近铝合金基准方案。针对热辊弯设计方案，完成了防撞梁与保险杠总成样件试制，验证了热辊弯工艺技术可行性。

关键词：热辊弯；保险杠；轻量化

Lightweight Application of Hot Roll Bending Process on Front Bumper Assembly

Zhang Jichao[1,2], Shi Lei[1,2], Yang Zhihui[1,2]

(1. Research Institute, Baoshan Iron & Steel Co., Ltd., Shanghai 201900, China; 2. State Key Laboratory of Development and Application Technology of Automotive Steels (Baosteel), Shanghai, 201900, China)

Abstract: Bumper is one of the most typical assemblies that lightweight materials and lightweight structures have been applied. In this paper, a lightweight design of the bumper assembly is presented. Lightweight of bumper beam is achieved through hot-roll-bending forming process that integrates high rigidity section design with ultra-high strength steel. Weight of bumper beam and assembly are increased by 8% and 14% respectively, while similar structure performance is maintained compared to Aluminum baseline design. Prototype of the bumper beam as well as other components of the assembly are completed, which validates the feasibility of hot roll bending process.

Key words: hot roll bending; bumper; lightweight

常化工艺对1.5%Si无取向硅钢组织和磁性的影响

余红春[1]，柳金龙[2]，毛一标[1]，周滨新[1]，吴圣杰[1]，马允敏[1]

（1. 沙钢集团有限公司质检处，江苏张家港　215625；
2. 东北大学材料工程学院材料研究所，辽宁沈阳　110006）

摘　要：本研究以1.5%Si无取向硅钢生产调试为基础，对比非常化、900℃及950℃常化工艺路线下的组织与磁性。950℃以上常化使得硅钢退火再结晶晶粒明显长大，且明显提高了α取向线上{110}和{100}织构的密度，提高磁性能。常化使得钢卷通卷的磁性能均匀化，最终得出走常化工艺路线的1.5%Si无取向硅钢的铁损$P_{15/50} \leq 3.4$ W/kg，磁感高于1.75 T，高于高效牌号$P_{1.5/50} \leq 3.60$ W/kg，磁感$B_{50} \geq 1.71$T的标准。

关键词：1.5%Si无取向硅钢；常化；组织；磁性能

Effect of Normalizing Process on Microstructure and Magnetic Properties of 1.5%Si Non-oriented Silicon Steel

Yu Hongchun[1], Liu Jinlong[2], Mao Yibiao[1], Zhou Binxin[1], Wu Shengjie[1], Ma Yunmin[1]

(1. Quality Inspection Office of Shagang Group Co., Ltd., Zhangjiagang 215625, China;

2. Material Research Institute of Institute of Materials of Northeast
University, Shenyang 110006, China)

Abstract: Based on the production and commissioning of 1.5%Si non-oriented silicon steel, compare with the microstructure and properties under non-normalized, 900℃, 950℃ normalized temperature. Above 950℃, the annealing recrystallized grains grow obviously, the density of {110} and {100} texture on the α line is obviously increased and the magnetic properties are improved. The properties of the coil are improved after normalizing process. Finally, after normalizing process, the iron loss $P_{15/50} \leqslant 3.4$ W/kg, magnetic induction $B_{50} \geqslant 1.75$T, higher than the standard $P_{1.5/50} \leqslant 3.60$ W/kg, $B_{50} \geqslant 1.71$T.

Key words: 1.5%Si non-oriented silicon steel; normalization; microstructure; magnetic properties

ML40Cr 马达轴断裂原因分析

王冬晨[1,2]，阮士朋[1,2]，王　欣[1,2]，杨　栋[2]

（1. 河北省线材工程技术研究中心，河北邢台　054027；
2. 邢台钢铁有限责任公司，河北邢台　054027）

摘　要： 客户使用某公司 ML40Cr 热轧盘条制作马达轴产品，在对成品检验时发生异常断裂。对断裂马达轴进行化学成分分析，满足 GB/T 3077 标准要求；利用电镜扫描（SEM）和金相显微镜分别对断口形貌和组织进行观察，断口边部为解理型断裂，心部存在较浅的韧窝，显微组织为铁素体+珠光体(球状+片层状)，轴向纵剖面观察发现在靠近心部位置存在微孔缺陷。结果表明：ML40Cr 马达轴的断裂原因是心部位置存在微孔缺陷，在受力时，裂纹由微孔处进行扩展直至发生断裂。

关键词： 马达轴；断裂；组织；微孔

Analysis on Cracking Reason of ML40Cr Motor Shaft

Wang Dongchen[1,2], Ruan Shipeng[1,2], Wang Xin[1,2], Yang Dong[2]

(1. Hebei Engineering Research Center for Wire Rod, Xingtai 054027, China;
2. Xingtai Iron & Steel Co., Ltd., Xingtai 054027, China)

Abstract: Customer used ML40Cr hot-rolled wire rod of a company to make motor shaft products, and abnormal fracture occurred during the inspection. The chemical composition of motor shaft was analyzed, it meets the requirement of GB/T 3077 standard; The morphology of the fracture and microstructure were observed by scanning electron microscopy (SEM) and metallographic microscope separately. The edge of the fracture was cleavage fracture, and there were shallow dimples in the heart. The microstructure consist of ferrite + pearlite (spheroid + lamellar). The microporous defects were found near the center by axial longitudinal section observation. The result show that the cracking reason of the ML40Cr motor shaft is the micropore defect at the core position, and when the force is applied, the crack propagates from the micropore until the fracture occurs.

Key words: motor shaft; fracture; microstructure; microporous

10B21紧固件低周疲劳和高周疲劳的断口分析

张 鹏[1,2]，李家杨[1,2]，阮士朋[1,2]，贾东涛[1,2]，王宁涛[1,2]

(1. 河北省线材工程技术研究中心，河北邢台 054027；
2. 邢台钢铁有限责任公司，河北邢台 054027)

摘 要：采用高频疲劳试验机测试了10B21材料制作的8.8级高强度紧固件在$10^4 \sim 10^6$周次范围内的疲劳性能。用扫描电子显微镜分析了疲劳断口形貌特征。结果表明，螺栓均在螺纹牙底发生疲劳开裂。低周疲劳具有多个疲劳源点。高周疲劳具有1个或少量几个疲劳源点。低周疲劳与高周疲劳样品中均有疲劳条带出现，但未观察到疲劳条带的断口亦可能为疲劳断口。

关键词：断口分析；高周疲劳；低周疲劳；8.8级紧固件；10B21

Fracture Analysis of the 10B21 Fasteners for Low Cycle Fatigue and High Cycle Fatigue

Zhang Peng[1,2], Li Jiayang[1,2], Ruan Shipeng[1,2], Jia Dongtao[1,2], Wang Ningtao[1,2]

(1. Hebei Engineering Research Center for Wire Rod, Xingtai 054027, China;
2. Xingtai Iron & Steel Co., Ltd., Xingtai 054027, China)

Abstract: Fatigue performance of grade 8.8 high strength fasteners made of 10B21 steel in the range of $10^4 \sim 10^6$ times was tested by high frequency fatigue testing machine. The fatigue fracture morphology was analyzed by scanning electron microscopy. The results show that the bolts are fatigued and cracked at the bottom of the thread. Low cycle fatigue (LCF) has multiple sources of fatigue. High cycle fatigue (HCF) has one or a few fatigue source points. Fatigue strips were observed in both low-cycle fatigue and high-cycle fatigue samples. But the fracture may also be fatigue fracture, in which fatigue strips were not be observed.

Key words: fracture analysis; high cycle fatigue; low cycle fatigue; grade 8.8 fastener; 10B21

6　先进钢铁材料

大会特邀报告

分会场特邀报告

炼铁与原料

炼钢与连铸

轧制与热处理

表面与涂镀

金属材料深加工

★ 先进钢铁材料

粉末冶金

能源、环保与资源利用

钢铁材料表征与评价

冶金设备与工程技术

冶金自动化与智能化

建筑诊治

其他

6.1 汽车用钢

低温环境专用汽车法兰用 00Cr12NiTi 钢的研制

王志军[1,2]

(1. 山西太钢不锈钢股份有限公司技术中心，山西太原 030003；2. 太原钢铁（集团）有限公司先进不锈钢材料国家重点实验室，山西太原 030003)

摘 要：不锈钢法兰作为汽车排气系统重要的密封连接件，在欧洲北部、北美等地区冬季环境温度大部分低于–32℃，这要求不锈钢法兰必须具有良好的低温冲击韧性。为此，我们研制了 00Cr12NiTi 钢以满足国内制造企业对于出口到欧洲、北美等地市场的法兰用钢需求。目前，我们已开发了 00Cr12NiTi 钢完整的工艺生产流程，可生产力学性能优良的卷切板。本钢种具有比 022Cr11Ti 相对较高的强度，特别是低温环境下具有良好的冲击韧性，在–40℃下测试冲击功值达到 50J 以上。

关键词：00Cr12NiTi；不锈钢；法兰；冲击功

Development of 00Cr12NiTi Steel for Automoblie Flange in Low Temperature Environment

Wang Zhijun[1,2]

(1. Technology Center, Shanxi Taiyuan Stainless Steel Co., Ltd., Taiyuan 030003, China;
2. State Key Laboratory of Advanced Stainless Steels Materials, Taiyuan Iron & Steel (Group) Co., Ltd., Taiyuan 030003, China)

Abstract: Stainless steel flange is an important sealing connector of automobile exhaust system. The main service environment is northern Europe and North America where the ambient temperature is mostly below –32℃ in winter, which requires the stainless steel flange must have good low temperature impact toughness.For this reason, we have developed 00Cr12NiTi steel to meet the demand of domestic manufacturers for flange steel exported to European and North American markets. At present, we have developed a complete process for the production of 00Cr12NiTi, which can produce coiled and cut boards with excellent mechanical properties. This steel has higher strength than 022Cr11Ti, especially good impact toughness at low temperature. The impact work value is more than 50J at –40℃.

Key words: 00Cr12NiTi; stainless steel; flange; impact energy

现代汽车板热处理工艺与热工炉设计要领探讨

许秀飞，王业科，夏强强

(中冶赛迪工程技术股份有限公司板带事业部，重庆 401122)

摘　要：本文通过对包括 QP 钢、DP 钢、TRIP 钢等高强钢在内的汽车板连续热处理工艺原理的分析，指出目前的带钢连续退火炉不能完全满足生产现代汽车板的需要，必须改进设计增加新功能的热工炉；在研究了 QP 钢热处理工艺流程、生产的关键工艺要点、两步法工艺路线和对热处理带来新要求的基础上，对现代汽车板连续热处理线、镀锌线和两用线热工炉设计的要领进行了探讨，设计出了典型的工艺流程布局，分别提出了加热段、保温段、缓冷段、快冷段的技术要求，特别是对各钢种在过时效段所发生的转变做了分析，提出了新型热工炉过时效段的结构和技术要求。

关键词：现代汽车板；高强钢热处理；QP 钢生产工艺；镀锌；连退；两用线

Discussion on Annealing Technology of Modern Automobile Sheet and Design Essentials of Annealing Furnace

Xu Xiufei, Wang Yeke, Xia Qiangqiang

(Department of Strip, CISDI Engineering Co., Ltd., Chongqing 401122, China)

Abstract: Based on the analysis of the principle of continuous annealing of automobile sheet, including QP steel, DP steel and TRIP steel, it is pointed out that the current strip continuous annealing furnace can not fully meet the needs of producing modern automobile sheet, and it is necessary to improve the annealing furnace designed to add new functions, and the annealing process of QP steel is studied. Based on the key technological points of production, the two-step process route and the new requirements for annealing, this paper discusses the essentials of the design of continuous annealing line, galvanized line and dual-use line annealing furnace of modern automobile sheet, and designs the typical process layout, and puts forward the technical requirements of heating section, soaking section, slow cooling section and fast cooling section respectively. In particular, the changes of various steels in the over-aging period are analyzed, and the structure and technical requirements of the over-aging section of the new annealing furnace over the aging section are put forward.

Key words: modern automobile sheet; annealing of high strength steel; production process of QP steel; galvanizing; continuous annealing; dual-purpose line

薄规格"以热代冷"热轧双相钢的研制与开发

王　成[1]，张　超[1]，刘永前[2]，赵　强[1]，李　波[1]，刘　斌[2]

（1. 武汉钢铁有限公司条材厂 CSP 分厂，湖北武汉　430083；
2. 宝钢股份中央研究院武汉分院，湖北武汉　430080）

摘　要：本文以一种低碳微合金钢为对象，基于 CSP 薄板坯连铸连轧工艺进行了汽车用热轧双相钢 DP780 的工业研制与开发。介绍了 DP780 的生产工艺流程、化学成分设计、控制轧制和基于超快冷设备的控制冷却工艺的确定原则，并对成品进行金相组织和力学性能检测分析。结果表明：在 CSP 薄板坯连铸连轧线采用 C-Mn-Cr-Nb-Ti 成分体系，成功开发出热轧双相钢 DP780，其室温组织为铁素体+马氏体，屈服强度为 480~560MPa，抗拉强度为 790~850MPa，伸长率为 18%~23%，屈强比 0.58~0.65。产品厚度为 1.2~2.0mm，具有低成本、高强度、薄规格以及低屈强比的特点，性能和质量达到冷轧同级别双相钢的标准要求，对推动汽车工业"以热代冷"进程具有重要的意义。

关键词：CSP；热轧双相钢；铁素体；马氏体

Development of Hot Rolled Thin Gauge Dual Phase Steel with "Cooling with Heat"

Wang Cheng[1], Zhang Chao[1], Liu Yongqian[2], Zhao Qiang[1], Li Bo[1], Liu Bin[2]

(1. CSP Branch of Long Product Plant of Wuhan Iron & Steel Co., Ltd., Wuhan 430083, China;
2. Wuhan Branch of Baosteel Center Research Institute, Wuhan 430080, China)

Abstract: The development of hot rolled steel DP780 for automobile on CSP was introduced using a low carbon microalloyed steel. The process flow, design of chemical composition, controlled rolling and controlled cooling with ultra fast cooling equipment for hot rolled high strength steel DP780 by CSP were introduced. Microstructure and properties of the hot rolled high strength steel DP780 were analyzed. The result shows that the microstructure of DP780 steel with C-Mn-Cr-Nb-Ti series are ferrite + martensite, the yield strength is 480~560MPa, the tensile strength is 790~850MPa, the total elongation is 18%~23%, the yield ratio is 0.58~0.65, the thickness is 1.2~2.0mm. The DP780 steel has low cost, high strength, thin gauge and low yield ratio. Performance and quality of this DP780 steel satisfied the request of cold rolled dual phase steel. The development of hot rolled high strength steel DP780 would promote the process of replacement of cold rolled products by hot rolled ones in automobile industry.

Key words: CSP; hot rolled dual phase steel; ferrite; martensite

SAPH400 热轧汽车用钢裂纹形成原因分析

王俊雄，时晓光，孙成钱，董　毅，张　宇，刘仁东，韩楚菲

（鞍钢集团钢铁研究院汽车与家电用钢研究中心，辽宁鞍山　114009）

摘　要： 本文以 SAPH400 热轧钢板为研究对象，简要介绍了该牌号热轧钢板的研制生产，并对开裂样件进行了缺陷分析。结果表明，钢板边部宏观可见"卷边"现象处微观形成明显组织流变，因此容易在后续变形时沿边部开裂。同时，基体组织主要为贝氏体，可能降低钢件整体塑性，同样容易导致钢板在加工变形时产生开裂失效。

关键词： SAPH400 钢；热轧；裂纹；夹杂物

Analysis of Causes of Crack Formation of Automobile Steel SAPH400

Wang Junxiong, Shi Xiaoguang, Sun Chengqian, Dong Yi,
Zhang Yu, Liu Rendong, Han Chufei

(Ansteel Group Iron and Steel Research Institute, Anshan 114009, China)

Abstract: Taking SAPH400 steel as research object, briefly introduced the development and production of this this hot rolled steel plate, defect analysis of the cracked samples was also performed. The result show that the macroscopically visible "rolling" phenomenon at the edge of the steel plate forms a microscopically obvious microstructure rheology, therefore, it is easy to crack along the edge during subsequent deformation. At the same time, the matrix structure is mainly

bainite, which may reduce the overall plasticity of the steel, and it is also easy to cause cracking failure of the steel sheet during processing deformation.
Key words: SAPH400 steel; hot rolling; crack; inclusion

500L-Z 热轧汽车用大梁钢裂纹形成原因分析

孙成钱，王俊雄，董 毅，时晓光，刘仁东，张 宇，韩楚菲

（鞍钢集团钢铁研究院，辽宁鞍山 114009）

摘 要：某汽车大梁用 500L-Z 钢在加工冲压过程中发生开裂，采用宏观检验、金相检验和扫描电镜检验对其冲压开裂的原因进行了分析。结果表明：冲压过程中在切边不平直处发生加工硬化导致开裂，同时试样中心位置存在珠光体偏析带，导致钢板断裂时出现断口分离现象，也是导致开裂的原因。

关键词：500L-Z 钢；热轧；裂纹；夹杂物

Analysis of Causes of Crack Formation of Automobile Beam Steel 500L-Z

Sun Chengqian, Wang Junxiong, Dong Yi, Shi Xiaoguang,
Liu Rendong, Zhang Yu, Han Chufei

(Ansteel Group Iron and Steel Research Institute, Anshan 114009, China)

Abstract: Cracks was found on 500L-Z steel automobile beam during stamping. Cracks of plate were analyzed by macrographic examination, metallographic examination and scanning electron microscope examination. The results show that the work-hardening phenomenon at the unflat trimming side of the steel plate results in the cracks defect during stamping. The pearlite segregation band in the middle of the sample is also a cause of the cracks.
Key words: 500L-Z steel; hot rolling; crack; inclusion

过程工艺对热镀锌合金化超高强钢强度的影响

李茫茫

（宝钢日铁汽车板有限公司技术质量管理部，上海 200941）

摘 要：热镀锌合金化超高强钢强度的稳定性直接受各生产过程工艺的影响。研究了抗拉强度 780MPa 级别以上的热镀锌合金化产品，通过炼钢成分、热卷冷却模式、冷轧变形量、退火和冷却温度及合金化过程的研究和试验，可以有针对性采取控制手段，以达到钢卷全长强度的稳定性，同时降低生产制造成本和借鉴到同类钢种等生产的目的。

关键词：过程工艺；热镀锌合金化；超高强钢；强度

# Effect of Process Technology on Strength of Hot-dip Galvanized and Alloyed Ultra-high Strength Steel

Li Mangmang

(Baosteel-Nippon Steel Automotive Steel Sheets Co., Ltd., Technical and Quality Control Division, Shanghai 200941, China)

Abstract: The strength stability of hot-dip galvanized and alloyed ultra-high strength steel is directly affected by the production process. The hot-dip galvanized products with tensile strength above 780 MPa were studied. Through the research and experiment of steelmaking composition, hot-coil cooling mode, cold rolling deformation, annealing and cooling temperature and alloying process, the control measures could be taken to achieve the stability of full-length strength of coil, and reduce the production cost and draw lessons from the production of the same kind of steel.

Key words: process technology; hot-dip galvanized and alloyed; ultra-high strength steel; strength

材料表面分析技术及其在锌基镀层热冲压钢开发中的应用

毕文珍[1,2], 洪继要[1,2]

（1. 宝山钢铁股份有限公司研究院，上海 201900；
2. 汽车用钢开发与应用技术国家重点实验室，上海 201900）

摘　要： 材料表面分析技术是通过分析探束或探针与材料表面发生作用产生的许多信息而研究表面的。主要分为表面形貌分析、表面组分分析和表面结构分析等几大部分。本文重点介绍了表面形貌分析技术，扫描电镜、透射电镜、扫描隧道显微镜、原子力显微镜等方法技术的原理、适用范围、特点。同时，以锌基镀层热冲压钢的开发为例，阐述了材料表面分析技术在产品开发上的应用。

关键词： 表面分析；SEM；AFM；AES；XRD；Zn基镀层热冲压钢

Material Surface Analysis Technology and Its Application in the Development of Zinc-coated Press Hardened Steel

Bi Wenzhen[1,2], Hong Jiyao[1,2]

(1. Research Institute, Baoshan Iron and Steel Co., Ltd., Shanghai 201900, China; 2. State Key Laboratory of Development and Application Technology of Automotive Steels, Shanghai 201900, China)

Abstract: The analytical technology of material surface worked on analysiny much information produced by reciprocity of probe and material surface. Geuerally speaking, surface analysis technology includes surface shape analysis, surface composition analysis, surface structure analysis and etc. This paper introduces the research progress of the surface shape analysis. Includes SEM, TEM, STM, AFM, etc. Besides, introduce the surface shape analysis on Znic-coated press hardened steel (PHS).

Key words: surface analysis; SEM; AFM; AES; XRD; Zinc-coated PHS

卷取后温度场对低碳微合金钢组织和性能的影响

张宇，时晓光，董毅，孙成钱，王俊雄，刘仁东

(鞍钢股份有限公司，辽宁鞍山 114009)

摘 要：通过在工业生产的热轧卷不同位置取样进行力学性能和显微组织分析，研究卷取过程中冷却速度的差异对低碳微合金钢组织性能的影响。结果表明：不同部位试样的显微组织均为形态各异的铁素体和 M/A 岛；随着取样部位由钢卷外部向内部变化，等轴铁素体比例增加，并且晶粒尺寸增大，M/A 岛减少；钢卷外部试样的(Nb, Ti)C 析出物尺寸小，数量较少，其力学性能较差。温度场模拟分析表明，沿热轧卷长度方向冷却速度变化较大，导致钢卷内部第二相粒子析出数量增加，因此随着取样位置由钢卷外部向内部变化，试验钢屈服强度和拉伸强度显著增加，伸长率变化不大。

关键词：低碳微合金钢；温度场；析出强化

Effect of Coiling Temperature Field on the Microstructure and Mechanical Properties of Low-carbon Micro-alloyed Steel

Zhang Yu, Shi Xiaoguang, Dong Yi, Sun Chengqian, Wang Junxiong, Liu Rendong

(Angang Steel Company Limited, Anshan 114009, China)

Abstract: The effect of cooling rate coiling process on the mechanical properties and microstructure of the tested low-carbon micro-alloyed steel plates, which are obtained from different parts of the industrial hot rolling coils, are studied in this paper. The results show that the microstructures are composed of various ferrites and M/A islands. As the sampling position changing from the external to the internal of the coil, the percentage of equiaxed ferrites increases obviously while the grain size becomes coarser and the quantity of M/A islands decreases. The external samples, which have (Nb, Ti) C precipitates with smaller size and less quantity, possess lower mechanical performance. Temperature field simulation analysis shows that the cooling rate of hot rolled coils along the length direction changes greatly. Therefore, a large number of nano size (Nb, Ti)C particles precipitate in the coil. It leads to the yield and tensile strength decreased gradually, while the elongation change small when the test samples obtained from the outer to the inner in the coil.

Key words: low-carbon micro alloyed steel; temperature field; precipitation strengthen

近年汽车用高强钢板生产和开发热点分析

郑瑞，代云红，魏丽艳

(首钢集团有限公司技术研究院，北京 100043)

摘 要：综述了 2018 年以来国内外重点钢铁企业汽车用高强钢板开发应用进展和产线新增、改造情况，分析了新建和改造产线的产品定位集中在 1000MPa 级以上超高强钢，着重介绍了汽车用超高强钢特别是冷成形用 1000MPa

级以上高延伸率超高强钢板、耐蚀性能良好的锌铝镁镀层钢板和铝硅镀层热成形钢等热点产品研发和应用方面的重要进展,并且对重点钢铁企业在印度汽车板市场开展的合作和产线布局进行了梳理。

关键词:汽车;高强钢;热镀锌;冷冲压;热成形;锌铝镁镀层;铝硅镀层

Hot Spot Analysis of Production and Development of High Strength Steel Sheet for Automobile in Recent Years

Zheng Rui, Dai Yunhong, Wei Liyan

(Shougang Research Institute of Technology, Beijing 100043, China)

Abstract: The development and application progress of automotive high strength steel sheet in key iron and steel enterprises at home and abroad since 2018, as well as the new addition and transformation of production line are summarized. It is pointed out that the product positioning of the new and modified production line is concentrated on ultra-high strength steel above 1000 MPa-grade.The important progress in development and application of hot-spot products such as ultra-high strength steels for automobiles, especially high elongation ultra-high strength steel sheets above 1000 MPa-grade for cold forming, Zn-Al-Mg coated steel sheets with good corrosion resistance and Al-Si coated hot stamping steel sheets is introduced emphatically. The cooperation and production line layout of key iron and steel enterprises in Indian automobile steel market are also sorted out.

Key words: automobile; high strength steel; hot galvanized; cold stamping; hot stamping; Zn-Al-Mg coating; Al-Si coating

汽车发动机连杆拉伸性能试验影响因素的探讨

刘运娜,郝彦英,刘献达

(石家庄钢铁有限责任公司技术中心,河北石家庄 050031)

摘 要:汽车发动机连杆拉伸性能检测时出现塑性指标偏低且数据不稳定现象,经现场调查,化学成分、拉伸断口、金相组织等分析,认为连杆拉伸性能试验时使用矩形拉伸试样工作长度部分的对称度以及试样夹持时与夹头对中不容易准确控制,会在拉伸时产生力偏心,产生附加弯曲应力,造成塑性指标偏低,且不稳定,如必须使用矩形试样,应该结合拉伸断口形貌判断检测指标的准确性,建议连杆拉伸试验最好使用圆形试样。

关键词:汽车发动机连杆;拉伸性能;拉伸断口;矩形试样;圆形试样

Investigation about the Influence Factors for Automotive Engine Connecting Rod Tensile Property Test Impact Property

Liu Yunna, Hao Yanying, Liu Xianda

(Shijiazhuang Iron & Steel Co., Ltd., Shijiazhuang 050031, China)

Abstract: Through tensile test, the plasticity index of connecting rod of automobile engine is low and unstable. Through chemical composition, tensile fracture and metallographic analysis, it is considered that the symmetry of working length of rectangular tensile specimen and the alignment between sample and clamp is not easy to control accurately in the tensile test, and force eccentricity will occur during tension, additional bending stress results in fluctuation of plasticity index. If rectangular specimen could be used, the accuracy of inspection index should be judged based on the morphology of tensile fracture. It is suggested that circular specimen should be used in the tensile test of connecting rod.

Key words: engine connecting rod; tensile property; tensile fracture; rectangular specimen; circular specimen

高强汽车板 DP980 在镀锌过程中的组织转变与相变

耿志宇，郭 健，宋海武

（唐山钢铁集团有限责任公司技术中心，河北唐山 063000）

摘 要： 通过 Gleeble3500 热模拟试验机对 DP980 钢的不同均热温度的镀锌工艺进行了模拟，并检测了镀锌过程中钢的相变以及镀锌后的组织和性能，结果表明，均热温度高于 740℃，DP980 组织中才开始出现马氏体，组织开始呈现双相钢的特征；均热温度在 740~780℃的范围内，随着均热温度升高，马氏体量逐渐增多，Ms 点逐渐降低，且屈强比变化不大。均热温度高于 780℃时，组织中开始出现贝氏体，且均热温度越高，贝氏体的量越多。贝氏体的出现使钢的屈强比升高，不利于冲压，因此 DP980 合适的镀锌工艺均热温度为 760~780℃。

关键词： DP980；镀锌；相变；双相钢

The Microstructure Change and Transformation of High Strength Automotive Plate DP980 during Galvanizing

Geng Zhiyu, Guo Jian, Song Haiwu

(Technology Center of Tangshan Iron and Steel Group Co., Ltd., Tangshan 063000, China)

Abstract: The galvanizing process of DP980 steel at different soaking temperatures was simulated by thermal simulator Gleeble3500, and the phase transformation, microstructure and properties of the steel after galvanizing were tested. The result shows that, when the soaking temperature is higher than 740℃, martensite appears in DP980, and Microstructure begins to take on the characteristics of dual-phase steel; When the soaking temperature ranges from 740 to 780℃, with the increase of soaking temperature, the amount of martensite increases gradually, the Ms point decreases gradually, and the yield-strength ratio changes little. When the soaking temperature is higher than 780℃, bainite begins to appear in the microstructure, and with the increase of soaking temperature, the amount of bainite increases. The appearance of bainite increases the yield ratio of steel, which is not conducive to stamping. Therefore, the suitable soaking temperature for DP980 galvanizing process is 760~780℃.

Key words: DP980; galvanizing; transfomation; DP steel

含 Nb 高强汽车用钢卷取前后冷却条件对组织性能的影响分析

苏振军[1]，杨建宽[1]，曹晓恩[1]，孔加维[1]，李超飞[2]

（1. 河钢集团邯钢公司技术中心，河北邯郸 056015；
2. 东北大学轧制技术及连轧自动化国家重点实验室，辽宁沈阳 110819）

摘　要：邯钢邯宝热轧厂生产含 Nb 高强钢时，受轧制、层流冷却、卷取后冷却工艺条件的不同，会得到不同的组织和性能，对于热轧中间环节，希望得到有利于酸轧轧制的软相组织，并且沿带钢长度方向均匀分布。由于热轧板组织对冷轧钢板组织具有遗传性，会导致后续冷轧产品厚度的波动和力学性能的差异，影响产品的使用性能。本文通过对比分析含 Nb 高强钢卷取前、后不同冷却条件下组织情况，制定了含 Nb 高强汽车用钢轧制冷却及卷取后冷却工艺，为冷轧提供均匀稳定的组织基料，为邯钢热轧高强钢稳定生产提供了基础保障。

关键词：卷取；高强钢；层流冷却；组织

Analysis of the Effect of Cooling Conditions before and after Coiling on Microstructure and Properties of Nb-contained High Strength Automobile Steel

Su Zhenjun[1], Yang Jiankuan[1], Cao Xiaoen[1], Kong Jiawei[1], Li Chaofei[2]

(1. HBIS Hansteel Technology Center, Handan 056015, China;
2. Northeastern University RAL, Shenyang 110819, China)

Abstract: The mechanical properties of Nb-contained ultra-high strength steel(UHSS.) are subject to rolling status such as laminar cooling mode and air cooling after coiling condition at Hanbao mill of Hansteel. Cold rolling process requires the upstream material featured with a strict even and soft mechanical properties through the whole coil of UHSS, as uneven properties of hot rolling material will be transported to cold rolling process. Severe thickness fluctuation of cold rolling sheets usually is not accepted by customers. This paper analysis how the cooling condition of both before and after coiling process affect mechanical property, offer proper solutions and provide a more even and stable mechanical properties coil for cold rolling process.

Key words: coiling; UHSS; laminar cooling; microstrure

轻量化耐疲劳热轧双相车轮钢 DP590 的开发

张志强，柳凤林，贾改风，裴庆涛，吕德文

（河钢股份有限公司邯郸分公司技术中心，河北邯郸　056015）

摘　要：采用 C-Si-Mn-Cr 的成分体系，通过密集冷却—空冷—常规冷却的三段式冷却和合理的轧制制度，在邯钢 2250mm 热连轧机组成功开发了热轧双相车轮钢 DP590，该钢种具有低屈强比、大延伸率及高加工硬化指数等特点。经某车轮生产企业试用，成型性能和焊接指标良好，制成的重卡车轮动态弯曲疲劳试验达到 190 万次，动态径向疲劳试验达到 150 万次，远高于标准要求。

关键词：轻量化；耐疲劳；热轧双相车轮钢

Development of Light Weight Fatigue Resistant Hot Rolled Dual Phase Wheel Steel DP590

Zhang Zhiqiang, Liu Fenglin, Jia Gaifeng, Pei Qingtao, Lv Dewen

(HBIS Group Handan Company, Handan 056015, China)

Abstract: Using the composition system of C-Si-Mn-Cr, through the three stage cooling of dense cooling, air cooling and dense cooling and reasonable rolling system, the hot-rolled dual phase wheel steel DP590 has been successfully developed in 2250mm hot rolling mill in Handan iron and steel company, which has the characteristics of low yield ratio, large elongation and high working hardening index. Through production trial of a wheel enterprise, the dynamic bending fatigue test of heavy truck wheel has reached $1.9×10^6$ times, and the dynamic radial fatigue test has reached $1.5×10^6$ times, which is far higher than the standard requirement.

Key words: light weight; fatigue resistant; hot-rolled dual phase wheel steel

淬火温度对 1000MPa 级 Q&P 钢组织性能的影响

王亚东，左海霞，陈虹宇

（本溪钢铁集团有限公司技术研究院，辽宁本溪　117000）

摘　要：针对 1000MPa 级 Q&P 钢，通过两相区退火结合一步淬火配分热处理工艺，采用扫描电镜和 X 射线衍射仪等手段，研究淬火温度对组织性能的影响。结果表明：实验钢组织为铁素体、马氏体和残余奥氏体，随着淬火温度的升高，屈服强度和抗拉强度均表现出下降趋势，而残余奥氏体含量和伸长率则先增加后减小，拉伸断口呈现典型的韧窝状形貌，当淬火温度为350℃时，得到最高的强塑积为24145MPa·%。

关键词：淬火温度；Q&P 钢；组织；性能

Effect of Quenching Temperature on Microstructure and Properties of 1000 MPa Grade Q&P Steel

Wang Yadong, Zuo Haixia, Chen Hongyu

(Technical Research Institute, Benxi Iron and Steel (Group) Co., Ltd., Benxi 117000, China)

Abstract: For 1000 MPa grade Q&P steel, the influence of quenching temperature on the microstructure and properties was studied by scanning electron microscopy, X-ray diffractometry and other means under intercritical annealing combined with one-step quenching process. The results show that the microstructure of the experimental steel is ferrite, martensite and retained austenite. With the increase of quenching temperature, the yield strength and tensile strength show a downward trend, while the retained austenite content and elongation are shown to increase first and then decrease. The morphology of fracture surface of appears a typical toughening nest shape. When the quenching temperature is 350°C, the product of strength and elongation could reach to 24145 MPa·%.

Key words: quenching temperature; Q&P steel; microstructure; properties

等温处理过程热轧 TRIP 钢过冷奥氏体的相变行为

王晓晖，康　健，袁　国，王国栋

（东北大学轧制技术及连轧自动化国家重点实验室，辽宁沈阳　110819）

摘　要：以不同铝含量热轧 TRIP 钢为研究对象，采用静态相变仪、动态相变仪、扫描电子显微镜和拉伸实验等实验方法，研究了贝氏体区等温处理过程中过冷奥氏体的相变行为。结果表明：对于不同铝含量的实验钢，随着等温时间的增加，实验钢均获得了稳定的残余奥氏体，随着实验钢铝含量的增加，获得稳定的残余奥氏体所需要的等温时间减少；随着等温时间的增加，实验钢的抗拉强度逐渐降低，断后延伸率和强塑积先升高后降低；随着实验钢铝含量的增加，铁素体开始转变温度升高，同时贝氏体相变速率加快。

关键词：过冷奥氏体；等温处理；显微组织；力学性能；相变行为

Transformation Behaviors of Undercooled Austenite in Hot Rolled TRIP Steels during Isothermal Treatments

Wang Xiaohui, Kang Jian, Yuan Guo, Wang Guodong

(State Key Laboratory of Rolling and Automation, Northeastern University, Shenyang 110819, China)

Abstract: Static dilatometer, dynamic dilatometer, scanning electron microscopy (SEM) and tensile tests were used to investigate transformation behaviors of the undercooled austenite in hot rolled TRIP steels with different Al content during isothermal treatments. The results show that for the tested steel with different Al content, the stable retained austenite are obtained with the isothermal time. When the Al content of the tested steel increases, the time required for the retained austenite obtainment decreases. The tensile strength of the tested steel decreases gradually with the isothermal time, whereas the total elongation and the strength ductility balance initially increases and then decreases. When the Al content of the tested steel increases, the starting transition temperature of ferrite increases, and the transformation rate of bainite is increased.

Key words: undercooled austenite; isothermal treatment; microstructures; mechanical properties; transformation behavior

热镀锌板高耐蚀性环保钝化试验研究

董学强[1]，郭太雄[1]，寸海红[2]，冉长荣[1]

(1. 攀钢集团研究院有限公司，钒钛资源综合利用国家重点实验室，四川攀枝花　617000；
2. 攀钢集团攀枝花钢钒有限公司，四川攀枝花　617000)

摘　要：为提高热镀锌板耐蚀性能，试验研究开发出了一种高耐蚀性三价铬钝化处理液。热镀锌板高耐蚀性钝化处理液采用铬酐被足量还原性有机物还原为三价铬后加入的胶体二氧化硅、含磷酸根物质等物质配制而成。该钝化处理液无六价铬，满足 RoHs 指令要求，钝化后的热镀锌板具有一定的环保性。此高耐蚀性钝化处理液可在 100℃、10～20s 内快速烘干，固化后钝化膜层表面光滑且致密，主要成分为 Cr、P、Si、C，除在板面低洼处膜层较厚外，膜层较为均匀，未出现发花、脱落等异常现象，具有良好的涂敷性能。在成膜过程中，新型热镀锌高耐蚀性环保钝化反应是电化学反应和沉淀反应及物理沉积的复合过程。在该钝化膜层中，以三价铬与磷酸根为主成膜物质，与涂层中的氢一起与镀锌板的 Zn 为反应成膜，使膜层与板面精密结合，而 SiO_2 和有机物起到填充作用，使膜层十分致密，可有效组织水及空气与镀锌板的接触，从而耐蚀性十分优异，其在中性盐雾试验条件下可达 120h 基本无锈蚀，216h 锈蚀面积小于 5%，远优于六价铬及无铬钝化液现有的 72h 锈蚀面积小于 5%。膜层主要成分为无机物，在 300℃空气中烘烤 20min 后膜层表面颜色无明显变化，耐热性良好。此外，热镀锌钝化板还具有优良的耐水性、耐指纹性，可满足用户的使用要求。

关键词：热镀锌板；高耐蚀；三价铬；钝化

Study on Environment-friendly Passivation for Hot-dip Galvanized Steel with High Corrosion Resistance

Dong Xueqiang[1], Guo Taixiong[1], Cun Haihong[2], Ran Changrong[1]

(1. Pangang Group Research Institute Co., Ltd., State Key Laboratory of Vanadium and Titanium Resources Comprehensive Utilization, Panzhihua 617000, China; 2. Pangang Group Panzhihua Steel & Vanadium Co., Ltd., Panzhihua 617000, China)

Abstract: In order to improve the corrosion resistance of hot-dip galvanized steel, experimental study developed a new trivalent chromium passivation film with high corrosion resistant. The preparation method of passivation coating with high corrosion resistant was that the CrO_3 was reduced to trivalent chromium by sufficient organic compound with reduction, then added colloidal SiO_2, the matter containing radical phosphate and other substances after reaction completely. The passivation coating did not contain hexavalent chromium, and can meet the requirement of RoHs directive, so the film on hot-dip galvanized steel after passivation was environment-friendly. After coated on the hot-dip galvanized steel, the coating could be fast drying at 100℃ about 10s to 20s. The film was uniform, and did not appear floating, falling and other abnormal phenomenon, so that the passivation coating had good coating performance. The passivation reaction of the new trivalent chromium coating on the hot-dip galvanized steel were compound processes of electrochemical reaction, precipitation reaction and physical deposition. The passivation film produced no corrosion basically after 120h in the neutral salt spray test according to the Chinese national standards GB/T 10125 and corroded area is less than 5% after 216 h, thus the corrosion resistance of the passivation film was excellent. In addition, the film has excellent heat resistance and fingerprint resistance, so it can meet the requirements of the user.

Key words: hot-dip galvanized steel; high corrosion resistance; Cr^{3+}; passivation

热轧集装箱用钢 SPA-H 薄规格的开发

田 鹏[1]，乔 俊[1]，梁静召[1]，刘 伟[1]，张振全[2]

（1. 河钢承钢板带事业部，河北承德 067102；2. 河钢承钢，河北承德 067102）

摘 要：通过对近年来世界集装箱需求量、我国出口量和出口金额的分析，预计2019年世界集装箱需求量约300万TEU，我国出口金额约90亿美元。结合集装箱行业的发展，认为1.5mm厚热轧SPA-H和1.2mm厚700MPa级高强钢将取代冷轧板，成为集装箱用钢的新方向。对1.5mm热轧集装箱用钢SPA-H的生产和使用进行研究，认为其表面质量和板形优良，力学性能稳定，满足用户的要求，并实现1.5mm SPA-H 的批量稳定供应,推动了热轧薄规格的技术进步。

关键词：热轧；集装箱用钢；SPA-H；薄规格

Development of Hot-rolled Thin Gauge SPA-H for Container Steel

Tian Peng[1], Qiao Jun[1], Liang Jingzhao[1], Liu Wei[1], Zhang Zhenquan[2]

(1. Plate and Strip Business Department, HBIS Chengsteel, Chengde 067102, China;

2. HBIS Chengsteel, Chengde 067102, China)

Abstract: With the analysis of the world's container demand, China's export volume and export value in recent years, it is estimated that the world container demand will be about 3 million TEU in 2019 and the China's export value will be about 9 billion dollars. Combined with the development of the container industry, it is believed that the hot-rolled SPA-H with 1.5 mm thickness and the 700 MPa grade high strength steel with 1.2 mm thickness will replace the cold rolled sheet and become the new direction of container steel. The production and use of hot-rolled container steel SPA-H with 1.5mm thickness is studied, it has excellent surface quality, strip shape and stable mechanical properties. These properties meet the users' requirements and can be achieved stably, the technology advancement of hot-rolled thin gauge is promoted.

Key words: hot-rolled; container steel; SPA-H; thin gauge

低碳钢铁素体轧制的力学性能和织构研究

刘宏博[1]，郝磊磊[1]，王建功[2]，夏银锋[2]，田　鹏[1]，康永林[1]

（1. 北京科技大学材料科学与工程学院，北京　100083；
2. 首钢京唐钢铁联合有限责任公司，河北唐山　063200）

摘　要：本文通过 XRD 检测方法对铁素体区轧制低碳钢热轧板和退火板的织构进行了分析，并对低碳钢退火板进行了力学性能测试和凸耳实验。结果表明：热轧板的织构整体较弱，强度在 1 到 2 之间，没有明显的择优取向；退火板中形成了较强且均匀的{111}<110>和{111}<112>织构，这有利于提高退火板的 r 值和降低 Δr 值。退火板的屈服强度为 216MPa，抗拉强度为 326MPa，伸长率为 42.5%、屈强比为 0.66、n 值为 0.18、r 值为 1.14、Δr 值为 0.07，最大凸耳率为 2.44%。

关键词：铁素体轧制；低碳钢；力学性能；织构；凸耳率

Study on Mechanical Properties and Texture of Low Carbon Steel under Ferritic Rolling

Liu Hongbo[1], Hao Leilei[1], Wang Jiangong[2], Xia Yinfeng[2],
Tian Peng[1], Kang Yonglin[1]

(1. School of Materials Science and Engineering, University of Science and Technology Beijing, Beijing 100083, China; 2. Shougang Jingtang United Iron and Steel Co., Ltd., Tangshan, 063200, China)

Abstract: In present paper, the texture of hot rolled and annealing plate of low carbon steel under ferritic rolling was analyzed by X-ray diffraction (XRD), and the mechanical properties and earing tests of annealing plate were carried out. According to the results, the texture of hot-plated is weak overall and the intensity ranges from 1 to 2, indicating that there is no strong preferred orientation. The strong and uniform textures of {111}<110> and {111}<112> were formed in the annealing plate, which is beneficial to improve the r value and reduce the Δr value. The yield strength, the tensile strength, the elongation, the yield ratio, the n value, the r value the Δr value, and the maximum earing rate of the annealing plate are 216MPa, 326MPa, 42.5%, 0.66, 0.18, 1.14, 0.07 and 2.44%, respectively.

Key words: ferritic rolling; low carbon steel; mechanical properties; texture; earing rate

铌微合金化对中碳冷镦钢组织和性能的影响

王利军[1,2]，李永超[1,2]，王　欣[1,2]，阮士朋[1,2]，王宁涛[1,2]

（1. 邢台钢铁有限责任公司，河北邢台　054027；
2. 河北省线材工程技术研究中心，河北邢台　054027）

摘　要：通过拉伸、硬度、末端淬火试验和热处理试验，研究铌微合金化对中碳冷镦钢组织和性能的影响。结果表明，中碳冷镦钢加入铌后，有利于改善中碳钢组织均匀性，细化了珠光体团尺寸且有细化珠光体片层间距和增加珠光体数量的趋势，提高了盘条的强度和塑性；铌微合金化细化了钢的奥氏体晶粒，热处理后具有更高的硬度；在本文末端淬火工艺下，固溶于奥氏体中的微量元素铌对淬透性的提高作用相比铌的碳氮化物降低淬透性的作用占优。

关键词：中碳冷镦钢；显微组织；力学性能；铌微合金化

Effect of Niobium Micro-alloying on Structure and Mechanical Properties of Medium-carbon Cold-heading Steel

Wang Lijun[1,2], Li Yongchao[1,2], Wang Xin[1,2], Ruan Shipeng[1,2], Wang Ningtao[1,2]

(1. Xingtai Iron and Steel Co., Ltd., Xingtai 054027, China;
2. Hebei Engineering Research Center for Wire Rod, Xingtai 054027, China)

Abstract: The effect of Nb micro-alloying on microstructure and mechanical properties of medium-carbon cold-heading steel was investigated by tensile, hardness, hardenability and heat treatment test. The results show that the addition of niobium in medium carbon cold heading steel is beneficial to improve the uniformity of the medium carbon steel, refines pearlite colony and intelamellar spacing, and there is a tendency to refine the pearlite lamellar spacing and increase the number of pearlites. As a result, the strength and plasticity of the wire rod are improved. Niobium micro-alloying refines the austenite grains of steel and has higher heat treatment hardness. At the end of the quenching process, the effect of the hardening of the niobium trace element in the austenite on the hardenability is superior to that of the carbonitride which reduces the hardenability.

Key words: medium-carbon cold-heading steel; microstructure; mechanical properties; niobium micro-alloying

高品质汽车齿轮钢关键控制技术

张　永

（石家庄钢铁有限责任公司技术中心，河北石家庄　050031）

摘　要：针对多数钢厂面临亟待解决的影响汽车齿轮高质量的原材纯净度、淬透性带宽、细晶粒这三方面关键因素，进行了高纯净度、窄淬透性带、细晶粒钢的技术研究，开发了纯净化和均质化的高品质汽车齿轮用钢，以满足高质量齿轮对制造材料越来越高的要求。

关键词：高品质；汽车齿轮；纯净度；淬透带

Key Technology of High Quality Automotive Gear Steel

Zhang Yong

(Shijiazhuang Iron & Steel Company, Technology Center, Shijiazhuang 050031, China)

Abstract: Aiming at the three key factors which affect the purity, hardenability bandwidth and fine grain of high quality automotive gear, which most steel mills are facing, the technical research of high purity, narrow hardenability band and fine grain steel is carried out. Pure purified and homogenized high quality automotive gear steel has been developed to meet the increasing demand of manufacturing materials for high quality gears.

Key words: high quality; automotive gear; purity; hardenability zone

6.2 船舶及海洋工程用钢

海洋平台用 620MPa 级超高强钢的热处理工艺研究和开发

周 成[1,2]，赵 坦[1,2]，金耀辉[1,2]，朱隆浩[1,2]，李家安[1,2]

（1. 海洋装备用金属材料及其应用国家重点实验室，辽宁鞍山 114009；
2. 鞍钢集团钢铁研究院，辽宁鞍山 114009）

摘　要： 针对 620MPa 超高强海洋工程用钢的性能要求，研究了淬火+回火热处理对实验钢显微组织和力学性能的影响。结果表明，两相区淬火+回火的组织内部有大量的铁素体板条束，当使前驱体调整为马氏体时，导致铁素体板条束尺寸缩短和增宽，强度下降，低温韧性变化不明显；完全奥氏体相区淬火+回火的组织为回火索氏体，调整前驱体为马氏体后，最终形成更加细小均匀的组织，强度变化不明显，低温韧性得到较大提升。

关键词： 淬火和回火；显微组织；力学性能；海洋工程用钢

Research and Development of Heat Treatment of 620MPa Grade High Strength Steel Plates for Offshore Platform

Zhou Cheng[1,2], Zhao Tan[1,2], Jin Yaohui[1,2], Zhu Longhao[1,2], Li Jiaan[1,2]

(1. State Key Laboratory of Metal Material for Marine Equipment and Application, Anshan 114009, China;
2. Iron & Steel Research Institute of Angang Group, Anshan 114009, China)

Abstract: To develop the 620MPa extra high strength marine engineering steel, the effect of quenching and tempering temperatures on microstructure and mechanical properties of the steel was studied. The results show that after intercritical quenching and tempering heat treatment, the microstructure consists of massive lathlike ferrite. The microstructure contains of much coarser and shorter lathlike ferrite can be obtained by the martensite precursor steel, which result in the reduced of strength, but low temperature impact toughness was unchanged. The tempered sorbite was formed by quenching after austenization and tempering. After quenching at austenite region and tempering process, the tested steel which has a martensite precursor has realized the excellent low temperature impact toughness, and much finer grains was obtain.

Key words: quenching and tempering; microstructure; mechanical property; marine engineering steel

Ti-Mg 氧化物冶金钢中夹杂物演变行为研究

娄号南，王丙兴，王昭东

(东北大学轧制技术及连轧自动化国家重点实验室，辽宁沈阳 110819)

摘 要： 本文对 Ti-Mg 氧化物处理钢中从 LF 炉到连铸过程夹杂物的演变行为进行了研究。结果显示经 LF 炉处理后，显微组织由块状铁素体逐渐变为针状铁素体与贝氏体。电子探针（EPMA）结果表明夹杂物平均尺寸细化至 0.34μm。然而，在 RH 炉处理后夹杂物面密度降低。另外对炼钢过程中不同元素的含量变化进行了分析，Ti 含量从 LF 炉到 RH 炉由于 Ca 的还原而降低。Mg 在 LF 炉添加后由于其化学性质含量也有所减少。夹杂物组成由 MnS-SiO$_3$ 向(Ti-Ca-Mg-Al-O)-MnS 变化。

关键词： Ti-Mg 氧化物冶金；夹杂物；炼钢；显微组织

Study on Inclusion Evolution Behaviors of Ti-Mg Oxide Metallurgy Steel

Lou Haonan, Wang Bingxing, Wang Zhaodong

(The State Key Laboratory of Rolling Automation, Northeastern University, Shenyang 110819, China)

Abstract: Evolution behaviors of inclusions from LF furnace to casting slab by Ti-Mg oxide metallurgy treatment were characterized. Results showed that microstructures changed from granular ferrite (GF) to acicular ferrite (AF) and bainite gradually after inclusion treatment in LF furnace. EPMA showed the average size of inclusions was refined to 0.34 μm. However, the area number density decreased after RH furnace treatment. In addition, the variation of different elements content in inclusion during the steelmaking process was analyzed by EDS. It was indicated that the content of Ti decreased from LF furnace to RH furnace because of reduction by Ca element. The content of Mg also decreased from the Mg addition in LF furnace because of its property. The composition of inclusions changed from MnS-SiO$_3$ to (Ti-Ca-Mg-Al-O)-MnS.

Key words: Ti-Mg oxide metallurgy; inclusion; steelmaking; microstructure

舞钢 LNG 船用低温高锰奥氏体钢的开发

莫德敏[1]，邓建军[2]，龙 杰[3]，张晨光[4]，赵燕青[5]，庞辉勇[3]，
杨 浩[5]，张俊凯[1]，王少义[4]，陈文杰[1]

(1. 河钢集团舞阳钢铁有限责任公司科技部军工钢研究室，河南舞钢 462500；

2. 河钢集团舞阳钢铁有限责任公司，河南舞钢　462500；

3. 河钢集团舞阳钢铁有限责任公司科技部，河南舞钢　462500；

4. 河钢集团舞阳钢铁有限责任公司第二轧钢厂，河南舞钢　462500；

5. 河钢集团钢研总院用户技术研究所，河北石家庄　050000）

摘　要：本文论述了舞钢 LNG 船用低温高锰奥氏体钢的开发，该钢板具有晶粒细小，夹杂物含量低，显微组织为等轴状准多边形晶粒组成的全奥氏体组织，晶粒内有孪晶，有韧窝状形貌，韧窝大小均匀且韧窝较深，为良好的强韧配比打下了基础，目前不同厚度规格的高锰钢板经检验塑韧性优良，性能指标远超 IMO 最新要求，达到国际先进水平，钢板焊后性能优良，舞钢完全具备了批量生产低温高锰奥氏体钢的生产技术。同时本文还对 LNG 船用低温高锰钢的市场前景进行了展望。

关键词：高锰奥氏体钢；超低温；LNG 储罐

Development of Low Temperature High Manganese Austenitic Steel for LNG Marine Use in Wugang

Mo Demin[1], Deng Jianjun[2], Long Jie[3], Zhang Chenguang[4], Zhao Yanqing[5], Pang Huiyong[3], Yang Hao[5], Zhang Junkai[1], Wang Shaoyi[4], Chen Wenjie[1]

(1. Reserch Institute of High Strength Steel for Millitary Application of Science and Technology Department of HBIS Group Wuyang Iron and Steel Co., Ltd., Wugang 462500, China;
2. HBIS Group Wuyang Iron and Steel Co., Ltd., Wugang 462500, China;
3. Science and Technology Department of HBIS Group Wuyang Iron and Steel Co., Ltd., Wugang 462500, China; 4. The Second Plant Mill of HBIS Group Wuyang Iron and Steel Co., Ltd., Wugang 462500, China; 5. Using Technology Research Institute of HBIS Group Technology Research Institute, Shijiazhuang 050000, China)

Abstract: This paper discuss the development of low temperature high manganese Austenite steel for LNG ship in Wugang. the steel plate has fine grain size and low inclusion content, microscopic structure is a full-austenitic structure of equiaxed quasi-polygonal grains, twins in grain, dimple morphology, uniform dimple size and deep dimple, which lays a foundation for good ratio of strength and toughness. At present, the Plasticity and toughness of high manganese steel plates with different thickness specifications have been tested to be excellent. The performance index far exceeds the latest requirements of IMO，and reaches the international advanced level. The steel plate has excellent post-welding properties. Wugang has the production technology of mass production of low temperature and high manganese Austenite steel. At the same time, The market prospect of LNG marine low-temperature high manganese steel is also discussed.

Key words: high manganese austenitic steel; ultralow temperature; LNG tank

高锰钢 TRIP/TWIP 效应对组织演变与应变硬化行为的影响规律研究

严　玲，张　鹏，王晓航，李文斌，李广龙

（海洋装备用金属材料及其应用国家重点实验室，鞍钢股份技术中心，辽宁鞍山　114009）

摘　要：研究了成分体系为 0.45C-20Mn-0.5Mo 的高锰钢经不同工艺热处理后，其显微组织演变以及在单向拉伸变形中的应变硬化行为变化规律。结果表明：高锰钢的塑性变形先后以滑移和孪生方式进行，亚稳态奥氏体组织在塑性变形中形成应变诱发马氏体和形变孪晶，位错滑移引起高应变硬化行为，而形变孪晶的产生对加工硬化性的提高起到阻滞作用，高锰钢的高强塑性是 TRIP/TWIP 两种效应综合作用的结果。

关键词：高锰钢；TRIP/TWIP 效应；应变硬化；层错能

Researched on Microstructures Evolution and Strain Hardening Behavior of TRIP/TWIP Effect of High Manganese Steel

Yan Ling, Zhang Peng, Wang Xiaohang, Li Wenbin, Li Guanglong

(State Key Laboratory of Metal Material for Marine Equipment and Application, Technology Center, Ansteel Co., Ltd., Anshan 114009, China)

Abstract: Studies on the microstructure evolution and in one direction tensile deformation the strain hardening behavior change law of high manganese steel with 0.45C-20Mn-0.5Mo after heat treatment with different processes .The results show that the plastic deformation of high manganese steel is carried out in the form of slip and twinning, The strain-induced martensite and deformation twin are formed in the process of Metastable Austenite structure plastic deformation, and the high strain hardening behavior is caused by dislocation slippage. The formation of deformation twin plays a retardant role in the improvement of machining hardening. The high strength plasticity of high manganese steel is the result of the combination of TRIP/TWIP.

Key words: high manganese steel; TRIP/TWIP effect; strain hardening; stacking fault energy

极地船用大线能量焊接用高强度船板的开发

张　朋，邓建军，龙　杰，王晓书

（河钢集团舞钢公司，河南舞钢　462500）

摘　要：本文采用微合金化的设计思路，钢板成分在低 C、低碳当量的基础上添加少量的 Nb、Ti 等微合金元素，同时在冶炼过程中采用氧化物冶金技术，在 TMCP 轧制过程中采用针对低碳当量的超快冷技术，最终在成品钢板中得到低碳贝氏体+铁素体混合组织，钢板强韧性匹配良好，-60℃冲击功达到 200J 以上。采用 100~250kJ/cm 的线能量进行焊接后，钢板焊缝及热影响区冲击韧性良好，能够满足大线能量焊接的需要，所开发钢板在国内首批极地船上取得批量应用。

关键词：大线能量焊接；NVE36；氧化物冶金；低碳贝氏体

The Development of Large Heat Input Welding High Strength Steel Plate used for Polar Ship

Zhang Peng, Deng Jianjun, Long Jie, Wang Xiaoshu

(HBIS Wuyang Iron and Steel Co., Ltd., Wugang 462500, China)

Abstract: In this paper, the composition of the steel with a small amount of Nb, Ti and large amount of Mn had been designed in micro-alloyed route. The content of C and the carbon equivalent were also designed to a low level. The technology of oxide metallurgy was used during the smelting process of the steel. The rolling technology of TMCP was controlled at a low rolling temperature and ultra-fast cooling technology was used, for the purpose of controlling the transformation of the microstructure. The microstructure of the steel plate was controlled to be the mixed microstructure of low carbon bainite and ferrite. Large amount of oxide particles dispersed in the microstructure of steel, which had a positive effect on the mechanical property and welding performance of the steel. The mechanical property of the steel plate was excellent and the value of longitudinal Akv at –60℃ is more than 200 J. The toughness of WM and HAZ were excellent after the steel plate was welded with a large heat input of 100~250 kJ/cm. The steel plate processed by mentioned above can meet the requirement of large heat input welding.

Key words: large heat input welding; NVE36; oxide metallurgy; low carbon bainite

6.3 电工钢

浅析硅钢连续退火炉炭套质量检验

何明生[1]，王雄奎[2]，张　敬[2]，龚学成[2]，杨　朝[2]，周旺枝[1]

（1. 武钢有限技术中心，湖北武汉　430080；2. 武钢有限硅钢部，湖北武汉　430080）

摘　要： 炭套是辊底式硅钢连续退火炉中用于钢带传输的重要部件，其质量和使用寿命对硅钢的生产效率和产品质量有重要影响。本文探讨了硅钢连续退火炉炭套的分类、质量评价理化性能指标，采购前后炭套质量检验指标以及硅钢生产现场的验收指标，并对目前炭套质量评价及检验中存在的问题进行了分析。这对硅钢生产企业来说，如何更科学、合理的使用和管理炭套有重要参考价值。

关键词： 炭套；质量检验；现场验收；硅钢；连续退火

Quality Inspection of Carbon Sleeve in Continuous Annealing Furnace for Silicon Steel Production

He Mingsheng[1], Wang Xiongkui[2], Zhang Jing[2],
Gong Xuecheng[2], Yang Chao[2], Zhou Wangzhi[1]

(1. R&D Center of Wuhan Iron & Steel Co., Ltd., Wuhan 430080, China;
2. Silicon Steel Division of Wuhan Iron & Steel Co., Ltd., Wuhan 430080, China)

Abstract: Carbon sleeve is the best hearth rollers to support and convey silicon steel strip in continuous annealing furnace. The quality and service life of carbon sleeve have an important influence on the production efficiency and product quality of the silicon steel. In this paper, the classification, quality evaluation physical and chemical performance index of carbon sleeve, quality inspection index before and after purchase and site acceptance index are discussed, and the problems in quality evaluation and inspection of carbon sleeve are analyzed. This provides important references for silicon steel production enterprises how to use and manage carbon sleeves more scientifically and reasonably.

Key words: carbon sleeve; quality evaluation; site acceptance; silicon steel; continuous annealing

氧化镁颗粒度检测的研究

齐 郁

（宝钢股份武钢有限公司质检中心，湖北武汉　430083）

摘　要：本文通过对激光粒度分析仪测量电工钢用氧化镁的分析条件进行优化，如分散介质、分散方式、样品预处理、仪器暗淡度等，开发出电工钢用 MgO 粒度范围测量重现性较好的试验方法，满足生产过程控制对电工钢用 MgO 粒度的要求。

关键词：氧化镁；粒径；激光粒度分析仪

Study on the Particle Size Measurement of Magnesium Oxide

Qi Yu

(Quality Inspect Center of Baosteel, Wuhan 430083, China)

Abstract: The analysis condition of grain oriented magnesium oxide based on the laser particle size analyzer were optimized, such as the dispersion medium, dispersion methods, sample preparation, and instrument obscuration, etc. Therefore, particle size measurement with good reproducibility for magnesium oxide is developed to meet the requirement of MgO size for control process.

Key words: magnesium oxide; particle size; laser particle size analyzer

薄板坯生产线无取向硅钢 50BW600 表面缺陷分析

李毅伟，李德君

（本钢技术研究院，辽宁本溪　117000）

摘　要：分析薄板坯连铸连轧生产线无取向硅钢 50BW600 产品表面缺陷种类及形成原因，找出化学成分、炼钢工艺、热轧工艺和冷轧工艺对表面质量的影响因素，提出优化措施，改善产品表面质量满足用户要求。

关键词：瓦楞缺陷；表面夹杂；叠装系数

Surface Quality Control of Non-oriented Silicon Steel 50BW600 in Thin Slab Production Line

Li Yiwei, Li Dejun

(Technology Research Institute of BX Steel, Benxi 117000, China)

Abstract: The types and causes of surface defects of 50BW600 in thin slab continuous casting and rolling production line are analyzed. The influencing factors of chemical composition, steelmaking process, hot rolling process and cold rolling process are found out. The optimization measures are put forward to improve the surface quality of products to meet user's requirements.
Key words: corrugated defects; surface inclusion; overlay coefficient

薄板坯连铸连轧工艺生产高牌号无取向硅钢研发情况

王 媛

(宝钢股份中央研究院武汉分院,湖北武汉 430080)

摘 要: 在介绍了高牌号无取向硅钢的基本情况的基础上,分析了利用薄板坯连铸连轧工艺生产高牌号无取向硅钢的优势和难点,最后介绍了国内各研究机构和企业在该方面的研发进展。
关键词: 高牌号;薄板坯连铸连轧;组织;缺陷

高牌号无取向硅钢在发电机中的应用浅研

高振宇[1],陈春梅[1],李亚东[1],刘文鹏[1],罗 理[2],赵 健[2],姜福健[1]

(1. 鞍钢集团钢铁研究院,辽宁鞍山 114009;2. 鞍钢股份有限公司,辽宁鞍山 114021)

摘 要: 本文针对高牌号硅钢产品在发电机领域的制备工艺及应用技术进行了分析研究;通过电机设计及制造工艺要求,指导了鞍钢高牌号硅钢 50AW310 产品的优化升级。电磁性能、磁化特性及产品综合质量的改进,进一步提升了鞍钢高牌号硅钢产品的生产技术水平;并以此明确了高牌号硅钢品种今后研制及发展的路径及方向。
关键词: 发电机;高牌号硅钢;冶金技术;涂层;应用研究

Abstract: The production and applying technology of high-performance silicon steel used in generator industry has been studied and analyzed. Its results are used to guild Ansteel's 50AW310 product's improvements of electromagnetic property, magnetization characteristics and overall quality. Through this the technological level of Ansteel's high-performance silicon steel is elevated, the path and direction of high-performance silicon steel's development are figured out.
Key words: generator; high-performance silicon steel; metallurgical technology; coating; applying research

提高无取向硅钢焊接性能的试验研究

姜福健[1],高振宇[1],张仁波[1],张本尊[2],王 铁[2],张智义[1],李文权[1]

(1. 鞍钢集团钢铁研究院,辽宁鞍山 114009;2. 鞍钢股份有限公司,辽宁鞍山 114021)

摘 要：焊接性是无取向硅钢的重要使用性能之一，采用焊接方法制作铁芯的用户非常关注此性能。焊接性主要与涂层中的有机成分含量和钢板基体中的碳硫成分含量有关。文章研究了硅钢焊接性的影响因素，对影响因素的影响程度进行了定量分析，提出了相关改进方法；并在应用中，使硅钢产品的焊接速度提高约 30%，可以达到约 100cm/min。

关键词：硅钢；涂层；焊接性

Experimental Research on Ameliorating Welding Properties of Non-oriented Silicon Steel

Jiang Fujian[1], Gao Zhenyu[1], Zhang Renbo[1], Zhang Benzun[2], Wang Tie[2], Zhang Zhiyi[1], Li Wenquan[1]

(1. Ansteel Group Iron and Steel Research Institute, Anshan 114009, China;
2. Ansteel Limited Liability Company, Anshan 114021, China)

Abstract: Weldability is one of the important properties of non-oriented silicon steel. This property is great concern to the users who make iron cores by welding. The weldability is mainly related to the content of organic components in the coating and carbon and sulfur components in the steel substrate. In this paper, the influencing factors of the weldability of silicon steel are studied. In the application, the welding speed of silicon steel products is increased by about 30%, which can reach about 100cm/min.

Key words: silicon steel; coating; weldability

常化工艺对 3.15% Si 硅钢冷加工性的影响

刘文鹏，高振宇，陈春梅，李亚东

（鞍钢集团钢铁研究院，辽宁鞍山 114009）

摘 要：本文针对常化工艺对 3.15% Si+0.85% Al 硅钢冷加工性的影响进行了生产实践探索，结果表明，常化板晶粒尺寸随常化温度降低而减小，冷加工性变好；在保证平均晶粒尺寸基本不变的前提下，通过少许降低常化机组速度和提升电阻带加热段温度、降低明火加热段温度，晶粒均匀性得到改善，冷加工性变好。

关键词：常化工艺；无取向硅钢；冷轧；晶粒尺寸

The Effect of Normalizing Parameters on the Cold Rollability of 3.15% Si Silicon Steel

Liu Wenpeng, Gao Zhenyu, Chen Chunmei, Li Yadong

(Iron and Steel Research Institutes of Ansteel Group Corporation, Anshan 114009, China)

Abstract: The effect of normalizing parameters on the cold rollability of "3.15% Si+0.85% Al" Si silicon steel is studied on production line. The results show that average grain size falls together with annealing temperature, then the cold rollability becomes better. While the average grain size is holden stable, as annealing time increases a little, or when the temperature of resistance wire furnace is elevated and the one of open fire annealing furnace is reduced, the homogeneity turns better and the cold rollability becomes better.

Key words: normalizing parameter; non-oriented silicon steel; cold rolling; grain size

C、Si、Mn 对 Hi-B 钢 γ-相影响规律的热力学研究

庞树芳[1]，游清雷[2]，李 莉[1]，贾志伟[1]，蒋奇武[2]

（1. 鞍钢集团钢铁研究院，辽宁鞍山 114009；2. 鞍钢股份产品发展部，辽宁鞍山 114044）

摘 要： 本文采用热力学软件 Thermo-Calc，研究 Hi-B 钢中 Si、C、Mn 元素与 γ-相相变温度以及相含量的关系。结果表明：Si 降低 γ-相相变温度，减少 γ-相最大含量；C 和 Mn 提高 γ-相相变温度、增加 γ-相最大含量、升高 γ-相含量最大时对应的温度。

关键词： Si；C；Mn；Hi-B 钢；奥氏体相

Thermodynamic Study on the Effect of C、Si and Mn on the Austenite Phase of Hi-B Steel

Pang Shufang[1], You Qinglei[2], Li Li[1], Jia Zhiwei[1], Jiang Qiwu[2]

(1. Iron and Steel Research Institutes of Ansteel Group Corporation, Anshan 114009, China;
2. Product development department of Ansteel Limited Company, Anshan 114044, China)

Abstract: In this paper, the relationships between Si、C and Mn Content in Hi-B steel and the phase transformation temperature and phase content of austenite phase were studied by thermodynamic software Thermo-Calc. Theconclusions show that Si decreases the phase transition temperature of austenite phase and the maximum content of austenite phase. C and Mn increase the phase transition temperature of austenite phase, the maximum content of austenite phase and the corresponding phase transitiontemperature.

Key words: Si; C; Mn; Hi-B steel; austenite phase

激光刻痕对取向硅钢铁损变化的影响规律

李 莉[1]，张 静[1]，贾志伟[1]，庞树芳[1]，蒋奇武[2]

（1. 鞍钢集团钢铁研究院，辽宁鞍山 114009；2. 鞍钢股份产品发展部，辽宁鞍山 114044）

摘 要： 通过激光刻痕磁畴细化降低铁损，是提高取向硅钢产品等级的有效方法。本文以鞍钢 27AG120 牌号取向硅钢为研究对象，分析了激光刻痕功率从 2W 提高到 18W 条件下，铁损和磁感的变化规律，进一步定量计算了刻痕前后的铁损分量并分析了铁损分量变化规律。实验结果表明，激光刻痕功率存在最优值，14W 激光功率下，铁损 $P_{17/50}$ 从 1.2W/kg 降低到 1.04W/kg，降幅达 12.8%，同时磁感 B_8 略有降低，降幅小于 0.5%。激光刻痕可显著降低取向硅钢反常损耗是铁损减小的主要原因，而激光功率存在最优值是由刻痕同时引起磁滞损耗和反常损耗变化所导致的。

关键词：取向硅钢；激光刻痕；细化磁畴；铁损分离

The Effect of Laser Scribing on the Core Loss of Grain-oriented Silicon Steel

Li Li[1], Zhang Jing[1], Jia Zhiwei[1], Pang Shufang[1], Jiang Qiwu[2]

(1. Iron and Steel Research Institutes of Ansteel Group Corporation, Anshan 114009, China;
2. Product Development Department of Ansteel Limited Company, Anshan 114044, China)

Abstract: Laser scribing is an effective method to reduce iron loss further and improve product grade by refining magnetic domains. In this paper, different laser scribing process parameters were used on 27AG120 high magnetic inductance grain-oriented silicon steel to find out the optimal value of laser scribing power to reduce iron loss. The domain refinement rate can reach 53% after indentation with optimized parameters. By comparing the iron loss before and after laser scribing, it can be seen that the magnetic hysteresis loss of grain oriented silicon steel is slightly increased and the abnormal loss is significantly reduced.

Key words: grain-oriented silicon steel; laser scribing; refinement of magnetic domains; iron loss seperation

6.4 建筑用钢

浅析一种新钢筋——精轧稀螺纹钢筋的科学合理性及应用前景

曹全兴

（宝武集团，上海 200940）

摘 要：精轧螺纹钢在钢筋连接上有极大的优势，它自带在任一截面上可拧入配套连接器的螺纹，实现了钢筋应力的无损传递。但由于螺纹密而粗，钢材浪费严重，使用成本大。过密的螺纹使石子难以进入螺纹间，流入的都是砂浆，降低了对螺纹的剪切力。现螺纹钢的各种连接方式，都有缺陷。

把此钢筋螺距放大一倍，即形成一种"精轧稀螺纹钢筋"，能有效取其优点，舍其不足，在建筑和高强钢筋的应用上，有广泛前景。螺纹减少后，并不需要加长连接器。

关键词：精轧螺纹钢；大量浪费钢材；影响握裹力；钢筋连接优势；精轧稀螺纹钢

Analysis of the Scientific Rationality and Application Prospect of Finish Rolling Thin Threar Reinforced Bar

Cao Quanxing

(Baowu Steel Group, Shanghai 200940, China)

Abstract: Finish rolling thread steel has great advantages in reinforced bar connection. It has threads that can be screwed into the connectors supported on any section, thus realizing the non-destructive transfer of reinforced bar stress. However, due to the dense and thick threads, steel waste is serious and the use cost is high. Over-dense threads make it difficult for stones to enter the threads. All the inflows are mortars that reduce the thread shearfoce, and all present connection methods of the thread steel have the defects.

Double the thread pitch of the reinforced bar, it forms a kind of "finish rolling thin thread reinforced bar", which can be applied in the construction and high-strength reinforced bar fields with board prospects based on effective use of its advantages and discard of its disadvantages. When threads are reduced, there is no need to lengthen the connector.

Key words: finely rolled thread steel; alarge amount of waste steel; inflaening the bonding force; advantage of reinforcing bar connection; finished rolled thin thread reinforced bar

建筑结构用低屈服点抗震钢板 LY225 的研制开发

王晓书，邓建军，张 朋，龙 杰，赵国昌

（河钢集团舞钢公司，河南舞钢 462500）

摘 要：本文采用低 C 低 Si 低 Mn 的成分设计思路，对比研究不同轧制工艺、不同正火工艺对钢板组织、晶粒度、力学性能的影响规律，表明合理的控制轧制可有效保证钢板具有较高的强度和韧性，在 920℃进行正火所得钢板内部组织均匀细小，各项力学性能完全满足标准要求。通过此工艺得到的强韧性适中的钢板满足标准要求，具备稳定批量生产的能力。

关键词：低屈服点；抗震；LY225；轧制；热处理

Research and Development of LY225 Low Yield Point Anti-seismic Steel Plate for Building Structure

Wang Xiaoshu, Deng Jianjun, Zhang Peng, Long Jie, Zhao Guochang

(HBIS Group Wusteel, Wugang 462500, China)

Abstract: The article adopts the chemical composition design concept of low C, low Si and low Mn content, compares the influencing law of different rolling process, different normalizing process on microstructure, grain size, mechanical properties. The results show that rational controlled rolling can effectively ensure high strength and toughness for steel plate. Refined microstructure and all items of mechanical properties gained after normalizing at 920℃ fully satisfy the standard requirements. The steel plate with proper strength and toughness produced with such process also meets the standard requirements, proving that the company has the capacity of stable batch production.

Key words: low yield point; anti-seismic; LY 225; rolling; heat treatment

宣钢 HRBE400 钢筋铌微合金化的探索

柳金瑞，王宏斌，张明海

（河钢宣钢一钢轧厂，河北张家口 075100）

摘　要：铌在钢铁中的应用及含铌钢的特点。宣钢一钢轧厂铌铁微合金化 HRB400E 钢筋的试验情况，研究了钢筋的成分、微合金化工艺及性能，对试制过程中出现的问题进行了分析和解决。

关键词：铌铁微合金化；HRB400ENb 高强度抗震钢筋；成本

氮在 T-5 CA 中的强化作用及其应用

孙超凡[1,2]，方　圆[1,2]，吴志国[1,2]，王雅晴[1,2]，刘　伟[3]

（1. 首钢集团有限公司技术研究院，北京　100043；2. 绿色可循环钢铁流程北京市重点实验室，北京　100043；3. 首钢京唐钢铁联合有限责任公司镀锡板事业部，河北唐山　063200）

摘　要：为了研究氮在 T-5 CA 中的强化作用及其效果，进而解决该钢种生产过程中硬度偏低的问题，采用物理化学相分析检测了钢中固溶氮和化合氮的含量，采用 TEM 分析了钢中 AlN 析出相的尺寸及其分布，通过相关材料强化理论计算公式定量分析了增氮对钢的强度的影响。理论计算和生产试验表明，T-5 CA 氮含量从 60ppm 提高到 80ppm 可使其屈服强度增加 12MPa，硬度增加 1.4 个单位。

关键词：镀锡板；增氮；强化；析出

Reinforcement and It's Application of Nitrogen in T-5 CA Steel

Sun Chaofan[1,2], Fang Yuan[1,2], Wu Zhiguo[1,2], Wang Yaqing[1,2], Liu Wei[3]

(1. Shougang Group Co., Ltd., Research Institute of Technology, Beijing 100043, China; 2. Beijing Key Laboratory of Green Recyclable Process for Iron & Steel Production Technology, Beijing 100043, China; 3. The Tinplate Department of Shougang Jingtang Iron and Steel Co., Ltd., Tangshan 063200, China)

Abstract: In order to study the reinforcement of Nitrogen in T-5 CA ,thus resolve the problem of low hardness of the steel during the production process, the amount of solid Nitrogen and combined Nitrogen were measured by physics-chemical phase analysis, the size and distribution of the AlN in the steel was studied by TEM. Based on the theory calculation formula of material strengthening method, the strengthening effect due to the increase of Nitrogen in the steel was analyzed quantitatively. The theoretical calculation and actual production show that with the Nitrogen amount in T-5 CA increasing from 60ppm to 80ppm, the yield strength of the steel would be promoted by 12MPa and the hardness of the steel would be increased by 1.4.

Key words: tinplate; nitrogen increasement; reinforce; precipitation

高层建筑用特厚高强结构钢的开发和应用

顾林豪[1,2,3]，路士平[1]，刘金刚[1]，刘春明[1]，初仁生[1]，张苏渊[1]

（1. 首钢集团有限公司技术研究院，北京　100043；2. 绿色可循环钢铁流程北京市重点实验室，北京　100043；3. 北京市能源用钢工程技术研究中心，北京　100043）

摘　要：本文涉及了高层建筑用 130mm 特厚欧标高强结构钢 S460G1-Z35 的成分设计和生产工艺的开发。通过

TMCP、淬火和回火工艺生产出具有高强度、高韧性、抗层状撕裂和良好焊接性能的特厚高强结构钢板。钢板屈服强度 Rel≥450MPa，抗拉强度 Rm≥550MPa，延伸率 A≥20%，−40℃低温冲击功均值 Akv≥250J，Z 向断面收缩率 Z≥65%。该高强结构钢板成功应用于澳门新濠天地酒店的工程建设。

关键词：高层建筑；高强结构钢；抗层状撕裂；焊接；特厚板

Research and Development of Ultra-thick High-strength Structural Steel for High-rise Buildings

Gu Linhao[1,2,3], Lu Shiping[1], Liu Jingang[1], Liu Chunming[1], Chu Rensheng[1], Zhang Suyuan[1]

(1. Shougang Research Institute of Technical, Beijing 100043, China; 2. Beijing Key Laboratory of Green Recyclable Process for Iron & Steel Production Technology, Beijing 100043, China; 3. Beijing Engineering Research Center of Energy Steel, Beijing 100043, China)

Abstract: This paper deals with the development of the composition design and production process of 130mm S460G1-Z35 for high-rise buildings. The ultra-thick and high-strength structural steel offers high strength, high toughness, lamellar tear resistance and good welding performance by using the TMCP, quenching and tempering process. For the super-thick high-strength structural steel, the yield strength is higher than 450MPa, the tensile strength is higher than 550MPa, the elongation is greater than 20%, the low temperature(−40℃) impact energy value is not less than 250J, the z-direction section shrinkage is more than 65%. The high-strength structural steel plate was successfully applied in the engineering construction of the city of dreams in Macau.

Key words: high-rise buildings; high-strength structural steel; lamellar tearing resistance; weldability; ultra-thick plate

建筑结构用高强抗震耐候钢的开发与应用研究

陈振业[1,2]，齐建军[1]，孙 力[1]，安会龙[1]，年保国[1]，姚纪坛[1]

（1. 河钢集团钢研总院，河北石家庄 050023；
2. 东北大学轧制技术及连轧自动化国家重点实验室，辽宁沈阳 110819）

摘 要：通过低碳及"Ni-Cr-Cu-Al-Nb"微合金化成分设计及控制轧制工艺的制定，成功开发出了高强度、低屈强比、耐候、易焊接的高建钢 Q460GJNH。该高强耐候钢组织由准多边形铁素体、贝氏体和珠光体组成，屈服强度 489~500MPa，抗拉强度 649~657 MPa，屈强比为 0.75~0.76，断后伸长率为 22.7%~23.1%，−40℃冲击功为 90~107J。所开发的 Q460GJNH 钢耐蚀性能显著优于 Q345B，同等条件下的腐蚀速率仅为 Q345B 的三分之一，焊接性能优良。试验钢除各项性能满足常规高建钢国标 GB/T 19879—2015《建筑结构用钢板》的要求外，还具有低屈强比、耐候及焊接性能良好的特性，为多功能高建钢的推广应用奠定了一定的基础。

关键词：高建钢；耐候；高强度；抗震；易焊接

Development and Application of High Strength Earthquake Resistant Weathering Steel for Building Structure

Chen Zhenye[1,2], Qi Jianjun[1], Sun Li[1], An Huilong[1], Nian Baoguo[1], Yao Jitan[1]

(1. The Centre Iron and Steel Technology Research Institute of HBIS, Shijiazhuang 050023, China;

2. State Key Laboratory of Rolling Technology and Automation,
Northeastern University, Shenyang 110819, China)

Abstract: The Q460GJNH steel for structural has been developed through Low carbon and "Ni-Cr-Cu-Al-Nb" microalloyed components design, controlled rolling and accelerated cooling process. The microstructure of Q460GJNH steel is composed of quasi-polygonal ferrite, bainite and pearlite. Besides, the yield strength is from 489MPa to 500MPa, and the tensile strength is from 649MPa to 657MPa. The yield ratio is 0.75~0.76, and the elongation after fracture is 22.7%~23.1%. The impact property below −40℃ keeps between 90J and 107J. The corrosion resistance of Q460GJNH steel is significantly better than that of Q345B, and the corrosion rate under the same conditions is only one third of that of Q345B, with excellent welding performance. In addition to meeting the requirements of GB/T 19879—2015 "steel plate for building structure", the test steel has low yield ratio, good weather resistance and welding performance, laying a certain foundation for the popularization and application of multifunctional high construction steel.

Key words: high construction steel; weather resistant; high strength; earthquake resistant; weldable

6.5 轨 道 用 钢

铝对 ML40Cr 冷镦钢的影响

高 航[1]，郭大勇[1]，王宏亮[2]，车 安[2]，王秉喜[1]，马立国[1]

（1. 鞍钢股份有限公司技术中心，辽宁鞍山 114001；
2. 鞍钢股份有限公司线材厂，辽宁鞍山 114001）

摘 要： 为进一步改善鞍钢合金冷镦钢盘条的使用性能，开展了 Al 元素对 ML40Cr 冷镦钢生产过程和使用性能的影响研究。结果表明，向钢中加入铝可以改善钢水的脱硫条件，在铁水脱硫条件一般时有利于冷镦钢中 S 含量的控制；加铝后生成的细小弥散的氮化铝会细化钢的组织，改善盘条的力学性能均匀性并提高盘条的塑性，但同时降低奥氏体稳定性，在加入 0.02% 的 Al 的情况下就可以明显降低钢的淬透性，因此应根据盘条规格及产品的最终用途选择合金冷镦钢的脱氧工艺。

关键词： 铝合金；冷镦钢；脱硫；力学性能；淬透性

Influence of Aluminum on ML40Cr Cold-heading Steel

Gao Hang[1], Guo Dayong[1], Wang Hongliang[2],
Che An[2], Wang Bingxi[1], Ma Liguo[1]

(1. Technology Center of Ansteel, Anshan 114001, China;
2. Wire Rod Plant of Ansteel, Anshan 114001, China)

Abstract: In order to improve the service performance of alloyed cold-heading steel wire rod of Ansteel, the effect of Aluminum on producing process and service performance of ML40Cr cold-heading steel was studied. The results show that adding of Aluminum can improve the desulfurization condition of molten steel, and it's beneficial to the control of S content in cold-heading steel when the desulfurization condition of molded iron is not so good; fine dispersed aluminum nitride particles produced in aluminum deoxidation can refine the micro-structure of steel, improve the uniformity of mechanical properties and rises the plasticity of wire rod, but at the same time reduce the stability of austenite, in case of adding 0.002% of aluminum can obviously reduce the hardenability, therefore should choose proper deoxidation process according to the diameter of wire rod and end use of alloyed cold-heading steel.

Key words: aluminum alloyed; cold-heading steel; desulfurization; mechanical property; hardenability

重轨开坯孔型系统优化及有限元模拟分析

付国龙[1,2]，赵利永[1]，刘小燕[1]，李聪颖[1]，王世杰[1]

（1. 河钢集团邯钢公司大型轧钢厂，河北邯郸 056000；
2. 东北大学轧制技术及连轧自动化国家重点实验室，辽宁沈阳 110004）

摘　要：对重轨BD区开坯轧辊孔型进行重新配置，对箱型孔参数及轧制规程进行优化设计，并利用Deform 3D软件对优化后的箱型孔轧制过程进行有限元模拟，验证了新的箱型孔参数及轧制规程的可行性。通过对重轨开坯孔型系统的优化，将BD区轧辊轧制量提高至6000t以上，将BD区总轧制道次减少至8道次，提高了轧制节奏，降低了辊耗成本。

关键词：重轨；箱型孔；有限元模拟

基于有限单元法的重轨在线余热淬火过程数值模拟研究

范　佳，易洪武，李钧正，韩志杰

（河钢集团邯钢公司技术中心，河北邯郸 056015）

摘　要：本文以邯钢生产的60kg/m U75V重轨为研究对象，基于有限单元法，对其水冷式在线余热淬火冷却过程开展数值模拟研究。通过将单位径向长度的重轨轨头冷却表面等效为相同面积大小的钢板，并对其不同机架冷却水流场以及作用到轨头等效表面后的温度场研究，得到不同重轨轨头温度下的等效换热系数。在此基础上，得到了重轨在余热淬火阶段轨头温度、踏面附近区域的金相组织及硬度的变化过程。研究结果表明：轨头温度、金相组织及硬度的变化与实际检测结果相吻合，从而为现场重轨淬火工艺和生产工艺的调整，提供了重要的参考依据。

关键词：重轨；余热淬火；温度场；换热系数；金相组织；硬度

Research on Numerical Simulation of Heavy Rail On-line Waste Heat Quenching Process based on Finite Element Method

Fan Jia, Yi Hongwu, Li Junzheng, Han Zhijie

(HBIS Group Hansteel Company Technology Center, Handan 056015, China)

Abstract: In this paper, the 60kg/m U75V heavy rail produced by Hansteel is taken as the research object. Based on the finite element method, the water-cooled on-line waste heat quenching cooling process is numerically simulated. The cooling surface of heavy rail head with unit radial length is equivalent to steel plate with the same area, and the cooling water field of different racks and the temperature field of the rail head are studied. The equivalent heat transfer coefficients at different temperature of heavy rail head are obtained. On this basis, the change process of rail head temperature, metallographic structure and hardness in the vicinity of tread during the waste heat quenching stage of heavy rail was obtained. The results show that the changes of rail head temperature, metallographic structure and hardness coincide with the actual test results, which provides an important reference for the adjustment of quenching process and production process of heavy rail in situ.

Key words: heavy rail; waste heat quench; temperature field; heat transfer coefficient; metallographic structure; hardness

R350HT 热处理钢轨的开发实践

王瑞敏，周剑华，费俊杰，欧阳珉路

（宝钢股份中央研究院武汉分院（武钢有限技术中心），湖北武汉　430080）

摘　要： 高强度热处理钢轨可显著改善钢轨耐磨和抗接触疲劳性能，尤其在小半径曲线上应用效果显著。通过开展成分优化与热处理工艺研究，成功开发了 R350HT 热处理钢轨，经批量生产，产品实物质量控制良好，各项性能均符合 EN13674.1 标准要求。钢轨的抗拉强度 $R_m \geq 1212$MPa，断后伸长率 $A \geq 11\%$，踏面硬度 358~375HB，横断面布氏硬度梯度分布满足判定方法，轨底残余应力 124~154MPa。

关键词： 钢轨；热处理；开发

Development of R350HT Heat-treated Rail

Wang Ruimin, Zhou Jianhua, Fei Junjie, Ouyang Minlu

（Wuhan Branch of Baosteel Central Research Institute(R and D Center of Wuhan Iron & Steel Co., Ltd.), Wuhan 430080, China）

Abstract: High strength heat treated rails can significantly improve the wear resistance and contact fatigue resistance of rails, especially on small radius curves. R350HT heat treatment rail has been successfully developed through the research of composition optimization and heat treatment process. After mass production, the product quality control is good, and all properties meet the requirements of EN13674.1 standard. Its mechanical properties is $R_m \geq 1212$MPa, $A \geq 11\%$, tread hardness 358~375HB, residual stress in rail foot is 124~54MPa, Brinell hardness in cross section meets the criterion.

Key words: rail; heat-treated; development

百米钢轨水淬工艺研究与应用

张海旺，李钧正，韩志杰

（河钢股份有限公司邯郸分公司技术中心，河北邯郸　056015）

摘　要： 百米钢轨全长在线水淬热处理工艺，其实质是用水对钢轨进行选择性软冷却，目的是冷却后在轨头硬化层内获得均匀的细片状珠光体组织。以水对钢轨进行均匀冷却时，冷却强度大，极易出现过冷组织，从而导致钢轨截面硬度及组织不均匀，淬火钢轨在使用过程中产生脆断或掉块，严重危及行车安全。通过研发水流形态技术、水轨换热技术、水压动态调节技术等相关技术，实现了水淬钢轨的批量高效稳定化生产。

关键词：百米钢轨；在线热处理；水淬工艺

Research and Application of Water Quenching Technology for 100m Rail

Zhang Haiwang, Li Junzheng, Han Zhijie

(Technology Center of Handan Branch of Hesteel Co., Ltd., Handan 056015, China)

Abstract: The on-line water quenching heat treatment process for 100-meter rail is essentially a selective soft cooling process with water to obtain uniform fine pearlite structure in the hardened layer of the rail head after cooling. When the rails are uniformly cooled by water, the cooling intensity is high and the supercooling structure is easy to appear, which results in the uneven hardness and structure of the rails. The quenched rails are brittle and fractured in use, which seriously endangers the safety of driving. Through the research and development of water flow morphology technology, rail heat transfer technology, water pressure dynamic regulation technology and other related technologies, the batch production of water quenched rail has been realized with high efficiency and stability.

Key words: 100m rail; on-line heat treatment; water quenching process

6.6 能源用钢

L485M 钢板的组织与性能

渠秀娟[1]，应传涛[2]，徐　烽[1]，陈军平[1]，惠兴伟[1]

(1. 鞍钢股份有限公司产品发展部，辽宁鞍山　114021；
2. 鞍钢股份有限公司鲅鱼圈分公司，辽宁营口　115007)

摘　要：本研究中，L485M 组织以贝氏体为主，低温韧性优良；钢板具有良好的抗酸性能。制管后，拉伸性能、夏比冲击性能与落锤撕裂性能均有不同程度的变化。结合现有数据与工程参数，利用断裂韧性计算公式分析钢管的服役状态，结果表明钢管的安全性良好。

关键词：煤制气；L485M；氢气；组织；性能

Microstructure and Property of L485M Plates

Qu Xiujuan[1], Ying Chuantao[2], Xu Feng[1], Chen Junping[1], Hui Xingwei[1]

(1. Department of Product Development of Angang Steel Co., Ltd., Anshan 114021, China;
2. Bayuquan Branch of Angang Steel Co., Ltd., Yingkou 115007, China)

Abstract: In this study, the L485M plates have microstructure of bainite mainly, which show excellent low temperature toughness and good anti-corrosion properties. After pipe-making process, mechanical properties of tensile, Charpy V-notched impact and drop weight tear test had evolved to some extent. Combined with the existing experimental data and the project parameters, the pipe safety was analyzed by the formula of fracture toughness. The calculation results showed the L485M line pipes are safe.

Key words: coal gas; L485M; H_2; microstructure; property

耐候钢板在转炉煤气柜的应用

郭 凌，王 河，李军红

（鞍钢股份有限公司能源管控中心燃气分厂，辽宁鞍山 114021）

摘　要： 鞍钢股份能源管控中心燃气分厂3#、4#8万立方米转炉煤气柜在近年来相继出现煤气柜壁板、底板、活塞支架根部煤气泄漏问题，通过分析转炉煤气冷凝水的成分、气柜壁板和活塞板及底板等泄漏位置情况，对转炉煤气成分进行分析，找出煤气柜泄漏原因，并根据煤气柜泄漏点的部位，采用耐候钢这一新型材料代替原有的普通碳钢制作气柜壁板和活塞板及底板。延长了煤气柜的使用寿命、消除了煤气泄漏的隐患。

关键词： 转炉煤气柜；泄漏；耐候钢；改进

大壁厚大管径 X65M 管线钢落锤性能优化

喻赛华[1]，史术华[2]

（1. 华菱湘钢销售部，湖南湘潭 411100；2. 华菱湘钢技术质量部，湖南湘潭 411100）

摘　要： 针对湘钢生产的1422mm管径25.4mm X65M管线钢出现落锤性能不合格的现象进行了金相组织分析，发现不合落锤试样显微组织为多边形铁素体+珠光体+贝氏体的混合组织，晶粒尺寸较大，主要是由于装炉和加热温度高，入水温度和冷却速率低造成。通过工艺优化，获得细小的针状铁素体和粒状贝氏体的混合组织，落锤性能指标大幅度提高。

关键词： 管线钢；X65M；落锤性能

高性能风电用 S420ML 宽厚钢板的研制与开发

潘中德，李 伟，胡其龙，高 飞

（南京钢铁股份有限公司板材事业部，江苏南京 210035）

摘　要： 随着风电装备大型化设计技术的发展，对钢板的高强度、低温断裂韧性和焊接性能提出了更为严格的要求。本文阐述了南钢通过低碳、低碳当量成分设计以及TMCP工艺设计，成功研发了高性能风电用S420ML宽厚钢板，经检测钢板各项性能指标满足 EN10025-4 标准要求，其中钢板屈服强度≥420MPa，-40℃低温冲击功平均值大于200J，钢板探伤质量满足 EN10160 标准 S2E3 级要求；钢板在焊接热输入量为 50 kJ/cm 情况下，焊接接头具有优异的综合性能，能够满足风力发电力设备的设计及使用要求。

关键词：S420ML；风电用钢；成分设计；工艺设计；焊接性能

Research and Development of S420ML Wide & Heavy Steel Plate for High Performance Wind Power

Pan Zhongde, Li Wei, Hu Qilong, Gao Fei

(Plate Business Unit of Nanjing Iron & Steel Co., Ltd., Nanjing 210035, China)

Abstract: With the development of large-scale design technology of wind power equipment, more stringent requirements are put forward for higher strength, low temperature fracture toughness and weldability. Through the composition design of low C and CEV, and process design of TMCP, NISCO successfully develops S420ML wide & heavy steel plate for high performance wind tower. Test results show that all property of the steel plate meet the standard of EN10025-4. The yield strength of steel plate is morn than 420MPa, and the value of average impact energy at temperature of -40℃ is over 200 J. The ultrasonic flaw detection conform to the stipulation of EN10160 standard class S2E3. When the welding heat input of steel plate is 50 kJ/cm, the welding joint has excellent comprehensive performance, and the plate can meet the design and use requirements of wind power equipment.

Key words: S420ML; wind power steel; composition design; process design; weldability

厚规格 X80 管线钢制管前后的力学性能变化分析

张豪臻，章传国，孙磊磊

（宝山钢铁股份有限公司中央研究院，上海 201900）

摘 要：研究了工业生产的 ϕ1422mm×25.7mm/30.8mm 规格的 JCOE/UOE X80 焊管在制管前后的拉伸性能、夏比冲击韧性以及 DWTT 性能，以分析不同厚度规格和制管工艺下板管力学性能的变化规律。总体上，钢管 DWTT 性能与钢板相比变化较小，钢管-10℃冲击功与-20℃钢板冲击功相比有所降低，而制管后屈服强度则出现不同程度的上升，抗拉强度相对变化较小，屈强比的变化趋势与屈服强度基本一致。UOE 与 JCOE 相比，UOE 制管后屈服强度以及屈强比的提升更为显著，且制管后的冲击性能相对较好，而 DWTT 性能变化的差异则相对较小。同样外径和制管方式下，对于较厚规格的钢板，其制管后冲击韧性相对较好，但拉伸及 DWTT 性能的变化趋势则没有明显差异。

关键词：X80 管线钢；力学性能；JCOE；UOE

Analysis on Mechanical Properties Variation of Thick-walled X80 Pipeline Steel before and after Pipe Manufacturing

Zhang Haozhen, Zhang Chuanguo, Sun Leilei

(Central Research Institute(R&D Center), Baoshan Iron & Steel Co., Ltd., Shanghai 201900, China)

Abstract: Investigated the tensile properties, charpy impact toughness and DWTT properties of JCOE/UOE X80 welded pipe(ϕ1422mm×25.7mm/30.8mm) before and after pipe manufacturing, to analyzed the mechanical properties of steel plates under different thickness and pipe-making processes. Generally, the DWTT performance of steel pipes has little change compared with steel plates, but the impact energy of steel pipes at −10℃ is lower than that of steel plates at −20℃, while the yield strength of steel pipes increases in varying degrees, the relative change of tensile strength is small, and the change trend of yield strength ratio is basically the same as that of yield strength. Compared with JCOE, the yield strength and yield-strength ratio are improved more significantly after UOE process, and the impact performance of UOE pipe is better, while the difference of DWTT performance is relatively small. Under the same external diameter and pipe making technology, the impact toughness of thicker steel plates after pipe making is relatively high, but there is no significant difference in the trend of tensile and DWTT properties.

Key words: X80 pipeline steel；mechanical properties；JCOE；UOE

高耐酸性管线钢的研制和性能研究

孔祥磊[1,2]，黄国建[1,2]，黄明浩[1,2]，王 杨[1,2]，张英慧[1,2]

（1. 海洋装备用金属材料及其应用国家重点实验室，辽宁鞍山 114009；
2. 鞍钢集团钢铁研究院，辽宁鞍山 114009）

摘 要： 为满足当前酸性油气田开发需要，针对当前管道工程对耐酸性管线钢要求的提高，鞍钢抓住市场机遇，成功开发了高耐酸性 X52MS/X60MS 管线钢热轧卷板，并稳定化生产了 9600t。技术上采用了低碳低锰微合金化设计、纯净钢冶炼技术和热机械轧制工艺技术，炼钢过程中严格控制杂质元素含量和夹杂物形貌、数量，连铸过程中着重抑制偏析，轧制过程中严格控制工艺窗口，使晶粒细小、组织均匀，并抑制珠光体带状组织产生。性能检验结果表明，产品具有优异的强度、韧性和耐酸性，特别是可耐加载实际屈服强度的 0.80 倍的 SSCC 检验，为国内开发优质耐酸性管线钢提供借鉴。

关键词： 酸性；管线钢；X52MS；X60MS；SSCC

Development and Properties Research of High Acid Resistance Pipeline Steel

Kong Xianglei[1,2], Huang Guojian[1,2], Huang Minghao[1,2], Wang Yang[1,2], Zhang Yinghui[1,2]

(1. State Key Laboratory of Metal Material for Marine Equipment and Application, Anshan, 114009, China;
2. Ansteel Group Iron & Steel Research Institute, Anshan, 114009, China)

Abstract: In order to meet the development needs of acidic oil and gas field and improvement of requirements for acidic pipeline steel in current pipeline projects, Ansteel seized the market opportunity and successfully developed high acidic X52MS/X60MS pipeline steel hot-rolled coils, and completed production of 9600 tons stabilized. Low carbon and low manganese, microalloying design, pure steel smelting technology and thermomechanical rolling technology are adopted. The content of impurity elements and the morphology and quantity of inclusions are strictly controlled in steelmaking process. Segregation is restrained in continuous casting process. The process window is strictly controlled in rolling process, so that the grain size is fine and the structure is uniform, and the production of pearlite banded structure is restrained. The test results show that the product has excellent strength, toughness and acid resistance, especially the SSCC test stressed with 0.80 of the actual yield strength, which provides a reference for the development of high-quality acid-resistant pipeline steel in China.

Key words: acid; pipeline steel; X52MS; X60MS; SSCC

低屈强比高强韧性海底管线钢控轧控冷工艺研究

李冠楠，贾改风，孙 毅

（河钢股份有限公司邯郸分公司技术中心，河北邯郸 056015）

摘 要： 为保证深海管线钢在恶劣的环境条件下能满足服役期的要求，本文重点进行低屈强比、高强韧性 X65MO 的开发。通过 Gleeble 试验机绘制出 CCT 曲线，结合不同冷速状态下金相组织的变化，对粗轧入口温度、精轧入口温度、卷取温度、冷速等参数进行了优化配置，实现产品强韧性的最优匹配；通过对卷取温度、冷却模式和中间坯厚度等工艺参数进行的正交试验数据分析，确定低温卷取，增加弛豫冷却时间，低中间坯厚度等方法可以有效降低材料的屈强比。最终产品–25℃落锤性能和–40℃冲击性能指标优异，屈强比平均在 0.83 左右，满足了工程需求。

关键词： 屈强比；强韧性；弛豫

Development of Low Yield Strength Ratio High Strength and Toughness Submarine Pipeline Steel

Li Guannan, Jia Gaifeng, Sun Yi

(Technique Center, Handan Branch Company of Hesteel Company limited, Handan 056015, China)

Abstract: In order to ensure that deep-sea pipeline steel can meet the requirements of service life in harsh environment, this paper focuses on the development of X65MO with low yield-strength ratio and high strength and toughness. CCT curves are drawn by Gleeble tester, and parameters such as roughing entry temperature, finishing entry temperature, coiling temperature and cooling speed are optimized according to the change of metallographic structure at different cooling rates, so as to achieve the optimal match of strength and toughness of products. The low temperature is determined by analyzing the orthogonal test data of coiling temperature, cooling mode and transfer bar thickness. Low coiling temperature, increasing relaxation cooling time and low thickness of transfer bar can effectively reduce the yield-strength ratio of the material. The final product has excellent DWTT property at − 25℃ and impact property at − 40℃, The average yield-strength ratio is about 0.83, which meets the engineering requirements.

Key words: yield-strength ratio; strength-toughness; relaxation

高等级 X90 管线钢奥氏体动态再结晶规律研究

张 海[1,2]，李少坡[1,2]，丁文华[1,2]

（1. 首钢集团有限公司技术研究院，北京 100041；
2. 北京市能源用钢工程技术研究中心，北京 100041）

摘 要： 在热模拟试验机上采用单道次压缩试验研究了 X90 管线钢在热变形过程中的动态再结晶行为。根据单道次压缩过程中的应力-应变曲线分析了 X90 管线钢在不同变形温度和变形速率条件下的动态再结晶规律,结果表明:高温低应变速率下,试验钢更容易发生动态再结晶,当变形温度足够低或者是变形速率足够大的时候,动态再结晶难以发生。采用回归方法确定了试验钢动态再结晶激活能和材料常数,最终建立了动态再结晶方程。同时根据试验钢的动态再结晶临界应变和稳态应变绘制了动态再结晶区域图。

关键词:X90 管线钢;热模拟;动态再结晶

Study on Dynamic Recrystallization Rule of Austenite for High Grade X90 Pipeline Steel

Zhang Hai[1,2], Li Shaopo[1,2], Ding Wenhua[1,2]

(1. Research Institute of Technology, Shougang Group Corporation, Beijing 100041, China;
2. Beijing Engineering Research Center of Energy Steel, Beijing 100041, China)

Abstract: The dynamic recrystallization behavior of X90 pipeline steel during hot deformation was studied by single pass compression test on a thermal simulator. The dynamic recrystallization rule of X90 pipeline under different deformation temperature and deformation rate was analyzed according to the stress-strain curve for the single pass compression process and it indicated that dynamic recrystallization was more likely to occur at high temperature and low strain rate, however, when the deformation temperature was low enough or deformation rate was high enough, dynamic recrystallization is difficult to occur. And the activation energy of dynamic recrystallization and the material parameters of the test steel were determined by regression method, and the dynamic recrystallizaiton equation was finally established. At the same time, the dynamic recrystallization diagram was drawn based on the dynamic recrystallization critical strain and steady-state strain.

Key words: X90 pipeline steel; thermal simulation; dynamic recrystallization

舞钢石油化工和煤化工用高性能容器钢板的生产

吴艳阳,龙 杰,李样兵

(舞阳钢铁有限责任公司,河南舞钢 462500)

摘 要:本文简要介绍了石油化工和煤化工用高性能容器钢板的产品结构,技术要求,生产关键工序控制点,同时介绍了舞钢生产石油化工和煤化工用高性能容器钢板与国外钢板的实物质量对比,根据对比可以看出,舞钢生产的产品实物性能与国外产品实物性能相当,部分指标甚至优于国外钢厂。

关键词:石油化工;煤化工;舞钢

The Production of High Performance Container Steel Plate for Wusteel Petrochemical and Coal Chemical Industry

Wu Yanyang, Long Jie, Li Yangbing

(Wuyang Iron and Steel Co., Ltd., Wugang 462500, China)

Abstract: This paper briefly introduces the product structure, technical requirements, production key process control points of high performance container steel plate for petrochemical and coal chemical industry. At the same time, the physical quality comparison between the high performance container steel plate produced by Wugang and foreign products is introduced. According to the comparison, it can be seen that the physical properties of the products produced by Wugang are equivalent to those of foreign products. Some indicators are even better than foreign steel mills.

Key words: petrochemical and coal chemical industry; coal chemical industry; Wugang

复合微合金化 J55 石油套管用钢生产实践

陆凤慧，邢俊芳，周少见，白宗奇

（河钢承钢板带事业部，河北承德 067102）

摘 要：河钢承钢 1780 生产线，采用在传统 16Mn 成分设计基础上，添加铌钛等微合金成分，采用低硫、低氮的洁净钢冶炼控制，控轧控冷的轧制技术，实现–20℃低温冲击性能要求的 J55 钢卷生产，满足客户需求。

关键词：石油套管；控轧控冷；微合金

Production Practice of Compound Microalloyed J55 Oil Casing Steel

Lu Fenghui, Xing Junfang, Zhou Shaojian, Bai Zongqi

(Plate and Strip Business Department, HBIS Chengsteel, Chengde 067102, China)

Abstract: On the basis of traditional 16Mn composition design, Niobium and Titanium microalloys were added to the 1780 production line of HBIS Chengsteel. The J55 coil with low sulfur and nitrogen content was produced by controlling rolling and cooling of clean steel to meet customers' demands.

Key words: oil casing; controlled rolling and cooling; microalloys

两种低温高强钢的比较研究

罗 毅

（宝钢股份中央研究院武汉分院（武钢有限技术中心），湖北武汉 430080）

摘 要：比较了相同工艺流程生产、厚度为 20mm 的两种低温高强钢板 LTS1 和 LTS2 的力学性能和焊接热影响区硬度，结果表明两者的屈服强度均在 600MPa 以上，LTS1 钢抗拉强度和–80℃、–100℃的冲击功均高于 LTS2 钢，而 LTS2 钢的焊接性能较好。通过组织观察发现，两钢组织均为回火马氏体+贝氏体，各自的表层和心部的组织相同，LTS1 钢的组织以回火马氏体为主，而 LTS2 钢的组织中贝氏体较多。两钢的强韧性均与其较高的合金元素相关，LTS1 钢较高的 C 和 Ni 含量使得其强度和低温冲击功更高，而 LTS2 钢中因含 Cu 产生的析出强化，使得其具有接近 LTS1 钢的强韧性。

关键词：低温高强钢板；性能；组织；焊接热影响区硬度

Comparative Study on Two Types of Low Temperature High Strength Steels

Luo Yi

(Central Research Institute of Baosteel Company, Wuhan Branch, Wuhan 430080, China)

Abstract: The mechanical properties and hardness in weld hot affect zone of the LTS1 and LTS2 low temperature high strength steel plates with thickness of 20 mm, and produced by the same process were compared. The results show that the yield strength of both LTS1 and LTS2 steel is above 600 MPa, and the tensile strength and impact energy at –80℃ and –100℃ of the LTS1 steel are higher than those of the LTS2 steel, while the weldability of the LTS2 steel is better. Microstructure observation shows that their microstructures at surface and core are tempered martensite + bainite. Tempered martensite is the main structure of the LTS1 steel, while bainite is more in the LTS2 steel. The strength and toughness of both steels are related to their higher alloying elements. The higher content of C and Ni in the LTS1 steel results in its higher strength and impact energy at low temperature. The precipitation strengthening of the LTS2 steel due to the addition of Cu makes its strength and toughness close to those of the LTS1 steel.

Key words: low temperature high strength steel plates; properties; microstructure; hardness in weld hot affect zone

6.7 工程机械用钢

500MPa级高耐磨焊管用钢的开发

惠亚军[1,2]，吴科敏[3]，刘锟[1]，许克好[3]，许斐范[1]，令狐克志[1]

（1. 首钢集团有限公司技术研究院薄板所，北京 100043；2. 绿色可循环钢铁流程北京市重点实验室，北京 100043；3. 北京首钢股份有限公司制造部，河北迁安 064404）

摘 要： 采用 OM、SEM 和 TEM 等仪器对 500MPa 级高耐磨焊管用钢的组织与性能进行了研究。结果表明，实验钢具有优良的强塑性，力学性能波动性较小，屈服强度在 535~551MPa，抗拉强度在 591~605MPa，断后伸长率均在 30.5%以上，屈强比在 0.91 左右。试验钢由铁素体和少量珠光体组成，铁素体晶粒均匀且细小，平均尺寸约为 2.5μm，珠光体球团尺寸约为 1.3μm，珠光体球团平均体积分数约为 3%；第二相析出物尺寸主要分布在 3~100nm 之间，平均尺寸约为 31nm，能谱分析显示其为 (Nb、Ti)CN。与 Q235B 相比，试验钢的耐磨性与耐蚀性更优，在 40N 载荷下磨损相对失重率在 70%以下，随着磨损时间的延长，相对失重率逐渐下降，耐磨性有所提升，这与在磨损过程中表层金属发生显著的加工硬化有关；试验钢的阻抗弧较大、自腐蚀电流较小，这与试验钢表面氧化铁皮中 Fe_3O_4 含量较高有关。

关键词： 500MPa 级；托辊用钢；低温韧性；耐磨性；耐蚀性

Development of 500MPa Grade High Wear-resistant Steel for Welded Tube

Hui Yajun[1,2], Wu Kemin[3], Liu Kun[1], Xu Kehao[3], Xu Feifan[1], Linghu Kezhi[1]

(1. Technology Institute of Shougang Group Co., Ltd., Sheet Metal Research Institute,

Beijing 100043, China; 2. Beijing Key Laboratory of Green Recyclable Process for Iron & Steel Production, Shougang Group, Beijing 100043, China; 3. Beijing Shougang Co., Ltd., Manufacturing Department, Qian'an 064404, China)

Abstract: The microstructure and properties of 500MPa grade high wear-resistant steel for welded tube were investigated by utilizing optical microscope, scanning electron microscopy and transmission electron microscopy. The results show that the tested steel has excellent mechanical properties, the yield strength is 535~551MPa, the tensile strength is 591~605MPa, the elongation is above 30.5%, and the yield ratio is about 0.91. The test steel consists of ferrite and a small amount of pearlite, the ferrite grains are uniform and fine, the average size is about 2.5 μm, the pearlite pellet size is about 1.3 μm, and the pearlite pellet average volume fraction is about 3%. The size of the second phase precipitates is mainly distributed between 3 and 100 nm, and the average size is about 31 nm, the energy spectrum analysis shows that it is (Nb, Ti)CN. Compared with Q235B, the tested steel has better wear resistance and corrosion resistance. Under 40N load, the relative weight loss rate is below 70%. With the increase of wear time, the relative weight loss rate gradually decreases, which is related to the significant work hardening of the surface metal during the wear process; the test steel has a large impedance arc and a small self-corrosion current, which is related to the high Fe_3O_4 content in the scale of the tested steel.
Key words: 500MPa grade; roller steel; low temperature toughness; wear resistance; corrosion resistance

模拟加速腐蚀条件下耐候螺栓的腐蚀行为研究

罗志俊[1,2]，王晓晨[1]，孙齐松[1,2]，徐士新[1]，刘厚权[4]，黄 耀[3]，田志红[1]

（1. 首钢集团有限公司技术研究院，北京 100043；2. 绿色可循环钢铁流程北京市重点实验室，北京 100043；3. 国家电网中国电力科学研究院，北京 100043；4. 首钢贵阳特殊钢有限公司，贵州贵阳 550299）

摘 要：利用实验室模拟工业大气环境加速腐蚀的试验方法，分析了镀锌螺栓锌层、耐候螺栓螺母及普碳螺栓螺母螺纹腐蚀行为。结果表明：在 0.01 mol·L^{-1} NaHSO$_3$ 溶液中加速腐蚀，锌层腐蚀深度逐渐增加，腐蚀 96h 时微裂纹扩展到基体交界处，但未腐蚀基体。在腐蚀 120~216h 时间内，锌层微裂纹深度、宽度逐渐增加变黑脱落，局部裂纹穿过交界腐蚀到基体，腐蚀形状呈树枝状，深度由 5μm 增加到 40μm，锌层失去防腐保护作用；在加速腐蚀 24~96h 时间范围内，锌层腐蚀深度、服役年限与腐蚀时间呈线性变化关系 H=7.5+0.52t，Y=26.76-0.22t。在普碳钢螺栓连接副产品加速腐蚀过程中，螺纹腐蚀深度达到全螺纹的 62.5%，而耐候螺栓连接副产品的螺纹腐蚀深度为 20%。耐候钢腐蚀过程中形成的致密稳定的锈层是腐蚀缓慢的主要原因。
关键词：0.01 mol·L^{-1} NaHSO$_3$ 溶液；加速腐蚀；耐候螺栓；螺纹腐蚀；锌层腐蚀

Study on Corrosion Behavior of Weathering Bolts under Simulated Accelerated Corrosion

Luo Zhijun[1,2], Wang Xiaochen[1], Sun Qisong[1,2], Xu Shixin[1],
Liu Houquan[4], Huang Yao[3], Tian Zhihong[1]

(1. Shougang Research Institute of Technology, Beijing 100043, China; 2. Key Laboratory of Environmental and Circulating Iron and Steel Manufacturing Process Beijing, Beijing 100043, China; 3. China Electric Power Research Institute, Beijing 100043, China; 4. Shougang Guiyang Special Steel Co., Ltd., Guiyang 550299, China)

Abstract: The corrosion behavior of galvanized bolt, weathering bolt nut and plain carbon steel bolt nut thread is analyzed by simulating accelerated corrosion test in industrial atmospheric environment. The result shows that the corrosion depth of the zinc layer gradually increases in the accelerated corrosion test in 0.01 mol·L^{-1} NaHSO$_3$ solution. When the corrosion test lasts for 96 hours, microcracks expand to the interface between zinc layer and matrix, but the matrix is not corroded. In the corrosion time range of 120~216h, the depth and the width of zinc layer microcracks gradually increases. Some part of the zinc layer turns black and falls off. Some microcracks corrod through the interface to the matrix, and the corrosion shape is dendritic. Moreover, the depth of the cracks increase from 5 μm to 40 μm and zinc layer loses anticorrosive protection effect. In the corrosion time range of 120~216h, the corrosion depth and service life of zinc layer show a linear relationship with corrosion time: H=7.5+0.52t, Y=26.76−0.22t. In the accelerated corrosion process of the plain carbon steel bolt, the corrosion depth reaches 62.5% of the whole thread, while the corrosion depth of the weathering steel bolt thread is 20%. The dense and stable rust layer formed in the corrosion process of weathering steel is the main reason of slow corrosion.

Key words: 0.01 mol·L^{-1} NaHSO$_3$ solution; accelerated corrosion; weather resistant bolt; thread corrosion; zinc corrosion

耐磨蚀钢的研制及应用

宋凤明，温东辉，杨阿娜

（宝山钢铁股份有限公司研究院，上海 201900）

摘　要： 为满足疏浚管道的耐磨蚀性能要求，开发了 450HBW 硬度级别的耐磨蚀钢板 BMS1400，并成功进行了钢板及钢管的试制。开发的钢板具有良好的成形性能、低温冲击韧性和耐磨蚀性能。制备的大口径厚规格疏浚管具有良好的平直度和椭圆度，性能指标均满足设计要求，钢管已经成功在疏浚管道上获得应用，耐磨蚀性能达到现有普碳钢的 2 倍。

关键词： 耐磨蚀；疏浚；管道；BMS1400

Development and Application of Erosion-corrosion Resistant Steel

Song Fengming, Wen Donghui, Yang Ana

(Research Institute, Baoshan Iron and Steel Co., Ltd., Shanghai 201900, China)

Abstract: Erosion-corrosion resistant steel BMS1400 with 450HBW hardness was developed in order to meet the erosion-corrosion requirement of dredging pipes, which has excellent formability, toughness in lower temperature and erosion-corrosion resistance, and the manufacture of steel plates and pipes were realized. Large diameter dredging pipes fabricated by BMS1400 plates with better ellipticity and flatness, and the erosion-corrosion resistance of which is two times as that of plain carbide steel pipe.

Key words: erosion-corrosion resistant; dredging; pipes; BMS1400

推土机铲斗用调质耐磨钢 27MnTiB 的开发

秦　坤，尹绍江，霍晶晶，李　行，任树洋，赵庆宇

（唐山中厚板材有限公司，河北唐山 063000）

摘　要：通过 Mn、Ti、B 等元素合金化、氮含量控制以及控轧工艺获得了理想的钢板性能，完全满足推土机铲斗用钢的性能需求。在试验生产过程中发现较低的终轧温度和较大的变形量能够得到部分针状铁素体+少量贝氏体组织，并通过调质热处理获得了较高的强度和冲击韧性，因而满足推土机铲斗用钢的性能需求。

关键词：B 含量；控制轧制；大变形量；调质热处理

Development of Tempered and Wear-resistant Steel 27MnTiB for Bulldozer Bucket

Qin Kun, Yin Shaojiang, Huo Jingjing, Li Hang, Ren Shuyang, Zhao Qingyu

(Tangshan Middle Thickness Plate Co., Ltd., Tangshan 063000, China)

Abstract: Through alloying of Mn, Ti, B, nitrogen content control and controlled rolling process, ideal steel sheet properties are obtained, which fully meet the performance requirements of bulldozer bucket steel. It was found that some acicular ferrite + a small amount of bainite can be obtained at lower finishing rolling temperature and larger deformation in the test production process, and higher strength and impact toughness can be obtained through quenching and tempering heat treatment, thus meeting the performance requirements of bulldozer bucket steel.

Key words: B content; controlled rolling; large deformation; quenching and tempering heat treatment

回火热处理对低碳贝氏体钢显微组织和冲击断裂过程的影响

黄少文，周　平，张学民，杨　恒，李长新

（山东钢铁集团有限公司研究院，山东济南　250100）

摘　要：研究了(610～650)℃×25min 回火热处理对低碳贝氏体钢显微组织和冲击断裂过程的影响。结果表明：淬火钢组织由板条贝氏体(LB)和马氏体/奥氏体(M/A)组元构成。回火试验钢保留了 LB 板条特征，细小 M/A 组元沿板条方向分布。回火温度升高，板条宽度增加，M/A 组元数量减少。610～650℃回火后，试验钢的屈服强度(R_{eL})由 796MPa 降低至 549～618MPa，抗拉强度(R_m)由 862MPa 降低至 611～684MPa，断后伸长率(A)由 17.5%提高至 20.5%～23.5%。−40℃的裂纹形成能(E_1)、脆性断裂扩展能(E_3)、脆性断裂扩展终止后的止裂能(E_4)和−60℃的韧性裂纹扩展能(E_2)、E_3、E_4 增加，综合力学性能得到改善。回火温度升高，试验钢的 R_{eL} 由 618MPa 降低至 549MPa，R_m 由 684MPa 降低至 611MPa，A 由 23.5%降低至 20.5%，−40℃的 E_2 和−60℃的 E_1 与淬火试验钢接近。630℃回火时，试验钢的 R_{eL} 为 609MPa，R_m 为 673MPa，A 为 21.5%，−60℃的冲击脆性断裂倾向较小，脆性断裂止裂性达到最佳。

关键词：低碳贝氏体钢；回火热处理；冲击韧性；显微组织

Effect of Tempering Heat Treatment Process on Microstructure and Impact Fracture Process of Low Carbon Bainite Steel

Huang Shaowen, Zhou Ping, Zhang Xuemin, Yang Heng, Li Changxin

(Research Institute of Shandong Iron and Steel Group Co., Ltd., Jinan 250100, China)

Abstract: Effects of quenching and (610～650)℃×25min tempering heat treatment on microstructure and Charpy V notch impact fracture process of low carbon bainite steel were investigated. Experimental results show that quenching steel consists of lath bainite (LB) and martensite/austenite (M/A) constituents. After (610~650)℃×25min tempering, the test steel retains LB lath features, the smal M/A constituents distribute along the bainite lath. The yield strength (R_{eL}) decreases from 796MPa to 549~618MPa, the tensile strength (R_m) decreases from 862MPa to 611~684MPa and the elongation (A) increases from 17.5% to 20.5%~23.5%. The crack formation energy(E_1), brittle fracture propagation energy(E_3), brittle crack arrest energy (E_4) of −40℃ and ductility fracture propagating energy (E_2), E_3, E_4 of −60℃ are improved by tempering. With tempering temperature's increasing, the lath of bainite is larger in width, the content of M/A constituents decrease. The test steel's R_{eL} reduces from 618MPa to 549MPa, R_m reduces from 684MPa to 611MPa, A reduces from 23.5% to 20.5%. The −40℃ E_2 and −60℃ E_1 during Charpy V notch impact fracture process is similar to the quenching steel. After 630℃ tempering, the test steel's R_{eL} is 609MPa, R_m is 673MPa, A is 21.5%, −60°C impact energy is 295J, the impact brittleness tendency is smaller, the test steel has better brittle crack arrest properties.

Key words: low carbon bainite steel; tempering heat treatment; impact toughness; microstructure

460MPa 级工程机械用钢表面边裂分析

张卫攀，吕德文，孙电强，刘红艳，张瑞超，冯俊鹏

（河钢集团邯钢公司技术中心，河北邯郸　056000）

摘　要： 某钢厂在生产460MPa级工程机械用钢时，发生了大批量表面边裂造成的降低改判。对边裂缺陷钢板取样，进行了宏观和微观的分析，以及成分和设备的排查，最终发现问题的原因在于结晶器铜板磨损，导致渗铜，造成钢板热塑性变差，出现了边部裂纹。同时采取了预防措施。

关键词： 工程机械；表面边裂；渗铜

Surface Edge Crack Analysis of Construction Machinery Steel of 460MPa Grade

Zhang Weipan, Lv Dewen, Sun Dianqing, Liu Hongyan,
Zhang Ruichao, Feng Junpeng

(Handan Iron and Steel Co., Ltd., Corporation，Hebei Steel and Iron Group, Handan 056000, China)

Abstract: A large number of defects of 460MPa grade steel for construction machinery occurred in a steel plant. Sample of edge crack was analyzed macroscopically and microscopically, and the composition and equipment were checked. Finally, it was found that the cause of the problem was the wear and tear of the copper plate in the mould, which led to copper infiltration, resulting in the deterioration of the thermoplasticity of the steel plate and the appearance of edge cracks. At the same time, preventive measures have been taken.

Key words: construction machinery; surface edge crack; copper infiltration

TMCP 工艺 Q550D 钢板及焊接组织和性能研究

孙电强，王会岭，张瑞超，王丽敏，张卫攀

（河钢集团邯郸分公司技术中心，河北邯郸　056015）

摘 要：利用低碳和 Nb-Ti 微合金化，适当添加 Cr、Mo 来增加钢的淬透性及利用合适的 TMCP 工艺设计，通过细化晶粒、析出强化和相变强化来实现钢板的高强度和高韧性，使 Q550D 钢板获得良好的综合力学性能，实现了 Q550D 的 TMCP 态直接交货。并对焊后组织和焊缝拉伸性能进行了研究，结果表明在非预热条件下，Q550D 钢板焊后性能良好，满足使用要求。

关键词：Q550D；焊接性能；贝氏体；热机械轧制

Research on Microstructure and Properties of TMCP Q550D Plate and Weld

Sun Dianqiang, Wang Huiling, Zhang Ruichao, Wang Limin, Zhang Weipan

(Handan Iron and Steel Co., Ltd., Corporation, Hebei Steel and Iron Group, Handan 056015, China)

Abstract: The fine mechanical properties were achieved through refining grain, precipitation strengthening and phase change, by using of low carbon and Nb-Ti microalloying and the adding of Cr-Mo. And the Q550D plate was realized TMCP delivery. The weld microstructure, hardness and tensile properties were researched in this paper. The results showed that Q550D weld properties were satisfied without preheating.

Key words: Q550D; welding performance; bainite; thermo-mechanical rolling

钛系热处理高强钢开发应用与强韧性机理研究

何亚元，宋育来，李利巍，陆在学，薛 欢

（宝钢股份中央研究院武汉分院，湖北武汉 430080）

摘 要：介绍了武钢热连轧产线基于超快冷工艺的钛系热处理高强钢开发情况，并针对典型实验钢，利用光学显微镜（OM）、透射电子显微镜（TEM）和 X 射线衍射法等手段检测并研究了热处理前后的组织和内应力演变规律。结果表明，钛系热处理高强钢回火后强度升高，塑韧性下降，显微组织未见明显变化，细小析出相是导致性能变化的关键。武钢钛系热处理高强钢强韧性匹配较好，内应力小而均匀，成功批量用于制造起重机吊臂等零件的建造，满足使用需求。

关键词：高强钢；内应力；析出相；回火

Development and Application of Titanium-based Heat-treated High Strength Steel and Study on Strength and Toughness Mechanism

He Yayuan, Song Yulai, Li Liwei, Lu Zaixue, Xue Huan

(Wuhan Branch of Baosteel Central Research Institute, Wuhan 430080, China)

Abstract: The development of titanium series heat treated high strength steel based on ultra-fast cooling technology in WISCO hot strip rolling line is introduced. For typical experimental steels, the microstructure and internal stress evolution before and after heat treatment are investigated by means of optical microscopy (OM), transmission electron microscopy (TEM) and X-ray diffraction. The results show that after tempering, the strength of high strength titanium heat treated steel increases, the plasticity and toughness decreases, and the microstructures do not change significantly. Fine precipitates are the key factors leading to the change of properties. WISCO Titanium series heat treated high strength steel has good strength and toughness matching, small and uniform internal stress. It has been successfully used in the construction of crane boom and other parts to meet the needs of use.

Key words: high strength steel; internal stress; precipitated phase; tempering

25%Mn 超低温用高锰钢熔敷金属的组织与性能

陈亚魁[1]，王红鸿[1]，孟 亮[1]，孙 超[2]，李东晖[2]

（1. 武汉科技大学，湖北武汉 430000；2. 南京钢铁集团有限公司研究院，江苏南京 210000）

摘 要：制备了三种不同工艺和成分的 25%Mn 钢熔敷金属，采用 OM、XRD、EBSD 等实验方法对三种熔敷金属进行组织类型以及相比例分析，结果显示三种熔敷金属组织均为全奥氏体组织，呈胞状树枝晶结构，在组织中发现了夹杂粒子。采用常温拉伸试验和低温冲击试验对熔敷金属的力学性能进行检测，检测结果显示三种熔敷金属的屈服强度分别为 441MPa、323MPa、359MPa，抗拉强度分别为 670MPa、600MPa、605MPa，在 -196℃下的平均冲击值分别为 67J、60J、67J。采用 SEM 对冲击断口和夹杂物粒子进行分析，分析结果显示三种熔敷金属低温冲击断裂方式均为延性断裂，韧窝中心存在夹杂物粒子，夹杂物粒子由 Al_2O_3 和 MnS 组成。

关键词：25%Mn 钢；熔敷金属；低温冲击试验；胞状树枝晶

Microstructure and Properties of Deposited Metal of 25%Mn Ultra-low Temperature Steel

Chen Yakui[1], Wang Honghong[1], Meng Liang[1], Sun Chao[2], Li Donghui[2]

(1. Wuhan University of Science and Technology, Wuhan 430000, China;
2. Institute of Nanjing Iron and Steel Group Co., Ltd., Nanjing 210000, China)

Abstract: Three 25%Mn deposited metals with different welding processes and compositions were designed. The microstructures of the three deposited metals were analyzed by OM, XRD, EBSD. The results show that the three deposited metal are full austenite with a cellular dendrites structures, and inclusions are found in the structure. The mechanical properties of the deposited metal were tested by normal temperature tensile test and low temperature impact test. The test results showed that the yield strengths of the three deposited metals were 441MPa, 323MPa and 359MPa, the tensile strengths were 670MPa, 600MPa and 605MPa, and the average impact values at -196℃ were 67J, 60J, and 67J. The impact fracture and inclusion particles were analyzed by SEM. The results show that the fracture mode of the three deposited metals are ductile fractures, inclusion particles are found in the center of the dimple, and inclusion particles composed of Al_2O_3 and MnS.

Key words: 25Mn steel; deposited metal; low temperature impact test; cellular dendrites

Nb 元素对 NM450 耐磨钢板组织和性能的影响

金 池[1]，邓想涛[1]，王昭东[1]，闫强军[2]，杨 柳[2]，张永青[3]

（1. 东北大学轧制技术及连轧自动化国家重点实验室，辽宁沈阳 110819；

2. 南京钢铁股份有限公司，江苏南京 210035；3. 中信金属股份有限公司，北京 100004）

摘　要：本文对比了经相同轧制工艺和热处理工艺处理后的含 Nb 量 0.045%和不含 Nb 元素耐磨钢板的组织演变规律和力学性能。实验结果表明，添加了质量分数为 0.045%的 Nb 元素钢板的抗拉强度和硬度，低温冲击韧性都得到了一定程度的提升。从材料组织决定力学性能的角度分析，钢板力学性能的提升主要是由于 Nb 元素的添加使钢板原始奥氏体晶粒细化导致的。

关键词：Nb 微合金化；耐磨钢；组织；力学性能

Effect of Niobium on the Microstructure and Mechanical Properties of the NM450 Wear Resistance Steel Plate

Jin Chi[1], Deng Xiangtao[1], Wang Zhaodong[1], Yan Qiangjun[2], Yang Liu[2], Zhang Yongqing[3]

(1. State Key Laboratory of Rolling Technology and Automation, Northeastern University, Shenyang 110819, China; 2. Nanjing Nangang Iron & Steel United Co., Ltd., Nanjing 210035, China; 3. CITIC Metal Group Limited, Beijing 100004, China)

Abstract: In this paper, the steel plate without Nb element and the steel plate with Nb content of 0.045% were used as the comparative experimental steel plate, and its microstructure and mechanical properties were studied. Under the same rolling process and heat treatment process, the influence of the addition of Nb on the microstructure of the steel plate was analyzed. The experimental results show that the tensile strength, hardness and low temperature impact toughness of the steel plate with 0.045% Nb element have been improved to a certain extent. From the perspective of the mechanical properties and material microstructure, the improvement of the mechanical properties of the steel plate is mainly caused by the addition of Nb element to the grain refinement of the prior austenite grain.

Key words: niobium microalloyed; wear resistant steel; microstructure; mechanical properties

新型超级耐磨钢板研制开发与工业化应用

邓想涛[1]，王昭东[1]，付天亮[1]，黄　龙[1]，梁　亮[2]，闫强军[3]

（1. 东北大学轧制技术及连轧自动化国家重点实验室，辽宁沈阳　110819；2. 湖南华菱涟源钢铁有限公司，湖南娄底　417009；3. 南京钢铁股份有限公司，江苏南京　210035）

摘　要：在常规低合金马氏体耐磨钢合金成分的基础上，添加一定量的 Ti 元素，通过冶炼连铸过程中形成大量微米、亚微米超硬 TiC 陶瓷颗粒，并结合控制轧制和控制热处理的工艺控制，使其弥散均匀分布在板条马氏体基体上，研发出一种新型连铸坯内生超硬 TiC 陶瓷颗粒增强耐磨性超级耐磨钢板，并在国内某钢厂进行了工业化生产；分析了连铸、热轧和离线热处理过程时实验钢中 TiC 的演变规律和组织性能的变化，并研究了其耐磨性能。结果表明，新型钢板中由于较多 Ti 元素的添加，在连铸凝固过程中形成仿晶界的微米、亚微米级的超硬 TiC 粒子，轧制和离线热处理过程中，仿晶界的 TiC 粒子在马氏体基体中弥散均匀分布；耐磨性测试表面，在同等硬度的条件下，新型耐磨钢板的耐磨性达到传统马氏体耐磨钢的 1.5~1.8 倍，展现出优异的耐磨性能。

关键词：耐磨钢；超硬 TiC 粒子；连铸；磨损

Development and Industrial Application of New Super Low Alloy Abrasion Resistant Steel

Deng Xiangtao[1], Wang Zhaodong[1], Fu Tianliang[1],
Huang Long[1], Liang Liang[2], Yan Qiangjun[3]

(1. The State Key Laboratory of Rolling and Automation, Northeastern University, Shenyang 110819, China; 2. Hunan Valin Lianyuan Iron and Steel Co., Ltd., Loudi 417009, China; 3. Nanjing Iron and Steel Co., Ltd., Nanjing 210035, China)

Abstract: We development a new super abrasion resistant steel which dependent on the super hard particles TiC formed during continuous casting, and the formation and evolution of TiC particles duiring casting, rolling and heat treatment were also studied. The industrial application mechanical properties and wear behavior of the new steel were tested. The results indicated that the new steel exhibits excellent combination of mechanical properties and wear resistance, the wear resistance of the new development steel was about 1.5 to 18 time for the same grade hardness steel.

Key words: abrasion resistant steel; super hard TiC particles; continuous casting; wear behavior

8.8 级紧固件用钢 SWRCH35K 盘条的研制开发

汪青山，贾元海，刘效云，王　林，李吉伟

（河钢集团承钢公司钒钛工程技术研究中心，河北承德　067002）

摘　要： 本文详细介绍了 8.8 级紧固件用钢 SWRCH35K 的开发。利用铁水预处理、保护浇注、LF 精炼、结晶器电磁搅拌、稳定拉速等先进控制手段进行铸坯生产，降低了钢中气体含量、夹杂物，提高了铸坯低倍质量；采用合理的控轧控冷工艺，确保钢具有良好的组织、高塑性及表面低硬度性能，避免了贝氏体等异常组织的产生。产品经用户使用，冷镦合格率达到 98% 以上，完全满足 8.8 级紧固件的技术要求。

关键词： 紧固件用钢；盘条；研制开发；控轧控冷

Development of Steel SWRCH35K Wire Rod for 8.8 Grade Fastener Steel

Wang Qingshan, Jia Yuanhai, Liu Xiaoyun, Wang Lin, Li Jiwei

(Vanadium-Titanium Engineering Technique Research Center, Chengde Iron and Steel Company, Hebei Iron and Steel Group, Chengde 067002, China)

Abstract: The development of grade 8.8 steel SWRCH35K for fasteners is introduced in detail in this paper. Advanced control means such as hot metal pretreatment, protective casting, LF refining, mould electromagnetic stirring and stable casting speed are used to produce billet, which reduces the gas content and inclusions in steel and improves the low-power quality of billet. Reasonable controlled rolling and cooling technology is adopted to ensure that the steel has good structure, high plasticity and low surface hardness. It can avoid the formation of abnormal structure such as bainite. The qualified rate of cold upsetting is over 98% after the product is used by users, which fully meets the technical requirements of 8.8 class fasteners.

Key words: fastener steel; wire rod; research and development; the controlled rolling control cold

10.9 级紧固件用钢 ML20MnTiB 盘条的研制开发

李吉伟，贾元海，王　林，刘效云

（河钢集团承钢公司钒钛工程技术研究中心，河北承德　067002）

摘　要：本文详细介绍了高强度紧固件用钢 ML20MnTiB 的开发。通过控制钢中的 Ti、B 的有效成分，保证了盘条的淬透性；采用合理的控轧控冷工艺，确保钢具有良好的组织、高塑性及表面低硬度性能，避免了贝氏体组织的产生。产品经用户使用，冷镦合格率达到98%以上，完全满足10.9级高强度紧固件的技术要求。

关键词：10.9级；紧固件；淬透性；控轧控冷

Development of Steel ML20MnTiB Wire Rod for 10.9 Class Fasteners

Li Jiwei, Jia Yuanhai, Wang Lin, Liu Xiaoyun

(Vanadium-Titanium Engineering Technique Research Center, Chengde Iron and Steel Company, Hebei Iron and Steel Group, Chengde 067002, China)

Abstract: The development of steel ML20MnTiB for high strength fasteners is introduced in detail in this paper. By controlling the active ingredients of Ti and B in steel, the hardenability of wire rod is ensured, and reasonable controlled rolling and cooling process is adopted to ensure that steel has good structure, high plasticity and low surface hardness, thus avoiding the formation of bainite structure. The qualified rate of cold heading is over 98% after the product is used by users, which fully meets the technical requirements of high strength fasteners of grade 10.9.

Key words: 10.9 grade; fastener; hardenability; the controlled rolling control cold

PC 钢棒用 40Si2Mn 盘条的研制与开发

王　林，梁　均，贾元海，李吉伟

（河钢集团承钢公司钒钛工程技术研究中心，河北承德　067002）

摘　要：依据 PC 钢棒用 40Si2Mn 的特点进行转炉冶炼、LF 精炼、连铸及轧制生产实践；生产出了纯净度高、夹杂物含量低、表面质量良好、通条性能均匀性好等质量优良的盘条。

关键词：PC 钢棒；40Si2Mn；盘条；开发

The Development of 40Si2Mn Wire Rod for PC Steel Bar

Wang Lin, Liang Jun, Jia Yuanhai, Li Jiwei

(Vanadium-Titanium Engineering Technique Research Center, Chengde Iron and Steel Company, Hebei Iron and Steel Group, Chengde 067002, China)

Abstract: According to the characteristics of 40Si2Mn used for PC steel bars, converter smelting, LF refining, continuous casting and rolling were carried out. High purity, low inclusion content, good surface quality and good uniformity of rod properties were produced.

Key words: PC steel bar; 40Si2Mn; wire rod; development

具有良好强韧性能高等级耐磨钢 NM500 生产工艺开发及应用

刘红艳[1]，王昭东[2]，吕德文[1]，邓想涛[2]，姚宙[1]

（1. 河钢集团邯郸分公司，河北邯郸　056015；
2. 东北大学轧制技术及连轧自动化国家重点实验室，辽宁沈阳　114001）

摘　要： 高等级耐磨钢 NM500 采用 Cr+Mo+B 成分体系，生产工艺主要包括炼钢、连铸、铸坯加热、双轧程轧制、淬火和回火工序。交叉热处理试验得出：随着淬火温度从 860℃升高到 920℃其布氏硬度值、–40℃低温冲击韧性值均呈现先上升再下降的的趋势；随着回火温度从 200℃升高到 400℃，硬度值呈缓慢下降趋势、–40℃低温冲击韧性值急剧下降，300℃回火时达到谷底。微观组织回火马氏体具有一定倾向性，即：随着回火温度从 200℃升高到 400℃，马氏体板条尺寸不断增大；碳化物析出量增加且尺寸不断增大和具有一定的球化趋势。淬火温度 890℃、回火温度 200℃时，NM500 高等级耐磨钢布氏硬度达到 515HBW、抗拉强度达到 1603MPa、延伸率达到 15.2%，–40℃低温冲击韧性达到 65J，高等级耐磨钢 NM500 的硬度、强度和低温冲击韧性呈现最优化匹配。

关键词： 强韧性能；高等级耐磨钢；NM500；热处理工艺

Development and Application of NM500 Production Process for High-grade Wear-resistant Steel with Good Strength and Toughness

Liu Hongyan[1], Wang Zhaodong[2], Lv Dewen[1], Deng Xiangtao[2], Yao Zhou[1]

(1. HBIS Group Hansteel Company, Handan 056015, China; 2. State Key Laboratory of Rolling Technology and Continuous Rolling Automation, Northeast University, Shenyang 114001, China)

Abstract: High-grade wear-resistant steel NM500 adopts Cr+Mo+B composition system, the production process mainly includes steelmaking, continuous casting, billet heating, double rolling, quenching and tempering. The results of cross-heat treatment test show that the Brinell hardness and –40°C low temperature impact toughness of the alloy increase first and then decrease with the quenching temperature rising from 860°C to 920°C, while the hardness decreases slowly, and the

−40℃ low temperature impact toughness decreases sharply with the tempering temperature rising from 200℃ to 400℃, the lowest is at 300℃. Microstructure tempered martensite has a certain tendency, as tempering temperature rises from 200℃ to 400℃, the size of martensite lath increases, the amount of carbide precipitation increases, the size increases and there is a certain tendency of spheroidization. The Brinell hardness of NM500 high-grade wear-resistant steel reaches 515HBW, tensile strength reaches 1603MPa, elongation reaches 15.2%, −40℃ low temperature impact toughness reaches 65J, the hardness, strength and low-temperature impact toughness of NM500 high-grade wear-resistant steel present the optimal match.

Key words: strength and toughness; high-grade wear-resistant steel; NM500; heat treatment process

6.8 其他用途用钢

BN 型易切削钢中夹杂物析出规律及对性能的影响

张 帆，任安超，夏艳花，丁礼权

（宝钢股份中央研究院武汉分院，湖北武汉 430080）

摘 要： 热力学计算表明，凝固过程中 B 的偏析程度远远超过 N，且低合金钢中 BN 在奥氏体温度区间开始析出；[Al]含量增加，硼氮平衡浓度积降低，过高的[Al]含量，会导致 AlN 优先于 BN 析出，当[Al]<0.03%时能有效避免 AlN 的形成。试验结果表明，N/B 比对试样钢力学性能和切削性能有重要影响，同时，与基础钢相比，BN 的添加对力学性能无害，并且能显著提高钢材的切削性能。

关键词： 易切削钢；BN；切削性能；N/B

Smelting and Performance Research of BN-type Automotive Cutting Steel

Zhang Fan, Ren Anchao, Xia Yanhua, Ding Liquan

(Wuhan Branch of Baosteel Central Research Institute, Wuhan 430080, China)

Abstract: Based on thermodynamic analysis, the paper shows that BN separate out at Austenitizing temperature, and along with [Al] increasing, the equilibrium concentration of [B][N] decrease, excess content of [Al] may cause AlN priority precipitation than BN, when [Al]<0.03%, there is no AlN precipitation in steel. Experiment results showed that, compared with foundation steel, BN does no harm to mechanical properties, and can significantly improve the machinability of steel.

Key words: free cutting steel; boron nitride; machinability; N/B

抗震阻尼器用极低屈服点钢的组织与性能调控研究

李昭东[1]，陈润农[1,2]，赵 刚[3]，张明亚[2]，杨忠民[1]，杨才福[1]

（1. 钢铁研究总院工程用钢研究所，北京 100081；2. 安徽工业大学冶金工程学院，

安徽马鞍山 243032；3. 鞍钢股份有限公司，辽宁鞍山 114021）

摘 要：本工作对 100MPa 级极低屈服点钢的组织、析出与力学性能进行了微合金化与软化热处理调控研究。对于 Nb 微合金化成分设计，进行 750~900℃软化热处理，粗化铁素体晶粒和 Nb(C,N)粒子，获得低温冲击韧性 KV_2（-20℃）超过 100J 的极低屈服点钢，其中 850℃软化热处理获得最佳综合力学性能。950℃软化热处理后该钢的强塑性优异，但因晶粒粗化至 90μm 以上导致韧性极低。对于 Ti 微合金化成分设计，将 S 含量提高至 80ppm 级，促使高温形成数量较多、尺寸较大的 $Ti_4C_2S_2$ 粒子，有利于降低屈服强度，但不显著损害塑韧性，为降低极低屈服点钢的微合金成本提供了可行的思路。

关键词：Nb 微合金化；Ti 微合金化；硫化物；强化机制；软化热处理

Study on the Controlling of Microstructure and Mechanical Properties of Ultra Low Yield Point Steel for Aseismic Damper

Li Zhaodong[1], Chen Runnong[1,2], Zhao Gang[3], Zhang Mingya[2],
Yang Zhongmin[1], Yang Caifu[1]

(1. Department of Structural Steels, Central Iron and Steel Research Institute, Beijing 100081, China;
2. School of Metallurgical Engineering, Anhui University of Technology, Maanshan 243032, China;
3. Angang Steel Co., Ltd., Anshan 114021, China)

Abstract: In this work, the microstructure, precipitation and mechanical properties of the 100MPa grade ultra-low yield point steel was controlled by microalloying and softening heat treatment. For the composition design of Nb microalloying, softening heat treatment at 750~900℃ was carried out, leading to the coarsening of ferrite grains and Nb (C, N) particles. The ultra low yield point steel with low impact toughness of KV_2 (-20℃) exceeding 100J was obtained, which has the best comprehensive mechanical properties after 850℃ softening heat treatment. However, excellent strength and plasticity was obtained after softening heat treatment at 950℃, but the toughness is extremely low due to grain coarsening to above 90μm. For the composition design of Ti microalloying, S content increased to 80ppm level could promote the formation of $Ti_4C_2S_2$ particles with large quantity and size at high temperature. It was beneficial to reduce yield strength, but did not significantly damage the plasticity and toughness. This was a feasible way to reduce the cost of microalloying for the ultra low yield point steel.

Key words: Nb microalloying; Ti microalloying; sulfide; strengthening mechanism; softening heat treatment

7　粉末冶金

大会特邀报告

分会场特邀报告

炼铁与原料

炼钢与连铸

轧制与热处理

表面与涂镀

金属材料深加工

先进钢铁材料

★ 粉末冶金

能源、环保与资源利用

钢铁材料表征与评价

冶金设备与工程技术

冶金自动化与智能化

建筑诊治

其他

气雾化高硅铁粉性能分析

康 伟[1,2]，廖相巍[1,2]，贾吉祥[1,2]，康 磊[1,2]

(1. 海洋装备金属材料及应用国家重点实验室，辽宁鞍山 114009；
2. 鞍钢钢铁集团研究院炼钢技术研究所，辽宁鞍山 114009)

摘 要：利用气雾化方法进行了高硅铁粉的制备，对两种方案所制备的硅铁粉进行了粒度、形貌、化学成分、物相分析、XPS 等分析，发现气雾化高硅铁粒径相对较大；铝含量高的硅铁粉形貌不规则，多呈棒状或块状，而铝含量较低的硅铁粉形貌较为规则，多呈圆球形分布；X 射线衍射显示两炉粉体的主要物相均为 Fe、$Fe_{0.9}Si_{0.1}$，从 XPS 检测结果看，1#样表面的单质硅少，而 SiO_2 多，Al_2O_3 也少，说明 1#样中成分铝少不能保护硅被氧化，2#样表面单质硅多，而 SiO_2 少，Al_2O_3 多，说明铝被氧化起到了保护硅的作用。

关键词：气雾化铁粉；高硅铁粉；高硅铁粉性能；高硅铁粉制备

Performance Analysis of High Silicon Iron Powder by Atomization Method

Kang Wei[1,2], Liao Xiangwei[1,2], Jia Jixiang[1,2], Kang Lei[1,2]

(1. Key Laboratory of Metal Materials for Marine Equipment and Application, Anshan 114009, China;
2. Ansteel Group Iron and Steel Research Institute, Anshan 114009, China)

Abstract: The atomization method of high silicon iron powder preparation, preparation of silicon nitride powder prepared with two kinds of schemes of the size, morphology, chemical composition, phase analysis, XPS analysis, found that the atomization of high silicon iron particle size is relatively large; the silicon nitride powder morphology of high aluminium content is irregular, rod or block, and ferrosilicon powder morphology with lower aluminum content is regular, is spherical shape distribution; X-ray diffraction showed that the main two furnace powder phases were Fe, $Fe_{0.9}Si_{0.1}$, XPS from the test results, the surface of silicon 1# less, and SiO_2, Al_2O_3, explained 1# like components in aluminum silicon is not less protective 2# like oxidation, surface elemental silicon, SiO_2 less, and Al_2O_3, that is to protect the aluminum oxide silicon.

Key words: atomized iron powder; high silicon iron powder; high silicon iron powder properties; preparation of high silicon iron powder

金属粉末做中间材热轧制备不锈钢复合板

康 磊[1,2]，廖相巍[1,2]，尚德礼[1,2]，贾吉祥[1,2]，李广帮[1,2]，康 伟[1,2]

(1. 海洋装备用金属材料及其应用国家重点实验室，辽宁鞍山 114009；
2. 鞍钢集团钢铁研究院，辽宁鞍山 114009)

摘 要：针对真空热轧制备不锈钢复合板工艺复杂和碳元素在复合界面扩散易形成碳化物影响结合强度的问题，进行了在低碳钢和不锈钢之间加入金属粉末的不锈钢/低碳钢非真空热轧试验研究。测量了复合板的结合强度，并对轧后的复合板进行了金相组织观察和扫描电镜元素分布的分析。结果表明，加入金属粉末中间层时，不锈钢和低碳钢容易达到良好的冶金结合，同时金属粉末的加入可以阻碍碳元素向复合界面处扩散，减少了碳铬化合物形成，有利于界面结合强度的提高。

关键词：不锈钢复合板；热轧复合；金属粉末中间层；结合强度

Stainless Steel Composite Plate Prepared by Hot-roll Bonding with Metal Powder as Intermediate Material

Kang Lei[1,2], Liao Xiangwei[1,2], Shang Deli[1,2],
Jia jixiang[1,2], Li Guangbang[1,2], Kang Wei[1,2]

(1. State Key Laboratory of Metal Materials for Marine Equipment and Application, Anshan 114009, China;
2. Iron & Steel Research Institutes of Ansteel Group Corporation, Anshan 114009, China)

Abstract: When stainless steel composites were produced by hot rolling under vacuum, the process of preparing composite slab was complex and diffusion of C to bonding surface are easy to form carbides that reduce the bonding strength during hot rolling. To attempt to solve these problem, metal powders as intermediate material were used to bond stainless steel and low-carbon steel by hot rolling under non-vacuum. The bonding strengths were measured, and the microstructure and the distribution of elements of the composites were examined. The results show that when metal powder is used, the stainless steel and the low-carbon steel have a metallurgical bonding. The diffusion of C to bonding surface can be hindered, it decreases the formation of carbon chromium compounds, and it is beneficial to improve the bonding strength.

Key words: stainless composite clad plate; hot-rolling bonding; metal powder intermediate material; bonding strength

WC-Ni 无磁硬质合金研究进展及应用

李金普，施瑜蕾，柳学全，姜丽娟

（北京安泰钢研超硬材料制品有限责任公司，北京　102200）

摘　要：WC-Ni 硬质合金和 WC-Co 硬质合金一样，硬度高、耐磨性好、抗弯强度高、导热系数大，但 WC-Ni 硬质合金耐腐蚀性较 WC-Co 好，且 WC-Ni 硬质合金无磁性。本文介绍了 Ni 作为粘结相的硬质合金替代 Co 作为粘结相硬质合金的必要性，综述了 WC-Ni 无磁硬质合金在国外国内的研究进展，最后介绍了无磁硬质合金在无磁模具、机械端面密封中的应用。

关键词：无磁；硬质合金；模具；密封

Research Development and Application on Non-magnetic Cemented Carbide

Li Jinpu, Shi Yulei, Liu Xuequan, Jiang Lijuan

(Beijing Gangyan Diamond Product Company, Beijing 102200, China)

Abstract: WC-Ni cemented carbide, as well as WC-Co cemented carbide has the physical properties of high hardness, good abrasion resistance, high bending strength and high thermal conductivity. However, the corrosion resistance of WC-Ni cemented carbide is better than that of WC-Co. This study shows the necessity of replacing Co with Ni as the bonding phase of cemented carbide, and reviewed the national and international research progress of WC-Ni non-magnetic cemented carbide. Finally, the application of non-magnetic cemented carbide in magnetic free die and mechanical section seal is introduced.

Key words: non-magnetic; cemented carbide; die; seal

Fe-2Cu-0.5C 粉末烧结钢高温拉伸流变应力及预测

李强[1]，郭彪[1]，吴辉[1]，敖进清[1,2]，宋欢[1]，敖逸博[1]

(1. 西华大学材料科学与工程学院，四川成都 610039；
2. 江苏开来钢管有限公司，江苏淮安 223001)

摘 要：在变形温度为 850~1000℃、应变速率为 0.01~10s^{-1} 条件下，采用 Gleeble-3500 热模拟机，研究了粉末烧结 Fe-2Cu-0.5C 钢高温拉伸时变形行为。结果表明：在变形初始阶段，流变应力随应变的增加迅速增大；之后，随着应变量增加，流变应力增速减缓，达到峰值后流变应力逐渐衰减直至试样断裂。当变形温度一定时，流变应力随应变速率上升而增高；当应变速率一定时，流变应力随变形温度的升高而降低。建立了具有较高预测精度的 Fe-2Cu-0.5C 粉末烧结钢高温拉伸流变应力本构方程。本构方程的流变应力预测值与实验值吻合较好，相关系数为 0.99637，平均绝对相对误差为 1.81%。

关键词：烧结钢；高温拉伸；流变应力；本构方程；预测

Flow Stress and Prediction of Powder Sintered Fe-2Cu-0.5C Steel during High Temperature Tensile

Li Qiang[1], Guo Biao[1], Wu Hui[1], Ao Jinqing[1,2], Song Huan[1], Ao Yibo[1]

(1. School of Materials Science and Engineering, Xihua University, Chengdu 610039, China;
2. Jiangsu Kailai Steel Pipe Company Ltd., Huaian 223001, China)

Abstract: Under the deformation temperature of 850~1000℃ and strain rate of 0.01~10s^{-1}, the hot deformation behaviors of powder sintered Fe-2Cu-0.5C steel during tensile was studied by Gleeble-3500 thermal simulator. The results show that in the initial stage of deformation, the flow stress increases rapidly with the increase of strain, and then the rate of flow stress rise decreases with the increase of strain. After the flow stress peaking, the flow stress gradually declines until the specimen fractures. Under the same deformation temperature, the flow stress rises with the increase of strain rate. When the strain rate is constant, the flow stress decreases with the increase of deformation temperature. The flow stress constitutive equation of powder sintered Fe-2Cu-0.5C steel during tensile tests at high temperature was established. The predicted stress values is in good agreement with the experimental stress values, the correlation coefficient was 0.99637, and the average absolute relative error is 1.81%.

Key words: sintered steel; high temperature tensile; flow stress; constitutive equation; prediction

不同碳含量粉末烧结钢镦粗流变致密化行为研究

吴 辉[1]，郭 彪[1]，李 强[1]，敖进清[1,2]，宋 欢[1]，敖逸博[1]

(1. 西华大学材料科学与工程学院，四川成都 610039；
2. 江苏开来钢管有限公司，江苏淮安 223001)

摘 要：在材料万能试验机上对碳含量为 0%、0.3%、0.6%、0.9%、1.2%（wt%）粉末烧结钢进行镦粗变形试验；利用 Hollomon 方程对流变应力数据进行非线性拟合，并结合显微组织观测，分析了烧结钢的镦粗流变致密化行为和孔隙、晶粒变形机制。结果表明：烧结钢的镦粗流变行为符合 Hollomon 方程且随着碳含量的增加，极限断裂应变量逐渐减小；在同一应变条件下，烧结钢加工硬化率随碳含量增加逐渐上升；随着应变的增加，烧结钢孔隙不断闭合，晶粒由等轴状变为纤维状；烧结钢的流变致密化过程在低应变下以致密化和致密化硬化为主，高应变下以变形和基体加工硬化为主。

关键词：烧结钢；镦粗；流变；致密化

Study on the Flow and Densification Behaviors of Powder Sintered Steel with Different Carbon Contents during Clod Upsetting

Wu Hui[1], Guo Biao[1], Li Qiang[1], Ao Jinqing[1,2], Song Huan[1], Ao Yibo[1]

(1. School of Materials Science and Engineering, Xihua University, Chengdu 610039, China;
2. Jiangsu Kailai Steel Pipe Company Ltd., Huaian 223001, China)

Abstract: The cold upsetting tests of powder sintered steel with 0%、0.3%、0.6%、0.9% and 1.2% (wt%) carbon contents were carried out on a universal material testing machine. The flow stress data were nonlinearly fitted by the Hollomon equation. The flow and densification behaviors, the pore and grain deformation mechanisms of sintered steel were analyzed based on the observation of microstructures. The results show that the upsetting flow behavior of sintered steel conforms to the Hollomon equation and the ultimate fracture strain decreases with the increase of carbon content. Under the same strain condition, the work hardening rate of sintered steel increases with the increase of carbon content. With the increase of strain, the pores of sintered steel gradually close and the grains change from equiaxed crystal into fibrous structure. During the flow and densification processes, the densification and densification hardening were dominant under low strains, while the deformation and matrix work hardening were dominant under high strains.

Key words: sintered steel; upsetting; low and deformation; densification

热处理对铜基粉末冶金刹车片的摩擦性能的影响

王 洋[1,2]，胡 铮[3]，张万昊[3]，张 坤[1,2]，魏炳忱[1,2]

(1. 中国科学院力学研究所国家微重力实验室，北京 100190；2. 中国科学院大学工程科学学院，北京 100049；3. 中国北方车辆研究所车辆传动重点实验室，北京 100072)

摘 要：本文对铜基粉末冶金刹车片进行热处理，首先利用销盘定速摩擦磨损试验机检验摩擦材料的短时摩擦学性能，同时通过 SAE2 台架试验研究实际工况下的长时接合摩擦性能。采用扫描电子显微镜（SEM）及维氏硬度实验分析了表面摩擦层热处理处理前后的微观结构和磨损形貌以及宏观硬度变化。结果表明，热处理试样摩擦区域表面相对光滑和平整，并未出现明显的犁沟状损伤；试样的宏观硬度增加，且硬度分布更加均匀。动摩擦因数显著提高，且随时间的变化趋势更加平稳。这说明热处理对铜基摩擦层的短时摩擦性能以及实际工况下的长时摩擦性能有一定的改善。

关键词：铜基粉末冶金刹车片；热处理；摩擦磨损

Effect of Heat Treatment on Friction Properties of the Cu-based Powder Metallurgy Brake Pads

Wang Yang[1,2], Hu Zheng[3], Zhang Wanhao[3], Zhang Kun[1,2], Wei Bingchen[1,2]

(1. Key Laboratory of Microgravity (National Microgravity Laboratory) Institute of Mechanics Chinese Academy of Sciences, Beijing 100190, China; 2. School of Engineering Science University of Chinese Academy of Sciences, Beijing 100049, China; 3. Science and Technology on Vehicle Transmission Laboratory China North Vehicle Research Institute, Beijing 100072, China)

Abstract: The effect of heat treatment on friction properties of Cu-based powder metallurgy brake pads was studied. Firstly, the short-term friction properties of friction layer are tested by pin-disc friction and wear test. Meanwhile, the long-term friction performance is investigated by SAE2 test. Scanning electron microscope (SEM) was used to characterize the microstructures and wear morphology of friction layer before and after heat treatment. The results show that the surface of the friction layer after treatment is more smooth and flat, without obvious furrow damage. Meanwhile, the macroscopic hardness of the friction layer after treatment increased obviously, and Weibull statistical analysis results also indicate that the friction layer is more homogeneous after treatment. Moreover, the dynamic friction coefficient of friction layer after treatment is significantly improved, and the change trend with time is more stable. The experimental results demonstrated that the short-term friction properties of Cu-based friction materials as well as the long-term friction performance of vehicle brake pads can be significantly improved by heat treatment.

Key words: Cu-based powder metallurgy brake pads; heat treatment; friction and wear

8 能源、环保与资源利用

大会特邀报告

分会场特邀报告

炼铁与原料

炼钢与连铸

轧制与热处理

表面与涂镀

金属材料深加工

先进钢铁材料

粉末冶金

★ 能源、环保与资源利用

钢铁材料表征与评价

冶金设备与工程技术

冶金自动化与智能化

建筑诊治

其他

8.1 能源与资源利用

热风炉热平衡软件开发与应用

陈冠军

（首钢集团有限公司技术研究院，北京 100043）

摘 要：热风炉通过蓄热原理加热鼓风，是高炉实现高风温的关键设备。热风炉热平衡计算以热力学第二定律为基础，结合热力学第一定律，实现热风炉各项收入热量与支出热量的计算。依据 GB/T 32287《高炉热风炉热平衡测定与计算方法》国家标准，开发了热风炉热平衡计算软件，可快速实现热风炉收入和支出热量的计算。通过热风炉热平衡模型和标准中的计算公式，结合燃烧和传热理论，详细介绍了热平衡各项收入热量、支出热量的计算，可实现的功能包括：混合气体比热、干湿成分转换、煤气热值计算，空气过剩系数、理论及实际空气量和烟气量等过程量的计算，热风炉本体及系统热效率的计算。通过实际案例的应用，分析了热风炉热平衡的主次项热量，指出了该案例中空气过剩系数偏大和热风管道散热过大的问题，提出降低空燃比和保温措施，为热风炉节能和改进操作提供参考。

关键词：热风炉；热平衡；软件开发

Development and Application of Heat Balance Software for Hot Blast Stove

Chen Guanjun

(Research Institute of Technology in Shougang Group, Beijing 100043, China)

Abstract: Hot blast stove is the key equipment to reach high blast temperature in blast furnace by heating blast with regenerative theory. Heat balance calculation in hot blast furnace is based on second law of thermodynamics with first law of thermodynamics, it can calculate the heat of each income and expenditure of the hot blast stove. According to the national standard of GB/T 32287 with "methods of determination and calculation of heat balance for hot blast stove in blast furnace", heat balance software for hot blast stove is developed, it can computer revenue and expenditure of heat in hot blast stove quickly. Though heat balance model and computing formula for hot blast stove, calculations of the heat of each income and expenditure are introduced with combustion and heat transfer theories, it includes functions such as dry and wet composition conversion, specific heat and calorific value of mixed gas, calculation of process variants such as excess coefficient and specific heat of air, theoretical and actual smoke volume and specific heat of smoke and calculation of self and system thermal efficiency. By the application of actual case of calculation of heat balance, primary and secondary items of heat balance in hot blast stove are analyzed, the problems of excess coefficient of per-heated air and excessive heat dissipation of hot blast pipeline are pointed out, measures to reduce air-fuel ratio and heat preservation are put forward in this case, it offers the references for energy-saving and operation improvement in hot blast stove.

Key words: hot blast stove; heat balance; software development

改性钢渣吸附 MTBE 污染废水的动力学研究

刘芳，胡绍伟，陈鹏，王飞，王永，徐伟

（鞍钢集团钢铁研究院环境与资源研究所，辽宁鞍山 114009）

摘　要：采用改性后的钢渣处理 MTBE 污染废水，优化了钢渣改性条件、吸附工艺运行参数并进行了动力学研究。实验结果表明：在钢渣与焦粉质量比 1∶3，马弗炉温度 400℃，焙烧时间 1h 条件下，改性后的钢渣对 MTBE 吸附效果最好；吸附 MTBE 的工艺优化参数为填料用量 0.5g，反应 pH 值 7.0，反应时间 1.5h；对于 MTBE 微污染废水（20mg/L），改性钢渣的吸附过程更适合用假二级动力学方程拟合，R^2 值在 0.99 以上。在上述工艺条件下，MTBE 去除率达到 70%以上。该工艺处理效果好，解决了钢渣任意堆放造成的资源浪费和污染环境的难题，以废治废，既经济又环保，大大降低了吸附剂成本，且易于工业化应用。

关键词：改性钢渣；吸附；MTBE；动力学

Dynamics Study on Adsorption of MTBE Wastewater by Modified Steel Slag

Liu Fang, Hu Shaowei, Chen Peng, Wang Fei, Wang Yong, Xu Wei

(Environment and Resource Institute, Ansteel Iron & Steel Research Institute, Anshan 114009, China)

Abstract: The modified steel slag was adopted to deal with MTBE wastewater. The steel slag modified conditions and the adsorption operational parameters were optimized. Then the adsorption dynamics were researched. The experimental results showed that the modified steel slag had the best adsorption effect in conditions of steel slag with coke powder mass ratio of 1∶3 and muffle furnace 400℃ roasting 1h. The optimum parameters of MTBE adsorption were filler dosage 0.5g, pH 7.0 and reaction time 1.5h. For MTBE wastewater(20mg/L), the reaction of MTBE solution with modified steel slag adsorption could be described by pseudo-second-order kinetics, the value of R^2 were above 0.99. Under these conditions, the removal rate of MTBE were above 70%. This process had a good effect and the problem of resource waste and environment pollution caused by steel slag piling up were solved. The research has the advantages of using waste treat waste, economic and environmental protection, greatly reducing the cost of adsorbent, and easy to industrial application.

Key words: modified steel slag; adsorption; MTBE; dynamics

苯加氢催化剂器外预硫化技术实践

陶帅江，董毅，杜建全，孙鸿君

（鞍钢集团化学科技有限公司，辽宁鞍山 114200）

摘　要：鞍钢集团化学科技有限公司苯加氢作业区自 2009 年开工投产以来，已经连续运行近 10 年，催化剂压差有升高趋势，反应器入口温度不断提高。为保证苯加氢装置能够稳定运行，公司购入一批新鲜催化剂。对主反应器和预反应的新鲜催化剂进行器外预硫化，预硫化催化剂在器内进行活化。装置开工后产品质量合格。本文重点介绍了器外预硫化催化剂技术指标、催化剂装载、器内进行活化的关键步骤，旨在为行业内催化剂器外预硫化和器内活化的实施提供参考。

关键词：苯加氢催化剂；器外预硫化；器内活化；装载

Practices of Ex-situ Pre-sulphurizing Process for Hydrogenation Catalysts

Tao Shuaijiang, Dong Yi, Du Jianquan, Sun Hongjun

(Angang Group Chemical Technology Co., Ltd., Anshan 114200, China)

Abstract: The process of crude benzene hydrogenation of Angang Group Chemical Technology Co., Ltd. has run 10 years since putted into operation in 2009.The pressure drop and the reactor inlet temperature of reactor continues to rise. In order to ensure the stable of benzene hydro-refining plant, the company purchased fresh catalysts which have to be sulphurized ex-situ and actived in-situ. The results showed the unit run smoothly and the product quality was good. The target of ex-situ pre-sulphurizing process and the key steps are mentioned in the paper, which intendes to provide a reference for the implementation of sulphurized ex-situ and actived in-situ for hydrogenation catalysts.

Key words: hydrogenation catalysts; ex-situ pre-sulphurizing; actived in-situ; loading

永磁调速技术在鞍钢生活水供应中的探索与实践

闫 涛，刘 星，邸光宇，魏 彬

（鞍钢股份能源管控中心给水分厂新水作业区，辽宁鞍山 114001）

摘 要：本文通过对鞍钢厂区生活水供水系统设备状况及运行的分析，结合永磁调速新技术的应用，对今后生活水系统的设备改造提出了建设性的意见。

关键词：永磁调速器；技术改造；经济节能；探索与实践

Exploration and Practice of Permanent Magnet Speed Regulation Technology in Life Water Supply of An steel

Yan Tao, Liu Xing, Di Guangyu, Wei Bin

(An Steel Energy Management and Control Center New Water Operation Area of Branch Factory, Anshan 114001, China)

Abstract: This paper analyses the equipment condition and operation of life water supply system in An Steel Factory and combines application of new technology of permanent magnet speed regulation. In this paper, constructive suggestions are put forward for the equipment renovation of life water system in the future.

Key words: permanent magnet governor; technological transformation; economic energy-saving; exploration and practice

煤化工企业蒸汽系统节能、优化的研究与应用

王 震,穆春丰,李冈亿

(鞍钢化学科技有限公司,辽宁鞍山 114000)

摘 要:针对鞍钢化学科技有限公司部分焦炉煤气净化、焦油加工等系统由于设计或历史原因,蒸汽系统存在的高质低用、泄漏直排、余热余能流失、制约生产管网参数频繁波动等造成蒸汽资源浪费及制约生产问题,通过对蒸汽使用及其疏水系统、供应管网系统等重新节能优化改进,实现了节约蒸汽使用,并消除由此带来的产品损失及生产安全隐患。

关键词:蒸汽系统;浪费;优化改进;节约蒸汽

Research and Application of Energy Saving and Optimization of Steam System in Coal Chemical Enterprises

Wang Zhen, Mu Chunfeng, Li Gangyi

(Angang Chemical Technology Co., Ltd., Anshan 114000, China)

Abstract: In view of the design or historical reasons of some coke oven gas purification and tar processing systems in Angang Chemical Technology Co., Ltd. The problems existing in steam system, such as high quality and low use, leakage and direct discharge, loss of waste heat and residual energy, restriction of frequent fluctuation of parameters of production pipe network, etc., result in waste of steam resources and restriction of production. Through the optimization and improvement of steam use, drainage system, supply pipe network system and so on, the steam saving is realized, and the product loss and production safety hidden danger caused by it are eliminated.

Key words: steam system; waste; optimization and improvement; steam saving

冶金企业保障输电线路稳定性相关措施

刘 鹏

(鞍钢集团公司能源管控中心,辽宁鞍山 114009)

摘 要:冶金企业为保障电网稳定性采取的相关方法,分架空线路和电力电缆两方面进行介绍。提出了一些目前正在使用的方式,并对未来检修方法提出了展望。

关键词:架空线路;电缆;检修;测温;GPS定位;隧道机器人

Metallurgical Enterprises to Ensure the Stability of Transmission Lines Related Measures

Liu Peng

(Energy Control Center of Anshan Iron & Steel Group Co., Ltd., Anshan 114009, China)

Abstract: The methods adopted by metallurgical enterprises to ensure the stability of power grid are introduced in two aspects: overhead lines and power cables. Some methods which are in use at present are put forward, and the prospect of future maintenance methods is put forward.

Key words: overhead line; cable; maintenance and repair; temperature measurement; GPS positioning; tunnel robot

论高炉煤气干湿除尘工艺及配套 TRT 运行优劣

李山虎，哈 君

（鞍钢股份能源管控中心，辽宁鞍山 114000）

摘 要： 文章结合高炉煤气干法除尘及 BXF 除尘工艺及配套 TRT 运行实践情况，综合从经济效益、生产运行、设备运行、安全环保运行四个方面进行论证评价，结合作者实际工作中涉及到的发电量指标、能量损失、安全生产、可靠运行、设备检修、环保风险多个角度开展论证，肯定了干法除尘工艺及配套 TRT 系统运行的优越性，并指出了该领域干法除尘工艺改造的发展趋势。

关键词： 高炉煤气；干法除尘；BXF 除尘；TRT；经济效益；生产运行；设备运行；安全环保运行

The Comparison of BFG Dry-type Dedusting and BXF Dedusting Technology with Related TRT

Li Shanhu, Ha Jun

(Energy Management and Control Center of Ansteel, Anshan, 114000, China)

Abstract: This paper combined the BFG dry-type dedusting and BXF dedusting technology with related TRT functioning practice. The demonstration evaluation was according to the economic performance, production operation, equipment operation, safety and environmental protection. Besides, the authors launched demonstration according to power generation index, energy loss, safety in production, reliable operation, equipment overhaul, environmental risk in practical work, confirming the advantages of dry-type dedusting and related TRT system operation. Besides, this paper also pointed out the development tendency of the reform of dry-type dedusting in its area.

Key words: blast furnace gas; dry-type dedusting; BXF dedusting; TRT; economic performance; production operation; equipment operation; safety and environmental protection

焦炉煤气制氢系统常见故障、影响及解决方案

李山虎，杨子壮，赵 冬

（鞍钢股份能源管控中心，辽宁 鞍山 114000）

摘 要：本文作者结合近年来实际工作经验，结合焦炉煤气制氢较为先进的PSA-6-3-2工艺，选取了包括压缩机、产品氢气质量、系统漏料、程控阀、换热器等多个方面存在的具有普遍指导性的故障及影响，并提出解决预防措施。本文提及缺陷针对性强，但未涵盖焦炉煤气制氢配套设备设施常规故障。

关键词：焦炉煤气制氢；压缩机；氢气质量；系统漏料；程控阀；换热器

Problem, Effect and Solution of Hydrogen Production from Coke Oven Gas

Li Shanhu, Yang Zizhuang, Zhao Dong

(Energy Management and Control Center of Ansteel, Anshan 114000, China)

Abstract: This paper combines the author's practical experience in recent years, according to the PSA-6-3-2 that is the advanced technology in hydrogen production. The problem covers compressor, hydrogen quality, leakage of adsorption tower, sequencing valve, heat exchanger and so on. The problems existed in hydrogen production units were analyzed, and the countermeasures were put forward. The defects this paper points out are highly targeted, but the paper does not cover all conventional failures in hydrogen production.

Key words: hydrogen production from coke oven gas; compressor; hydrogen quality; leakage of adsorption tower; sequencing valve; heat exchanger

一种适用于轧钢加热炉的煤气除湿技术

徐言东，谢仲豪，刘 涛，余 伟，程知松，魏付豪

（北京科技大学工程技术研究院，北京 100083）

摘 要：轧钢厂加热炉大多使用转炉煤气加热并用鼓风机系统鼓风助燃，根据工艺要求，可在加热炉前设置煤气除湿装置，既能降低加热炉内气氛的湿度，提高煤气热值，又能进一步减少钢坯脱碳，提高产品质量，消除造成棒线、板材扭曲、开裂等内在质量问题。目前此技术在钢铁企业具有广泛的应用前景和较高的使用价值。

关键词：轧钢加热炉；煤气；除湿

A Gas Dehumidification Technology for Steel Rolling Heating Furnace

Xu Yandong, Xie Zhonghao, Liu Tao, Yu Wei, Cheng Zhisong, Wei Fuhao

(Engineering and Technology Research Institute, University of Science and Technology Beijing, Beijing 100083, China)

Abstract: Most heating furnaces in rolling mills use converter gas to heat and use blower system to blast combustion. According to technological requirements, a gas dehumidification device can be installed in front of the heating furnace, which can not only reduce the humidity of the atmosphere in the heating furnace, increase the calorific value of the gas, but also further reduce the decarbonization of steel billets, improve product quality and eliminate the internal quality problems such as twisting and cracking of rods and plates. At present, this technology has broad application prospects and high application value in iron and steel enterprises.

Key words: steel rolling heating furnace; gas; dehumidification

鞍钢高炉工序 CO_2 排放水平及减排展望

孟凡双，李建军，曾　宇，李晓春

（鞍钢股份有限公司炼铁总厂，辽宁鞍山　114021）

摘　要： 计算鞍钢高炉工序 CO_2 排放量情况，分析目前高炉工序 CO_2 排放水平及分布，进一步提高现有高炉的能源利用效率，并研究并应用新技术大幅度降低能源消耗，可减少高炉工序 CO_2 排放量，展望鞍钢下一步高炉工序 CO_2 减排工作方向。

关键词： 高炉；二氧化碳；减排

鞍钢低成本有效提高热风温度技术

刘德军，袁　玲，赵爱华

（鞍钢集团钢铁研究院环境与资源研究所，辽宁鞍山　114009）

摘　要： 核心介绍了鞍钢高炉热风炉高风温及其相应的节能技术的进步。重点就鞍钢热风炉长期使用低热值煤气烧炉的特点，介绍了鞍钢梯次实施的热风炉结构形式的改造和热风炉自预热、前置炉及辅助热风炉等根本性改造；继而开展了针对热风炉的板换替代管换实施双预热、送风换炉技术优化、富氧烧炉、复合涂料的使用、送风系统关键部位预制预警技术等多项综合节能技术的研究与应用，实现了热风温度的大幅提高和热风炉烧炉煤气消耗的大幅降低，取得了良好的效果，极大地推动了鞍钢高炉热风炉技术的进步。

关键词： 高炉；热风炉；技术；低成本

Low Cost Effective Hot Air Temperature Technology in Angang

Liu Dejun, Yuan Ling, Zhao Aihua

(Environment and Resource Institute, Iron & Steel Research Institutes of Ansteel Group Corporation, Anshan 114009, China)

Abstract: The progress of high blast temperature and corresponding energy saving technology in Anshan iron and steel co. In this paper, the features of low-calorific value gas fired furnace in Anshan iron and steel co, this paper introduces the structural transformation of hot blast furnace implemented, and self-preheating, front furnace and auxiliary heating furnace and other fundamental transformation by steps in Anshan iron and steel group co. Then conducted for hot blast stove plate for replacement tube in the implementation of double preheating, the air distribution in the furnace technology optimization, the rich oxygen burning furnace, the use of the composite coating, the air supply system key parts prefabrication early warning technology and so on many research and application of comprehensive energy saving technology, realize the large increase in the hot blast temperature and hot blast stove burning furnace gas consumption greatly reduced, and achieved good effect, greatly promoted the Anshan iron and steel blast furnace hot blast stove technology progress.

Key words: blast furnace; hot blast stove; technology; lowcost

冷轧热镀锌退火炉测试分析与节能

高军，马光宇，刘常鹏，张天赋，赵俣

（鞍钢集团钢铁研究院，辽宁鞍山 114009）

摘　要： 针对某冷轧镀锌线退火炉煤气单耗较高的问题，采用热工测试手段，分析退火炉各加热段热工操作及炉体保温等存在的问题，制定解决方案，提高退火炉热效率，实现降低煤气消耗的目的。

关键词： 热镀锌退火炉；工艺改进；设备完善；节能

The Energy Saving and Thermal Measurements on Annealing Furnace in Cold Rolled Hot Dip Galvanizing Line

Gao Jun, Ma Guangyu, Liu Changpeng, Zhang Tianfu, Zhao Yu

(Ansteel Iron & Steel Research Institutes, Anshan 114009, China)

Abstract: For the issue of higher energy intensity on annealing furnace in cold rolled dip galvanizing line, thermal measurements and the analysis on issues such as thermal operation and furnace surface insulation have been made. The solution is put forward and the aim of improving furnace thermal efficiency and lowering the gas consumption is achieved.

Key words: annealing furnace; process improvement; equipment perfect; energy saving

鞍钢高炉热风炉燃烧控制系统应用实践

李建军，孟凡双，曾　宇，李晓春

（鞍钢股份有限公司炼铁总厂，辽宁鞍山　114021）

摘　要：针对热风炉现场已有的最基本测控仪表和执行机构，采用寻找动态最佳空燃比的自寻优算法，以及智能软测量技术、趋势优化控制技术、扰动观测器和智能控制技术，实现热风炉燃烧过程的自动优化控制，进而提高送风风温，降低煤气消耗。鞍钢 5 高炉热风炉应用智能燃烧优化技术，提升了热风炉自动控制水平，节能效益 400 万元以上。

关键词：热风炉；燃烧优化；自动控制

Application Practice of Combustion Control System for Hot Blast Furnace of Angang

Li Jianjun, Meng Fanshuang, Zeng Yu, Li Xiaochun

(Angang Iron and Steel Co., Ltd., Anshan 114021, China)

Abstract: Aiming at the most basic measurement and control instruments and actuators existing in the field of hot blast stove, this paper adopts the self-optimization algorithm to find the dynamic optimal air-fuel ratio, as well as intelligent soft sensing technology, trend optimization control technology, disturbance observer and intelligent control technology to realize the automatic optimization control of hot blast furnace combustion process, so as to increase the air supply temperature and reduce the gas consumption. The intelligent combustion optimization technology is applied to the hot blast furnace of Anshan Iron and Steel Co., Ltd., which improves the automatic control level of the hot blast furnace, and the energy saving benefit is more than 4 million yuan.

Key words: hot blast furnace; combustion optimization; automatic control

红土镍矿资源特点及矿物改性研究现状

武兵强[1,2]，齐渊洪[1,3]，周和敏[1,3]，高建军[1,3]，邹宗树[2]

（1. 钢铁研究总院先进钢铁流程及材料国家重点实验室，北京　100081；2. 东北大学冶金学院，辽宁沈阳　110819；3. 钢研晟华科技股份有限公司，北京　100081）

摘　要：随着硫化镍矿资源的不断大量消耗和镍需求量的持续增长，从红土镍矿资源中提取镍以及其他有价元素成为当前研究的热点，红土镍矿资源的高效开发和综合利用具有极其重要的现实意义。本文简要阐述了红土镍矿的资源特点，包括红土镍矿资源的分布、红土镍矿的分类和矿物组成等三个方面，综述了红土镍矿矿物改性的研究现状，并探讨了红土镍矿改性的研究方向和应用前景，为充分利用红土镍矿资源提供参考。

关键词：红土镍矿；特点；矿物改性；现状

Characteristics and Research Status of Mineral Modification of Laterite Nickel Ore

Wu Bingqiang[1,2], Qi Yuanhong[1,3], Zhou Hemin[1,3], Gao Jianjun[1,3], Zou Zongshu[2]

(1. State Key Laboratory of Advanced Steel Processes and Products, Central Iron and Steel Research Institute, Beijing 100081, China; 2. School of Metallurgy, Northeastern University, Shenyang 110819, China; 3.CISRI Sunward Technology Co., Ltd., Beijing 100081, China)

Abstract: With the continuous consumption of nickel sulfide ore resources and the continuous growth of nickel demand, the extraction of nickel and other valuable elements from laterite nickel ore resources has become a hot research topic. It is of great practical significance for the efficient exploitation and comprehensive utilization of laterite nickel ore resources. Characteristics of laterite nickel ore resources, including distribution, classification and mineral composition, were briefly expounded. Research status of mineral modification of laterite nickel ore was summarized. Research direction and application prospect of mineral modification of laterite nickel ore were discussed. These provide reference for making full use of laterite nickel ore resources.

Key words: laterite nickel ore; characteristic; mineral modification; status

鞍钢 2000m³ 焦炉煤气制氢项目探析

舒 畅，杨子壮

（鞍钢股份有限公司能源管控中心，辽宁鞍山 114021）

摘 要：在总结鞍钢本部五套制氢装置建设及运营的基础上，新建 2000m³/h 焦炉煤气制氢装置项目体现了工艺先进完善、设备选型合理、施工组织有序等优点，同时分析了项目的不足，从而对相关的项目建设具有借鉴意义。

关键词：焦炉煤气；PSA6-3-2 工艺；压缩机

Study on the 2000m³ COG to H₂ Project in Ansteel

Shu Chang, Yang Zizhuang

(Energy Control Centre of Ansteel Co., Ltd., Anshan 114021, China)

Abstract: Based on construction and operation of 5 COG to H₂ projects, we build the new 2000m³/h COG to H₂ project, it refldcted with advanced and perfect technology, reasonable equipment selection and built in orderly, in addition the stortcomings the project are analyzed in order to be used for refercence for the construction of the related projects.

Key words: coke oven gas; technique of PSA6-3-2; compressor

碳纤维复合芯铝绞线在钢铁企业的应用

韩 帅,李 达

(鞍钢股份有限公司能源管控中心,辽宁鞍山 114003)

摘 要:碳纤维复合芯铝绞线作为一种节能环保的新型导线,成为目前传统钢芯铝铰线输电线路增容改造的理想选择。本文结合鞍钢股份有限公司能源管控中心 66kV 电北线改造的典型案例,分别从国内外应用情况、导线的特点、线路输送能力、技术经济性等方面对碳纤维芯铝绞线进行分析,证明采用碳纤维复合材料芯导线的可行性,以及所取得的经济、社会效益及推广价值。

关键词:输电;改造;导线;碳纤维

Application of Carbon Fiber Composite Core Aluminum Strand in Iron and Steel Enterprises

Han Shuai, Li Da

(Energy Control Center of Anshan Iron & Steel Group Co., Ltd., Anshan 114003, China)

Abstract: Carbon fiber composite core aluminum strand as a new energy-saving and environmental protection conductor. It has become an ideal choice for capacity-increasing and retrofitting of traditional steel-core aluminium hinged transmission lines at present. This paper combines the typical case of the renovation of 66kV electricity north line in the energy management and control center of Anshan Iron and Steel Co., Ltd., Carbon fiber reinforced aluminium strand is analyzed in terms of its application at home and abroad, characteristics of conductors, transmission capacity and technical economy.The feasibility of using carbon fiber composite core conductor is proved, and the economy achieved, social benefits and promotional value.

Key words: transmission; reform; wireway; carbon fibre

厚板加热炉自动炉温设定方法初探

张敏文

(湛江钢铁厚板厂,广东湛江 524076)

摘 要:为实现厚板加热炉炉温自动设定,利用大样本生产实绩,区分不同装入温度、抽出温度,基于传热解析对炉温设定方法做初步研究。

关键词:厚板;加热炉;炉温;自动设定

Research for the Automatic Furnace Temperature Setting of Heavy Plate Mill

Zhang Minwen

(Heavy Plate Mill, Zhanjiang Iron and Steel, Zhanjiang 524076, China)

Abstract: For automatic furnace temperature setting of heavy plate mill, based on heat transfer theory and large-sample production data, the furnace temperature setting way is initially designed with different values of charging temperature and discharging temperature.

Key words: heavy plate mill; reheating furnace; furnace temperature; automatic setting

冷轧浓油废水油水分离废油的资源化分析

张国志[1]，尹婷婷[2]

（1. 宝钢股份冷轧厂，上海 201900；2. 宝钢中央研究院能源与环境研究所，上海 201900）

摘 要： 冷轧工艺需要，使用大量的乳化油、防锈油等油相试剂，经过轧制或碱洗等工艺后，产生大量的浓含油废水，需先进行除油预处理后再进行生化、高级氧化有机降解手段，实现达标排放。除油环节通过油水分离会产生大量废油，废油作为水处理的副产物，它的资源化再生具有重要经济意义，也是衡量除油工艺经济性的重要考察标准。本文针对聚结分离技术所产生的废油副产物，对含水率、燃烧性能、毒性等资源化重要影响因子进行分析，探索其资源化利用的途径，估算其经济价值。

关键词： 冷轧；浓油废水；废油；资源化

Reclamation Analysis of Oil-Water Separation Waste Oil from Cold Rolling Heavy Oil Wastewater

Zhang Guozhi[1], Yin Tingting[2]

(1. Baosteel Cold Rolling Plant, Shanghai 201900, China; 2. Baosteel Institute for Energy and Environmental Research, Shanghai 201900, China)

Abstract: Because of the processing requirement, a large number of oil-phase reagents, such as emulsified oil and rust-proof oil, are used in cold rolling mills. After rolling or alkali washing, a large number of wastewater containing heavy oil is produced. These heavy oil wastewaters are necessary to carry out deoiling pretreatment before biochemical and advanced oxidative organic degradation to achieve standard discharge. Oil-water separation will produce a large amount of waste oil. As a by-product of water treatment, the recycling of waste oil is of great economic significance, and it is also an important criterion to measure the economy of oil-removal process. Aiming at the by-products of waste oil produced by coalescence and separation technology, this paper analyses the important influencing factors of water content, combustion performance, toxicity and so on, explores the ways of resource utilization, and estimates its economic value.

Key words: cold rolling; heavy oil wastewater; waste oil; reclamation

TRT 旁通管变工况下的仿真及设计优化

潘 宏[1], 李星星[2], 刘丹瑶[1], 胡学羽[1]

(1. 中冶南方工程技术有限公司,湖北武汉　430223;
2. 中核武汉核电运行技术股份有限公司,湖北武汉　430223)

摘　要: 本文在对实际工程 TRT 系统配置的旁通管进行流动及换热分析的基础上, 计算变工况条件下旁通管的临界截面及出口截面参数,并分析了不同压力工况下旁通快开调节阀开度变化情况, 为冶金工程 TRT 旁通管的设计优化及调节阀的选型提供了理论依据。

关键词: TRT; 旁通管; 临界截面; 出口截面

Simulation and Design Optimization of TRT Bypass Pipeline under Variable Conditions

Pan Hong[1], Li Xingxing[2], Liu Danyao[1], Hu Xueyu[1]

(1. WISDRI Engineering & Research Incorporation Limited, Wuhan 430223, China;
2. China Nuclear Power Operation Technology Corporation, Ltd., Wuhan 430223, China)

Abstract: Based on the analysis of the flow and heat transfer of by-pass pipeline in TRT system of practical engineering. The critical section and outlet section and parameters of bypass pipeline under variable operating conditions are calculated. The variation of by-pass quick-opening control valve opening under different pressure conditions is analyzed. It provides a theoretical basis for the design optimization of TRT bypass pipe and the selection of control valve in metallurgical engineering.

Key words: TRT; by-pass pipeline; critical section; exit section

鱼雷罐加金属化球团对铁水温度的影响

谢俊波, 牛长胜, 金　奕

(宝钢湛江钢铁有限公司制造管理部, 广东湛江　524000)

摘　要: 在倒完铁水的空鱼雷罐中加入金属化球团, 鱼雷罐再次受铁过程中, 在铁水的动力学冲击和热力学作用下, 金属化球团中的氧化铁与铁水中的碳元素和硅元素发生还原反应, 实现了金属化球团的无危害返生产利用。研究了金属化球团对鱼雷罐中铁水温度的影响, 从理论分析和生产实践两个方面, 证实了金属化球团的加入不会降低铁水温度。

关键词: 金属化球团; 返生产; 鱼雷罐; 铁水温度

Effect of Taking Metallized Pellets into Torpedo Tank on the Temperature of Molten Iron

Xie Junbo, Niu Changsheng, Jin Yi

(Baosteel Zhanjiang Iron & Steel Co., Ltd. Manufacturing Management Department,
Zhanjiang 524000, China)

Abstract: After adding metallized pellets to the torpedo tank with molten iron, in the process of iron, under the dynamic impact and thermodynamics of molten iron, the iron oxide in the metallized pellet and the carbon and silica in the molten iron have a reduction reaction, and the non-hazardous return production and utilization of metallized pellets are realized. The influence of metallized pellets on the temperature of molten iron in torpedo cans is studied, and from two aspects of theoretical analysis and production practice, it is proved that the addition of metallized pellets will not reduce the temperature of molten iron.

Key words: metallized pellets; return to production; torpedo tank; molten iron temperature

板框压滤机在含酸废水处理改进工艺中的应用

杨 跃

(鞍钢股份有限公司冷轧厂，辽宁鞍山 114021)

摘 要： 本文对针对含酸废水传统处理工艺进行改进，充分利用板框压滤机的过滤作用对废水进行过滤处理，消除因 pH 值波动变化对絮凝沉降效果的影响，保证工艺稳定连续运行，处理后废水达标排放。

Application of Plate and Frame Filter Press in Improvement Process of Acid Wastewater Treatment

Yang Yue

(Cold Rolling Mill, Anshan Iron and Steel Co., Ltd., Anshan, 114021, China)

Abstract: In this paper, the traditional treatment process for acid containing wastewater is improved with the filtration effect of the plate frame filter press be used to filter the wastewater. The flocculating settlement effect of the pH fluctuation is eliminated, the stable and continuous operation of the process is ensured, and the wastewater treatment reaches the standard discharge.

梅钢高炉水渣水处理工艺优化

翟玉龙

(宝武集团上海梅山钢铁股份有限公司炼铁厂，江苏南京 210039)

摘　要：介绍了梅钢高炉水渣水处理工艺流程，分析制渣水处理工艺组成和系统管网及其存在的问题，分析大高炉区域工业废水来源并检测其化学成分，对水渣水处理工艺优化改进，消除转鼓冲扫水的非正常消耗，把原排放的水冷钢槽冷却水改为循环封闭冷却水，在高炉水渣生产区域增加配置水回收处理系统，防止水渣热水池等溢流水外排，工艺优化后，高炉水渣系统水处理实现封闭循环利用，实现了高炉区域生产废水零外排的目标，已取得明显效果，环境效益明显。

关键词：水处理；INBA 制渣；工艺优化；高炉

The Water Treatment Process Optimization for Blast Furnace Slag in Meishan Steel

Zhai Yulong

(Baowu Group Shanghai Meishan Iron & Steel Co., Ltd., Nanjing 210039, China)

Abstract: Introduces water treatment process flow of the blast furnace slag in Meishan Steel, analysis problems of slag water treatment process and system network, and analysis of industrial waste water sources of blast furnace and testing its chemical composition, to improve water slag water treatment process, eliminate the abnormal water consumption used for drum flush, changed the original water emissions of water-cooling steel tank to loop cooling water, configure water recycling system, to prevent the effluent water in water slag heat sink overflow and let out, after process optimization, blast furnace slag water treatment system achieved closed cycle water use, and achieved the goal of the production waste-water no emissions, environmental benefits are obvious.

Key words: water treatment; INBA technique; improve process; blast furnace

无焰富氧燃烧技术在马钢板坯加热炉上的应用

丁　毅，田　俊，范满仓，周劲军，翟　炜

（马鞍山钢铁股份有限公司，安徽马鞍山　243000）

摘　要：为解决热轧线产能不匹配问题，马钢采用无焰富氧燃烧技术对 1580 热轧加热炉燃烧控制系统进行了改造。实施效果表明，加热产能提高 15% 以上，燃耗下降 15% 以上，烧损下降 18% 以上，钢坯加热温度均匀性明显改善，烟气 NO_x 排放没有明显变化，达到了预期目标。

关键词：无焰富氧燃烧；板坯加热炉；应用

Application of Flameless Oxy-fuel Burning Technology in Slab Heating Furnace of Masteel

Ding Yi, Tian Jun, Fan Mancang, Zhou Jinjun, Zhai Wei

(Maanshan Iron & Steel Co., Ltd., Maanshan 243000, China)

Abstract: In order to solve the problem of mismatching production capacity of hot rolling line, masteel has transformed the combustion control system of 1580 rolling reheating furnace by means of flameless oxy-fuel burning technology. The implementation results show that the heating productivity is increased by more than 15%, the fuel consumption is reduced by more than 15%, the rate of burn loss is decreased by 15%, the heating temperature uniformity of steel billet is obviously improved, and the NO_x emission of flue gas is not significantly changed, reaching the expected goal.

Key words: flameless oxy-fuel burning; slab heating furnace; application

CCPP 机组夏季燃料水喷射降温加湿技术及应用

薛晓金

（鞍钢股份鲅鱼圈钢铁分公司能源动力部，辽宁营口　115007）

摘　要：文章主要对如何提高 CCPP 机组夏季负荷进行介绍，通过对煤气压缩机出口的高温高压煤气进行喷水降温加湿，使得燃机燃烧温度降低，增加燃机可燃烧煤气量，进而提高机组负荷。对水喷射工艺进行介绍，水喷射相关设备的工况以及控制方式，水喷射投运时空气、煤气温度对其影响，以及水喷射投运时对机组关键参数的影响进行了介绍。

关键词：水喷射；降温；负荷

Technology and Application of Cooling and Humidifying by Fuel Water Injection in CCPP Unit in Summert

Xue Xiaojin

(The Energy and Power Department of Ansteel Bayuquan Iron & Steel Subsidiary, Yingkou, 115007, China)

Abstract: This paper mainly introduces how to increase the summer load of CCPP unit. By spraying water to cool and humidify the high temperature and high pressure gas at the outlet of gas compressor, the combustion temperature of the gas turbine is reduced, the combustible gas quantity of the gas turbine is increased, and the load of the unit is increased. This paper introduces the water jet technology, the working conditions and control methods of water jet related equipment, the influence of gas and gas temperature on water jet operation, and the influence of water jet operation on the key parameters of the unit.

Key words: water injection; cooling; load

蒸汽管网安全运行的分析与事故防治

黄显保，何　嵩，刘柏寒

（鞍钢股份能源管控中心，辽宁鞍山　114021）

摘　要：鞍钢股份能源管控中心负责鞍钢股份本部生产生活的蒸汽供应，管线长，范围广。用户蒸汽压力使用等级繁多，蒸汽管网压损大，温降大，极易发生管道汽水冲击事故，严重影响蒸汽管网的安全运行。本文主要是从生产运行管理、设备维护、施工管理等方面进行剖析，制定措施，防止事故的发生。

关键词：蒸汽管网；管道汽水冲击

新型接地选线装置的应用及案例分析

李达，韩帅，王林松，唐巍，沙金

（鞍钢股份有限公司能源管控中心，辽宁鞍山 114000）

摘 要：钢铁企业电网结构较为复杂，同时由于设备运行环境较为恶劣，电气事故发生频次较高。尤其是接地事故时有发生，传统的处理程序是通过接地选线装置及倒母线或轮停的方式确定接地线路。由于选线装置准确率不高，所以通常根据运行方式特点依靠倒闸操作及特定顺序停电的办法进行选接地。但是在选接地操作过程中（如合切母联及其他断路器）及间歇性弧光接地经常出现电气设备（电压互感器及过电压保护器）短路放炮事故的发生，使单相接地演变为短路停电事故，给公司的生产稳顺带来极大影响。根据公司最大变电站运行特点，结合目前行业出现的转移型接地选线装置工作原理，探索了一套适合本系统的接地选线装置及其运行管理办法。

关键词：小电流接地系统；转移型接地选线装置；电压互感器；过电压保护器

Application and Case Analysis of New Earth Line Selection Device

Li Da, Han Shuai, Wang Linsong, Tang Wei, Sha Jin

(Enegy Manage and Control Center, Ansteel Company Limited, Anshan 114000, China)

Abstract: The structure of iron and steel power grid is more complex. At the same time, due to the poor operating environment of equipment, the frequency of electrical accidents is high. In particular, grounding accidents occur from time to time. The traditional processing program determines the grounding line by means of grounding line selection device and inverted master line or wheel stop. Due to the low accuracy of the wire selection device, it is usually based on the operation of the backgate and the method of power outage in a specific order. However, during the ground selection operation(such as syncopation bus and other circuit breakers) and intermittent arc light grounding, there are often short-circuit firing accidents in electrical equipment(voltage transformer and overvoltage protector), which makes single-phase connection the ground evolved into a short-circuit power outage accident. It has a great impact on the stability of the company's production. According to the operation characteristics of the company's largest substation and the working principle of the transfer type grounding line selection device that appears in the current industry, a set of grounding line selection device suitable for the system and its operation management methods are explored.

Key words: little currentearth line select system; distract type of earth line select equipment; VT; overvoltage protector

焦化废水处理工艺氨氮指标的控制措施

李国庆

（河北钢铁集团宣钢公司焦化厂，河北宣化 075105）

摘　要：通过对焦化废水 A/O 生物脱氮处理工艺的运行机理和影响出水氨氮指标的各种因素进行分析，总结出焦化废水生化处理系统出水氨氮指标的控制措施，在运行中采取针对性措施及时调节，保证系统正常运行。

关键词：焦化废水；氨氮；硝化；反硝化

The Control Measures of NH$_3$-N Index in Coking Wastewater Treatment System

Li Guoqing

(Coking Plant of HBIS Xuan Steel, Xuanhua 075105, China)

Abstract: By analyzing the operation mechanism of coking wastewater A/O biological denitrification process and various factors affecting the NH$_3$-N index of effluent, the control measures of NH$_3$-N index in effluent of coking wastewater biochemical treatment system were summarized. Take targeted measures in the operation of timely adjustment to ensure the normal operation of the system.

Key words: coking wastewater; NH$_3$-N; nitrating; denitrating

钢化联合企业生产能源一体化智能管控系统探讨

刘书文，欧　燕，叶理德

（中冶南方工程技术有限公司钢铁公司技术研究院，湖北武汉　430023）

摘　要：本文探讨了利用钢铁副产煤气进行化工生产的钢化联合企业，在生产与能源高度耦合的情况下，如何基于生产计划、检修计划，获取突发的生产异常事件，使用能源动态预测、动态平衡与优化调度等手段，组建有效的生产能源一体化智能管控系统，在保证能源平衡的前提下，给出生产调度和能源调度的指导性策略，提高能源综合利用效率，最终实现调度周期内总的能源利用效益最高的目的。

关键词：钢化联合；能源管理；能源平衡与调度；能源动态预测；生产能源一体化

换热器在步进式加热炉生产中的技术进步

李国辉

（鞍钢股份有限公司大型厂小型分厂，辽宁鞍山　114021）

摘　要：轧钢加热炉的排烟温度很高，烟气带走的热量很大，对加热炉进行余热回收是提高加热炉热效率，节约能源的有效途径。换热器是轧钢加热炉余热回收的主要设备，对换热器的节能运行是一项综合的系统工程，它与炉型、燃烧器型式、供风系统、排烟系统等都有关系。生产上对换热器的要求是在最小的能耗情况下获得最大的强化传热效果，也就是在能够获得较高的传热系数的前提下尽量缩小换热器的体积，降低换热器的阻力。本文对金属对流换热器材质的提升，换热器系统控制方式中的新技术，以及在生产中确定空气预热温度时应考虑的问题进行了综合论述。

关键词：换热器；传热系数；发热值；烟气温度

酒钢矿棉加工可行性探析

赵贵清[1]，陈亚团[2]

（1. 酒钢技术中心，甘肃嘉峪关 735100；2. 酒钢钢铁研究院，甘肃嘉峪关 735100）

摘 要：介绍了矿棉技术的发展概况，提出了酒钢矿物渣高效利用的具体方案，并着重就酒钢实现矿棉生产的可行性进行了详细探讨。

关键词：矿物渣；矿棉；可行性；探析

Feasibility Investigation of Mineral Wool Produced in Jiusteel

Zhao Guiqing[1], Chen Yatuan[2]

(1. Technology Center of Jiusteel, Jiayuguan 735100, China;
2. Iron and Steel Institute of Jiusteel, Jiayuguan 735100, China)

Abstract: The development of mineral wool technology were introduced. A plan of the efficient utilization of metallurgical slag of Jiusteel was suggested.And the feasibility of mineral wool produced in detail was discussed, including process method and the market prospect.

Key words: metallurgical slag; mineral wool; feasibility; investigate

鞍钢第二发电厂 CCPP 厂用电系统经济运行方案分析

吴 猛[1]，孙德俊[1]，丁伟伟[2]

（1. 鞍山钢铁集团有限公司第二发电厂，辽宁鞍山 114012；
2. 鞍钢矿业公司动力总厂，辽宁鞍山 114012）

摘 要：发电厂电气系统如何做到经济运行，主要从以下几个方面来考虑：第一，变压器在合理区间运行；第二，采用合理的运行方式；第三，根据负荷情况调节变压器的分接头；第四，备用变在备用时应不带电压。

关键词：厂用电；节能降耗；经济运行

Analysis on the Economic Operation Scheme of CCPP Power Plant in Angang Second Power Plant

Wu Meng[1], Sun Dejun[1], Ding Weiwei[2]

(1. Anshan Iron and Steel Group Co., Ltd., Second Power Plant, Anshan 114012, China;

2. Angang Mining Company Power Plant, Anshan 114012, China)

Abstract: How to realize the economic operation of power plant electrical system is mainly considered from the following aspects: First, the transformer operates within a reasonable range; Second, adopt reasonable operation mode; Third, according to the load condition, adjust the transformer junction; Fourth, the standby change should not have voltage when standby.

Key words: plant electricity; energy conservation and consumption reduction; economic functioning

钢铁厂粉尘处理及二次产物综合利用

孙利军[1]，张　晨[2]，张丽俊[3]

（1. 陕钢集团汉中钢铁有限责任公司炼铁厂，陕西汉中　723000；
2. 中钢集团工程设计研究院，北京　100083；3. 西安建筑科技大学，西安　710000）

摘　要：目前以冶金粉尘为原料，采用转底炉工艺生产出的预还原球团，无法达到炼钢使用要求。日本等国将其作为高炉炉料使用。国内外生产实践证明，高炉使用金属化炉料能够提高产量，降低焦比。实验结果表明：预还原球团具有良好的冶金性能，无低温还原粉化、无高温还原膨胀、软熔性能优良。但是其物理性能差，抗压强度仅有800N，影响块状带透气性，是其大量使用的制约条件。根据物料平衡计算，预还原球团在炉料中仅占到约5%，总体而言利大于弊，可以作为高炉炉料使用。

关键词：转底炉；预还原球团；抗压强度；冶金性能

Dust Treatment and Comprehensive Utilization of Secondary Products in Iron and Steel Works

Sun Lijun[1], Zhang Chen[2], Zhang Lijun[3]

(1. Shaanxi Steel Group Hanzhong Iron and Steel Co., Ltd., Hanzhong 723000, China;
2. Sinosteel Engineering Design & Research Institute Co., Ltd., Beijing, 100083, China;
3. Xi'an University of Architecture and Technology, Xi'an, 710000, China)

Abstract: At present, prereduction pellets produced by rotary hearth furnace with metallurgical dust as raw material cannot meet the requirement of steel-making. Thus Japan and other countries use it as blast furnace material. Production practice both at home and abroad has proved that the use of metallized pellets as burden in blast furnace can improve output and reduce coke ratio. The experimental results illustrate that the prereduction pellets have good metallurgical properties, such as no pulvetization during low temperature reduction, no expansion during high temperature reduction and favorable soft melting. However, its physical performance is poor, its compressive strength being only ≥800N. It also affects the permeability of the block zone, which constrains its extensive use. Calculation of material balance, The prereduction pellets account for only ~ 5% of the all burden entering the furnace. Generally speaking, the advantages outweigh the disadvantages and prereduction pellets can be used as blast furnace material.

Key words: rotary hearth furnace; prereduction pellets; compressive strength; metallurgical properties

梅钢烧结矿竖冷炉内物料运动特性的模拟研究

孙俊杰[1],程乃良[1],徐 骥[2]

(1. 宝钢股份研究院梅钢技术中心,江苏南京 210039;
2. 中国科学院过程工程研究所多相复杂系统国家重点实验室,北京 100190)

摘 要:竖式冷却回收烧结矿显热工艺是一种新型余热回收技术,其本质上是气固逆流换热散料床,梅钢从环保和节能两方面考虑引入该技术代替传统的环冷机来对其烧结矿余热进行回收利用。本文以烧结矿竖冷炉为研究对象,利用离散单元法(DEM)模拟烧结矿颗粒的运动情况。计算结果显示:(1)在竖冷炉内部,烧结矿颗粒的分布呈现小颗粒聚集在进料管周边,大颗粒集中在炉墙边缘的规律;(2)烧结矿石的偏析与颗粒的落点有关,与竖冷炉内颗粒的堆积高度无关。

关键词:烧结矿;竖冷炉;离散单元法;颗粒偏析

Simulation of Particle Motion Characteristics in Sinter Shaft Cooler of Meisteel

Sun Junjie[1], Cheng Nailiang[1], Xu Ji[2]

(1. R&D Center for Meisteel of Baosteel Research Institute, Nanjing 210039, China;
2. State Key Laboratory of Multiphase Complex Systems, Institute of Process Engineering, Chinese Academy of Sciences, Beijing 100190, China)

Abstract: The process of recycling the sensible heat of sinter by using sinter shaft cooler is a new technology. The shaft cooler is a typical powder bed with gas-solid countercurrent heat transfer. Considering environmental protection and energy conservation, the process of sinter shaft cooler was introduced into Meishan steel to replace the traditional ring cooler. Through focusing on the motion law of particles in the shaft cooler, the DEM (Discrete Element Method) is used on an industrial scale in this paper. The result of simulation indicates two conclusions. Firstly, the distribution of sinter particles shows a law that small particles gather around the feeding pipe and large particles concentrate at the edge of furnace wall. Secondly, the segregation of sinter particles is related to the drop point of particles and has nothing to do with the accumulation height of particles in the shaft cooler.

Key words: sinter; shaft cooler; DEM; segregation of particle

燃气锅炉燃烧自动控制优化

任 珲,降万勇

(宝武集团广东韶关钢铁有限公司能源环保部,广东韶关 512123)

摘 要：通过采用大数据、人工智能、机器深度学习技术优化锅炉燃料燃烧控制，实现对燃气锅炉以及辅机系统的自控监控和安全检测功能，使得锅炉能更安全可靠运行，提升锅炉燃料效率。

关键词：锅炉；自动控制优化；汽包水位；炉膛负压；减温水；氧量；煤气量

Optimization of Automatic Combustion Control for Gas-fired Boilers

Ren Hui, Jiang Wanyong

(Baowu Group Guangdong Shaoguan Iron and Steel Co., Ltd., Shaoguan 512123, China)

Abstract: The automatic monitoring and safety detection functions of gas-fired boilers and auxiliary systems can be realized through the use of big data, artificial intelligence and machine deep learning technology to optimize the fuel combustion control of boiler, so that the boiler can operate more safely and reliably, and the fuel efficiency of the boiler can be improved.

Key words: boiler; automatic control optimization; drum water level; hearth negative pressure; temperature reducing water; oxygen; gas volume

节能减排装置"高炉煤气透平机组"故障分析及解决方法

谢福成

（宝武集团广东韶关钢铁有限公司能源环保部，广东韶关 512123）

摘 要：大型管道的安装好坏直接影响 TRT 机组的正常运行，从另外一个角度看，管道安装不当将引起机组振动超标，破坏密封系统，引发煤气泄漏，造成人员伤亡的重大安全事故，对此，我们根据现场振动的实际情况，查找分析故障，并提出管道改造方案。

关键词：TRT；管道；支座；传感器

Fault Analysis and Solution of "Top Gas Turbine Unit" for Energy Saving and Emission Reduction Device

Xie Fucheng

(Baowu Group Guangdong Shaoguan Iron and Steel Co., Ltd., Shaoguan 512123, China)

Abstract: The installation of large-scale pipelines directly affects the normal operation of TRT units. From another point of view, improper installation of pipelines will cause excessive vibration of units, damage the sealing system, cause gas leakage, and cause major safety accidents of casualties. For this, the faults are found and analyzed according to the actual situation of on-site vibration. The scheme of pipeline rebuild is put forward.

Key words: TRT; pipeline; support; sensor

电场处理对渣金反应元素迁移的影响研究

廖直友[1,2]，王博文[1]，王海川[1,2]，尹振兴[1,2]

（1. 安徽工业大学冶金工程学院，安徽马鞍山 243032；
2. 冶金减排与资源综合利用教育部重点实验室，安徽马鞍山 243002）

摘　要：通过在渣-金反应间添加脉冲电场，来使渣中的杂质元素进行定向迁移，进入铁水中，使热态钢渣得以回收利用，同时高磷硫的铁水也可以加以利用。实验发现增大脉冲电压能够显著提高磷和硫元素在渣金间的迁移率，同样，降低钢渣的黏度，也能够有利于磷和硫从渣相向金属相迁移；钢渣的黏度越小，即渣中 CaF_2 含量越高，磷和硫元素的迁移率越高，钢渣中磷和硫含量降低直至趋于稳定值。

关键词：渣金反应；脉冲电场；杂质元素；迁移

Effect of Electric Field Treatment on the Transport of Reactive Elements in Slag-metal Reaction

Liao Zhiyou[1,2], Wang Bowen[1], Wang Haichuan[1,2], Yin Zhenxing[1,2]

(1. School of Metallurgy Engineering, AHUT, Maanshan, 243032, China; 2. Key Laboratory of Metallurgical Emission Reduction & Resources Recycling (AHUT), Ministry of Education, Maanshan 243002, China)

Abstract: By adding pulsed electric field to the reaction between slag and metal, the impurity elements in slag can be directionally migrated into molten iron, so that the molten slag can be recycled, and high phosphorus and sulfur molten iron can also be used. Experiments show that increasing pulse voltage can significantly improve the mobility of phosphorus and sulfur between slags. Similarly, reducing the viscosity of steel slag can also facilitate the migration of phosphorus and sulfur from slag to molten iron. The lower the viscosity of steel slag, which means a higher content of CaF_2 in slag, the higher the mobility of phosphorus and sulfur. And the content of phosphorus and sulfur in steel slag would reduce to a stable value gradually.

Key words: slag-metal reaction; pulsed electric field; impurity elements; migration

钢铁工业能耗现状和节能潜力分析

王维兴

（中国金属学会，北京 100711）

摘　要：近年钢铁工业能耗有所下降，各企业之间发展不平衡，能耗水平差距较大；我国钢铁工业总体上与国际先进水平相比尚有 10%左右的节能潜力。行业节能工作的重点在炼铁系统，特别是高炉工序；要努力提高高炉入炉铁品位、热风温度、改善焦炭质量，提高操作水平等；二次能源回收利用水平进展不大，尚有一定的节能潜力。

关键词：钢铁能耗现状；节能潜力

The Status of Energy Consumption and Analysis of Energy Saving Potential in Iron and Steel Industry

Wang Weixing

(The Chinese Society for Metals, Beijing 100711, China)

Abstract: In recent years, the energy consumption of the steel industry has declined, the development between enterprises has been uneven and the gap in energy consumption has been raised. In general, there is still about 5% energy saving potential compared with the international advanced level. The iron making system is the focus of energy saving work in the steel industry, especially the blast furnace process. Efforts should be made to improve the operating level of the BF, We should increase the iron grade of the ore, increase the hot air temperature, and improve the quality of the furnace charge. Although the level of recycling of secondary energy has not progressed heavily, there is still a certain potential for energy conservation.

Key words: steel energy consumption; energy saving potential

转炉烟气中低温余热回收工艺研究

周 涛，侯祥松，谢 建，董茂林

（中冶赛迪技术研究中心有限公司，重庆 401122）

摘 要：通过对炼钢转炉烟气余热回收和利用的现状分析，结合转炉烟气产生的特点和目前相关技术应用，介绍一种转炉烟气中低温余热回收工艺的原理和特点，全面回收转炉炼钢的烟气余热并加以充分利用，实现节能降耗，进一步提高转炉负能炼钢水平。

关键词：转炉烟气；余热回收；节能降耗

Abstract: Based on the analysis of the status quo of waste heat recovery and utilization of steelmaking converter flue gas, combined with the characteristics of converter flue gas generation and the application of related technologies, this paper introduces the principle and characteristics of a low-temperature waste heat recovery process in converter flue gas. The residual heat of the flue gas of steelmaking is fully utilized to achieve energy saving and consumption reduction, and further improve the level of negative energy steelmaking in the converter.

Key words: converter flue gas; waste heat recovery; energy saving

8.2 冶金环保

承压一体化冶金污水净化装置在连铸浊环水系统中的应用实例

李建康，李 杰，梁思懿

(中冶京诚工程技术有限公司，北京 100176)

摘 要：本文介绍了某钢铁企业对其连铸浊环水供水系统进行升级改造的工程实例。采用"一次铁皮坑→承压一体化冶金污水净化装置→冷却塔"短流程作为核心工艺，取代了传统平流沉淀池、化学除油器、高效澄清器、稀土磁盘、过滤器等设施，具有节省投资、节约占地、节约能源等特点。通过实际运行表明，改造后系统出水 SS<20mg/L，矿物油<5mg/L，满足循环水回用要求，具有良好的示范效果和推广价值。

关键词：钢铁企业；浊环水；一体化；承压

Application of Pressure Type Integration Metallurgy Sewage Purification Treatment Device in Turbid Circulating Water System of Continuous Casting

Li Jiankang, Li Jie, Liang Siyi

(MCC Capital Engineering & Research Incorporation Limited, Beijing 100176, China)

Abstract: This paper introduces an engineering example of upgrading and renovation of turbid water supply system for continuous casting in an iron and steel enterprise. The short process of "primary Iron pit→pressure type integration metallurgy sewage purification treatment device→cooling tower" is adopted as the core process, which replaces the traditional advection sedimentation tank, chemical deoiler, high efficiency clarifier, rare earth magnetic disk, filter and other facilities. It has the characteristics of saving investment, occupying land and energy. The actual operation shows that the effluent SS<20mg/L and Oil<5mg/L of the reformed system can meet the requirements of recycling water reuse and have good demonstration effect and popularization value.

Key words: iron and steel enterprises; turbid circulating water; integration; pressure type

吹扫捕集/气相色谱-质谱法和气相色谱法同步测定水中挥发性有机物和挥发性石油烃

凌 冰，高巨鹏，黄 晓

(上海宝钢工业技术服务有限公司，上海 201900)

摘　要：本文建立了吹扫捕集-气相色谱／质谱法和气相色谱法同步检测水中挥发性有机物和挥发性石油烃的方法，优化了试验条件，该方法的回收率在86%~108%之间，挥发性石油烃的检出限为10.0 μg/L，15种挥发性有机物的检出限为0.1~0.3μg/L，均满足国家相关标准要求，通过对实际样品的测定表明，本方法能够满足水样分析的要求。

关键词：吹扫捕集；挥发性有机物；挥发性石油烃

Simultaneous Determination of Volatile Organic Compounds and Volatile Petroleum Hydrocarbons in Water by Purge and Trap/Gas Chromatography-mass Spectrometry and Gas Chromatography

Ling Bing, Gao Jupeng, Huang Xiao

(Shanghai Baosteel Industrial Technology Service Co., Ltd., Shanghai 201900, China)

Abstract: A method for simultaneous determination of volatile organic compounds and petroleum hydrocarbons in water by purge-trap/gas chromatography-mass spectrometry and gas chromatography was established. The recovery of this method was between 86% and 108%, the method detection limit of volatile petroleum hydrocarbons was 10.0 μg/L, and the method detection limit of 15 kinds of volatile organic compounds was between 0.1~0.3 μg/L. The results of actual samples determination shows the method can meet the demand of water analysis.

Key words: purge and trap; volatile organic compounds; volatile petroleum hydrocarbons

冶金除尘灰科学利用技术集成与应用

刘德军[1]，张延辉[3]，唐继忠[2]，袁　玲[1]

（1. 鞍钢集团钢铁研究院，辽宁鞍山　114009；2. 鞍钢股份有限公司鲅鱼圈炼铁部，辽宁营口　115007；3. 鞍钢股份有限公司炼铁总厂，辽宁鞍山　114021）

摘　要：通过对当前冶金除尘灰各种再利用方式的特点进行详细分析与甄别，阐明了"除尘灰与粉煤同喷"这种与"熔融还原"巧妙结合利用方式的科学性；明确了冶金粉尘再资源化科学利用的重要性；同时就冶金粉尘高炉综合喷吹技术对冶金粉尘的种类及量等因素的普适性和最佳性、含铁除尘灰等冷料下料量的工程级精准配料、高炉喷吹含铁除尘灰回旋区热补偿实现手段、消除喷吹除尘灰输送管道易堵塞技术以及整个工艺过程无二次扬尘绿色技术等进行了详细的论述；分析了冶金粉尘高炉综合喷吹技术在鞍钢鲅鱼圈炼铁部的应用状况。

关键词：冶金除尘灰；科学利用；再资源化；集成技术

Technology Integration and Application of Scientific Utilization of Metallurgical Dust

Liu Dejun[1], Zhang Yanhui[3], Tang Jizhong[2], Yuan Ling[1]

(1. Environment and Resource Institute, Iron & Steel Research Institutes of Ansteel Group Corporation, Anshan 114009, China; 2. Bayuquan Branch of Angang Steel Co., Ltd., Yingkou 115007,

Abstract: It's stated that it's important to utilize scientifically metallurgical dusts recycling. After analyzing and distinguishing features of different kinds of metallurgical dusts, we concluded that it's scientificity of the technology that injecting metallurgical dust and pulverized coal into burst furnace together. At the same time, it's expounded on 1) universality and optimality, about the technology to the kinds and amount of dusts injected into BF, are confirmed; 2) the accurately mixture in engineering of the materials such as dusts with iron injected into BF; 3)thermal compensation of dusts with iron in the blast furnace raceway; 4) avoidance of blocking the tunnel as injecting into the BF and 5) the green environment technology without second generate dust in the whole process. This paper introduces the influence of the application of the metallurgical dust blast furnace comprehensive injection technology to the BF in the of Ansteel Bayuquan Iron-making Department.

Key words: metallurgical dust; science utilization; recycling; integration technology

一种芽孢杆菌在降解焦化废水中苯酚的作用研究

王 永，袁 玲，陈 鹏，胡绍伟，王 飞，刘 芳

（鞍钢股份有限公司技术中心，辽宁鞍山 114009）

摘 要：以某焦化厂好氧池进水为研究对象，针对该出水苯酚指标较高的特点，向好氧池生化池中加入人工筛选出的菌剂，强化苯酚的去除效果，在不改变整个工艺的情况下提高COD降解能力。结果表明，添加菌剂以后，对比池中的苯酚浓度去除率提高了51.2%，COD去除率提高了40.2%，降低后续的处理焦化废水的费用。

关键词：芽孢杆菌；焦化废水；苯酚；降解

Study on the Role of a Bacillus in the Degradation of Phenol in Coking Wastewater

Wang Yong, Yuan Ling, Chen Peng, Hu Shaowei, Wang Fei, Liu Fang

(Technology Center of Angang Steel Co., Ltd., Anshan 114009, China)

Abstract: Taking the aerobic tank influent in a coking plant as research target, based on the characteristic of the high content of phenol in effluent, adding the selected bacteria agent, keep the whole process unchanged, the effect of removal phenol is intensified and biodegradation capacity of COD is improved. The results show that, after the adding of bacteria agent, in the compared tank, the removal rate of phenol and COD is improved by 51.2% and 40.2% respectively and the cost for coking wastewater treatment is reduced.

Key words: bacillus; coking wastewater; phenol; degradation

$321m^2$ 带式焙烧机头电除尘器的选型与实践

金勇成，李小丽

（鞍钢股份有限公司炼铁总厂，辽宁鞍山 114021）

摘　要： 鞍钢股份有限公司炼铁总厂一台321m²带式焙烧机，原配备2×130m²双室两电场电除尘，不能满足排放标准要求。2018 年带式焙烧机大修，机头电除尘同步进行升级改造。本文论述鞍钢股份炼铁总厂球团机头电除尘的选型及结构形式，运行后达到99%的除尘效率与最佳的性价比。

关键词： 机头电除尘；选型；运行效率

Type Selection and Practice of 321m² Belt - type Electric Dust Collector

Jin Yongcheng, Li Xiaoli

(General Iron Plant, Anshan Iron & Steel Company Limited, Anshan 114021, China)

Abstract: Angang Steel Company Limited ironmaking general factory has a 321m² belt calciner, which was originally equipped with 2×130m² double-chamber electric dust removal, which cannot meet the emission standards. In 2018, the belt calciner will be overhauled and the head electric dust removal will be upgraded simultaneously. This paper discusses the selection and structure of pelletizing head electric dust removal in angang iron smelting general plant.

Key words: head electric dust removal; selection; operation efficiency

焦炉烟囱二氧化硫排放浓度分析与治理

张其峰，张允东，周晓锋，孔　弢，李富鑫

（鞍钢股份有限公司炼焦总厂，辽宁鞍山　114000）

摘　要： 本文依据鞍钢股份炼焦总厂 9#、10#焦炉当前生产条件，对 9#、10#焦炉烟囱二氧化硫来源及排放浓度曲线进行分析，得出焦炉烟囱二氧化硫超标排放的原因。为此，鞍钢股份炼焦总厂采取多项整改措施，以保证 9#、10#焦炉烟囱二氧化硫排放浓度达到 GB 16171—2012《炼焦化学工业污染物排放标准》中表 5 规定的排放要求，对焦炉清洁生产、改善大气环境有重要意义。

关键词： 焦炉烟囱；二氧化硫；排放标准；整改措施

Analysis and Control of Sulfur Dioxide Emission from Coke Oven Chimney

Zhang Qifeng, Zhang Yundong, Zhou Xiaofeng, Kong Tao, Li Fuxin

(Angang Co., Ltd., Coking Plant, Anshan 114000, China)

Abstract: According to the current production conditions of 9# and 10# coking furnaces in Angang coking plant, the source and emission concentration curves of sulfur dioxide from 9# and 10# coke oven chimneys were analyzed, it is concluded that the cause of the coke oven chimney excess emissions of sulfur dioxide. To this end, Angang coking plant to take a number of rectification measures, in order to ensure that the sulfur dioxide emission concentration of 9# and 10# coke oven chimney reaches the emission requirements specified in table 5 of GB 16171—2012 "Emission standard of pollutants for coking chemical industry", it is of great significance to the clean production of coke oven and the improvement of atmospheric environment.

Key words: coke oven chimney; sulfur dioxide; emission standard; rectification measures

电吸附除盐技术用于梅钢再生水中试研究

邹正东，初　明，胡小龙

（梅山钢铁公司能源环保部，江苏南京　210039）

摘　要：采用电吸附除盐技术，对梅山钢铁再生水厂出水进行除盐试验，处理水量为 $0.5m^3/h$，回收率为85%，24h 不间断运行。结果表明，电吸附除盐工艺对氯离子、硫酸根、总硬度和总离子的平均去除率分别为82%、80%、93% 和81%，对 COD 也有一定的去除率，电耗为 $1.208kW·h/m^3$；该工艺自控程度高、运行操作简单、出水水质稳定、除盐效果明显且直接运行成本较低，用于梅钢再生水的深度处理是可行的。

关键词：钢铁企业废水；电吸附；除盐

Pilot Study on Desalination of Recycled Water by Electro-absorption Technique in Meigang

Zou Zhengdong, Chu Ming, Hu Xiaolong

(Energy and Environment Protection Department of Meishan Iron & Steel Co., Ltd., Nanjing 210039, China)

Abstract: Desalination was performed to effluent from waste water treatment plant of meigang by electro-absorption technique with treating capacity of $0.5 m^3/h$, The results showed that the average removal rate of chloride ions, sulfate, total hardness and total ions was 82%, 80%, 93% and 81%, respectively, and there was a certain removal rate of COD, and the power consumption is $1.208kW·h/m^3$. The process has high automation, simple operation, stable effluent quality, obvious desalination effect and low operation cost and it is feasible for advanced treatment of Steel companies reclaimed water.

Key words: iron and steel waste water; electric adsorption; desalination

冶金物料清洁高效转运新技术的研究与探讨

毕　琳，王得刚，徐培万

（中冶京诚工程技术有限公司，北京　100176）

摘　要：本文首先简要介绍了冶金原料场的散状物料输送系统在转运环节的几种主要结构形式和转运特点。其次，阐述了物料转运环节存在的粉尘、冲击、跑偏、堵料和撒料问题。通过问题分析对症入手，提出改进型清洁转运设备结构和特制型高效原料转运系统的技术措施，并重点介绍了该新技术研究的具体内容和技术特点。最后，进一步探讨了如何研究物料转运影响因素及开展优化设计，更好地实现绿色输送目标。

关键词：物料；输送；转运；设计；清洁；高效

Research and Discussion on the new Technology of Clean and Efficient Transport of Metallurgical Materials

Bi Lin, Wang Degang, Xu Peiwan

(Capital Engineering & Research Incorporation Limited, Beijing 100176, China)

Abstract: This paper first briefly introduces several main structural forms and transport characteristics of the loose material transportation system in the metallurgical raw material field. Secondly, it expounds the problems of dust, impact, run-off, blocking and sowing in the material transportation link. Through the problem analysis, the technical measures of improving the structure of the clean transport equipment and the special high-efficiency raw material transfer system are put forward, and the specific content and technical characteristics of the new technology research are highlighted. Finally, it further discusses how to study the factors of material transportation and carry out optimized design to better achieve the goal of green transportation.

Key words: material; convey; transfer; design; clean; high effective

含有氯化亚铁酸性废水出水超标的原因分析

胡金良

（河钢集团邯钢公司冷轧厂，河北邯郸 056015）

摘 要： 含有氯化亚铁的酸性废水处理过程中 SS、COD 超标的原因进行了分析，采取了提前提高 pH 值、废水多次循环、延长停留时间和提高溶解氧等措施，基本上解决了出水水质超标的难题。

关键词： 氧化；溶解氧；沉淀效果

Causal Analysis of Acidic Waste Water Containing Ferrous Chloride Beyond-standard Discharging

Hu Jinliang

(HBIS Group Handan Iron and Steel Cold Rolling Plant, Handan 056015, China)

Abstract: In the process of waste water treatment containing ferrous chloride, The discharge water quality (suspended solids, chemical oxygen demand) also is beyond standard. raising pH value of the waste water in advance and continue recycling the waste water and prolonging the hydraulic retention time and increasing dissolved oxygen, those measures are taken in the process of waste water treatment. so problems in beyond-standard discharge are be solved basically.

Key words: oxidize; dissolved oxygen; the precipitation effect

硅钢退火机组废气中氮氧化物含量的控制

汪剑飞,卢禹龙,范建军,毕 军,马允敏

(江苏沙钢集团冷轧硅钢车间,江苏张家港 215600)

摘 要:随着环保形势的日益严峻,国家生态环境部计划在2022年底前逐步将氮氧化物排放标准调整到150mg/Nm³以下。目前硅钢连续退火炉尾气排放中压力最大的是如何确保氮氧化物的达标排放;事实证明在不改变现有烧嘴结构的情况下暂时只能通过调整烧嘴空燃比,来确保烟气中的氮氧化物在300mg/Nm³以下;若需达到超低排放标准150mg/Nm³,则需进一步改变现有烧嘴的结构。

关键词:氮氧化物;焦炉煤气;烧嘴;空气;火焰温度;控制系统

Control of Nitrogen Oxide Content in Waste Gas of Annealing Unit of Silicon Steel

Wang Jianfei, Lu Yulong, Fan Jianjun, Bi Jun, Ma Yunmin

(Jiangsu Shagang Group Co., Ltd., Zhangjiagang 215600, China)

Abstract: With the increasingly serious environmental protection situation, the ministry of ecological environment of China plans to gradually adjust the emission standard of nitrogen oxides to below 150mg/Nm³ by the end of 2022. At present, the biggest pressure in exhaust gas emission of continuous annealing furnace for silicon steel is how to ensure the emission of nitrogen oxides up to the standard. It has been proved that the structure of the existing burner can only be changed temporarily by adjusting the air-fuel ratio of the burner to ensure that the nitrogen oxides in flue gas are below 300mg/Nm³ without changing the structure of the burner. If the exhaust gas emission standard of 150mg/Nm³ is to be reached, the structure of the existing burner needs to be further changed.

Key words: nitrogen oxide; coke ovengas; burner; air; flam temperature; control system

生态环境强约束下的京津冀钢铁节能环保产业发展战略研究

汪 浩[1],潘崇超[1,2],刘育松[2],张立明[1],王一帆[1],赵喜彬[2]

(1. 北京科技大学能源与环境工程学院,北京 100083;
2. 北京鼎鑫钢联科技协同创新研究院,北京 100083)

摘 要:京津冀区域生态环境随着一系列政策的发布实施,空气质量改善明显。钢铁行业在京津冀地区第二产业中起着支撑作用,同时随着"生态文明建设"的推进,京津冀钢铁节能环保产业面临前所未有的压力和机遇。为了推进经济新常态下钢铁节能环保产业的发展,推动钢铁企业结构优化和转型升级,本文采用PEST分析和SWOT分析模型从多个维度对京津冀钢铁节能环保产业发展进行了系统研究。研究结果表明钢铁节能环保产业发展应京津两地的技术优势结合企业的自身优势发展高效节能环保技术和装备，强化节能环保第三方治理；同时应发展城市钢厂的钢铁全流程能源利用和环保治理技术及方案，促进钢铁节能环保升级。

关键词：京津冀；钢铁节能环保；PEST 分析；SWOT 分析

Research on the Development Strategy of Beijing-Tianjin-Hebei Steel Energy-saving and Environmental Protection Industry under the Strong Ecological Environment

Wang Hao[1], Pan Chongchao[1,2], Liu Yusong[2], Zhang Liming[1],
Wang Yifan[1], Zhao Xibin[2]

(1. School of Energy and Environment Engineering, University of Science and Technology Beijing, Beijing 100083, China; 2. Dingxing Steel Union Collaborative Innovation of Science and Technology Research Institute Co., Ltd, Beijing 100083, China)

Abstract: The regional ecological environment of Beijing-Tianjin-Hebei has been improved with the release of a series of policies. The steel industry plays a supporting role in the secondary industry in the Beijing-Tianjin-Hebei region. At the same time, with the advancement of "ecological civilization construction", the Beijing-Tianjin-Hebei steel energy-saving and environmental protection industry faces unprecedented pressures and opportunities. In order to promote the development of steel energy-saving and environmental protection industry under the new economic normalization and promote the structural optimization and transformation and upgrading of iron and steel enterprises, this paper uses PEST analysis and SWOT analysis model to systematically study the development of Beijing-Tianjin-Hebei steel energy-saving and environmental protection industry from various dimensions. The research results show that the development of steel energy-saving and environmental protection industry should combine the technological advantages of Beijing and Tianjin with the advantages of enterprises to develop high-efficiency energy-saving technologies and equipment, and strengthen the third-party governance of energy conservation and environmental protection. At the same time, it should develop the whole process of steel utilization and environmental protection of urban steel mills. Governance technology and programs to promote steel energy conservation and environmental protection upgrades.

Key words: Beijing-Tianjin-Hebei; energy saving and environmental protection of iron and steel; PEST analysis; SWOT analysis

热电厂 5 号 6 号锅炉烟气在线监控系统矩阵流速仪改造的实践

卢 亮

（上海梅山钢铁股份有限公司，江苏南京 210039）

摘 要：本文重点介绍了钢铁行业锅炉废气烟道在线监控系统中的流量测量装置的安装工控条件、应用现状及存在问题，根据现状及问题分析流量测量不准确产生的原因及整改方案，通过分析矩阵流速仪的测量原理，介绍安装矩阵流速仪的改造方案，以及投用后取得的应用效果，并提出问题与建议。

钢铁企业 CO_2 排放计算方法探讨

王维兴

(中国金属学会,北京 100081)

摘 要:我们不能用国外 CO_2 排放的计算方法,要根据本企业的生产条件去计算。钢铁工业 CO_2 排放,重点在炼铁、烧结工序;降低生产过程中燃煤的消耗有降低 CO_2 排放的效果。炼铁要降低燃料比,烧结降低固体燃耗,可以降低 CO_2 排放。

关键词:CO_2;排放;计算方法

Discussion on the Calculation Method of CO_2 Emission in Iron and Steel Enterprises

Wang Weixing

(The Chinese Society for Metals, Beijing 100081, China)

Abstract: Emission of CO_2 is one of the important causes of climate warming. CO_2 emission from the iron and steel industry is account for 11% in China, especially on the ironmaking and sintering process. Reducing the consumption of coal during the production process has the effect of reducing CO_2 emission. In order to reduce CO_2 emission, the fuel ratio should be reducd in ironmaking process and the solid fuel consumption should also be reduced in the sintering process.

Key words: CO_2; emission; calculation method

9 钢铁材料表征与评价

大会特邀报告

分会场特邀报告

炼铁与原料

炼钢与连铸

轧制与热处理

表面与涂镀

金属材料深加工

先进钢铁材料

粉末冶金

能源、环保与资源利用

★ 钢铁材料表征与评价

冶金设备与工程技术

冶金自动化与智能化

建筑诊治

其他

基于大数据和 XGBoost 的热轧板带力学性能预测模型

陈金香,赵 峰,孙彦广,张 琳,尹一岚

(中国钢研冶金自动化研究设计院,混合流程工业自动化及装备技术国家重点实验室,北京 100081)

摘 要:针对现有热轧板带力学性能预测模型所存在的精准性不足、通用性和经济性弱、知识获取量不足、无联想推演和自学习能力等缺点,基于大数据分析与机器学习方法,提出了一种基于 XGBoost 的热轧板带力学性能预测算法,并以热轧带钢的力学性能预测作为研究背景,以某钢厂的 17710 组热轧带钢过程数据作为样本数据集,其中,90%的数据作为训练样本,10%的数据作为测试样本,对所提出模型进行了验证。结果表明该模型对抗拉强度、屈服强度和伸长率的预测精度(R^2 值)分别为 0.99895、0.99576、0.96260,均优于 BP 神经网络模型。应用 XGBoost 模型预测抗拉强度的绝对误差为 8MPa,相对误差为 0.02%;预测屈服强度的绝对误差为 18MPa,相对误差为 0.05%;预测伸长率的绝对误差为 4%,相对误差为 0.1%,远低于生产现场要求的误差范围。本研究为热轧板带力学性能的获知,提供一种"模型预测"代替"局部样本实验测试"的新思路。

关键词:热轧带钢;力学性能;工艺参数;智能预测;XGBoost;机器学习

A Prediction Model based on Big Data and XGBoost for the Mechanical Properties of Hot-rolled Strips

Chen Jinxiang, Zhao Feng, Sun Yanguang, Zhang Lin, Yin Yilan

(State Key Laboratory of Hybrid Process Industry Automation Systems and Equipment Technology, Automation Research and Design Institute of Metallurgical Industry, China Iron & Steel Research Institute Group, Beijing 100081, China)

Abstract: An XGBoost prediction model based on the data analysis and machine learning methods is presented to predict the mechanical properties of hot-rolled strips, which can solve the problems of the existing prediction methods, for example, the low precision, weak versatility and economy, insufficient knowledge information and low self-learning ability, etc. The provided model is verified by using the 17710 samples data of the chemical composition, processing parameters and mechanical properties of hot-rolled steel strips under the prediction for the mechanical properties and process parameters of hot rolled steel strips are looked as the application background. The results show that the proposed XGBoost model has high prediction accuracy, so that the prediction accuracies (R^2 value) of the tensile strength, yield strength and elongation are 0.99895, 0.99576, 0.96260, respectively, which are better than the results by using BP neural network. The obtained absolute error and relative error by using the XGBoost model to predict tensile strength are 8MPa and 0.02%, respectively. The absolute error and relative error to predict the yield strength are 18MPa and 0.05%, respectively. The absolute error and relative error to predict the elongation are 4% and 0.1%, respectively. These errors are lower than the target ranges required by production. This study can provide a new idea to obtain the mechanical properties of hot-rolled strips, which is the approach to replace "the experiment testing for local strips" with "model prediction".

Key words: hot rolled steel strips; mechanical properties; process parameters; intelligent prediction; XGBoost; machine learning

船运皮带自动取样设备选型及应用

罗 湘

（华菱湘钢技术质量部，湖南湘潭 411100）

摘 要：根据湘钢自身物流条件、硬件情况、人员安排等各方面因素，对船运皮带自动取样设备进行设计计算后选型。试样结果接近商检结果，试样代表性强，完全满足国家质检标准，确保质检公正和权威性。

关键词：自动取样；皮带运输机；取样器

38MnVS6 活塞销座孔内壁掉肉原因分析

李富强，林晏民，罗新中，章玉成，朱祥睿，杨明梅

（宝武集团广东韶关钢铁有限公司检测中心，广东韶关 512123）

摘 要：用户对 38MnVS6 活塞进行精车时，销座孔内壁出现掉肉现象，采用光学显微镜和扫描电镜等分析手段，对 38MnVS6 活塞销座孔内壁掉肉原因进行分析。分析结果表明：活塞销座孔内壁掉肉是由于连铸圆钢中存在气孔缺陷。

关键词：活塞；掉肉；气孔

Cause Analysis of Flesh out on the Inner Wall of 38MnVS6 Piston Pin Hole

Li Fuqiang, Lin Yanmin, Luo Xinzhong, Zhang Yucheng, Zhu Xiangrui, Yang Mingmei

(Baowu Steel Group Guangdong Shaoguan Iron and Steel Co., Ltd., Shaoguan 512123, China)

Abstract: When the user finishes the 38MnVS6 piston, there is flesh out on the inner wall of the Piston Pin Hole. The optical microscope and scanning electron microscope are used to analyze the cause of flesh out on the inner wall of the 38MnVS6 piston pin hole. The analysis results show that the flesh out on the inner wall of the piston pin hole is due to the existence of porosity defects in the continuous casting round steel.

Key words: pistons; flesh out; Stoma

二安替吡啉甲烷光度法测定钛铁中钛含量

邓军华，高 品，王 莹，刘冬杰，李 颖

（鞍钢集团钢铁研究院，辽宁鞍山 114009）

摘 要：本文建立了二安替吡啉甲烷分光光度法测定钛铁中钛含量的快速分析方法。采用稀硫酸和过硫酸铵溶解试样，以1.8mol/L盐酸为显色酸介质，利用二安替吡啉甲烷与四价钛能形成稳定1:3的黄色络合物[Ti(DAPM)$_3$]$^{4+}$的特性，通过测定被测元素的吸光度值，建立钛含量与吸光度之间的线性拟合，回归曲线方程为A=2.6618×w+0.0033，线性相关系数为0.9998。方法适用于钛铁中钛含量范围12%~50%的测定，标准样品测定结果与认定值一致，其相对标准偏差（RSD）小于0.45%，分析速度快，操作简单，值得推广。

关键词：钛铁；钛；二安替吡啉甲烷；分光光度法

Spectrophotometric Determination of Titanium Content in Titanium Ferroalloy with Diantipyrylmethane

Deng Junhua, Gao Pin, Wang Ying, Liu Dongjie, Li Ying

(Iron & Steel Research Institutes of Ansteel Group Corporation, Anshan 114009, China)

Abstract: Using dilute sulfuric acid and ammonium persulfate to dissolve the sample, with 1.8mol/L hydrochloric acid as the determination medium, the characteristic of diantipyrylmethane and tetravalent titanium can form stable yellow complex[Ti(DAPM)$_3$]$^{4+}$ of 1:3 was used, by increasing the absorbance value to reduce the measurement error, the method for determination of titanium content in titanium iron with diantipyrylmethane photometric method was established. A linear fit between the titanium content and the absorbance was established by measuring the absorbance value of the measured element. The regression curve equation was A=2.6618×w+0.0033, and the linear correlation coefficient was 0.9998. The method is suitable for the determination of titanium content in titanium iron from 12% to 50%. The determination result of standard sample is consistent with the certified value. The relative standard deviation (RSD) is less than 1.0%, the analysis speed is fast, the operation is simple, and it is worthy of promotion.

Key words: titanium ferroalloy; titanium; diantipyrylmethane; spectrophotometry

SWRCH22A冷镦钢线材拉拔断裂原因分析

李富强，罗新中，孙福猛，杨明梅，章玉成，朱祥睿

（宝武集团广东韶关钢铁有限公司检测中心，广东韶关 512123）

摘 要：针对SWRCH22A线材拉拔过程中出现断裂的质量问题，采用光学显微镜和扫描电镜等分析手段，对拉拔钢丝断裂原因进行分析。分析结果表明：SWRCH22A线材拉拔断裂的原因有拉拔润滑不良、表面增碳、心部大尺

寸不变形非金属夹杂物、组织异常等。

关键词：SWRCH22A；拉拔断裂；润滑不良；非金属夹杂物；组织异常

Analysis on Causes of Fracture Occurred in Drawing of SWRCH22A Cold Heading Wire Rod

Li Fuqiang, Luo Xinzhong, Sun Fumeng, Yang Mingmei,
Zhang Yucheng, Zhu Xiangrui

(Baowu Steel Group Guangdong Shaoguan Iron and Steel Co., Ltd., Shaoguan 512123, China)

Abstract: In view of the quality problems that there is fracture during the SWRCH22A wire drawing, optical microscopy and scanning electron microscopy were used to analyze the causes of fracture during drawing. The analysis results show that the causes of fracture during SWRCH22A wire drawing fracture are poor drawing lubrication, surface carbonization, non-metallic inclusions of large size in the heart and tissue abnormalities.

Key words: SWRCH22A; drawing fracture; poor lubrication; non-metallic inclusions; tissue abnormalities

调质处理对油井管的微观组织和耐蚀性能的影响

钟　彬[1,2]，陈义庆[1,2]，高　鹏[1,2]，李　琳[1,2]，
艾芳芳[1,2]，伞宏宇[1,2]，肖　宇[1,2]

（1. 海洋装备用金属材料及其应用国家重点实验室，辽宁鞍山　114009；
2. 鞍钢集团钢铁研究院，辽宁鞍山　114009）

摘　要：利用 OM、TEM 及电化学等分析手段研究调质处理对油井管的微观组织、析出相、力学性能和耐蚀性能的影响关系。结果表明：调质处理后油井管的金相组织为回火索氏体，随着调质处理次数的增加，晶粒和析出相尺寸都是先变小后长大，强度和耐蚀性能先升高后降低，经过两次热处理的油井管晶粒细化，微观组织均匀，析出相细小弥散，使得油井管的屈服强度和抗拉强度最高，极化电阻最大，耐蚀性能最好。

关键词：油井管；微观组织；析出相；耐蚀性能

Effects of Quenching and Tempering Treatment on Microstructure and Corrosion Resistance of Oil Well Pipes

Zhong Bin[1,2], Chen Yiqing[1,2], Gao Peng[1,2], Li Lin[1,2],
Ai Fangfang[1,2], San Hongyu[1,2], Xiao Yu[1,2]

(1. State Key Laboratory of Metal Material for Marine Equipment and Application, Anshan 114009, China;
2. Anshan Iron & Steel Institute, Anshan 114009, China)

Abstract: The effects of quenching and tempering treatment on microstructure, precipitated phase, mechanical properties and corrosion resistance of oil well pipes were studied by OM, TEM and electrochemical methods. The results show that the metallographic structure of the quenched and tempered oil well pipes were tempered sorbite. With the increase of times of quenching and tempering treatment, the size of crystalline grain and precipitated phase reduced firstly and then increased, the strength and corrosion resistance increased firstly and then decreased. After being treated with quenching and tempering treatment twice, the crystalline grain of oil well pipes became fine, the microstructure was homogeneous, precipitated phase was fine and dispersed, which made the oil well pipe have the highest yield strength and tensile strength, the largest polarization resistance and the best corrosion resistance.

Key words: oil well pipe; microstructure; precipitated phase; corrosion resistance

冷轧薄钢板 DC04 表面波纹度测试及分析

郭 晶，孟庆刚，王鲲鹏

（本钢板材股份有限公司技术研究院，辽宁本溪 117000）

摘 要： 利用白光干涉的三维光学轮廓仪，测试分析冷轧薄钢板 DC04 表面波纹度。依据相关标准，确定表面波纹度参数 Wsa_{1-5} 及 $Wa_{0.8}$ 的评估方法，测量冷轧薄钢板 DC04，厚度为 0.65 mm 的表面波纹度，并对不同变形程度的冷轧薄钢板进行表面波纹度参数 $Wa_{0.8}$ 和 Wsa_{1-5} 的测试。整个测试过程中，随着变形程度 0～10%增加，变形程度与表面波纹度参数 Wsa_{1-5} 和 $Wa_{0.8}$ 存在正相关性。但是变形程度对两个波纹参数的影响程度不同：较小变形程度 0～5%时，变形程度对表面波纹度参数 Wsa_{1-5} 影响较大，而较大变形程度 5%～10%时，变形程度对表面波纹度参数 $Wa_{0.8}$ 影响较大。

关键词： 白光干涉；表面轮廓；表面波纹度；冷轧薄钢板；变形程度；$Wa_{0.8}$；Wsa_{1-5}

Measurement and Analysis of Surface Waviness of DC04 Cold Rolled Sheet Steel

Guo Jing, Meng Qinggang, Wang Kunpeng

(Technology Research Institute of Benxi Steel Plate Co., Ltd., Benxi 117000, China)

Abstract: To test and analyze the surface Waviness of DC04 cold rolled sheet steel, the evaluation methods of surface Waviness parameters Wsa_{1-5} and $Wa_{0.8}$ were determined. By using a white light interferometer 3D optical profiler, the surface Waviness of DC04 cold rolled steel sheet with a thickness of 0.65 mm was measured. During the whole testing process, with the increase of deformation degree from 0 to 10%, the deformation degree is positively correlated with surface Waviness parameters Wsa_{1-5} and $Wa_{0.8}$. However, the degree of deformation has a different influence on the two waviness parameters. When the degree of deformation is 0~5%, the degree of deformation has a greater influence on the surface Waviness parameter Wsa_{1-5}, while when the degree of deformation is 5%~10%, the degree of deformation has a greater influence on the surface waviness parameter $Wa_{0.8}$.

Key words: white light interference; surface profile; surface waviness; cold rolled sheet steel; degree of deformation; $Wa_{0.8}$; Wsa_{1-5}

激光切割技术在制备钢铁中氮元素含量分析样品中的应用

王 鹏，刘步婷，杨 琳，刘 俊

（武钢有限质量检验中心，湖北武汉 430083）

摘 要：将激光切割技术应用于钢铁中氮元素含量分析的试样加工过程中。分别使用激光切割法与传统的冲孔法加工钢铁样品，并比对加工效果的属性，发现激光切割法在一次加工合格率，试样形貌，加工钢种的通用性方面均优于传统的冲孔法。用激光切割法加工出的试样，其氮元素含量分析结果的准确度和精密度均能够满足生产需要。因此，激光切割法能够用于制备钢铁中氮元素含量分析的试样。

关键词：激光切割；钢中氮元素含量；试样加工；冲孔；准确度；精密度

Application of Laser Cutting Technology in Preparing Nitrogen Content Analysis Samples in Steel

Wang Peng, Liu Buting, Yang Lin, Liu Jun

(Quality Inspection Center of WISCO Limited Company, Wuhan 430083, China)

Abstract: Laser cutting technology was applied to sample processing of nitrogen content analysis in steel. Laser cutting method and traditional punching method are used to process steel samples, and the attributes of the processing effect are compared. It is found that laser cutting method is superior to traditional punching method in one-time processing qualification rate, sample morphology and generality. The accuracy and precision of the nitrogen content analysis results of the samples processed by laser cutting method can meet the production needs. Therefore, laser cutting method can be used to prepare samples for nitrogen content analysis in steel.

Key words: laser cutting; nitrogen content in steel; sample preparation; punching; accuracy; precision

40Cr 连接螺栓断裂失效分析

王德宝，杨 峥，牟祖茂，徐 辉，徐 雁，宋鑫晶

（马鞍山钢铁股份公司技术中心，安徽马鞍山 243000）

摘 要：针对 40Cr 连接螺栓在使用过程中发生断裂事故，采用化学分析、金相、硬度、扫描电镜等实验手段对失效件进行了检验分析。结果表明：螺栓群在服役过程中受力不均匀是造成螺栓断裂的主要原因，同时显微组织中上贝氏体组织以及 B 类非金属夹杂物也促进了螺栓的断裂。在以上分析的基础上提出了相应的改进措施和建议。

关键词：连接螺栓；受力不均匀；疲劳断裂；失效分析

Failure Analysis of 40Cr Steel Connecting Bolt

Wang Debao, Yang Zheng, Mou Zumao, Xu Hui, Xu Yan, Song Xinjing

(Technology Center of Maanshan Iron and Steel Co., Ltd., Maanshan 243000, China)

Abstract: Be aimed at the crack on the 40Cr bolt in the using process, the failed bolt were analysis by means of chemical composition, hardness, metallographic and SEM examination. The results showed that the reason 40Cr bolts breakage were the uneven load of the bolts during service, At the same time, the upper bainite structure and the Non-Metallic Inclusion inclusions in the microstructure also promote the bolt breakage. Based on the above analysis, corresponding improvement measures and suggestions were proposed.

Key words: connecting bolt; uneven load; fracture; failure analysis

冷轧稀碱含油废水中挥发性有机污染物特征研究

洪 涛，何晓蕾，张 毅，李恩超

（宝山钢铁股份有限公司中央研究院，上海 201900）

摘 要： 研究了冷轧稀碱含油废水中 VOCs 的组成特征，考察了生化处理工艺对 VOCs 的处理能力，并利用气质联用法定量分析了其中卤代烃和苯系物的含量。结果表明生化处理前共检出 VOCs 39 种，包含的种类是卤代烃、苯系物、氯苯类、多环芳烃、酚类、胺类、含硫化合物、杂环化合物。生化处理后 VOCs 种类减少至 22 种，与处理前相比减少了接近一半，其中胺类被完全去除，其余化合物都有不同程度的减少。定量分析结果表明苯系物的去除率在 82%～90%之间。部分氯代烃被完全去除但也有部分氯代烃含量升高，同时有新生成的溴代烃，说明生化处理过程既消减了一些卤代烃，同时也产生了一些新的污染物。

关键词： 稀碱含油废水；挥发性有机污染物；特征；种类

Study on Characteristics of VOCs in Cold Mill Dilute Alkali Oily Wastewater

Hong Tao, He Xiaolei, Zhang Yi, Li Enchao

(Central Research Institute, Baoshan Iron & Steel Co., Ltd., Shanghai 201900, China)

Abstract: The composition characteristics of VOCs in cold mill dilute alkali oily wastewater were studied. The treatment capacity of VOCs in biochemical treatment process was investigated. The content of halogenated hydrocarbons and benzene series were quantitatively analyzed by GC-MS. The results showed that 39 kinds of VOCs were detected before biochemical treatment, including halohydrocarbons, benzene series, chlorobenzene, polycyclic aromatic hydrocarbons, phenols, amines, sulfur compounds and heterocyclic compounds. After biochemical treatment, the number of VOCs decreased to 22, which was reduced by nearly half. Amines were completely removed, and other compounds were reduced to varying degrees. Quantitative analysis showed that the removal rate of benzene homologues ranged from 82% to 90%.

Some chlorinated hydrocarbons were completely removed, but some are elevated, and there are newly formed brominated hydrocarbons, indicating that the biochemical treatment process not only reduces some halohydrocarbons, but also produces some new pollutants.

Key words: dilute alkali oily wastewate; volatile organic compounds; characteristic; species

波长色散 X 射线荧光光谱法测定渣类化学成分

刘青桥，郭登元

（湖南华菱湘潭钢铁有限公司技术质量部，湖南湘潭　411100）

摘　要：采用渣类标准样品按常规称量，灼烧氧化，熔融法制样，X 射线荧光光谱法测量，制作校准曲线，可直接测定保护渣、精炼渣等渣类中的 CaO、MgO、SiO_2 等 7 种化学成分。本方法准确、快速，完全可满足常规检测生产的需求。

关键词：预氧化；熔融法；X 射线荧光光谱法；渣类

原子吸收分光光度计火焰法测铅检出限结果不确定度评定

张启发，吴　琨

（宝武集团八钢公司制造管理部理化检验中心，乌鲁木齐　830022）

摘　要：通过原子吸收分光光度计火焰法对铅检出限的测量过程，分析了不确定度的来源，并根据实际数据计算得到了各标准不确定度分量。

关键词：原子吸收分光光度计；火焰法；铅；检出限；不确定度

Evaluation of Uncertainty in Results of Lead Detection Limit by Flame Atomic Absorption Spectrophotometer

Zhang Qifa, Wu Kun

(Baowu Group Ba Steel Company Manufacturing Management Department Physical and Chemical Testing Center, Urumchi 830022, China)

Abstract: Through the measurement process of lead detection limit by flame method of atomic absorption spectrophotometer, the sources of uncertainty are analyzed, and the uncertainty components of each standard are calculated according to actual data.

Key words: atomic absorption spectrophotometer; flame method; lead; the detection limit; uncertainty

汽运铁精粉检验频次优化的探索实践

王 浩，吴 琨，张启发

(宝武集团八钢公司制造管理部理化检验中心，新疆乌鲁木齐 830022)

摘 要：汽运铁精粉车车检测不仅增加检化验工作强度，有时还会造成过度检测。为此，将根据以往数据和判级依据运用统计学数学模型来探索优化铁精粉检验频次，达到提升卸车效率、节省人力、有效检测和有效防范风险的目的。

关键词：铁精粉；检验频次；优化；质量

Exploration and Practice of Frequency Optimization of Steam Transport Iron Powder Inspection

Wang Hao, Wu Kun, Zhang Qifa

(Baowu Group Basteel Company Manufacturing Management Department Physical and Chemical Testing Center, Urumchi 830022, China)

Abstract: The testing of the truck is not only to increase the strength of testing and testing, but also cause excessive testing. Therefore, the statistical mathematical model will be applied to optimize the inspection frequency of iron fine powder according to the previous data and grade basis, so as to improve the unloading efficiency, save manpower, effectively detect and effectively prevent risks.

Key words: iron powder; inspection frequency; optimization; quality

焊接工艺对耐候钢焊接接头疲劳性能的影响

邱 宇，蔡 宁，杨永达，王泽阳，鞠新华

(首钢集团有限公司技术研究院，北京 100041)

摘 要：为了选取合适的焊接工艺，从而获得具有最优抗疲劳性能的焊接接头，结构设计人员需要对不同焊接工艺下的焊接接头进行疲劳性能评估。本文采用应力比为 0 的拉向高周疲劳试验，对 30kJ/cm 埋弧焊 1、30kJ/cm 埋弧焊 2、20kJ/cm 气保焊接以及 20kJ/cm 气保大间隙焊接的焊接接头的有限寿命区 S-N 曲线和无限寿命区的疲劳强度进行了分析，发现采用 20kJ/cm 气保焊接工艺的焊接接头的抗疲劳性能最好，采用 30kJ/cm 埋弧焊 2 焊接工艺的焊接接头的抗疲劳性能最差。

关键词：耐候钢；焊接工艺；焊接接头；疲劳性能

Influence of Welding Processes on the Fatigue Properties of Welding Joints for Corten Steel

Qiu Yu, Cai Ning, Yang Yongda, Wang Zeyang, Ju Xinhua

(Shougang Group Co., Ltd., Research Institute of Technology, Beijing 100041, China)

Abstract: To obtain the optimum fatigue properties of welding joint with an appropriate welding process, the mechanical designers need to evaluate fatigue properties of joints welded with different welding processes. In this paper, the pulling high-cycle fatigue (HCF) tests were carried out to analyze the S-N curves in the finite life zone and the fatigue strengths in the infinite life zone for the welding joints. In the HCF tests, the stress ratios were 0. Welding joints were welded with four different welding processes including the submerged arc welding 1 (SAW1) with 30kJ/cm, the SAW2 with 30kJ/cm, the gas shielded arc welding with 20kJ/cm and the gas shielded wide gap arc welding with 20kJ/cm. It is found that the welding joint with the gas shielded arc welding with 20kJ/cm has the optimum fatigue properties, while that with the SAW2 with 30kJ/cm has the worst fatigue properties.

Key words: corten steel; welding process; welding joints; fatigue properties

电阻应变片在高速拉伸试验中的应用

鹿宪宝，邱 宇，孙 博，张清水，鞠新华

（首钢集团有限公司技术研究院，北京 100041）

摘 要： 为了更好地获得金属材料在高应变速率下的拉伸性能，模拟其在服役中真实应变速率下的变形行为，需要进行高速拉伸试验。本文比较了高速拉伸试验机压电传感器和电阻应变片所测得的力值曲线，结果表明在低应变速率下，通过高速拉伸试验机压电传感器所测得的力值曲线足够光滑，无需采用电阻应变片测量力值曲线；在高应变速率下，通过电阻应变片测量力值曲线的方法可以有效地减小通过高速拉伸试验机压电传感器所测得的力值曲线的波动程度。

关键词： 应变片；高速拉伸；力值曲线；波动

Application of Resistance Strain Gage in the High Speed Tensile Test

Lu Xianbao, Qiu Yu, Sun Bo, Zhang Qingshui, Ju Xinhua

(Shougang Group Co., Ltd., Research Institute of Technology, Beijing 100041, China)

Abstract: To obtain the tensile properties of metal materials at high strain speed, simulating their deformation behaviors at true strain rates under the service. In this paper, the comparison between the results of the force curves measured with the resistance strain gage and the piezoelectric sensor of the high speed tensile machine was carried out, the results show that at the low strain rate, the force curve obtained with piezoelectric sensors of the high speed tensile machine is smooth enough that there is no need to choose the resistance strain gage to measure the force, but at high strain rate, adopting the resistance strain gage to measure the force can help to reduce the fluctuation range of the force curves measured with the piezoelectric sensor of the high speed tensile machine.

Key words: resistance strain gage; high speed tensile test; force curve; fluctuation

河钢石钢钢材宏观纯净度检测技术

王殿峰，周立波，丁 辉，李双居

（河钢集团石家庄钢铁有限责任公司技术中心，河北石家庄 050031）

摘 要：水浸高频超声检测设备可以检测和评估产品宏观夹杂物，由于检测速度快、检测体积大，所以具有快速、准确评估产品纯净度的能力，检测结果可直观反映钢中缺陷的形状、位置和分布，并可通过计算量化纯净度指数，进而得到一段时间内纯净度变化的趋势，指导炼钢工艺完善和技术进步。河钢石钢利用水浸高频超声设备对产品进行了长期的跟踪检测和总结分析，为河钢石钢产品迈向高端架起了一座桥梁。本文通过对河钢石钢高端钢纯净度检测分析，指出影响产品纯净度的主要夹杂物类型，为进一步改善部分产品的纯净度，提出了工艺改进措施并验证实施效果。同时探讨通过高频超声设备检测微观夹杂或较小宏观夹杂物的可能性。

关键词：纯净度；水浸高频超声；评估；纯净度指数；趋势

Detection Technology of Steel Macroscopic Cleanness Rate

Wang Dianfeng, Zhou Libo, Ding Hui, Li Shuangju

(Hesteel Group Shijiazhuang Iron and Steel Co., Ltd., Technology Center, Shijiazhuang 050031, China)

Abstract: Water immersion high-frequency ultrasonic testing machine can test and evaluate the macroscopic inclusion, the testing speed is fast, testing volume is large, the testing have the fast and accurate ability to evaluate the product's cleanness, testing results will directly show the defect profile, position and distribution, and after calculation will get the cleanness index, collecting some period cleanness index data will get the cleanness change trend, this help to improve the steel making process and technical progress. Hesteel Shisteel use this testing machine made the long terms trace test, summary and analysis, build up a bridge for Hesteel Shisteel products go to high level. In this thesis, we use the cleanness index data, along with the inclusion type, give the instruction steel making process to promoting some product's cleanness and verify its results, meanwhile make a discussion for the possibility to use this testing machine to check the microscopic inclusion or the small macroscopic inclusion.

Key words: cleanness; high-frequency ultrasonic; evaluation; cleanness index; trend

汽车外板冲压后表面轮廓变化对2C1B涂装质量的影响

胡燕慧[1]，张 浩[1]，刘李斌[1]，滕华湘[1]，李蔚然[2]

（1. 首钢集团有限公司技术研究院薄板所，北京 100043；
2. 北京首钢股份有限公司营销中心汽车板服务室，北京 100043）

摘 要：通过研究汽车外板冲压后表面轮廓变化和表面轮廓参数对2C1B涂装质量的影响，探讨从钢板表面质量控制方面改善 2C1B 涂装质量的可能性。对 IF 钢连退汽车外板进行冲压实验，分析冲压过程中表面轮廓的变化；对冲压后的零件进行 2C1B 涂装实验，分析表面轮廓在涂装过程中的遗传规律，并分析冲压后表面轮廓参数对涂装后表面质量的影响。结果表明：冲压后钢板表面粗糙度 R_a 有所提高，峰值数 R_{Pc} 稍有下降，波纹度 W_{sa} 有明显提高。电泳后，R_a 和 R_{Pc} 大幅衰减，W_{sa} 衰减不明显。R_a 和 R_{Pc} 对桔皮和鲜映性 DOI 未见明显影响，随着 W_{sa} 提高，涂装后长波数提高，DOI 下降。冲压过程中 W_{sa} 提高是表面塑形变形不均匀引起的，为了得到良好的涂装后表面质量，应控制冲压后表面波纹度 W_{sa} 在 0.4μm 以下。

关键词：汽车外板；IF 钢；2C1B 涂装工艺；波纹度；桔皮；鲜映性

Effect of Surface Profile Change during Forming Process of Automobile Outer Panel on Painting Quality for 2C1B Coating Process

Hu Yanhui[1], Zhang Hao[1], Liu Libin[1], Teng Huaxiang[1], Li Weiran[2]

(1. Strip and Sheet Department, Shougang Group Co., Ltd., Research Institute of Technology, Beijing 100043,China; 2. Automotive Steel Service Department, Beijing Shougang Co., Ltd., Sales Company, Beijing 100043, China)

Abstract: The automotive 2C1B coating process reduces the coating process, the thickness of the paint film and the hiding effecton the automotive steel sheets, so that the surface quality of the automobile outer panels has more significant effect on the painting quality. The study on the surface profile change and the influence of surface profile parameters on the 2C1B painting quality was carried out and the ways to improve painting quality were discussed. The stamping forming tests on continuous annealing cold-rolled interstitial-free steel sheets and the 2C1B coating experiments were carried out to study the change of surface profile during the forming process. Then the heritability of the surface profile during the coating processand the effect of profile parameters after forming on the surface quality were analyzed. The results show that the roughness average R_a increased, the peak counts R_{Pc} decreased slightly, and the waviness W_{sa} increased significantly after stamping. After electrophoresis process, R_a and R_{Pc} greatly decreased, and W_{sa} decreased slightly. R_a and R_{Pc} had no significant effect on orange peel and distinctness of Image(DOI). With the increase of W_{sa}, the long wave numbers increased and the DOI decreased after painting. The increase of W_{sa} during forming is caused by inhomogeneous surface deformation. In order to improve the painting quality, the surface waviness W_{sa} after forming should be controlled to be less than 0.4μm.

Key words: automobile outer panel; interstitial-free steel; 2C1B coating process; waviness; orange peel; distinctness of image

钢中铌原子晶界偏聚的研究现状及发展

吕俊南，魏　然，王红鸿

（武汉科技大学国际钢铁研究院，湖北武汉　430081）

摘　要：铌微合金化钢是先进钢铁材料研发的重要方向。铌不仅是强碳化物形成元素，而且在钢中具有强烈的晶界偏聚倾向，该特性对组织性能控制有重要作用。本文介绍了钢中 Nb 的晶界偏聚对相变及组织控制的影响，综述了钢中 Nb 晶界偏聚的实验及理论研究，根据晶界偏聚理论及 Nb 晶界偏聚特点提出了钢中 Nb 晶界偏聚研究中亟待解决的问题以及介绍了作者的部分探索工作。

关键词：铌微合金化钢；晶界偏聚；动力学；溶质-空位复合体

Research of Niobium Segregation at Grain Boundary in Steel

Lv Junnan, Wei Ran, Wang Honghong

(Wuhan University of Science and Technology, Wuhan 430081, China)

Abstract: Niobium atom has a strong grain boundary segregation tendency in steel which plays an important role in controlling phase transformation, microstructure and mechanical properties. Nb segregation at grain boundary influences phase transformation in steel. The research work about Nb segregation at grain boundary in the steel are reviewed experimentally and theoretically. The key points of Nb segregation on which should be focused are described based upon theory of grain boundary and the part of the exploration work are also introduced.

Key words: Nb micro-alloyed steel; grain boundary segregation; kinectics; solute-vacancy comple

试样加工对冷轧汽车板拉伸性能检验的影响

张宏岭，叶 筱，金 鑫，甘 霖，秦 胜，江亚平

（武钢有限质检中心，湖北武汉 430083）

摘 要： 本文从试样制取和试样制备两大方面分析了试样加工对冷轧汽车板拉伸性能的影响。结果表明：取样方向、取样位置、加工方式、加工精度等不同程度地影响了冷轧汽车板的拉伸性能。在实际生产检验中，应正确认识这些因素对拉伸试验结果的影响，实验前确保选择正确的取样方向及取样部位以加工成符合标准规定的试样。此外，制取冷轧汽车板拉伸试样时还应避免产生加工硬化，并尽量提高加工精度。

关键词： 冷轧汽车板；试样加工；拉伸性能

Influence of Sample Processing on Tensile Property Test of Cold Rolled Automobile Plate

Zhang Hongling, Ye Xiao, Jin Xin, Gan Lin, Qin Sheng, Jiang Yaping

(Quality Inspect Center of Baosteel, Wuhan 430083, China)

Abstract: In this paper, the influence of sample processing on tensile properties of cold-rolled automobile plate was analyzed from two aspects: sample manufacture and sample preparation. The results show that the tensile properties of cold rolled automobile plate was affected by sampling direction, sampling position, machining mode and machining accuracy. In the actual production inspection, the influences of these factors on the tensile test results should be correctly understood. Before the experiment, the correct sampling direction and sampling position should be selected to process the samples in line with the standards. In addition, the working hardening should be avoided and the machining accuracy should be improved.

Key words: cold rolling utomobile; plate sample processing; tensile properties

铝渣球金属铝化学分析方法探讨

张启发，吴 琨

（宝武集团八钢公司制造管理部理化检验中心，乌鲁木齐 830022）

摘　要：实验探讨铝渣球金属铝测定方法。设计实验方案，通过正交实验法，分别采用氧化还原法和碱融法计算回收率。实验数据显示碱融法回收率更高，结果更可靠。

关键词：金属铝；正交试验；回收率

Discussion on Chemical Analysis Method of Aluminum Slag Ball Metal Aluminum

Zhang Qifa, Wu Kun

(Baowu Group Ba Steel Company Manufacturing Management Department Physical and Chemical Testing Center, Urumchi, 830022, China)

Abstract: The experimental method of aluminum aluminum slag metal aluminum determination was designed.the experimental scheme was designed. The recovery rate was calculated by the orthogonal experiment method using redox method and alkali fusion method. the experimental data show that the alkali-melting method has higher recovery rate and the result is more reliable.

Key words: solid aluminum; orthogonal test; recovery rate

比表面积分析仪校准方法的探讨

毕经亮

（钢研纳克检测技术股份有限公司，北京 100081）

摘　要：鉴于目前国家没有颁布比表面积分析仪相关的计量技术规范，而在实验室活动中此类设备又有较多的计量需求，本文探讨了利用多种类比表面积标准物质整机校准的方法，给出了相关的计量性能要求，校准项目等，并评定了校准结果的不确定度，为仪器测量方法的验证、测量结果的比对及相关计量技术规范的制定等提供了参考。

关键词：比表面积；示值误差；重复性；不确定度评定

Discussion on Calibration Method of Specific Surface Integral Analyses

Bi Jingliang

(NCS Testing Technology Co., Ltd., Beijing 100081, China)

Abstract: In view of the fact that the country has not promulgated the technical specifications of specific surface area analyzer and there are many measuring requirements for this kind of equipment in laboratory activities, this paper discusses the calibration method of the whole machine using various analogous surface area reference materials, gives the relevant measurement performance requirements, calibration items, and evaluates the uncertainty of the calibration results, so as to measure the instrument. The validation of the method, the comparison of measurement results and the formulation of relevant measurement technical specifications provide a reference.

Key words: specific surface area; indicative error; repeatability; uncertainty evaluation

电感耦合等离子体原子发射光谱测定磷化液中3种主要元素

叶筱，魏静，熊立波，张宏岭

（武钢有限质检中心，湖北武汉 430080）

摘 要：目前，磷化板广泛应用含镍、锰的三元锌系磷化工艺前处理。为了满足对磷化液中 Zn、Ni、Mn 化学成分的准确测定要求，建立了以电感耦合等离子体原子发射光谱法(ICP-AES)测定磷化液中 Zn、Ni、Mn 含量的方法。试样采用盐酸溶解，选择 Mn 259.373nm、Ni 231.604nm、Zn 202.548nm 作为分析线，以校准溶液制作校准曲线，在功率为 1150 W、雾化器压力为 0.7L/min、辅助气流量为 0.5L/min 和泵速为 50r/min 的条件下测定，测定结果的相对标准偏差≤1.29%，加标回收率在 100.67%～124.67%之间。

关键词：电感耦合等离子体原子发射光谱法（ICP-AES）；磷化液；Mn；Ni；Zn

Determination of Three Main Elements in Phosphating Liquid by Inductively Coupled Plasma Atomic Emission Spectrometry

Ye Xiao, Wei Jing, Xiong Libo, Zhang Hongling

(Quality Inspect Center of Baosteel, Wuhan 430080, China)

Abstract: At present, the three-way zinc phosphating process containing nickel and manganese is widely used in the phosphating steel. In order to meet the requirements for the accurate determination of Zn, Ni and Mn in phosphating liquid, an ICP-AES method for the determination of Zn, Ni and Mn in phosphating liquid has been established. The samples were dissolved by hydrochloric acid, and Mn 259.373nm, Ni 231.604nm and Zn 202.548nm were selected as analysis lines. The calibration curve was prepared with the calibration solution. The power was 1150 W, the atomizer pressure was 0.7L/min,

the auxiliary gas flow was 0.5L/min and the pump speed was 50r/min, the relative standard deviation of the measured results was ≤1.29%, and the standard recovery rate was between 100.67% and 124.67%.
Key words: inductively coupled plasma atomic emission spectrometry (ICP-AES); phosphating liquld, Mn; Ni; Zn

对于低合金钢化学成分的不同测定方法比对

孙璧瑶[1]，高宏斌[2]，王 蓬[2]，冯浩州[1]，卢毓华[1]，蔺 菲[2]

（1. 钢铁研究总院，北京 100081；2. 钢研纳克检测技术股份有限公司，北京 100081）

摘 要： 本文针对低合金钢中化学成分的测定，通过比较不同测定方法的理论与实际应用部分，得出不同方法在使用时有不同的侧重点。在理论部分，比较了测定方法的测量范围、重复性限与再现性限的异同之处；关于实际应用，分析了测定方法的合格率以及测定方法的使用趋势变化。分析结果表明，火花放电原子发射光谱法适用于常规检测，而燃烧红外吸收法与电感耦合等离子体原子发射光谱法更加适用更加严格，有科研技术要求的项目。
关键词： 低合金钢；能力验证；标准；统计结果；测定方法

Comparison of Different Determination Methods for Chemical Composition of Low Alloy Steel

Sun Biyao[1], Gao Hongbin[2], Wang Peng[2], Feng Haozhou[1], Lu Yuhua[1], Lin Fei[2]

(1. Central Iron & Steel Research Institute, Beijing 100081, China;
2. NCS Testing Technology Co., Ltd., Beijing 100081, China)

Abstract: In view of the determination of chemical composition in low alloy steel, by comparing the theoretical and practical application parts of different determination methods, it is concluded that different methods have different focuses when used. In the theoretical part, the similarities and differences between the measurement range, repeatability limit and reproducibility limit of the measurement method are compared. Regarding the practical application, the qualification rate of the measurement method and the change trend of the use method of the measurement method are analyzed. The analysis results show that spark discharge atomic emission spectrometry is suitable for routine detection, while the infrared absorption method after combustion in an induction furnace method and inductively coupled plasma atomic emission spectrometric method are more suitable for more stringent projects with scientific research requirements.
Key words: low alloy steel; proficiency testing; standard; statistical results; determination method

ICP-AEC 法测定锰铁中铬锑铅锡砷含量

何志明，陈 溪

（上海梅山钢铁股份有限公司制造管理部，江苏南京 210039）

摘 要： 建立了电感耦合等离子体原子发射光谱法(ICP-AES 法)同时测定锰铁中微量铬锑铅锡砷的分析方法。采用硝酸和氢氟酸加热溶解样品，运用基体匹配法来校正基体效应，通过优化仪器测定参数，选择分析线为Cr283.5nm、Sb217.5nm、Pb220.3nm、Sn189.9nm、As189.0nm测定锰铁中微量铬锑铅锡砷的含量。研究结果表明：铬、锑、铅、锡、砷校准曲线的线性相关系数在0.999以上，检出限（质量分数）分别为0.0005%、0.001%、0.0004%、0.0002%、0.0002%。标准样品的测定值与标准值一致，相对标准偏差（$n=5$）均小于6%，加标回收率为96%~108%。该方法分析周期短，精密度和准确度能满足测量要求。

关键词：电感耦合等离子体原子发射光谱法(ICP-AES法)；铬；锑；铅；锡；砷

Determination of Chromium, Antimony, Lead, Tin and Arsenic in Ferromanganese by ICP-AEC

He Zhiming, Chen Xi

(Manufacturing Management Department of Shanghai Meishan Iron and Steel Co., Ltd., Nanjing 210039, China)

Abstract: A method for the simultaneous determination of trace chromium, antimony, lead, tin and arsenic in ferromanganese by inductively coupled plasma atomic emission spectrometry (ICP-AES) was established. The matrix effect was corrected by heating the sample with nitric acid and hydrofluoric acid. By optimizing the instrument parameters, the analytical lines were selected as Cr283.5 nm, Sb217.5 nm, Pb220.3 nm, Sn189.9 nm and As 189.0 nm for the determination of trace chromium, antimony, lead, tin and arsenic in ferromanganese. The results show that the linear correlation coefficients of the calibration curves of chromium, antimony, lead, tin and arsenic are above 0.999, and the detection limits (mass fraction) are 0.000 5%, 0.001%, 0.0004%, 0.0002% and 0.0002%, respectively. The relative standard deviation ($n=5$) of the standard sample is less than 6%, and the recovery of standard addition is 96%~108%. This method has a short analysis period, and its precision and accuracy can meet the measurement requirements.

Key words: inductively coupled plasma atomic emission spectrometry (ICP-AES); chromium; antimony; lead; tin; arsenic

晶粒度对奥氏体不锈钢321低周疲劳影响机理的研究

张玉刚[1]，何国求[2]，林 媛[1]，佘 萌[2]，张怀征[3]，张文康[1]

（1. 山西太钢不锈钢股份有限公司技术中心，山西太原 030003；2. 同济大学材料科学与工程学院，上海 201804；3.西门子中国研究院燃气轮机设计工程中心，上海 310112）

摘 要：通过对321采用固溶处理的方法，得到粗晶粒奥氏体不锈钢321；在室温及高温下，对其进行一系列的低周疲劳试验并将结果与原321材料的疲劳实验结果分别进行对比；利用XRD、金相显微镜及扫描电镜（SEM）相结合的方式，对材料的微观组织演变和断口形貌进行了研究和分析，揭示了晶粒度这一因素对奥氏体不锈钢321的低周疲劳影响机理。结果表明：在室温时，在应变幅值为0.5%及以下时，粗晶粒材料的低周疲劳寿命要低于细晶粒材料，应变幅值高于0.5%时相反；在高温下，粗晶粒材料在各个应变幅值下的低周疲劳寿命均高于细晶粒材料；细晶粒材料在各应变幅值下疲劳裂纹均先于微孔处形核，粗晶粒材料在低应变幅值（小于0.5%）下裂纹萌生于晶界，在高应变幅值（大于0.5%）下断口形貌出现了准解理面，且微孔处产生了二次裂纹。

关键词：晶粒度；奥氏体不锈钢；低周疲劳；应变幅值

Research on the Influence Mechanism of Grain Size on the LCF of Austenitic Stainless Steel 321

Zhang Yugang[1], He Guoqiu[2], Lin Yuan[1], She Meng[2],
Zhang Huaizheng[3], Zhang Wenkang[1]

(1. Technical Center, Taiyuan Iron and Steel (Group) Co., Ltd., Taiyuan 030003, China;
2. School of Material Science and Engineering, Tongji University, Shanghai 201804, China;
3. Siemens CT Gas Turbine Engineering Hub, Shanghai, 310112, China)

Abstract: The coarse grained austenitic stainless steel 321 was obtained in the way of solution treatment; a series of LCF tests were conducted on coarse grained 321 and the original ones both at room temperature and elevated temperature and the results were compared; through the method of using XRD, macroscopic and SEM, the microstructure evolution and fracture morphologies were analyzed, and the influence mechanism of grain size on austenitic stainless steel 321 was revealed. The results are listed as follows: At room temperature, LCF life of coarse grained materials is short when the strain amplitude is under 0.5%, which is reversed when higher than 0.5%. However, the fatigue life of coarse grained material is higher at elevated temperature for any strain amplitude; the nucleation of fatigue crack for fine grained materials is primarily occurred at microviods, while the crack of coarse grained materials at low strain amplitudes (under 0.5%) is initiated at grain boundaries, and the cleavage facets are formed on fracture at high strain amplitudes (higher than 0.5%), also the secondary cracks are appeared at the microviods.

Key words: grain size; austenitic stainless steel; low-cycle fatigue; strain amplitude

汽车结构用钢 SAPH440 疲劳寿命预测研究

韩 丹，陈虹宇

（本钢技术研究院，辽宁本溪 117000）

摘 要：对汽车结构用钢 SAPH440 进行低周疲劳试验，试验采用等幅应变控制方式，应变比 $R=-1$，在 MTS Landmark 液压伺服疲劳试验机上进行，测定了应力-应变迟滞回线，循环应力响应特征曲线，拟合出应变-寿命曲线，采用 Zwick SUPRA55 扫描电镜对疲劳断口形貌观察，分析了裂纹萌生区、扩展区及瞬断区的特点，并给出了疲劳断裂机理。试验结果表明：SAPH440 具有较好的塑性变形能力，循环应力响应特征为循环软化，根据 Manson-Coffin 公式拟合疲劳数据，获得了 SAPH440 疲劳寿命预测公式，疲劳断裂方式为韧性断裂。SAPH440 低周疲劳性能的研究为高强钢疲劳行为分析和寿命预测提供了理论依据。

关键词：低周疲劳；高强钢；迟滞回线；裂纹萌生

Research on Prediction of Fatigue Life of SAPH440 Steel for Automobile Structure

Han Dan, Chen Hongyu

(Technology Research Institute of Benxi Steel Plate Co., Ltd., Benxi 117000, China)

Abstract: The low cycle fatigue test of SAPH440 steel for automotive structure was carried out. The test was carried out on

MTS testing machine by strain control method, strain ratio was $R=-1$. The stress-strain hysteresis loop and cyclic stress response characteristic curve were measured, then the strain-life curve was fitted. The fatigue fracture was characterized by SEM, analyzed the crack initiation area, crack extension area and fracture area, and the fatigue fracture mechanism was given. The results showed that SAPH440 has good plastic deformation ability, and the cyclic stress response was characterized by cyclic softening. According to Manson-Coffin formula, the fatigue life prediction formula of SAPH440 was obtained, and the fatigue fracture was ductile fracture. The study of low cycle fatigue properties of SAPH440 provided a theoretical basis for fatigue behavior analysis and life prediction of high strength steel.

Key words: low cycle fatigue; high strength steel; fatigue hysteresis loop; crack initiation

熔融制样-X 射线荧光光谱法测定磷铁中磷、硅、锰和钛

宋祖峰，英江霞，王忠乐，郭士光，付万云

（马鞍山钢铁股份有限公司技术中心，安徽马鞍山 243000）

摘 要： 采用陶瓷坩埚石墨垫底低温预氧化后，高温熔融制样，建立了 X 射线荧光光谱（XRF）测定磷铁中磷、硅、锰和钛含量的检测方法。选用磷铁标准样品，按照一定的比例合成及在磷铁标准样品中加入标准溶液方式，配制成一定梯度的磷铁校准样品，拓宽了校准曲线的含量范围。以碳酸锂和过氧化钡复合氧化剂，从 400℃缓慢升温至 800℃，对磷铁样品进行预氧化，避免了熔融过程中对铂金坩埚的腐蚀。实验结果表明，以四硼酸锂为熔剂，溴化铵为脱模剂，稀释比 40∶1，于 1100℃下熔融 20min，制得的玻璃熔片均匀稳定。各元素的检出限在 24.56～60.23μg·g^{-1}之间。在最佳实验条件下对磷铁标准样品进行测定，各元素测定结果的相对标准偏差（RSD，n=10）在 0.18%～1.03%之间。实验方法应用于磷铁实际样品的测定，测定值与其他方法测得结果相吻合。

关键词： 熔融制样；陶瓷坩埚；石墨垫底；X 射线荧光光谱法；磷铁；复合氧化剂

Determination of Phosphorus, Silicon, Manganese and Titanium in Ferrophosphorus by X-ray Fluorescence Spectrometry with Fusion Sample Preparation

Song Zufeng, Jia Jiangxia, Wang Zhongle, Guo Shiguang, Fu Wanyun

(Technology Center of Maanshan Iron & Steel Co., Ltd., Maanshan 243000, China)

Abstract: A method for determination of the content of phosphorus, silicon, manganese and titanium in ferrophosphorus by X-ray fluorescence spectroscopy (XRF) was established by low temperature pre-oxidation of ceramic ruthenium graphite bottom. The standards sample of ferrophosphorus were selected, synthesized according to a certain ratio and a standard solution was added to the standard samples of phosphorusiron to prepare a gradient phosphorite calibration samples, broadening the content range of the calibration curve. The ferrophosphorus samples were pre-oxidized by slowly increasing the temperature from 400℃ to 800℃ with lithium carbonate and ruthenium peroxide composite oxidant to avoid corrosion of platinum ruthenium during the melting process. The experimental results showed that the lithium tetraborate flux and ammonium bromide were the release agents, and the dilution ratio was 40∶1. At 1100℃, the melting was 20 min, and the glass piece obtained was evenly and stably. The detection limit of each elements was between 24.56μg·g^{-1} and 60.23μg·g^{-1}. The standard sample of ferrophosphorus was measured under the optimal experimental conditions, and the relative standard deviation (RSD, n=10) of each elements was between 0.18% and 1.03%. The proposed method was applied to the

激光光谱原位分析技术对铝合金焊缝的成分分布分析

史孝侠，崔飞鹏，徐 鹏，李晓鹏，郭飞飞

（钢研纳克检测技术股份有限公司，北京 100081）

摘　要：采用激光光谱原位统计分布分析技术对两个不同焊接工艺的铝合金焊接样品进行了成分分布面分析，得到铝合金焊接后母材区及熔合区域的合金元素的分布及变化情况。实验建立了铝合金中 Zn、Mg 元素含量面分布的激光诱导击穿光谱原位统计分布分析方法（LIBSOPA），绘制了匹配的铝合金的定量分析工作曲线。该方法应用于铝合金焊缝及母材的重要元素含量分布及变化的原位分布分析，可帮助探索元素含量分布对焊缝组织性能的影响，并为焊接工艺提供有效评估手段。

关键词：激光诱导击穿光谱法；原位统计分布分析；铝合金；焊缝；成分分布

Components Distribution Analysis in Aluminium Alloy Welding Joint by LIBSOPA

Shi Xiaoxia, Cui Feipeng, Xu Peng, Li Xiaopeng, Guo Feifei

(NCS Testing Technology Co., Ltd., Beijing 100081, China)

Abstract: Composition distribution of two aluminium alloy welding samples with different welding processes were determined by laser-induced breakdown spectroscopy-original position statistic distribution analysis （LIBSOPA）. Elements change in the base welding joint were obtained. Distribution method of Zn and Mg elements in aluminum alloy was established, and their work curves was drawn. This method is applied to the analysis of important elements in aluminum alloy welding joint and base material. It can help to explore the influence of the distribution of elements on the properties of weld structure, and can provide an effective assessment method for welding process.

Key words: laser-induced breakdown spectroscopy; original position statistic distribution analysis; aluminium alloy; welding joint; Components distribution

一种大气压下液体阴极辉光放电装置的研制

史孝侠，李晓鹏，崔飞鹏，徐 鹏

（钢研纳克检测技术股份有限公司，北京 100094）

摘　要：建立了一种基于大气压液体阴极辉光放电(Solution Cathode—Glow Discharge，SC-GD)的用于水体中金属离子检测的装置，该装置包括液体阴极、金属阳极、液体进样单元、高压直流电源及辉光检测单元。该装置的优点是在常温常压下即可产生辉光放电，无需真空系统、辅助气体及复杂的样品前处理，在现场及在线检测方面有很大的应用前景。基于该装置，通过优化硬件参数及实验条件得到了水体中 Mg、Cr、Ni 及 V 元素光谱信号，具有一定的实用价值和科研价值。

关键词：辉光放电；液体阴极；大气压；发射光谱；金属离子检测

Development of a Solution Cathode—Glow Discharge (SC-GD) Device at Atmospheric Pressure

Shi Xiaoxia, Li Xiaopeng, Cui Feipeng, Xu Peng

(Beijing NCS Analytical Instruments Co., Ltd., Beijing 100094, China)

Abstract: A Solution Cathode—Glow Discharge Device (SC-GD) has been developed in this paper, which can be used in detecting of metal element in Aqueous Solutions. The device consists of solution cathode metal anode, liquid sampling, a high voltage supply and detector of Glow Discharge (GD). It can generate GD in Atmospheric Pressure and normal temperature with this device averting vacuum system, auxiliary gas and complicated pretreatment of sample. With this device, the emission spectrum of Mg、Cr、Ni and V are detected respectively. The result provides an available method for further research of metal detecting in Aqueous Solutions by a Solution Cathode—Glow Discharge Device (SC-GD).

Key words: glow discharge; solution cathode; atmospheric pressure; optical emission spectrometry; detection of metals

以钛合金锻件为例的不同标准体系对比分析

王矛，陈鸣

（钢研纳克检测技术股份有限公司材料试验标准部，北京 100081）

摘 要：目前，市场上存在许多标准体系，各种标准体系间交叉重叠，内容繁杂，标准水平参差不齐，对标准的选择和使用造成了许多障碍。本文以 TC4 钛合金锻件为例，通过对比其关键指标试验方法（化学成分分析，室温拉伸，冲击功，断裂韧性）在国标，航标和 ASTM 体系中的差异，对各标准体系的特点进行总结，加强对标准科学性的认识，以期对我国标准体系的改进升级工作有所帮助。

关键词：TC4 钛合金锻件；国标；航标；ASTM；标准对比

Comparison of Different Standard Systems based on Titanium Alloy Forgings

Wang Mao, Chen Ming

Abstract: Nowadays, many standard systems contain considerable overlaps and repetitions and the contents are usually miscellaneous and complicated, resulting in multiple problems and difficulties during standard selections and applications. In this paper, TC4 titanium alloy forgings are taken as an example, the testing methods of its key index (chemical composition analysis, tension testing at room temperature, notched bar impact testing and fracture toughness testing) in GB code, HB code and ASTM are compared to show the differences. Through the comparison, the characteristics of these standard systems are summarized and a better realization about the scientific nature of standards is achieved. This work is expected to be helpful for improving and upgrading Chinese standard systems.

Key words: TC4 titanium alloy forgings; GB code; HB code; ASTM; standards comparison

10 冶金设备与工程技术

大会特邀报告

分会场特邀报告

炼铁与原料

炼钢与连铸

轧制与热处理

表面与涂镀

金属材料深加工

先进钢铁材料

粉末冶金

能源、环保与资源利用

钢铁材料表征与评价

★ 冶金设备与工程技术

冶金自动化与智能化

建筑诊治

其他

一种开口机凿岩机油管运行装置

邓振月,陈文彬,王仲民,何润平,刘 斌,冯 伟,康大鹏

(首钢股份公司炼铁作业部,河北迁安 064400)

摘 要:本课题重点阐述了高炉炉前液压开口机的工作原理和结构特点,分析了开口机三路角及油管在使用中出现的故障,这些故障虽然不直接影响生产,但对维修来说成为了炉前维护量最大的工作,通过深入分析故障的原因,结合炉前工况环境进行改变开口机凿岩机油管运行方式,避免了出铁时铁渣的喷溅,极大地延长了油管的使用寿命,同时本装置的应用可最大限度地减少凿岩机油管的烧损,延长高压油管的使用寿命,为高炉的顺稳生产提供了有力的保障。

关键词:高炉;开口机;油管运行装置

Oil Pipe Operation Device of Rock Drill Machine with Opening Machine

Deng Zhenyue, Chen Wenbin, Wang Zhongmin, He Runping,
Liu Bin, Feng Wei, Kang Dapeng

(Beijing Shougang Co., Ltd., Qian'an 064400, China)

Abstract: This paper focuses on the working principle and structural characteristics of the hydraulic opening machine in front of the furnace, and analyzes the three-way angle of the opening machine and the faults in the use of the tubing. Although these faults do not directly affect the production, they have become the work with the largest amount of maintenance in front of the furnace. Through in-depth analysis of the causes of the failure, combined with the working condition environment in front of the furnace, the operation mode of the tubing of the drilling machine is changed. The spatter of iron slag is avoided and the service life of tubing is greatly prolonged. At the same time, the application of this device can minimize the burning loss of drill pipe and prolong the service life of high pressure tubing, which provides a powerful guarantee for the smooth and stable production of blast furnace.

Key words: blast furnaces; opening machine; tubing operation device

基于AMESim的双缸液压同步仿真分析

刘秀军

(中冶京诚工程技术有限公司冶金工程事业部,北京 100176)

摘 要:双缸液压回路在轧钢生产线很普遍,是典型的同步回路。本文运用AMESim软件中的液压模块建立了双

阀控缸液压系统模型，探讨了双缸同步模型的参数设计，并进行了动态特性仿真分析，得出仿真结果，具有一定的参考价值。

关键词：AMESim；双缸同步；仿真

Simulation Analysis of Hydraulic Synchronization System with Double Cylinders based on AIMSim

Liu Xiujun

(Metallurgical Engineering Division, Capital Engineering & Research Incorporation Limited, Beijing 100176, China)

Abstract: The hydraulic system of two-cylinder is common in the Steel Rolling Production Line. Two-cylinder synchronous circuit is very typical synchronous control. Based on AMESim, this paper establishes a valve controlling cylinder model, discusses the parameter design of the two-cylinder synchronous circuit, and carries out simulation analysis of dynamic characteristics to obtain simulation results. This paper has certain reference value.

Key words: AMESim; two-cylinder synchronous; simulation

卧式活套套量设计计算

惠升谋

（宝钢工程技术集团有限公司，上海 201900）

摘　要：本文结合卧式活套的设计分析结果及其控制方式，对常用的活套设计提出了有关的计算公式。具体阐述了该机组活套的工艺、设备选型、机组控制联锁和设备结构设计应考虑的技术要点，对常见活套的特点进行分析比较，分析了活套基本原理和结构类别，结合具体的工程对活套进行了优化设计，阐述活套设计计算过程。

关键词：入口活套；活套设计；PL-TCM

Design and Calculation of Horizontal

Hui Shengmou

(Baosteel Engineering & Technology Group Co., Ltd., Shanghai 201900, China)

Abstract: According to the design and control method of Zhanjiang 1550mm PL-TCM entry looper, the common looper design related calculation formula are put forward. The main technical points of looping process, equipment type selection, control interlock and equipment structure design which should be consider were specifically elaborated. The characteristics of common looperwere analyzed, the basic principle of loop structure and categories were researched. Combined with specific engineering for theoper optimized design, the looper design calculation process was introduced.

Key words: entry looper; looper design; PL-TCM

冷轧酸洗除尘效果提升技术

武明明

（鞍钢股份有限公司冷轧厂，辽宁鞍山 114000）

摘　要：通过对冷连轧酸洗除尘设备功能和工作原理的了解，结合现场实际生产情况，氧化铁皮产生的重点部位进行分析，现场的铁皮粉尘不仅恶化了污染了工作环境，还对工人的身心健康造成了损伤。通过对冷轧酸洗段机组原有的除尘设备进行技术改造，除尘效率显著提升，现场污染情况得到了有效治理。

关键词：冷轧；带钢；氧化铁皮；除尘

Abstract: Based on the understanding of the function and working principle of the acid-cleaning and dust removal equipment for cold continuous rolling, the key parts of the oxidized iron sheet were analyzed according to the actual production situation. The tin dust on the site not only deteriorated the working environment, but also caused damage to the physical and mental health of the workers. Through the technical transformation of the original dedusting equipment of the cold rolling pickling section, the dedusting efficiency is greatly improved and the field pollution is effectively controlled.

Key words: cold rolling; steel belt; oxidized iron sheet; dust removal

红外吸收法测定汽车板钢中氧含量的不确定度评定

刘步婷，刘　俊，王　鹏，彭　涛

（武钢有限质检中心，湖北武汉 430083）

摘　要：本文通过脉冲加热熔融-红外吸收法对汽车板钢中氧含量的不确定度来源进行了分析并建立数学模型，分别对测量不确定度的来源，诸如测量值的重复性、标准物质校准仪器的变动性、标准物质标准值的不确定度、天平称量的不确定度、仪器变动性、显示分辨力等进行了评定，得出汽车板钢中氧含量的测量不确定度，为提高氧的测量可靠性提供了分析依据。

关键词：红外吸收法；不确定度评定；氧；不确定度；汽车板钢

Assess of Uncertainty for Determination of Oxygen Content in Automobile Plate Steel by Infrared Absorption Method

Liu Buting, Liu Jun, Wang Peng, Peng Tao

(Quality Inspection Center of WISCO, Wuhan 430083, China)

Abstract: In this paper, the source of uncertainty of oxygen content in automobile plate steel is analyzed by pulsed heating, melting and infrared absorption method, and a mathematical model is established to measure the source of uncertainty. For example, the repeatability of the measurement value, the variability of the standard material calibration instrument, the uncertainty of the standard value of the standard material, the uncertainty of the balance weight, the variability of the instrument, and the display resolution were assessed. The uncertainty of the measurement of oxygen content in automobile plate steel is obtained, which provides a basis for improving the reliability of oxygen measurement.

Key words: infrared absorption method; uncertainty assessment; oxygen; uncertainty; automobile plate steel

一种冷轧带钢内径擦伤的原因分析和解决方案

李 黎

（鞍钢股份有限公司冷轧厂，辽宁鞍山 114021）

摘 要：本文以鞍钢冷轧厂二分厂电解清洗机组的带钢内径擦伤缺陷为例，对鞍钢冷轧电解清洗机组的工艺布局进行了简单介绍，对清洗机组内径擦伤的缺陷内容进行了简单描述，并且重点论述了该种类型内径擦伤产生的设备和控制系统方面的原因，以及如何通过对控制系统的调整和控制方法的优化抑制该类型内径擦伤缺陷产生的方法和实现过程。

关键词：外延型内径擦伤；张力调节；涨径压力；内核张力控制

Cause Analysis and Solution of Inner Diameter Scratch of Cold Rolling Strip

Li Li

(Cold Rolling Department of Ansteel, Anshan 114021, China)

Abstract: This paper takes the inner diameter scratch of strip steel in ECL(The Electronic Cleaning Line) of the second factory of Angang Cold Rolling as an example. The process layout of ECL(The Electronic Cleaning Line) of Angang Cold Rolling is briefly introduced. In this paper, the defect of the inner diameter scratch of strip steel of the ECL is briefly described. It is discussed about the causes of the equipment and the control system by this type of the inner diameter scratch of strip steel. And how to control the generation and realization process of the inner diameter scratch of strip steel through the adjustment of control method of the control system.

Key words: the extension type of the inner diameter scratch of strip steel ; tension adjust; the pressure of expand the roll; tension control of the inner core

提高棒材650t冷剪机生产能力的技术改造

刘 宏[1]，戴江波[2]，金培革[1]，张业华[1]

（1. 武钢集团襄阳重型装备材料有限公司，湖北襄阳 441100；
2. 中冶南方武汉钢铁设计研究院有限公司，湖北武汉 430080）

摘 要：冷剪机运行状态的好坏，直接关系到棒材生产线的正常生产及产品质量。本文对650t冷剪机运行过程中出现的故障进行详细分析，提出了有效的技术改造方法，同时相应的提出了在改造后应做好的设备预防维护和检修工作，减少了设备故障停机率，增加了冷剪设备生产能力超出原设计能力的20%，满足了棒材线高产稳产的需要。

关键词：冷剪；生产能力；改造

Technical Reform of Increasing Production Capacity of 650t Bar Cold Shearing Machine

Liu Hong[1], Dai Jiangbo[2], Jin Peige[1], Zhang Yehua[1]

(1. Baowu Group Xiangyang Steel Co., Ltd., Xiangyang 441100, China;
2. WISDRI Wuhan Iron and Steel Design Institute Co., Ltd., Wuhan 430080, China)

Abstract: The running state of cold shear is directly related to the normal production of bar production line and product quality. This paper makes a detailed analysis of the faults occurring during the operation of 650t cold shear, and puts forward effective technical transformation methods. At the same time, it puts forward corresponding preventive maintenance and overhaul work of equipment after transformation, which reduces the failure rate of equipment, increases the production capacity of cold shear equipment by 20% exceeding the original design capacity, and meets the need of high and stable production of bar wire.

Key words: cold shear; scale of production; reform

储煤筒仓防自燃因素分析及设计

向甘乾

（中冶南方武汉钢铁设计研究院有限公司，湖北武汉 430081）

摘　要：筒仓具有占地小、空间利用率高、无粉尘污染、自动化程度高且易于配煤等优点，但极易发生自燃事故，对安全生产造成重大影响。本文简述了筒仓储煤工艺，通过对筒仓储煤自燃原因的分析，充分考虑生产中可能出现的问题，提出防止筒仓储煤自燃的关键是解决下煤堵塞问题的观点。并从设计角度出发阐述防堵要点。

关键词：筒仓储煤；防自燃；防堵；设计

Analysis of Influencing Factors and Design for Anti-self-ingition of Coal Storage Silos

Xiang Gan'gan

(WISDRI Wugang Engineering Co., Ltd., Wuhan 430081, China)

Abstract: The silo has so many advantages, such as covering a small area, high utilization rate of space, reducing the dust pollution to zero, high degree of automation, simple distribution of coal and etc. But on the other side, the spontaneous combustion accidents tend to take place easily in the silo, which takes negative impact on the safety in production. This paper briefly describes the silo storage technology and proposes the key point that the prevention of spontaneous combustion for coal is to handle the problem of congestion when the coal falls down the silo by analyzing the cause of spontaneous combustion in the silo and fully taking all kinds of situation that may occur in the production into consideration.

Finally setting forth the key point of the prevention of congestion from the perspective of design would end up this paper.

Key words: coal storage silos; anti- self-ignition; anti-clogging; design

湛江钢铁厚板二切线传动系统改造

谢清新

(宝钢湛江钢铁有限公司厚板厂，广东湛江 524072)

摘 要： 湛江钢铁厚板二切设备均搬迁利用罗泾4200mm厚板厂原有设备，由于辊道的传动方式采用的是悬臂式齿轮电机，给设备维护和生产组织带来了各种问题。本次改造内容包括电机选型及传动方式，低压供配电设施，电线、电缆的敷设方式，机械、土建及辅供。通过技术改造后，联轴器传动的电机能有效的抵抗平直度较差的板材在运送过程中带来的振动和冲击，相应的各个配套系统运行稳定。改造后二切线生产作业率提高到96%以上，检修率和检修费用明显降低。同时提升了生产效率，实现250℃以上板坯热装热送率达40%。

关键词： 二切；传动；电机；宽厚板

Cutting Line Drive System Transformation in Zhanjiang Steel Heavy Plate Plant

Xie Qingxin

(Heavy Plate Plant, Baosteel Zhanjiang Steel Co., Ltd., Zhanjiang 524072, China)

Abstract: The original equipment of Luojing was used in zhanjiang iron and steel heavy plate plant cutting line. As the transmission mode of roller table adopts the cantilever gear motor, it brings various problems to equipment maintenance and production organization. This transformation includes motor selection and transmission mode, low-voltage power supply and distribution facilities, wire and cable laying mode, machinery, civil construction and auxiliary supply. After the technical transformation, the motor driven by the coupling can effectively resist the vibration and impact of the plate with poor flatness in the transportation process, and the corresponding supporting systems operate stably. After the transformation, the operation rate of the cutting line production increased to more than 96%, and the maintenance rate and cost were significantly reduced. At the same time, the production efficiency has been improved, and the hot charging and hot delivery rate of slabs above 250℃ has reached 40%.

Key words: double cut; transmission; motor; heavy plate

一种热轧厂SSP曲轴拆装技术

曹叶飞

(上海宝钢工业技术服务有限公司，上海 201999)

摘 要： 本文主要介绍了热轧厂SSP定宽侧压机关键部件曲轴拆装修复面临的难题，重点分析了齿接手和大型轴承无损的拆装过程，并介绍了专用工装的设计及强度校核过程。根据在SSP曲轴拆装项目过程中遇到的项目难题、技术攻关手段,对其拆装过程进行了总结,形成本论文所述曲轴拆装技术。

关键词：SSP 曲轴；拆装工装；加热温度；装拆工艺

A Crankshaft Disassembly and Assembly Technology of SSP in Hot Rolling Mill

Cao Yefei

(Shanghai Baosteel Industry Technological Service Co., Ltd., Shanghai 201999, China)

Abstract: This paper mainly introduces the difficult problems of the crankshaft disassembly, assembly and repair of SSP in hot rolling mill. In this paper, we mostly analysis the non-destructive disassembly and assembly process of teeth joint and large bearing, and introduces the design of special tooling and the process of strength checking. According to the project difficulties and technical means encountered in the process of the SP crankshaft disassembly and assembly project, the disassembly and assembly process is summarized, and the crankshaft disassembly and assembly technology described in this paper is formed.

Key words: SSP crankshaft; disassembly and assembly tooling; heating temperature; assembly and disassembly process

混铁车电气故障诊断与处理

牟德新

(安徽马钢工程技术集团有限公司钢结构分公司,安徽马鞍山 243000)

摘 要：混铁车是用来将高炉铁水运送到钢厂进行冶炼的专用车辆,其运行时出故障将会影响钢厂正常受铁水时间,而混铁车故障基本上都是电气故障。为了缩短维修人员电气故障检修时间,大幅提高检修效率,本文通过对混铁车电气设备及线路等的详细介绍,重点分析了混铁车常见电气故障诊断与处理方法。

关键词：混铁车；电气故障；故障诊断；故障处理

Electrical Fault Diagnosis and Treatment of Mixed Iron Furnace Type Ladle Car

Mou Dexin

(Anhui Masteel Engineering Technology Group Ltd., Steel Structure Branch Company, Maanshan 243000, China)

Abstract: The mixed iron furnace type ladle car is a special vehicle used to send the blast furnace iron water transport to the steel mill for smelting. The failure of the operation will affect the normal iron water time of the steel mill, and the failure of the mixed iron furnace type ladle car is basically an electrical fault. In order to shorten the time of maintenance personnel's electrical fault and improve the efficiency of maintenance, this paper analyzes the common electrical fault diagnosis and treatment method of the hybrid truck through the detailed introduction of electric equipment and circuit.

Key words: the mixed iron furnace type ladle car; electrical failure; fault diagnosis; fault handling

基于定位跟踪的自动抛丸控制技术

赵进友

(湛江钢铁厚板厂设备室，广东湛江 524000)

摘　要：通过检测元件的跟踪定位技术、Wincc 人机交换界面和 LAD 编辑语言等工具对抛丸机的钢板位置信息不准、抛后钢板表面质量存在缺陷等问题进行分析，合理布置跟踪元件位置，优化改进丸料平衡分配的控制逻辑。结果表明，改进后的钢板位置跟踪准确，表面质量得到极大改善，原本需抛板 2~3 次才能符合工艺质量要求的钢板，现抛板 1 次已满足工艺要求，节约能耗、改进钢板质量和提高生产节奏。

关键词：跟踪定位技术；Wincc 人机交换界面；LAD 编辑语言；表面质量；控制逻辑；节约能耗；生产节奏

Automatic Shot Blasting Control Technology based on Positioning and Tracking

Zhao Jinyou

(Equipment Management Room of Thick Plate Plant, Zhanjiang 524000, China)

Abstract: By means of detecting components'tracking and positioning technology, Wincc man machine Exchange interface and Lad editing language, the paper analyzes the problems of the plate position information of the shot blasting machine and the defects in the surface quality of the steel plate, and arranges the position of the tracking components reasonably, and optimizes the control logic of the balance distribution of the pellets. Results the surface of the improved steel plate position tracking accuracy, the surface quality has been greatly improved, the original need to cast 2~3 times to meet the quality requirements of the steel plate, now cast 1 times has met the process requirements, save energy consumption, improve steel quality and improve production rhythm.

Key words: tracking and positioning technology; Wincc man machine exchange interface; ladder logic programming language; surface quality; control logic; save energy consumption; production rhythy

机旁库备件二维码的应用

陈敏华

(上海宝信软件股份有限公司信息化事业本部，上海 201900)

摘　要：随着钢铁业的迅速发展，要求钢铁企业的安全稳定运行水平也不断提高，而完善的备件供应是保证钢铁企业安全、可靠、持续运行的关键因素。钢铁厂备件种类繁多，管理不够专业、智能，将直接影响机组安全、高效运行，人力、物力的浪费也将最终反映于经济效益的损失。为了解决管理过程中的不足，宝山钢铁开发了一套机旁库备件智能管理系统，以计算机网络技术与二维码技术为依托，结合移动端 APP 的使用，依托设备管理信息系统的物料管理、用二维码和移动端 APP 来实现快速实现机旁备件的管理。文章详细介绍了系统设计的要点与理念，以及系统开发建设的流程，实现了机旁备件的入库、出库、移库、调拨、台账等全过程信息化闭环管理、信息查询，有效地解决了管理过程中效率低、账不符实等多项问题。

关键词：机旁库备件；设备管理系统；二维码；移动APP

Application of Two-dimensional Code for Spare Parts in Machine Side Storage

Chen Minhua

(Information Business Headquarters of Shanghai Baosight Software Co., Ltd., Shanghai 201900, China)

Abstract: With the rapid development of iron and steel industry, it is required that the level of safe and stable operation of iron and steel enterprises is constantly improved. Perfect spare parts supply is the key factor to ensure the safe, reliable and sustainable operation of iron and steel enterprises. There are many kinds of spare parts in iron and steel plant, and the management is not professional and intelligent enough. It will directly affect the safe and efficient operation of the unit. The waste of manpower and material resources will ultimately be reflected in the loss of economic benefits. In order to solve the shortcomings in the management process, Baoshan Iron and Steel has developed an intelligent management system for spare parts of machine-side warehouse. Based on computer network technology and two-dimensional code technology, combined with the use of mobile App, relying on material management of equipment management information system, using two-dimensional code and mobile APP to achieve rapid input. This paper introduces the main points and concepts of the system design and the process of system development and construction in detail. It realizes the closed-loop information management and information inquiry of all the spare parts beside the machine, such as warehousing, outgoing, moving, transferring and managing the accounts. It effectively solves many problems such as low efficiency in the process of management and accounting compliance.

Key words: spare parts for hangar equipment; management system; two-dimensional; mobile APP

LECO CS844 测定铌铁合金中碳元素含量

郭士光，宋祖峰，英江霞，王忠乐，张　洁，龙如成

（马鞍山钢铁股份有限公司技术中心，安徽马鞍山　243000）

摘　要：使用燃烧红外吸收法代替 GB/T 223.71—1997 中燃烧重量法测定铌铁碳量。最佳分析条件为：纯铁和钨做助熔剂，0.5g 纯铁助熔剂+2.0g 钨助熔剂，0.2g 称样量 SL28-04、SL28-02 两种标准物质的相对标准偏差分别为 0.77%、1.32%。该方法测定铌铁中碳元素含量可以获得准确、稳定的结果。

关键词：铌铁；燃烧；红外；助熔剂；碳含量

Determination of Carbon Content in Ferroniobium by LECO CS844

Guo Shiguang, Song Zufeng, Jia Jiangxia, Wang Zhongle, Zhang Jie, Long Rucheng

(Technical Center of Masteel, Maanshan 243000, China)

Abstract: High frequency combustion infrared absorption method can be used to instead combustion weight method in

GB/T 223.71—1997 for determination of carbon content in ferroniobium.The best analytical conditions are: using pure iron and tungsten as flux, 0.5g pure iron +2.0g tungsten as flux , sampling amount is 0.2g .The relative standard deviations of the two standard substances SL28-04 and SL28-02 are 0.77% and 1.32%, respectively. The method is accurate and stable in determine carbon content of ferroniobium.

Key words: ferroniobium;combustion;infrared; flux;carbon content

35kV 变压器非电量保护故障分析与改进

叶 恩

（宝钢湛江钢铁有限公司厚板厂设备管理室，广东湛江 524000）

摘 要：本文分析并论述了非电量保护系统在大型电力变压器应用中的重要性，介绍了湛江钢铁厚板厂 35kV 变压器非电量保护装置整定值设置失误引发系统故障的典型案例及其改进方案。

关键词：变压器；非电量保护；故障；改进

The Analysis and Improvement of Fault of Non-electric Protection for 35kV Transformer

Ye En

(Baosteel Zhanjiang Iron and Steel Co., Ltd., Equipment Management Department of Heavy Plate Plant, Zhanjiang 524000, China)

Abstract: This paper analyses and discusses the importance of non-electric protection system in the application of large-scale power transformer. Typical cases of system faults caused by setting errors of 35kV transformer non-electric protection devices in Zhanjiang Heavy Plate Plant and the improvement schemes are introduced.

Key words: transformers; non-electric protection; malfunction; improvement

无功补偿装置在湛江厚板厂的应用与优化

叶 恩

（宝钢湛江钢铁有限公司厚板厂设备管理室，广东湛江 524000）

摘 要：分析湛江钢铁厚板厂 35kV SVC 静止型动态无功补偿装置的安装必要性与作用，介绍其在厚板供电系统中的实际应用与运行效果。并针对前期 SVC 出现的问题隐患，做到在不更换新电容器的情况下，通过电容值的测量与计算，对电容位置进行重新匹配的优化，使不平衡度维持在正常范围内，实现补偿系统的平衡。在节省备件损耗的同时有效地降低设备故障发生的几率，达到降本增效的效果。

关键词：无功补偿；SVC；滤波；电容；计算；优化

The Application and Optimization of Reactive Compensation Device in Zhanjiang Heavy Plate Mill

Ye En

(Baosteel Zhanjiang Iron and Steel Co., Ltd., Heavy Plate Mill Equipment Management Room, Zhanjiang 524000, China)

Abstract: This paper analyzes the necessity and function of 35kV SVC static dynamic reactive power compensation device in Heavy plate mill of Baosteel Zhanjiang Iron and Steel Co., Ltd., and introduces its practical application and operation effect of power supply system in Heavy plate mill. In view of the potential problems of SVC in the early stage, the position of capacitor can be rematched and optimized by measuring and calculating the capacitance value without replacing new capacitor, so as to maintain the unbalance degree within the normal range and realize the balance of the compensation system. While saving the loss of spare parts, it can effectively reduce the probability of equipment failure and achieve the effect of cost reduction and efficiency increase.

Key words: reactive compensation; SVC; filter; capacitance; calculation; majorization

鞍钢大连铸双线供钢铁路改造项目组织研究

苟 涛[1]，刘彦栋[1]，李新颖[1]，李荣升[2]，林传山[1]

（1. 鞍山钢铁集团有限公司铁路运输分公司，辽宁鞍山 114021；
2. 鞍山钢铁集团有限公司，辽宁鞍山 114021）

摘 要： 鞍钢大连铸双线供钢铁路改造项目是立足于解决鞍钢大连铸钢水运输过程中的安全隐患和效率低下问题，本文介绍了在钢水运输生产正常的条件下，通过采取有效方法和措施，完成大连铸双线供钢铁路改造施工任务，为企业既有铁路站场改造积累了经验。

关键词： 连铸；铁路；施工；改造

Study on Reconstruction Project of Steel Provision Double-track Railway

Xun Tao[1], Liu Yandong[1], Li Xinying[1], Li Rongsheng[2], Lin Chuanshan[1]

(1. Railway Transportation Company of Anshan Iron & Steel Group Co., Ltd., Anshan 114021, China;
2. Anshan Iron & Steel Group Co., Ltd., Anshan 114021, China)

Abstract: The reconstruction project of steel provision double-track railway aims to solve the problems of potential safety hazard and low efficiency in molten steel transportation. In this paper, effective methods and measures are taken to complete the project of steel provision railway reconstruction and construction under the normal production conditions. It also provides valuable experience for the reconstruction of existing railway.

Key words: continuous casting; railway; construction; reconstruction

鞍钢 3200m³ 高炉煤气除尘升级改造研究与应用

李艳[1]，卢楠[2]，李志远[1]，李洪宇[2]，袁冬海[3]

（1. 鞍钢工程技术有限公司，辽宁鞍山 114003；2. 鞍钢股份公司能源管控中心，辽宁鞍山 114003；北京建筑大学，北京 100044）

摘 要：文章从荒煤气系统设施的配备、煤气降温方案的选择、双排稀相气力输灰系统配置、卸灰机的选择等方面确定了改造工艺方法，尤其是在 3200m³ 大高炉上首次实现了除尘器一排布置工艺，解决了旧有厂区受空间限制双排布置无法实现的技术难题，并给出了项目改造一年后的运行效果。

关键词：高炉；湿法除尘；干法除尘；一排布置；TRT

Research and Application of Upgrading and Reforming of Gas Dust for 3200m³ Blast Furnace in Ansteel

Li Yan[1], Lu Nan[2], Li Zhiyuan[1], Li Hongyu[2], Yuan Donghai[3]

(1. Angang Engineering Technology Co., Ltd., Anshan 114003, China;
2. Energy Control Center of Anshan Iron and Steel Co., Ltd., Anshan 114003, China;
3. Beijing University of Civil Engineering and Architecture, Beijing 100044, China)

Abstract: The dry dedusting transformation process is put forward from the aspects of equipment of raw gas system, selection of gas cooling scheme, configuration of dual-row dilute-phase pneumatic ash conveying system and selection of ash unloader, etc. Especially, the single-row layout process of dust collector has been realized for the first time on 3200 m³ blast furnace, which solves the problem of double-row layout restricted by space in old plant area. The technical difficulties are solved and the operation effect after one year of the project transformation is given.

Key words: blast furnace; wet dedusting; dry dedusting; row layout; TRT

鞍钢西区焦炉煤气加压站变频系统改造

尚士鑫，牟柳春，杨大海，陕卫东，芦刚

（鞍钢能源管控中心，辽宁鞍山 114021）

摘 要：鞍钢能源管控中心燃气分厂西区焦炉煤气加压站设备运行多年，变频器元器件老化严重，故障频出，已严重影响用户的生产顺行。对变频器系统进行改造，彻底解决了设备运行隐患，降低了工人劳动强度，降低了机组停机机率，达到了生产工艺的要求，为保障生产顺行做好支撑，实现经济效益最大化，为能源管控中心实现自动化集中管控打下坚实的基础。

关键词：高压变频；低压变频；PLC；转换；低电压；偏差

The Reform on Frequency Converter System in Coke Coal Gas Pressure Station of West Zone Anshan Iron-steel Group

Shang Shixin, Mou Liuchun, Yang Dahai, Shan Weidong, Lu Gang

(Energy Sources Manage and Control Center of Anshan Iron and Steel Group, Anshan 114021, China)

Abstract: The equipment has worked many years in coke coal gas pressure station of west zone Anshan Iron-steel Group. The frequency converter component aged heavily, and users production has been effected due to equipment faults. The reform on frequency converter system can solve the problem, lower workers' labor intensity, reduce downtime ratio of pressure unit, and reach the need of production. Economy benefit maximum can be realized also. Solid basis for automation centralized control can be turn out in advance.

Key words: high voltage frequency converter; low voltage frequency converter; PLC; convert; low voltage; difference

棒材生产线轧后工序的智能改造研究与技术开发

徐言东[1]，张华鑫[1]，程知松[1]，郭新文[2]，吴 杰[2]，徐 平[3]

（1. 北京科技大学工程技术研究院，北京 100083；2. 首钢长治钢铁有限公司，山西长治 046031；3. 江苏沙钢集团淮钢特钢股份有限公司，江苏张家港 223002）

摘 要： 目前棒材生产线存在产品超差、计数不准确、打捆质量差、天车运输管理混乱、码垛不规范、出库效率低下等问题，严重阻碍了产品质量的提升，用户异议不断，成本居高不下。本文提出棒材生产线轧后工序要增加配置棒材自动测径、自动表面缺陷检测跟踪、高速冷床、精整系统智能化升级改造、无人天车系统等项目，为实现棒材生产线无人化运行打下基础，逐步解决存在问题。

关键词： 棒材；测径仪；表面检测；无人天车；智能改造

Intelligent Transformation Research and Development for the Bar Finishing System

Xu Yandong[1], Zhang Huaxin[1], Cheng Zhisong[1], Guo Xinwen[2], Wu Jie[2], Xu Ping[3]

(1. Institute of Engineering Technology, USTB, Beijing 100083, China;
2. Shougang Changzhi Iron and Steel Co., Ltd., Changzhi 046031, China;
3. Jiangsu Shagang Group Huaigang Special Steel Co., Ltd., Zhangjiagang 223002, China)

Abstract: Problems existing in current bar plants, such as low accuracy in products and the chaos in rhythm of production, have seriously hindered the improvement of product quality, which not only caused the dispute on product quality, but also increased the cost. Formation of these problems can be attributed to the low accuracy in product measuring, disorder in crane transporting, and also the low efficiency on product delivery. Here we propose a method to upgrade the bar plants, which involves the addition of automatic diameter measuring system and surface defect detection system. Moreover, addition of high-speed cooling bed, intelligent finishing system, and the unmanned crane system, can help realizing the high-quality production as well as the intelligent management, further improving the competitiveness of products.

Key words: bar; caliper; surface inspection; unmanned crane; intelligent transformation

山钢日照 2050mm 热连轧工程设计中采用的先进技术

王 永，吴兆军

（山东省冶金设计院股份有限公司轧钢室，山东济南 250101）

摘 要： 山钢日照 2050mm 热连轧工程设计中采用了多项先进技术，整体工艺技术装备达到国际先进水平，本工程在近年来国内外建设的热连轧工程基础上，充分发挥后发优势，按照产品定位—工艺需求—工程设计的思路来保证整体的工程质量。本文依次介绍工程设计过程的产品定位、工艺流程、工艺需求、采用的先进技术和高架式设计特点。

关键词： 热连轧；热轧带钢；先进技术；高架式

Advanced Technologies of 2050mm Hot Strip Rolling Line at Shansteel Rizhao

Wang Yong, Wu Zhaojun

(Shandong Province Metallurgical Engineering Co., Ltd., Jinan 250101, China)

Abstract: Shandong Steel Rizhao 2050mm hot strip rolling project has adopted many advanced technologies, and the overall process technology and equipment reach the international advanced level. This project has taken advantage of the late-model hot strip rolling projects built in the world in recent years. Product positioning-process requirements-engineering design ideas has been carried out to ensure the whole project quality. This article mainly introduces the project's product schedule, process flow, process requirements, adopted advanced technology and elevated type design features.

Key words: hot rolling; hot strip rolling; advanced technology; elevated type

奥氏体不锈钢水冷隔热罩焊接裂纹分析及对策

秦书清，赵守林，殷 栋，贾士忠

（山东钢铁集团日照有限公司，山东日照 276805）

摘 要： 通过对 0Cr18Ni9 奥氏体耐热不锈钢水冷隔热罩裂纹原因分析，指出了晶间腐蚀和应力腐蚀开裂是造成裂纹产生的主要原因。从而在生产实践中采取有针对性的措施，包括焊材选择、焊接工艺参数确定、焊后处理等技术，便可获得优良的焊缝，提高隔热罩的使用寿命。

关键词： 奥氏体不锈钢；焊接工艺；晶间腐蚀；应力腐蚀

Analysis and Countermeasures of Crack Welding of Austenitic Stainless Steel Water-cooled Heat Cover

Qin Shuqing, Zhao Shoulin, Yin Dong, Jia Shizhong

(Shandong Iron and Steel Group Rizhao Co., Ltd., Rizhao 276805, China)

Abstract: By analyzing the reasons of cracks in the water-cooled thermal insulation cover of 0Cr18Ni9 austenitic stainless steel, pointed out that intercrystalline corrosion and stress corrosion cracking are the main causes of cracks. Thus, in the production practice, adopting targeted measures, including welding material selection, welding process parameters determination, post-welding treatment and other technologies, can obtain excellent weld, improve the service life of the heat shield.

Key words: austenitic stainless steel; welding procedure; intercrystalline corrosion; stress corrosion

高效的真空感应熔炼炉上加料装置

钱红兵

（应达工业（上海）有限公司，上海 201203）

摘 要： 按生产方式,真空感应熔炼炉分为周期炉和半连续炉，小型真空感应熔炼炉采用周期式的生产方式有独特的优势，大型真空感应熔炼炉采用半连续生产方式生产效率更高。移动式上加料室的腔体可水平移动，采用双向密封插板阀加隔热挡板阀的设计，能满足正常生产的要求，但生产效率偏低。Consarc 上加料室为固定加料室腔体，设备故障率低，可用性高。旋转真空隔离阀的压紧环结构，在熔炼室内压力的任何波动都不可能冲开阀板。旋转真空隔离阀内部上下的圆筒状的保护罩，避免密封圈受到热辐射、金属飞溅的伤害，或在加料、小颗粒炉料的损害。

关键词： 真空感应熔炼；上加料；生产率；真空隔离阀；保护环

High Efficiency Overmelt Charging Chamber of VIM

Qian Hongbing

(Inductotherm (Shanghai) Industrial Company, Shanghai 201203, China)

Abstract: Vacuum induction smelting (VIM) furnace is divided into periodic furnace and semi-continuous furnace according to production mode. Small VIM furnace has unique advantages in using periodic production mode. Large VIM furnace using semi-continuous production mode has higher productivity. The movable charging chamber can move horizontally. The design of two-side sealed slide gate valve and heat insulation baffle valve can meet the requirements of normal production, but the productivity is low. Overmelt feeding chamber from Consarc is a static charging chamber with low failure rate and high usability. Because of the pressure ring of swing vacuum isolation valve, the valve plate can not be opened in any condition of pressure fluctuation in the smelting chamber. The cylindrical protective shield inside the swing vacuum isolation valve prevents the sealing ring from being damaged by thermal radiation, metal spattering, or small grain burden.

Key words: VIM; overmelt charging; productivity; vacuum isolation valve; protection ring

倾斜摄影在三维厂区可视化中的应用

左春雷，刘九阳，马亚敏，韦书剑，王海旭，王东明

（中冶沈勘工程技术有限公司测绘地理信息公司，辽宁沈阳 110169）

摘　要：随着无人机技术的快速发展，倾斜摄影技术应用越来越广泛，将倾斜摄影技术与 BIM 技术结合在三维可视化应用是一个趋势，本文通过采用 context capture 软件进行三维模型生产，采用第三方软件进行单体化处理及管道建模，将重建三维模型在 skyline 三维可视化软件上进行展示，对三维模型进行可视化操作及分析，对生产和管理具有重要意义，对今后推动多元技术结合在三维可视化应用具有重要影响。

关键词：倾斜摄影；实景三维模型；BIM；三维可视化

The Application of Oblique Photography in Three-dimensional Visualization Plant Area

Zuo Chunlei, Liu Jiuyang, Ma Yamin, Wei Shujian, Wang Haixu, Wang Dongming

(Shenkan Engineering & Technology Corporation. MCC, Shenyang 110169, China)

Abstract: With the rapid development of UAV technology, the application of oblique photography technology is more and more widely. It is a trend to combine oblique photography technology with BIM technology in 3D visualization application. In this paper, we use context capture software to produce 3D model, use third-party software to process and model pipelines, and reconstruct 3D model on skyline 3D visualization software. Display, visualization operation and analysis of three-dimensional model is of great significance to production and management, and will have an important impact on promoting the combination of multi-technology in the application of three-dimensional visualization in the future.

Key words: oblique photography; real-time 3D model; BIM; three-dimensional visualization

智慧检修的思考与探索

时元海，张鸿元，李玉凤

（莱芜钢铁集团有限公司，山东济南 271200）

摘　要：当今社会随着互联网的快速发展，互联网+、智能、数字、智慧等词语成了最火热与亮眼的名片。"中国制造 2025" 国家战略计划的推出也给传统制造业的智能转型加上助推器，与制造业相关配套产业也迅速向智能化、智慧型产业转型。智能制造、智慧检修已成为社会发展的新趋势。本文拟从检修业务拓展转型角度，立足提升钢铁主业检修质效进行分析，并提出智慧检修建设目标和意见建议，以期为建立智慧检修体系提供参考。

关键词：智慧检修；思考；探索

Thinking and Exploration of Smart Overhaul

Shi Yuanhai, Zhang Hongyuan, Li Yufeng

(Laiwu Steel Group Co., Ltd., Jinan 271200, China)

Abstract: With the rapid development of the Internet in today's society, words such as "Internet +", "intelligence", "data" and "smart" have become the hottest and most eye-catching business cards. The launch of the "made in China 2025" national strategic plan also gives a booster to the smart transformation of traditional manufacturing industry, and the supporting industries related to manufacturing industry are also rapidly transforming to smart manufacturing and smart overhaul industries. Smart manufacturing or smart overhaul has become a new trend of social development. From the perspective of maintenance business expansion and transformation, this paper analyzes the maintenance quality and efficiency of the core area of steel industry, and puts forward the goals and suggestions of smart overhaul, construction, so as to provide references for the establishment of smart overhaul system.

Key words: smart overhaul; thinking; exploration

石钢公司轴流压缩机 EPU 应用实践

李 绅,曹俊卿,时永海

(河钢集团石家庄钢铁有限责任公司炼铁厂,河北石家庄 050000)

摘 要:高炉鼓风用轴流压缩机是炼铁过程中的核心动力设备,其工作性能与稳定性对高炉生产至关重要。在高炉风机控制策略中,采用科学、先进的控制方法,不仅可以更有效地保证机组的安全和稳定,同时可以使风机性能范围有所扩大,在提高风量的同时尽最大减少不必要的放风能耗,对高炉安全稳定和节能降耗起到显著的促进作用。研究 EPU(Economy & Performance Upgrade)对高炉风机喘振控制实践,实现了对高炉鼓风机安全稳定运行,减少了鼓风机放风量,促进风机节能降耗,改善了铁前成本电耗指标。

关键词:轴流压缩机;EPU;防喘线;喘振控制;节能降耗

Application Practice of EPU in Axial Compressor of Shisteel Company

Li Shen, Cao Junqing, Shi Yonghai

(Shisteel Company, Iron Works, Shijiazhuang 050000, China)

Abstract: The axial compressor for blast furnace blast is the core power equipment in the iron making process, and its working performance and stability are crucial for blast furnace production. In the blast furnace fan control strategy, scientific and advanced control methods can not only ensure the safety and stability of the unit more effectively, but also expand the performance range of the fan, and reduce the unnecessary air release energy while increasing the air volume. Consumption, play a significant role in promoting the safety and stability of the blast furnace and energy saving. Study EPU

(Economy & Performance Upgrade) on the surge control practice of blast furnace fan, realize the safe and stable operation of the blast furnace blower, reduce the air volume of the blower, promote the energy saving and consumption reduction of the fan, and improve the cost and power consumption index before the iron.

Key words: axial compressor; EPU (economy & performance upgrade); anti-asthmatic line; surge control; energy saving

调速器因素导致游车故障的原因分析及处理方法

邓朝强

（华菱湘钢运输部，湖南湘潭 411101）

摘 要：介绍湘钢 GKIC 型内燃机车近年来在所用 302Y-Z 调速器时，因为调速器因素造成机车游车故障的判断办法，并对各种产生的故障原因进行了全面分析，提出了相应故障的处理方法和改进措施。

关键词：302Y-Z；调速器；因素；游车故障；原因分析；处理方法

Analysis of Vehicle Failure Caused by 302Y-Z Governor and Countermeasures

Deng Zhaoqiang

(Transportation Department of Xiangtan Steel, Xiangtan 411101, China)

Abstract: This paper introduces the judgment method of vehicle failure caused by 302Y-Z speed generator in Xiangtan Steel's recent years of application of GKIC diesel train. And a comprehensive analysis was done about the various causes of the failure. Corresponding fault treatment methods and improvement measures are put forwarded.

Key words: 302Y-Z; speed governor; factors; vehicle failure; analysis of causes; countermeasures

冷轧厂镀锌线卸卷小车升降定位不准问题的解决与思考

胡增雨，冯志新，王 军

（鞍钢股份有限公司冷轧厂，辽宁鞍山 114000）

摘 要：卸卷小车是镀锌线生产的最后一道工序，它能否顺利卸卷是关系到机组能否正常生产的关键。但卸卷小车定位不准造成钢卷内径"抽芯"问题一直困扰镀锌线机组成材率的提高并产生机组非正常停车。本文介绍了关于卸卷小车定位不准问题的解决方法以及关于对其液压控制的一些思考。

关键词：卸卷小车；钢卷；电磁换向阀

钢包渣线砖在 RH 精炼终点后增碳影响研究

王洛，汪雷，王俊北，樊明宇

(马鞍山钢铁股份有限公司技术中心，安徽马鞍山 243000)

摘 要：通过跟踪统计钢厂超低碳钢产品的增碳情况，分析了钢包渣线镁碳砖的侵蚀情况，计算了钢包渣线砖对钢水增碳幅度，研究了钢包渣线镁碳砖从 RH 出站到连铸中间包期间对超低碳钢的增碳影响。结果表明：第 5 层和第 4 层渣线镁碳砖失碳的主要方式是氧化和渣蚀，不进入钢水；第 1 层~第 3 层渣线镁碳砖除第一次使用前烘烤稍有失碳外，其余渣线砖侵蚀的[C]若全部溶解进入钢水，据此计算钢包渣线砖导致钢水增碳幅度为 3.73 ppm。

关键词：超低碳钢；侵蚀；渣线镁碳砖；增碳

Research on Carburization of Ladle Slag Lining from Refining End

Wang Luo, Wang Lei, Wang Junbei, Fan Mingyu

(Technology Center of Maanshan Iron and Steel Co., Ltd., Maanshan 243000, China)

Abstract: The carburization of ultra-low carbon steel has been studied by statistic data. Through the use of corrosion on MgO-C brick used as refining ladle slag line, the carburization effects of ladle slag lining on the ultra-low carbon steel were analyzed from refining end to tundish process. Results show that the damage causes of the fifth layer and fourth layer MgO-C bricks is oxidation and slag corrosion, with no carbon element entering steel; Besides minor decarburization of the first layer to the third layer MgO-C bricks during drying before the first use, the rest carbon loss is dissolved into molten steel completely. Based on these, the ladle slag line brick on steel recarburization from RH refining end is about 3.73 ppm.

Key words: ultra-low carbon steel; corrosion; MgO-C brick; carburization

机床丝杠断裂原因分析

赵亮，樊一丁

(河钢集团石钢公司技术中心开发部，河北石家庄 050031)

摘 要：从化学成分、硬度、断口形貌、显微组织等方面对校直过程中出现的断裂机床丝杠进行了分析。结果表明机床丝杠断裂是由于用户本身加工不当造成的表面加工缺陷，在校直过程中产生应力集中，在校直力作用下丝杠表面沿缺陷位置发生脆性断裂，裂纹沿表面向中心扩展导致丝杠完全断裂。最终确定机床丝杠断裂的原因与原材料无关，建议用户结合自身的加工工艺查找原因。

关键词：机床丝杠；断裂；断口分析；表面缺陷

Fracture Analysis of Machine Tool Lead Screw

Zhao Liang, Fan Yiding

(Shisteel Co., Ltd., Hegang Group Technology Center Development, Shijiazhuang 050031, China)

Abstract: The fracture screw of machine tool during straightening was analyzed in terms of chemical composition, hardness, fracture morphology and microstructure. The results show that the lead screw fracture is the surface machining defect caused by improper machining of the user, and the stress concentration occurs during the straightening process. Under the action of straightening force, brittle fracture occurs along the defect position on the lead screw surface, and the crack extends along the surface to the center, leading to the complete fracture of the lead screw. Finally, it is determined that the reason of machine lead screw fracture has nothing to do with raw materials.

Key words: machine tool screw; fracture; fracture analyzed; surface defect

210t 铁水罐出铁状态一体化安全校核

赵 阳[1]，董现春[1]，刘新垚[1]，姜 猛[2]，张彦丰[2]，汪海东[2]

（1. 首钢集团有限公司技术研究院用户技术研究所，北京 100043；
2. 迁安首钢设备结构有限公司，河北迁安 064404）

摘 要： 铁水罐出铁过程是一个复杂的热、流体、动态耦合过程，长期以来是一个仿真分析的难题。由此造成铁水罐、拉杆等零部件受力工况不明确，必须采用过安全设计的方式保证设备安全。本文通过 Workbench 有限元软件，借鉴前人热仿真分析和力学理论等结果，采用铁水罐及拉杆一体化模型，将铁水罐出铁过程简化为准静态有限元模型，采用本模型，可以计算出铁水罐出铁状态下最危险位置各零部件的受力情况，尤其是拉杆部件在出铁状态下的复杂受力工况，通过该模型分析，可以确定焊接拉杆可以满足安全生产需求，可以替代铸造拉杆。通过本模型校核的 210t 某铁水罐已安全使用，焊接拉杆替代铸造拉杆为生产制造节约了工期的同时也降低了生产成本。

关键词： 铁水罐；有限元分析；一体化模型；安全校核

Integrated Safety Check of a 210t Iron Ladle in the Tapping State

Zhao Yang[1], Dong Xianchun[1], Liu Xinyao[1], Jiang Meng[2], Zhang Yanfeng[2], Wang Haidong[2]

(1. Research Institute of Technology of Shougang Group Co., Ltd., Beijing 100043, China;
2. Qianan Shougang Equipment Structure Co., Ltd., Qianan 064404, China)

Abstract: The iron tapping process is a complex heat, fluid, dynamic coupling process. While, it's also a difficult problem of simulation analysis. As a result, the working conditions of parts such as hot metal ladle and tie rods are not clear. In order to ensure the safety of the equipment, an over-safe design must be adopted. By using an integrated model of hot metal ladle and tie rods, This paper simplify the multi-coupling dynamic process into a quasi-static finite element model. Using this model, it is possible to calculate the force of each component in the most dangerous position of the hot metal ladle. The 210t hot metal ladle checked by this model has been safely used, which indicated the correctness of the model.

Key words: hot metal ladle; finite element analysis; integrated model; safety check

梅钢公司回用水厂V型滤池运行状况分析及控制技术研究

初 明

(宝武集团梅山钢铁公司能源环保部，江苏南京 210039)

摘 要：梅钢回用水厂建于2006年9月，主要处理梅钢公司工业污水和处理后的生活污水。该回用水厂采用法国得利满技术，主要采用物化法的处理工艺。其中V型滤池是很重要的一个处理设施，我们对V型滤池的运行控制进行了研究及改进，使得处理后的废水出水水质稳定达标，为下一步的生产使用打下了良好的基础。

关键词：梅钢公司；V型滤池；运行控制研究及改进

Study on the Operation and Control Technology of V-type Filter in the Rewater Plant of Meigang Company

Chu Ming

(Baowu Group Meishan Iron and Steel Company Energy and Environmental Protection Department, Nanjing 210039, China)

Abstract: Established in September 2006, Meigang Recycle Water Plant mainly deals with industrial sewage and domestic sewage after treatment. The recycling water plant adopts French Delium technology and mainly adopts the physicochemical treatment process. Among them, V-shaped filter is an important treatment facility. We have studied and improved the operation control of V-shaped filter, so that the treated wastewater effluent quality reaches the standard stably, laying a good foundation for the next production and use.

Key words: Meishan Iron and Steel Company; V-shaped filter; studied and improved the operation control of V-shaped filter

新型高炉旋风除尘器仿真分析

胡 伟[1,2]，耿云梅[1]，陈玉敏[1]，李 欣[1,2]，章启夫[1]

(1. 北京首钢国际工程技术有限公司，北京 100043；
2. 北京市冶金三维仿真设计工程技术研究中心，北京 100043)

摘 要：传统的重力除尘器除尘效率低，进入干法除尘系统的煤气灰量大，布袋负荷大，管道和阀门的磨损严重。为有效降低进入干法除尘系统的煤气含尘量，需开发研究高效的旋风除尘器。结合迁钢及京唐等工程，对旋风除尘器的本体结构、内部耐磨材料及结构、流场分布等方面进行研究。结果表明，旋风除尘器内部最大速度为20 m/s，气体压力损失约853Pa；导流板上靠近水平出气管一侧磨损较为严重，在磨损严重位置需要加装锆莫来石陶瓷衬板。

通过对计算结果的分析，为设计过程中的设备选型与优化提供参考。

关键词：高炉；旋风除尘器；计算流体力学；优化设计

Simulation and Analysis of New Type Cyclone Dust Collector for Blast Furnace

Hu Wei[1,2], Geng Yunmei[1], Chen Yumin[1], Li Xin[1,2], Zhang Qifu[1]

(1. Beijing Shougang International Engineering Technology Co., Ltd., Beijing 100043, China;
2. Beijing Metallurgical Three-Dimensional Simulation Design Engineering Technology Research Center, Beijing 100043, China)

Abstract: The traditional gravity dust collector has low dust removal efficiency, a large amount of gas ash entering the dry dust removal system, a large load of bags, and serious wear and tear of pipes and valves. In order to effectively reduce the dust content of the gas entering the dry dust removal system, it is necessary to develop and research an efficient cyclone dust collector. Combining with Qiangang and Jingtang projects, the main structure, internal wear-resistant materials, structure and flow field distribution of cyclone dust collector were studied. The results show that the maximum internal velocity of cyclone dust collector is 20m/s, and the gas pressure loss is about 853Pa. The wear on the side of the guide plate near the horizontal outlet pipe is serious, and the zirconium mullite ceramic liner is needed to be installed in the place where the wear is serious. Through the analysis of the calculation results, it can provide reference for equipment selection and Optimization in the design process.

Key words: blast furnace; cyclone dust collectors; computational fluid dynamic; optimal design

两轴转向架在 130t 保温平车上的应用

项克舜，严 峰，张海滨

（武汉钢铁有限公司运输部，湖北武汉 430083）

摘 要：本文总结了既有武钢 130t 保温平车所用的三轴转向架在运用过程中所暴露出的问题，分析了问题产生的原因，并提出了采用 45t 轴重两轴转向架替换原有三轴转向架的改进方案。同时，简要介绍了 45t 轴重两轴转向架的主要结构及技术参数，样车的运用考验情况及技术经济效益。进一步地，分析了首批改造的 10 辆车运用过程中出现的问题、产生的问题及改进措施。最后，根据 1 年多的小批量运用考核效果，提出了在其他车辆上继续推广使用的建议。

关键词：130t 保温平车；转向架；运用

Application of Two Axle Bogies in 130t Insulated Flat Wagon

Xiang Keshun, Yan Feng, Zhang Haibin

(Transportation Department of Wuhan Iron and Steel Co., Ltd., Wuhan 430083, China)

Abstract: This paper summarizes the problems exposed in the use process of the 3 axles bogies on the existing 130t insulation flat wagon owned by Wuhan Iron and Steel Group, analyzes its causes, and raise up the improvement to replace the original 3 axles bogies with 45t 2 axles bogies. It briefly introduces the main structure and technical parameters of the 45tons 2 axles bogies, the application and technical and economic benefit of the prototype at the same time. It analyzes the problem in the use process of the first 10 modified wagons the causes, and the improvements further. Lastly, it proposes the suggestion to use the 2 axles bogies widely on other wagons according to the use effect of the small batch in more than one year.

Key words: 130t insulation flat wagon; bogie; application

钢管超声波检测可靠性的探讨

曾海滨，王 伟，陈 杰，李 阳

（宝山钢铁股份有限公司钢管条钢事业部，上海 201900）

摘 要：本文针对无缝钢管产品，介绍了相关的无损检测标准，用户对钢管产品质量的要求，无损检测超声波方法的优点和局限性，钢管厂自动化无损检测设备能力分析，提高无损检测可靠性方面进行探讨。

关键词：钢管；超声波；检测；可靠性

Discussion on Reliability of Ultrasonic Testing of Steel Pipe

Zeng Haibin, Wang Wei, Chen Jie, Li Yang

(Baoshan Iron and Steel Co., Ltd., Shanghai 201900, China)

Abstract: This paper introduces the relevant nondestructive testing standards for seamless steel pipe products, the requirements of users for the quality of steel pipe products, the advantages and limitations of ultrasonic nondestructive testing methods, the analysis of the capability of automatic nondestructive testing equipment in steel pipe plants, and the discussion on improving the reliability of nondestructive testing.

Key words: steel pipe;ultrasonic wave;testing;reliability

自力式调节阀原理及其在连退炉氮氢混合站的应用

张贵春，周 涛，胡剑斌，陈 璐

（新余钢铁集团有限公司，江西新余 338001）

摘 要：自力式压力调节阀是一种新型阀门，在我国工业领域得到了越来越广泛的应用。文章重点介绍了RMG430（Inc pilot RMG610）自力式压力调节阀的工作原理和结构特点，以便于自力式调节阀在生产中更好地使用和维护。同时给出了自力式压力调节阀在连退炉氮氢混合站的具体应用实例。

关键词：自力式压力调节阀；工作原理；主阀；指挥器；调节级；负载限制级

Principle and Application of Self-acting Pressure Regulating Valve in Continuous Annealing Furnace N₂H₂ Mixing Station

Zhang Guichun, Zhou Tao, Hu Jianbin, Chen Lu

(Xinyu Iron & Steel Group Co., Ltd., Xinyu 338001, China)

Abstract: Self-acting pressure regulating valve is a new style valve, which is more and more widely used in the industry. This paper emphases introduce the working principle and structure characteristic of RMG430 (Inc pilot RMG610) self-acting pressure regulating valve, in order to facilitate self-acting pressure regulating valve in production can be better used and maintained. Simultaneously provide its application in continuous annealing furnace N_2H_2 mixing station.

Key words: self-acting pressure regulating valve; working principle; main valve; pilot; regulating stage; load limiting stage

最新一代热轧钢卷运输技术及其工程应用

韦富强[1,2,3]，郑江涛[2,3]，杨建立[2,3]，刘天柱[2,3]

（1. 北京首钢云翔工业科技有限责任公司，北京　100043；
2. 北京市冶金三维仿真设计工程技术研究中心，北京　100043；
3. 北京首钢国际工程技术有限公司智能运输装备研究所，北京　100043）

摘　要： 本文概述了最新一代钢卷运输技术，即智能化新能源钢卷运输技术，介绍了某热轧工程采用的智能化新能源钢卷运输系统工艺、设备构成及电气自动化功能特点，描述了该运输系统运行一年多来的应用情况。该技术全面满足智能制造要求，为智能化冶金工厂的建设提供了一种可靠的界面技术解决方案，是最新一代的热轧钢卷运输技术。

关键词： 智能化；新一代；重载；钢卷运输；工程；应用

Latest Generation of Hot Strip Coil Conveying Technology and Its Engineering Application

Wei Fuqiang[1,2,3], Zheng Jiangtao[2,3], Yang Jianli[2,3], Liu Tianzhu[2,3]

(1. Beijing Shougang Yunxiang Industrial Technology Co., Ltd., Beijing 100043, China;
2. Beijing Metallurgical 3-D Simulation Design Engineering Technology Research Center, Beijing 100043, China; 3. Institute of Intelligent Transportation Equipment of BSIET Co., Ltd., Beijing 100043, China)

Abstract: This paper outlines the latest generation of hot strip coil conveying technology, that is intelligent new energy coil transportation technology, introduces the process description, the equipment composition and electrical automation features of the intelligent new energy coil system used in a real plant, and describes the application of the transportation system which has been in operation for more than one year. This technology fully meets the requirements of intelligent manufacturing and provides a reliable interface technology solution for the construction of intelligent metallurgical plant. It is the latest generation of hot strip coil conveying technology.

Key words: intelligent; new generation; heavy load; steel coil transportation; engineering; application

钢铁企业应用自动化立体仓库的分析与探讨

毕 琳，李洪森

（中冶京诚工程技术有限公司冶金工程事业部，北京 100176）

摘 要：本文首先简要介绍了自动化立体仓库的主要组成系统和各自作业特点。然后，阐述了钢铁企业内综合物资的仓库类型、仓储现状和存在的问题。重点针对钢铁企业备件和资材等仓储物资的种类、重量、存储特点，对自动化立体仓库的设计应用进行了初步分析与研究，给出了布置方式、设备选择和控制管理等方面的设计方案，并探讨了自动化立体仓库在钢铁企业进一步推广应用的方向。

关键词：自动化；立体仓库；钢铁；物资；仓储；设计

Analysis and Discussion of the Application of Automated Storage and Retrieval System in Steel Works

Bi Lin, Li Hongsen

(Capital Engineering & Research Incorporation Limited Metallurgical Division, Beijing 100176, China)

Abstract: This paper first briefly introduces the main component system and operational characteristics of the automated storage and retrieval system, and expounds the type of warehouse, storage status and existing problems of steel works .Then, focusing on the types, weight and storage characteristics of storage materials such as spare parts, a preliminary study was made on the application of automated storage and retrieval system in steel works, and the design scheme of layout, equipment selection and control management is given, and the development direction of further the application of automated storage and retrieval system in steel works was discussed.

Key words: automated; storage and retrieval system; iron and steel; goods and materials; storage; design

电涡流振动位移传感器自动检定技术方案研究

郑建忠

（宝武集团韶关钢铁设备管理部，广东韶关 512122）

摘 要：根据JJG 644—2003《振动位移传感器检定规程》研究电涡流振动位移传感器自动检定对测量数据、技术指标要求，分析比较常规电涡流传感器校验装置测试方法与JJG 644—2003在数据采集及指标计算上的差别。按照规程技术要求，制定电涡流振动位移传感器静、动态自动检定量值溯源和传递系统图，在常规校验校准装置上开展自动检定的技术方案，对该自动检定计量标准装置的不确定度进行评定。

关键词：电涡流位移传感器；自动检定；灵敏度；不确定度

Study on Automatic Verification Technology for Eddy-current Vibration Displacement Sensor

Zheng Jianzhong

(Shaoguan Steel Equipment Management Department Baowu Steel Group Corporation Limited, Shaoguan 512122, China)

Abstract: This study examines the requirement of data measurement and technical specifications for automatic verification of eddy-current vibration displacement sensor based on JJG 644—2003, the Vibration Displacement Sensor Verification Regulation. It further compares and analyzes the difference between the conventional eddy-current sensor calibration device test method and JJG 644—2003 in data acquisition and index calculation. The traceability and transmission system diagram of both static and dynamic automatic verification of eddy-current vibration displacement sensor is formulated according to the technical requirements of the regulation. Lastly, the study applies the automatic verification technology on the conventional calibration device and evaluates its uncertainty.

Key words: eddy current displacement sensor; automatic verification; sensitivity; uncertainty

Φ250mm 4340 钢棒的超声波 C 扫描检测研究

万 策

（钢研纳克检测技术股份有限公司，北京 100081）

摘 要： 对 Φ250 mm 的 4340 钢棒进行超声波 C 扫描检测，包括了适用的范围、检测的方法原理、设备检测参数的选取和调整及缺陷检测结果的评定与分析。根据锻轧钢棒超声检测方法的国家标准，针对直径范围为 250mm 的钢棒材，实现超声 C 扫描检测，从而保证了钢棒材产品质量检测结果的数字化和可视性。

关键词： 钢棒；超声波检测；C 扫描；水浸聚焦

Research on Ultrasonic C-scan Method for Φ250mm 4340 Steel Rod

Wan Ce

(NCS Testing Technology Co., Ltd., Beijing 100081, China)

Abstract: Ultrasonic C-scan to the Φ250 mm 4340 steel rod was performed in this paper, including the applicable scope, method principle of detection, selection and adjustment of equipment inspection parameters, and evaluation and analysis of the defect detection results. According to the national standard of ultrasonic testing method of forged and rolled steel rod, the ultrasonic C scanning is realized for steel rods with the diameter range of 250mm, which ensures the digitalization and visibility of the product quality inspection result of steel rods.

Key words: steel rod; ultrasonic detection; C-scan; immersion focusing

油膜轴承在韶钢宽厚板轧机的应用及改进初探

陈功彬，王 萍

（宝武集团广东韶关钢铁有限公司特轧厂，广东韶关 512123）

摘 要：简要介绍了韶钢松山股份有限公司板材厂宽厚板轧机的油膜轴承结构，重点阐述了油膜轴承在应用中遇到的问题，主要是油品质量明显异常，同时油膜轴承密封使用寿命大幅缩短，进一步检查发现轧机辊系窜动较大。针对发现的问题，通过调整轧机牌坊滑板间隙，控制了辊系窜动，同时对油膜轴承结构改良，将原单止推油膜轴承，改良为双止推油膜轴承。改进措施实施后，取得了明显效果。本文对轧机油膜轴承日常设备维护、油膜轴承密封国产化、改进措施的实施效果以及取得的经验进行了总结，以供业内其他宽厚板厂参考和借鉴。

关键词：油膜轴承；工作原理；改进措施

Application and Improvement Practice of Oil Film Bearing for Wide and Heavy Plate Rolling Mill at SGIS

Chen Gongbin, Wang Ping

(Baowu Group Guangdong Shaoguan Iron and Steel Co.,Ltd., Shaoguan 512123, China)

Abstract: A brief introduction on the oil film bearing structure of wide and heavy plate rolling mill at SGIS is given.The problems encountered in practical application are mainly described, that is the oil quality is obviously abnormal, at the same time, the service life of the oil film bearing seal is shortened greatly. Further examination shows that roll axial shifting is high. In view of the problems found, by adjusting the mill housing slide clearance, the roll axial shifting has been controlled, and at the same time, the improvement on oil film bearing structure has been done, by replacing the single thrust bearings with double thrust bearings. Obvious effect has been obtained since the implementation of improvement measures. The oil film bearing in rolling mill equipment daily maintenance, oil film bearing seal effect of localization, the implementation of improvement measures and experience are summarized.

Key words: oil film bearing; working principle; improvement measures

镶嵌式限位支架在真空槽合金翻板阀中的应用

何肇恒

（宝钢湛江钢铁有限公司炼钢厂，广东湛江 524000）

摘 要：为了减少RH合金翻板限位更换作业时工作人员身体所受热辐射的伤害、提高真空槽上下线的效率以及限位拆装作业的安全性，作者自行设计一种镶嵌式的限位支架，以镶嵌的限位安装方式代替原来拧螺丝的限位安装方式。实践表明使用镶嵌式限位支架可降低真空槽合金翻板限位拆装的工作难度和工作强度，减少劳动者工作过程所受热辐射的时间，可有效保护劳动者的身体健康。对于安装在温度高、操作困难、工作环境差的限位可使用镶嵌式限位支架以提高工作效率，降低工作人员的劳动强度。

关键词：合金翻板阀；限位支架；镶嵌式；热辐射

Application of Inlaid Limit Bracket in Vacuum Tank Alloy Flap Valve

He Zhaoheng

(Baosteel Zhanjiang Iron & Steel Making Plant, Zhanjiang 524000, China)

Abstract: In order to reduce the damage of heat radiation to the health of the workers when the limit of RH alloy is changed, and to improve the efficiency of the vacuum tank and the safety of the limit disassembly, the author has designed a set of limit switch bracket, and the original way of screw limit switch installation would be replaced. The practice shows that the use of the inlaid limit switch bracket can reduce the difficulty and intensity of the disassembly and assembly of the alloy plate with vacuum tank. Also, it can reduce the time of heat radiation to the working process of the laborers, and effectively protect the physical health of the laborers. As a result, the inlaid bracket can be used especially in the limit switch installation in high temperature, difficult operation, poor working environment so as to improve work efficiency, reduce the labor intensity of the staff.

Key words: alloy plat; limit switch bracket; inlaid; heat radiation

一种缓冲阀的设计和应用

肖争光，李松磊，丛津功，高轶桐

（鞍钢股份鲅鱼圈钢铁分公司厚板部，辽宁营口 115000）

摘 要：在液压与气动设备中，工作缸在刚启动和刚停止的短时间内会有压力和流量的突变，对工作缸、管路及运动执行机构造成较大冲击，需要进行缓冲处理。常用的缓冲结构存在一些弊端，本文旨在提出一种成本小、耦合性低、缓冲行程易改变、布置灵活、维修更换简便的新型缓冲阀。该缓冲阀可在多种场合替代其他缓冲元件或结构，提高液压及气动系统设计和使用的经济效益。

关键词：缓冲；阀；液压；气动；经济效益

Design and Application of a New Type of Buffer Valve

Xiao Zhengguang, Li Songlei, Cong Jingong, Gao Yitong

(The Heavy Plate Department of Ansteel Bayuquan Iron & Steel Subsidiary, Yingkou 115000, China)

Abstract: In hydraulic and pneumatic equipments, the working cylinder will have sudden changes in pressure and flow within a short period of starting and stopping, which will cause a great impact on the working cylinder, pipeline and motion actuator, and need to be buffered. The common buffer structure has some disadvantages. This paper presents a new type of buffer valve with low cost, low coupling, flexible arrangement and easy maintenance and replacement. The buffer valve can replace other buffer elements or structures in many occasions to improve economic benefits in the design and use of hydraulic and pneumatic systems.

Key words: buffer; valve; hydraulic; pneumatic; economic benefits

UCM 冷轧机力学行为及板形调控性能研究

贾生晖[1]，陶　浩[2]，韩国民[2]，李洪波[2]

（1. 天津市新宇彩板有限公司，天津　300382；2. 北京科技大学机械工程学院，北京　100083）

摘　要：为分析某厂 UCM 六辊冷轧机的力学行为及其对带钢的板形调控性能，结合现场实际工艺参数，利用 ANSYS 有限元软件建立了 UCM 轧机四分之一辊系静力学仿真模型。通过仿真计算得到了单位轧制力、弯辊力和窜辊量这些因素作用下的承载辊缝形状，其反映了轧机板形调控性能的强弱；并得出了这些因素对轧机辊间接触压力的影响规律，可以发现辊间接触压力峰值一般出现在接触区域边部，且随着单位轧制力的增大，辊间接触压力整体增大。所建模型与分析结果将为 UCM 冷轧机的现场板形控制提供理论依据，同时也为改善轧辊的辊间磨损甚至剥落提供一定的参考。

关键词：UCM 冷轧机；有限元仿真；辊间接触压力；板形调控

Analysis of Mechanical Behavior and Flatness Control Characteristics for UCM Cold Rolling Mill

Jia Shenghui[1], Tao Hao[2], Han Guomin[2], Li Hongbo[2]

(1. Tianjin Xinyu Color Coated Board Co., Ltd., Tianjin 300382, China; 2. School of Mechanical Engineering, University of Science and Technology Beijing, Beijing 100083, China)

Abstract: In order to analyze the mechanical behavior and the flatness control characteristics for a certain UCM cold rolling mill, combined with the actual parameters in the field, the ANSYS finite element software was used to establish a static simulation model of the quarter roll systems of the UCM mill. Through the simulation calculation, the roll gap shapes under the factors of the unit rolling force, the bending force and the roll shift value are obtained, which reflects the flatness control characteristics; and the influences of these factors on the roll contact pressure are obtained. It can be found that the roll contact pressure peaks generally appear at the edge of the contact area, and with the unit rolling force increasing, the roll contact pressure increases on the whole. Ultimately, the simulation model and the analysis results will not only provide a theoretical basis for the flatness control of the UCM cold rolling mill but also provide a reference for improving the roll wear even the spalling.

Key words: UCM cold rolling mill; finite element simulation; roll contact pressure; flatness control

轧钢加热炉垫块的选型应用及发展趋势

赵　俣，马光宇，刘常鹏，张天赋

（鞍钢集团钢铁研究院，辽宁鞍山　114009）

摘　要：介绍了轧钢加热炉垫块选型应用的发展现状，结合现场实际应用情况提出了传统垫块的材质及其安装方式存在的问题，对新型垫块的发展趋势进行了展望。

关键词：加热炉；垫块；安装方式；金属玻璃；发展趋势

Selection Application and Development Trend of the Skid in Steel Rolling Heating Furnace

Zhao Yu, Ma Guangyu, Liu Changpeng, Zhang Tianfu

(Ansteel Group Iron and Steel Research Institute, Anshan 114009, China)

Abstract: Introducing the current situation of skid selection and application in reheating furnace, combining the situation in production process, the problem related to material and installation mode on traditional skid have been put forward, and the development trend on newly skid is made.

Key words: heating furnace; skid; installation mode; metallic glass; development trend

冷轧带钢激光拼焊板焊缝过拉矫机力学行为仿真研究

陈 兵，张海涛，唐晓垒，李晋鹏

（北京科技大学机械工程学院，北京 100083）

摘 要：本文针对某钢企酸轧线拉矫机为提高机械破鳞效率及提高带钢板带表面质量，实现拉矫工艺的降本增效之目的，对拉矫机工艺参数优化展开技术攻关的实际需求。随着机械破鳞效率的提升，导致板带焊缝在过拉矫工艺时发生裂纹萌生甚至断带等事故发生概率增大，为保证生产顺行，急需对板带焊缝部位过拉矫工艺段时，焊缝部位与拉矫工艺参数之间关系展开研究，探究宽幅薄板带激光焊接部位过拉矫机的力学行为与拉矫工艺参数间作用规律。在试验测试获取板带物性参数基础上，使用有限元的方法对带钢焊缝区域过拉矫时的动态受力状态进行仿真模拟。通过设置典型仿真工况，改变拉矫机的插入深度和拉矫张力组合，探究拉矫机工艺参数变化时焊缝不同区域的受力变化情况，分析拉矫工艺参数对带钢缺陷产生的影响，找到带钢焊缝区域易发生断裂的原因，减少焊缝处断带的可能性，为带钢焊缝过拉矫机工作辊时工作参数的选择提供理论指导。

关键词：冷轧；拉矫机；焊缝；力学行为；有限单元法；仿真

Simulation Study on Mechanical Behavior of Welding Seam Tension Correction Machine for Laser Welding Plate of Cold Rolled Strip Steel

Chen Bing, Zhang Haitao, Tang Xiaolei, Li Jinpeng

(School of Mechanical Engineering, University of Science and Technology Beijing, Beijing 100083, China)

Abstract: Aiming at the purpose of improving the mechanical scale-breaking efficiency and surface quality of steel sheet and strip, this paper presents the practical requirements of technical optimization of the tension levator for the purpose of reducing cost and increasing efficiency of the tension levator. As to promote the efficiency of mechanical breaker, which leads to the plate with weld crack initiation in a masking process during the accident probability increase, such as strip break even in order to ensure the production line, need to strip weld parts through proper process segment, the weld area and researches on the relationship between process parameters correction, explore the wide sheet with laser welding parts have pull machine mechanical behavior and the proper role law between the process parameters. On the basis of the physical properties of strip obtained by test, the dynamic stress state of strip weld area under tension correction is simulated by finite element method. By setting the typical simulation condition, change pull machine insertion depth and proper tension, explore pull machine process parameters change in different areas of the weld stress changes of analysis, the impact of process parameters on the strip defect correction, find the reason of strip steel weld area prone to fracture, reduced the likelihood of seam broken belt, in strip steel weld machine work roll during the working parameters of choice to provide theoretical guidance.

Key words: cold rolling; tension leveler; weld seam; mechanical behavior; finite element method; simulation

成品筛激振器的发热分析及改进实践

程明森

（宝钢股份宝山基地炼铁厂，上海 201900）

摘 要：成品筛作为烧结工艺中成品矿的筛分设备，对于成品矿的粒度控制、成品率、粉矿率等有着重要的影响。目前成品筛激振器存在寿命短，尤其是 4DL 成品一次筛激振器在夏季高温天气极易出现激振器发热、轴承损坏的故障。本文通过对目前成品筛一次筛激振器故障情况进行汇总分析，得出成品筛激振器劣化的各因素，从理论和实践上给出可行的改进方案，提高激振器的寿命。

关键词：烧结成品筛；激振器；轴承；故障分析；提升寿命；改进

Heating Analysis and Improvement Practice of Fine Screen Exciter

Cheng Mingsen

(Baoshan Iron & Steel Co., Ltd., Baoshan Base Ironmaking Plant, Shanghai 201900, China)

Abstract: As a screening equipment for finished ore in sintering process, finished product screening has an important influence on the granularity control, finished product rate and powder ore rate of finished ore. At present, the service life of the finished vibrator is short. Especially in the high temperature weather in summer, the 4DL primary vibrator is prone to the failure of the vibrator heating and bearing damage. By summarizing and analyzing the fault of the primary screen vibrator of the finished product screen at present, the factors of the deterioration of the screen vibrator of the finished product are obtained, and the feasible improvement scheme is given in theory and practice, so as to improve the life of the vibration exciter.

Key words: vibrating screen; exciter; bearing; failure analysis; prolong life span; improvement

11　冶金自动化与智能化

大会特邀报告

分会场特邀报告

炼铁与原料

炼钢与连铸

轧制与热处理

表面与涂镀

金属材料深加工

先进钢铁材料

粉末冶金

能源、环保与资源利用

钢铁材料表征与评价

冶金设备与工程技术

★ 冶金自动化与智能化

建筑诊治

其他

绝对值编码器常用算法缺陷及解决策略

孙 抗

（首钢京唐钢铁联合有限责任公司冷轧部，河北唐山 063200）

摘 要：绝对值编码器是工业生产中常用的位置检测元件，通过其反馈的码值计算实际位置。本文简要介绍了绝对值编码器的工作原理，常用算法。以笔者经历的事故为例，提出常用算法的两类主要缺陷：数据溢出与编码器码值溢出。针对第一类缺陷给出了解决方案，针对第二类缺陷的两种形式给出补偿算法，并给出注意事项。

关键词：绝对值编码器；位置检测；常用算法缺陷；数据溢出；码值溢出

Common Algorithmic Deficiencies and Solutions for Absolute Encoders

Sun Kang

(Cold Metal Dept., Shougang Jingtang Iron and Steel
United Corporation in Limited, Tangshan 063200, China)

Abstract: The absolute encoder is a position detecting component commonly used in industrial production, and the actual position is calculated by the code value of the feedback. This article briefly introduces the working principle of the absolute encoder and the commonly used algorithms. The author takes the accidents that occurred in his own production line as an example, and proposes two defects of main types of common algorithms: compensation of data overflow caused by inappropriate data types, and encoder code value overflow. A solution is given for the first type of defects, and a detailed compensation algorithm is given for the two forms of the second type of defects, and many considerations for practical use are given.

Key words: absolute encoder; position detecting; defects of common algorithms; data overflow; code value overflow

热轧轧辊磨损及补偿控制

陈 晨，孙建林，谭耘宇

（上海梅山钢铁股份有限公司，江苏南京 210039）

摘 要：热轧带钢板型的控制是多方面的，影响因素主要有轧制力、弯辊力、串辊位置、轧辊热膨胀、轧辊轮廓、轧辊磨损。这几个因素相互影响制约，共同控制着带钢板型的好坏。在众多因素中，轧辊磨损因素较为不可控。支撑辊表面往往是不均匀的、长期的磨损，并且支撑辊属于基础性的承载轧制力，所以对支撑辊磨损模型的与计算是十分必要的。在现有的板型控制模型中加入了补偿控制方法，产品的质量也产生了非常可观的提升。

关键词：轧辊磨损；模型；控制

Wear and Compensation Control of Hot Rolling

Chen Chen, Sun Jianlin, Tan Yunyu

(Shanghai Meishan Iron and Steel Co., Ltd., Nanjing 210039, China)

Abstract: The control of hot-rolling strip steel is various, and the main influencing factors include force, bending force, rolling position, roll thermal expansion, roll prof wear. These factors influence each others and control the quality of steel together. Roller wear is the most uncontrollable factor among these factors. The surface of supporting rolling is often uneven, long-term wear and support the basic force.So it is very necessary to calculate the roll wear model.The correct control method is added into the existing control model, and the product quality is great.

Key words: roll wear; model; control

钢卷库无人天车系统的开发及应用

丁 辉[1]，邹永生[2]，赵 平[2]

（1. 首钢智新迁安电磁材料有限公司，河北迁安 064400；
2. 北京首钢股份有限公司，河北迁安 064400）

摘 要：近年来，随着自动化技术的不断进步，很多钢铁企业把物流和信息化有机结合，形成智能仓储物流系统，满足企业物流自动化、专业化、信息化、系统化、智能化的需求。本文针对大型钢厂钢卷库内存在的用工人数多、效率低、安全风险大的客观特点研发了无人天车系统，该系统实现了现场无人看管值守，自动吊运货物的功能，为企业节约了用工成本。

关键词：钢卷库；无人天车；控制系统；PLC

Development and Application of Unmanned Crane System for Steel Coil Warehouse

Ding Hui[1], Zou Yongsheng[2], Zhao Ping[2]

(1. Shougang Zhixin Qianan Electromagnetic Material Co., Ltd., Qianan 064400, China;
2. Beijing Shougang Co., Ltd., Qianan 064400, China)

Abstract: In recent years, with the continuous progress of automation technology, many steel companies organically integrate logistics and information technology to form an intelligent warehouse logistics system to meet the needs of enterprise logistics automation, specialization, informationization, systemization, and intelligence. In this paper, the unmanned crane system is developed for the objective characteristics of large number of workers, low efficiency and high safety risk in the steel coil warehouse of large steel mills. The system realizes the function of unattended on-site inspection and automatic lifting of goods. Enterprises save labor costs.

Key words: steel coil library; unmanned crane; control system; PLC

基于机器学习实现冷轧产品的动态工艺调整

张振宇，彭晶晶，梁燕燕

（宝钢湛江钢铁有限公司，广东湛江 524000）

摘 要：为了实现热镀锌产品在炼钢、热轧工艺波动时，冷轧工序可以自动做出相应的调整，而不是按照原设计的目标值控制，提升产品性能的稳定性。采用机器学习算法随机森林作为预测模型，设计损失函数，利用梯度下降算法计算工艺调整的最优解，使最终产品性能波动降低。

关键词：机器学习；热镀锌产品；590DP；动态工艺调整；损失函数

Dynamic Process Adjustment of Cold Rolled Products based on Machine Learning

Zhang Zhenyu, Peng Jingjing, Liang Yanyan

(Baosteel Zhanjiang Iron and Steel Co., Ltd., Zhanjiang 524000, China)

Abstract: In order to realize 590DP hot-dip galvanizing product in steelmaking and hot rolling process fluctuation, the cold rolling process can be adjusted automatically, instead of controlling according to the original design target value, so as to improve the stability of product performance. Machine learning algorithm stochastic forest is used as prediction model, loss function is designed, and gradient descent algorithm is used to calculate the optimal solution of process adjustment, which reduces the fluctuation of final product performance.

Key words: machine learning; hot galvanized products; 590DP; dynamic process adjustment; loss function

焦炉电机车自动化应用

孔弢，张允东，甘秀石，赵明，高薇，张其峰

（鞍钢股份有限公司炼焦总厂，辽宁鞍山 114000）

摘 要：根据岗位每日往复频繁操作电机车实际完成的接红焦，运输焦罐，提升机对位的动作易出错，造成事故的问题，通过智能化读码器读取各车码牌，反馈到自身车辆的走行PLC，无线通信传输到平台确定相对位置关系，驱动车辆完成各种动作，提高机械效率，使整体焦炉车辆系统的自动化水平提高，节约人力，减少人工劳动强度，而且由于各车辆准确有效的生产秩序对环保效果和焦炉机械设备维护均有改进。

关键词：焦炉；自动化；电机车

Automation Application of Coke Oven Locomotive

Kong Tao, Zhang Yundong, Gan Xiushi, Zhao Ming, Gao Wei, Zhang Qifeng

(Angang Co., Ltd., Coking Plant, Anshan 114000, China)

Abstract: According to the fact that the position frequently operates the electric locomotive daily to and fro to and fro, the action of conveying the coke tank and hoist alignment is easy to make mistakes, which causes accidents. Intelligent reader reads the license plates and feeds back to the running PLC of its own vehicle. It determines the relative position relationship through the vehicle wireless communication transmission to the platform, drives the vehicle to complete various actions, improves the mechanical efficiency and makes the whole vehicle complete. The automation level of body coke oven vehicle system is improved, manpower is saved and labor intensity is reduced. Moreover, the environmental protection effect and the maintenance of coke oven machinery and equipment are improved due to the accurate and effective production order of each vehicle.

Key words: coke oven; automation; locomotive

MES 系统在冶金产线的应用与发展

范海龙

（酒钢（集团）宏兴股份公司宏兴股份公司不锈钢分公司，甘肃嘉峪关 735100）

摘 要：目的： 研究以 MES 系统为代表的智能化、信息化制造技术在冶金、加工产线的典型应用及发展前景。**方法：** 从企业全产线订单排产、物料跟踪、生产过程实绩采集、质量管理、产品运输管理等方面，对比采用 MES 系统管理前后的生产经营情况与指标。**结果：** MES 系统能够贴合并指导实际生产运营，同时实现所有产品全生命周期跟踪管理，规范了业务流程与生产操作，系统运行稳定，已经为企业带来现实以及潜在的经济和社会效益。**结论：** 利用 MES 项目可以使生产过程控制系统与 ERP 系统衔接，实现企业从生产模式、机构设置到生产运营管理业务流程的规范化和标准化，从而全方位提高企业的生产经营和信息化管理水平。同时能够对企业内部实施成本控制，极大提高生产效率、产品质量和服务水平。

关键词： 冶金产线；MES 系统信息化；标准化

Application and Development of MES System in Metallurgical Production Line

Fan Hailong

(Hongxing Iron & Steel Co., Ltd., Jiuquan Iron and Steel Group Corporation, Jiayuguan 735100, China)

Abstract: Objective: To study the application and development prospect of intelligent manufacturing and information technology represented by MES system in typical steel manufacturing lines. **Methods:** From the aspects of order management, material management, production process tracking, quality management and product outward tracking, the situation before and after the implementation of MES system was compared and analyzed. **Result:** MES system can well adapt to the actual production situation. It can also track all products in production and standardize the business operation process. The production data in the system conforms to the actual situation and runs steadily. It has brought realistic and potential economic benefits to enterprises. **Conclusion:** The implementation of MES project can make the production centralized control system and SAP system link up well, realize the standardization and standardization of enterprise's production mode, organization setup and business process, so as to improve enterprise's production management and information management level in an all-round way. Implementing accurate control of the internal cost of the enterprise greatly improves the production efficiency, product quality and service level.

Key words: metallurgical production line; MES system informatization; standardization

基于 PCA 算法的高炉参数规则指标排序研究

张秀春，宋 政，周道付，范 晨

(飞马智科信息股份有限公司，安徽马鞍山 243000)

摘 要：本文提出一种基于 PCA 算法的高炉参数指标排序评价方法。以马钢 3 号高炉历史参数指标数据为基础，通过 PCA 算法对原有参数指标进行降维提取，得到新的高炉综合指标；根据实验结果对指标的优先级别进行排序；最后，结合本文算法对高炉冶炼状态进行总体评价。

关键词：PCA 算法；指标排序；高炉参数

Research on Index Ranking of Blast Furnace Parameter Rules based on PCA Algorithms

Zhang Xiuchun, Song Zheng, Zhou Daofu, Fan Chen

(PHIMA Intelligence Technology Co., Ltd., Maanshan 243000, China)

Abstract: The paper, a method of ranking and evaluating blast furnace parameters based on PCA algorithm is presented. Based on the Historical Parameter Index Data of No.3 BF in Masteel, Using PCA algorithm to extract dimension reduction of original parameters, Obtaining New Comprehensive Index of Blast Furnace. Priority ranking of indicators based on experimental results. Finally, the overall evaluation of BF smelting status is carried out with the algorithm proposed in this paper.

Key words: PCA algorithm; index ranking; blast furnace parameters

iSCADA 系统在 CSP 厂系统迁移项目中的应用

徐 重

(宝钢股份武汉钢铁有限公司条材厂，湖北武汉 430080)

摘 要：介绍了 iSCADA 技术原理和特点，iSCADA 与传统服务器架构方式下的优势，重点叙述了 CSP 厂系统迁移项目的背景，以及 iSCADA 在 CSP 厂系统迁移项目应用的可行性，iSCADA 系统采用一体化云平台作为整个系统的管理平台，整个系统分为就地控制层、网络通讯层及控制管理层。就地控制层采用具有以太网通讯功能的可编程控制器完成对生产工艺数据的采集及对设备的控制，网络通讯层实现现场与控制管理层之间的数据传输，通讯方式全部统一改为以太网通讯，控制管理层实现对生产工艺数据的监视、调度及控制，控制室根据用户需求分成地面系统中控室；用户可通过管理员授权在内网覆盖的地方登入平台，监控和管理整个系统的运行情况。

关键词：iSCADA；云计算；分布式架构；虚拟机

Application of iSCADA System in System Migration Project of CSP Plant

Xu Zhong

(Bar Mill of Wuhan Iron and Steel Co., Ltd., Baosteel, Wuhan 430080, China)

Abstract: This paper introduces the principle and characteristics of iSCADA technology, the advantages of iSCADA and traditional server architecture, and focuses on the background of system migration project in CSP plant, as well as the feasibility of application of iSCADA in system migration project in CSP plant. iSCADA system uses integrated cloud platform as the management platform of the whole system. The whole system is divided into local control layer, network communication layer and control management layer. Local control layer uses programmable controller with Ethernet communication function to collect production process data and control equipment. Network communication layer realizes data transmission between field and control management layer. Communication mode is changed to Ethernet communication. Control management layer realizes monitoring, scheduling and control of production process data. Control room divides production process data according to user's needs. Users can log in to the platform where the intranet is covered by administrator authorization to monitor and manage the operation of the whole system.

Key words: iSCADA; cloud computing; distributed architecture; virtual machine

Python 数据分析工具在 CSP 厂信息化开发中的应用

徐 重

（宝钢股份武汉钢铁有限公司条材厂，湖北武汉 430083）

摘 要： 信息技术的飞速发展导致大量的数据生成，为了从这些海量的数据中提取有用的关键信息，各个企业投入了大量的人力物力去进行数据的整理筛选和报表开发，往往分析的结果却差强人意，现如今电子商务、CRM、MES 等信息技术越来越发达，我们身处数字信息大爆炸的时代，只有通过采用先进的数据信息处理产品，才能改变钢铁行业信息化技术发展落后、保守的现状，促使钢铁企业在发展的道路上越来越顺畅。本篇论文介绍了 Python 数据分析工具在 CSP 信息化开发中的应用实践经验。

关键词： Python；钢铁企业；数据分析；信息化

Application of Python Data Analysis Tool in Information Development of CSP Plant

Xu Zhong

(Bar Mill of Wuhan Iron and Steel Co., Ltd. Baosteel, Wuhan 430083, China)

Abstract: The rapid development of information technology has led to a large number of data generation. In order to extract useful key information from these massive data, various enterprises have invested a lot of manpower and material resources in data sorting, screening and report development. The results of analysis are often unsatisfactory. Nowadays, e-commerce,

CRM, MES and other information technologies are more and more developed, and we are in a big explosion of digital information. In the era, only through the use of advanced data information processing products, can we change the backward and conservative status of information technology development in iron and steel industry, and promote iron and steel enterprises on the road of development more and more smoothly. This paper introduces the application experience of Python data analysis tool in CSP information development.

Key words: Python; iron and steel enterprises; data analysis; informatization

炼钢数字化单元的集成实践与思考

李立勋，杨晓江，康书广，王　欣，张道良

（河钢乐亭钢铁有限公司，河北唐山　063000）

摘　要： 当前，新一代信息技术与制造业的深度融合正在引发影响深远的产业变革，智能制造已成为主要制造强国的竞争制高点。钢铁工业发展智能制造是行业提质增效、提高有效供给水平的一条重要途径。钢铁企业的智能制造工作是庞杂的系统工程，在推进过程中既要有顶层格局又要有细致的支撑步骤。全面数字化、高端模型化、深度信息化是拉动冶金企业走向中高端的三架马车。扎实做好钢铁流程每个工序和车间单元的数字化、模型化、信息化工作筑牢底层基础的同时预留网络协同制造系统的数据接口，通过工业网络和分布式工业ICT系统将各基础单元组网感知与协同运转，用高质量数据、高端冶金工艺模型、深度网络自感知与广域工序网络协同驱动冶金工业高质量发展。

关键词： 钢铁工业；数字化；模型化；智能制造；网络协同

Practice and Consideration of Steelmaking Digital Unit Integration

Li Lixun, Yang Xiaojiang, Kang Shuguang, Wang Xin, Zhang Daoliang

(Laoting Iron & Steel Co., Ltd. of HBIS Group, Tangshan 063000, China)

Abstract: At present, the deep integration of new generation information technology and manufacturing industry is triggering far-reaching industrial changes. Intelligent manufacturing has become the competitive highlight of major manufacturing powers. The development of Intelligent Manufacturing in iron and steel industry is an important way to improve the quality and efficiency of the industry and to improve the level of effective supply. Intelligent manufacturing in iron and steel enterprises is complex system engineering. In the process of advancing, both pattern and detailed steps should be taken. Comprehensive digitalization, advanced process control and deep informatization are the three carriages driving metallurgical enterprises to the middle and high-end. The data interface of network cooperative manufacturing system is reserved while the digitization, modulation and informatization of each process and workshop of manufacturing process are well done. Through industrial network and distributed industrial ICT system, the basic units are networked, perceived and coordinated, the whole metallurgical process driven by high-quality data, advanced metallurgical process model, deep network self-perception and wide area process network-cooperation system.

Key words: iron and steel industry; digital; modulation; intelligent manufacturing; network-cooperation

基于 Spark 的大数据网络日志采集分析和预警模型研究

易可可，钱哲怡，王 威

（上海宝信软件股份有限公司，上海 201203）

摘 要：随着互联网的深入应用，电商系统中相关网络日志信息规模爆发式增长，为了应对各种情况的监控，我们运用最新的大数据技术，研究一种基于 Spark 技术进行网络日志大数据的采集分析和预警模型，并对其进行总体架构设计和实现，详细地阐述了系统的底层日志原始文件日志采集、存储、逻辑处理、分析结果存储和展现。该模型不仅能实时、高效的分析日志信息，计算统计出分析结果实时展现，而且针对历史日志数据进行快速存储和离线计算，帮助运维人员快速定位实际发生的问题。

关键词：大数据 Spark；离线计算；实时计算；日志分析

Research on Big Data Network Log Collection Analysis and Early Warning Model based on Spark

Yi Keke, Qian Zheyi, Wang Wei

(Shanghai Baosight Software Co., Ltd., Shanghai 201203, China)

Abstract: With the development of the Internet, network system log information scale explosive growth, in order to deal with all kinds of condition monitoring, we use the latest technology of big data, research a Spark technology based on network log large data collection and analysis and early warning model, and carries on the overall architecture design and implementation, in detail elaborated the system of the underlying original log file log collection storage logic processing storage and analysis results show the model can not only real-time Efficient analysis of log information, calculation and statistics to show the analysis results in real time, and for the historical log data for fast storage and off-line calculation, to help the operation and maintenance personnel to quickly locate the actual problems.

Key words: big data Spark, off-line computation; real-time computation; log analysis

武钢 CSP 厂 1 号 RH 真空系统 S4a、S4b 喷嘴腐蚀失效故障的分析判断及改进

张 云，王南生，李文晓，徐 重

（武钢条材总厂 CSP 分厂，湖北武汉 430080）

摘 要：运用真空泵系统工作原理对武钢 CSP 厂 1 号 RH 真空系统 S4a、S4b 喷嘴腐蚀失效的故障进行分析判断，查出影响真空度的原因，有效地解决设备故障，并在此基础上提出一些相应的改进措施。

关键词：真空系统原理；内循环；真空度

Failure Analysis and Improvement of RH Vacuum System S4a, S4b Nozzle Corrosion

Zhang Yun, Wang Nansheng, Li Wenxiao, Xu Zhong

(CSP Branch of General Wire Rod Mill of WISCO, Wuhan 430080, China)

Abstract: The system works using a vacuum pump for WISCO CSP plant 1 # RH vacuum system S4a, S4b nozzle corrosion failure analysis to determine fault, to identify factors affecting the degree of vacuum, effective solution to equipment failure, corresponding well to make some improvements on this basis measures.

Key words: vacuum system principles; inner loop; vacuum degree

PRINCE2 方法在 CSP 一键式炼钢模型项目的应用

徐 重，张 云

（宝钢股份武汉钢铁有限公司条材厂，湖北武汉 430080）

摘 要：目的： 整合资源，形成合力，完成 CSP 一键式炼钢模型项目系统建设。**方法：** 采用全球流行的 PRINCE2 项目管理方法对项目实施进行管理，以保证自动炼钢这项影响 CSP 产品结构调整的信息化建设达到预期效果。**结果：** 经过一年的时间，项目进展顺利，从项目论证到工程的完成，实现了预期目标。**结论：** PRINCE2 是一种开放的项目管理方法论，通过标准化的项目实施方法确保了自动炼钢项目的完成，标准的管理改变了靠经验和能力决定项目成败的传统做法，提升了信息建设的管理能力，同时也为 PRINCE2 方法论在冶金制造行业的落地提供了范例。

关键词： 项目管理；自动炼钢；模型；PRINCE2

Application of PRINCE2 Method in CSP Fully Automatic Steelmaking Model Project

Xu Zhong, Zhang Yun

(Bar Mill of Wuhan Iron and Steel Co., Ltd., Baosteel, Wuhan 430080, China)

Abstract: Objective: To integrate resources and form joint efforts to complete the construction of CSP fully automatic steelmaking model project system. **Methods:** The project implementation was managed by the global popular PRINCE2 project management method to ensure that the information construction of automatic steelmaking, which affects the structural adjustment of CSP products, achieves the expected results. Result: After one year, the project progressed smoothly, from project demonstration to project completion, and achieved the expected goal. **Conclusion:** PRINCE2 is an open project management methodology, which ensures the completion of automatic steelmaking projects through standardized project implementation methods. Standard management has changed the traditional way of deciding the success or failure of projects by experience and ability, improved the management ability of information construction, and provided an example for the landing of PRINCE2 methodology in metallurgical manufacturing industry.

Key words: project management; automated steelmaking; model; PRINCE2

高炉炉衬监测技术发展和应用探讨

张 伟，朱建伟，张立国，任 伟，李金莲

（鞍钢股份有限公司技术中心，辽宁鞍山 114009）

摘 要：高炉安全长寿一直是炼铁工作者重点关注的课题，炉衬侵蚀状态的监测意义重大，对分别基于热、电、声等反应变化判断炉衬状态的几类监测技术进行了阐述，重点介绍了热流强度推断法、电阻法、超声波法、冲击回波法等典型技术的基本原理和特点，通过分析不同监测技术的优势和不足，对企业未来应用高炉炉衬监测技术提出了建议。

关键词：高炉；炉衬；厚度检测；冲击回波法；炭砖

Application and Its Development for Blast Furnace Lining Thickness Detection Technology

Zhang Wei, Zhu Jianwei, Zhang Liguo, Ren Wei, Li Jinlian

(Technology Center, Ansteel Limited Company, Anshan 114009, China)

Abstract: The long campaign life of blast furnace operation are always the focus for iron-making researchers, and the BF lining erosion became more and more important.the review for lining thickness detecting technology based on heat/elcetricity/wave has been conducted, and highlight the principle and character of the heat detection method, resistance method, ultrosonic wave method,impact clastic wave method. Finally some advice and suggestion has been given for these detecting technology via the advantage and dis-advantage analyse.

Key words: blast furnace; lining; testing thickness; impact-echo method; carbon brick

冷轧镀锌线减少焊缝过光整机张力波动的优化研究

崔伦凯，郭俊明，张 军，张振勇，李凤惠

（北京首钢冷轧薄板有限公司，北京 101300）

摘 要：冷轧镀锌机组通常配置有光整机，通常采用延伸率控制，焊缝过光整机时会采用减压过焊缝的方式。这样在减压时，由于压下量的剧烈变化，导致张力的大幅波动，从而造成焊缝前后带钢表面的斜纹缺陷，也会影响到内部的力学性能，降低了成材率。本文针对首钢冷轧镀锌线焊缝过光整机张力波动的问题进行分析,通过在焊缝到达光整机前，光整机停止延伸率控制，设定轧制力降低到减压轧制力，过光整机后，再提升至新带钢的L2设定轧制力。在设定轧制力变化时，光整机的设定速度同步发生变化，以补偿轧制力变化带来的秒流量变化，从而提升了张力的稳定性，提高了表面质量。

关键词：镀锌线；光整机；焊缝；张力

The Optimization of Tension Control on the Skin Pass Mill

Cui Lunkai, Guo Junming, Zhang Jun, Zhang Zhenyong, Li Fenghui

(Beijing Shougang Cold Rolling Co., Ltd., Beijing 101300, China)

Abstract: The cold galvanizing line usually equips with skin pass mill (SPM), which is used to improve surface quality and mechanical properties of the coating. SPM always uses elongation control, and the force will decrease greatly when the weldseam passes mill. In this way, the drastic change of reduction leads to a large fluctuation of tension, resulting in surface defects of strip steel, which also affects the internal mechanical properties and reduces the yield. This paper analyses the problem appeared in the SPM's application of Shougang CGL. By disabling the SPM elongation control before the weldseam reaches the mill, the rolling force setpoint is reduced to the reducing rolling force, and it is raised to the L2 rolling force setpoint of the new strip after weldseam passed the mill. At the meantime, the speed of the SPM changes synchronously to compensate for the change of the mass flow caused by the change of rolling force, thus improving the stability of tension and surface quality.

Key words: cold galvanizing line; skin pass mill; weldseam; tension

一种环境在线监测数据采集系统

楼 纬

(山信软件股份有限公司,山东济南 250101)

摘 要: 文章介绍了山钢集团日照有限公司新上线的环境在线监测数据采集系统,首先,介绍了采集系统的整个架构,然后详细介绍了系统的软硬件组成和工作原理,最后介绍了系统上线后在日照有限公司的应用情况和所创造的效益。

关键词: 数据采集;网关;环境;CitectSCADA2016;Modbus-RTU

Environment Online Monitoring Data Acquisition System

Lou Wei

(Shanxin Software Co., Ltd., Jinan 250101, China)

Abstract: This paper introduces a new online environment online monitoring data acquisition system in Shandong Iron & Steel Group Rizhao Co., Ltd., first of all, this paper introduces the whole structure of this collection system, then introduced the system hardware and software composition and working principle, at last, the application effectiveness and beneficial result are introduced after this project go to use in Rizhao Co., Ltd.

Key words: data collection; gateway; environment; CitectSCADA2016; Modbus-RTU

污水处理的自控系统设计

王 猛

(山信软件莱芜自动化分公司,山东济南 271104)

摘 要:本文介绍了污水处理的自控系统的设计。自控系统是整个污水处理设备的核心部分。全厂的自动化运行控制应不依赖于一个控制装置或系统,以提高控制系统的整体可靠性。局部故障的影响范围应被限制在单体内部,即使中央控制室因故停止运行,各单体控制系统仍可独立运行。

关键词:污水处理;自控系统;管理监视;数据报表

Design of Automatic Control System for Sewage Treatment

Wang Meng

(Shanxin Software Laiwu Automation Branch, Jinan 271104, China)

Abstract: This paper introduces the design of automatic control system for sewage treatment. The automation system is the core part of the entire sewage treatment equipment. The automatic operation control of the whole plant should not depend on a control device or system to improve the overall reliability of the control system. The impact of local faults should be limited to the internal single unit. Even if the central control room stops operating for some reason, each single unit control system can still operate independently.

Key words: sewage treatment; automatic control systems; manage surveillance; data statements

关键皮带故障原因分析与技术改进

朱庆庙[1],庞克亮[1],张允东[2],甘秀石[2],侯士彬[2],马银华[3]

(1. 鞍钢集团钢铁研究院,辽宁鞍山 114009;2. 鞍钢股份有限公司炼焦总厂,
辽宁鞍山 114021;3.鞍钢股份有限公司鲅鱼圈钢铁分公司,辽宁营口 115007)

摘 要:确立炼焦生产系统中的关键皮带,对关键皮带系统故障、耦合器窜油故障原因进行系统分析,阐述了耦合器、易熔塞的工作原理,最终通过对关键皮带负载电流进行监测、控制进而控制皮带输送焦炭量,确立了当关键皮带电流达到80A并超过40s时整个皮带系统自动停机,改进了关键皮带上游第一条皮带堵溜器的位置及型号,并增加了堵溜器负重,避免由于皮带超负荷运行造成耦合器窜油故障,避免关键皮带故障造成的焦炉紧急停产事故的发生。

关键词:关键皮带;耦合器;易熔塞;电流;自动控制;研究;应用

Analysis and Technical Improvement of Key Belt Fault

Zhu Qingmiao[1], Pang Keliang[1], Zhang Yundong[2], Gan Xiushi[2], Hou Shibin[2], Ma Yinhua[3]

(1. Technology Center of Angang Steel Co., Ltd., Anshan 114009, China;

2. Ansteel Stock Company Coking Plant, Anshan 114021, China;
3. Bayuquan Branch of Angang Steel Co., Ltd., Yingkou 115007, China)

Abstract: Key belt in coking production system is established, the key to the belt system fault coupler oil system analysis the cause of the problem, this paper expounds the working principle of coupler soluble plug, finally through to the key monitoring control and control of belt conveyor belt load current amount of coke, establish key belt when current of 80A and more than 40 s automatic stop when the belt system, improved the upstream first key belt belt plugging position and type of slip, and increases the wall slip load and avoid couplers oil failure due to belt overload operation, to avoid the key fault belt caused production of coke oven emergency accidents.

Key words: key belt; coupler; fusible plug; electricity; automation; research; application

焦炉机车信息化管控系统研究与实践

于庆泉，周 鹏，马富刚，崔金林，孙 强

（鞍钢股份有限公司鲅鱼圈钢铁分公司，辽宁营口 115007）

摘 要：随着信息化与工业化的融合发展，对工业设备的信息化管控已成为工业生产提质增效的迫切要求。为了提高炼焦生产中焦炉机车的管理效率和控制效果，鞍钢股份鲅鱼圈钢铁分公司炼焦部建设了涵盖全部焦炉机车的无线通信网络，搭建了用于焦炉机车信息化管控的硬件平台和软件平台，开发了多种管理功能和控制功能，形成了一种焦炉机车信息化管控系统。实现了焦炉机车在信息化、自动化、智能化等方面的全面提升。

关键词：焦炉机车；信息化；管控

Research and Practice of Informationalized Control System for Coking Locomotive

Yu Qingquan, Zhou Peng, Ma Fugang, Cui Jinlin, Sun Qiang

(Ansteel Company Bayuquan Iron & Steel Subsidiary, Yingkou 115007, China)

Abstract: With the integration and development of informatization and industrialization, the information management and control of industrial equipment has become an urgent requirement for improving the quality and efficiency of industrial production. In order to improve the management efficiency and control effect of coking locomotives in coking production, the coking Department of Angang Bayuquan Iron and Steel Company has built a wireless communication network covering all coking locomotives, built a hardware platform and software platform for information management and control of coking locomotives, developed a variety of management functions and control functions, and formed an information management and control system of coking locomotives. The comprehensive improvement of coking locomotive in information, automation and intellectualization has been realized.

Key words: coking locomotive; informationize; management and control

基于产销协同处理的钢铁资源分配模型

王丽婷，刘志军

（北京首钢自动化信息技术有限公司信息事业部，北京 100041）

摘 要：为了满足钢铁企业生产、销售计划管理的实际业务，加强信息化系统在实际业务中的应用，在调研明确某大型钢铁股份有限公司硅钢用户需求的基础上，本文围绕产品价格和客户订货需求，综合考虑客户资源预报结果、资源预测、市场情况、价格政策、产能计划建立资源分配模型且根据一定规则实现资源分配，同时启用交期算法将客户交期精确至 5 天之内。从而在提高客户满意度的基础上保证了企业的效益，实现了产销无缝衔接。

关键词：资源预报；资源预测池；备料；销售计划；资源平衡

Steel Resource Allocation Model based on Production and Marketing Co Processing

Wang Liting, Liu Zhijun

(Information Business Department of Beijing Shougang Automation Information Technology Co., Ltd., Beijing 100041, China)

Abstract: In order to meet the actual business of production and sales plan management in iron and steel enterprises and strengthen the application of information system in actual business, we are focusing on the product price and the demand of ordering, and takes into account the results of customer resource forecast, the resource prediction, the market situation, the price policy and the capacity planning to establish a resource allocation model on the basis of investigating and expliciting the demand of Silicon Steel customer which is in a big steel company Limited by Share Ltd . It realizes the resource allocation according to the certain rules and takes advantage of the intersection algorithm to compute the delivery date of every customer at the same time. The model controls the delivery date within 5 days accurately. It improves the satisfaction of customers, ensures the efficiency of enterprises and realizes seamless connection between production and marketing.

Key words: resource forecast; resource forecast pool; stock preparation; sales plan; resource balance

超大规模项目全生命周期管理信息系统研究

王晓蕾，牛春波

（山钢集团日照有限公司信息计量部，山东日照 276800）

摘 要：钢铁企业工程建设项目投资大、耗时长、采购范围广，工程建设期间的项目管理工作、招标采购管理和财务核算管理工作由于涉及部门多，业务流程长，管理难度特别大，因此迫切需要信息系统的支撑，以实现项目建设过程中的规范化、透明化和精益化管理。本文研究了利用信息系统，在钢铁企业工程建设背景下项目全生命周期的管理工作，对信息化时代的工程建设管理工作具有指导意义。

关键词：项目管理；核算；工程建设；钢铁企业

Research on Life Cycle Management Information System of Ultra Large-scale Project

Wang Xiaolei, Niu Chunbo

(SD Steel Rizhao Co., Ltd., Rizhao 276800, China)

Abstract: Iron and steel enterprises have large investment, long time-consuming and wide purchasing scope. Project management, bidding and purchasing management and financial accounting management during the construction period need the support of information system urgently in order to achieve standardization, transparency and lean management in the process of project construction, because they involve many departments, have long business processes and are particularly difficult to manage. This paper studies the management of the whole life cycle of the project under the background of the construction of iron and steel enterprises by using information system, which has guiding significance for the construction management in the information age.

Key words: project management; accounting; engineering construction; steel enterprises

智能化管控信息系统的设计与应用

李 健，陈向阳，刘 杰

（山东钢铁集团日照有限公司，山东日照 276800）

摘 要：工业化的发展离不开信息化的支撑。随着业务快速增长，公司的生产、运营对信息化系统的依赖越来越强，对信息化系统的稳定性、可靠性提出了更高的要求。同时，业务的多样化也使信息系统技术和管理日渐复杂，需要有稳定、可靠、智能化的信息化服务支撑。本文针对新模式下信息化管控模式，综合运用数据集中采集、在线状态监控、数据深度挖掘、二次开发利用等技术，完成了对信息化设备与系统运行状况的问题预知预判、故障自诊断、知识沉淀、专家知识库等智能化功能，实现了信息化设备与系统运行状态全方位、全天候、全覆盖的管控，提升了公司管控信息化水平。

关键词：信息化管控；知识沉淀；专家知识库；智能管理

Design and Application of Intelligent Information System Control

Li Jian, Chen Xiangyang, Liu Jie

(Shandong Iron and Steel Group Rizhao Co., Ltd., Rizhao 276800, China)

Abstract: The development of industrialization can not be separated from the support of information technology. With the rapid growth of business, the company's production and operation of the information system is more and more dependent on the stability and reliability of the information system put forward higher requirements. At the same time, the diversification of business has also made information system technology and management increasingly complex, and it needs to be supported by stable, reliable and intelligent information services. Under the new model, this paper comprehensively uses the techniques of centralized data collection, online status monitoring, data depth mining, and secondary development and utilization. It has completed intelligent functions such as forecasting problems, self-diagnosis of failures, knowledge precipitation, and expert knowledge bases for the operation of information-based equipment and systems, and has achieved omni-directional, all-weather, and full-coverage control over the operation of information-based equipment and systems. It has improved the level of corporate information management.

Key words: informatization control; knowledge precipitation; expert knowledge base; intelligent management

大包回转台准确定位法在异形坯连铸机中的应用

魏晨溪

（莱芜钢铁集团有限公司，山东济南 271104）

摘 要：大包浇铸位置的准确定位，是确保滑动水口正常开关的关键。否则会导致大包水口无法正常打开，钢水无法顺利浇注，造成连铸机降拉速、停浇等事故。本文提出的大包回转台准确定位的方法，是通过PLC系统实时采集回转台旋转时变频器转速信号，计算出大包回转台的实际旋转的位置。对大包回转台旋转位置进行精确定位，避免了位置检测元件不准导致的水口无法正常打开的弊端，确保了连铸机生产顺行。

关键词：大包回转台；变频器转速；旋转的位置；准确定位法；连铸机

Application of the Method of Precise Positioning for Ladle Turret on Beam Blank Continuous Caster

Wei Chenxi

(Laiwu Steel Group Co., Ltd., Jinan 271104, China)

Abstract: The precise positioning of ladle casting location is the key to ensure the sliding nozzle normally open and close. Otherwise, it will result in the failure of the opening of ladle nozzle and the smooth pouring of molten steel, resulting in the drawdown speed of continuous caster and the failure of pouring. In this paper, the method of precise positioning of the ladle turret is to calculate the actual rotation position of the ladle turret by collecting the speed signal of the frequency converter in real time when the ladle turret rotates through the PLC system. The rotary position of the ladle turret was precisely positioned to avoid the defects that the nozzle could not be opened normally due to inaccurate position detection elements and ensure the production of continuous caster smoothly.

Key words: ladle turret; speed signal of the frequency converter; rotation position; the method of precise positioning; continuous caster

大型钢铁企业产销一体化信息系统设计与实现

徐守新，牛春波，杨金光

（山东钢铁集团日照有限公司，山东日照 276800）

摘 要：本文结合山钢日照公司实际案例，探讨了大型钢铁企业产销一体化信息系统的设计与实现，介绍了系统设计背景、设计思路及技术方案。产销系统一体化信息系统借鉴吸收国际先进的管理理念，实现了IT与业务融合创新，为营销、研发、生产、成本、质量管理等提供有效支撑，实现生产经营管理流程的智能化、简约化、扁平化，实现生产经营物流、资金流、信息流的集成控制。

关键词：产销；B/S；订单；质量设计；生产调度；质量判定；管理流程

Design and Implementation of Integrated Information System for Sales and Production in Large Iron and Steel Enterprises

Xu Shouxin, Niu Chunbo, Yang Jinguang

(Shandong Iron and Steel Group Rizhao Co., Ltd., Rizhao 276800, China)

Abstract: Based on a practical case of Shandong Iron and Steel Group Rizhao Co., Ltd, this paper discusses the design and implementation of the integrated marketing and production information system for a large iron and steel enterprise, and introduces the design background, design ideas and technical scheme of the system. The integrated information system of production and marketing system draws lessons from advanced international management concepts, realizes the integration and innovation of IT and business, provides effective support for marketing, R&D, production, cost and quality management, realizes the intellectualization, simplification and flattening of production and operation management process, and realizes the integrated control of production and operation logistics, capital flow and information flow.

Key words: production and marketing; B/S; order; quality design; production scheduling; quality judgment; management process

智能视频监控绊线穿越检测在河钢石钢的研究与应用

刘 霞

（河钢石钢信息物流中心，河北石家庄 050031）

摘 要： 近年来，智能视频监控[1]已经在很多领域得到广泛的应用。它是一个多个学科交叉的研究领域，也是当前计算机视觉、图像处理方向的研究热点。目前，社会各界对物流安全监控的投入力度越来越大，在某些特殊的工作场合更是如此。智能视频监控加上智能分析很好的解决了"人的工作注意力不能长时间集中在传统的视频监控画面上"这个问题，且系统无需人工干预就能在及时报警的同时留下影像，以备事后查证。河钢石钢成功的应用了视频监控的绊线穿越检测技术，在车辆违规转弯控制上取得新突破。

关键词： 智能视频监控；智能分析；绊线报警；车辆转弯布控

Research and Application of Intelligent Video Monitoring Trigger Crossing Detection in HBIS Shisteel

Liu Xia

(HBIS Shisteel Information Logistics Center, Shijiazhuang 050031, China)

Abstract: In recent years, intelligent video surveillance has been widely used in many fieids.It is an interdisciplinary field, and also a hot research topic in the direction of computer vision and image processing. At present, all sectors of society are investing more and more in logistics secutity monitoring,especially in some special work occasions.Intelligent video surveillance and intelligent analysis can solve the problem of "people's working attention can not be focused on the traditional video surveillance screen for a long time", and the system can alarm in time without manual intervention while leaving images for later verification. HBIS SHISTEEL has successfully applied video surveillance tripping detection technology, and made a new breakthrough in vehicle illegal tutning control.

Key words: intelligent video surveillance; intelligent analysis; tripping wire alarm; vehicle turning control

轧机主电机负载平衡控制的研究应用

王 博

(宝钢湛江钢铁有限公司厚板厂,广东湛江 524000)

摘 要： 轧机上下辊电机转矩平衡控制是轧机主传动控制系统不可或缺的一部分，对生产工艺具有重大意义，本文针对厚板厂轧机主电机原负载平衡控制功能导致的速度波动问题，介绍了一种解决在负载平衡控制投入后导致速度波动大的问题。在粗、精轧主传动负载平衡控制功能经调试投入运行后，效果良好。它的使用可以有效地平衡上下辊电机的输出转矩，可以有效地解决原负载平衡控制功能导致的电机运行速度不稳的问题，防止速度不稳对钢板质量产生影响，对厚板生产轧制工艺具有重要意义。

关键词： 主传动；负载平衡；转矩偏差；附加速度

Research and Application of Load Balancing Control for Main Motor of Rolling Mill

Wang Bo

(Heavy Plate Plant of Baosteel Zhanjiang Iron and Steel Co., Ltd., Zhanjiang 524000, China)

Abstract: Torque balance control of upper and lower roll motors is an indispensable part of the main drive control system of rolling mill, which is of great significance to the production process. In this paper, the problem of speed fluctuation caused by the original load balance control function of the main motor of rolling mill in heavy plate mill is introduced, which solves the problem of speed fluctuation caused by the input of load balance control. After debugging and putting into operation, the load balancing control function of the main drive of roughing and finishing rolling has achieved good results. Its use can effectively balance the output torque of the upper and lower roll motor, effectively solve the problem of motor speed instability caused by the original load balancing control function, and prevent the speed instability from affecting the quality of steel plate, which is of great significance to the rolling process of thick plate production.

Key words: main drive; load balancing; torque deviation; additional speed

炼焦煤质量预警系统的实现和应用

卢 瑜，杜 屏，赵华涛

(江苏省沙钢钢铁研究院,江苏张家港 215625)

摘 要：炼焦煤的质量是影响焦炭质量的重要因素。某 7.63m 焦炉使用的炼焦煤品种多、种类复杂、变化大，造成炼焦配煤难度大、配比变动大、焦炭质量稳定性差。针对该问题，基于公司级原辅料检化验平台，开发了一套炼焦煤质量自动预警系统。系统基于炼焦煤质量大数据库，实现炼焦煤质量自动预警和炼焦煤质量趋势显示等功能。系统的上线，解决了炼焦煤质量数据分散的问题，能对质量异常来煤进行大数据分析，对异常项目进行自动预警，并能根据用户需求个性化显示炼焦煤质量趋势，减少了因为来煤质量异常造成的焦炭质量脱标事故，使焦炭质量趋于稳定。
关键词：炼焦煤；焦炭质量；质量预警；大数据；配煤

Realization and Application of Coking Coal Quality Early Warning System

Lu Yu, Du Ping, Zhao Huatao

(Institute of Research of Iron and Steel of Jiangsu Province (Shasteel), Zhangjiagang 215625, China)

Abstract: The quality of coking coal is an important factor affecting the quality of coke. There are many types of coking coal used in a 7.63m coke oven, the types are complex, and the changes are large, which makes the coking coal blending personnel difficult to blend coal, the ratio of the mix is large, and the coke quality stability is poor. In response to this problem, based on the company-level raw and auxiliary materials inspection and testing platform, an automatic warning system for coking coal quality was developed. The system includes a large database of coking coal quality, automatic warning of coking coal quality and a trend chart of coking coal quality. The on-line of the system solves the problem of dispersing the quality data of coking coal. It can analyze the big data of abnormal coals and automatically warn the abnormal items, and can personally display the quality trend of coking coal according to the needs of users, and reduced the quality of coke quality due to abnormal coal quality, and coke quality tends to be stable.
Key words: coking coal; coke quality; quality warning; big data; blend of coal

基于大数据分析的可视化智能运维系统在钢铁企业的应用

杨 恒[1]，周 平[1]，范 鹍[2]，黄少文[1]，李长新[1]

（1. 山东钢铁集团有限公司研究院，山东济南 250100；
2. 山东钢铁集团日照有限公司信息计量部，山东日照 276800）

摘 要：设备高质量运行是钢铁企业提升产品质量和企业竞争力的基本要素，基于大数据分析的可视化智能运维系统的应用，有效的解决了钢铁企业设备管理的共性问题，实现了关键设备的全生命周期管理，提升了企业设备管理水平。
关键词：大数据分析；可视化；智能运维系统；全生命周期管理

Application of Visual Intelligent Operation and Maintenance System based on Big Data Analysis in Iron and Steel Enterprise

Yang Heng[1], Zhou Ping[1], Fan Kun[2], Huang Shaowen[1], Li Changxin[1]

(1. Research Institute of Shandong Iron and Steel Group Co., Ltd., Jinan 250100, China;
2. Shandong Iron and Steel Group Rizhao Co., Ltd., Rizhao 276800, China)

Abstract: The high-quality operation of equipment is the basic factor of improving product quality and competitiveness of iron and steel enterprises. The application of visual intelligent operation and maintenance system based on big data analysis effectively solves the common problems of equipment management in iron and steel enterprises, realizes the life cycle management of key equipment and improves the level of equipment management in enterprise.

Key words: big data analysis; isualization; intelligent operation and maintenance system; life cycle management

钢铁流程工序界面信息一体化融合实践

谢 晖[1]，周 平[1]，范 鹍[2]，王成镇[1]，杨 恒[1]

(1. 山东钢铁集团有限公司研究院，山东济南 250100；
2. 山东钢铁集团日照有限公司信息计量部，山东日照 276805)

摘 要： 山东钢铁集团日照有限公司利用物联网、大数据、移动互联等先进信息技术，在原料、炼铁、炼钢、热轧、冷轧、成品等关键钢铁流程工序实施了系列"界面技术"，实现生产过程中工序间柔性衔接、合理匹配，促进生产流程整体运行的稳定协调和连续化、高效化，从而提高钢铁生产流程的资源效率、能源效率，实现钢铁生产流程的整体优化。

关键词： 钢铁流程；工序界面；一体化融合；智能制造

Practice of Integrative Information Fusion of Process Interface in Iron and Steel Process

Xie Hui[1], Zhou Ping[1], Fan Kun[2], Wang Chengzhen[1], Yang Heng[1]

(1. Research Institute of Shandong Iron and Steel Group Co., Ltd., Jinan 250100, China; 2. Informatization & Meterlogy Dept. of Shandong Iron and Steel Group Company Rizhao Co., Ltd., Rizhao 276805, China)

Abstract: Rizhao Co., Ltd. of Shandong Iron and Steel Group has implemented a series of "interface technologies" in key steel process processes such as raw materials, ironmaking, steelmaking, hot rolling, cold rolling and finished products by using advanced information technology such as Internet of Things, big data and mobile interconnection, so as to realize flexible connection and reasonable matching between processes in production process, and to promote stable coordination, continuity and high efficiency of the overall operation of production process. In order to improve the resource efficiency and energy efficiency of iron and steel production process and realize the overall optimization of iron and steel production process.

Key words: iron and steel process; process interface; integrated integration; intelligent manufacturing

基于钢板轮廓精确检测的智能优化剪切方法的应用

吴昆鹏，杨朝霖，石 杰

(北京科技大学高效轧制国家工程研究中心，北京 100083)

摘 要：中厚板平整工艺主要包括切头、切尾、切双边和定尺等，其中的关键在于尺寸的定位。由于人工剪切过程大多依赖经验目测，容易因头部过剪切而导致板长不足的问题。本文提出了基于轮廓精确提取的优化剪切方法，首先，针对轮廓提取的精度优化问题，分别提出通过高精度编码器控制线阵相机速率来优化长度方向精度，以及通过多相机排布横向角度偏移方法来优化宽度方向精度。然后，针对常见的板形缺陷镰刀弯，提出两个指标对其数据进行量化处理。最后提出优化剪切方法，获取钢板最大有效长度、头尾剪切余量等参数指导剪切。本方法应用于现场在宽度方向测量精度可达到±2mm，长度方向测量精度在±20mm，可实现智能剪切的目的，提高钢板成材率。

关键词：图像拼接；轮廓检测；镰刀弯检测；优化剪切

Application of Intelligent Optimization Shearing Method based on Accurate Detection of Steel Plate Contour

Wu Kunpeng, Yang Chaolin, Shi Jie

(National Engineering Research Center for Advanced Rolling Technology, University of Science and Technology Beijing, Beijing 100083, China)

Abstract: The plate flattening process mainly includes cutting head, tail cutting, cutting bilateral and fixed length, etc., and the key is the positioning of the size. Since the manual shearing process mostly relies on empirical visual inspection, it is easy to cause the problem of insufficient plate length due to excessive shearing of the head. In this paper, an optimal shearing method based on contour extraction is proposed. Firstly, for the accuracy optimization problem of contour extraction, the high-speed encoder is used to control the linear camera speed to optimize the length direction accuracy, and the horizontal angle is shifted by multiple cameras. Move the method to optimize the width direction accuracy. Then, for the common plate shape defect, two indicators are proposed to quantify the data. Finally, an optimized shearing method is proposed to obtain the parameters such as the maximum effective length of the steel plate and the head-to-tail shear allowance. The method is applied to the field to measure the accuracy in the width direction to ±2mm, and the measurement accuracy in the length direction is ±20mm, which can achieve the purpose of intelligent shearing and improve the yield of the steel plate.

Key words: image stitching; contour detection; sickle bending detection; optimized cutting

六辊轧机厚控系统精度问题分析及应对措施

李 胤

（宝钢股份武钢有限硅钢部，湖北武汉 430083）

摘 要：厚控系统是以厚度作为被控制量的自动控制系统，生产过程中，厚控系统异常可能导致产品厚度波动超标，影响产品的质量产量及新产品的开发，厚控系统的正常投用及控制精度对轧机的稳定轧制和产品质量具有重要意义。本文介绍了厚控系统的基本原理和前馈、反馈、秒流量厚度控制方式原理，通过分析厚控系统电气控制程序和典型问题iba曲线，结合现场生产实际，对厚控系统精度问题产生的原因和各个要素进行了详细的分析和探讨研究，并从生产工艺、操作、零级、一级等方面提出了切实可行的应对措施。现场实践证明，改进后效果良好，可以确保厚控系统正常投用及产品控制精度，满足生产需求。

关键词：厚控系统；前馈；反馈；秒流量；控制精度

Analysis and Countermeasure of Accuracy Problem of Thickness Control System of Six-high Mill

Li Yin

(Baosteel Co., Ltd., WISCO Silicon Steel Department, Wuhan 430083, China)

Abstract: Thickness control system is an automatic control system with thickness as the controlled quantity. In the production process, abnormal thickness control system may lead to excessive fluctuation of product thickness, affecting the quality and output of products and the development of new products. The normal operation and control accuracy of thickness control system are of great significance to the stable rolling of rolling mill and product quality. This paper introduces the basic principle of the thickness control system and the principle of feedforward, feedback and second flow thickness control mode. By analyzing the electrical control program of the thickness control system and the IBA curve of typical problems, and combining with the actual production in the field, the causes of the accuracy problems of the thickness control system and the various elements are analyzed and studied in detail. Practical countermeasures are put forward in terms of production process, operation, zero-level and first-level. Field practice has proved that the effect of the improvement is good, which can ensure the normal commissioning of the thickness control system and the accuracy of product control, and meet the production needs.

Key words: thickness control system; feedforward; feedback; mass flow; control accuracy

设备全生命周期管理系统建设与应用

安丰涛，宋振兴，苏　威，石　江

（河钢集团宣钢公司，河北张家口　075100）

摘　要： 大型钢铁企业的设备管理中，基于信息化系统下的设备全生命周期管理技术应用，能为企业的设备的综合管理提升新高度。在设备前期、中期、后期的各种业务应用场景中，通过信息化技术，进行流程化、规范化管理。以企业现有信息化系统为基础，参照两化融合体系管理标准，依据现行组织管理体系，建立一套规范、实用、可靠、高效、可扩展的设备全生命周期管理信息系统。

关键词： 设备；全生命周期；信息化技术；ERP

Construction and Application of Equipment Life Cycle Management System

An Fengtao, Song Zhenxing, Su Wei, Shi Jiang

(Xuan Steel Company of Hebei Iron and Steel Group, Zhangjiakou 075100, China)

Abstract: In the equipment management of large iron and steel enterprises, the application of equipment life cycle management technology based on information system can improve the comprehensive management of equipment of enterprises to a new height. In all kinds of business application scenarios in the early, middle and late stages of equipment, process and standardized management is carried out through information technology. Based on the existing enterprise information system, referring to the management standard of the integration system of two modernizations, and according to the current organizational management system, a set of standardized, practical, reliable, efficient and scalable equipment life cycle management information system is established.

Key words: equipment; life cycle; information technology; ERP

宣钢高炉 EPU 防喘控制系统技术应用

李蕴华

（河北钢铁集团宣钢公司设备能源部，河北张家口 075100）

摘　要：节能降耗是钢铁行业重要的经济指标之一，本文针对冶金行业高炉鼓风机实际运行工况，分析了高炉与鼓风机对应风量的性能关系，并基于实际案例采用了 EPU 防喘振性能技术，实现了节能降耗，提高了经济效益，具有较高的应用推广价值。

关键词：高炉鼓风机；防喘控制；动力装置

宣钢 2500m³ 高炉炉顶设备最优控制

王春元，闫新卓，蔡　宇，薄　巍，倪力鑫，周建军

（河北钢铁集团宣钢公司机电公司，河北宣化 075100）

摘　要：本文重点介绍了宣钢 2500m³ 高炉通过提高检测精度、控制程序优化等方式，提高了高炉炉顶设备的控制精度，对于高炉利用系数、降低焦比奠定了基础，在同类型高炉的炉顶设备自动化检测控制方面具有一定的推广和应用价值。

关键词：无料钟炉顶；料面高度检测；料流阀控制；倾动控制

横河 DCS 在 435m² 烧结机自控系统中的开发及应用

霍迎科，申存斌，宋林昊

（河钢集团邯钢公司自动化部，河北邯郸 056015）

摘　要：435m² 烧结机是邯钢东区产业升级改造中的一个重要项目。435m² 烧结机的工艺构成主要有原燃料上料系统、配料系统、烧冷系统、筛分等各部分组成；DCS 控制系统根据工艺生产设备分布情况由三套 CPU 控制系统组成，主要论述了横河 DCS 控制系统的结构、组成及控制程序的控制功能及应用。

关键词：DCS；ESB 总线；ES；OS；冗余

Development and Application of YOKOGAWA DCS Automation System in the 435m² Sintering Machine

Huo Yingke, Shen Cunbin, Song Linhao

(Hebei Iron and Steel Group Handan Iron and Steel Co., Ltd., Handan 056015, China)

Abstract: 435m² sintering machine is an important project in the industrial upgrading and transformation of Eastern District of Handan Iron and Steel Co., Ltd. The process composition of 435m² sintering machine is mainly composed of raw fuel feeding system and so on. DCS control system is composed of three sets of CPU control system according to the distribution of process production equipment. The structure, compostion and control function and application of the control system of the YOKOGAWA DCS are discussed.

Key words: DCS, ESB BUS, ES, OS, redundancy

基于模式识别的电机系统在线监测研究

黄晨晨，郭光辉，郝春辉，张勇军

（北京科技大学工程技术研究院，高效轧制国家工程研究中心，北京　100083）

摘　要： 电机是钢铁冶金工业生产中不可或缺的电气设备，其正常运转对流程化大规模生产的稳定性和连续性至关重要。电机在运行过程中的任何异常故障都可能会导致整个生产线中断运行，造成巨大的经济损失甚至危及设备安全。电机运行状态在线实时监测技术可以有效地保证电机系统的运行安全，本文通过智能传感器采用无接触式采集得到表征电机健康状态的数据，并尝试利用小波包和BP神经网络分析运行电机的实时关键参数，根据电机的当前和历史运行状态，提前对可能出现的故障状态做出预警，实现对电机运行状态的在线监测。通过对基于该方法自主开发的电机在线监测装置进行测试，结果表明在线监测系统能够实时对电机运行状态进行状态判定，有助于对生产线中的在线运行电机进行不停机检测与预测性维护。

关键词： 电动机；在线监测；预警；状态识别

Research on Motor Online Monitoring System based on Pattern Recognition

Huang Chenchen, Guo Guanghui, Hao Chunhui, Zhang Yongjun

(Institute of Engineer Technology, University of Science and Technology Beijing, National Engineering Research Center for Advanced Rolling Technology, Beijing 100083, China)

Abstract: The electric machine is an indispensable electrical equipment in the production of iron and steel metallurgy industry, and its normal operation is essential for the stability and continuity of large-scale process production. Any abnormal faults of the motor during operation may lead to the interruption of the entire production line, causing huge economic losses and even jeopardizing equipment safety. The on-line real-time monitoring technology of motor running state can effectively ensure the safety of the motor system. This paper uses the non-contact acquisition of intelligent sensors to obtain the data characterizing the health of the motor, and attempts to analyze the real-time key parameters of the running motor by using wavelet packet and BP neural network. According to the current and historical operating state of the motor, an early warning of possible fault conditions is made in advance to realize online monitoring of the running state of the motor. By testing the on-line monitoring device of the motor independently developed based on the method, the results show that the online monitoring system can judge the running state of motor in real time, which is helpful for non-stop detection and predictive maintenance of the online running motor in the production line.

Key words: motor; online monitoring; early warning; state recognition

在线实时成分检测技术在炼铁行业应用前景

马洪斌

（北京宇宏泰测控技术有限公司技术研发中心，北京　100043）

摘　要： 在线实时成分检测技术能够在线连续检测各种元素含量，以"在线-实时-数字化"解决了荧光和化学分析方法"离线-滞后-人工化"的弊病，适用于钢铁企业原料采购、混匀料场、球团车间、烧结车间、高炉车间、焦化车间、喷煤车间生产过程自动控制。蒂森克虏伯 Duisburg 厂 2008 年，柳钢烧结厂 2014 年，宝钢股份炼铁部烧结分厂 2016 年，鞍钢东鞍山烧结厂 2017 年安装在线实时成分检测设备，另外在线实时成分检测在钢铁行业前端的矿山行业已经广泛应用于来料分级、料堆质量管理、原料配料、细粒、中粒、大块产品监测等。炼铁行业应用方案，一类是提供在线成分检测结果，另一类是通过在线成分检测结果，进行闭环自动成分调整，稳定原料成分，稳定生产过程和生产指标。在线实时成分检测技术已经度过了其在炼铁行业的适应期，硬件、软件针对炼铁行业工况已经能够很好的适应，自动配料系统经过各项参数选择、摸索、试验后，已经能够平稳的运行，并取得了很好的生产指标。

关键词： 炼铁；在线实时成分检测；自动配料；中子活化

加热炉二级优化控制系统研究与开发

蒋国强，张宝华，陈建洲

（广东韶钢松山股份有限公司，广东韶关　512123）

摘　要： 针对广东韶钢松山股份有限公司大棒 1#加热炉二级优化控制系统开展研究工作，开发了钢坯加热过程数学模型，实现了大棒 1#加热炉物料及温度跟踪、工艺查询及修改、历史数据查询和钢坯详细温度记录等功能模块，所开发的二级优化控制系统可对钢坯加热过程进行有效控制，为大棒 1#加热炉钢坯加热质量提高及节能降耗做出贡献。

关键词： 大棒加热炉；二级优化控制系统；钢坯加热过程

Research and Development of the Secondary Optimization Control System for the Heating Furnace

Jiang Guoqiang, Zhang Baohua, Chen Jianzhou

(Guangdong Shaoguan Songshan Co., Ltd., Shaoguan 512123 China)

Abstract: Research on the secondary optimization control system of the No. 1 heatingfurnace of Shaogang's large bar mill was carried out. A mathematical model of billet heatingprocess was developed, realizing function modules of material and temperature tracking, processinquiry and modification, history data inquiry and detailed recording of billet temperature.The newly developed secondary optimization control system can perform effective controlof billet heating process, to improve billet heating quality and achieve energy saving andconsumption reduction.

Key words: large bar heating furnace; secondary optimization control system; billetheating process

钢铁企业数据应用创造价值的实践

熊 鑫，王里程，白 雪

（鞍钢集团自动化有限公司，辽宁鞍山 114021）

摘 要：工业互联网相关技术逐步成熟与深入应用，为钢铁企业工业互联网平台建设提供了技术手段，提升了企业挖掘数据应用创造价值的能力。针对钢铁企业全流程价值创造的需求，结合钢铁企业的生产组织特点，通过数据有效治理，提升数据成为资产的转化率，通过数据应用服务的典型案例，实现为企业带来经济效益，助推钢铁企业数字化转型，引领行业高质量发展。

关键词：钢铁企业；工业互联网；数据资产；全流程；价值创造

Practice of Creating Value by Data Application in Iron and Steel Enterprises

Xiong Xin, Wang Licheng, Bai Xue

(Information Industry Co., Ltd., Anshan Iron and Steel Group, Anshan 114021, China)

Abstract: Industrial Internet technology has gradually matured and applied in depth, which provides a technical means for the construction of industrial Internet platform in iron and steel enterprises, and improves the ability of enterprises to mine data applications and create value. In view of the demand of value creation in the whole process of iron and steel enterprises, combined with the characteristics of production organization of iron and steel enterprises, the conversion rate of data into assets can be improved through effective data management. Through the typical cases of data application service, it can bring economic benefits to enterprises, promote the digitalized transformation of iron and steel enterprises, and lead the high-quality development of the industry.

Key words: iron and steel enterprises; industrial internet; data assets; whole process; value creation

立式连续退火炉炉辊温度模型研究

李洋龙[1]，任伟超[2]，陈 飞[1]，王 慧[1]

（1. 首钢集团有限公司技术研究院冶金过程研究所，北京 100043；
2. 首钢京唐钢铁联合有限责任公司冷轧部，河北唐山 064000）

摘　要：根据炉辊不同位置的传热方式，对炉辊进行区域划分，基于传热学原理建立了立式连续退火炉炉辊温度模型，用于预测炉辊长度方向的温度分布，研究了带钢宽度、带钢温度和辐射管温度对炉辊温度分布的影响。带钢宽度增加使炉辊温度分布趋向于带钢温度，随着带钢温度和辐射管温度的升高，炉辊整体温度逐渐增加，当辐射管温度明显大于带钢温度时，炉辊边缘区的温度大于炉辊中心区温度，削弱了炉辊正凸度；当辐射管温度明显小于带钢温度时，炉辊边缘区的温度小于炉辊中心区温度，增强了炉辊正凸度。利用立式连续退火炉炉辊温度模型，提高了立式连续退火炉2级系统中对带钢发生瓢曲和跑偏现象的判断能力，降低了带钢发生瓢曲和跑偏现象对生产的不利影响。

关键词：连续退火炉；炉辊；带钢；温度模型

Study on Temperature Model of Furnace Roller in Vertical Continuous Annealing Furnace

Li Yanglong[1], Ren Weichao[2], Chen Fei[1], Wang Hui[1]

(1. Department of Metallurgical Process, Research Institute of Technology of Shougang Group Co., Ltd., Beijing 100043, China; 2. Cold Rolling Department, Shougang Jingtang United Iron & Steel Co., Ltd., Tangshan 064000, China)

Abstract: According to the different heat transfer ways at different positions of the furnace roller, the temperature model of furnace roll in the vertical continuous annealing line is established based on the heat transfer principle, which is used to predict the temperature distribution in the length direction of the furnace roller. The influence of the strip width, the strip temperature and the radiant tube temperature on the temperature distribution of the furnace roller is studied. The increase of strip width makes the temperature distribution of the roller tends to the strip temperature. With the increase of the strip temperature and radiant tube temperature, the overall temperature of the furnace roller increases gradually. When the radiant tube temperature is obviously larger than the strip temperature, the temperature at the edge of the furnace roller is greater than that at the center of the roll, which weakens the positive crown of the roller. When the radiant tube temperature is less than the strip temperature, the temperature at the edge of the roller is smaller than that in the central area of the roller, which enhances the positive crown of the furnace roll. Based on the temperature model of furnace roller in vertical continuous annealing furnace, the judgment of strip buckling and running deviation in the lever 2 system for the continuous annealing furnace is improved, and the adverse effects of strip buckling and running deviation on production are reduced.

Key words: continuous annealing furnace; furnace roller; strip; temperature model

AI技术在冷轧连退平整机控制模型中的应用

袁文振[1]，黄华钦[2]，司华春[2]，高新建[2]，谢　谦[2]

（1. 宝山钢铁股份有限公司冷轧厂，上海　201900；
2. 安徽工业大学冶金工程学院，安徽马鞍山　243002）

摘　要：随着AI技术的创新发展，人工智能与神经网络技术的应用也日趋广泛。目前，连退及热镀锌机组平整机的控制模式仍为静态表控制，如何获取准确的参数设定值，提高产品质量，成为制约平整机控制效果的瓶颈问题。本文通过建立AI神经网络预测模型，将共轭梯度算法作为反向传播训练算法，对现场积累的大量生产数据进行智能分析，实现准确的轧制力及配套张力的设定，为平整机轧制模型与控制系统研究提供帮助。

关键词：冷轧工艺；平整机；人工智能；神经网络；控制系统

Application of AI Technology in Control of Cold Rolling Continuous Leveling Mill

Yuan Wenzhen[1], Huang Huaqin[2], Si Huachun[2], Gao Xinjian[2], Xie Qian[2]

(1. Baoshan Iron and Steel Co., Ltd., Division of Cold Rolling, Shanghai 201900, China;
2. Anhui University of Technology, School of Metallurgical Engineering, Maanshan, 243002, China)

Abstract: With the innovation and development of AI technology, the application of artificial intelligence and neural network technology is becoming more and more important. The control mode of leveling mill of continuous retreat and hot dip galvanizing unit is still static table control. How to obtain accurate parameter setting values and improve product quality has become the bottleneck problem that impede the improving of leveling control effect. In this paper, the AI neural network prediction model is established to control the leveling process, and the conjugate gradient algorithm is used as the back-propagation training algorithm. The large amount of production data accumulated in the field are analyzed intelligently. The accurate setting of rolling force and matching tension can be achieved, which are helpful for the research of rolling model and control system of temper mill.

Key words: cold rolling process; temper mill; artificial intelligence; neural network; control system

物资远程智能计量系统发展状况

郑建忠

(宝武集团韶关钢铁设备管理部,广东韶关 512122)

摘　要: 介绍钢铁企业物资远程智能计量系统发展过程,分析比较本地称量单机版称量系统的功能、操作流程、优缺点,计量数据采集系统基本架构、通信方式,计量质量一体化系统发展与应用特点,无人值守远程集中计量系统总体结构、业务流程、系统配置、系统功能,及汽车衡、轨道衡、铁水衡、平台秤、皮带秤等衡器实施物资远程集中称量的操作模式、操作要求、防止作弊功能等,以韶钢为例介绍从远程集中计量到智能化、现场无人化计量的改造实践。

关键词: 物资；计量；远程；无人值守；智能

The Development Status of Material Remote Intelligent Measurement Systems

Zheng Jianzhong

(Shaoguan Steel Equipment Management Department Baowu Steel Group
Corporation Limited, Shaoguan 512122, China)

Abstract: This paper presents the development of remote intelligent material metering system in steel industry. It reviews the functions, operation procedures, and pros and cons of local single-machine weighing system and introduces the basic structure of metering data acquisition, communication, and the application of integrated metering quality system, and unattended remote centralized metering system. The study explores the overall structure, business process, system configuration and function of the remote centralized material weighing system, such as truck scale, track scale, molten iron scale, platform scale and belt scale. It also investigates its operation mode, operation requirement and anti-cheating function. The practice of Shaogang is used as an example for the remote centralized weighing and intelligent on-site unmanned weighing.

Key words: materials; measurement; remote; unattended; intelligence

钢铁制造流程智能制造与智能设计的研究

颉建新[1,3]，张福明[2,3]

（1. 北京首钢国际工程技术有限公司战略技术部，北京 100043；2. 首钢集团有限公司总工室，北京 100041；3. 北京市冶金三维仿真设计工程技术研究中心，北京 100043）

摘 要： 本文分析了钢铁制造流程智能制造和智能设计的主要内容。钢铁制造流程智能制造包括：智能化流程设计、智能化流程生产运行、智能化流程管理、智能化流程供应链、智能化流程服务体系。钢铁制造流程智能设计包括：基础科学研究，采用CAE三维仿真设计分析计算技术；技术科学研究，采用机械三维仿真设计技术；工程科学研究，采用数字化三维仿真工厂设计技术。实现钢铁制造流程智能制造，智能设计既是智能制造的关键环节，又发挥着先导和引领作用，没有先进的智能设计，很难有先进的智能制造，必须高度重视钢铁制造流程智能设计的作用和意义，智能设计的先进程度决定着智能制造的先进水平与未来。

关键词： 钢铁制造流程；智能制造；智能设计；研究

Research of Intelligent Manufacturing and Intelligent Design on Iron and Steel Manufacturing Process

Xie Jianxin[1,3], Zhang Fuming[2,3]

(1. Strategy and Technology Department of Beijing Shougang International Engineering Co., Ltd., Beijing 100043, China; 2. Chief Engineer Department of Shougang Group Co., Ltd., Beijing 100041, China; 3. Beijing Metallurgy Three-dimensional Simulation Design Engineering Technology Research Center, Beijing 100043, China)

Abstract: The main content of intelligent manufacturing and intelligent design on iron and steel manufacturing process are analyzed in this paper. The intelligent manufacturing on iron and steel manufacturing process includes: the smart process design, the smart process production run, the smart process management, the smart process supply chain, the smart process service system. The intelligent design on iron and steel manufacturing process includes: the basic scientific research, adopting CAE three-dimensional simulation design analysis computing technology; the technical science research, adopting mechanical engineering three-dimensional simulation design technique; the engineering science research, adopting digitization three-dimensional simulation plant layout technique; realizing intelligent manufacturing on iron and steel manufacturing process, the intelligent design is not only key link of intelligent manufacturing, but also give play to forerunner and guide affect. Without having advanced intelligent design, there will be no advanced intelligent manufacturing, so we must pay high attention to the affect and significance of intelligent design on iron and steel manufacturing process. The advanced degree of intelligent design decides the advanced level and future of intelligent manufacturing.

Key words: steel manufacturing process; intelligent manufacturing; intelligent design; research

控冷装备控制精度的自动测定与提高

曹育盛

(宝山钢铁股份有限公司厚板部生产技术室，上海 200941)

摘　要：分析了影响控冷装备精度的影响因素，并通过长期实践摸索出测定的标准和方法，并实现对基础数据的自动计算后输出精度周报。利用周报相关数据的解读，可以合理的揭露设备劣化倾向并提出改进建议。

关键词：控冷；控冷工艺；控冷装备；控制精度；精度测定

Auto-evaluating and Improving the Controlling Accuracy of Control Cooling Equipment

Cao Yusheng

(Heavy Plate Plant, Baosteel, Shanghai 200941, China)

Abstract: To analyze the factors that affect the accuracy of the control cooling equipment, and through long-term practice test standards and methods, and to realize the automatic calculation on the basis of data output precision. Use weekly data interpretation, reveal equipment deterioration can be reasonably and make recommendations for improvement

Key words: control cooling; cooling process; the fast cooling equipment; control accuracy; precision measurement

韶钢原燃料汽车运输智能采样系统的设计与实现

肖命冬，欧连发，吴红兵，石志钢

(宝武集团广东韶关钢铁有限公司检测中心，广东韶关 512123)

摘　要：本文详细介绍了汽车采样机改造过程运用网络技术实现远程智能采样控制技术的功能，分析了智能采样控制系统完成控制流程、系统架构和数据库的设计，并对系统开发运用的车牌IC卡读取技术、采样数字图像生成处理技术、三维技术、PLC控制技术、连锁保护技术、自动打包技术、样品编码解码技术等关键技术进行了详细阐述，提出了符合韶钢原燃料运输汽车进厂检验的管理模式。采样过程全流程自动检测不需要人工干预，将信息处理、过程节点的分析判断、操纵控制等智能化技术完美的运用，解决了制约韶钢原燃材料采样管理困难的问题。改造完成后堵塞了管理漏洞，消除了管理盲区，规范了检测工作管理，最大限度地解放和优化了人力资源，提升了韶钢在智慧制造的钢铁行业中技术运用水平和设备管理水平，为网络钢厂建设提供条件，增强了企业竞争力。

关键词：远程；智能；采样系统；设计

Design and Application of Remote Auto-sampling System for Trucks at SGIS

Xiao Mingdong, Ou Lianfa, Wu Hongbing, Shi Zhigang

(Inspecting & Testing Center, Baowu Guangdong Shaoguan Iron and Steel Co., Ltd., Shaoguan 512123, China)

Abstract: The network technique which was used for the reform of a truck sampler for achieving the function of remote sampling control is described in detail. The design of control flow of collecting control system, system architecture and data base is analyzed. The system, how to achieve the application ability of key techniques, including IC card reading technique of license plates, sampling digital image generation processing technology, 3D technology, PLC control technology, auto-packing technology, sample coding and decoding technology and full automatic application effect, is expounded in detail. The management mode which will meet the inspecting and testing of receiving trucks at SGIS is suggested. The full flow auto-inspecting and testing of sampling does not need manual intervention. The perfect application of automation technology such as information processing, analysis and determination of process nodes and control eliminates the bottleneck of difficulty of raw material sampling management at SGIS. After the reform, it eliminates the management defects, avoids the insufficient management, and maximizes the liberation and optimization of human resources which boosts the technical application level in iron and steel industry of intelligent manufacturing and strengthens the competing edge of SGIS.

Key words: remote; intelligence; sampling system; design

基于连续域蚁群的钢坯加热炉操作优化

周 毅[1]，罗国盛[2,3]，王小毛[1]，邱奕敏[3]

（1. 武汉科技大学冶金自动化与检测技术教育部工程研究中心，湖北武汉 430081；
2. 中船重工第712研究所，湖北武汉 430064；
3. 武汉科技大学信息科学与工程学院，湖北武汉 430081）

摘 要： 钢铁工业生产中，加热炉能耗巨大，对其进行有效的优化控制，是实现节能减排的关键。本文针对加热炉生产过程进行建模以解决其操作优化问题。首先建立了加热机理模型以通过离散化计算预测钢坯温度分布，然后基于钢坯温度预报，采用了连续域蚁群建模求解优化问题。仿真实验表明，本文所提算法能较快收敛，并且可以有效求得优化的炉温设定值，具有很好的应用前景。

关键词： 蚁群算法；钢坯加热炉；操作优化；连续域优化

Continuous Ant Colony Optimization based Operational Optimization for Slab Reheating Furnace

Zhou Yi[1], Luo Guosheng[2,3], Wang Xiaomao[1], Qiu Yimin[3]

(1. Wuhan University of Science and Technology, Engineering Research Center of

Metallurgical Automation and Measurement Technology, Wuhan 430081, China;
2. China Shipbuilding Industry Corporation, 712 Research Institute, Wuhan 430064, China;
3. Wuhan University of Science and Technology, School of Information Science and Engineering, Wuhan 430081, China)

Abstract: In iron and steel industry, slab reheating furnace (SRF) is of high energy consumption. The effective control of the furnace is significant to the energy conservation and emission reduction of the production process. In this work, we modeling the SRF process to solve the operational optimization problem. Firstly, we establish the mechanism model of heat transfer and exchange process and solve it using finite difference method to predict the temperature distribution of a slab. Secondly, based on the temperature prediction module, we solve the optimization problem by continuous Ant Colony Optimization. The simulation result demonstrates that the algorithm can quickly converge to sub-optimal solution and the performance is promising.

Key words: ant colony optimization; slab reheating furnace; operational optimization; optimization for continuous domain

基于现代信息化平台的炼钢调度在沙钢的运用

陈 超

（江苏沙钢集团有限公司智能制造项目办，江苏张家港 215625）

摘 要： 炼钢-连铸是钢铁生产的瓶颈环节，其过程是多阶段、半连续的，工序间呈现顺序加工关系，且前后工序紧密衔接，存在着物质与能量的转换与传递。炼钢连铸生产计划调度系统是钢铁企业制造执行系统 MES 的重要组成部分，在企业的生产管理中起着承上启下的作用。通过它来决定作业在炼钢连铸区域的加工顺序和作业时间。

关键词： 智能调度；MES；数据采集；系统接口

Application of Steelmaking Scheduling based on Modern Information Platform in Shagang

Chen Chao

(Jiangsu Shagang Group Co., Ltd., Zhangjiagang 215625, China)

Abstract: Steelmaking-continuous casting is the bottleneck link of steel production. The process is multi-stage and semi-continuous, with sequential processing relationship between processes and close connection between the former and the latter, and the conversion and transfer of matter and energy. Steelmaking and continuous casting production scheduling system is an important component of MES, which plays a connecting role in production management. It can determine the processing sequence and operation time in the area of steelmaking and continuous casting.

Key words: intelligent scheduling; MES; the data collection; the system interface

全自动控制在钢包车上的应用

王雪明，黄春东，施军贤，陆网军

（中国沙钢集团有限公司电炉炼钢厂三车间，江苏张家港 215625）

摘　要：为了提高钢包车设备自动化控制水平，减少操作安全事故发生、减少操作工劳动强度、减少设备故障、减少能耗，提高钢水质量，中国沙钢集团有限公司电炉炼钢厂三车间对钢包车设备进行了系统的改进，由原先的继电器控制改为变频控制、增设氩气自动插拔接头、钢包车增设激光测距仪控制、增设数字高清摄像头、增设氩气智能控制系统，使钢包车控制真正实现了全自动控制，取得了良好使用效果。

关键词：钢包车；智能控制；激光测距；自动插拔

Full-automation Control Use in Ladle Car

Wang Xueming, Huang Chundong, Shi Junxian, Lu Wangjun

(Electric Furnace Steel Plant Third Workshop, Shagang Group Co., Ltd., Zhangjiagang 215625, China)

Abstract: In order to increase ladle car equipment automation control lelel, reduce safety accidents of operation happen、reduce labor intensity or operator、reduce equipment fault happen、reduce energy consumption, increase molten steel quality, Electric furnace steel plant third workshop, Shagang Group Co., Ltd. improve ladle car, for relay control is changed to frequency converter control、add argon Automatic insertion and withdrawal connecto、add laser range findei、add Digital HD camera、add argon intelligence control system，the full-automation control of ladle car is realized，goodresults have been achieved.

Key words: ladle car; full-automation control; laser ranging; Automatic insertion and withdrawal

板坯长度跟踪在 CSP 连铸机上的应用及优化

雷希璋，张　超，何金平，徐　重

（武钢有限条材厂 CSP 分厂，湖北武汉　430080）

摘　要：主要介绍了 CSP 连铸机板坯长度跟踪的原理及武钢有限条材厂 CSP 分厂在板坯长度跟踪的应用及优化。

关键词：CSP 连铸机；板坯长度跟踪；应用优化

Application and Optimization of Slab Length Tracking in CSP Caster Machine

Lei Xizhang, Zhang Chao, He Jinping, Xu Zhong

(Plants of Long Product & CSP, Wuhan Iron & Steel Co., Ltd., Wuhan 430080, China)

Abstract: The introduce principles of CSP caster machine slab length tracking system ,As well as application and optimization of slab length tracking in the plants of long product & CSP, Wuhan Iron & Steel Co., Ltd.

Key words: CSP caster machine; slab length tracking; application & optimization

退休管理网上办公系统在企业中的应用

蔺 飞，李彩虹，杨雅娟，赵媛媛

（河钢集团邯钢公司自动化部，河北邯郸　056015）

摘　要：随着信息技术的飞速发展和办公数据的增多，企业已不再满足于独立、零散的办公自动化应用，企业需要的是协同工作、综合、集成化的解决方案。而网络是解决由于物理距离造成的信息交流不畅、协商沟通不便的办公瓶颈问题的最佳方式。企业网上办公自通过对各办公自动化要素的闭环整合，实现了工作流、信息流、知识流和办公系统的整合管理，提供了一个科学、开放、先进的信息化办公平台，实现办公自动化，并进行远程办公或在家办公。本系统采用结构化与原型法结合的系统开发方法，基于公司本身内部管理及业务发展需求，应用计算机技术，Internet 技术，实现企业办公管理信息化。

关键词：网络；管理信息化；退休管理

The Application of Retirement Management Information System

Lin Fei, Li Caihong, Yang Yajuan, Zhao Yuanyuan

Abstract: With the rapid development of information technology and the increasing of business request, enterprises are no longer satisfied with the office application independent, fragmented, enterprises need is a solution of collaborative work, comprehensive, integrated. The network is the best way to solve the bottleneck problem of information exchange office due to physical distance caused by the poor, the consultative communication inconvenient. The enterprise online office automation through closed-loop integration of the office automation elements, to achieve the integration of workflow management, information flow, knowledge flow and office automation system, provides a scientific, open, advanced information office platform, and remote office or home office. The system development method the system uses a combination of structured and prototyping, their own internal management and business development based on the demand, the application of computer technology, Internet technology, enterprise office automation, information management.

Key words: network; management information system; retirement

八钢 150t 精炼炉冶炼变压器调压绕组烧损故障分析及处理

刘兴海

（宝钢集团八钢公司炼钢厂，新疆乌鲁木齐　830022）

摘　要：文章分析了八钢炼钢 150t 精炼炉冶炼变压器调压绕组烧损故障的原因，介绍了处理过程及应对措施，同时对该变压器的维护保养提出了建议。

关键词：变压器；调压绕组；分接开关

Analysis and Treatment of Burning Fault of Voltage Regulating Winding of Smelting Transformer in 150t Refining Furnace of Bagang

Liu Xinghai

(Baosteel Group Bagang Co., Ltd., Urumchi 830022, China)

Abstract: This paper analyses the causes of burnout of voltage regulating winding of transformer in 150t refining furnace of Bagang, introduces the treatment process and countermeasures, and puts forward some suggestions for maintenance of transformer.

Key words: transformer; voltage regulating winding; tap changer

12　建筑诊治

大会特邀报告

分会场特邀报告

炼铁与原料

炼钢与连铸

轧制与热处理

表面与涂镀

金属材料深加工

先进钢铁材料

粉末冶金

能源、环保与资源利用

钢铁材料表征与评价

冶金设备与工程技术

冶金自动化与智能化

★ 建筑诊治

其他

带大开间巨型框架结构抗震性能研究

胡光林,吴志华

(安徽马钢工程技术集团有限公司钢结构工程分公司,安徽马鞍山 243000)

摘　要:以钢筋混凝土巨型框架结构为例,基于 SAP2000 有限元软件通过模态分析、反应谱分析、时程分析研究大开间对结构周期、质量参与系数、最大楼层位移、层间位移角、时程位移的影响。结果表明:带大开间巨型框架与普通巨型框架相比,周期最大增幅达 10.78%,结构刚度较弱;最大楼层位移和层间位移角的最大增幅分别是 3.16%、11.05%;层间位移角极大值出现在巨型框架结构中间层,层间位移角极小值出现在巨型框架梁所在层;基于 EL-Centro 波、Taft 波、人工波作用下结构最大时程位移增幅分别为 17.4%、14.3%、3.4%,最大楼层位移增幅分别是 12.8%、21.7%、3.4%。

关键词:巨型框架;最大楼层位移;层间位移角;周期;时程位移;质量参与系数

The Research on the Seismic Performance of Mega-frame Structures with a Large Bay

Hu Guanglin, Wu Zhihua

(Steel Structure Engineering Company, Anhui Masteel Engineering & Technology Group Co., Ltd., Maanshan 243000, China)

Abstract: It set the reinforced concrete mega-frame structures as an example. It used the finite element software SAP2000 to carry out modal analysis, response spectrum analysis and time history analysis, which studied the large bay had an influence on period, mass participation, maximal floor displacement, story drift angle, time-history displacement. The results indicated that compared mega-frame with a large bay to conventional mega-frame the max increase of period was 10.78%, the stiffness of the structure was weaker. The largest increase of the maximum floor displacement and story drift angle is respectively 3.16%, 11.05%. The story drift angle of maximum value appeared in the middle of the mega-frame structures. The story drift angle of minimum value appeared in the main beam of the mega-frame structures. Based on the EL-Centro wave, Taft wave and artificial wave increase of maximum displacement was respectively 17.4%, 14.3%, 3.4%. The increase of maximum floor displacement was respectively 12.8%, 21.7%, 3.4%.

Key words: mega-frame structures; maximal floor displacement; story drift angle; period; time-history displacement; mass participation ratio

提高建筑工程抗震性能的探究

杨　倩

(莱芜钢铁集团建筑安装工程有限公司锅炉安装工程部,山东济南 271104)

摘　要:随着建筑工程的建设与发展,建筑物本身性能的要求逐渐提高,我国是一个地震多发的国家,地震区域分布广泛，很多城市位于地震区，70%以上的城市都应该进行抗震设防。在进行建筑设计的初期应该充分考虑到建筑结构的抗震性能设计，提高建筑就够的抗震性能与承载力。本文通过对当前建筑工程抗震设计问题的分析，提出提升建筑工程抗震性能的有效措施。

关键词：建筑工程；抗震性能；抗震设计；措施

Research on Improving Seismic Performance of Building Engineering

Yang Qian

(Laiwu Iron & Steel Group Construction and Installation Engineering Co., Ltd., Jinan 271104, China)

Abstract: With the construction and development of construction engineering, building itself performance requirements gradually improve, our country is an earthquake-prone country, earthquake area are widely distributed, many cities in earthquake zones, more than 70% of the city should be seismic fortification in the early stages of the architectural design should fully consider the construction of the structure of the seismic performance design, improve the seismic performance and bearing capacity of the building is in this article, through the analysis of the current seismic design of building engineering problems, put forward the effective measures to improve seismic performance of construction project.

Key words: building engineering; seismic performance; seismic design; measures

建筑抗震性能分析及提升技术

付洪坤

（山东莱钢建设有限公司建筑安装分公司，山东莱芜　271104）

摘　要：近几年，随着我国经济的快速发展，人们对建筑的需求也越来越多，若建筑结构的抗震性能相对较低，一旦有强烈等级的地震发生，将会造成重大的人员伤亡和财产损失。为此,提高建筑结构抗震能力，是建筑结构设计和施工时均值得深入思考的问题。主要针对目前主流的房屋建筑结构抗震性能技术进行阐述，正确地进行结构抗震设计，掌握好结构设计的抗震机理，不断提升房屋建筑的抗震性能，保障人民的生命和财产安全，可以使我国房屋结构设计中的抗震技术走得更远。再次，从提高建筑物结构规则性、增强建筑物的稳定性、保证建筑物的刚度以及到建筑物的施工管理等层面来说明如何提高建筑物的抗震性能。

关键词：建筑物结构；抗震性能；结构设计；选择场地；施工管理及质量

钢结构无损检测技术以及质量控制研究

刘冠

（莱芜钢铁集团有限公司，山东济南　271104）

摘　要：随着科学技术的不断发展，人们的生活质量得到了有效的改善，同时工程建设领域也出现了越来越多的新材料和新技术，随着钢材在建筑工程当中的应用，建筑工程也得到了颠覆性的变化。钢结构具有强度高、质量轻、耐腐蚀、性能稳定等特点，在建筑工程当中得到了非常广泛的应用。钢结构工程的检测也成为了工程建设中非常重要的组成部分，其中无损检测就是目前最为常用的检测方法和技术。基于此，本文主要在对当前的主要的无损检测技术进行总结的基础上，对不同的钢结构无损检测就无损检测质量控制进行了分析。

关键词：钢结构；无损检测技术；检测质量

吊车梁辅助桁架与柱连接破坏情况分析

王九根，郑　钧

（上海宝钢工业技术服务有限公司，上海　201900）

摘　要：吊车梁辅助桁架在吊车荷载反复作用下，出现辅助桁架与上柱连接处拉接破坏。本文对吊车梁辅助桁架上弦与柱连接的节点板以及节点板之间的焊缝有多处出现裂纹以及或者断裂的情况从强度计算上、吊车荷载作用下连接出现疲劳破坏。

关键词：吊车梁；辅助桁架；疲劳；破坏

Failure Analysis of Connection between Auxiliary Truss and Column of Crane Girder

Wang Jiugen, Zheng Jun

(Shanghai Baosteel Industrial Technology Service Limited Company, Shanghai 201900, China)

Abstract: Under repeated crane loads, the auxiliary truss of crane girder fails at the connection between the auxiliary truss and the upper column. In this article, the fatigue failure of the joints between the upper chord and the column of the auxiliary truss of crane girder and the weld seam between the joints and the joints is studied in terms of strength calculation and crane load.

Key words: crane girder; auxiliary truss; fatigue; destruction

某高炉出铁场除尘管道凹陷变形原因分析

王建强，庄继勇，郑　钧，王华丹

（上海宝钢工业技术服务有限公司，上海　201900）

摘　要：除尘管道在压力的作用下容易出现凹陷变形，本文结合工程实例，对某高炉出铁场负压除尘管道凹陷变形后进行了详细检测，并对凹陷变形过程进行了详细的探讨，依据除尘管道凹陷后的残余变形状况，并根据现场构件变形测量情况对管道凹陷变形原因及过程进行了分析，提出了相关设计改进措施，具有较强的实用性，可供类似工程参考。

关键词：负压除尘管道；检测；稳定性；加强环板

Deformation Cause Analysis of Dust Removal Pipeline Depression in a Blast Furnace Discharge Yard

Wang Jianqiang, Zhuang Jiyong, Zheng Jun, Wang Huadan

(Shanghai Baosteel Industrial Technology Service Limited Company, Shanghai 201900, China)

Abstract: Dust removal pipeline is prone to depression deformation under the action of pressure. In this paper, combined with an actual example, the depression deformation of negative pressure dust removal pipeline in a blast furnace tapping yard is detected in detail, and the process of depression deformation is discussed in detail. According to the residual deformation after depression of dust removal pipeline, analyzing the field component deformation measurement base on the process of pipeline depression deformation, we proposed the improvement measure about design, which has strong practicability and can be used for similar projects as reference.

Key words: dust removal pipeline; inspection; stability; strengthening ring plate

某厂房抗震鉴定

陆明磊

(上海宝钢工业技术服务有限公司,上海 201900)

摘 要:本文通过对厂房的房屋质量检测结果,以及对其抗震能力进行研究、分析,给出了相应的加固处理措施。

关键词:检测;抗震能力;加固

The Seismic Appraiser of an Workshop

Lu Minglei

(Shanghai Baosteel Industrial Technology Service Limited Company, Shanghai 201900, China)

Abstract: The survey result of an workshop was present in this paper, having analyzed the aseismic ability of the building, some necessary reinforce method were given.

Key words: survey; aseismic ability; reinforce

浅析建筑结构抗震技术的应用

郭之亮

(莱芜钢铁集团建筑安装工程有限公司,山东济南 271104)

摘 要:在这个与时俱进的新时代里,人们对建筑结构的抗震技术提出了更高的要求。房屋抗震标准已成为购房者置业参考的一个重要因素。《中华人民共和国防震减灾法》规定：新建、扩建、改建建设工程，应当达到抗震设防要求。应当按照地震烈度区划图或者地震动参数区划图所确定的抗震设防要求进行抗震设防。对于医院、学校、消防站等一些人员密集场所的房屋建设工程，应当按照高于当地房屋建筑的抗震设防要求进行设计和施工，同时采取有效措施，增强抗震设防能力。为此，本文主要对建筑框架式结构抗震性能特点、技术应用及措施进行了分析与探究。框架结构是由许多梁和柱共同组成的框架来承受房屋全部荷载的结构。高层的民用建筑和多层的工业厂房，砖墙承重已不能适应荷重较大的要求，往往采用框架作为承重结构；结构抗震性能是指在地震作用下，结构构件的承载能力、变形能力、耗能能力、刚度及破坏形态的变化和发展。

关键词：建筑框架式结构；抗震性能；抗震方式

工业建筑钢结构疲劳损伤检测、评估及加固关键技术研究

幸坤涛[1,2]，赵晓青[2]，郭小华[1,2]，惠云玲[2]，杨建平[2]

(1. 中冶建筑研究总院有限公司，北京 100088；
2. 国家工业建构筑物质量安全监督检验中心，北京 100088)

摘　要：我国工业建筑中钢结构占很大比重，而且很多处于动载、重载、腐蚀等不利条件下。许多工业建筑钢结构，尤其是重级工作制吊车梁，未达到设计使用年限就发生早期破坏。据统计工业建筑钢结构破坏80%是疲劳破坏，工业建筑钢结构疲劳破坏90%以上发生在重级工作制的钢吊车梁系统，我国重级工作制钢吊车梁系统使用15~20年会出现疲劳开裂甚至于断裂。承受动荷载的钢结构吊车梁疲劳问题显得越来越突出，如何对既有钢结构工业建筑进行检测鉴定、安全评估和加固修复成为迫切需要解决的问题。本文介绍了"工业建筑钢结构疲劳损伤检测评估及加固修复关键技术研究"的项目研究背景、项目研究过程及主要内容、主要研究成果及应用推广情况。

关键词：工业建筑；钢结构；疲劳；检测鉴定；疲劳寿命评估；疲劳可靠性；加固修复

Research on Key Technologies of Fatigue Damage Detection, Assessment and Reinforcement of Steel Structures in Industrial Buildings

Xing Kuntao[1,2], Zhao Xiaoqing[2], Guo Xiaohua[1,2], Hui Yunling[2], Yang Jianping[2]

(1. Central Research Institute of Building and Construction Co., Ltd., MCC, Beijing 100088, China; 2. National Industrial Construction Quality and Safety Supervision and Inspection Center, Beijing 100088, China)

Abstract: Steel structure accounts for a large proportion of industrial buildings in China, and many of them are under unfavorable conditions such as dynamic load, heavy load and corrosion. Many steel structures of industrial buildings, especially heavy working crane girders, fail to reach the designed service life and cause early damage. According to statistics, 80% of the damage of steel structures in industrial buildings is fatigue damage. More than 90% of the fatigue damage of steel structures in industrial buildings occurs in the heavy-duty steel crane girder system. Fatigue cracking or even fracture will occur in the heavy-duty steel crane girder system in China during 15~20 years of service. The fatigue problem of steel crane girders under dynamic loads is becoming more and more prominent. How to inspect and identify existing steel industrial buildings, safety assessment and reinforcement and repair has become an urgent problem to be solved. This paper introduces the research background, research process and main contents, main research results and application popularization of the project "Key Technologies for Fatigue Damage Detection, Assessment and Reinforcement and Repair of Steel Structures in Industrial Buildings".

Key words: industrial buildings; steel structures; fatigue; detection and assessment; fatigue life assessment; fatigue reliability; reinforcement and repair

钢框架-中心支撑结构的结构抗震鉴定

郑 钧

（上海宝钢工业技术服务有限公司，上海 201900）

摘 要： 某房屋主体结构为钢框架-中心支撑结构，屋盖为钢网架。业主拟对房屋重新装饰装修，楼层荷载增加，为保证房屋后续的使用安全，并为后续改造提供依据，故对房屋进行抗震鉴定。按照现行规范对房屋改造方案进行验算，评估房屋改造方案是否可行，并提出合理加固处理方案。

关键词： 钢框架；网架结构；结构测绘；抗震鉴定

Seismic Appraisal of Steel Frame-central Braced Structure

Zheng Jun

(Shanghai Baosteel Industrial Technology Service Limited Company, Shanghai 201900, China)

Abstract: A steel frame-center support and steel net frame roof located, needs to be rebuilt. the current status of the building is surveyed and seismic proofed.according to the current regulations. In order to ensure the follow-up use safety of the house and provide the basis for the follow-up renovation, the seismic appraisal of the house is carried out. Checking the housing reform scheme in accordance with the current standards .The plan carries out the check calculation,, evaluates whether the house renovation plan is feasible, and proposes a reasonable reinforcement treatment plan.

Key words: steel frame; grid structure; structural mapping; anti-earthquake evaluation

其他

- 大会特邀报告
- 分会场特邀报告
- 炼铁与原料
- 炼钢与连铸
- 轧制与热处理
- 表面与涂镀
- 金属材料深加工
- 先进钢铁材料
- 粉末冶金
- 能源、环保与资源利用
- 钢铁材料表征与评价
- 冶金设备与工程技术
- 冶金自动化与智能化
- 建筑诊治
- ★ 其他

应用新型工业化应力退火技术研究矩形纳米晶磁芯的磁性能

李雪松[1]，薛志勇[1]，Sajad Sohrabi[1,2]，汪卫华[1,2]

(1. 先进材料研究院，华北电力大学，北京 102206；
2. 物理研究所，中国科学院大学，北京 100190)

摘　要：人为施加机械应力或外加磁场的条件下对非晶和纳米晶磁环进行退火处理，可以改变材料的磁各向异性，显著影响磁化曲线的形状。这一特点使得非晶纳米晶磁环具有广阔的应用前景：从具有高饱和磁感应强度和稳定的、低的磁导率（具有较大磁各向异性）的扼流圈和电抗器，到需要高磁导率（具有低磁各向异性）的共模扼流圈。为了诱导磁各向异性，传统应力退火的方法是先对退火的过程中的带材施加应力，之后再卷绕成磁芯，但这种方法并不适用于具有直角曲率的矩形磁芯。本研究介绍了一种新型工业规模应力退火技术，先将带材卷绕成磁芯，对矩形磁芯预先施加应力后再进行常规热处理，改变矩形磁芯的磁各向异性。与无应力退火相比，应力退火后的磁芯具有较大的磁各向异性和较低的磁导率，随着外加应力的增加，磁各向异性逐渐增大，而磁导率逐渐减小。

关键词：纳米晶软磁合金；应力退火；磁各向异性

Tailoring Magnetic Properties of Rectangular-Shaped Magnetic Cores Using a Novel Industrial-scale Stress Annealing Technique

Li Xuesong[1], Xue Zhiyong[1], Sajad Sohrabi[1,2], Wang Weihua[1,2]

(1. Institute of Advance Materials, North China Electric Power University, Beijing 102206, China;
2. Institute of Physics, Chinese Academy of Sciences, Beijing 100190, China)

Abstract: Annealing of amorphous and nanocrystalline soft magnetic cores under mechanical stress or a magnetic field can remarkably influence the shape of magnetization curves by controlling the magnetic anisotropy of treated ribbons. This enables producing soft ferromagnetic cores covering a wide range of applications: from choke coils and reactors with large saturation and consistent, low permeability (cores with large magnetic anisotropy and sheared B-H hysteresis loops), to common-mode chokes where high permeability (cores with low magnetic anisotropy and squared B-H hysteresis loops) is desirable. The common stress annealing method for inducing magnetic anisotropy consists of passing the ribbon through a furnace under stress prior to core winding, but this method is not adequate for the production of rectangular-shaped cores where right angle curvatures are required. In this study, we introduce a novel large-scale production technique for stress annealing of this type of cores and show that the magnetic anisotropy can be induced in rectangular-shaped FINEMET cores by pre-straining of the tape wound core through a mandrel during conventional furnace annealing. The treated cores under stress annealing have larger magnetic anisotropy and lower permeability compared to the cores annealed without stress. The extent of the increase in anisotropy and the decrease in permeability in treated cores is proportional to the extent of the imposed stress on the cores.

Key words: nanocrystalline soft magnetic alloys; stress annealing; magnetic anisotropy

冶金行业三维工厂设计与二维设计的分析比较

陈 馨,郭彩虹

(中冶京诚工程技术有限公司,北京 100176)

摘 要:本文以台塑项目为例,具体介绍了三维工厂设计在冶金行业的应用,通过分析比较三维设计与二维设计的区别及优势,可以看出未来三维设计将逐步取代二维设计,成为冶金行业以及复杂工程建设的主流设计方法。本文为设计人员选择设计方法提供了参考意见。

关键词:冶金行业;工厂设计;三维设计;二维设计;协同

Analysis and Comparison between 3d and 2d Design in Metallurgical Industry

Chen Xin, Guo Caihong

(Capital Engineering & Research Incorporation Limited, Beijing 100176, China)

Abstract: It is illustrated by case of FHS project, introduced the application of 3d plant design in the metallurgical industry. According to the difference and advantages between the 3d and 2d, 3d will instead of 2d in the future and become popular design method in the metallurgical industry and complex project construct. It will provide some advice for designer on choosing design method.

Key words: metallurgical industry; plant design; 3d design; 2d design; coordination

草酸盐从钒渣酸浸液中除锰试验研究

杨 晓,李道玉

(攀钢集团研究院有限公司钒钛资源综合利用国家重点实验室,四川攀枝花 617000)

摘 要:针对目前钒渣钙化焙烧-酸浸提钒工艺所得酸浸液中锰含量高,给后续沉钒及产品纯度造成不利影响,提出酸浸液预先除锰净化工艺。采用草酸盐为除锰剂对高锰酸浸液进行了除锰研究,考察了除锰剂种类、除锰剂用量、反应pH值、反应温度及反应时间对除锰的影响。在最佳试验条件下,除锰率达92%以上,钒的损失率在0.5%以内。该除锰方法简单,对原有提钒工艺无需做大的调整,易于进行工业化。

关键词:钒渣;酸浸液;除锰;草酸钠

Experimental Study on Removal of Manganese from Acid Leaching Solution of Vanadium Slag by Oxalate

Yang Xiao, Li Daoyu

(Pangang Group Research Institute Co., Ltd., State Key Laboratory of Vanadium and Titanium Resources Comprehensive Utilization, Panzhihua 617000, China)

Abstract: In view of the high manganese content in the acid leaching solution obtained by the calcination of vanadium slag and the adverse influence on the subsequent vanadium precipitation and low product purity, a purification process of manganese removal from the acid leaching solution is put forward. Manganese removal in high manganese contented leaching solution was studied by using oxalate as manganese removal agent. The effects of kinds of manganese removal agent, amount of manganese removal agent, reaction pH value, reaction temperature and reaction time on manganese removal were investigated. Under the best experimental conditions, the removal rate of manganese is more than 92%, and the loss rate of vanadium is less than 0.5%. The method of removing manganese is simple, and it is easy to industrialize.

Key words: vanadium slag; axid leaching solution; manganese removal; oxalate

Cu 表面超疏水薄膜的制备及耐腐蚀性的研究

汪建，蔡浩，张强，龙剑平

（成都理工大学材料与化学化工学院材料科学与工程系，四川成都 610059）

摘 要： 本文采用自组装法在 Cu 表面制备超疏水膜。研究了刻蚀液中 NaOH 溶液和硬脂酸-乙醇溶液的浓度对超疏水薄膜的影响，观察了薄膜在不同环境介质中的腐蚀行为。

研究表明，铜表面的黑色膜层为沿着特定晶面(-1 1 1)/(0 0 2)和(1 1 1)生长的 CuO 膜。在酸性介质中，超疏水膜层会直接消解而达不到电化学稳定状态。而在碱性或盐类介质中，空白样的耐腐蚀性会随着电解质浓度的增大而增大。在碱性介质中，薄膜的缓蚀率较低，随着电解质浓度的增大而逐渐增大。而在盐类电解质中，薄膜的缓蚀率较高，同样随着电解质浓度的增大而增大。研究表明，涂覆有超疏水表面的铜箔片在盐溶液中腐蚀最为严重，表面结构完全消失。

关键词： 铜箔片；自组装；超疏水薄膜；耐蚀性；缓蚀率

Preparation and Corrosion Resistance of Superhydrophobic Films on Cu Foil

Wang Jian, Cai Hao, Zhang Qiang, Long Jianping

(Department of Materials and Engineering, College of Materials and Chemistry & Chemical Engineering, Chengdu University of Technology, Chengdu 610059, China)

Abstract: In this paper, a superhydrophobic surface is prepared on the surface of copper foil by self-assembly method. The effect of NaOH concentration in the etching solution and the concentration of stearic acid-ethanol solution on the superhydrophobic film was studied, and the corrosion behavior under different environmental media was observed.

The results show that the black film on the copper surface is CuO grown along a specific crystal plane (−1 1 1)/(0 0 2) and (1 1 1). In an acidic medium, the superhydrophobic membrane layer will directly digest and fail to reach an electrochemically stable state. Whether in alkaline or salt media, the corrosion resistance of the blank increases with increasing electrolyte concentration. It was found that the corrosion inhibiting rate of the superhydrophobic sample in the alkaline medium was relatively low, and gradually increased with the increasing electrolyte concentration. In the salt electrolyte, the corrosion inhibiting rate of the superhydrophobic sample is relatively high, and also increases with the increasing electrolyte concentration. It was found that the copper foil coated with the superhydrophobic film was the most severely corroded in the salt medium, and the surface structure completely disappeared.

Key words: copper foil; self-assembly; superhydrophobic film; corrosion resistance; corrosion inhibiting rate

TMEIC 激光焊机 X 轴激光焦点系统分析与维护

邓承龙

（攀钢集团西昌钢钒公司板材厂设备室，四川西昌　615000）

摘　要：以攀钢西昌钢钒有限公司板材厂 酸轧机组配备的 TMEIC 激光焊机为研究对象，介绍了 X 轴激光焦点系统在 TMEIC 激光焊机中的应用，并重点分析了焊机 X 轴激光焦点系统的结构组成、焦点偏移影响因素、焦点位置计算调整及零位标定，掌握这些调整及标定方法对保证焊机 X 轴激光焦点系统的稳定精准运行至关重要。

关键词：激光焦点；X 轴；激光焊机；零位调整

Analysis and Maintenance of Laser Focus System of the X-ray Axis of TMEIC Laser Welder

Deng Chenglong

(The Plate Factory Equipment Room of Pangang Xichang, Xichang, 615000, China)

Abstract: This paper takes TMEIC laser welder as the research object, and introduce the application of X-ray axis laser focus system in TMEIC laser welder, focus analysis the structure of the X-ray axis laser focus system、focus offset Influencing factors、focus position calculation adjustment and zero position calibration, It is very important to master these adjustment and calibration methods to ensure the stable and accurate operation of the X-ray axis laser focus system.

Key words: laser focus; X-ray axis; laser welder; zero position calibration